Systems Biogeochemistry
of Major Marine Biomes

Systems Biogeochemistry of Major Marine Biomes

Edited by

Aninda Mazumdar
CSIR–National Institute of Oceanography
Goa, India

Wriddhiman Ghosh
Bose Institute
Kolkata, India

WILEY

This edition first published 2022

Registered Office
John Wiley & Sons, Inc., 111 River Street, Hoboken, NJ 07030, USA

Editorial Office
111 River Street, Hoboken, NJ 07030, USA

For details of our global editorial offices, customer services, and more information about Wiley products visit us at www.wiley.com.

Wiley also publishes its books in a variety of electronic formats and by print-on-demand. Some content that appears in standard print versions of this book may not be available in other formats.

Library of Congress Cataloging-in-Publication Data

Names: Mazumdar, Aninda, editor | Ghosh, Wriddhiman, editor.
Title: Systems biogeochemistry of major marine biomes / edited by Aninda Mazumdar, CSIR-National Institute of Oceanography, Goa, India, Wriddhiman Ghosh, Bose Institute, Kolkata, India.
Description: First edition. | Hoboken, NJ : Wiley, 2022. | Includes bibliographical references and index.
Identifiers: LCCN 2021048467 (print) | LCCN 2021048468 (ebook) | ISBN 9781119554387 (cloth) | ISBN 9781119554370 (adobe pdf) | ISBN 9781119554363 (epub)
Subjects: LCSH: Biogeochemistry. | Chemical oceanography.
Classification: LCC QH343.7 .S97 2022 (print) | LCC QH343.7 (ebook) | DDC 577/.14–dc23/eng/20211103
LC record available at https://lccn.loc.gov/2021048467
LC ebook record available at https://lccn.loc.gov/2021048468

Cover image: © Wriddhiman Ghosh
Cover design by Wiley

Set in 10/12pt TimesNewRomanMTStd by Straive, Pondicherry, India

CONTENTS

Biome IV Extreme Environments

LIST OF CONTRIBUTORS

Felix Bast
Department of Botany
Central University of Punjab
Bathinda, India

Bhaskar Bhadra
Reliance Technology Group, Reliance Industries Limited
Navi Mumbai, India

Sabyasachi Bhattacharya
Department of Microbiology
Bose Institute
Kolkata, India

Moumita Bhowmik
Department of Life Sciences
Presidency University
Kolkata, India

Sumit Chatterjee
Department of Microbiology
Bose Institute
Kolkata, India

Kalyan De
CSIR-National Institute of Oceanography
Goa, India

Aniket Desai
CSIR-National Institute of Oceanography
Goa, India;
and
Academy of Scientific and Innovative Research
Goa, India

Svetlana Fernandes
CSIR-National Institute of Oceanography
Goa, India

Lina L. Fernandes
CSIR-National Institute of Oceanography
Goa, India

Sheryl Oliveira Fernandes
ESSO-National Centre for Polar and
Ocean Research
Ministry of Earth Sciences
Goa, India

Wriddhiman Ghosh
Department of Microbiology
Bose Institute
Kolkata, India

Baban Ingole
ESSO-National Centre for Polar and Ocean Research
Ministry of Earth Sciences
Goa, India

Pratima M. Kessarkar
CSIR-National Institute of Oceanography
Goa, India

Amrutha Karayakath
Department of Geology
Central University of Punjab
Bathinda, India

K. P. Kottekkatu
ESSO-National Centre for Polar and Ocean Research
Ministry of Earth Sciences
Goa, India

Pushpendu Kundu
Centre for Biosciences
Central University of Punjab
Bathinda, India

John P. Kurian
ESSO-National Centre for Polar and Ocean Research
Ministry of Earth Sciences
Goa, India

Siby Kurian
CSIR-National Institute of Oceanography
Goa, India

L. Surya Prakash
ESSO-National Centre for Polar and Ocean Research
Ministry of Earth Sciences
Goa, India

Subhrangshu Mandal
Department of Microbiology
Bose Institute
Kolkata, India

Sumit Mandal
Department of Life Sciences
Presidency University
Kolkata, India

Jittu Mathai
CSIR-National Institute of Oceanography
Goa, India

Aninda Mazumdar
CSIR-National Institute of Oceanography
Goa, India

Patrick Meister
Department of Geodynamics and Sedimentology
University of Vienna
Vienna, Austria

Rahul Mohan
ESSO-National Centre for Polar and
Ocean Research (NCPOR)
Ministry of Earth Sciences
Goa, India

Nibendu Mondal
Department of Microbiology
Bose Institute
Kolkata, India

Sambhaji Mote
CSIR-National Institute of Oceanography
Goa, India

Mandar Nanajkar
CSIR-National Institute of Oceanography
Goa, India

Vidya P. Jayapalan
ESSO-National Centre for Polar and Ocean Research (NCPOR)
Ministry of Earth Sciences
Goa, India

Sabu Prabhakaran
ESSO-National Centre for Polar and Ocean Research (NCPOR)
Ministry of Earth Sciences
Goa, India

Shramik Patil
ESSO-National Centre for Polar and Ocean Research (NCPOR)
Ministry of Earth Sciences
Goa, India

Jitendra K. Pattanaik
Department of Geology
Central University of Punjab
Bathinda, India

Aditya Peketi
CSIR-National Institute of Oceanography
Goa, India

Venigalla P. Rao
CSIR-National Institute of Oceanography
Goa, India;
and
Department of Civil Engineering
Vignan's University
Guntur, India

Carolina Reyes
Department of Environmental Geosciences
University of Vienna
Vienna, Austria

Khem C. Saini
Centre for Biosciences
Central University of Punjab
Bathinda, India

Jagannath Sarkar
Department of Microbiology
Bose Institute
Kolkata, India

Sabyasachi Sautya
Regional Centre, CSIR-National Institute of
Oceanography
Mumbai, India

Suhas S. Shetye
CSIR-National Institute of Oceanography
Goa, India

Rupesh K. Sinha
ESSO-National Centre for Polar and Ocean Research (NCPOR)
Ministry of Earth Sciences
Goa, India

Kalyani Sivan
Academy of Scientific and Innovative Research (AcSIR)
Ghaziabad, India;
and
CSIR-National Institute of Oceanography
Goa, India

Melena Soares
ESSO-National Centre for Polar and Ocean Research (NCPOR)
Ministry of Earth Sciences
Goa, India

Herald Strauss
Institut für Geologie und Paläontologie
Westfälische Wilhelms-Universität Münster
Münster, Germany

Sarat C. Tripathy
ESSO-National Centre for Polar and Ocean Research (NCPOR)
Ministry of Earth Sciences
Goa, India

PREFACE

Biological processes have profound roles in the chemical transformations of the marine realm. The nature and extent of biological activities, in turn, are controlled by a number of physical and chemical phenomena such as interaction of the ocean with the atmosphere and earth's crust, oceanic upwelling, oxygenation, sediment accumulation, and inorganic and organic matter fluxes in dissolved and particulate forms from the land and the deep earth. The overall biogeochemical dynamics of the water column and sediment fluid also exert feedback influences on the structure and function of the biota, and, in the entire process, the system evolves holistically in time and space. In this context, a marine biogeochemical system can be thought of as a geographically and geologically defined physiography that transcends latitudes, longitudes and water-depths of the global ocean, and has its characteristic geobiological framework, overarching all the spatiotemporal dynamics of its structural (e.g. chemical and biological species) and functional (e.g. reactions and transformation) components.

Comprehensive understanding of marine biogeochemical regimes entails systemic knowledge of the biomes. Historically, studies of inorganic and organic geochemistry for water-columns and sediments (including both past and present records) contributed the most in deciphering biogeochemical processes within the marine realm. In recent times, remarkable advances in high-throughput meta-omics-based microbiological research, and molecular-taxonomy-based understanding of benthic ecology, have afforded increasing knowledge interfaces between preserved geochemical records and ecosystem structure/function. Such cross-talks of disciplines have led to the discovery of a number of novel (and often cryptic) biogeochemical processes within the global ocean that were impossible to have been revealed previously from preserved chemical records alone. At the same time, increasing cross-disciplinary interfaces have also given rise to biogeochemical enigmas, of which some may well be the potential seeds of future paradigm changes in marine biogeochemistry. In this intellectual landscape, where our understanding of *in situ* biological processes, their geological manifestations, and cybernetic controls, are evolving faster than ever, this book adopts a systems-based approach to integrate and update the information available on the geochemistry, geophysics, biology and ecology of four major categories of marine physiography: continental margins, ocean depths, polar oceans, and biophysically extreme environments. The idea, at large, is to envision holistic pictures of all biogeochemical happenings within distinct systems of these oceanic provinces.

While some of the biogeochemical regimes mulled over hold crucial implications for planetary health and biosphere sustainability (by virtue of being the bellwethers of perturbations brought about by anthropogenic interference, pollution, global warming and/or climate change), others are central to the understanding of the Earth's early biosphere, so their geobiologies hold critical implications for life-detection during space-exploration missions and potential habitability of differently constituted oceanic systems in extraterrestrial locations.

We expect the biome-based systemic perspective to complement the traditional way of looking at marine biogeochemistry from the standpoint of individual elemental cycles. Systemic consideration of biogeochemical provinces of the global ocean would not only provide an intellectual baseline for integrating chemical records with biome structures/functions but also lay the foundation of holistic policy frameworks for sustainable utilization of marine resources, risk avoidance, and disaster management.

Aninda Mazumdar
Geological Oceanography
CSIR-National Institute of Oceanography, India

Wriddhiman Ghosh
Department of Microbiology
Bose Institute, India

Biome I
Continental Margins

1

Biogeochemistry of Marine Oxygen Minimum Zones with Special Emphasis on the Northern Indian Ocean

Svetlana Fernandes[1,†], Subhrangshu Mandal[2,†], Kalyani Sivan[1,3], Aditya Peketi[1], and Aninda Mazumdar[1*]

ABSTRACT

Oxygen minimum zones (OMZs) enclose O_2 depleted subsurface water masses in the global ocean extending approximately 150 to 1200 m below sea level. The most pronounced OMZs occur in Eastern Tropical North Pacific off Mexico and California (ETNP), Eastern Tropical South Pacific off Peru and Chile (ETSP), and the Arabian Sea (AS) defined by secondary nitrite maxima attributed to intense denitrification in the water column. These OMZs sites are critical for biogeochemical processes that control the biodiversity and primary productivity of the ocean. The preservation of organic carbon is efficient within the sediments underlying oxygen-depleted waters as a result of incomplete decomposition as it sinks through the water column and diminished bioturbation activity. The partially degraded (reactive) organic matter fuels microbe-mediated biogeochemical processes in the anoxic marine sediments where sulfate reduction is a significant remineralization pathway. The OMZs exert a strong influence on the abundance, diversity, and composition of microbial communities. Recent geochemical and environmental genomic studies identified the prevalence of C, N, and S cycles in the OMZs. Here, we review the progress and current understanding of the C–S–N cycle in the OMZ sediments with regard to its biogeochemistry and microbial ecology, and present a brief account of the mechanism of the formation of OMZs in the northern Indian Ocean.

1.1. INTRODUCTION

Oxygen minimum zones (OMZs) are oxygen-depleted intermediate-depth water masses in the global ocean, usually between 150 and 1200 m below sea level (mbsl). The upper threshold of dissolved oxygen (DO) concentrations defining the OMZs is ~20 μM (Figure 1.1; Paulmier and Ruiz-Pino, 2009; Ruvalcaba Baroni et al., 2020), while the lower concentration of DO detected in an OMZ may be <2 nM O_2 (Revsbech et al., 2009), amounting to functionally anoxic conditions. Oceanic regions generally classified as OMZs include (1) the Eastern Tropical North Pacific off Mexico and California (ETNP), (2) the Eastern Tropical South Pacific off Peru and Chile (ETSP), (3) the Bay of Bengal (BoB), and (4) the Arabian Sea (AS). Of these OMZs, the ETNP, ETSP, and the AS have also been classified as anoxic marine zones (Ulloa et al., 2012), owing to the buildup of secondary nitrite maxima (SNM) within the water column. The SNM forms because of denitrification in the water column, which occurs when DO concentrations drop to <5 mM (Anderson et al., 2007; Banse et al., 2017). The functionally anoxic parts of these OMZs occupy only ~0.8% of the world ocean but are responsible for ~35% of the production of N_2 through denitrification (Ward et al., 2009).

[1]*CSIR-National Institute of Oceanography, Dona Paula, Goa 403004, India*

[2]*Department of Microbiology, Bose Institute, P-1/12 CIT Scheme VIIM, Kolkata 700054 , West Bengal, India*

[3]*Academy of Scientific and Innovative Research (AcSIR), Ghaziabad 201002, India*

**Corresponding Author: Aninda Mazumdar (maninda@nio.org) ORCiD code: 0000-0002-7897-1646*

† *Equal contribution*

Systems Biogeochemistry of Major Marine Biomes, First Edition. Edited by Aninda Mazumdar and Wriddhiman Ghosh.

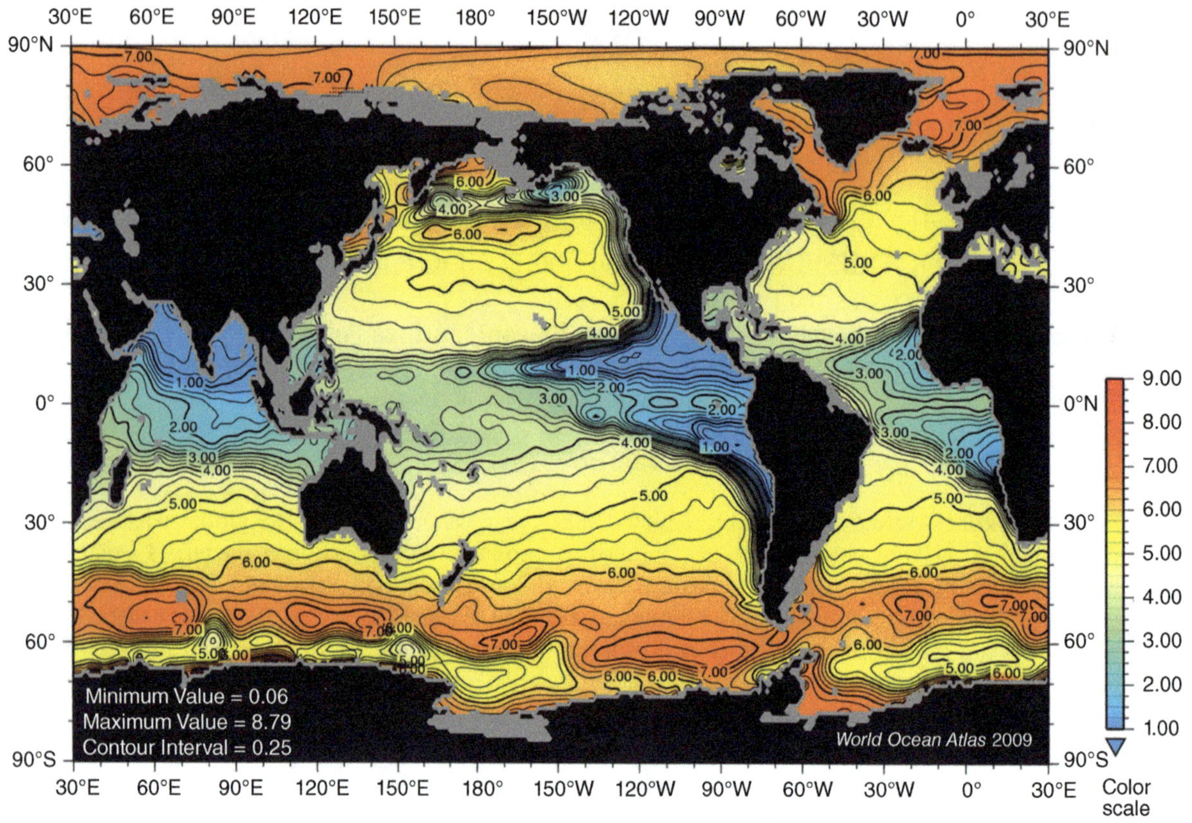

Figure 1.1 Global annual oxygen concentration (ml l^{-1}) at 150 mbsl. *Source:* World Ocean Atlas 2009.

The buildup of OMZs is controlled by the interplay of physical and biological processes coupled with regional geography. Inadequate ventilation of the oxygenated water masses (e.g. the northern Indian Ocean), upwelling of nutrient-rich deep waters (leading to high productivity in the euphotic zone), and subsequent microbial respiration of organic particulates (phytodetritus, fecal pellets, dead organisms, etc.) lead to a depletion of DO in the water column, resulting in the formation of OMZs (Behrenfeld et al., 2006; Regaudie-de-Gioux and Duarte, 2012). OMZs are high in nutrient concentrations and support highly productive fishing regions (Garçon et al., 2019).

In the northern Indian Ocean, OMZs occur in both the AS and the BoB (Figure 1.2) because of their closed northern geographical boundaries and monsoon-driven seasonal upwelling (Naqvi, 1991; Stramma et al., 2008). The DO concentration within the Arabian Sea (ASOMZ) reflects a balance between biological O_2 consumption and O_2 replenishment. The primary source of intermediate water (Indian Ocean Central Water: ICW) (Stramma et al., 1996; Schott and McCreary, 2001; Rixen et al., 2020) in the northern Indian Ocean include contributions from the Antarctic Intermediate Water (AAIW), Subantarctic Mode Water (SAMW) and the Indonesian intermediate waters (Lachkar et al., 2019). The dense and highly saline Persian Gulf water (PGW) and relatively less saline but denser Red Sea water (RSW) are the lateral sources of intermediate water in the AS (Bower et al., 2005).

1.1.1. The Arabian Sea Oxygen Minimum Zone

The AS is believed to contain the thickest and most intense of the OMZs (Agnihotri et al., 2003; Naqvi et al., 2010a; Prakash et al., 2012; Banse et al., 2014; Acharya and Panigrahi, 2016), extending over the entire sea, with its upper boundary occurring at 100–150 m, and its lower boundary at 1000-1200 m (Wyrtki, 1971; Bange et al., 2005; Banse et al., 2014), impinging upon a large area of the continental slope (Naqvi et al., 2006). The benthic area of the upper slope underlying hypoxic water in the AS is computed to be 230 440 km^2, making it responsible for over one-quarter of the world's naturally hypoxic seafloor (Global estimate = 764 000 km^2; Helly and Levin, 2004). The ASOMZ arises from its landlocked geography, mainly owing to the presence of the Asian landmass that restricts the flow of oxygenated water from the north. Monsoonal upwelling, wind-driven mixing, Ekman pumping during summer (Kumar et al., 2009), convective mixing during winter (Madhupratap et al., 1996;

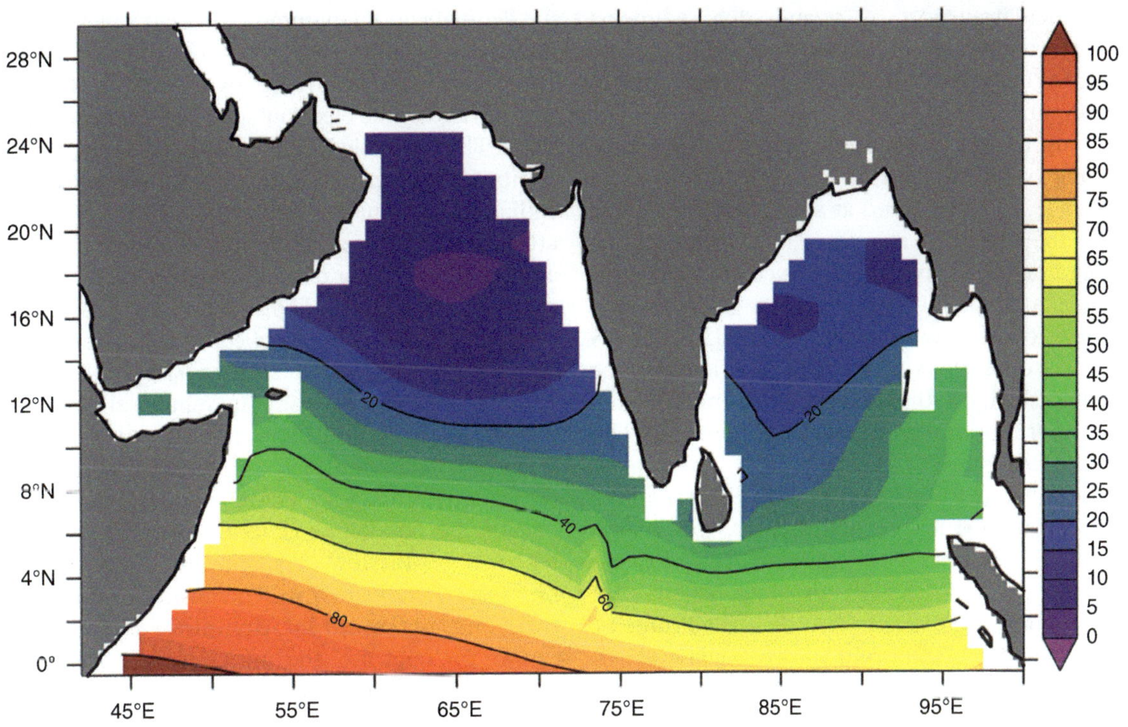

Figure 1.2 Annual oxygen concentration (µmol l^{-1}) in the water column of the Indian Ocean. The data is obtained from the World Ocean Atlas 2013 (Boyer et al., 2013). The map was produced using generic mapping tool software.

Prasanna Kumar et al., 2001; Kumar et al., 2009), and advection of NO_3^- rich upwelled water mass off the Oman, Yemen, and Somalia margins leads to large-scale fertilization of the euphotic zone (Naqvi et al., 2006; 2010b). The productivity in this region is controlled by the NO_3^- and Fe (aeolian) limitations which also show seasonal variation (Naqvi et al., 2010b; Banerjee and Kumar, 2014).

The ASOMZ shows an east–west variation in structure, where the upper part (400 m) is located in the central/eastern basin, and the lower part (below 400 m) extends to the Omani coast, indicating a northward intensification of the ASOMZ (McCreary et al., 2013). The PGW water enters the ASOMZ from the northwest at shallow depths of – 300 m and spreads around the perimeter of the basin and southward along the Omani coast (Prasad et al., 2001), and the denser RSW encroaches to depths of 600–1000 m and spreads across the basin (Shankar et al., 2005; Shenoi et al., 2005).

A poleward undercurrent (West India Undercurrent, WIUC) carries the ICW northwards into the eastern Arabian Sea (EAS) up to 16°N at a water depth of – 500 m (Shenoy et al., 2020; Schmidt et al., 2020a). In the EAS, the OMZ expands southwards during the SW monsoon (~ 9°N) and retreats northwards during the NE monsoon (11°N) (Shenoy et al., 2020). Although the productivity across the EAS is significantly lower compared with that of the Western Arabian Sea (WAS), the OMZ is more intense in the central and eastern AS (Naqvi, 1991). This observation may be attributed to one or more hypotheses, including:

water column O_2 consumption (via respiration of organic matter) during eastward transit of ICW from Somali coast to the west coast of India (McCreary et al., 2013; Acharyya and Panigrahi, 2016; Rixen et al., 2020);

rapid sinking of the large phytoplankton detritus in the upwelling regions of the WAS (Naqvi et al., 2010b) results in minimum respiration, alternatively, the slow sinking rate of organic matter (Hood et al., 2009) produced in the WAS and subsequent eastwardly advection of the organic particulates leads to DO depletion in central and eastern AS;

effective renewal of subsurface water along the western boundary of the AS, by advection from the south, from the Red Sea and from the Persian Gulf (Naqvi et al., 2006). The prolonged ventilation times 7–8 years for RSW and (2–3 years) for PGW in the EAS compared with the ventilation time of (3–4 years and 1–2 years, respectively, in WAS lead to more significant DO depletion in the EAS (Schmidt et al., 2020a)

physical processes such as mesoscale eddies prevailing in the WAS are expected to contribute towards effective oxygen renewal in the water column (McCreary et al., 2013; Schmidt et al., 2020b);

the cross-shelf transport of organic-rich sediments from the western continental shelf of India (Somayajulu et al., 1996; Sarma et al., 2020) and subsequent respiration, causing suboxia in the EAS.

the warming of the PGW. Recent observations have seen warming of the PGW (shallow semi-enclosed sea with an average depth of 35 m) at a rate of two to three times faster than the global average rate, increasing its buoyancy and may lead to poor, intermediate water ventilation (Lachkar et al., 2019).

In addition to the perennial OMZ (pOMZ), the AS experiences seasonal suboxia and denitrification over the inner and mid-shelf off the west coast of India during and shortly after the southwest monsoon (Sharma, 1978; Shetye et al., 1990; Naqvi et al., 2000, 2006). This coastal oxygen-deficient zone is separated from the pOMZ in the central AS by the presence of slightly more oxygenated waters of the WIUC that flow poleward (Naqvi et al., 2006). During the peak of upwelling season (in September), almost the entire Indian shelf (and some of the Pakistan shelf) is severely hypoxic having an O_2 concentration <0.5 ml l^{-1} covering an area of about 180 000 km^2 (Naqvi et al., 2000). The seasonal suboxia (even anoxia) reported in this region occurs because of coastal upwelling occurring along the western Indian shelf during June to November. Upwelling begins in April, along the southwest coast of India (along with Kerala), and gradually moves northward. The intense DO depletion in both the seasonal OMZ (sOMZ) and pOMZ enhances the deposition of organic matter in the underlying sediments as a result of a combination of high (upwelling-driven) primary productivity and inefficient degradation during sinking in the oxygen-depleted water column (Paropkari et al., 1992, 1993; Cowie, 2005). The intense DO depletion has a profound influence on the underlying sediments, with respect to the redox conditions, microbial community, nature and activity of benthic communities, pore water redox processes, and consequently on the diagenetic pathways (e.g. Cowie, 2005).

1.1.2. The Bay of Bengal Oxygen Minimum Zone

The BoB, in the northern Indian Ocean, is the largest bay in the world, bordered by India, Bangladesh, Sri Lanka, and Myanmar. The BoB receives a large amount of freshwater from several river systems, which drain into it. The rivers Ganga-Brahmaputra (G-B), Irrawaddy, Godavari, Mahanadi, Krishna, and Kaveri contribute 60% of the total freshwater received by the BoB. Riverine water flux of 2.95×10^{12} m^3 $year^{-1}$ (Sengupta et al., 2006) combined with an excess of precipitation over evaporation results in a salinity stratified water column (Prasanna Kumar et al., 2007). The BoB is one of the major global OMZs. Although the conditions appear to be conducive for denitrification/anammox to occur, there has been no indication of nitrogen loss via denitrification in the BoB. Anammox reaction involves oxidation of NH_4^+ to N_2 gas using NO_2^- as the electron acceptor under anoxic conditions. The relatively less intense OMZ in the BoB may be attributed to weaker upwelling along the east coast of India than in Oman and Somalia. McCreary et al. (2013) attributed lower respiration in the BoB to the absence of an advective flux of additional organic matter, as observed in the EAS. The aggregation of organic matter with mineral particles supplied by high river discharge in the BoB increases the sinking speed of the organic detritus, thereby decreasing its remineralization rate and impacting on the OMZ structure and intensity (Al Azhar et al., 2017). In contrast, Sarma et al. (2018) suggested the activities of cyclonic and anticyclonic eddies to be one of the possible reasons for excess DO in the BoB reported by Bristow et al. (2017). The cyclonic eddies bring up nutrient-rich water, causing enhanced productivity and subsequent DO drawdown as a result of respiration. In contrast, the anticyclonic eddies cause the downwelling of DO-rich surface water to deeper layers causing the weakening of the OMZ. In aAdition, Sarma and Bhaskar (2018) suggested westward advection of oxygenated waters within the 150– 300 mbsl depth zone driven by anticyclonic eddies originating in the Andaman waters. Recently, however, Bristow et al. (2017) observed extremely low levels of DO in the BoB and hypothesized that future increases in the deposition of nutrients from atmospheric sources combined with the varying intensity of the summer monsoon could enhance organic matter decomposition and O_2 consumption leading to anammox and nitrogen loss. The vertical distribution of O_2 and NO_2^- in the AS (Lam et al., 2011) and BoB (Bristow et al., 2017) is plotted in Figure 1.3.

1.2. PRESERVATION OF ORGANIC MATTER AND SEDIMENT BIOGEOCHEMISTRY

The preservation of organic carbon (OC) is efficient within the sediments underlying oxygen-depleted waters (Paropkari et al., 1993; Eglinton et al., 1994; Van der Weijden et al., 1999; Hartnett et al., 1998; Hartnett and Devol, 2003; Böning et al., 2004; Jessen et al., 2017, More et al., 2018) because of incomplete breakdown while sinking through the water column as well as diminished bioturbation activity. A more significant proportion of labile organic matter escapes degradation while sinking through hypoxic or oxygen minimum regions of the modern ocean than oxic zones (Van Mooy et al., 2002). The partly degraded (reactive) organic matter fuels microbe-mediated sediment biogeochemical processes during burial, which results in modification of the pore fluid composition and precipitation/dissolution

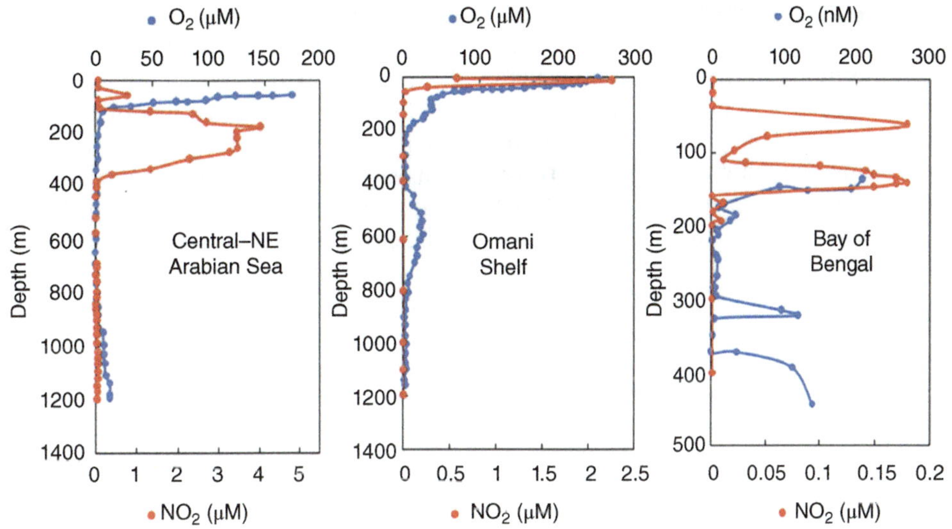

Figure 1.3 Vertical distribution of oxygen (solid blue circle) and nitrite concentrations (solid red circle) over the Central–northeast Arabian Sea, Omani Shelf and the Bay of Bengal. Redrawn from Lam et al. (2011) and Bristow et al. (2017).

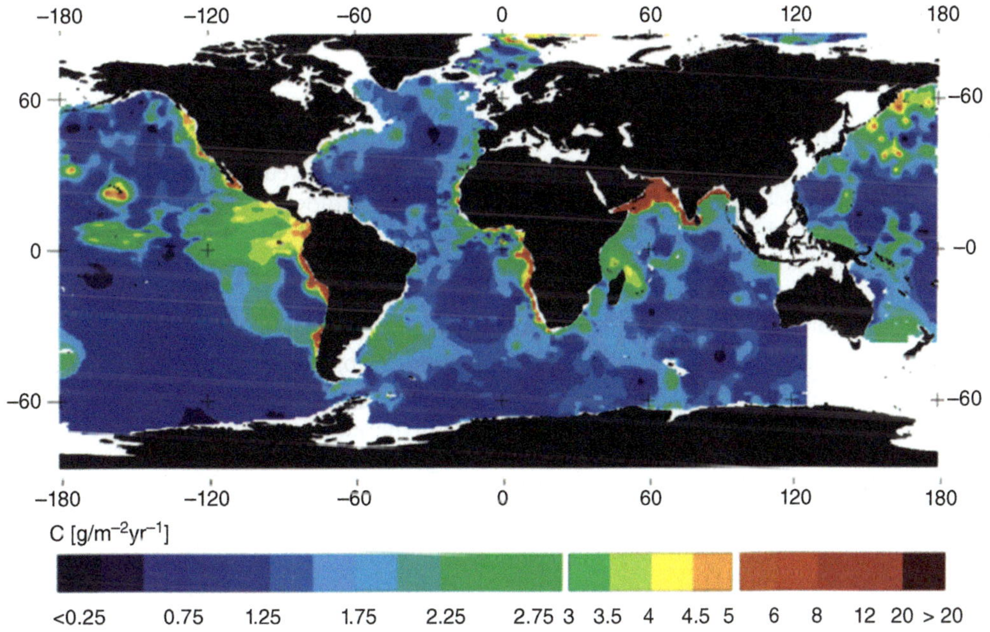

Figure 1.4 Global distribution pattern of particulate carbon flux to the seafloor. *Source:* Reimers (2007) with permission from American Chemical Society.

of inorganic mineral species (Madigan et al., 2000; Middelburg and Levin, 2009) Fernandes et al., 2018). The global distribution pattern of particulate organic carbon flux (Reimers, 2007) is depicted in Figure 1.4.

Remineralization of the partly degraded organic matter is strongly influenced by benthic biotic activity controlled by sediment–water interface conditions such as DO concentration, diffusion of CH_4/H_2S, sedimentation rate, organic matter content, and sediment grain size. The upper and lower boundaries of the ASOMZ have a lower TOC content than the OMZ core, which is attributed to lack of bioturbation and high organic matter flux from the DO-depleted water column in the latter (Fernandes et al., 2018). In contrast, the upper and lower edges of the OMZ are associated with a relatively high remineralization rate due to higher DO availability.

1.3. PORE FLUID GEOCHEMISTRY

The chemical compositions of sediment pore-waters are altered from their original seawater-like compositions by various microbially mediated biogeochemical reactions taking place at and below the sediment–water interface. The biogeochemical reactions are associated with the remineralization of organic matter. Thermodynamically, O_2 is the most favorable terminal electron acceptor for the remineralization of organic matter. When the aerobic respiration entirely consumes O_2, other electron acceptors (in order of decreasing energy gain) such as nitrate (NO_3^-), manganese oxides (Mn (IV)), iron oxides (Fe (III)), and sulfate (SO_4^{2-}) are used, depending on their concentration in solution and their thermodynamic efficiency (Froelich et al., 1979; Pattan et al., 2013; Turchyn and Schrag, 2006, Turchyn et al., 2016; Müller, 2018). Thus, there is a geochemical zonation in the sediments, with redox conditions ranging from oxic to methanic. The sediment layer where NO_3^-, Mn (IV) and Fe (III) reduction controls organic matter remineralization is termed the suboxic zone (Froelich et al., 1979). Although energetically more favorable than SO_4^{2-}, electron acceptors in the suboxic zone are usually limited in supply, which makes them biogeochemically less significant (Aller, 1994; Thamdrup et al., 1994; Jørgensen and Kasten, 2006; Seitaj et al., 2015). Microbial sulfate reduction is the major pathway of organic carbon oxidation in the anoxic zone as the high concentration of SO_4^{2-} in seawater makes it a dominant electron acceptor (Heinrich and Reeburgh, 1987; Widdel et al., 2007). The global estimate of marine sulfate reduction is 11.3 teramoles of sulfate per year, which accounts for the oxidation of 12-29% of organic carbon that reaches the sea floor (Bowles et al., 2014). The organoclastic sulfate reduction (OSR) (Froelich et al., 1979; Treude et al., 2005; Jørgensen and Kasten, 2006; Riedinger et al., 2017) and anaerobic oxidation of methane (AOM) (Froelich et al., 1979; Reeburgh, 1980; Valentine and Reeburgh, 2000; Boetius et al., 2000) are the two pathways of microbial sulfate reduction in marine sediments, represented by equations (1) and (2).

$$(CH_2O)_{106}(NH_3)_{16}(H_3PO_4) + 53SO_4^{2-} \rightarrow 106HCO_3^- + 53HS^- + 16NH_3 + H_3PO_4 + 53H^+ \quad (1)$$

$$CH_4 + SO_4^{2-} \rightarrow HCO_3^- + HS^- + H_2O \quad (2)$$

The AOM involves a syntrophic consortium of sulfate-reducing bacteria and methanotrophic archaea (Boetius et al., 2000; Knittel and Boetius, 2009). The AOM process results in the formation of SMTZ, which is a redox interface in the sediment where porewater SO_4^{2-} and CH_4 intersect and are depleted to near-zero levels (Treude et al., 2005; Sultan et al., 2016). The depth of SMTZ depends on the organic matter availability, oxygen exposure time (OET) of organic matter, flux of CH_4 and SO_4^{2-}, and the activity of sulfate-reducing bacteria and methanotrophic archaea and bioturbation (Niewöhner et al., 1998; Hong et al., 2014; Komada et al., 2016; Lin et al., 2016; Hu et al., 2017). The AOM process at the SMTZ leads to high pore water alkalinity owing to bicarbonate (HCO_3^-) and hydrogen sulfide (HS^-) production. The sulfate reduction rates in marine sediment pore waters are controlled by the availability of labile organic matter, SO_4^{2-} concentration, bottom water temperature (OET), bioturbation, bacterial distribution, and abundance (Aller, 1994; Kristensen, 2000; Hedges and Keil, 1995; Nierop et al., 2017; Jessen et al., 2017). Depth integrated sulfate reduction rates have been reported to be maximum around continental margins (Bowles et al., 2014), apparently linked to the higher deposition/preservation of organic carbon in these areas of the global ocean (Dean et al., 1994; Littke et al., 1997; Van der Weijden et al., 1999; Böning et al., 2004; Seiter et al., 2004, 2005; Rasiq et al., 2016; Dale et al., 2015; Fernandes et al., 2018).

Bottom water DO depletion and high preservation of reactive organic matter within the OMZ sediment led to considerably higher sulfate reduction rates. The higher sulfate reduction rate is associated with higher porewater NH_4^+ and dissolved inorganic carbon (DIC) concentrations and shallow SMTZ (Mazumdar et al., 2009, 2012; Fernandes et al., 2018; Fernandes et al., 2020).

Studies from the Oman margin sediments by Pedersen and Shimmield (1991) show depletion of porewater SO_4^{2-} concentration (<1 mM) within 50 to 100 meters below seafloor (mbsf). Furthermore, Passier et al. (1997) (from Oman margin), Schenau et al. (2002) (from Pakistan and Oman margins), and Van der Weijden et al. (1999) (Oman margin) reported low sulfate reduction in the sediments. Lower sulfate reduction in the Oman margin sediment is also supported by a very low concentration of reduced sulfur compounds (Fe-sulfide) relative to total organic carbon content (Emeis et al., 1991). The low sulfate reduction rate in the ASOMZ sediments may result from the incorporation of sulfur into the organic matter, which renders the organic matter refractory (Passier et al., 1997; Lückge et al., 1999; Schenau et al., 2002; Law et al., 2009). In contrast, Law et al. (2009) attributed low sulfate reduction rates (0–0.45 mmol m^{-2} d^{-1}) from the Pakistan margin sediments to the dominance of other metabolic pathways (Fe cycling) or lack of reactive organic matter. The low sulfate reduction rates in the ASOMZ sediments compared with other upwelling regions (e.g. Namibian and Peruvian margins) is attributed to the low availability of labile organic matter and supply of reactive Fe in the sediments (Lückge et al., 1999; Law et al., 2009; Naik

Table 1.1 Summarizes the calculated depth-integrated sedimentary sulfate reduction rates (SRRs) from different oxygen minimum zones (OMZs) regions of the global ocean.

Area	OMZ water depth (mbsl)	SRR (mmol cm^{-2} yr^{-1})	Reference
Eastern Arabian Sea OMZ	200–1200	0.0008–0.0113	Fernandes et al. (2018)
Pakistan margin	200–1200	0.00036–0.01642	Law et al. (2009)
Oman margin	200–1000	0.000035–0.00011	Pedersen and Shimmield (1991)
Off central Chile	70–450	0.3504–1.09135	Ferdelman et al., (1997), Thamdrup and Canfield (1996)
Off central Peru	50–650	0.3212–1.25195	Fossing (1990)
Off NW Mexico	150–800	0.011315–0.06862	Hartnett and Devol (2003)

et al., 2017). See Table 1.1 for depth-integrated sedimentary SRRs from different OMZ regions of the global Ocean.

1.4. SEDIMENTARY SULFIDIZATION AND SULFURIZATION

Hydrogen sulfide produced during sulfate reduction in sediments is diffused from the microbial cell into the surrounding pore water and is trapped in the sediment as Fe sulfide minerals (sulfidization) and organosulfur compounds (sulfurization). A significant fraction of H_2S may diffuse out of the sediment–water interface and be oxidized to SO_4^{2-} or elemental sulfur (S^0). Porewater H_2S is oxidized to intermediate species such as S^0, thiosulfate (polysulfides: S_x^{2-}, HS_x^{2-}) and sulfoxyanoins (sulfite, thiosulfate, and polythionates) (Chen and Morris, 1972; Berner and Westrich, 1985; Thamdrup et al., 1994; Canfield et al., 2005). Sulfide oxidation can be microbe-mediated (e.g. Nielsen et al., 2010; Eckert et al., 2011; Rao et al., 2016) or abiotic, by reaction with oxides such as MnO_2 or FeO(OH) (Berner and Westrich, 1985; Thamdrup et al., 1993; Yao and Millero, 1993, 1995, 1996; Canfield et al., 2005). The intermediate sulfur species produced from sulfide oxidation can undergo disproportionation reactions where the sulfur intermediates are simultaneously transformed to both H_2S and SO_4^{2-}; these reactions are often referred to as inorganic fermentation because they do not involve any other electron donor or acceptor (Bak and Cypionka, 1987, Canfield and Thamdrup, 1994; Habicht et al., 1998; Böttcher et al., 2001).The burial of Fe sulfide minerals and sulfurized organic matter depends on the availability of reactive Fe (Yucel et al., 2010; Zhu et al., 2016), sedimentation rate, reactive organic matter flux (Berner, 1985; Raiswell and Berner, 1985; Wilkin and Barnes, 1997; Werne et al., 2003; Markovic et al., 2015), and bottom water DO concentration. The degree of pyritization (DOP) values in sediments underlying the oxygenated waters are < 0.45, whereas, under dysoxic and euxinic waters the values range between 0.46 and 0.75 and > 0.75, respectively (Raiswell and Berner, 1985; Raiswell and Canfield, 2012). In the ASOMZ, pyritization is limited by the availability of reactive Fe oxides attributed to the reductive dissolution of Fe in the water column or at the sediment–water interface as a result of low DO concentration (Scheneau et al., 2002). The bottom water DO concentrations and detrital organic matter availability controls the spatial extent and the degree of bioturbation and bioirrigation. The activity of benthic organisms not only modifies the porewater concentration profiles it also plays an important role by oxidizing Fe-sulfide minerals close to the sediment–water interface. The sediments underlying OMZs show the minimum influence of bioturbation (Levin, 2003, Cowie and Levin, 2009). The absence of burrowing/irrigation would minimize the exposure of early diagenetic Fe-sulfides to oxidants such as NO_3^-/NO_2^- thereby enhancing the preservation of early sulfidization. The relative significance of organic matter sulfurization depends on the availability of labile organic molecules and the extent and rate of reactive Fe consumption (Zaback and Pratt, 1992; Passier et al., 1997; Werne et al., 2004). Loss of sedimentary reactive Fe through reductive dissolution and enhanced preservation of highly labile organic matter in the sediments underlying OMZs would promote enhanced organic matter sulfurization (Lückge et al., 2002; Schneau et al., 2002). The OBS constitutes an estimated 50% of the total sedimentary sulfur in the Peru margin following the polysulfide pathway (Mossmann et al., 1991). The incorporation of sulfur in organic molecules (sulfurization) occurs either intramolecularly (as cyclo sulfo groups such as thiolane, thiane, and thiophene) or intermolecularly through S^{2-} or S_x^{2-} bonds between larger molecules (Sinninghe Damsté and de Leeuw, 1990; Abdulla et al., 2020). Lallier-Verges et al. (1993) found pyrite infillings in pore spaces and pyrite framboids in sediment rich in autochthonous organic matter from the Oman margin. Furthermore, they suggested that 50% of organic matter degradation in organic poor sediments is supported by sulfate reduction and attributed it to the deposition of zooplankton debris (in addition to phytoplankton) with mineral skeletons, which increase porosity and enhance sulfate reduction. While in organic-rich sediments, only

20 % of the organic matter degradation is supported by sulfate reduction. This may be caused by the deposition of purely organic phytoplankton with low porosity.

1.5. BENTHIC BIOLOGY

Oxygen Minimum Zones, which impinge on sediments of continental margins, have a strong influence on the abundance, diversity, and composition of benthic fauna (Levin, 2003). Low bottom-water DO concentrations limit O_2 penetration into the sediment, which creates unfavorable conditions for most benthic organisms to thrive. The abundance and diversity of benthic macrofaunal communities within the OMZs are typically low, and only a few tolerant benthic species are known to survive in these regions (Levin and Gage, 1998). However, small organisms (microbes, metazoan meiofauna, and foraminifera) may have a higher density in OMZs because of the abundance of detrital food and the absence of predators (e.g. Levin et al., 1991; Neira et al., 2001, 2018).

If sulfide is present in the upper sedimentary layers of OMZs, it inhibits aerobic respiration and can also react with trace amounts of O_2 to form hydrogen peroxide (H_2O_2), which can cause cell damage (Bernhard and Bowser, 2008). Many benthic organisms living in OMZ environments develop morphological adaptations to cope with the O_2 limitation. Low O_2 availability in OMZs leads to the prevalence of hypoxia-tolerant fauna such as nematodes, polychaetes, and some calcareous foraminifera. Most calcifying invertebrates tend to disappear in suboxic environments. However, some exceptions, such as the snail *Alia permodesta* and the mussel *Amygdallum politum*, have thin shells in suboxic or anoxic sediments (Levin, 2003; Moffitt et al., 2015). Some of the other commonly observed adaptations in benthic fauna colonizing OMZs include an increase in gill surface area in some crustaceans and polychaetes (Levin, 2003), elongation of branchiae in polychaetes (Lamont and Gage, 2000; Levin et al., 2009, 2010), test flattening of some foraminiferal species, and reduced body size in metazoan macrofauna (Gooday et al., 2010).

The lower transition zones of OMZs, where O_2 concentrations begin to increase, are often regions of intense benthic biological activity characterized by a dramatic change in the population density of several benthic species (Levin, 2003; Gooday et al., 2009), an observation commonly termed as the 'edge effect' (Mullins et al., 1985; Levin, 2003). A zonation of faunal communities across O_2 gradients has been observed on OMZ-impinged continental margins, indicating the existence of substantial O_2 tolerance thresholds for different benthic biota. These gradients are formed as a result of the specific O_2 tolerances of each species and potentially also the absence of larger predators (Levin, 2003). Seasonal changes in oxygenation have also been observed to cause a compositional change in benthic communities (e.g. Sellanes and Neira, 2006; Woulds et al., 2007; Gutiérrez et al., 2008).

Bioturbation (particle mixing) in OMZ sediments is generally diminished owing to the reduction in species diversity and body size of benthic fauna. This usually leads to the formation of a generation of laminated or varved sediments underlying OMZs (e.g. Levin et al., 2009; Schimmelmann et al., 2016). However, some gutless symbiont taxa are capable of bioturbating OMZ sediments (Levin, 2003). Bioturbation generally leads to the churning of organic matter and particles deeper into the deposit, contributing to carbon burial and can fuel subseafloor microbial processes. Laminated anoxic sediments typically contribute less to nutrient recycling. Thus, more labile organic material remains unused (Levin and Gallo, 2019).

1.6. MICROBIAL METABOLISM IN THE MARINE OXYGEN MINIMUM ZONE WATER COLUMN AND SEDIMENT

Marine phototrophic and chemolithoautotrophic microorganisms produce massive quantities of organic compounds that, in conjunction with other inorganic compounds, go through intricate circuits of biogeochemical transformations involving a wide variety of consumers and decomposers. These chemoorganoheterotrophic microorganisms, in their turn, form and sequester an equally wide variety of dissolved and particulate matter in marine waters and sediments. These biogeochemical transformations and recycling networks are central to the productivity, biodiversity, ecological balance, and resourcefulness of the oceanic waters and sediments. However, major microbe-assisted functional metabolic processes identified so far in the water column of OMZ, include oxidative-reductive cycling of sulfur compounds, methanogenesis, and N transformations.

1.7. NITROGEN METABOLISM IN THE MARINE MARINE OXYGEN MINIMUM ZONE WATER COLUMN

Bilogical nitrogen fixation (BNF) is the process by which N_2 is converted to NH_3 and subsequently into biomass by N_2-fixing bacteria and archaea (Sohm et al., 2011). The biomass produced during BNF and assimilation may then release N compounds into the ocean. The N loss through denitrification and anammox processes are dominant in the water column of the OMZs and the sediment underlying the OMZ (Capone and Knapp, 2007). For the last few decades, a substantial increase in oceanography and omics-based data has dramatically improved our understanding of the marine N cycle, particularly in the OMZs. However, a contrasting influence of O_2 concentrations on anammox and denitrification may be observed. On one hand, some experimental data shows 200 and 900 nM concentrations of DO cause a 50% drop in N_2 production by the microbial community through denitrification and anammox metabolism, respectively (Dalsgaard et al., 2014). On the other hand, a separate study reports active denitrification and anammox metabolism in the presence of micromolar levels of DO in OMZ off Conceptión, Chile (Bristow et al., 2016).

Several literature reports show the selective prevalence of anammox group in the marine OMZ water column off northern Chile and ETSP ocean, compared with other types (Molina et al., 2005; Galán et al., 2009; Lam et al., 2009; Kalvelage et al., 2013), which may be attributed to release of NH_4^+ during decomposition of sinking organic matter in the water column and benthic NH_4^+ flux (Kalvelage et al., 2013).

Although some studies have revealed anammox (OMZs) off Namibia, Peru, and Chile (Thamdrup et al., 2006; Hamersley et al., 2007) as the predominant pathway of fixed N loss, data from the ASOMZ revealed dominance of denitrification metabolism over anammox activity in the water column (Ward et al., 2009; Bulow et al., 2010). Some recent studies have revealed the presence of *Candidatus* Scalindua clades 2 (predicted to have anammox activity) sequences from the ASOMZ (Woebken et al., 2008). The aerobic ammonia-oxidizing archaea (AOA) and anaerobic ammonia-oxidizing (anammox) bacteria show a unique depth distribution and niche segregation pattern (Pitcher et al., 2011) in the water column of ASOMZ, where the prevalence of AOA was found at a water depth of 170 m while anammox retained their highest abundance at a water depth of 450–750 m. However, archaeal and bacterial mediated ammonia oxidation also has been reported in OMZs of ETNP (Peng et al., 2015) (see Table 1.2 for detailed information). However, some recent reports also show nitrite oxidation in the OMZs of the ETNP and Namibian sea region (Füssel et al., 2012; Beman et al., 2013). The co-existence of nitrite oxidation and anammox activity has been observed in the OMZ of the Namibian sea region,; indicating competition between the microbial population with each other for a common substrate, namely NO_2^- (Füssel et al., 2012). Both *Nitrococcus* and *Nitrospina* were found to be the prevalent bacterial groups (9% of the total microbial community) in the OMZ of the Namibian sea region, whereas only the latter, was dominant for efficient nitrite oxidation in the OMZ of ETNP (Füssel et al., 2012; Beman et al., 2013). Moreover, functionally active microbial groups involved in OMZ water column nitrogen transformations identified to date through different omics-based, culture-based, and molecular approach includes Nitrosopumilus, Nitrosopelagicus, Nitrosospira, Nitrospina, Nitrococcus, Sagittula, *Candidatus* Scalindua, SAR11 group, SUP05/ARCTIC96BD-19, Planctomycetes, among others (Molina et al., 2005; Beman et al., 2013; Bristow et al., 2016; Bertagnolli and Stewart, 2018). Furthermore, high-throughput genetic and proteome-level investigation (functional marker gene analysis, metatranscriptome, metaproteome) also highlighted the presence and functionality of microbes capable of nitrogen compound metabolism by the identification of respective enzymes from the OMZ waters such as ammonia monooxygenase enzyme active subunit (*amoA*), nitrous oxide reductase (*nosZ*), nitrite reductase (*nirS*), nitric oxide reductase (*norB*), nitrate reductase (*narG*), hydrazine oxidoreductase (*hzo*), nitrite reductase (*nirK*), nitrite oxidase (*NXR*) (Molina et al., 2005; Lam et al., 2009; Dalsgaard et al., 2014).

1.8. MICROBIOLOGICAL PERSPECTIVE OF SULFUR METABOLISM IN THE MARINE OXYGEN MINIMUM ZONE WATER COLUMN

Pioneering work by Canfield et al. (2010) first brought to the fore the role of sulfur species in the anaerobic sulfur cycling of Chilean OMZs. However, before this study, detection and isolation of several sulfur oxidizing–reducing, bacterial members from geographically distinct OMZ (off the coast of Peru and Chile) water also gave enough hints about their probable ecological relevance (Stevens and Ulloa, 2008; Finster and Kjeldsen, 2010). In general, sulfide produced in anoxic organic-rich marine sediments via bacterial sulfate reduction is reoxidized within the sediment, resulting in minimal fluxes of the sulfide to the water column (Brüchert et al., 2003). Furthermore, retention sulfide in the water column also depends on biological or chemical oxidation: by precipitation as Fe sulfides and organic-bound sulfur (OBS) formation (Brüchert, 1998). Nonetheless, some of the literature showed the presence of specific sulfide oxidizingand nitrate respiring or storing microbes (*Thiomargarita* sp.; *Beggiatoa* spp.) at the juncture of the sediment–water interface, although their ecological relevance in OMZ water column is not yet clearly understood (Schulz et al., 1999). However, for the last few years, advanced molecular-omics and biogeochemical approaches have unravelled complex S-cycling in the OMZ water column. For eample, a maximum abundance of metagenomic reads (individual DNA fragment sequenced) affiliated to sulfide oxidizing (6.3 to 16.2%) and sulfate reducing (2.1 to 2.4%) taxa have been detected in the Chilean OMZ water column (Canfield et al., 2010). Moreover, another bacterial group, Marinimicrobia clade SHBH1141, was also abundant in anoxic and anoxic–sulfidic OMZ waters, and can oxidize sulfide to polysulfide

Table 1.2 Microbiological features of the well-studied oxygen minimum zones (OMZs) water-column of the global ocean.

Sl. no	Name of the OMZ	O_2 (μM)	*In situ* microbial processes detected so far (nitrogen metabolism)	Microbiology or molecular technique used for decrypting biogeochemical process	Taxonomic groups responsible for *in situ* metabolic process	References
1	OMZ off northern Chile	2–12	Anammox activity	16S rRNA gene and partial genes sequencing, CARD-FISH, ^{15}N labelling incubations	Planctomycetes, Scalindua spp.	Galán et al. (2009)
2	Off the coast of Concepcion, Chile (36°30′85 S, 73°07′75 W)	0.005–0	Ammonium and nitrite oxidation	16S rRNA gene and partial genes sequencing, ^{15}N labelling incubations	Nitrosopumilus, Nitrosopelagicus	Bristow et al. (2016)
3	Off the coast of Peru and northern Chile	<10	Anammox activity	16S rRNA gene and partial genes sequencing, ^{15}N labeling incubations, FISH	Anammox-related 16S ribosomal ribonucleic acid gene sequences	Hamersley et al. (2007)
4	Eastern Tropical South Pacific (ETSP)	<10	Anammox activity. dissimilatory reduction of nitrate to ammonia (DNRA)	Functional gene analysis, metagenome, ^{15}N labelling incubations	*Candidatus* Scalindua sp. T23	Lam et al. (2009), Ulloa et al. (2012)
5	Eastern South Pacific off northern Chile	<10	Aerobic ammonium oxidation	16S rRNA gene and partial genes sequencing, Functional gene analysis	*Nitrosospira*-like βAOB	Molina et al. (2007)
6	Eastern tropical North Pacific Ocean	<1	Ammonia and Nitrite oxidation, DNRA, Anammox activity	16S rRNA gene and partial genes sequencing, ^{15}N labeling incubations, Functional gene analysis	*Nitrospina*, archaeal and β-proteobacterial groups	Beman et al. (2013), Peng et al. (2015), Pajares et al. (2019)
7	Namibian sea region	<10	Nitrite oxidation	FISH, ^{15}N labelling incubations	*Nitrococcus*, *Nitrospina*	Füssel et al. (2012)
8	Arabian Sea OMZ	<2	Denitrification, ammonia-oxidation, anammox activity	16S rRNA gene and partial genes sequencing, N tracer incubations, Functional gene analysis	Ammonia-oxidizing archaea (AOA) and anaerobic ammonia-oxidizing (anammox) bacteria	Ward et al. (2009), Bulow et al. (2010), Pitcher et al. (2011)
Sulfur metabolism						
9	Northern Chilean coast	<13	Sulfide oxidation, Sulfate reduction	Metagenome, Functional gene analysis	SUP05 group, ARTIC96BD lineage of the gamma-proteobacteria, *Desulfatibacillum*, *Desulfobacterium*, *Desulfococcus*, *Syntrophobacter*, and *Desulfovibrio*	Canfield et al. (2010), Crowe et al. (2018)

10	Eastern Tropical South Pacific (ETSP)	<10	Sulfide oxidation	16S rRNA gene and partial genes sequencing, metagenomics and genome binning CARD-FISH, ^{15}N labeling incubations	SUP05, *Candidatus* Thioglobus autotrophicus	Schunck et al. (2013), Callbeck et al. (2018)
11	North eastern subarctic Pacific (NESAP)	<10	Sulfide oxidation	Metagenome, Functional gene analysis, genome binning	Marinimicrobia clades	Hawley et al. (2017)
Methane metabolism						
12	Eastern tropical North Pacific Ocean	<10	Anaerobic methane oxidation	16S rRNA gene and partial genes sequencing, Metagenome, Functional gene analysis, genome binning, $^{3}H\text{-}CH_4$ and $^{14}C\text{-}CH_4$ labeling experiment	NC10 bacterial clade, clade OPU3	Padilla et al. (2016, 2017), Chronopoulou et al. (2017), Thamdrup et al. (2019)

CARD-FISH, catalyzed reporter deposition–fluorescence *in situ* hybridization; βAOB, Betaproteobacteria ammonia–oxidizing archaea; DNRA, dissimilatory nitrate reduction to ammonia.

(polyS), that is ultimately stored and later regenerated to H_2S. It also has the genetic potential for nitrous oxide reduction to N_2 and is considered to have roles in both S and N cycles (Hawley et al., 2017; Bertagnolli and Stewart, 2018). However, as far as the sulfate-reducing bacterial community in OMZ water is concerned, significant proportions of metagenomic reads ascribable to the genera *Desulfatibacillum*, *Desulfobacterium*, *Desulfococcus*, *Syntrophobacter*, and *Desulfovibrio* species have been found (Canfield et al., 2010; Bertagnolli and Stewart, 2018) (see Table 1.2 for detailed information). Furthermore, a survey of functional marker genes related to sulfur metabolism identified in the Chilean OMZ water column includes gene clusters for dissimilatory sulfite reductase enzyme (*dsr*), the *sox* gene complex, and the adenosine 5-phosphosulfate (APS) reductase gene (*apr*) (Canfield et al., 2010). In the context of coupling between the S and N cycle, it is noteworthy that sulfate reduction may also contribute to the NH_4^+ requirements for anammox bacteria because there is ample evidence suggesting the inadequate liberation of NH_4^+ during the heterotrophic denitrification that is necessary to drive anammox activity in many OMZ waters (Thamdrup et al., 2006; Lam et al., 2009).

Another study, based on the Chilean OMZ water column, also revealed the existence of active sulfide oxidizing bacterial community (SUP05/ARTIC96BD lineage of the gammaproteobacteria) that have a high affinity for sulfide and have the ability to oxidize it at very low concentration (<100 nM) (see Table 1.2 for detailed information; Crowe et al., 2018). This phenomenon indicates that such an anaerobic sulfide oxidizing bacterial community (having high affinity for sulfide) is likely to maintain vanishingly low sulfide concentrations in OMZs water-column, thereby keeping the S cycling cryptic (rapid oxidation of sulfide into sulfate). Furthermore, another geomicrobiological exploration of ETSP region in sulfide-poor offshore OMZ waters off the coast of Peru revealed the abundance and activity of sulfide oxidizing–denitrifying bacteria (SUP05, *Candidatus* Thioglobus autotrophicus, belonging to SUP05 bacterial clade) having a role in driving the S cycling; via the continued oxidation of co-transported elemental sulfur (Schunck et al., 2013; Callbeck et al., 2018). Considering the present state of information, there was no apparent reason to presume that such sulfide-based cryptic S cycling would be operational in geographically distinct, nitrite and sulfide rich OMZ water columns with similar geomicrobiological features. In the light of the present topic, notably, a recent study based on two ~3 m long sediment cores (collected from 530 m and 580 m below sea level) located within the pOMZ of the EAS off the west coast of India revealed active tetrathionate-based biogeochemical S cycling in anoxic marine sediment horizon. It highlighted the role of microbial redox metabolisms in preempting the accumulation of this highly reactive polythionate in sediment pore-fluids, over and above the known abiotic mechanisms of tetrathionate scavenging by *in situ* sulfide (Mandal et al., 2020). Notably, in-depth genome data of SUP05 also revealed that this organism has the genetic potential for the utilization of thiosulfate. Thus, althoughcurrent knowledge highlights the role of sulfide, the role of other intermediate reactive redox species of S cannot be ruled out in the S cycle of the OMZ water column.

1.9. MICROBIOLOGY OF METHANE CYCLING IN THE OXYGEN MINIMUM ZONE WATER COLUMN

High-resolution water-column studies revealed the existence of a 300 m thick layer with elevated methane concentrations (20–105 nM) in the anoxic core of ETNP (Chronopoulou et al., 2017). Another geochemical investigation also revealed presence of CH_4 in the eastern ETNP water column (Pack et al., 2015) (for concentrations and its oxidation rates see Table 1.2). Several omics-based investigations revealed the presence of genes (particulate methane monooxygenase: *pMMO*) and transcripts (16S rRNA and other relevant functional genes) belonging to a unique group of denitrifying methanotrophs in the candidate bacterial division NC10 from Eastern Pacific OMZs (Padilla et al., 2016). This bacterial group has the genetic potential to perform nitric oxide dismutation along with oxygen production, thus holding significant importance in coupling the NO_2^- and CH_4 cycle. However, successive high-throughput genome binning experiments from the aforementioned ecosystem recovered near-complete (95%) draft genome representing another methanotroph clade OPU3 (having genomic potential for partial denitrification) that forms a maximum abundance of (4%) of the total microbial community sequenced (Padilla et al., 2017). Although metagenomic studies on the ASOMZ water column showed the existence of CH_4 cycling, the active functionality of this cycle needs further biogeochemical and microbiological substantiation (Lüke et al., 2016).

1.10. MICROBIAL METABOLISM IN MARINE OXYGEN MINIMUM ZONE SEDIMENTS

In general, OMZ sediments are characterized by two predominant microbial processes: sulfate reduction and methanogenesis. However, most of the well-characterized OMZ sediments exhibited distinct zonation patterns depending on these two metabolic processes. It has generally been observed that sulfate reducing bacteria and methanogenic archaea compete for the common substrates such as hydrogen and acetate. Owing to higher substrate affinity, sulfate reducing bacteria have the ability to outcompete methanogenic archaea as long as SO_4^{2-} ions are present in the sediment porewaters to act as electon acceptor (Treude et al., 2014; Maltby et al., 2016). As seawater is the main source of SO_4^{2-} in the marine sediment, a downward replenishing of the SO_4^{2-} concentration in the sediment-pore-water influences the sulfate reduction rate. However, microbial methanogenesis only starts when the SO_4^{2-} in the sediment is nearly or entirely exhausted. The overlapping zone between these two is generally known as a sulfate methane transition zone (SMTZ), where both the SO_4^{2-} and CH_4 coexist (Treude et al., 2014; Maltby et al., 2016). Irrespective of the depth at which they occur, SMTZs are commonly found in all anoxic sediments where SO_4^{2-} is transported from above and CH_4 is transported from the bottom. Thus, SMTZ is considered a hot spot for anaerobic oxidation of methane via the syntrophic relationship between sulfate reducers and methanotrophs (Knittel and Boetius, 2009).

Sulfate reduction is generally carried out by both bacteria and archaeal groups. They typically couple the oxidation of organic compounds or molecular hydrogen to the reduction of SO_4^{2-} to H_2S to obtain their energy. Sulfate reducing bacteria are also able to utilize a variety of low molecular mass organic compounds such as monocarboxylic and dicarboxylic aliphatic acids, alcohols, polar aromatic compounds, and hydrocarbons as electron donors (Rabus et al., 2006). In contrast, methanogenesis, is restricted only to archaea. They are obligate CH_4 producers: i.e. they obtain all of their required energy by producing CH_4 only. Methanogenic archaea can use a restricted number of substrates for CH_4 production, such as CO_2 and H_2 (for the hydrogenotrophic group) along with acetate (acetoclastic group) and methanol, methylamine, etc. (methylotrophic group) (Hedderich and Whitman, 2006). Substrates for both sulfate reduction and methanogenesis are formed as the end products of biodegradation and fermentation of organic macromolecules. While hydrogenotrophic and acetoclastic methanogenesis depends on common substrates with sulfate reduction, methylotrophic methanogenesis uses distinct substrates. Having a higher affinity towards their substrates, sulfate reducers always outcompete hydrogenotrophic and acetoclastic methanogens.

However, methylotrophic methanogens can escape this competition with sulfate reducers and operate methanogenesis metabolism simultaneously with sulfate reduction in the SO_4^{2-} containing sediment zone. However, the co-occurrence of sulfate reduction and methanogenesis is also possible and mostly observed in organic-rich sediments. Anaerobic methanotrophic archaea (with AOM potential) are also found in syntrophic association with sulfate reducing bacteria. So far as their taxonomic affiliations are concerned, methanotrophic archaea are formed three distinct clusters related to orders *Methanosarcinales* and *Methanomicrobiales* under the phylum Euryarchaeota (Knittel and Boetius 2009). Moreover, AOM enriches the sediment with HCO_3^- and HS^- and selectively increases the ^{12}C in the *in situ* DIC.

Notably, an abundance of sulfate reducing bacteria along with high aerial sulfate reduction rate, high porewater sulfide concentration, and shallowing of the SMTZ depth has been detected in the ASOMZ sediments located in the center of the vertical expanse of the OMZ (Fernandes et al., 2018). The AOM was also identified in the same sediment cores from the signature of significant ^{13}C depletion in the DIC content (Fernandes et al., 2018). However, another recent study based on seasonal oxygen minimum zones (sOMZ) and pOMZ across the western Indian shelf; revealed the abundance of methanogens/anaerobic-methane-oxidizers/sulfate-reducers/acetogens that heightened in the topmost sediment layer and then declined via synchronized fluctuations until the SMTZ was reached (Bhattacharya et al., 2021). Furthermore, another recent study based on the same ASOMZ sediment system also revealed the functionally active aerobic bacterial community that belongs to a diverse bacterial group (sulfur chemolithotrophs; methylotrophs etc.) (Bhattacharya et al., 2020) (see Table 1.3 for detailed information). The functional genomics and transcriptomics approach detected several genes responsible for aerobic respiration and oxidation of methane, ammonia, alcohol, thiosulfate,sulfite, and organosulfur compounds, thereby indicating the existence of diverse metabolic groups distinct from prototypical sulfate reducers or methanogens. Notably, in this context, another work by Mandal et al. (2020) also revealed the dynamics of functionally active tetrathionate metabolizing bacterial community along with decrypting the role of tetrathionate in the sedimentary S cycle of ASOMZ sediments. Conversely, although some other geochemical exploration based on OMZs of the Pakistan margin of the AS revealed sedimentary Mn, P, Fe, and S cycling, microbial involvement for those metabolism has still to be addressed (Law et al., 2009; Kraal et al., 2012).

Another study revealed the co-occurrence of methanogenesis and sulfate reduction in the sediment surface of the upwelling region off the Peruvian coast, which is considered to be one of the most intense OMZs (Maltby et al., 2016). This study indirectly indicated the use of non-competitive substrates for methanogenesis in the sulfate reducing zone and a higher availability of organic carbon in the sediments as the potential cause of this unusual phenomenon. Furthermore, active benthic N cycling in the Peruvian OMZ was also reported by Dale et al. (2016), who also elucidated the role of *Thioploca* community dynamics

Table 1.3 Microbiological features of the well-studied oxygen minimum zones (OMZs) sediment of the global ocean.

Sl. no	Name of OMZ	O_2 (μM)	*In situ* microbial processes detected so far (nitrogen metabolism)	Microbiology or molecular technique used for decrypting biogeochemical process	Taxonomic groups responsible for *in situ* metabolic process	References
1	Eastern Tropical SouthPacific (ETSP)	<detection limit	Dissimilatory reduction of nitrate to ammonia (DNRA)	Seafloor imaging	Thioploca	Sommer et al., 2016; Bohlen et al., 2011
Sulfur metabolism						
2	Arabian Sea OMZ	<detection limit	sulfate reduction	16S rRNA gene and partial genes sequencing	*Desulfurococcus, Desulfurolobus,*	Fernandes et al., 2018
3	Arabian Sea OMZ	<detection limit	Sulfur oxidation	Metagenome, Functional gene analysis, whole genome, pure culture isolation	*Cereibacter, Guyparkeria, Halomonas, Methylophaga, Pseudomonas, Sulfitobacter Thioploca*	Bhattacharya et al., 2020; Schmaljohann et al., 2001
Methane metabolism						
4	Arabian Sea OMZ	<detection limit	Anaerobic methane oxidation	16S rRNA gene and partial genes sequencing	–	Fernandes et al., 2018, Bhattacharya et al., 2021

as a NO_3^- reservoir and an effective barrier to H_2S emissions to the water column. Nevertheless, taxon-specific meta-omics exploration from physiographically distinct marine OMZ identified a diverse bacterial group that thus far includes Deltaproteobacteria (Syntrophobacterales and Desulfovibrionales), Chloroflexi, Planctomycetes, Spirochetes, Firmicutes, Acidobacteria, and Verrucomicrobia, etc.(Divya et al., 2010; Podlaska et al., 2012; Fernandes et al., 2018).

1.11. OXYGEN MINIMUM ZONE EXPANSION

Several studies have shown that oceanic OMZs are currently expanding as a result of global warming (e.g. Stramma et al., 2008; Schmidtko et al., 2017; Breitburg et al., 2018). Climate change caused by anthropogenic emissions of CO_2 leads to an increase in surface seawater temperature, thereby decreasing the solubility of O_2 in surface waters, leading to enhanced stratification (Bopp et al., 2002; Stramma et al., 2009) that could further prevent ventilation of the interior of the ocean.

Deutsch et al. (2011) predicted that an increase in surface ocean temperatures by 1°C could result in a threefold increase in seawater volume containing <5 μM O_2. Expansion of OMZs could have a significant impact on the biogeochemical cycles of major elements such as Fe, P, and N in the ocean (Deutsch et al., 2011), ultimately influencing the burial of organic carbon (Ruvalcaba Baroni et al., 2020)

1.12. CONCLUSION

Marine oxygen minimum zones (OMZs) are typical biogeochemical milieus that play crucial roles in regulating the global ocean's productivity and organic matter burial. The OMZ sediments are key sites for C–S–N cycling characterized by high labile organic matter preservation and diminished bioturbation activity, thereby exerting a strong influence on the benthic abundance, diversity, and composition biota, and gas and metal fluxes across the sediment–water interface. The present review has emphasized the significance of biogeochemical cycling in the OMZ sediments with respect to the pore fluid and solid phase geochemistry and microbial ecology, along with a brief account of the mechanism of formation of OMZ in the northern Indian Ocean. The OMZ in the northern Indian Ocean (AS and BoB) arises from high monsoon-driven primary productivity, high respiratory O_2 demand, and poor intermediate water ventilation owing to its landlocked geography. The AS in the northwestern Indian Ocean contains the thickest and most intense perennially DO depleted water mass in the world associated with SNM attributed to denitrification. In contrast, the OMZ in the BoB is less intense and thinner than the AS. The low DO concentration sufficient to support nitrite oxidation in the BoB inhibits denitrification to some extent.

Oxygen minimum zones host unique microbial communities that depend on different electron acceptors for metabolism, leading to fixed N loss and production of greenhouse gases. The high abundance and diversity of small benthic organisms (meiofauna and microbes) over macrofauna are observed in OMZs as a result of decreased predation and food competition. The prevalence of anammox over denitrification as the predominant N loss pathway is reported from the ETSP while the opposite is the case in the AS. Other than microbes capable of N compound metabolism, recent metagenomic and geochemical studies also identified sulfur oxidizing and reducing bacterial communities from the OMZ water column, suggesting its importance as a biochemical pathway in oxygen-depleted conditions. Microbial sulfate reduction and methanogenesis are important biogeochemical processes in anoxic marine sediments. The prevalence and activity of sulfate reducing/sulfide oxidizing bacterial communities and methanogens/methanotrophs are also reported from the OMZ sediments.

The DO depleted water mass globally is increasing due to global warming and anthropogenic input of nutrients leading to eutrophication. The expansion of OMZs would also affect the intensity of biogeochemical processes in the underlying sediments, influencing the benthic fluxes across the sediment–water interface. High sulfate reduction rates and shallowing of SMTZ can lead to seepage of H_2S into the water column, which has a deleterious effect on the benthic community. This effect has been reported from OMZs of Peru and Chile.

Perturbations in the homeostasis of the OMZs would have a momentous impact on the biogeochemical cycling of nutrients (particularly the C–N–S cycles) and the maintenance of ecological balance in the global ocean. Nonetheless, it is worthy of mention that in the future, new discoveries and research focusing on a deeper understanding of OMZ microbiology and geochemistry will unravel novel metabolic strategies or pathways in this critical ecosystem than have not been anticipated to date.

ACKNOWLEDGMENT

We acknowledge Director-NIO and funding from CSIR for the research program. We thank the CSIR-NIO publication committee for carrying out the internal review.

REFERENCES

Abdulla, H.A., Burdige, D.J. and Komada, T. (2020). Abiotic formation of dissolved organic sulfur in anoxic sediments of Santa Barbara Basin. *Organic Geochemistry, 139*, 103879. https://doi.org/10.1016/j.orggeochem.2019.05.009

Acharya, S.S. and Panigrahi, M.K. (2016). Eastward shift and maintenance of Arabian Sea oxygen minimum zone: Understanding the paradox. *Deep Sea Research Part I: Oceanographic Research Papers 115*: 240–252. https://doi.org/10.1016/j.dsr.2016.07.004

Agnihotri, R., Bhattacharya, S.K., Sarin, M.M. et al. (2003). Changes in surface productivity and subsurface denitrification during the Holocene: a multiproxy study from the eastern Arabian Sea. *The Holocene 13* (5): 701–71. https://doi.org/10.1191/0959683603hl656rp

Al Azhar, M., Lachkar, Z., Lévy, M. et al. (2017). Oxygen minimum zone contrasts between the Arabian Sea and the Bay of Bengal implied by differences in remineralization depth. *Geophysical Research Letters 44* (21): 11–106. https://doi.org/10.1002/2017GL075157

Aller, R.C. (1994). Bioturbation and remineralization of sedimentary organic matter: effects of redox oscillation. *Chemical Geology 114* (3–4): 331–345. https://doi.org/10.1016/0009-2541(94)90062-0

Anderson, T.A., Ryabchenko, V.A., Fasham, M.J.R. et al (2007). Denitrification in the Arabian Sea: a 3D ecosystem modelling study. *Deep-Sea Research I 54*: 2082–2119. https://doi.org/10.1016/j.dsr.2007.09.005

Bak, F. and Cypionka, H. (1987). A novel type of energy metabolism involving fermentation of inorganic sulphur compounds. *Nature 326*: 891–892. https://doi.org/10.1016/j.dsr.2007.09.005

Banerjee, P. and Kumar, S.P. (2014). Dust-induced episodic phytoplankton blooms in the Arabian Sea during winter monsoon. *Journal of Geophysical Research: Oceans 119* (10): 7123–7138. https://doi.org/10.1002/2014JC010304

Bange, H.W., Naqvi, S.W.A. and Codispoti, L.A. (2005). The nitrogen cycle in the Arabian Sea. *Progress in Oceanography 65*: 145–158. https://doi.org/10.1016/j.pocean.2005.03.002

Banse, K., Naqvi, S.W.A. and Narvekar, P. (2014). Oxygen minimum zone of the open Arabian Sea: variability of oxygen and nitrite from daily to decadal timescales. *Biogeosciences 11* (8): 2237–2261. https://doi.org/10.5194/bg-11-2237-2014

Banse, K., Naqvi, S.W.A. and Postel, J.R. (2017). A zona incognita surrounds the secondary nitrite maximum in open-ocean oxygen minimum zones. *Deep Sea Research Part I: Oceanographic Research Papers 127*: 111–113. https://doi.org/10.1016/j.dsr.2017.07.004

Behrenfeld, M.J., O'Malley, R.T., Siegel, D.A. et al. (2006). Climate-driven trends in contemporary ocean productivity. *Nature 444* (7120): 752–755. https://doi.org/10.1038/nature05317

Beman, J.M., Shih, J.L. and Popp, B.N. (2013). Nitrite oxidation in the upper water column and oxygen minimum zone of the eastern tropical North Pacific Ocean. *The ISME Journal 7* (11): 2192–2205. https://doi.org/10.1038/ismej.2013.96

Berner, R.A. (1985). Sulphate reduction, organic matter decomposition and pyrite formation. *Philosophical Transactions of the Royal Society of London. Series A, Mathematical and Physical Sciences 315* (1531): 25–38. https://doi.org/10.1098/rsta.1985.0027

Berner, R.A. and Westrich, J.T. (1985). Bioturbation and the early diagenesis of carbon and sulfur: *American Journal of Science 285*: 193–206. https://doi.org/10.2475/ajs.285.3.193

Bernhard, J.M. and Bowser, S.S. (2008). Peroxisome proliferation in foraminifera inhabiting the chemocline: an adaptation to reactive oxygen species exposure? 1. *Journal of Eukaryotic Microbiology 55* (3): 135–144. https://doi.org/10.1111/j.1550-7408.2008.0031

Bertagnolli, A.D. and Stewart, F.J. (2018). Microbial niches in marine oxygen minimum zones. *Nature Reviews Microbiology 16* (12): 723–729. https://doi.org/10.1038/s41579-018-0087-z

Bhattacharya, S., Mapder, T., Fernandes, S. et al. (2021). Sedimentation rate and organic matter dynamics shape microbiomes across a continental margin. Biogeosciences, 18, 5203–5222. https://doi.org/10.5194/bg-18-5203-2021

Bhattacharya, S., Roy, C., Mandal, S. et al. (2020) Aerobic microbial communities in the sediments of a marine 910 oxygen minimum zone. *FEMS Microbiolology Letters 367*: fnaa157, 2020. doi: 10.1093/femsle/fnaa157

Boetius, A., Ravenschlag, K., Schubert, C.J. et al. (2000). A marine microbial consortium apparently mediating anaerobic oxidation of methane. *Nature 407* (6804): 623–626. https://doi.org/10.1038/35036572

Bohlen, L., Dale, A.W., Sommer, S. et al. (2011). Benthic nitrogen cycling traversing the Peruvian oxygen minimum zone. *Geochimica et Cosmochimica Acta 75* (20): 6094–6111. https://doi.org/10.1016/j.gca.2011.08.010

Böning, P., Brumsack, H.J., Böttcher, M.E. et al. (2004). Geochemistry of Peruvian near-surface sediments. *Geochimica et Cosmochimica Acta 68* (21): 4429–4451. https://doi.org/10.1016/j.gca.2004.04.027

Bopp, L., Le Quéré, C., Heimann, M. et al. (2002). Climate-induced oceanic oxygen fluxes: Implications for the contemporary carbon budget. *Global Biogeochemical Cycles 16* (2): 6–1. https://doi.org/10.1029/2001GB001445

Böttcher, M.E., Thamdrup, B. and Vennemann, T.W. (2001). Oxygen and sulfur isotope fractionation during anaerobic bacterial disproportionation of elemental sulfur. *Geochimica et Cosmochimica Acta 65*: 1601–1609. https://doi.org/10.1016/S0016-7037(00)00628-1

Bower, A.S., Johns, W.E., Fratantoni, D.M. et al. (2005). Equilibration and circulation of Red Sea outflow water in the Western Gulf of Aden. *Journal of Physical Oceanography 35* (11): 1963–1985. https://doi.org/10.1175/JPO2787.1

Bowles, M.W., Mogollón, J.M., Kasten, S. et al. (2014). Global rates of marine sulfate reduction and implications for sub-sea-floor metabolic activities. *Science 344* (6186): 889–891. https://doi.org/10.1126/science.1249213

Boyer, T.P., Antonov, J.I., Baranova, O.K. et al. (2013). *World Ocean Database 2013*. Silver Spring: NOAA.

Breitburg, D., f, L.A., Oschlies, A., Grégoire, M. et al. (2018). Declining oxygen in the global ocean and coastal waters. *Science, 359* (6371): eaam7240. https://doi.org/10.1126/science. aam7240

Bristow, L.A., Callbeck, C.M., Larsen, M. et al. (2017). N_2 production rates limited by nitrite availability in the Bay of Bengal oxygen minimum zone. *Nature Geoscience 10* (1): 24–29. https://doi.org/10.1038/NGEO2847

Bristow, L.A., Dalsgaard, T., Tiano, L. et al. (2016). Ammonium and nitrite oxidation at nanomolar oxygen concentrations in oxygen minimum zone waters. *Proceedings of the National Academy of Sciences 113* (38): 10601–10606. https://doi.org/10.1073/pnas.1600359113

Brüchert, V. (1998). Early diagenesis of sulfur in estuarine sediments: the role of sedimentary humic and fulvic acids. *Geochimica et Cosmochimica Acta 62* (9): 1567–1586. https://doi.org/10.1016/S0016-7037(98)00089-1

Brüchert, V., Jørgensen, B.B., Neumann, K. et al. (2003). Regulation of bacterial sulfate reduction and hydrogen sulfide fluxes in the central Namibian coastal upwelling zone. *Geochimica et Cosmochimica Acta 67* (23): 4505–4518. https://doi.org/10.1016/S0016-7037(98)00089-1

Bulow, S.E., Rich, J.J., Naik, H.S. et al. (2010). Denitrification exceeds anammox as a nitrogen loss pathway in the Arabian Sea oxygen minimum zone. *Deep Sea Research Part I: Oceanographic Research Papers 57* (3): 384–393. https://doi.org/10.1016/j.dsr.2009.10.014

Callbeck, C.M., Lavik, G., Ferdelman, T.G. et al. (2018). Oxygen minimum zone cryptic sulfur cycling sustained by offshore transport of key sulfur oxidizing bacteria. *Nature Communications 9* (1): 1–11. https://doi.org/10.1038/s41467-018-04041-x

Canfield, D.E. and Thamdrup, B. (1994). The production of ^{34}S-depleted sulfide during bacterial disproportionation of elemental sulfur. *Science 266*: 1973–1975. https://doi.org/10.1126/science.11540246

Canfield, D.E., Kristensen, E. and Thamdrup, B. (2005). The sulfur cycle. *Advances in Marine Biology 48*: 313–381. https://doi.org/10.1016/S0065-2881(05)48009-8

Canfield, D.E., Stewart, F.J., Thamdrup, B. et al. (2010). A cryptic sulfur cycle in oxygen-minimum zone waters off the Chilean coast. *Science 330* (6009): 1375–1378. https://doi.org/10.1126/science.1196889

Capone, D.G. and Knapp, A.N. (2007). Oceanography: a marine nitrogen cycle fix? *Nature 445* (7124): 159. https://doi.org/10.1038/445159a

Chen, K.Y. and Morris, J.C. (1972). Kinetics of oxidation of aqueous sulfide by O_2. *Environmental Science and Technology 6* (6): 529–537. https://doi.org/10.1021/es60065a008

Chronopoulou, P.M., Shelley, F., Pritchard, W.J. et al. (2017). Origin and fate of methane in the Eastern Tropical North Pacific oxygen minimum zone. *The ISME Journal 11* (6): 1386–1399. https://doi.org/10.1038/ismej.2017.6

Cowie, G. (2005). The biogeochemistry of Arabian Sea surficial sediments: a review of recent studies. *Progress in Oceanography 65* (2 –4): 260 –289. https://doi.org/10.1016/j.pocean.2005.03.003

Cowie, G.L. and Levin, L.A. (2009). Benthic biological and biogeochemical patterns and processes across an oxygen minimum zone (Pakistan margin, NE Arabian Sea). *Deep Sea Research Part II: Topical Studies in Oceanography 56* (6–7): 261–270. https://doi.org/10.1016/j.dsr2.2008.10.001

Crowe, S.A., Cox, R.P., Jones, C. et al. (2018). Decrypting the sulfur cycle in oceanic oxygen minimum zones. *The ISME Journal 12* (9): 2322–2329. https://doi.org/10.1038/s41396-018-0149-2

Dale, A.W., Sommer, S., Lomnitz, U. et al. (2015). Organic carbon production, mineralisation and preservation on the Peruvian margin. *Biogeosciences 12* (5): 1537–1559. https://doi.org/10.5194/bg-12-1537-2015

Dale, A.W., Sommer, S., Lomnitz, U. et al. (2016). Biological nitrate transport in sediments on the Peruvian margin mitigates benthic sulfide emissions and drives pelagic N loss during stagnation events. *Deep Sea Research Part I: Oceanographic Research Papers 112*: 123–136. https://doi.org/10.1016/j.dsr.2016.02.013

Dalsgaard, T., Stewart, F.J., Thamdrup, B., et al. and (2014). Oxygen at nanomolar levels reversibly suppresses process rates and gene expression in anammox and denitrification in the oxygen minimum zone off northern Chile. *Marine Biology 5* (6): https://doi.org/10.1128/mBio.01966-14

Dean, W.E., Gardner, J.V. and Anderson, R.Y. (1994). Geochemical evidence for enhanced preservation of organic matter in the oxygen minimum zone of the continental margin of northern California during the late Pleistocene. *Paleoceanography 9* (1): 47–61. https://doi.org/10.1029/93PA02829

Deutsch, C., Brix, H., Ito, T., Frenzel, H. et al. (2011). Climate-forced variability of ocean hypoxia. *Science 333* (6040): 336–339. https://doi.org/10.1126/science.1202422

Divya, B., Soumya, K.V. and Nair, S. (2010). 16SrRNA and enzymatic diversity of culturable bacteria from the sediments of oxygen minimum zone in the Arabian Sea. *Antonie van Leeuwenhoek 98* (1): 9–18. https://doi.org/10.1007/s10482-010-9423-7

Eckert, T., Brunner, B., Edwards, E.A. et al. (2011). Microbially mediated reoxidation of sulfide during dissimilatory sulfate reduction by *Desulfobacter latus*. *Geochimica et Cosmochimica Acta 75*: 3469–3485. https://doi.org/10.1016/j.gca.2011.03.034

Eglinton, T.I., Irvine, J.E., Vairavamurthy, A. et al. (1994). Formation and diagenesis of macromolecular organic sulfur in Peru margin sediments. *Organic Geochemistry 22*: 781–799. https://doi.org/10.1016/0146-6380(94)90139-2

Emeis, K.C., Morse, J.W. and Mays, L.L. (1991). Organic carbon, reduced sulfur and iron in Miocene to Holocene upwelling sediments from the Oman and Benguela upwelling systems. In: *Proceedings of the Ocean Drilling Program, Scientific Results 117*: 517–527. College Station, TX: Ocean Drilling Program https://doi.org/10.2973/odp. proc.sr.117.155.1991

Ferdelman, T.G., Lee, C., Pantoja, S. et al. (1997). Sulfate reduction and methanogenesis in a *Thioploca*-dominated sediment off the coast of Chile. *Geochimica et Cosmochimica Acta 61* (15): 3065–3079. https://doi.org/10.1016/S0016-7037(97)00158-0

Fernandes, S., Mazumdar, A., Bhattacharya, S. et al. (2018). Enhanced carbon-sulfur cycling in the sediments of Arabian Sea oxygen minimum zone center. *Scientific Reports 8* (1): 1–15. https://doi.org/10.1038/s41598-018-27002-2

Fernandes, S., Mazumdar, A., Peketi, A. et al. (2020). Sulfidization processes in seasonally hypoxic shelf sediments: a study off the West coast of India. *Marine and Petroleum Geology 117*: 104353. https://doi.org/10.1016/j.marpetgeo.2020.104353

Finster, K.W. and Kjeldsen, K.U. (2010). *Desulfovibrio oceani* subsp. *oceani* sp. nov., subsp. nov. and *Desulfovibrio oceani* subsp. *galateae* subsp. nov., novel sulfate-reducing bacteria isolated from the oxygen minimum zone off the coast of Peru. *Antonie van Leeuwenhoek 97* (3): 221–229. https://doi.org/10.1007/s10482-009-9403-y

Fossing, H. (1990). Sulfate reduction in shelf sediments in the upwelling region off Central Peru. *Continental Shelf Research 10* (4): 355–67. https://doi.org/10.1016/0278-4343(90)90056-R

Froelich, P.N., Klinkhammer, G.P., Bender, M.L. andet al. (1979). Early oxidation of organic matter in pelagic sediments of the eastern equatorial Atlantic: suboxic diagenesis. *Geochimica et Cosmochimica Acta 43*: 1075–1090. https://doi.org/10.1016/0016-7037(79)90095-4

Füssel, J., Lam, P., Lavik, G. andet al. (2012). Nitrite oxidation in the Namibian oxygen minimum zone. *The ISME Journal 6* (6): 1200–1209. https://doi.org/10.1038/ismej.2011.178

Galán, A., Molina, V., Thamdrup, B. et al. (2009). Anammox bacteria and the anaerobic oxidation of ammonium in the oxygen minimum zone off northern Chile. *Deep Sea Research Part II: Topical Studies in Oceanography 56* (16): 1021–1031. https://doi.org/10.1016/j.dsr2.2008.09.016

Garçon, V., Karstensen, J., Palacz, A. et al. (2019). Multidisciplinary observing in the world ocean's oxygen minimum zone regions: from climate to fish – the VOICE Initiative. *Frontiers in Marine Science 6*:722. https://doi.org/10.3389/fmars.2019.00722

Gooday, A.J., Bett, B.J., Escobar, E. et al. (2010). Habitat heterogeneity and its influence on benthic biodiversity in oxygen minimum zones. *Marine Ecology 31* (1): 125–147. https://doi.org/10.1111/j.1439-0485.2009.0034

Gooday, A.J., Levin, L.A., da Silva, A.A. et al. (2009). Faunal responses to oxygen gradients on the Pakistan margin: a comparison of foraminiferans, macrofauna and megafauna. *Deep Sea Research Part II: Topical Studies in Oceanography 56* (6–7): 488–502. https://doi.org/10.1016/j.dsr2.2008.10.003

Gutiérrez, D., Enríquez, E., Purca, S. et al. (2008). Oxygenation episodes on the continental shelf of central Peru: remote forcing and benthic ecosystem response. *Progress in Oceanography 79* (2–4): 177–189. https://doi.org/10.1016/j.pocean.2008.10.025

Habicht, K., Canfield, D.E. and Rethmeier, J. (1998). Sulfur isotope fractionation during bacterial reduction and disproportionation of thiosulfate and sulfite. *Geochimica et Cosmochimica Acta 62*: 2585–2595. https://doi.org/10.1016/S0016-7037(98)00167-7

Hamersley, M.R., Lavik, G., Woebken, D. et al. (2007). Anaerobic ammonium oxidation in the Peruvian oxygen minimum zone. *Limnology and Oceanography 52* (3): 923–933. https://doi.org/10.1016/S0016-7037(98)00167-7

Hartnett, H.E. and Devol, A.H. (2003). The role of a strong oxygen deficient zone in the preservation and degradation of organic matter: a carbon budget for the continental margins of NW Mexico and Washington. *Geochimica et Cosmochimica Acta 67* (2): 247–264. https://doi.org/10.1016/S0016-7037(02)01076-1

Hartnett, H.E., Keil, R.G., Hedges, J.I. et al. (1998). Influence of oxygen exposure time on organic carbon preservation in continental margin sediments. *Nature 391*: 2–4. https://doi.org/10.1038/35351

Hawley, A.K., Nobu, M.K., Wright, J.J. et al. (2017). Diverse Marinimicrobia bacteria may mediate coupled biogeochemical cycles along eco-thermodynamic gradients. *Nature Communications 8* (1): 1–10. https://doi.org/10.1038/s41467-017-01376-9

Hedderich, R. and Whitman, W.B. (2006). Physiology and biochemistry of the methane-producing Archaea. *The Prokaryotes 2*: 1050–1079. https://doi.org/10.1007/0-387-30742-7_34

Hedges, J.I. and Keil, R.G. (1995). Sedimentary organic matter preservation: an assessment and speculative synthesis. *Marine Chemistry 49* (2–3): 81–115.https://doi.org/10.1016/0304-4203(95)00008-F

Heinrich, S.M. and Reeburgh, W.S. (1987). Anaerobic mineralization of marine sediment organic matter: rates and the role of anaerobic processes in the oceanic carbon economy. *Geomicrobiology Journal 5*: 191–237. https://doi.org/10.1080/01490458709385971

Helly, J. and Levin, L.A. (2004). Global distribution of naturally occurring marine hypoxia on continental margins. *Deep Sea Research I 51*: 1159–1168. https://doi.org/10.1016/j.dsr.2004.03.009

Hong, W.L., Torres, M.E., Kim, J.H. et al. (2014). Towards quantifying the reaction network around the sulfate–methane-transition-zone in the Ulleung Basin, East Sea, with a kinetic modeling approach. *Geochimica et Cosmochimica Acta 140*: 127–141. https://doi.org/10.1016/j.gca.2014.05.032

Hood, R.R., Wiggert, J.D. and Naqvi, S.W.A. (2009). Indian Ocean research: opportunities and challenges. *Geophysical Monograph Series 185*: 409–428. https://doi.org/10.1029/2007GM000714

Hu, C.Y., Yang, T.F., Burr, G.S. et al. (2017). Biogeochemical cycles at the sulfate-methane transition zone (SMTZ) and geochemical characteristics of the pore fluids offshore southwestern Taiwan. *Journal of Asian Earth Sciences 149*: 172–183. https://doi.org/10.1016/j.jseaes.2017.07.002

Jessen, G.L., Lichtschlag, A., Ramette, A. et al. (2017). Hypoxia causes preservation of labile organic matter and changes seafloor microbial community composition (Black Sea). *Science Advances 3* (2): e1601897. https://doi.org/10.1126/sciadv.1601897

Jørgensen, B.B. and Kasten, S. (2006). Sulfur cycling and methane oxidation. In: *Marine Geochemistry*, 2e (eds H.D. Schulz and M. Zabel), 271–309. Berlin: Springer.

Kalvelage, T., Lavik, G., Lam, P. et al. (2013). Nitrogen cycling driven by organic matter export in the South Pacific oxygen minimum zone. *Nature Geosciences*, *6* (3), 228–234. https://doi.org/10.1038/ngeo1739

Knittel, K. and Boetius, A. (2009). Anaerobic oxidation of methane: progress with an unknown process. *Annual Review of Microbiology 63*: 311–334. https://doi.org/10.1146/annurev. micro.61.080706.093130

Komada, T., Burdige, D.J., Li, H.L. et al. (2016). Organic matter cycling across the sulfate–methane transition zone of the Santa Barbara Basin, California Borderland. *Geochimica et Cosmochimica Acta 176*: 259–278. https://doi.org/10.1016/j.gca.2015.12.022

Kraal, P., Slomp, C.P., Reed, D.C. et al. (2012). Sedimentary phosphorus and iron cycling in and below the oxygen minimum zone of the northern Arabian Sea. *Biogeosciences 9* (7): 2603–2624. https://doi.org/10.1016/j.gca.2015.12.022

Kristensen, E. (2000). Organic matter diagenesis at the oxic/anoxic interface in coastal marine sediments, with emphasis on the role of burrowing animals. In: *Life at Interfaces and Under Extreme Conditions, Developments in Hydrobiology*, Vol. *151* (eds. G. Liebezeit, S. Dittmann and I. Kröncke), 1–4. Dordrecht: Springer. https://doi.org/10.1007/978-94-011-4148-2_1

Kumar, S.P., Roshin, R.P., Narvekar, J. et al. (2009). Response of the Arabian Sea to global warming and associated regional climate shift. *Marine Environment Research 68* (5): 217–222. https://doi.org/10.1016/j.marenvres.2009.06.010

Lachkar, Z., Lévy, M. and Smith, K.S. (2019). Strong intensification of the Arabian Sea oxygen minimum zone in response to Arabian Gulf warming. *Geophysical Research Letters 46* (10): 5420–5429. https://doi.org/10.1029/2018GL081631

Lallier-Verges, E., Bertrand, P. and Desprairies, A. (1993). Organic matter composition and sulfate reduction intensity in Oman Margin sediments. *Marine Geology 112* (1–4), 57–69. https://doi.org/10.1016/0025-3227 (93)90161-N

Lam, P., Lavik, G., Jensen, M.M. et al. (2009). Revising the nitrogen cycle in the Peruvian oxygen minimum zone. *Proceedings of the National Academy of Sciences 106* (12): 4752–4757. https://doi.org/10.1073/pnas.0812444106

Lam, P., Jensen, M.M., Kock, A. et al. (2011). Origin and fate of the secondary nitrite maximum in the Arabian Sea. *Biogeosciences 8* (6): 1565–1577. https://doi.org/10.5194/bg-8-1565-2011

Lamont, P.A. and Gage, J.D. (2000). Morphological responses of macrobenthic polychaetes to low oxygen on the Oman continental slope, NW Arabian Sea. *Deep Sea Research Part II: Topical Studies in Oceanography 47* (1–2): 9–24. https://doi.org/10.1016/S0967-0645 (99)00102-2

Law, G.T., Shimmield, T.M., Shimmield, G.B. et al. (2009). Manganese, iron and sulphur cycling on the Pakistan margin. *Deep Sea Research Part II: Topical Studies in Oceanography 56* (6–7): 305–323. https://doi.org/10.1016/S0967-0645(99)00102-2

Levin, L.A. (2003). Oxygen minimum zone benthos: adaptation and community response to hypoxia. *Oceanography and Marine Biology: an Annual Review 41*: 1–45.

Levin, L.A. and Gage, J.D. (1998). Relationships between oxygen, organic matter and the diversity of bathyal macrofauna. *Deep Sea Research Part II: Topical Studies in Oceanography 45* (1–3): 129–163. https://doi.org/10.1016/S0967-0645 (97)00085-4

Levin, L.A. and Gallo, N.D. (2019). The significance of ocean deoxygenation for continental margin benthic and demersal biota. In:, *Ocean Deoxygenation: Everyone's Problem. Causes, Impacts, Consequences and Solutions* (eds. D. Laffoley and J.M. Baxter), 341–361). IUCN.

Levin, L.A., Huggett, C.L. and Wishner, K.F. (1991). Control of deep-sea benthic community structure by oxygen and organic-matter gradients in the eastern Pacific Ocean. *Journal of Marine Research, 49* (4), 763–800. https://doi.org/10.1357/002224091784995756

Levin, L.A., Ekau, W., Gooday, A.J., et al. (2009). Effects of natural and human-induced hypoxia on coastal benthos. *Biogeosciences 6* (10): 2063–2098. https://doi.org/10.5194/bg-6-2063-2009 and Levin, L.A., Mendoza, G.F., Gonzalez, J.P., andet al. (2010). Diversity of bathyal macrofauna on the northeastern Pacific margin: the influence of methane seeps and oxygen minimum zones. *Marine Ecology* 31 (1): 94–110. https://doi.org/10.1111/j.1439-0485.2009.0033

Lin, Q., Wang, J., Algeo, T.J. et al. (2016). Enhanced framboidal pyrite formation related to anaerobic oxidation of methane in the sulfate-methane transition zone of the northern South China Sea. *Marine Geology 379*: 100–108. https://doi.org/10.1016/j.margeo.2016.05.016

Littke, R., Lückge, A. and Welte, D.H. (1997). Quantification of organic matter degradation by microbial sulphate reduction for Quaternary sediments from the northern Arabian Sea. *Naturwissenschaften 84* (7): 312–315. https://doi.org/10.1007/s001140050402

Lückge, A., Ercegovac, M., Strauss, H. et al. (1999). Early diagenetic alteration of organic matter by sulfate reduction in Quaternary sediments from the northeastern Arabian Sea. *Marine Geology 158* (1–4): 1–13. https://doi.org/10.1016/S0025-3227 (98)00191-1

Lückge, A., Horsfield, B., Littke, R. et al. (2002). Organic matter preservation and sulfur uptake in sediments from the continental margin off Pakistan. *Organic Geochemistry 33* (4). 477 488. https://doi.org/10.1016/S0146-6380 (01) 00171-1

Lüke, C., Speth, D.R., Kox, M.A. et al. (2016). Metagenomic analysis of nitrogen and methane cycling in the Arabian Sea oxygen minimum zone. *PeerJ 4*: p.e1924. https://doi.org/10.7717/peerj.1924

Madhupratap, M., Prasanna Kumar S., Bhattathiri P.M.A. et al. (1996). Mechanism of the biological response to winter cooling in the northeastern Arabian Sea. *Nature 384*: 549–552. https://doi.org/10.1038/384549a0

Madigan, M.T., Martinko J.M. and Parker, J. (2000). *Brock Biology of Microorganisms*. Hoboken, NJ: Prentice Hall, Inc.

Maltby, J., Sommer, S., Dale, A.W. et al. (2016). Microbial methanogenesis in the sulfate-reducing zone of surface sediments traversing the Peruvian margin. *Biogeosciences 13* (1): 283–299. https://doi.org/10.5194/bg-13-283-2016

Mandal, S., Bhattacharya, S., Roy, C. et al. (2020) Cryptic roles of tetrathionate in the sulfur cycle of marine sediments: microbial drivers and indicators, *Biogeosciences 17*: 4611–4631, https://doi.org/10.5194/bg-17-4611-2020, 2020

Markovic, S., Paytan, A. and Wortmann, U.G. (2015). Pleistocene sediment offloading and the global sulfur cycle. *Biogeosciences, 12* (10), 3043–3060. https://doi.org/10.5194/bg-12-3043-2015

Mazumdar, A., Peketi, A., Dewangan, P. et al. (2009). Shallow gas charged sediments off the Indian west coast: Genesis and distribution. *Marine Geology 267* (1–2): 71–85. https://doi.org/10.1016/j.margeo.2009.09.005

Mazumdar, A., Peketi, A., Joao, H. et al. (2012). Sulfidization in a shallow coastal depositional setting: diagenetic and palaeoclimatic implications. *Chemical Geology 322*: 68–78. https://doi.org/10.1016/j.chemgeo.2012.06.005

McCreary, Jr, J.P., Yu, Z., Hood, R.R. andet al. (2013). Dynamics of the Indian-Ocean oxygen minimum zones. *Progress in Oceanography 112*: 15–37. https://doi.org/10.1016/j.pocean.2013.03.002

Middelburg, J.J. and Levin, L.A. (2009). Coastal hypoxia and sediment biogeochemistry. *Biogeosciences 6* (7): 1273–1293. https://doi.org/10.5194/bg-6-1273-2009

Moffitt, S.E., Hill, T.M., Roopnarine, P.D. et al. (2015). Response of seafloor ecosystems to abrupt global climate change. *Proceedings of the National Academy of Sciences 112* (15): 4684–4689. https://doi.org/10.1073/pnas.1417130112

Molina, V., Farías, L., Eissler, Y. et al. (2005). Ammonium cycling under a strong oxygen gradient associated with the Oxygen Minimum Zone off northern chile (~ 23 S). *Marine Ecology Progress Series 288*: 35–43. https://doi.org/10.3354/meps288035

Molina, V., Ulloa, O., Farías, L. et al. and (2007). Ammonia-oxidizing β-proteobacteria from the oxygen minimum zone off northern Chile. *Applied and Environmental Microbiology 73* (11): 3547–3555. https://doi.org/10.1128/AEM.02275-06

More, K.D., Orsi, W.D., Galy, V. et al and. (2018). A 43 kyr record of protist communities and their response to oxygen minimum zone variability in the Northeastern Arabian Sea. *Earth and Planetary Science Letters 496*: 248–256. https://doi.org/10.1016/j.epsl.2018.05.045

Mossmann, J.R., Aplin, A.C., Curtis, C.D. et al. (1991). Geochemistry of inorganic and organic sulphur in organic-rich sediments from the Peru Margin. *Geochimica et Cosmochimica Acta 55* (12): 3581–3595. https://doi.org/10.1016/0016-7037(91)90057-C

Müller, F.L. (2018). Exploring the potential role of terrestrially derived humic substances in the marine biogeochemistry of iron. *Frontiers in Earth Science 6*: 159. https://doi.org/10.3389/feart.2018.00159

Mullins, H.T., Thompson, J.B., McDougall, K. et al. (1985). Oxygen-minimum zone edge effects: evidence from the central California coastal upwelling system. *Geology 13* (7): 491–494. https://doi.org/10.1130/0091-7613 (1985)13<491:OZEEEF>2.O;2

Naik, R., Naqvi, S.W.A. and Araujo, J. (2017). Anaerobic carbon mineralisation through sulphate reduction in the inner shelf sediments of eastern Arabian Sea. *Estuaries and Coasts 40* (1): 134–144. https://doi.org/10.1007/s12237-016-0130-0

Naqvi, S.W.A. (1991). Geographical extent of denitrification in the Arabian Sea in relation to physical processes. *Oceanologica Acta 14*: 281–290.

Naqvi, S.W.A., Jayakumar, D.A., Narvekar, P.V. et al. (2000). Increased marine production of N_2O due to intensifying anoxia on the Indian continental shelf. *Nature 408*: 346–349. https://doi.org/10.1038/35042551

Naqvi, S.W.A., Naik, H., Pratihary, A. et al. (2006). Coastal versus open-ocean denitrification in the Arabian Sea, *Biogeosciences 3*: 621–633. https://doi.org/10.5194/bg-3-621-200

Naqvi, S.W.A., Naik, H., Jayakumar, A. et al. (2009). Seasonal anoxia over the western Indian continental shelf. *Indian Ocean Biogeochemical Processes and Ecological Variability 185*: 333–345. https://doi.org/10.1029/2008GM000745

Naqvi, S.W.A., Bange, H.W., Farias, L. et al. (2010a). Marine hypoxia/anoxia as a source of CH_4 and N_2O. *Biogeosciences 7*: 2159–219. https://doi.org/10.5194/bg-7-2159-2010

Naqvi, S.W.A., Moffett, J.W., Gauns, M.U. et al. (2010b). The Arabian Sea as a high-nutrient, low-chlorophyll region during the late Southwest Monsoon. *Biogeosciences 7* (7): 2091–2100. https://doi.org/10.5194/bg-7-2091-2010

and Neira, C., Sellanes, J., Levin, L.A. et al. (2001). Meiofaunal distributions on the Peru margin: relationship to oxygen and organic matter availability. *Deep Sea Research Part I: Oceanographic Research Papers 48* (11): 2453–2472. https://doi.org/10.1016/S0967-0637 (01)00018-8

Neira, C., Ingels, J., Mendoza, G., et al. (2018). Distribution of meiofauna in bathyal sediments influenced by the oxygen minimum zone off Costa Rica. *Frontiers in Marine Science 5*: 448. https://doi.org/10.3389/fmars.2018.00448

Nielsen, L.P., Risgaard-Petersen, N., Fossing, H. et al. (2010). Electric currents couple spatially separated biogeochemical processes in marine sediment. *Nature 463*: 1071–1074. https://doi.org/10.1038/nature08790

Nierop, K.G., Reichart, G.J., Veld, H. et al. (2017). The influence of oxygen exposure time on the composition of macromolecular organic matter as revealed by surface sediments on the Murray Ridge (Arabian Sea). *Geochimica et Cosmochimica Acta 206*: 40–56. https://doi.org/10.1016/j.gca.2017.02.032

Niewöhner, C., Hensen, C., Kasten, S. et al. (1998). Deep sulfate reduction completely mediated by anaerobic methane oxidation in sediments of the upwelling area off Namibia. *Geochimica et cosmochimica Acta 62* (3): 455–464. https://doi.org/10.1016/S0016-7037 (98)00055-6

Pack, M.A., Heintz, M.B., Reeburgh, W.S., et al. (2015). Methane oxidation in the eastern tropical North Pacific Ocean water column. *Journal of Geophysical Research: Biogeosciences 120* (6): 1078–1092. https://doi.org/10.1002/2014JG002900

Padilla, C.C., Bristow, L.A., Sarode, N. et al. (2016). NC10 bacteria in marine oxygen minimum zones. *The ISME Journa, 10* (8): 2067–2071. https://doi.org/10.1038/ismej.2015.262

Padilla, C.C., Bertagnolli, A.D., Bristow, L.A. et al. (2017). Metagenomic binning recovers a transcriptionally active Gammaproteobacterium linking methanotrophy to partial denitrification in an anoxic oxygen minimum zone. *Frontiers in Marine Science 4*: 23. https://doi.org/10.3389/fmars.2017.00023

Pajares, S., Soto-Jiménez, M.F. and Merino-Ibarra, M. (2019). Molecular and isotopic evidence of the distribution of nitrogen-cycling microbial communities in the oxygen minimum zone of the Tropical Mexican Pacific. *FEMS Microbiology Ecology 95* (10): p.fiz143. https://doi.org/10.1093/femsec/fiz143

Paropkari, A.L., Babu, C.P. and Mascarenhas, A. (1992). A critical evaluation of depositional parameters controlling the variability of organic carbon in Arabian Sea sediments. *Marine Geology 107*: 213–226. https://doi.org/10.1016/0025-3227 (92)90168-H

Paropkari, A.L., Babu, C.P. and Mascarenhas, A. (1993). New evidence for enhanced preservation of organic-carbon in contact with oxygen minimum zone on the western continental-slope of India. *Marine Geology 111* (1–2): 7–13. https://doi.org/10.1016/0025-3227 (93)90185-X

Passier, H.F., Luther, III, G.W. and De Lange, G.J. (1997). Early diagenesis and sulphur speciation in sediments of the Oman Margin, northwestern Arabian Sea. *Deep Sea Research Part II: Topical Studies in Oceanography 44* (6–7): 1361–1380. https://doi.org/10.1016/S0967-0645 (97)00014-3

Pattan, J.N., Mir, I.A., Parthiban, G. et al. (2013). Coupling between suboxic condition in sediments of the western Bay of Bengal and southwest monsoon intensification: a geochemical study. *Chemical Geology 343*: 55–66. https://doi.org/10.1016/j.chemgeo.2013.02.011

Paulmier, A. and Ruiz-Pino, D. (2009). Oxygen minimum zones (OMZs) in the modern ocean. *Progress in Oceanography 80*: 113–128. https://doi.org/10.1016/j.pocean.2008.08.001

Pedersen, T.F. and Shimmield, G.B. (1991). Interstitial water chemistry, Leg 117: contrasts with the Peru Margin. In: *Proceedings of the Ocean Drilling Program, Scientific Results*, Vol. 117, (W.L. Prell, N. Niitsuma, et al.), 499–513). https://doi.org/10.2973/odp. proc. ir.117.1989

Peng, X., Fuchsman, C.A., Jayakumar, A., andet al. (2015). Ammonia and nitrite oxidation in the Eastern Tropical North Pacific. *Global Biogeochemical Cycles 29* (12): 2034–2049. https://doi.org/10.1002/2015GB005278

Pitcher, A., Villanueva, L., Hopmans, E.C., et al. (2011). Niche segregation of ammonia-oxidizing archaea and anammox bacteria in the Arabian Sea oxygen minimum zone. *The ISME Journal 5* (12): 1896–1904. https://doi.org/10.1038/ismej.2011.60

Podlaska, A., Wakeham, S.G., Fanning, K.A. et al. (2012). Microbial community structure and productivity in the oxygen minimum zone of the eastern tropical North Pacific. *Deep Sea Research Part I: Oceanographic Research Papers 66*: 77–89. https://doi.org/10.1016/j.dsr.2012.04.002

Prakash, S., Balakrishnan Nair, T.M., Udaya Bhaskar, T.et al. (2012). Oxycline variability in the central Arabian Sea: an Argo-oxygen study. *Journal of Sea Research 71*: 1–8. https://doi.org/10.1016/j.seares.2012.03.003

Prasad, T.G., Ikeda, M. and Kumar, S.P. (2001). Seasonal spreading of the Persian Gulf Water mass in the Arabian Sea. *Journal of Geophysical Research: Oceans, 106* (C8), 17059–17071. https://doi.org/10.1029/2000JC000480

Prasanna Kumar, S., Ramaiah, N., Gauns, M. et al. (2001). Physical forcing of biological productivity in the Northern Arabian Sea during the Northeast Monsoon. *Deep-Sea Research II 48*: 1115–1126. https://doi.org/10.1016/S0967-0645 (00)00133-8

Prasanna Kumar, S, Nuncio, M., Ramaiah, N. et al. (2007). Eddy-mediated biological productivity in the Bay of Bengal during fall and spring intermonsoons. *Deep Sea Research*

Part I: Oceanographic Research Papers 54 (9): 1619–1640. https://doi.org/10.1016/j.dsr.2007.06.002

Rabus, R., Hansen, T.A. and Widdel, F. (2006) Dissimilatory sulfate- and sulfur-reducing prokaryotes. In: *The Prokaryotes* (eds. M. Dworkin, S. Falkow, E. Rosenberg, t et al.), 659–768. New York, NY: Springer. https://doi.org/10.1007/0-387-30742-7_22

Raiswell, R. and Berner, R.A. (1985). Pyrite formation in euxinic and semi-euxinic sediments. *American Journal of Science 285* (8): 710–724. https://doi.org/10.2475/ajs.285.8.710

Raiswell, R. and Canfield, D.E. (2012). The iron biogeochemical cycle past and present. *Geochemical Perspectives 1* (1): 1–2. https://doi.org/10.7185/geochempersp.1.1

Rao, A.M.F., Malkin, S.Y., Hidalgo-Martinez, S. et al. (2016). The impact of electrogenic sulfide oxidation on elemental cycling and solute fluxes in coastal sediment. *Geochimica et Cosmochimica Acta 172*: 265–286. https://doi.org/10.1016/j.gca.2015.09.014

Rasiq, K.T., Kurian, S., Karapurkar, S.G. et al. (2016). Sedimentary pigments and nature of organic matter within the oxygen minimum zone (OMZ) of the Eastern Arabian Sea (Indian margin). *Estuarine, Coastal and Shelf Science 176*: 91–101. https://doi.org/10.1016/j.gca.2015.09.014

Reeburgh, W.S. (1980). Anaerobic methane oxidation: rate depth distributions in Skan Bay sediments. *Earth and Planetary Science Letters 47* (3): 345–352. https://doi.org/10.1016/0012-821X (80)90021-7

Regaudie-de-Gioux, A. and Duarte, C.M. (2012). Temperature dependence of planktonic metabolism in the ocean. *Global Biogeochemical Cycles 26* (1). https://doi.org/10.1029/2010GB003907

Reimers, C.E. (2007). Applications of microelectrodes to problems in chemical oceanography. *Chemical Reviews 107* (2): 590–600. https://doi.org/10.1021/cr050363n

Revsbech, N.P., Larsen, L.H., Gundersen, J. et al. (2009). Determination of ultra-low oxygen concentrations in oxygen minimum zones by the STOX sensor, *Limnology and Oceanography Methods 7*: 371–381. https://doi.org/10.4319/lom.2009.7.371

Riedinger, N., Brunner, B., Krastel, S., et al. (2017). Sulfur cycling in an iron oxide-dominated, dynamic marine depositional system: the Argentine continental margin. *Frontiers in Earth Science*, *5*, 33. https://doi.org/10.3389/feart.2017.00033

Rixen, T., Cowie, G., Gaye, B., et al. (2020). Present past and future of the OMZ in the northern Indian Ocean. *Biogeoscience Discussions 10*. https://doi.org/10.5194/bg-2020-82

Ruvalcaba Baroni, I., Palastanga, V. and Slomp, C.P. (2020). Enhanced organic carbon burial in sediments of oxygen minimum zones upon ocean deoxygenation. *Frontiers in Marine Science*, *6*, 839. https://doi.org/10.3389/fmars.2019.00839

Sarma, V.V.S.S. and UdayaBhaskar, T.V.S. (2018). Ventilation of oxygen to oxygen minimum zone due to anticyclonic eddies in the Bay of Bengal. *Journal of Geophysical Research: Biogeosciences 123* (7): 2145–2153. https://doi.org/10.1029/2018JG004447

Sarma, V., Jagadeesan, L., Dalabehera, H. et al. (2018). Role of eddies on intensity of oxygen minimum zone in the Bay of Bengal. *Continental Shelf Research 168*: 48–53. https://doi.org/10.1016/j.csr.2018.09.008

Sarma, V.V.S.S., Bhaskar, T.U., Kumar, J.P. et al. (2020). Potential mechanisms responsible for occurrence of core oxygen minimum zone in the north-eastern Arabian Sea. *Deep Sea Research Part I: Oceanographic Research Papers 165*: 103393. https://doi.org/10.1016/j.dsr.2020.103393

Schenau, S.J., Passier, H.F., Reichart, G.J. et al. (2002). Sedimentary pyrite formation in the Arabian Sea. *Marine Geology 185* (3–4): 393–402. https://doi.org/10.1016/S0025-3227 (02)00183-4

Schimmelmann, A., Lange, C.B., Schieber, J. et al. (2016). Varves in marine sediments: a review. *Earth-Science Reviews 159*: 215–246. https://doi.org/10.1016/j.earscirev.2016.04.009

Schmaljohann, R., Drews, M., Walter, S. et al. (2001). Oxygen-minimum zone sediments in the northeastern Arabian Sea off Pakistan: a habitat for the bacterium *Thioploca*. *Marine Ecology Progress Series 211*: 27–42. https://doi.org/10.3354/meps211027

Schmidt, H., Czeschel, R. and Visbeck, M. (2020a). Seasonal variability of the circulation in the Arabian Sea at intermediate depth and its link to the Oxygen Minimum Zone. *Ocean Science Discussions 1–31*. https://doi.org/10.5194/os-2020-9.

Schmidt, H., Czeschel, R. and Visbeck, M. (2020b). Seasonal variability of the Arabian Sea intermediate circulation and its impact on seasonal changes of the upper oxygen minimum zone. *Ocean Science 16* (6): 1459–1474. https://doi.org/10.5194/os-16-1459-2020

Schmidtko, S., Stramma, L. and Visbeck, M. (2017). Decline in global oceanic oxygen content during the past five decades. *Nature 542* (7641): 335–339. https://doi.org/10.1038/nature21399

Schott, F.A. and McCreary, Jr, J.P. (2001). The monsoon circulation of the Indian Ocean. *Progress in Oceanography 51* (1): 1–123. https://doi.org/10.1016/S0079-6611 (01)00083-0

Schulz, H.N., Brinkhoff, T., Ferdelman, T.G. et al. (1999). Dense populations of a giant sulfur bacterium in Namibian shelf sediments. *Science 284* (5413): 493–495. https://doi.org/10.1126/science.284.5413.493

Schunck, H., Lavik, G., Desai, D.K. et al. (2013). Giant hydrogen sulfide plume in the oxygen minimum zone off Peru supports chemolithoautotrophy. *PloS One 8* (8): p.e68661. https://doi.org/10.1371/journal. pone.0068661

Seitaj, D., Schauer, R., Sulu-Gambari, F. et al. (2015). Cable bacteria generate a firewall against euxinia in seasonally hypoxic basins. *Proceedings of the National Academy of Sciences 112*(43): 13278–13283. https://doi.org/10.1073/pnas.1510152112

Seiter, K., Hensen, C., Schröter, J. et al. (2004). Organic carbon content in surface sediments – defining regional provinces. *Deep-Sea Research Part I 51*: 2001 2026. https://doi.org/10.1016/j.dsr.2004.06.014

Seiter, K., Hensen, C. and Zabel, M. (2005). Benthic carbon mineralization on a global scale. *Global Biogeochemical Cycles 19*: 1–26. https://doi.org/10.1029/2004GB002225

Sellanes, J. and Neira, C. (2006). ENSO as a natural experiment to understand environmental control of meiofaunal community structure. *Marine Ecology 27* (1): 31–43. https://doi.org/10.1111/j.1439-0485.2005.0006

Sengupta, D., Bharath Raj, G.N. and Shenoi, S.S.C. (2006). Surface freshwater from Bay of Bengal runoff and Indonesian throughflow in the tropical Indian Ocean. *Geophysical Research Letters 33* (22). https://doi.org/10.1029/2006GL027573

Shankar, D., Shenoi, S.S.C., Nayak, R.K., et al. (2005). Hydrography of the eastern Arabian Sea during summer monsoon 2002. *Journal of Earth System Science 114* (5): 459–474. https://doi.org/10.1029/2006GL027573

Sharma, G.S. (1978). Upwelling off the southwest coast of India. *Indian Journal of Marine Sciences 7* (4): 209–218.

Shenoi, S.S.C., Shankar, D., Michael, G.S., et al. (2005). Hydrography and water masses in the southeastern Arabian Sea during March–June 2003. *Journal of Earth System Science 114* (5): 475–491. https://doi.org/10.1007/BF02702024

Shenoy, D.M., Suresh, I., Uskaikar, H., et al. (2020). Variability of dissolved oxygen in the Arabian Sea Oxygen Minimum Zone and its driving mechanisms. *Journal of Marine Systems 204*: 103310. https://doi.org/10.1016/j.jmarsys.2020.103310

Shetye, S.R., Gouveia, A.D., Shenoi, S.S.C. et al. (1990). Hydrography and circulation off the west coast of India during the southwest monsoon 1987. *Journal of Marine Research 48* (2): 359–378. https://doi.org/10.1357/002224090784988809

Sinninghe Damsté, J.S. and De Leeuw, J.W. (1990). Organic sulfur compounds and other biomarkers as indicators of palaeosalinity. In: *Geochemistry of Sulfur in Fossil Fuels. ACS Symposium Series*, Vol. *429*, 417–443. Washington DC: American Chemical Society. https://doi.org/10.1021/bk-1990-042h024

Sohm, J.A., Webb, E.A. and Capone, D.G. (2011). Emerging patterns of marine nitrogen fixation. *Nature Reviews Microbiology 9* (7): 499–508. https://doi.org/10.1038/nrmicro2594

Somayajulu, B.L.K., Sarin, M.M. and Ramesh, R. (1996). Denitrification in the eastern Arabian Sea: evaluation of the role of continental margins using Ra isotopes. *Deep Sea Research Part II: Topical Studies in Oceanography 43* (1): 111–117. https://doi.org/10.1016/0967-0645 (95)00079-8

Sommer, S., Gier, J., Treude, T. et al. (2016). Depletion of oxygen, nitrate and nitrite in the Peruvian oxygen minimum zone cause an imbalance of benthic nitrogen fluxes. *Deep Sea Research Part I: Oceanographic Research Papers 112*: 113–122. https://doi.org/10.1016/j.dsr.2016.03.001

Stevens, H. and Ulloa, O. (2008). Bacterial diversity in the oxygen minimum zone of the eastern tropical South Pacific. *Environmental Microbiology 10* (5): 1244–1259. https://doi.org/10.1111/j.1462-2920.2007.0153

Stramma, L., Fischer, J. and Schott, F. (1996). The flow field off southwest India at 8N during the southwest monsoon of August 1993. *Journal of Marine Research 54* (1): 55–72. https://doi.org/10.1357/0022240963213448

Stramma, L., Johnson, G.C., Sprintall, J. et al. (2008). Expanding oxygen-minimum zones in the tropical oceans. *Science 320*: 655–658. https://doi.org/10.1126/science.1153847

Stramma, L., Visbeck, M., Brandt, P., andet al. (2009). Deoxygenation in the oxygen minimum zone of the eastern tropical North Atlantic. *Geophysical Research Letters, 36* (20): L20607. https://doi.org/10.1029/2009GL039593

Sultan, N., Garziglia, S. and Ruffine, L. (2016). New insights into the transport processes controlling the sulfate-methane-transition-zone near methane vents. *Scientific Reports, 6* (1), 1–9. https://doi.org/10.1038/srep26701

Thamdrup, B.O. and Canfield, D.E. (1996). Pathways of carbon oxidation in continental margin sediments off central Chile. *Limnology and oceanography, 41* (8), 1629–1650. https://doi.org/10.4319/lo.1996.41.8.1629

and Thamdrup, B., Finster, K., Hansen, W. et al. (1993). Bacterial disproportionation of elemental sulfur coupled to chemical reduction of iron and manganese. *Applied and Environmental Microbiology 59* (1): 101–108. https://doi.org/10.1128/AEM.59.1.101-108.1993

Thamdrup, B., Fossing, H. and Jorgensen B.B. (1994). Manganese, iron and sulfur cycling in a coastal marine sediment, Aarhus Bay, Denmark. *Geochimica et Cosmochimica Acta 58*: 5115–5129. https://doi.org/10.1016/0016-7037 (94)90298-4

Thamdrup, B., Dalsgaard, T., Jensen, M.M. et al. (2006). Anaerobic ammonium oxidation in the oxygen-deficient waters off northern Chile. *Limnology and Oceanography, 51* (5), 2145–2156. https://doi.org/10.4319/lo.2006.51.5.2145

Thamdrup, B., Steinsdóttir, H.G., Bertagnolli, A.D., et al. (2019). Anaerobic methane oxidation is an important sink for methane in the ocean's largest oxygen minimum zone. *Limnology and Oceanography 64* (6): 2569–2585. https://doi.org/10.1002/lno.11235

Treude, T., Niggemann, J., Kallmeyer, J., et al. (2005). Anaerobic oxidation of methane and sulfate reduction along the Chilean continental margin. *Geochimica et Cosmochimica Acta 69* (11): 2767–2779. https://doi.org/10.1016/j.gca.2005.01.002

Treude, T., Krause, S., Maltby, J. et al. (2014). Sulfate reduction and methane oxidation activity below the sulfate-methane transition zone in Alaskan Beaufort Sea continental margin sediments: implications for deep sulfur cycling. *Geochimica et Cosmochimica Acta 144*: 217–237. https://doi.org/10.1016/j.gca.2014.08.018

Turchyn, A.V. and Schrag, D.P. (2006). Cenozoic evolution of the sulfur cycle: insight from oxygen isotopes in marine sulfate. *Earth and Planetary Science Letters 241* (3–4): 763–779. https://doi.org/10.1016/j.epsl.2005.11.007

Turchyn, A.V., Antler, G., Byrne, D. et al. (2016). Microbial sulfur metabolism evidenced from pore fluid isotope geochemistry at Site U1385. *Global and Planetary Change 141*: 82–90. https://doi.org/10.1016/j.gloplacha.2016.03.004

Ulloa, O., Canfield, D.E., DeLong, E.F. et al. (2012). Microbial oceanography of anoxic oxygen minimum zones. *Proceedings of the National Academy of Sciences, 109* (40): 15996–16003. https://doi.org/10.1073/pnas.1205009109

Valentine, D.L. and Reeburgh, W.S. (2000). New perspectives on anaerobic methane oxidation: minireview. *Environmental Microbiology 2* (5): 477–484. https://doi.org/10.1046/j.1462-2920.2000.0013

Van der Weijden, C.H., Reichart, G.J. and Visser, H.J. (1999). Enhanced preservation of organic matter in sediments deposited within the oxygen minimum zone in the northeastern Arabian Sea. *Deep Sea Research Part I. Oceanography Research Papers 46*: 807–830. https://doi.org/10.1016/S0967-0637 (98)00093-4

Van Mooy, B.A.S., Keil, R.G. and Devol, A.H. (2002). Impact of suboxia on sinking particulate organic carbon: enhanced carbon flux and preferential degradation of amino acids via denitrification. *Geochimica et Cosmochimica Acta 66*: 457–465. https://doi.org/10.1016/S0016-7037 (01)00787-6

Ward, B.B., Devol, A.H., Rich, J.J. et al. (2009). Denitrification as the dominant nitrogen loss process in the Arabian Sea. *Nature 461* (7260): 78–81. https://doi.org/10.1038/nature08276

Werne, J.P., Lyons, T.W., Hollander, D.J. et al. (2003). Reduced sulfur in euxinic sediments of the Cariaco Basin: sulfur isotope constraints on organic sulfur formation. *Chemical Geology 195* (1–4): 159–179. https://doi.org/10.1016/S0009-2541 (02)00393-5

Werne, J.P., Hollander, D.J., Lyons, T.W. et al. (2004). Organic sulfur biogeochemistry: recent advances and future research directions. *Geological Society of America Special Papers 379*: 135–150.

Widdel, F., Musat, F., Knittel, K. and Galushko, A. (2007). Anaerobic degradation of hydrocarbons with sulphate as electron acceptor. In: *Sulphate-Reducing Bacteria: Environmental and Engineered Systems* (eds. L.L. Barton and W.A. Hamilton). Cambridge, Cambridge University Press. https://doi.org/10.1017/CBO9780511541490.010

Wilkin, R.T. and Barnes, H.L. (1997). Formation processes of framboidal pyrite. *Geochimica et Cosmochimica Acta 61* (2): 323–339. https://doi.org/10.1016/S0016-7037 (96)00320-1

Woebken, D., Lam, P., Kuypers, M.M. et al. (2008). A microdiversity study of anammox bacteria reveals a novel *Candidatus* Scalindua phylotype in marine oxygen minimum zones. *Environmental Microbiology 10* (11): 3106–3119. https://doi.org/10.1111/j.1462-2920.2008.0164

Woulds, C., Cowie, G.L., Levin, L.A. et al. (2007). Oxygen as a control on sea floor biological communities and their roles in sedimentary carbon cycling. *Limnology and Oceanography 52* (4): 1698–1709. https://doi.org/10.4319/lo.2007.52.4.1698

Wyrtki, K. (1971). *Oceanographic Atlas of the International Indian Ocean Expedition*, Vol. *531*. Washington, DC: National Science Foundation.

Yao, W. and Millero, F.J. (1993). The rate of sulfide oxidation by MnO2 in seawater. *Geochimica et Cosmochimica Acta 57*: 3359–3365. https://doi.org/10.1016/0016-7037 (93)90544-7

Yao, W. and Millero, F.J. (1995). Oxidation of hydrogen sulfide by Mn (IV) and Fe (III) (hydr)oxides in seawater. *ACS Symposium Series 612*: 260–279. https://doi.org/10.1021/bk-1995-061h014

Yao, W. and Millero, F.J. (1996). Oxidation of hydrogen sulfide by hydrous Fe (III) oxides in seawater. *Marine Chemistry 52*: 1–16. https://doi.org/10.1016/0304-4203 (95)00072-0

Yucel, M., Konovalov, S.K., Moore, T.S. andet al. (2010). Sulfur speciation in the upper Black Sea sediments. *Chemical Geology 269* (3–4): 364–375. https://doi.org/10.1016/j.chemgeo.2009.10.010

Zaback, D.A. and Pratt, L.M. (1992). Isotopic composition and speciation of sulfur in the Miocene Monterey Formation: reevaluation of sulfur reactions during early diagenesis in marine environments. *Geochimica et Cosmochimica Acta 56* (2): 763–774. https://doi.org/10.1016/0016-7037 (92)90096-2

Zhu, M.X., Chen, K.K., Yang, G.P., et al. (2016). Sulfur and iron diagenesis in temperate unsteady sediments of the East China Sea inner shelf and a comparison with tropical mobile mud belts (MMBs). *Journal of Geophysical Research: Biogeosciences 121* (11): 2811–2828. https://doi.org/10.1002/2016JG003391

2

Sedimentary Records of Present and Past Marine Sulfur Cycling

Herald Strauss

ABSTRACT

The sulfur isotopic composition of marine sediments and sedimentary rocks reveals a wealth of information about sulfur cycling on different spatial and temporal scales. Sulfate minerals such as gypsum/anhydrite, barite or carbonate-associated sulfate provide a temporal record of seawater evolution that reflects secular changes in the global sulfur cycle. Dissolved porewater sulfate and sulfide, but more so sedimentary iron sulfides and/or organic sulfur reveal details about the diagenetic evolution of marine sediments. Mass-independent sulfur isotope anomalies have proven to be a prime recorder for the atmospheric oxygen abundance in the first half of Earth's history.

2.1. INTRODUCTION

Sulfur is present in the ocean-atmosphere system in multiple forms, ranging in oxidation states from +6 to –2. It is an essential element of life and participates in numerous microbially driven redox processes. In the modern ocean, dissolved sulfate is the most abundant form of sulfur with a concentration of 29 mM (Canfield, 2001a). Only a few oceanic basins, most prominently the Black Sea, exhibit a permanent stratification of the water body characterized by an anoxic lower part of the water column where dissolved sulfide is stable; such a water body is termed euxinic. Modern marine sediments contain dissolved sulfate and sulfide in their porewaters, the latter resulting from microbially driven sulfate reduction. Dissolved sulfide generally combines with reduced iron to form sedimentary iron sulfide (Rickard and Luther, 2007). In ancient sedimentary rocks, sulfur is most prominently present as sulfate mineral (gypsum, anhydrite, barite), as carbonate-associated sulfate (CAS), as sedimentary iron sulfide (pyrite), or as organic-bound sulfur.

Institut für Geologie und Paläontologie, Westfälische Wilhelms-Universität Münster, Correnstraße 24, 48149 Münster, Germany (hstrauss@uni-muenster.de)

ORCiD code: 0000-0003-2639-1225

In our quest to unravel the global sulfur cycle, stable sulfur isotopes have been successfully utilized to identify potential source(s) and (microbial) processing of sulfur in the sedimentary realm. Traditionally, evaporitic sulfate minerals (i.e. gypsum and anhydrite) were considered as recorders of seawater evolution (Holser and Kaplan, 1966; Claypool et al., 1980; Strauss, 2004). Subsequently, the analytical approach was extended to barite (Paytan et al., 2004) and carbonate-associated sulfate (Kampschulte and Strauss, 2004). In contrast, sedimentary pyrite was generally studied to investigating the presence and temporal evolution of microbial sulfate reduction, being an ancient process of microbially driven sulfur cycling (Strauss, 1997, 1999; Canfield, 2001a; Shen et al., 2001). Only in more recent years has there been an increase in sulfur isotope studies of organic-bound sulfur, recognizing that sulfurization of organic matter is one way of increasing the preservation of sedimentary organic matter, while also acknowledging the fact that organic-bound sulfur reflects part of the sedimentary sulfur budget that is largely neglected in global mass balance calculations (Werne et al., 2004). In the past 20 years, the analysis of all four stable sulfur isotopes have been instrumental in reconstructing the temporal evolution of atmospheric oxygen abundance during the early part of Earth's

Systems Biogeochemistry of Major Marine Biomes, First Edition. Edited by Aninda Mazumdar and Wriddhiman Ghosh.

history, specifically defining the first major rise in atmospheric oxygen abundance some 2.3 billion years ago (Farquhar et al., 2000). Another application for such multiple sulfur isotope approach has been the differentiation of microbial processes of sulfur cycling (Johnston et al., 2005; Zerkle et al., 2009).

This review will address key aspects of the global sulfur cycle, as recorded by the sulfur isotopic composition and archived in marine sediments. Starting with the modern ocean, different aspects of sulfur cycling on different spatial and temporal scales archived in the marine sedimentary sulfur isotope record will be explored.

2.2. STABLE SULFUR ISOTOPES - A TOOL FOR RECONSTRUCTING SPATIAL AND TEMPORAL CHANGES IN SULFUR CYCLING

Sulfur has four stable isotopes, ^{32}S, ^{33}S, ^{34}S, and ^{36}S, with average natural abundances of 94.93%, 0.76%, 4.29%, 0.02%, respectively (Coplen et al., 2002). Results are commonly expressed in the delta notation, placing the ratio of the two major isotopes ^{32}S and ^{34}S in a sample in relation to this ratio in a reference material and normalizing it to the Vienna Canon Diablo Troilite standard (VCDT; Krouse and Coplen, 1997):

$$\delta^{34}S(‰) = \left[\left({}^{34}S/{}^{32}S_{Sa}/{}^{34}S/{}^{32}S_{St}\right)-1\right]\times 1000$$

Despite early studies recording all four stable sulfur isotopes (Hulston and Thode, 1965), only the last 20 years has there been an increasing number of multiple sulfur isotope studies that also recorded the minor ^{33}S and ^{36}S isotopes. Acknowledging the fact that modern day physical, chemical and biologically mediated reactions are generally associated with a mass-dependent fractionation in sulfur isotopes, these publications express their results as deviation from the calculated mass-dependent isotope fractionation (Farquhar et al., 2000):

$$\Delta^{33}S(‰) = \delta^{33}S - 1000\times\left[\left(1+\delta^{34}S/1000\right)^{0.515}-1\right]$$

$$\Delta^{36}S(‰) = \delta^{36}S - 1000\times\left[\left(1+\delta^{34}S/1000\right)^{1.90}-1\right].$$

Early sulfur isotope measurements in marine sediments were published by, for example, Thode et al. (1949, 1953) and Szabo et al. (1950) and provided the basis for the application of sulfur isotopes in earth and life sciences.

2.3. THE MODERN MARINE REALM

Today's ocean is the largest modern sedimentary reservoir, with sulfur present as dissolved sulfate. A sulfate concentration of 29 mM is distributed homogeneously in the lateral and vertical dimension because of a residence time (8.7×10^6 years) that is substantially longer than the oceanic mixing time (around 10^3 years) (Holland, 1984). Dissolved oceanic sulfate exhibits a stable sulfur isotopic composition of +21‰ ($\delta^{34}S$ vs. VCDT; Rees et al., 1978) and shows no discernible variations with depth (Figure 2.1) based on a compilation of data from several ocean basins (Böttcher et al., 2007). More recent accounts for the sulfur isotopic composition of seawater sulfate were presented by Tostevin et al. (2014), who reported an average $\delta^{34}S$ value of +21.24 ± 0.44‰ and a $\Delta^{33}S$ value of 0.050 ± 0.014‰, based on seawater sulfate samples from the eastern margin of the Pacific Ocean, offshore from the California and Peru coastlines, and by Johnston et al. (2014), who reported an average $\delta^{34}S$ value of +21.15 ± 0.15‰ and a $\Delta^{33}S$ value of 0.048 ± 0.006‰ for water samples from oxygen minimum zones in the Pacific and Atlantic Ocean. Significant deviations from the very homogeneous $\delta^{34}S$ value of +21‰ exist solely in marginal seas as a result of the in-mixing of isotopically variable but generally less ^{34}S-enriched riverine sulfate, showing a global average $\delta^{34}S$ value of 4.4 ± 4.5‰ (Burke et al., 2018).

Depth profiles through the upper part of the marine sedimentary column indicate that the dissolved sulfate

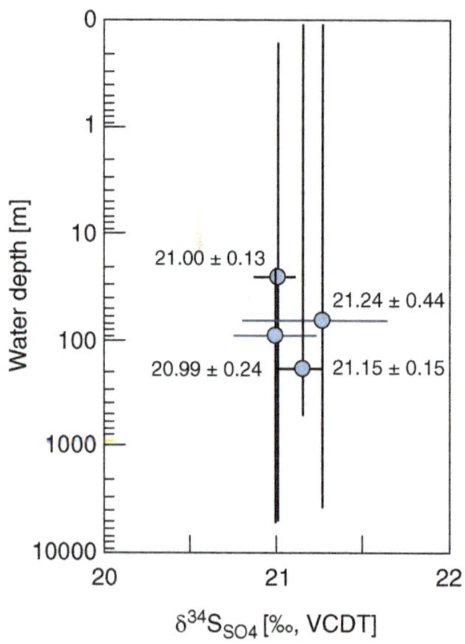

Figure 2.1 Homogeneous sulfur isotopic composition of modern seawater sulfate as compiled by Rees et al. (1978), Böttcher et al. (2007), Tostevin et al. (2014), and Johnston et al. (2014). Vertical lines represent depth ranges of samples studied. VCDT: Vienna Canon Diablo Troilite standard.

load contained in the porewater provides the principal electron acceptor for the microbially mediated mineralization of sedimentary organic matter via organoclastic sulfate reduction, once dissolved oxygen has been fully consumed (Jørgensen, 1982). At depths, ranging from millimeter to meter scale, the concentration of porewater sulfate becomes limited because microbial sulfate consumption exceeds sulfate supply via diffusion. The decrease in sulfate concentration with depth is accompanied by an increase in its $\delta^{34}S$ value (Figure 2.2a). The residual porewater sulfate becomes progressively enriched in ^{34}S due to the preferential utilization of sulfate containing the light ^{32}S isotope during microbial sulfate reduction. In addition to organoclastic sulfate reduction, numerous studies (Jørgensen et al., 2004; Peketi et al., 2012; Lin et al., 2016) have revealed the complete consumption of porewater sulfate at depth when upwards diffusing (biogenic) methane is oxidized at the expense of sulfate. A consortium of methanotrophic archaea and sulfate reducing bacteria (Boetius et al., 2000) pursues this reaction. Termed sulfate-driven anaerobic oxidation of methane (SO_4-AOM), this reaction frequently defines a distinct reaction zone at depth termed the sulfate–methane transition zone (SMTZ; Borowski et al., 1996). Thereby, the methane flux determines the depth of the SMTZ (Figure 2.2b). In the sedimentary column SO_4-AOM is recorded by abundant sedimentary (biogenic) pyrite that exhibits distinctly positive sulfur isotope values (Borowski et al., 2013; Lin et al., 2016; for further details, see text below).

Modern marine sediments, i.e. sediments deposited from a bottom water containing dissolved oxygen, contain on average 0.6 weight per cent of sulfur (Goldhaber, 2003), generally present as sedimentary pyrite and attributed to microbial sulfate reduction and subsequent precipitation as iron sulfide (Canfield, 2001a; Rickard and Luther, 2007). Microbial sulfate reduction, more specifically organoclastic sulfate reduction, is associated with a distinct isotopic fractionation of up to 70‰, generally displaying a $\delta^{34}S$ value for the resulting sulfide that is ^{34}S-depleted compared with the parental sulfate. The magnitude in isotopic fractionation is determined by a multitude of factors including the availability and reactivity of sulfate and organic substrate as well as physicochemical boundary conditions, such as temperature. Milestones in our understanding in this respect were published by Kaplan and Rittenberg (1964), Canfield (2001b), Detmers et al. (2001), and more recently by Johnston et al. (2007), Sim et al. (2011), Leavitt et al. (2013), and Wing and Halevy (2014).

Questions in studying marine sedimentary pyrite with sulfur isotopes center on identifying the principle processes related to its formation. Depth records for dissolved sulfide in porewater but more so for iron monosulfides and disulfides in ancient sedimentary rocks (and elemental and organic sulfur, if preserved) provide a detailed record of sediment diagenesis. Important supplementary information includes detailed petrographic examination of pyrite morphology, distinguishing early diagenetic framboidal pyrite from late diagenetic overgrowth or late-stage euhedral crystals (Figure 2.3), as well as additional sediment geochemical data (e.g. organic carbon content, iron speciation).

In a series of publications, Lin et al. (2016, 2017) convincingly argued for a spatio-temporal sequence of diagenetic sulfur cycling in marine sediments involving both organoclastic sulfate reduction and sulfate-driven anaerobic methane oxidation. Conclusions are based on

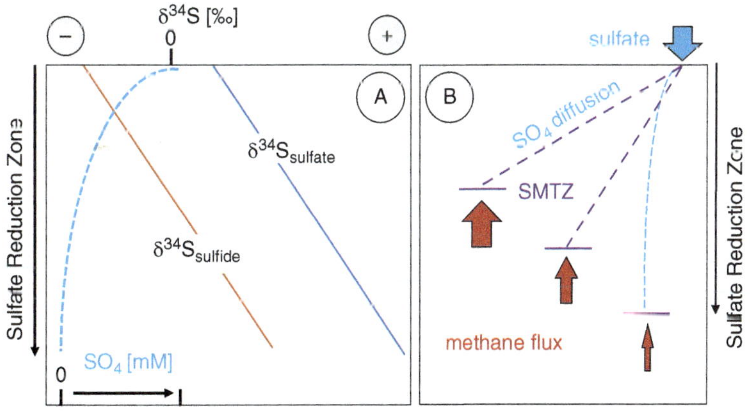

Figure 2.2 (a) Simplified depth distribution of changes in porewater sulfate concentration and sulfur isotopic composition of dissolved sulfate and sulfide as a consequence of progressive microbial sulfate reduction. (b) Depth distribution of sulfate as a consequence of an upwards methane flux of difference intensity. SMTZ: sulfate methane transition zone (Modified from Borowski et al., 1996).

Figure 2.3 Different generations of sedimentary pyrite with early diagenetic framboidal pyrite overgrown by late(r) diagenetic pyrite (SEM image courtesy of Zhiyong Lin).

respective evidence from geochemical and isotopic data (Figure 2.4). Porewater sulfate concentration decreases with increasing depth,. Sedimentary pyrite that was chemically extracted from bulk rock (chromium reducible sulfur: CRS) exhibits a strongly negative sulfur isotopic composition that clearly indicates a biological origin via organoclastic sulfate reduction (OSR). At greater depth, the decrease in dissolved sulfate concentration overlaps with an increase in methane concentration. This overlap defines the SMTZ that is frequently characterized by a sudden increase in sedimentary pyrite abundance. More importantly, however, strongly ^{34}S-enriched sulfur isotope values characterize this pyrite, suggesting rapid sulfate consumption through SO_4-AOM, thereby exceeding sulfate replenishment via diffusion. In addition, different degrees of ^{34}S-enrichment became apparent when comparing the sulfur isotope results from different analytical approaches (chromium reducible sulfur extracted from bulk rock, hand-picked pyrite, different generations of pyrite ranging from early diagenetic framboids to late diagenetic pyrite overgrowth studied via secondary ion microprobe analysis, SIMS), including some of the most

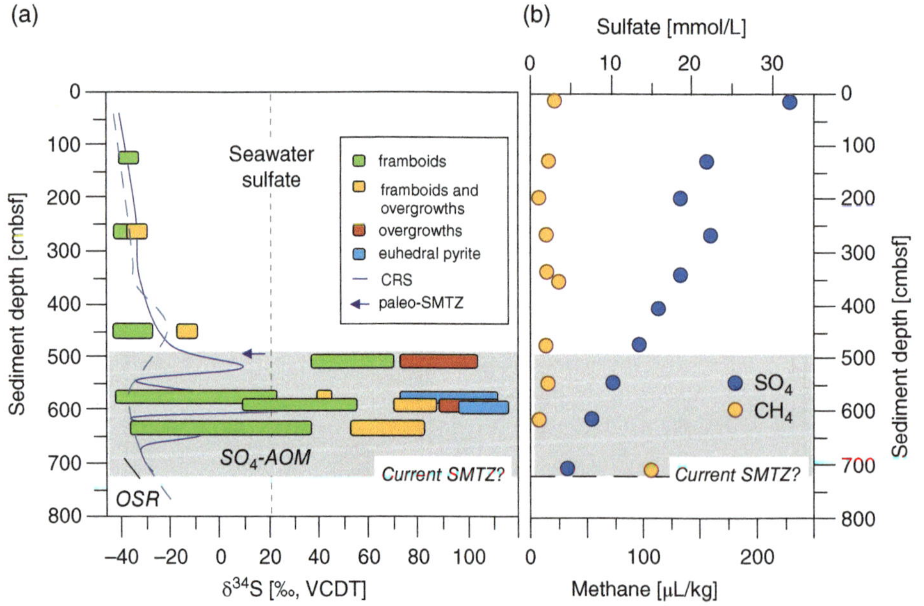

Figure 2.4 (a) Variations in $\delta^{34}S$ for chromium reducible sulfur from bulk rock extraction (CRS blue line) and differentiated for early to late stage diagenetic pyrite exhibiting different morphologies. The dashed line separates a zone to the left which is suggested to be dominated by organoclastic sulfate reduction (OSR) and a zone to the right suggested to be dominated by sulfate-driven anaerobic oxidation of methane (SO_4-AOM). (b) Depth profiles for dissolved sulfate and methane concentrations from porewater. The gray field indicates the depth, which exhibits sulfate-driven anaerobic oxidation of methane. (Modified from Lin et al., 2016.)

positive $\delta^{34}S$ values ever reported for sedimentary pyrite (up to +115‰). Comparable results, i.e. a substantial sulfur isotopic variation at the (sub)millimeter scale of up to 75‰, was reported from modern microbial mat systems (Fike et al., 2009) as well as for Proterozoic (Present et al., 2018) and even Archean sediments (Marin-Carbonne et al., 2019).

By including the minor sulfur isotopes ^{33}S and ^{36}S (with results expressed as $\Delta^{33}S$ and $\Delta^{36}S$ values), Lin et al. (2017) were able to clearly distinguish the involvement of different microbially driven processes, i.e. sulfate reduction, disproportionation of sulfur intermediates and (partial) sulfide oxidation. Moreover, distinctly different data populations became apparent in a cross-plot of $\Delta^{33}S_{py} - \Delta^{33}S_{SO4}$ vs. $\delta^{34}S_{py} - \delta^{34}S_{SO4}$ (Figure 2.5), allowing to distinguish sedimentary pyrite that formed via OSR vs. pyrite that formed as a result of SO_4-AOM. Most recently, Jørgensen et al. (2019) further advanced our understanding about spatiotemporal variations in the importance of organoclastic sulfate reduction vs. sulfate-driven anaerobic oxidation of methane.

Studies such as those by Lin et al. (2016, 2017) clearly indicate the sometimes substantial complexity in sediment diagenesis that is recorded by sulfur isotopes. In terms of its application, the high variability in $\delta^{34}S$ and additional information obtained from $\Delta^{33}S$ offers a strong potential for unraveling this complexity. Yet, how detailed a respective (hi)story unveils strongly depends on the detail of the analytical approach.

2.4. OCEANIC SULFATE AND ITS EVOLUTION THROUGH TIME

Seawater sulfate is our prime recorder of secular changes in global sulfur cycling. Abundance and sulfur isotopic composition of oceanic sulfate reflect the delicate balance between inputs and outputs. Principal input parameters are the riverine delivery of sulfate from oxidative continental weathering of sulfide and the dissolution of sulfate as well as the interaction of seawater and ocean crustal rocks, most prominently exhibited at submarine hydrothermal vent sites along mid-ocean ridges, island arcs and back-arc settings. Dissolved sulfate leaves the ocean via precipitation of sulfate minerals or through microbially driven sulfate reduction and subsequent precipitation of the resulting hydrogen sulfide as iron sulfide or its incorporation into organic matter. In a simplified isotope mass balance

$$\delta^{34}S_{input} = f_{sulfide} \times \delta^{34}S_{sulfide} + (1-f)_{sulfide} \times \delta^{34}S_{sulfate}$$

these two principal output functions (i.e. sulfate precipitation and biogenic sulfide formation) are balanced against an overall input function that supposedly reflects average crustal sulfur and that is generally considered to be invariable through time (Holser et al., 1988; for recent discussions, see Halevy et al., 2012 and Canfield, 2013).

As the precipitation of a sulfate mineral is not associated with a substantial sulfur isotope fractionation (Raab

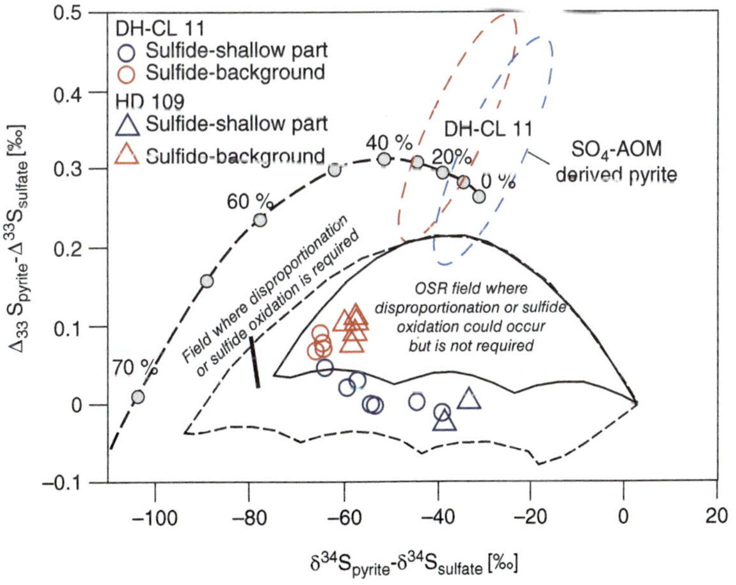

Figure 2.5 Differentiation of different forms of microbial sulfur cycling as demonstrated by multiple sulfur isotopes. (Redrawn after Lin et al., 2017). OSR: organoclastic sulfate reduction; SO_4-AOM: sulfate-driven anaerobic oxidation of methane; DH-CL 11 and HD 109 are drill cores.

and Spiro, 1991), the sulfur isotopic composition of marine evaporitic sulfates faithfully records the isotopic composition of dissolved oceanic sulfate. In considering the simplified sulfur isotope mass balance, the temporal evolution in $\delta^{34}S_{sulfate}$ has been interpreted as a reflection of secular changes in the balance between the principal output functions from the oceanic reservoir.

Early accounts of the sulfur isotopic composition of oceanic sulfate through time were published by Nielsen (1965) and Holser and Kaplan (1966), followed by the seminal paper from Claypool et al. (1980). These studies used evaporitic sulfate minerals (generally gypsum or anhydrite) for reconstructing the temporal changes in $\delta^{34}S_{sulfate}$, thereby acknowledging the absence of a substantial isotope effect during mineral precipitation. Two observations were most apparent from these studies: the sulfur isotopic composition of oceanic sulfate was not homogeneous through time and the temporal record was not continuous, with gaps that reflect the nondeposition of evaporites or their loss during weathering (Strauss, 1997). Barite was considered as an alternative early on (Goodfellow and Johansson, 1984), and our present understanding about secular changes in the sulfur isotopic composition of oceanic sulfate during the past 130 million years is indeed based on a high-resolution record of marine barite (Paytan et al., 2004).

The most promising and widely utilized alternative, however, is carbonate-associated sulfate (CAS). During carbonate precipitation, sulfate is incorporated into the crystal lattice in concentrations of tens to thousands of parts per million (Busenberg and Plummer, 1985; Staudt and Schoonen, 1995), and this sulfate archives the sulfur isotopic composition of the dissolved sulfate at the time of carbonate precipitation. As with true sulfate minerals, it is important to independently constrain the marine nature of carbonates used, such as marine biogenic carbonates. Similar to the modern ocean, limited variability in $\delta^{34}S_{CAS}$ can be expected for multiple time-equivalent carbonate samples, as carbonate associated sulfate would reflect the homogeneous nature of the seawater sulfate sulfur isotopic composition at any given time in Earth history. However, recent studies have reported significant differences in $\delta^{34}S_{CAS}$ for different sedimentary carbonate components within an individual stratigraphic unit (Present et al., 2015), indicating that diagenesis can exert control on the $\delta^{34}S_{CAS}$ value of carbonates (Present et al., 2019). Consequently, a prerequisite for successfully applying $\delta^{34}S_{CAS}$ is a thorough evaluation of the diagenetic history of the carbonate (Fichtner et al., 2017), with a rigid analytical procedure (Wotte et al., 2012) being mandatory.

Kampschulte and Strauss (2004) published the first CAS-based sulfur isotope time series of oceanic sulfate for the Phanerozoic. Clear differences were discernible such as high $\delta^{34}S_{sulfate}$ values in the early Paleozoic and minimum $\delta^{34}S_{sulfate}$ values during Permian times. While the overall temporal trend was comparable to observations published earlier and based on evaporite minerals, this new time series of the sulfur isotopic composition of Phanerozoic seawater sulfate showed substantially more internal structure because of its significantly better temporal resolution. Numerous studies on $\delta^{34}S_{sulfate}$, published in the past 15 years, have provided a detailed view of the secular variations in the sulfur isotopic composition through time; these have most recently been compiled by Crockford et al (2019) extending the record back into the Proterozoic (Figure 2.6).

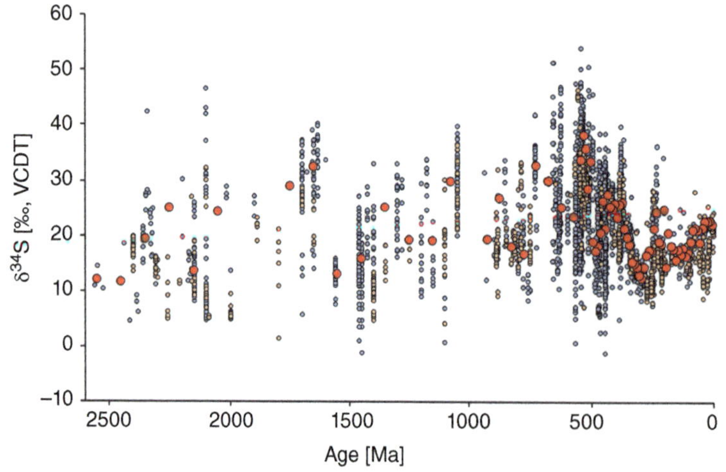

Figure 2.6 Secular variations in $\delta^{34}S$ for carbonate-associated sulfate (CAS: blue circles) and evaporitic sulfate (orange circles). Red circles indicate 10 million year average values (0–500 Ma), 50 million-year average values (500–1000 Ma) and 100 million-year average values (1100–2600 Ma) calculated from $\delta^{34}S_{CAS}$. (Modified from Crockford et al., 2019.)

A first-order interpretation, based on isotope mass balance considerations, views these temporal variations as a reflection of changes in pyrite burial ($f_{sulfide}$), with high rates in the early Paleozoic, minimum rates in the late Paleozoic and its transition into the Mesozoic, and an evolution towards modern values. However, additional information can be gathered from a detailed multiple sulfur isotope record through time.

Based on complimentary records of $\delta^{34}S_{sulfate}$ and $\Delta^{33}S_{sulfate}$ for Paleozoic and Mesozoic carbonate-associated sulfate, and considering the temporal record of $\delta^{34}S_{sulfide}$ through time, Wu et al. (2010, 2014) proposed significant changes in the operational mode of the global sulfur cycle through the Phanerozoic. Most notably, these authors highlighted a change in the magnitude of isotopic fractionation between sulfate and sulfide from lower values in the Paleozoic to higher values in the Cenozoic. Moreover, they argue for a change in the isotopic composition of the input function ($\delta^{34}S_{in}$) to the global sulfur cycle (i.e. the continental weathering signature), as previously discussed by Canfield (2013). Wu et al. (2014) attribute this to changes in the Earth surface sulfur pool, notably a rapid recycling of newly formed sulfate minerals. Discernible shorter-term fluctuations (on the tens of million years scale) in the temporal records are interpreted as reflecting changes in the intensity of sulfide oxidation during cycling of sulfur and/or by rapid changes in sulfur influx to the oceans and its associated sulfur isotopes. Recently, Crockford et al. (2019) compiled and substantially extended the time series of $\delta^{34}S_{sulfate}$ and $\Delta^{33}S_{sulfate}$ for seawater sulfate (and $\delta^{18}O_{sulfate}$ and $\Delta^{17}O_{sulfate}$) back in time into the late Archean, with new data mostly derived from gypsum or anhydrite. As in other previous studies, the observed large scale secular variations in the sulfur isotopic composition of seawater sulfate are attributed to temporal changes in burial/weathering of sedimentary sulfide. Moreover, these time series substantiate earlier suggestions (Melezhik et al., 2005) that the operational mode of the sulfur cycle as we know it today and, in particular, the continental weathering and riverine delivery of sulfate to the ocean only commenced in the early Proterozoic, postdating the first significant rise in atmospheric oxygen.

Two additional aspects related to the study of carbonate-associated sulfate are noteworthy as they likely reflect on future research. An increasing number of studies have focused on critical time boundaries such as the Permian–Triassic transition (Schobben et al., 2015, 2017), and careful work has resulted in high-resolution profiles across prominent rock successions allowing for a renewed view on the causes and consequences of short-term perturbations of global sulfur cycling. The second aspect pertains to an analytical improvement, notably the measurement of sulfur isotopes at very low sulfate concentrations (nanomole level) using multi-collector induced coupled plasma mass-spectrometry (MC-ICP-MS) (Craddock et al., 2008; Paris et al., 2013). Applications of the latter will allow for an even higher temporal resolution for rock successions across critical time boundaries than presently available.

2.5. PYRITE AND ORGANIC-BOUND SULFUR AS RECORDERS OF MICROBIAL SULFUR CYCLING IN THE PAST

Being a key element of life, sulfur is taken up by plants and microorganisms via assimilatory reduction of inorganic sulfate with little sulfur isotope fractionation associated. In contrast, dissimilatory sulfate reduction is an energy-yielding process associated with the release of hydrogen sulfide. Kaplan and Rittenberg (1964) in their seminal paper on isotopic fractionation associated with dissimilatory sulfate reduction established the foundation for the application of sulfur isotopes as recorder of microbial activities in the geological past. Hartmann and Nielsen (1969) were the first to apply this new understanding that dissimilatory sulfate reduction is generally associated with a substantial isotopic discrimination against the heavy ^{34}S isotope in their study of marine sediments. Subsequently, a negative $\delta^{34}S$ value measured for a sedimentary pyrite or at least a sizeable apparent isotopic fractionation between the parental sulfate (preserved in the rock record as evaporite) and resulting sedimentary (iron) sulfide, as observed in natural marine settings as well as in the early laboratory experiments, were considered as evidence for the biogenicity of a sedimentary sulfide: it provided the basis for tracing microbial sulfur cycling through time.

Since the 1970s, numerous studies explored the antiquity of microbial sulfur cycling by studying sedimentary rocks as far back as 3.8 billion years ago (Monster et al., 1979). Time series of $\delta^{34}S$ values for pyrite in sedimentary rocks were presented, among others, by Schidlowski et al. (1983), Hayes et al. (1992), Strauss (1999), and Canfield (2001a). Reviewing sulfur isotope research targeting sedimentary sulfides with the objective of identifying microbial sulfur cycling and trace it through time two milestone discoveries in the past 20 years by Sim et al. (2011) and Pellerin et al. (2019) are most notable.

In 1964, Kaplan and Rittenberg reported that a maximum sulfur isotopic fractionation of 46‰ associated with dissimilatory sulfate reduction, modern, and ancient marine sediments and sedimentary rocks frequently yielded a larger isotopic difference between a measured sedimentary pyrite and the coeval seawater sulfate (Strauss, 1997). This enigmatic difference between nature and experiment persisted for nearly 30 years until a new type of microbial sulfur metabolism termed disproportionation of sulfur intermediates resulting from sulfide oxidation (such as thiosulfate or elemental sulfur) was identified as a major microbially mediated process of sulfur cycling in the marine realm (Jørgensen, 1990). Subsequent sulfur isotope studies of this type of microbial sulfur cycling (Cypionka et al., 1998; Habicht et al., 1998) revealed sizeable cumulative sulfur isotope effects that were greater than the canonical 46‰ defined by earlier studies for dissimilatory sulfate reduction. Consequently, large isotopic fractionations observed for marine sediments and sedimentary rocks above the canonical 46‰ (cf. Kaplan and Rittenberg, 1964) were attributed to disproportionation of intermediate sulfur compounds. Analyzing the temporal record of $\delta^{34}S_{pyrite}$ available at that time, Canfield and Teske (1996) concluded that the onset of such sizeable sulfur isotopic fractionations could be attributed to sulfide oxidation. Linking the temporal evolution of the global sulfur and oxygen cycles, these authors concluded that disproportionation of sulfur intermediates commenced around 850 million years ago as a reflection of the Neoproterozoic rise in atmospheric oxygen abundance in the dawn of animal life (Knoll et al., 1986). More recent work, however, places the onset of microbial sulfur disproportionation at around 635 million years, post-dating the Marinoan glaciation (Kunzmann et al., 2017).

Recently, Sim et al. (2011) reported a sulfur isotopic fractionation of nearly up to 70‰ for microbial sulfate reduction, both in a marine sediment on the New Jersey Shelf (USA) and in their cultivation experiments. The discovery of such a high magnitude of sulfur isotopic fractionation exerted solely by a sulfate-reducing bacterium finally terminated the long-standing discrepancy between sulfur isotope data from cultivation experiment and observational results from marine sediments and, thus, a long-time paradigm of a canonical isotopic fractionation associated with microbial sulfate reduction of 46‰.

Most recent work by Pellerin et al. (2019) challenged the assumption that only microbial sulfate reduction and disproportionation of sulfur intermediates are associated with sizeable isotopic fractionations. These authors reported a large sulfur isotope fractionation associated with bacterial sulfide oxidation. However, the implications of this new discovery for interpreting the deep time record of $\delta^{34}S_{pyrite}$ will have to be determined in future studies.

Looking back on more than 50 years of sulfur isotope studies of sedimentary iron sulfides, the major aim of respective research has always been to constrain pyrite formation with a strong focus on identifying the biogenic nature of the pyrite under study. This holds true in particular for Precambrian sedimentary rocks, but a characterization of ancient microbial sulfur cycling in the marine realm under variable environmental conditions was commonly the major driver behind respective studies of the geological past.

While most of the scientific focus was placed on the sulfur isotopic composition of sedimentary pyrite as a proxy for reconstructing sediment diagenesis or the temporal record of sulfur metabolism on Earth, organic sulfur and its sulfur isotopic composition were given comparatively little attention.

Seawater sulfate is also the principal source of organic sulfur in the marine realm. Incorporation of sulfur into organic matter via assimilatory reduction of sulfate (biosynthetic organic sulfur) represents one pathway, albeit of minor importance considering that sulfur represents <1% of marine biomass (Canfield, 2001a). Of much greater importance is the dissimilatory reduction of sulfate, resulting in highly reactive hydrogen sulfide. This can either react with iron and form a sedimentary iron sulfide or be oxidized to sulfur phases of intermediate valence (such as elemental sulfur, polysulfide, thiosulfate), or to sulfate, or be incorporated into organic matter. Thereby, an intermolecular reaction between sulfur and organic matter, resulting in organosulfur macromolecules, is distinguished from an intramolecular mechanism where sulfur becomes part of an organic molecule forming ring structures (Sinninghe-Damsté et al., 1989). Prerequisites for the formation of organic sulfur include the abundance of reduced sulfur species, the presence of organic matter with functional sites for the reaction with sulfur, and a low concentration of metals, such as iron (Amrani, 2014; Shawar et al., 2018). The last of these is assumed to exert a strong influence because the reaction of hydrogen sulfide with iron is kinetically favored compared with the sulfurization of organic matter (Mossmann et al., 1991; Hartgers et al., 1997). However, several studies have reported the coformation of organic sulfur and pyrite (Brüchert and Pratt, 1996; Hennecke et al., 1997; Urban et al., 1999; Filley et al., 2002).

The sulfur isotopic composition of organic sulfur is dependent upon the isotope signature of the sulfur source, the isotopic fractionation associated with microbial sulfur metabolism resulting in a sulfur phase that is reactive towards organic matter (a possible isotope effect associated with the incorporation of sulfur into the organic matter), and a possible isotope effect related to the progressing maturation of the organic matter (Siedenberg et al., 2018). In 1963, Kaplan et al. had already noted a difference in $\delta^{34}S$ between organic sulfur from marine animals and algae that was close to the isotopic composition of seawater sulfate and organic sulfur that displays a ^{34}S-depleted sulfur isotope signature. While the former suggests the assimilatory reduction/incorporation of sulfate into the biomass, the latter sulfur isotope signature suggests the incorporation of reduced sulfur likely resulting from microbial sulfate reduction. Reviews by Anderson and Pratt (1995) or Werne et al. (2004) indicated that organic sulfur appeared to always be ^{34}S enriched (by 10–30‰) compared with sedimentary sulfide. The later timing of organosulfur compound formation, when progressive sulfate reduction had already resulted in sulfate limitation and an associated enrichment in ^{34}S, was proposed as an explanation. Recent analytical advances, specifically compound-specific sulfur isotope analysis (Werne et al., 2008) allowed for a significant advancement in our understanding of organic sulfur and its isotopic composition, as reviewed by Amrani (2014). Most notably, the formation of organic sulfur is much more complex than previously assumed. In addition, fractionations associated with the formation of organic sulfur, with the thermal maturation of organic matter as well as with preservation and mixing – all processes within the realm of progressing sediment diagenesis – exert a shift in the sulfur isotopic composition towards ^{34}S-enriched values of the bulk organic sulfur. Consequently, a substantial variability in isotopic composition for organic sulfur is to be expected, both in relation to coexisting iron sulfide as well as between different individual organic-sulfur compounds. For the latter, compound-specific sulfur isotope analyses revealed a variability in $\delta^{34}S$ of as much as 50‰ (Amrani, 2014; Shawar et al., 2020). Two aspects became apparent: organic sulfur is an integral part of the diagenetic history of a sediment/sedimentary rock and organic sulfur as an important component of the total sedimentary sulfur pool is generally underestimated.

2.6. MASS-INDEPENDENTLY FRACTIONATED SULFUR ISOTOPES – A RECORD OF EARTH'S OXYGENATION

Another milestone discovery in respect of the sulfur isotopic composition of sedimentary sulfur and its evolution through time was the observation of mass-independently fractionated sulfur isotopes preserved in sedimentary sulfides and sulfates. This observation revolutionized the field of sulfur isotope research as much as it extended the importance of sulfur isotopes as a recorder of Earth System Evolution.

It started with the discovery of mass-independently fractionated sulfur isotopes (MIF-S) archived in pyrite and barite in sedimentary rocks older than 2.4 billion years (Farquhar et al., 2000). Respective sulfur phases were characterized by a $\Delta^{33}S$ value more positive/negative than 0.3‰ and followed a regression line between $\Delta^{33}S$ and $\Delta^{36}S$ with a slope near −1, later called the Archean array. Inspired by mass-independently fractionated oxygen isotopes recorded in ozone (Thiemens and Heidenreich, 1983), Farquhar et al. (2000) proposed UV-driven photochemical disproportionation of volcanogenic SO_2 as the principal reaction behind these sulfur isotope anomalies. Already early on, the exclusive occurrence of MIF-S in rocks older than 2.4 billion years suggested at least a temporal relationship to the abundance of atmospheric oxygen on the early Earth. Following our understanding of an anoxic atmosphere during the early part of Earth history (Kasting, 1993) and a first significant rise in atmospheric oxygen abundance between 2.4 and 2.2 billion years ago, termed the Great Oxidation Event (Holland, 2006), not only a temporal but also a causal relationship between atmospheric pO_2 and MIF-S was proposed (Farquhar et al., 2000). Subsequent experimental and modeling approaches revealed that the formation of MIF-S and their preservation in sediments deposited at the Earth's surface required that the atmospheric oxygen abundance was below 10^{-5} present atmospheric level (PAL; Pavlov and Kasting, 2002; but see also Zahnle et al., 2006).

Since the discovery paper by Farquhar et al. (2000), numerous research articles have added to a now sizeable record of $\Delta^{33}S$ values for the Precambrian sedimentary record (Figure 2.7). This record displays clear temporal variations in the magnitude of mass-independently fractionated sulfur isotopes. In line with the overall mechanism initially proposed for generating MIF-S (i.e. the photochemistry of sulfur dioxide), respective variations in the magnitude of isotopic fractionation were related to photochemical reactions at different wavelengths (Johnston, 2011). However, the more important second feature that had already been noted in the discovery paper and is prominently apparent from the MIF-S record available now, is the termination of mass-independent sulfur isotopic fractionation in a narrow time interval between 2.4 and 2.2 billion years ago. Neither pyrite nor sedimentary sulfate minerals or carbonate-associated sulfate in sedimentary rocks younger than 2.2 billion years display the two prominent features recognized as MIF-S (i.e. a deviation of the $\Delta^{33}S$ value from zero by more than 0.3‰ and a negative correlation between $\Delta^{33}S$ and $\Delta^{36}S$ following the Archean array). The demise of MIF-S is clearly observed in stratigraphic successions from South Africa (Bekker et al., 2004; Guo et al., 2009), Canada (Papineau et al., 2007) and Russian Fennoscandia (Reuschel et al., 2013). This termination of MIF-S in the sedimentary record is considered as the 'smoking gun' for the first rise in atmospheric oxygen abundance at that time in the early Paleoproterozoic.

Owing to the causal relationship between the presence of MIF-S and atmospheric oxygen abundance, including a mechanistic understanding about formation of this signature as well as its transfer and preservation in the sedimentary rock record (Johnston., 2011), MIF-S became a

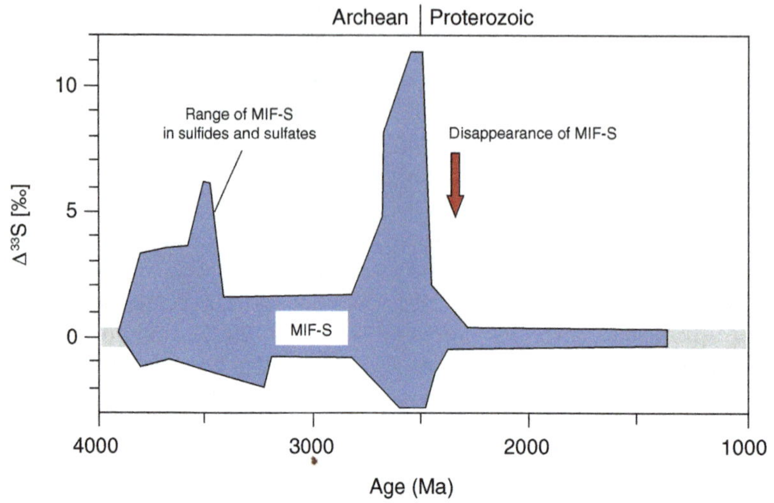

Figure 2.7 Range in $\Delta^{33}S$ for sedimentary sulfide and sulfate exhibiting mass-independently fractionated sulfur isotopes in the Archean and early Paleoproterozoic and its disappearance around 2300 million years ago. (Modified from Reuschel et al., 2013.)

recorder of early atmospheric evolution. The latter always centered on the evolution of atmospheric oxygen abundance and with it questions related to the advent of oxygenic photosynthesis (Kurzweil et al., 2013). More recent work questions the synchronous disappearance of the MIF-S-signature as a clear sign for atmospheric oxygenation of the Earth, arguing for a re-evaluation of placing the Great Oxidation Event at the last occurrence of the MIF-S signature (Philippot et al., 2018). This argument follows an earlier modeling approach addressing the aspect of a possible long-term sedimentary recycling of the MIF-S signature (Reinhard et al, 2013).

Recent work on other redox-sensitive elements and their isotopic composition (e.g., Mo, Cr, U) has added substantially to this discussion, resulting in a refinement in our understanding and challenging in particular the timing of the first significant rise in atmospheric oxygen abundance.

2.7. SUMMARY AND DIRECTION OF FUTURE RESEARCH

It is without doubt that the sedimentary records of sulfide and sulfate have archived a vast range of information in respect to the evolution of our Earth System. Geological processes at the surface of our planet (e.g., deposition and weathering) as well as at depth leave their traces as much as microbially mediated processes of sulfur cycling. The sulfur isotopic composition has been and continues to be the most important source of information in respect to these processes, despite the fact that diagenesis tends to obscure to some extent any primary signal. Moreover, through the intimate relationship between the geochemical cycles of sulfur, carbon and oxygen, sulfur isotopes have become a prime proxy signal for the overall evolution of Earth's redox state. Future efforts should continue to explore these possibilities in our quest to quantitatively understanding Earth System Evolution.

ACKNOWLEDGMENTS

I thank the editors for inviting me to contribute this review. Financial support for my research on sulfur isotope geochemistry by the Deutsche Forschungsgemeinschaft (DFG) throughout the years is gratefully acknowledged. Ted Present and two anonymous reviewers are thanked for their constructive comments that improved the clarity of this manuscript.

REFERENCES

Amrani, A. (2014). Organosulfur compounds: molecular and isotopic evolution from biota to oil and gas. *Annual Review of Earth and Planetary Sciences 42*: 733–768.

Anderson, T.F. and Pratt, L.M. (1995). Isotopic evidence for the origin of organic sulfur ad elemental sulfur in marine sediments. In: *Geochemical Transformations of Sedimentary Sulfur* (eds. M.A. Vairamurthy and M.A.A. Schoonen). Washington DC, *American Chemical Society Symposium Series 612*: 378–396.

Bekker, A., Holland, H.D., Wang, P.L. et al. (2004). Dating the rise of atmospheric oxygen. *Nature 427*: 117–120.

Boetius A., Ravenschlag K., Schubert C. J. et al. (2000). A marine microbial consortium apparently mediating anaerobic oxidation of methane. *Nature 407*: 623–626.

Böttcher, M.E., Brumsack, H.-J. and Dürselen, C-D. (2007). The isotopic composition of modern seawater sulfate: I. Coastal waters with special regard to the North Sea. *Journal of Marine Systems 67*: 73–82.

Borowski, W.S., Paull, C.K. and Ussler III, W. (1996). Marine pore-water sulfate profiles indicate *in situ* methane flux from underlying gas hydrate. *Geology 24*, 655–658.

Borowski, W.S., Rodriguez, N.M., Paull, C.K. et al. (2013). Are ^{34}S-enriched authigenic sulfide minerals a proxy for elevated methane flux and gas hydrates in the geologic record? *Marine Petroleum Geology 43*:, 381–395.

Brüchert, V. and Pratt, L.M. (1996). Contemporaneous early diagenetic formation of organic and inorganic sulfur in estuarine sediments from St. Andrews Bay, Florida, USA. *Geochimica et Cosmochimica Acta 60*: 2325–2332.

Busenberg, E. and Plummer, L.N. (1985). Kinetic and thermodynamic factors controlling the distribution of SO_4^{2-} and Na^+ in calcites and selected aragonites. *Geochimica et Cosmochimica Acta 49*: 713–725.

Burke, A., Present, T.M., Paris, G. et al. (2018). Sulfur isotopes in rivers: insights into global weathering budgets, pyrite oxidation and the modern sulfur cycle. *Earth and Planetary Science Letters 496*: 168–177.

Canfield, D.E. (2001a). Biogeochemistry of Sulfur Isotopes. *Reviews in Mineralogy and Geochemistry 43*: 607–635.

Canfield, D.E. (2001b). Isotope fractionation by natural populations of sulfate-reducing bacteria. *Geochimica et Cosmochimica Acta 65*: 1117–1124.

Canfield, D.E. (2013). Sulfur isotopes in coal constrain the evolution of the Phanerozoic sulfur cycle. *Proceedings of the National Academy of Sciences of the United States of America 110*: 8443–8446.

Canfield, D.E. and Teske, A. (1996). Late Proterozoic rise in atmospheric oxygen concentration inferred from phylogenetic and sulphur-isotope studies. *Nature 382*: 127–132.

Claypool, G.E., Holser, W.T., Kaplan, I.R. et al. (1980). The age curves of sulfur and oxygen isotopes in marine sulfate and their mutual interpretation. *Chemical Geology 28*, 199–260.

Coplen, T.B., Bohlke, J.K., De Bievre, P. et al. (2002). Isotope-abundance variations of selected elements (IUPAC Technical Report). *Pure and Applied Chemistry 74*: 1987–2017.

Craddock, P.R., Rouxel, O.J., Ball, L.A. et al. (2008). Sulfur isotope measurement of sulfate and sulfide by high-resolution MC-ICP-MS. *Chemical Geology 253*: 102–133.

Crockford, P.W., Kunzmann, M., Bekker, A., et al. (2019). Claypool continued: extending the isotopic record of sedimentary sulfate. *Chemical Geology 513*: 200–225.

Cypionka, H., Smock, A.M. and Böttcher, M.E. (1998). A combined pathway of sulfur compound disproportionation in *Desulfovibrio desulfuricans*. *FEMS Microbiology Letters 166*: 181–186.

Detmers, J., Brüchert, V., Habicht, K.S. et al. (2001). Diversity of sulfur isotope fractionations by sulfate-reducing prokaryotes. *Applied and Environmental Microbiology 67*: 888–894.

Farquhar, J., Bao, H.M. and Thiemens, M. (2000). Atmospheric influence of Earth's earliest sulfur cycle. *Science 289*: 756–759.

Fichtner, V., Strauss, H., Immenhauser, A. et al. (2017). Diagenesis of carbonate associated sulfate. *Chemical Geology 463*: 61–75.

Fike, D.A., Finke, N., Zha, J. et al. (2009). The effect of sulfate concentration on (sub)millimeter-scale sulfide δ^{34}S in hypersaline cyanobacterial mats over the diurnal cycle. *Geochimica et Cosmochimica Acta 73*: 6187–6204.

Filley, T.R., Freeman, K.H., Wilkin, R.T. et al. (2002). Biogeochemical controls on reaction of sedimentary organic matter and aqueous sulfides in Holocene sediments of Mud Lake, Florida. *Geochimica et Cosmochimica Acta 66*: 937–954.

Goldhaber, M.B. (2003). Sulfur-rich sediments. In: *Sediments, Diagenesis and Sedimentary Rocks* (ed F.T. Mackenzie), 257–288. Elsevier Ltd.

Goodfellow, W.D. and Jonasson, I.R. (1984). Ocean stagnation and ventilation defined by ^{34}S secular trends in pyrite and barite, Selwyn basin, Yukon. *Geology 12*: 583–586.

Guo, Q.J., Strauss, H., Kaufman, A.J. et al., (2009). Reconstructing Earth's surface oxidation across the Archean–Proterozoic transition. *Geology 37*: 399–402.

Habicht, K.S., Canfield, D.E. and Rethmeier, J. (1998). Sulfur isotope fractionation during bacterial reduction and disproportionation of thiosulfate and sulfite. *Geochimica et Cosmochimica Acta 62*: 2585–2595.

Halevy, I., Peters, S.E., Fischer, W.W. (2012). Sulfate burial constraints on the Phanerozoic sulfur cycle. *Science 337*: 331–334.

Hartgers, W.A., Lopez, J.F., Sinnighe-Damsté, J.S. et al. (1997). Sulfur-binding in recent environments: II. Speciation of sulfur and iron and implications for the occurrence of organo-sulfur compounds. *Geochimica et Cosmochimica Acta 61*: 4769–4788.

Hartmann, M. and Nielsen, H. (1969). δ^{34}S-Werte in rezenten Meeressedimenten und ihre Deutung am Beispiel einiger Sedimentprofile aus der westlichen Ostsee. *Geologische Rundschau 58*: 621–655.

Hayes, J.M., Lambert, I.B. and Strauss, H. (1992). The sulfur-isotopic record. In: *The Proterozoic Biosphere: A Multidisciplinary Study* (eds. J.W. Schopf and C. Klein), 129–132. Cambridge: Cambridge University Press.

Hennecke, E., Luther, G.W., de Lange, G.J. et al. (1997). Sulphur speciation in anoxic hypersaline sediments from the eastern Mediterranean Sea. *Geochimica et Cosmochimica Acta 61*, 307–321.

Holland, H.D. (1984). *The Chemical Evolution of the Atmosphere and Oceans*. Princeton, NJ: Princeton University Press.

Holland, H.D. (2006). The oxygenation of the atmosphere and oceans. *Philosophical Transactions of the Royal Society Series B 361*: 904–916.

Holser, W.T. and Kaplan, I.R. (1966). Isotope geochemistry of sedimentary sulfates. *Chemical Geology 1*: 93–35.

Holser, W.T., Schidlowski, M., Mackenzie, F.T. et al. (1988). Geochemical cycles of carbon and sulfur. In: *Chemical Cycles in the Evolution of the Earth* (eds. C.B. Gregor, R.M. Garrels, F.T. MacKenzie, et al.), 105–173. New York: Wiley.

Hulston, J.R. and Thode, H.G. (1965). Variations in the S^{33}, S^{34} and S^{36} contents of meteorites and their relation to chemical and nuclear effects. *Journal of Geophysical Research 70*: 3475–3484.

Jørgensen, B.B. (1982). Mineralization of organic matter in the sea bed: the role of sulfate reduction. *Nature 296*, 643–645.

Jørgensen, B.B. (1990). A thiosulfate shunt in the sulfur cycle of marine sediments. *Science 249*: 152–154.

Jørgensen, B.B., Böttcher, M.E., Lüschen, H. et al. (2004). Anaerobic methane oxidation and a deep H_2S sink generate isotopically heavy sulfides in Black Sea sediments. *Geochimica et Cosmochimica Acta 68*: 2095–2118.

Jørgensen, B.B., Beulig, F., Egger, M. et al. (2019). Organoclastic sulfate reduction in the sulfate-methane transition of marine sediments. *Geochimica et Cosmochimica Acta 254*: 231–245.

Johnston, D.T. (2011). Multiple sulfur isotopes and the evolution of Earth's surface sulfur cycle. *Earth-Science Reviews 106*: 161–183.

Johnston, D.T., Farquhar, J. and Canfield, D.E. (2007). Sulfur isotope insights into microbial sulfate reduction: when microbes meet models. *Geochimica et Cosmochimica Acta 71*: 3929–3947.

Johnston, D.T., Farquhar, J., Wing, B.A. et al. (2005). Multiple sulfur isotope fractionations in biological systems: a case study with sulfate reducers and sulfur disproportionators. *American Journal of Sciences 305*: 645–660.

Johnston, D.T., Gill, B.C., Masterson, A. et al. (2014). Placing an upper limit on the cryptic marine sulphur cycling. *Nature 513*: 530–533.

Kampschulte, A. and Strauss, H. (2004). The sulfur isotopic evolution of Phanerozoic seawater based on the analysis of structurally substituted sulfate in carbonates. *Chemical Geology 204*: 255–286.

Kaplan, I.R. and Rittenberg, S.C. (1964). Microbiological fractionation of sulphur isotopes. *Journal of General Microbiology 34*: 195–212.

Kaplan, I.R., Emery, K.O. and Rittenberg, S.C. (1963). The distribution and isotopic abundance of sulphur in recent marine sediments off southern California. *Geochimica et Cosmochimica Acta 27*: 297–331.

Kasting, J.F. (1993). Earth's early atmosphere. *Science 259*: 920–926.

Knoll, A.H., Hayes, J.M., Kaufman, A.J. et al. (1986). Secular variation in carbon isotope ratios from upper Proterozoic successions of Svalbard and East Greenland. *Nature 321*: 832–838.

Krouse, H.R. and Coplen, T.B. (1997). Reporting of relative sulfur isotope-ratio data. *Pure and Applied Chemistry 69*, 293–295.

Kunzmann, M., Bui, T.H., Crockford, P.W. et al. (2017). Bacterial sulfur disproportionation constrains timing of Neoproterozoic oxygenation. *Geology 45*: 207–210.

Kurzweil, F., Claire, M., Thomazo, C. et al. (2013). Atmospheric sulfur rearrangement 2.7 billion years ago: evidence for oxygenic photosynthesis. *Earth and Planetary Science Letters 366*: 17–26.

Leavitt, W.D., Halevy, I., Bradley, A.S. et al. (2013). Influence of sulfate reduction rates on the Phanerozoic sulfur isotope record. *Proceedings of the National Academy of Sciences of the United States of America 110*: 11244–11249.

Lin, Z.Y., Sun, X.M., Peckmann, J. et al. (2016). How sulfate-driven anaerobic oxidation of methane affects the sulfur isotopic composition of pyrite: a SIMS study from the South China Sea. *Chemical Geology 440*: 26–41.

Lin, Z., Sun, X., Strauss, H. et al. (2017). Multiple sulfur isotope constraints on sulfate-driven anaerobic oxidation of methane: evidence from authigenic pyrite in seepage areas of the South China Sea. *Geochimica et Cosmochimica Acta 211*: 153–173.

Marin-Carbonne, J., Busigny, V., Miot, J. et al. (2019). *In situ* Fe and S isotope analyses in pyrite from the 3.2 Ga Mendon Formation (Barberton Greenstone Belt, South Africa): evidence for early microbial iron reduction. *Geobiology* 2020. DOI:10.1111/gbi.12385.

Melezhik, V.A., Fallick, A.E., Hanski, E.J. et al. (2005). Emergence of the aerobic biosphere during the Archean–Proterozoic transition: challenges of future research. *GSA Today 15*: 4–11.

Monster, J., Appel, P.W.U., Thode, H.G. et al. (1979). Sulfur isotope studies in early Archaean sediments from Isua, West Greenland: implications for the antiquity of bacterial sulfate reduction. *Geochimica et Cosmochimica Acta 43*: 405–413.

Mossmann, J.-R., Aplin, A.C., Curtis, C.D. and Coleman, M.L. (1991). Geochemistry of inorganic and organic sulphur in organic-rich sediments from the Peru Margin. *Geochimica et Cosmochimica Acta 55*: 3581–3595.

Nielsen, H. (1965). Schwefelisotope im marinen Kreislauf und das $\delta^{34}S$ der früheren Meere. *Geologische Rundschau 55*: 160–172.

Papineau, D., Mojzsis, S.J. and Schmitt, A.K. (2007). Multiple sulfur isotopes from Paleoproterozoic Huronian interglacial sediments and the rise of atmospheric oxygen. *Earth and Planetary Science Letters 255*: 188–212.

Paris, G., Sessions, A.L., Subhas, A.V. et al. (2013). MC-ICP-MS measurement of $\delta^{34}S$ and $\Delta^{33}S$ in small amounts of dissolved sulfate. *Chemical Geology 345*: 50–61

Paytan, A., Kastner, M., Campbell, D. and Thiemens, M.H. (2004). Seawater sulfur isotope fluctuations in the Cretaceous. *Science 304*: 1663–1665.

Pavlov, A.A. and Kasting, J.F. (2002). Mass-independent fractionation of sulfur isotopes in Archean sediments: Strong evidence for an anoxic Archean atmosphere. *Astrobiology 2*, 27–41.

Peketi, A., Mazumdar, A., Joshi, R. et al. (2012). Tracing the paleo sulfate-methane transition zones and H_2S seepage events in marine sediments: an application of C-S-Mo systematics. *Geochemistry Geophysics Geosystems 13*: Q10007. http://dx.doi.org/10.1029/2012GC004288.

Pellerin, A., Antler, G., Holm, S.A. et al. (2019). Large sulfur isotope fractionation by bacterial sulfide oxidation. *Science Advances 5*: 1480–1486.

Philippot, P., Ávila, J.N., Killingsworth, B.A. et al. (2018). Globally asynchronous sulphur isotope signals require re-definition of the Great Oxidation Event. *Nature Communications 9*: 2245–2255.

Present, T.M., Paris, G., Burke, A. et al. (2015). Large carbonate associated sulfate isotopic variability between brachiopods, micrite and other sedimentary components in Late Ordovician strata. *Earth and Planetary Science Letters 432*: 187–198.

Present, T.M., Bergmann, K.D., Myers, C. et al. (2018). Pyrite-walled tube structures in a Mesoproterozoic sediment-hosted metal sulfide deposit. *Geological Society of America Bulletin 130*: 598–616.

Present, T.M., Gutierrez, M., Paris, G. et al. (2019). Diagenetic controls on the isotopic composition of carbonate-associated sulphate in the Permian Capitan Reef Complex, West Texas. *Sedimentology 66*: 2605–2626.

Raab, M. and Spiro, B. (1991). Sulfur isotopic variations during seawater evaporation with fractional crystallization. *Chemical Geology Isotope Geoscience 86*: 323–333.

Rees, C.E., Jenkins, W.J. and Monster, J. (1978). The sulphur isotopic composition of ocean water sulphate. *Geochimica et Cosmochimica Acta 42*: 377–381.

Reinhard, C.T., Planavsky, N.J. and Lyons, T.W. (2013). Long-term sedimentary recycling of rare sulphur isotope anomalies. *Nature 497*: 100–103.

Reuschel, M., Strauss, H., Lepland, A. et al. (2013). The end of mass-independent fractionation of sulphur isotopes. In: *Reading the Archive of Earth's Oxygenation. Volume 3: Global Events and the Fennoscandian Arctic Russia – Drilling Early Earth Project*, Chapter 7.1 (eds V.A., Melezhik, A.R. Prave, A.E. Fallick et al.), 1049–1058. Dordrecht: Springer.

Rickard, D. and Luther III, G.W. (2007). Chemistry of iron sulfides. *Chemical Reviews 107*: 514–562.

Schidlowksi, M., Hayes, J.M. and Kaplan, I.R. (1983). Isotopic inferences of ancient biochemistries: carbon, sulfur, hydrogen and nitrogen. In: *Earth's Earliest Biosphere – Its Origin and Evolution* (ed. J.W. Schopf), 149–186. Princeton, NJ: Princeton UniversityPress.

Schobben, M., Stebbins, A., Ghaderi, A. et al. (2015). Flourishing ocean drives the end-Permian marine mass-extinction. *Proceedings of the National Academy of Sciences of the United States of America 112*: 10298–10303.

Schobben, M., Stebbins, A., Algeo, T.J. et al. (2017). Volatile Early Triassic sulfur cycle: a consequence of persistent low seawater sulfate concentrations and a high sulfur cycle turn-over rate? *Palaeogeography, Palaeoclimatology, Palaeoecology 486*: 74–85.

Shawar, L., Halevy, I., Said-Ahmad, W. et al. (2018). Dynamics of pyrite formation and organic matter sulfurization in organic-rich carbonate sediments. *Geochimica et Cosmochimica Acta 241*: 219–239.

Shawar, L., Said-Ahmad, W., Ellis, G.S. and Amrani, A. (2020). Sulfur isotope composition of individual compounds in immature organic-rich rocks and possible geochemical implications. *Geochimica et Cosmochimica Acta 274*: 20–44.

Shen, Y., Buick, R. and Canfield, D.E. (2001). Isotopic evidence for microbial sulphate reduction in the early Archaean era. *Nature 410*: 77–81.

Siedenberg, K., Strauss, H., Podlaha, O.G. et al. (2018). Multiple sulfur isotopes ($\delta^{34}S$, $\Delta^{33}S$) of organic sulfur and pyrite from Late Cretaceous to Early Eocene oil shales in Jordan. *Organic Geochemistry 125*: 29–40.

Sim, M.S., Bosak, T. and Ono, S. (2011). Large sulfur isotope fractionation does not require disproportionation. *Science 333*: 74–77.

Sinninghe-Damsté, J.S., Rijpstra, W.I.C., Kock-van-Dalen, A. et al. (1989). Quenching of labile functionalized lipids by inorganic sulphur species: evidence for the formation of sedimentary organic sulphur compounds at the early stages of diagenesis. *Geochimica et Cosmochimica Acta 53*: 1343–1355.

Staudt, W.J. and Schoonen, M.A.A. (1995). Sulfate incorporation into sedimentary carbonates. In: *Geochemical Transformations of Sedimentary Sulfur* (eds. M.A. Vairavamurthy and M.A.A. Schoonen), 332–345. Washington DC: American Chemical Society.

Strauss, H. (1997). The isotopic composition of sedimentary sulfur through time. *Palaeogeography, Palaeoclimatology, Palaeoecology 132*: 97–118.

Strauss, H. (1999). Geological evolution from isotope proxy signals: sulfur. *Chemical Geology 161*: 89–101.

Strauss, H. (2004). 4 Ga of seawater evolution: evidence from the sulfur isotopic composition of sulfate. In: *Sulfur Biogeochemistry: Past and Present. Special Paper 379* (eds. J.P. Amend, K.J. Edwards and T.W. Lyons), 195–205. Geological Society of America.

Szabo, A., Tudge, A., Macnamara J. et al. (1950). The distribution of S^{34} in nature and the sulfur cycle. *Science 111*: 464–465.

Thiemens, M.H. and Heidenreich, J.E. (1983). The mass-independent fractionation of oxygen – a novel isotope effect and its possible cosmochemical implications. *Science 219*: 1073–1075.

Thode, H.G., MacNamara, J. and Collins, C.B. (1949). Natural variations in the isotopic content of sulphur and their significance. *Canadian Journal of Research B 27*: 361–73.

Thode, H.G., Macnamara, J. and Fleming, W.H. (1953). Sulfur isotope fractionation in nature and geological and biological time scales. *Geochimica et Cosmochimica Acta 3*: 235–243.

Tostevin, R., Turchyn, A.V., Farquhar et al. (2014). Multiple sulfur isotope constraints on the modern sulfur cycle. *Earth and Planetary Sciences Letters 396*: 14–21.

Urban, N.R., Ernst, K. and Bernasconi, S. (1999). Addition of sulfur to organic matter during early diagenesis of lake sediments. *Geochimica et Cosmochimica Acta 63*: 837–853.

Werne, J.P., Hollander, D.J., Lyons, T.W. et al. (2004). Organic sulfur biogeochemistry: recent advances and future research directions. In: *Sulfur Biogeochemistry: Past and Present. Special Paper 379* (eds. J.P. Amend K.J. Edwards and T.W. Lyons), 135–149. Geological Society of America.

Werne, J.P., Lyons, T.W., Hollander, D.J. et al. (2008). Investigating pathways of diagenetic organic matter sulfurization using compound-specific sulfur isotope analysis. *Geochimica et Cosmochimica Acta 72*: 3489–3502.

Wing, B.A. and Halevy, I. (2014). Intracellular metabolite levels shape sulfur isotope fractionation during microbial sulfate reduction. *Proceedings of the National Academy of Sciences 111*: 18116–18125.

Wotte, T., Shields-Zhou, G.A. and Strauss, H. (2012). Carbonate-associated sulfate: Experimental comparisons of common extraction methods and recommendations toward a standard analytical protocol. *Chemical Geology 326–327*: 132–144.

Wu, N., Farquhar, J., Strauss, H., Kim, S.T. and Canfield, D.E. (2010). Evaluating the S-isotope fractionation associated with Phanerozoic pyrite burial. *Geochimica et Cosmochimica Acta 74*, 2053–2071.

Wu, N., Farquhar, J. and Strauss, H. (2014). $\delta^{34}S$ and $\Delta^{33}S$ records of Paleozoic seawater sulfate based on the analysis of carbonate associated sulfate. *Earth and Planetary Science Letters 399*: 44–51.

Zahnle, K., Claire, M. and Catling, D. (2006). The loss of mass-independent fractionation in sulfur due to a Palaeoproterozoic collapse of atmospheric methane. *Geobiology 4*: 271–283.

Zerkle, A.L., Farquhar, J., Johnston, D.T. et al. (2009). Fractionation of multiple sulfur isotopes during phototrophic oxidation of sulfide and elemental sulfur by a green sulfur bacterium. *Geochimica et Cosmochimica Acta 73*: 291–306.

3

The Role of Microorganisms in Iron Reduction in Marine Sediments

Carolina Reyes[1] and Patrick Meister[2]

ABSTRACT

Iron-reducing microbial activity may occur in marine sediments in zones where it is not outcompeted by abiotic reaction of three-valent (ferric) iron with sulfide. Because near neutral pH ferric iron prevails in solid form, microorganisms have developed different strategies for extracellular electron transfer. Microbial pili, nanowires, and electron shuttles have been observed in two microorganisms, *Shewanella* and *Geobacter*, to transfer electrons to iron oxide mineral surfaces. Alternatively, iron may be chelated and transported into the cell. However, other than in these two model microorganisms, the pathways of electron transfer are poorly understood. Previous diversity studies showed that a number of phylogenetic groups of both Bacteria and Archaea, capable of using iron as an electron acceptor, occur in marine sediments, some of which are versatile, capable of reducing various electron acceptors. These groups seem to be associated with zones of iron reduction, such as suboxic zones and sulfide-free methanogenic zones, but their pathways of iron reduction are largely unknown. We highlight the necessity to further elucidate electron transport pathways including extracellular systems, using molecular genetic studies, to understand their ecological role in natural systems such as in marine sediments.

3.1. INTRODUCTION

Iron is often reduced and oxidized within closely spaced zones in the marine sedimentary biome (Froelich et al. 1979), and iron oxides often represent the most abundant solid phase electron acceptor. Although three-valent iron [iron(III)] does not spontaneously react with organic electron donors, such reactions may be mediated by microbial enzymatic pathways, providing a source of energy to the respective organisms residing in a particular location in a redox gradient. In contrast, iron(III) is abiotically reduced by sulfide. Under anoxic marine conditions, various reduced sulfur species may be available to react with iron (Canfield 1989; Rickard and Luther III, 2007). It is often difficult to assess whether iron cycling is driven by abiotic reaction with sulfide in a prevailing redox gradient, or whether it is the result of a microbial process. Nevertheless, evidence in a number of studies reviewed herein show that bacteria in marine environments are indeed using iron as electron acceptor in their metabolic pathways.

Microbially driven iron reduction most likely occurs in locations where reactions are not outcompeted by abiotic reduction of iron. Ocean margins, including shoreline, continental shelf, and slope regions, are abundant in terrigenous-clastic sediments, which are generally iron-rich but also rich in organic matter. In contrast, open-ocean sediments generally receive limited iron from the continent but large amounts of iron are delivered from mid-ocean ridge hydrothermal activity (not further discussed here). Sedimentary organic matter can be delivered from land to the ocean by river discharge or become deposited from

[1]Department of Environmental Geosciences, University of Vienna, Centre for Microbiology and Environmental Systems Science, Althanstrasse 14, 1090, Vienna, Austria

[2]Department of Geodynamics and Sedimentology, University of Vienna, Faculty of Earth Sciences, Geography and Astronomy, Althanstrasse 14, 1090, Vienna, Austria https://orcid.org0000-0002-7413-607X

Corresponding Author: Carolina Reyes (creyes6@gmail.com), Patrick Meister (patrick.meister@univie.ac.at)

ORCiD code: 0000-0003-3623-6456

Systems Biogeochemistry of Major Marine Biomes, First Edition. Edited by Aninda Mazumdar and Wriddhiman Ghosh.

marine primary production (de Haas et al. 2002). As a result, coastal upwelling zones, estuaries, and stagnant shelf basins may show anoxic conditions in bottom water, further enhancing the preservation of organic matter, giving rise to predominantly sulfidic conditions in the sediment. In turn, in shallow coastal areas water energy is often high (above the wave base), resulting in better ventilated, more oxic conditions at the sediment surface and suboxic conditions within the sediment, supporting iron cycling. Deep-sea sediments also show suboxic conditions in some areas (e.g. Meister et al., 2009). Overall, different environmental factors support microbial iron cycling, but may also support reactions via competing pathways. It is still not entirely clear which factors control microbial iron reduction in marine sediments and, therefore, it is difficult to assess which pathways the organisms use to access this source of electron acceptors.

The metabolic pathways involved in iron reduction are explored in detail in only two model organisms, *Shewanella* and *Geobacter*, but only the latter was suggested to be responsible for iron reduction in marine sediments. However, as shown by Reyes et al. (2016), several representatives of other phylogenetic groups with members capable of performing iron reduction were detected in Baltic Sea sediments, using DNA sequencing. Among these organisms, several groups are sulfate reducers or versatiles, possibly switching to iron reduction if the conditions are favorable. Furthermore, as iron itself plays a fundamental role as a reactive center in many electron-transporting enzymes and cofactors, the availability of iron at trace amounts is a requirement for all organisms. Thus, understanding which organism can reduce iron and what strategies they use to access solid-phase iron(III) would be fundamental to understanding the ecological role of iron cycling in marine systems.

While current literature on this topic is rather limited, this chapter will provide a brief overview of current knowledge and highlight the role of iron-reducing organisms in marine sediments. This chapter focuses on the question of where in the sediment iron reduction occurs and what microbial pathways are potentially involved. It then discusses which microorganisms are involved and how diverse these communities are. In addition, a compilation and a phylogenetic tree of known iron-reducing microorganisms in the marine environment is presented and their functional diversity is discussed.

3.2. THE REDOX ZONES IN SHALLOW MARINE SEDIMENTS

Dissolved iron(III) [Fe^{3+}] under neutral to slightly alkaline pH prevailing in ocean margin sediments has a redox potential around 0 V, which enables this ion to serve as an electron acceptor for most anaerobic metabolic pathways, except for the oxidation of Mn(II) to Mn(IV) and NH_4^+ to NO_3^- (Schulz and Zabel, 2006, p. 187). Accordingly, an iron reduction zone commonly occurs below the oxic/anoxic boundary, starting at the depth where nitrate and dissolved iron overlap (Froelich et al. 1979; König et al. 1999; Meister et al. 2014). The iron zone is also limited by the presence of dissolved sulfide, which reacts spontaneously with ferric iron to form insoluble iron sulfides (Schulz and Zabel, 2006, p. 247). In fact, the iron zone may be missing in cases where the iron content is low and sulfide production is overwhelming the production of dissolved iron(II) [Fe^{2+}], which is generally the case in organic carbon-rich sediments showing strong reducing conditions. In these cases, sulfide may even reach the seafloor, as shown, for example, by Calvert and Karlin (1991), or may be linked to nitrate reduction by *Thioploca* (Otte et al. 1999, pp. 3148–3157).

The depth of the iron zone in the sequence of downward decreasing electron potential is, however, not just the result of spontaneous interaction of the ions. Froelich et al. (1979) demonstrated that a redox zonation arises due to a competitive advantage of the organisms capable of using a stronger electron acceptor coupled with the oxidation of a limited pool of carbon sources, while organisms using an electron acceptor with a lesser redox potential are outcompeted. The redox zonation is thus a result of microbial ecology, favouring organisms that can gain more energy and accordingly produce more adenosine triphosphate (ATP) per electron transferred. In sediments, the competitive advantage of iron reducers thus also depends on organic carbon supply, reactivity of organic molecules, sediment properties, and sedimentation rate, interacting with the microbial community.

One factor that may modify the sequence of electron acceptors in marine sediments is the poor solubility of iron and manganese oxides, which prevail as minerals under neutral pH. Two different effects play a role for the availability of iron as an electron acceptor for microbial metabolisms:

1. Dissolution may be accelerated by chelating agents in the solution. Chelators are capable of maintaining iron in an oxidized state in solution, and they can be transported to organisms. However, chelation also strongly depends on the reactivity of the mineral.
2. Reduction of the iron may occur while the iron atom is still at its site in the crystal lattice, and it is then released as Fe^{2+} to the surrounding solution. Reductive dissolution may occur abiotically via sulfide oxidation or biotically using extracellular electron transport systems (see below).

According to Afonso and Stumm (1992), iron reduction by reaction with sulfide essentially depends on the surface area of the mineral. Canfield (1989) showed that reactivity towards sulfide strongly depends on the mineralogy of the iron phase. Poorly ordered iron oxides, but also iron oxides such as hematite and magnetite, are rather rapidly reduced, while some sheet silicates, such as illite or mica, are essentially insoluble over millions of years. This has also been observed in the ratio of pyrite over reactive iron content, showing a steep increase in the uppermost few metres below seafloor but only a minor increase below this depth in sediments of the Peruvian continental shelf (Meister et al. 2019). However, in sediments where sulfide is depleted, reactive iron may survive over a longer period if the sediment becomes rapidly buried into the methanogenic zone, or if microbial activity is generally too low to produce significant amounts of sulfide (suboxic zones). In methanogenic zones, iron (III) can persist over geological timescales (Riedinger et al., 2017) and thereby give rise to cryptic sulfur cycling, if sulfur becomes available (Holmkvist et al. 2011) or it can serve as an electron acceptor driving the anaerobic oxidation of methane (Beal et al. 2009; Wankel et al. 2012). Indeed, methanogenic zones can occur at very shallow depth (ca. 1 m, Thang et al. 2013; or even less: Oni et al. 2015), in which case solid iron phases are likely to bypass the sulfate reduction zone.

However, even within the zone where free iron is present in solution, it cannot be excluded that reductive dissolution via sulfide is driving iron reduction (Picard et al. 2016). Although the most soluble forms of iron sulfide, such as mackinawite or finely dispersed and poorly crystalline FeS in the sediment, allow for sulfide concentrations only in the nanomolar level in the presence of Fe^{2+} (Figure 3.1 shows the sulfide concentrations and activities for any given iron concentration /activity at saturation with respect to sedimentary and crystalline FeS), reduced sulfur species may always be available as polysulfide or organosulfur complexes. Thus, a zone with iron in solution, as shown for example in Figure 3.2a, is not necessarily an indication for microbially mediated iron reduction. Hence, in these zones it is difficult to reveal whether or not microbial iron reduction truly takes place. Evidence that could support that iron reduction is unrelated to sulfur cycling is provided by the oxygen isotope

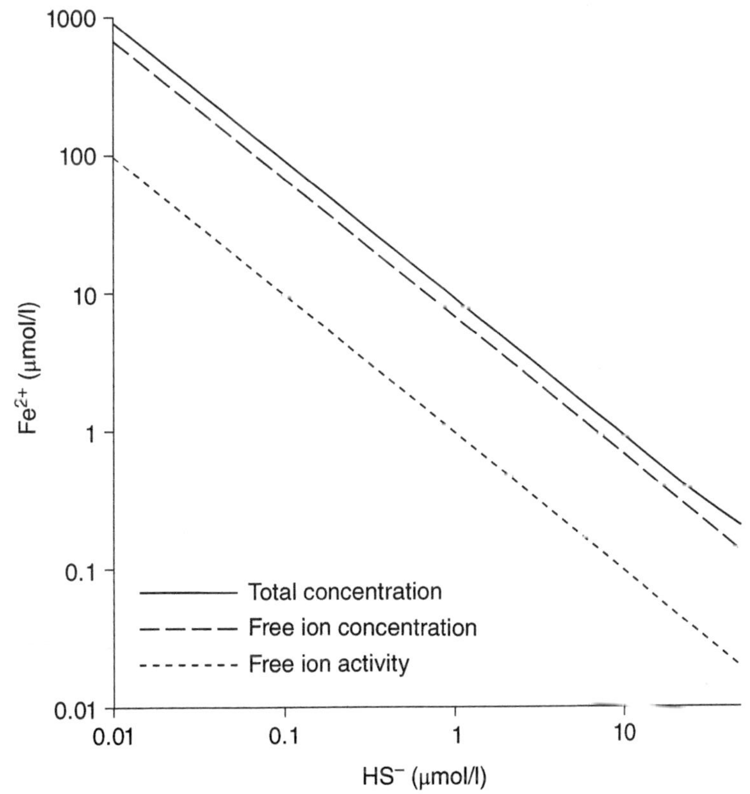

Figure 3.1 Concentrations and activities of reduced iron and sulfide in seawater in equilibrium with sedimentary FeS precipitate [log (IAP/Ks) = 0]. The solid line indicates concentrations of total dissolved Fe^{2+} and ΣH_2S. The dashed line indicates the concentrations of free Fe^{2+} and HS^-. The dotted line indicates the activities of free Fe^{2+} and HS^-. Values are corrected for ion activity using the extended Debye–Hückel method and using parameters from the wateq4f.dat database in PhreeqC (Parkhurst and Appelo, 2013). Note that the solubilities of mackinawite (FeS), greigite (Fe_3S_4), and pyrite (FeS_2) are magnitudes lower than bulk sedimentary FeS.

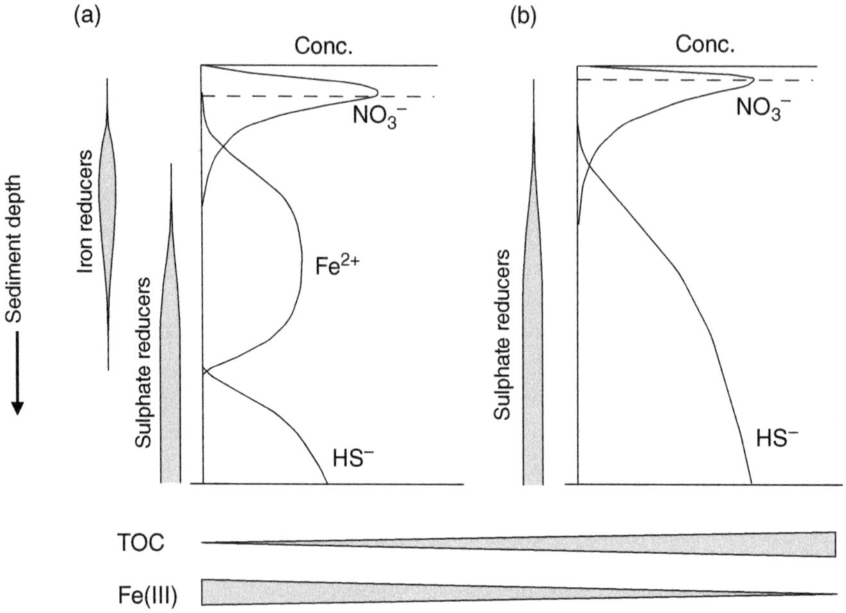

Figure 3.2 Schematic porewater profiles and dominant microbial groups through the uppermost sediment showing two different types of redox zonation: (a) With low and/or poorly reactive organic matter (TOC) and high reactive ferric iron content, a suboxic zone with free Fe^{2+} in the porewater is established. Within the suboxic zone, organisms may increasingly use sulfur metabolism as iron becomes more and more limited. (b) If organic matter is more abundant and/or reactive iron content is lower, dissimilatory iron reduction cannot keep up with sulfate reduction and the suboxic zone disappears. The dashed line represents the oxic front.

value ($\delta^{18}O$) of sulfate. The $\delta^{18}O$ in sulfate in seawater (~10‰ relative to standard mean ocean water) is not in equilibrium with seawater, and equilibration occurs upon reoxidation of sulfide, as new S–O bonds are formed (Wortmann et al. 2007). Therefore, $\delta^{18}O$ in sulfate is a good indicator for sulfur cycling, or the lack of it, in a zone of iron reduction. Furthermore, iron reduction has been linked to the consumption of several different organic substrates in incubation experiments using sediments from Skagerrak (Baltic Sea; Jensen et al. 2003). A quantitative consumption of organic substrate with iron as electron acceptor has also been shown by Canfield et al. (1993) in the Skagerrak by mass balance calculation.

Although the above-mentioned studies cannot entirely exclude the contribution of sulfur cycling, Reyes et al. (2016) discussed a hypothesis for how microbial iron reduction is controlled. Iron reduction is limited by the reactivity of the mineral surface, as shown by Afonso and Stumm (1992), and depends on the mineralogy of the iron oxide. Iron reduction has a competitive advantage over sulfate reduction, as it provides more energy (sensu Froelich et al. 1979), but it is limited by the reactivity of the solid phase. Hence, iron reduction would only outcompete sulfate reduction to the level at which the reactivity of the iron mineral can keep up with the rate of sulfate reduction. In surface sediments with abundant reactive organic matter, sulfate reduction would outcompete iron reduction (Figure 3.2b). However, in somewhat less productive areas, or perhaps in zones with more refractory organic matter (e.g. organic matter that has already experienced partial degradation under aerobic conditions, or organic matter derived from land plant material) decomposing more sluggishly, at a level that iron reduction rates can keep up, iron reduction may dominate over sulfate reduction. The ecological role of iron reducers in the marine biome may thus depend on a delicate balance of factors, such as reactivity of iron minerals, reactivity of organic matter, availability of sulfur species and the microbe's ability to mediate each of the different processes.

3.3. BIOCHEMICAL PATHWAYS OF IRON REDUCTION

3.3.1. The Specificity of Microbial Pathways with Respect to Iron

Coupling of iron reduction to the oxidation of specific organic molecules does not occur spontaneously under Earth surface conditions. Microbes can catalyse these reactions by means of specific enzymatic pathways and, therefore, take advantage of organic molecules as energy sources. As discussed earlier, such reactions can be bypassed abiotically if other microbes produce reduced sulfur species. The overall difference in redox potential from the substrate

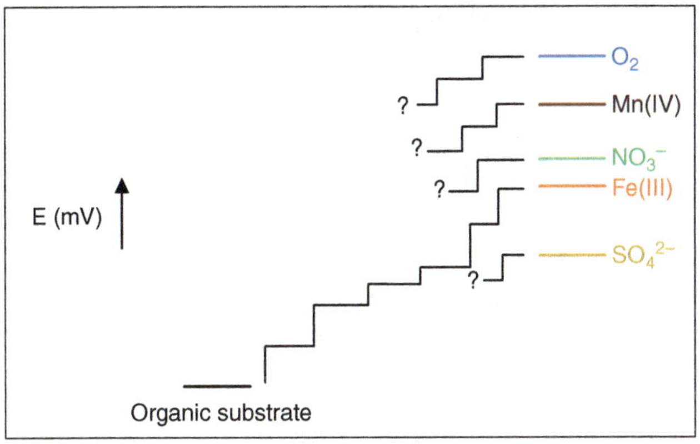

Figure 3.3 Schematic diagram of redox potentials of enzymes in the cascade of an electron transport chain, linking the organic substrate with iron oxide as terminal electron acceptor. Specificity in the pathway with respect to iron may arise from its capacity to yield the maximum amount of ATP or other energy equivalents. This pathway would not yield the maximum energy equivalents if linked to any of the other terminal electron acceptors.

to the electron acceptor must match the sum of differences in redox potential of all steps in a pathway (as shown schematically in Figure 3.3), in order to yield the maximum amount of ATP or other energy equivalents for the organism to develop its full competitive potential. Thereby each enzyme, and each reactive site within the enzymes (e.g. heme groups in a cytochrome; cf. Santos et al. 2015), should represent a step in the a cascade of increasing redox potential through the electron transport chain from the electron donor to the terminal electron acceptor.

What is best understood about extracellular electron transport is the general pathway of the electron from the cytoplasm to ferric iron (Figure 3.4). The passage of electrons to iron, residing on the cell exterior, happens by way of enzymes and molecules that traverse the inner and outer membranes. In the bacterial strains *Shewanella* MR-1 and ANA-3, a series of multi-heme cytochrome proteins can shuttle electrons from inside the cytoplasmic membrane to the outer membrane (El-Naggar et al. 2010; Clarke et al. 2011; Reyes et al. 2012; Watanabe et al. 2017).

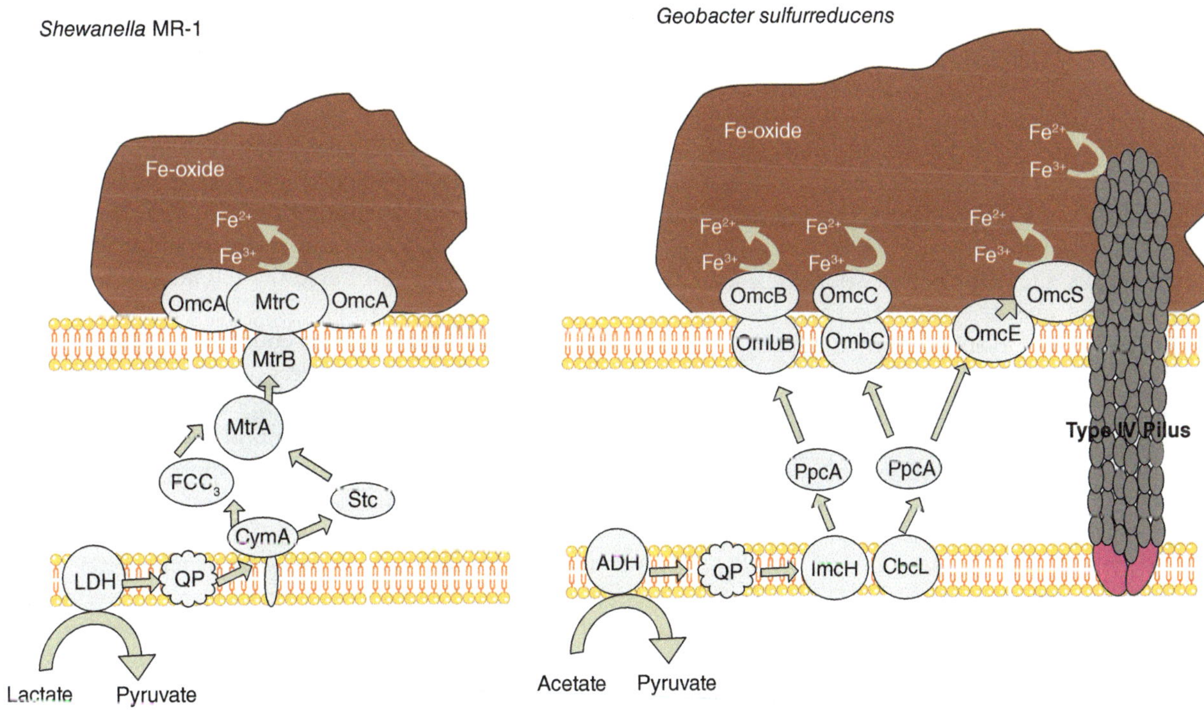

Figure 3.4 Schematic of multiheme cytochromes involved in transporting electrons from the organic substrate within the cell to the surface of an iron oxide mineral in (A) Shewanella and (B) Geobacter, adapted from Shi et al. (2007), Santos et al. (2015), and Shi et al. (2016). (A) When MR-1 oxidizes a carbon substrate such as lactate by a dehydrogenase, electrons from this reaction are transported to a quinone/quinol pool. Electrons are passed onto cytoplasmic and periplasmic multiheme cytochromes and to the outermembrane multiheme cytochromes. These cytochromes can then reduce the iron oxide surface directly or further pass electrons via an extracellular transport system. (B) In Geobacter sulfurreducens, electrons derived from the oxidation of a carbon substrate by a dehydrogenase can also be passed onto a quinone/quinol pool and then onto cytoplasmic (e.g. ImH, CbcL) and periplasmic multiheme cytochromes (e.g. PpcA), from which electrons can be delivered to outer membrane multiheme cytochromes (e.g. OmbB and OmbC). Alternatively, electrons can be transferred to outermembrane cytochromes (e.g. OmcE and OmcS), which are associated with the pili, further passing the electrons to the iron oxide. The pili themselves are also capable of extracellular electron transfer.

During iron reduction (e.g. ferrihydrite or iron(III)-citrate) electrons are donated from a reduced substrate in the cytoplasm (e.g. lactate) to a dehydrogenase in the inner membrane (Burns 2010). Proton translocation from the cytoplasm to the periplasm is presumed to occur as electrons proceed via a quinone pool (quinones and quinols embedded in the membrane) to the periplasmic cytochrome CymA (Shi et al. 2016). From CymA, electrons are passed to the periplasmic cytochromes Fcc3 and STC (Shi et al. 2016). These cytochromes in turn transfer electrons to MtrA, a decaheme cytochrome in the periplasm (Courselle et al. 2010), and MtrA transfers the electrons to the outer membrane cytochromes MtrC and OmcA via the porin protein MtrB (Shi et al. 2016).

Hypotheses exist regarding how electrons could be transferred from certain hemes of the outer membrane cytochrome MtrF in MR-1 (Clarke et al. 2011) to iron oxide (see below), but whether the electron transfer per se is specific to particular electron acceptors remains unclear. To date, there is only one study, involving *Shewanella oneidensis* MR-1 that has provided protein-crystallographic evidence of how iron as iron(III)-citrate and iron(III)-NTA bind to specific sites of the outer membrane cytochrome UndA before reduction (Edwards et al. 2012). Single frozen crystals of UndA, soaked in chelated iron, were analysed by synchrotron radiation. While binding sites near heme 7 of the undecaheme cytochrome UndA of MR-1 appear to specifically bind chelated iron(III), they do not appear to differentiate between the type of chelator, e.g. NTA or citrate (Edwards et al. 2012). Thus, sites near heme 7 could in general bind different types of chelated iron(III). These results imply that at least for this cytochrome there are specific sites for soluble iron reduction.

Geobacter sulfurreducens possesses at least 30 outer membrane cytochromes (Costa et al. 2018), although they are not structurally characterized to the extent of those of MR-1, and it is not known which hemes interact with different electron acceptors. Otero et al. (2018) have shown that multiple pathways of iron-oxide and iron(III)-citrate reduction are possible in *G. sulfurreducens* involving different combinations of cytochromes, supporting results from earlier studies. One proposed pathway for the reduction of iron oxides involves electron transfer from a quinone pool to multiheme cytochromes ImcH (Levar et al. 2014) and CbcL (Zacharoff et al. 2015) in the inner membrane. These in turn could transfer electrons to the periplasmic cytochrome PpcA (Lloyd et al. 2003) and from PpcA to the outer membrane cytochromes OmcBC via the porin proteins OmaBC (Shi et al. 2016). Other outer membrane cytochromes OmcE and OmcS were shown to be important for reduction of iron oxides but not of iron(III)-citrate (Mehta et al. 2005).

3.3.2. Microbial Strategies to Reduce Solid Iron Phases

In addition to the fact that the electrons have to be channeled from the substrate to the electron acceptor, it is particularly challenging for microorganisms that iron is usually not dissolved under marine, circumneutral pH conditions. Despite these challenges, several strategies still allow bacteria to deliver electrons to a solid-phase electron acceptor. Biochemical methods have allowed for whole cell interactions with metal oxides (Lower et al. 2007) and electrodes (Coursolle et al. 2010) to be determined and abiotic (Lower et al. 2007, 2008; Clarke et al. 2011; Breuer et al. 2015) and biotic (Lower et al. 2007) binding of specific multiheme cytochromes with different iron oxide mineral surfaces to be resolved. However, the mechanism of electron transport from the hemes to these solid-phase electron acceptors is still unknown.

Many studies of electron transfer from MR-1 cell suspensions to the surface of electrodes in microbial fuel cells and in cyclic voltammetry have been made using different types of electrodes including glassy carbon (Kim et al. 1999, p. 128; Kim et al. 2002), graphite (Bretschger et al. 2007; Wang et al. 2011), and hematite (Meitl et al. 2009). In all cases under specific experimental conditions with specific redox potentials, the cells have been able to deliver electrons to these various types of electrodes. Thus, it is irrelevant that the electron acceptor is iron, as electrons would be equally transported to any electrode with a particular electron potential. What matters is that the microorganism has the molecular pathway to transfer electrons from a substrate in the cell interior to an extracellular electron acceptor.

Theoretically, the bacteria could be growing directly on the surface of the iron mineral, which is indeed observed in laboratory experiments (Jeong et al. 2006). However, this is not always possible in natural sediments. Organisms have developed strategies to transport electrons to the mineral surface either via secreted organic molecules that act as electron shuttles, such as flavins, or through extensions of the outer membrane, such as nanowires, capable of transporting electrons. Alternatively, there are also strategies to transport Fe^{3+} ions from the mineral to the cell, using chelators (i.e. siderophores). In any case, the communities have to be highly organized, as otherwise the energetic costs of manufacturing such structures would not be compensated by the energy gain of reducing the extracellular electron acceptor.

All known *Shewanella* strains have been shown to secrete flavin mononucleotide (FMN), a molecule that can shuttle one or two electrons from its alloxazine ring structure to the extracellular electron acceptors (Marsili et al. 2008; von Canstein et al. 2008; Coursolle et al. 2010). Three out of the 10 hemes of the multiheme cytochrome MtrF, an MtrC homolog, have been shown to directly reduce FMN (–150 mV). En route to reduce FMN, electrons move from MtrF hemes with a low redox potential to MtrF hemes with a higher redox potential (Watanabe et al. 2017).

Another way in which MR-1 can reduce solid phase electron acceptors is via nanowires. Nanowires in MR-1 extend out of its outer-membrane and include the periplasm (Gorby et al. 2006; Pirbadian et al. 2014). These outer-membrane extensions bud out of the cell and form into vesicles. Thus, a nanowire is not a smooth filament, it is in fact a chain of outer membrane vesicles connected to each other with a periplasmic interior and can be differentiated from flagella and pili by electron cryotomography (see Figures 3, 4, and 7 in Subramanian et al. 2018). The outside of the vesicle is coated with outer membrane cytochromes (e.g. MtrC) and the inside (periplasm) is coated with periplasmic cytochromes (e.g. MtrA; Subramanian et al. 2018). These vesicles along the nanowire in turn are divided by gap junctions, which keep vesicles apart (Subramanian et al. 2018). These vesicles are also dynamic and can grow and shrink in size or transition from multiple vesicles into a single giant vesicle and vice versa (see Figure 5 in Subramanian et al. 2018). When multiple vesicles are connected to each other, it is hypothesized that electrons hop between the vesicles from cytochrome complex to cytochrome complex when heme groups are in close range. When these complexes are not in close range, it is hypothesized that cytochrome complexes or electron shuttles physically diffuse between vesicles until reaching another cytochrome complex thereby passing on the electrons (Subramanian et al. 2018).

It is known that *G. sulfurreducens* has two different types of filaments, which are implicated to participate in iron oxide reduction. Their three-nanometer-wide pili (termed e-pili) are mostly composed of the PilA protein with the hexaheme cytochrome OmcS distributed along the pili (Costa et al. 2018). In *Geobacter*, the e-pili are extensions of the cytoplasmic membrane (Reguera et al. 2005) that protrudes out of a channel protein embedded in the outer membrane (Shi et al. 2007). Studies by Malvankar et al. (2011, 2015) showed that the *Geobacter* e-pili are essential for reducing iron oxide. These e-pili have also been shown to transfer electrons between *Geobacter* and other species (Lovley, 2017). It is hypothesized that electrons could travel along the e-pili, from pili protein to pili protein, until reaching the solid phase electron acceptor (Feliciano et al. 2012; Lampa-Pastirk et al. 2016). Vargas et al. (2013) showed that overlapping π–π orbitals of aromatic moieties and aromatic amino acids that make up the pili protein PilA give the pili their conductive property. Recently, Filman et al. (2018) showed through cryo-electron microscopy (cryo-EM) and biochemical modeling that a second type of conductive filament is produced by *G. sulfurreducens* (termed OmcS-filaments), which could be involved in long-range electron transfer to iron minerals and other microorganisms. Unlike the outer membrane cytochrome vesicles of the MR-1 nanowire, the OmcS cytochromes are packed close enough along the length of the filament to allow for direct electron transfer between the heme groups of these cytochromes (Filman et al. 2018). A recent publication by Walker et al. (2019) shows that the archaellum of the archaeon *Methanospirillum* is conductively similar to the e-pili of *Geobacter*, demonstrating that these structures are not limited to bacteria.

Furthermore, siderophores, which are low molecular weight chelators (400–1000 kDa; Saha et al. 2012), are produced by organisms under iron-limited conditions in order to acquire and take up iron that would otherwise be bound to iron oxides or organic matter. There are various types of siderophores, which are classified based on the type of ligand that binds the iron. The ligands include moieties, such as catecholates, phenolates, hydroxamates, carboxylates, and mixed types (Saha et al. 2012), where the oxygen groups of the ligands form a bond to the iron center. Siderophores are excreted out of a bacterial cell by a transport protein or a pump (e.g. ATPase) and form a bond with Fe^{3+} ions (Saha et al. 2012). Once captured, either the Fe^{3+} ion enters the cell after being released from the siderophore, or the entire siderophore-iron complex enters the cell (Saha et al. 2012). While siderophores have a high affinity for iron, most siderophores also chelate other metals albeit with a lower affinity (Kraemer 2004). For example, *Pseudomonas aeruginosa* secretes the siderophore pyochelin, which in addition to Fe^{3+} also chelates other cations (Braud et al. 2009). While the chelation is thus not precisely specific for iron, specificity arises at the stage where pyochelin bound to iron(III) is taken into the cell by a specific transporter at a higher rate than three other pyochelin metal complexes, implying that siderophore-iron uptake by its transporter is selective in this organism.

A further strategy that is employed by uncultivated members of the *Desulfobulbaceae* family (filament-forming cable bacteria; Nielsen et al. 2010; Pfeffer et al. 2012; Schauer et al. 2014, pp.1314–1322) is cell-to-cell conductive electron transport, possibly via c-type cytochromes which are present in the envelope of single cable bacterial filaments (Figure S5, Kjeldsen et al. 2019) and show a redox gradient along the filament (Kjeldsen et al. 2019). However, this has yet to be experimentally verified. Cells attach to one another in a long chain to form a bacterial "cable". The outer membrane of cable bacteria is rigid and has tubular channels that connect the periplasm of neighboring cable bacteria to each other (Reguera 2018). The periplasm of these cells appears to be packed with c-type cytochromes (Bjerg et al. 2018).

Using minicores of shallow marine sediment from Aarhus Bay, Risgaard-Petersen et al. (2012) observed that sulfate can be oxidized 2 cm away from the reduction of oxygen, and that an electron gradient was present in the pore fluid. This was interpreted as indicative of microbial communities performing long-distance electron transport. Pfeffer et al. (2012) observed that sulfate could be reduced by cable bacteria at the bottom of long filaments within the top 2 cm of marine sediments. Subsequent in-situ studies of cable bacteria, have also observed them within the top few millimeters of coastal marine sediments (Otte et al. 2018).

While more details about the molecular pathways of electron transport are being discovered, it is still poorly understood how the different strategies of accessing solid electron acceptors affect the rates of dissolution of different iron mineral phases. Future studies using nondestructive in vivo techniques, such as interferometry (Waters et al. 2009) may provide possible routes to better understand and quantify mineral surface reactions in the presence of living microbes.

3.3.3. Uptake of Iron as a Nutrient

In addition to acting as an electron acceptor, iron is an important nutrient for all living organisms. Most importantly, iron plays a major role as reactive centers in electron shuttling enzymes, e.g. cytochrome, ferredoxin, and iron-sulfur proteins. Hence, it is critical that a sufficient level of iron is always available in order to drive the essential catabolic functions in the cell. Even under strongly sulfidic conditions, apparently there is still enough iron available despite the low solubility of iron sulfides (cf. Figure 3.1).

MR-1 in particular requires a large number of heme-containing cytochromes for its respiratory versatility. In a hypothesized model, one way iron could enter MR-1, is via undefined porins on the cell surface (Bennett et al. 2018). Under iron-deficient conditions MR-1 can secrete siderophores, such as putrebactin, to acquire Fe^{3+} and transport the Fe^{3+}-siderophore complex via a TonB protein complex on the outer membrane (Bennett et al. 2018). Once inside the periplasm, the protein FeoB could transport Fe^{2+} into the cytoplasm (Bennett et al. 2018). MR-1 could also use siderophores secreted by other microorganisms, transporting these into the cell via various transporters (Liu et al. 2018; Bennett et al. 2018). Interestingly, small alpha-hydroxy acids such as lactate could help to transport iron(III) into MR-1 via undefined porins in a fashion similar to siderophores (Bennett et al. 2018). Once in the cell, Fe^{3+} could be reduced to Fe^{2+} and, if concentrations of Fe^{2+} are high, imported from the periplasm into the cytoplasm with the help of Feo and FicI protein importers (Bennett et al. 2018). Otherwise, if concentrations are low, Feo alone could suffice for importing iron.

Similar to MR-1, in *G. sulfurreducens* cytoplasmic transporters FeoB and FeoB2 transport Fe^{2+} from the periplasm into the cytoplasm (Cartron et al. 2006, Embree et al. 2014). However, siderophore uptake mechanisms are not yet as well understood as in MR-1. Furthermore, microbes can also take up heme groups from outside the cell, produced by other bacteria in the community, and they can use special ATP-binding cassette (ABC) transporters to bring iron into the cell (Li et al. 2014, Toulza et al. 2012).

The strategy that siderophores are produced if iron becomes limiting, may in fact be related to the importance of iron as a catalyst – as a cofactor in many enzymes – rather than an electron acceptor (Andrews et al. 2003). If these siderophores are capable of accessing iron from more insoluble silicates, which are prevalent in most ocean margin sediments, such strategies may become more efficient (perhaps even outcompeting the reaction with sulfide). As reactive iron centers, such as in heme groups or iron-sulfur clusters, play an important role in all organisms of the phylogenetic tree, it is understandable that such pathways of iron uptake must have occurred in the earliest organisms on Earth, which may indeed have occurred in marine or brackish environments on the margins of early continents (cf. Sleep 2018), and which would have encountered some challenges in accessing insoluble iron after the onset of an oxic atmosphere and a complete sulfur cycle.

3.4. DIVERSITY OF POTENTIAL IRON-REDUCING AND IRON-OXIDIZING ORGANISMS

3.4.1. Correlation of Phylogenetic Abundances with Porewater Chemistry Data

Before discussing correlations of known or potential iron reducers with geochemical data, it is necessary to raise a note of caution. It has become a widely used practice to correlate phylogenetic abundance with all kinds of geochemical parameters, in particular with porewater concentration of potential metabolites, using multivariate statistics (principal component analysis, PCA; Ramette 2007). However, such correlations can be misleading, as the highest solute concentrations do not necessarily equate to highest activity with respect to a particular metabolite. Ramette (2007) specifically warns not to overinterpret the correlations by relying on unjustified causality. Depth profiles in porewater concentrations are mainly subject to diffusive transport along concentration gradients. A linearly increasing or decreasing gradient with depth essentially means that no consumption or production occurs. Instead the second derivative of the concentration after depth, which represents the curvature of the profile according to Fick's second law of diffusion (Eq. 1; Fick 1855), is an indication of the rate of production or consumption of a particular solute.

$$\frac{\partial c}{\partial t} = D\frac{\partial^2 c}{\partial z^2} \pm s(z) \tag{1}$$

c = concentration of a solute (mmol/l)
t = time
D = diffusion coefficient (m^2/s)
z = sediment depth
s = source or sink of solute

According to Eq. 1, the source or sink of a solute is equal to the second derivative after depth, if the profile is at steady state, i.e. $\partial c/\partial t = 0$. In theory, the second derivative of concentration after depth could be used in PCA analyses instead, which could potentially provide a more meaningful outcome. However, such an approach would have to be used with caution, as nonsteady state conditions may result in incorrect rate determinations. Ideally, phylogenetic data should be compared with geochemistry using depth profiles. In this review, we do not consider interpretations that are based on the correlation of solute concentrations in PCA analyses alone.

3.4.2. Diversity of Iron Reducers in Suboxic Zones

Several studies have attempted to find and cultivate iron reducers from the iron-reduction zone in the uppermost centimetres of marine sediments. Stapleton et al. (2005) isolated seven bacteria closely related to different *Shewanella* genera (91-98% similarity) from the iron-reduction zone (0-10 cm) of sediments collected from a transect off the coast of Washington State (USA) and Puget Sound (USA).

In the Scheldt Estuary (Netherlands, Lin et al. 2007), DNA was extracted from five layers of a 15-cm long sediment core (Site Waard), and the 16S rRNA genes of *Geobacteraceae*, *Shewanella*, *Anaeromyxobacter*, and *Geothrix* were amplified and quantified using a most probable number polymerase chain reaction (MPN-PCR) method. *Shewanella* and *Geobacteraceae* as well as the dissimilatory iron reducers *Anaeromyxobacter* and *Geothrix* were detected at all depths. Both *Shewanella* and *Geobacteraceae* were relatively abundant (between 0.1 and 1%) compared with the other two dissimilatory iron reducers (0.001%). Lin et al. (2007) also made enrichment cultures using the 5–15 cm depth interval of the sediment supplemented with various iron oxides and acetate and lactate as substrates. Microorganisms in the enrichments were identified by first amplifying the 16S rRNA gene followed by denaturing gradient gel electrophoresis (DGGE). In these enrichments, the fermenters *Ralstonia* and *Clostridium*, which are both capable of reducing iron, were the most abundant community members, while *Shewanella* was not as dominant (≤1%; see Figure 5 in Lin et al. 2007). Overall, Lin et al. (2007) found that the classical iron reducers and other dissimilatory iron reducers were not the predominant iron reducers in these suboxic sediments. Instead, fermenters that are known to donate their electrons to Fe(III) (Hamman and Ottow 1974; Dobbin et al. 1999; Park et al. 2001) were the most abundant.

Several other bacteria and archaea that have been isolated from marine sediment environments and can reduce iron are listed in Table 3.1 and include: *Shewanella loihica*, isolated from iron-rich mats near hydrothermal vents (Stapleton et al. 2005; Gao et al. 2006), and *Geothermobacter* sp. str. HR-1, isolated from Lōʻihi Seamount, Hawaiʻi (Smith et al. 2018).

In-situ diversity studies using 16S rRNA sequencing to detect potential iron reducers in marine surface sediments have been performed at several sites, such as Brown Bay in the Antarctic (Powell et al. 2003), the Baltic Sea at Askö (Sweden; Edlund et al. 2008), and the Archipelago Sea located at the intersection of the Gulf of Bothnia and the Gulf of Finland with the Baltic Sea (Sinnko et al. 2011, see Table S5 in Reyes et al. 2016). Although in some of these studies multivariate statistics were used to link microbial communities to zones of iron reduction, it is worth mentioning the fact that, independent of statistical analysis, these groups were identified as being closely related to known iron reducers.

Using next generation sequencing techniques, potential iron reducers were detected within the zone of iron reduction in the top few centimetres of the sediment of Skagerrak and Bothnian Bay (Figure 3.5; Reyes et al. 2016). These communities were later shown by quantitative PCR (qPCR) to be potentially active in situ and in incubations with marine sediments (Reyes et al. 2017). Microorganisms that are known to carry out dissimilatory iron reduction, such as *Desulfobacteraceae* (Lovley et al. 1993), *Desulfobulbaceae* (Lovley et al. 1993; Knoblauch et al. 1999), *Desulfurmonadaceae* (Roden and Lovley 1993; Liesack and Finster 1994; Coates et al. 1995; Krumholz et al. 1996; Lonergan et al. 1996), *Geobacteraceae* (Lovley and Phillips 1998; Caccavo et al. 1994), and *Pelobaceraceae* (Lovley et al. 1995; Lonergan et al. 1996), are present and possibly active in the iron reduction zone. As mentioned above, these bacteria are versatiles capable of reducing both iron and sulfate. While the abundance of potential iron reducers decreases with depth, the abundance of sulfate reducers *Desulfoccocus* and *Desulfobacteria* increases with depth. To date, pure cultures of *Desulfococcus* have not been shown to reduce iron.

Table 3.1 List of iron-reducing bacteria and archaea isolated from marine sediments.

Taxa	References
Archaea	
Methanothermococcus thermolithotrophicus str. DSM 2095	Huber (1982), Weber et al. (2006), Lovley et al. (2004)
Methanosarcina barkeri culture str. DSM 800	Sivan et al. (2016)
Ferroglobus placidus	Kashefi et al. (2002)
Geoglobus acetivorans str. SBH6	Slobodinkina et al. (2009)
Methanopyrus kandleri str. AV19	Kurr et al. (1991), Weber et al. (2006)
Pyrodictium abyssi str. DSM 6158	Pley et al. (1993), Vargas et al. (1998)
Pyrodictium delaneyi str. Su06	Lin et al. (2016)
Bacteria	
Thermatoga	
Thermatoga maritima	Vargas et al. (1998)
Defferibacteres	
Deferribacter thermophilus	Lovley (2013)
Deferribacter autotrophicus strain str. SL50	Slobodinkina et al. (2009)
Alphaproteobacteria	
Sinirhodobacter ferrireducens str. SgZ-3	Yang et al. (2013)
Betaproteobacteria	
Rhodoferax ferrireducens str. T118	Finneran et al. (2003)
Gammaproteobacteria	
Shewanella sp. str. ANA-3	Saltikov et al. (2003)
Shewanella oneidensis str. MR-1	Lovley (2013)
Shewanella loihica str. PV-4	Gao et al. (2006)
Shewanella arctica str. IR12	Kim et al. (2012)
Shewanella piezotolerans str. WP3	Wang et al. (2007)
Shewanella colwelliana	Laufer et al. (2015)
Shewanella japonica	Ivanova et al. (2001), Lee et al. (2013)
Shewanella gelidimarina str. ACAM 456	Bowman (1997)
Shewanella frigidimarina str. ACAM 591T	Bowman (1997)
Shewanella profunda str. LT13a	Picard et al. (2015)
Shewanella alga str. BrY	Turick et al. (2002)
Shewanella baltica str. NCTC10735	Satomi (2014)
Shewanella livingstonensis str. LMG19866	Bosal et al. (2002)
Shewanella pealeona	Zhao et al. (2006)
Shewanella xiamensis	Huang et al. (2010)
Shewanella amazonensis	Venkateswaran et al. (1998)
Enterobacter sp. L6	Liu and Wang (2016)
Deltaproteobacteria	
Desulfovibrio profundus	Weber et al. 2006, Bale et al. (1997)
Desufobacterium autotrophicum str. DSM 3382	Lovley et al. (1993)
Desulfobacter postgatei str. DSM 2034	Widdel and Pfenning (1981), Lovley et al. (1993)
Desulfobulbus propianicus	Lovley et al. (1993), Holmes et al. (2004a)
Desulfotalea arctica str. LSv514	Knoblauch et al. (1999)
Desulfotalea psychrophilia str. LSv54	Knoblauch et al. (1999)
Desulfuromnas acetoxidans str. DSM 684	Lovley (2013)
Desulfuromonas palmitatis	Lovley (2013)
Pelobacter carbinolicus str. DSM 2380	Lovley (2013)
Desulfuromonas acetexigens str. 2873	Kashefi et al. (2003)

Table 3.1 (Continued)

Taxa	References
Geopsychrobacter electrodiphilus	Holmes et al. (2004b)
Desulfuromusa succinoxidans str. Gylac	Lovley (2013)
Desulfuromusa bakii str. Gyprop	Lovley (2013)
Geobacter sulfurreducens str. PCA	Lovley (2013), Bond and Lovley (2003), Call and Logan (2011), Sajana et al. (2017)
Geobacter chapellei str. 172	Lovley (2013)
Geothermobacter str. HR-1	Smith et al. (2018)
Geothermobacter str. EPR-M	Tully et al. (2017)
Firmicutes	
Bacillus 29	Castro and Ehrlich (1970)
Renamed *Bacillus pumilus* DSM str. 766	
Clostridium sp. str. EDB2	Bhushan et al. (2004), Bhushan et al. (2006)
Caloranerobacter ferrireducens sp. str. DY22619	Zheng et al. (2015)
Anoxybacter fermentans str. DY22613	Li et al. (2019)
Orenia metallireducens str. Z6	Dong et al. (2016)
Desulfotomaculum reducens str. MI-1	Tebo et al. (1998)

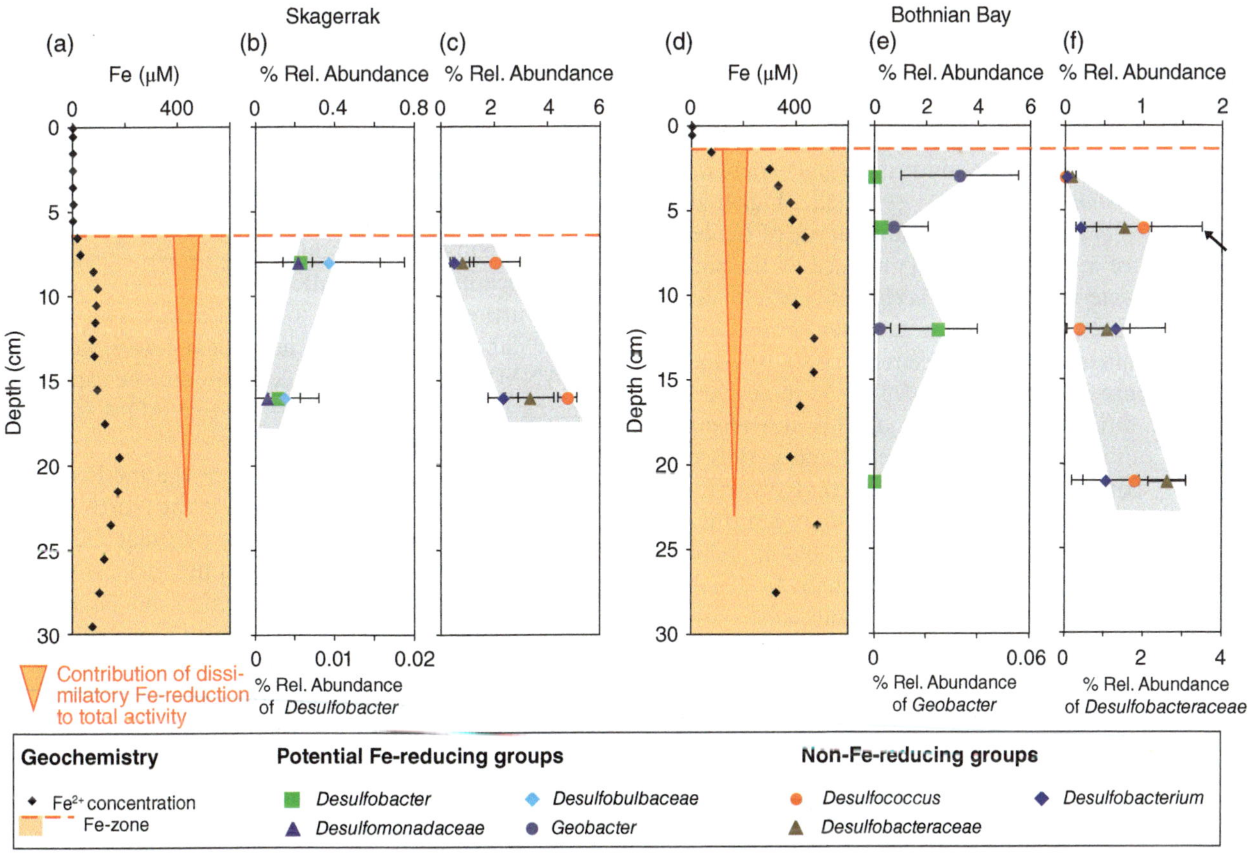

Figure 3.5 Dissolved iron concentrations and depth stratification in community composition in Baltic Sea sediments: (a) Skagerrak; (b) Bothnian Bay. Decreasing relative contribution of iron reduction to overall dissimilatory microbial activity is schematically indicated. Relative abundance of potential iron-reducing groups (b and e) tend to decrease with depth, while other groups (c and f) increase. A deviation of the trend occurs at the Bothnian Bay site, at 5 cm, which is probably due to a layer of freshly deposited marine organic matter. Modified from Reyes et al. (2016) and reproduced with permission from Oxford University Press.

The observed depth-zonation of the microbial community can be explained in the geochemical context. In sediments from Site Geo2a in the Skagerrak, containing moderate amounts of reactive organic matter, dissolved iron(II) is present in the porewater (Figure 3.5; Reyes et al. 2016). Sulfate concentrations are also constant with depth, while constant $\delta^{34}S$ and $\delta^{18}O$ values of sulfate suggest that sulfide is not re-oxidized to sulfate. It is possible that, due to limitation in organic substrates, iron(III) is initially the dominating electron acceptor, but diminished availability of iron(III) may lead to a shift from iron reducers to sulfate reducers with depth in these sediments. Fermenters that are known to donate their electrons to iron(III) are also observed in the iron-reduction zone, including *Clostridiaceae*, *Bacillaceae*, and *Rhodobacteraceae*. Their abundance remains constant with depth, matching the persistent availability of organic carbon.

A similar distribution of iron reducers with depth was observed in sediments from Site At4 in the Bothnian Bay (Reyes et al. 2016). Dissimilatory iron reducers, including *Desulfobacter*, *Geobacter*, *Desulfobulbaceae*, *Desulfuromonadaceae*, and *Pelobacteraceae* had a higher abundance near the surface of the sediment and their abundances decreased with depth (Figure 3.5). In addition, *Clostridiaceae* and *Bacillaceae* abundances decreased with depth. In contrast, *Desulfobacteraceae*, *Desulfobacterium,* and *Desulfococcus* had a higher abundance at greater depths. The high abundance of *Geobacter* at the surface and the higher abundance of the sulfate-reducing bacterium *Desulfococcus* at greater depths suggests that near the top of the sediment iron reduction is the dominant process, whereas sulfate reduction becomes more important with depth. As indicated by a high C/N ratio, organic matter at this site is mainly derived from terrestrial plant material and is thus more refractory compared with the Skagerrak. However, near the surface, a layer of marine derived organic matter occurs, probably as a result of an algal bloom, which is evidently capable of fuelling sulfate reduction. At the same time iron is abundant (~1 wt%) and most probably reactive, so that iron reduction and sulfate reduction may overlap. The decrease in the fermenters with depth reflects the decrease in organic matter reactivity.

Several other studies used next generation sequencing to identify active iron-cycling microorganisms in marine surface sediments. Otte et al. (2018) detected *Desulfobulbus* (1–4%), *Desulfobacterium* (0.1–0.2%), *Geobacter* (0.1–0.2%), and *Shewanella* (0.1–1%) in sediments of Norsminde Fjord and Kalø Vig (Denmark), but these organisms occurred at all depths and their presence was not related to the geochemical zones. In the study by Li et al. (2019), *Shewanella* was highly abundant (4.4%) in a layer where iron(III) and sulfate reduction co-occurred (at 330 cm). Nevertheless, the latter studies proposed only a hypothetical involvement of observed organisms in iron reduction without relation to the geochemical context.

3.4.3. Methanogenic Zones

In the methanogenic zone, where both sulfate and sulfide are depleted, AOM may be linked to the reduction of iron, a process termed Fe-AOM. This was suggested based on experiments with flow-through reactors using sediments recovered from Juan de Fuca Ridge (Wankel et al. 2012). The presence of green rust and other mixed-valance iron oxides in these high-temperature incubations at 90°C suggested that AOM could be coupled to iron reduction. Based on pyrosequencing and qPCR, members of the archaeal group ANME-1, including ANME-1c, were abundant and, thus, were likely to contribute to methane oxidation. Egger et al. (2015) observed enhanced oxidation of ^{13}C-labelled methane in the presence of iron(III) nanoparticles in incubations with sediments from a methanogenic zone of the Bothnian Sea, suggesting that iron-reduction was linked to methane oxidation. Ettwig et al. (2016) showed that enrichment cultures of a new group within the *Methanosarcinales* termed AOM-associated archaea (AAA), and sometimes referred to as ANME-2d, can couple the oxidation of methane to the reduction of iron and manganese oxides. Control incubations without methane did not produce Fe^{2+}, indicating that no alternative carbon source was used. Furthermore, based on incubation experiments with ^{13}C-labelled methane, Scheller et al. (2016) suggested that ANME-2a and ANME-2c could also be capable of reducing insoluble iron oxides and that AOM is decoupled from sulfate reduction.

An in-situ study in sediment from the methanogenic zone of the Helgoland mudflat area in the North Sea was performed by Oni et al. (2015), using pyroseqeuncing and 16S rRNA qPCR. This site showed high concentrations of dissolved Fe^{2+} in the methanogenic zone. High abundances of the bacterial candidate division JS1 (SB45 lineage) and *Methanohalobium* ANME-3 archaea were detected in the methanogenic zone with free Fe^{2+} in the porewater, suggesting that they could be involved in iron reduction (Oni et al. 2015). Alternatively also *Desulfuromonadales* can reduce iron, but they were present also outside the iron zone.

3.4.4. The Phylogenetic Tree of Marine Iron Reducers

Iron-reducing microorganisms belong to many different phyla and include Bacteria and Archaea (Figure 3.6). Most of the iron reducers characterized by cultivation to

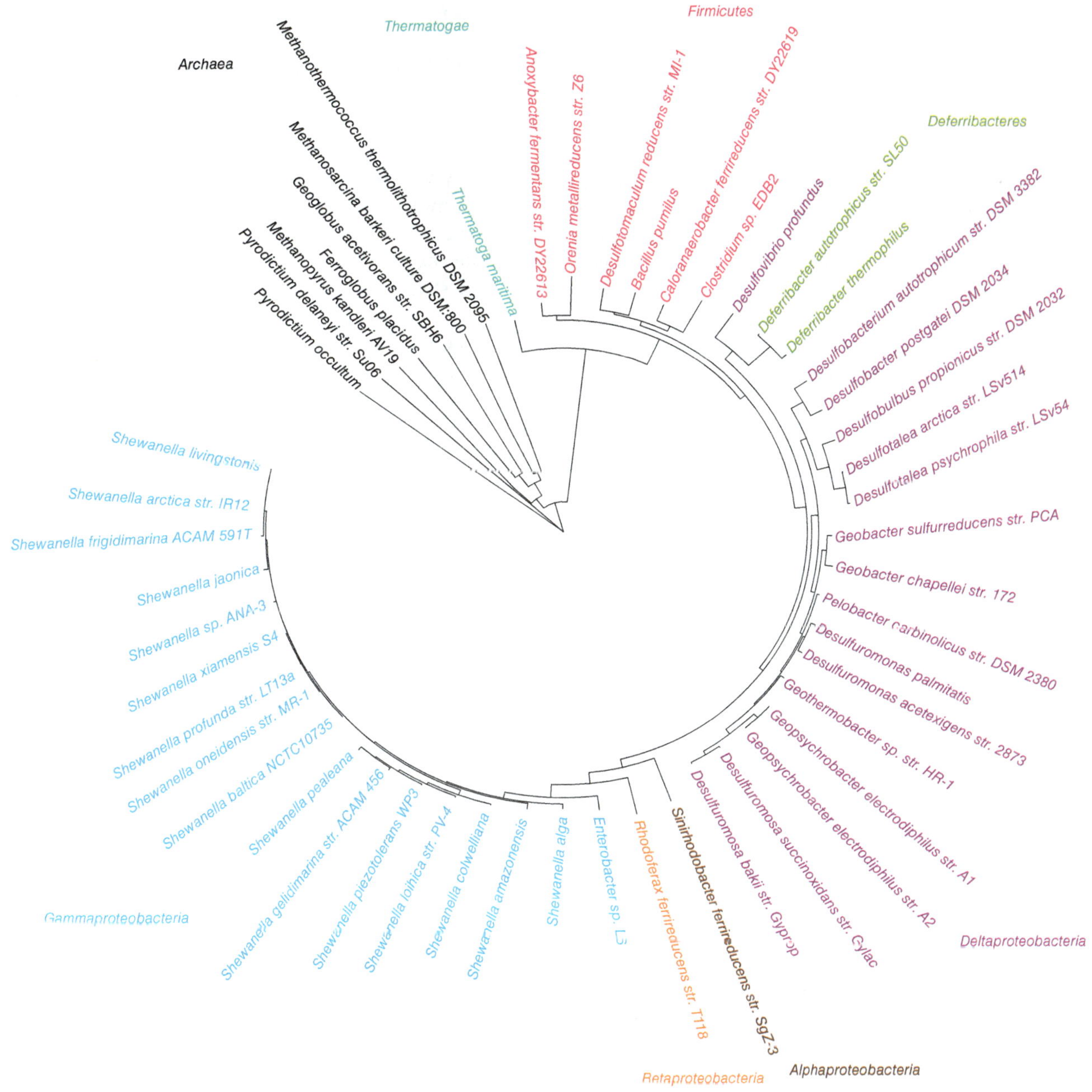

Figure 3.6 Phylogenetic tree of iron-reducing bacteria and archaea isolated from marine sediments. The unrooted phylogenetic tree was adapted from Weber (2006) and was based on nearly full-length 16S ribosomal DNA sequences compiled from the National Center for Biotechnology Information database. Sequences were aligned using Clustal Omega 1.2.4.1 and curated using Gblocks 0.91.1 (default settings) accessed from the website Phylogeny.fr (https://ngphylogeny.fr) (Lemonie et al. 2019) on October 22, 2019. The tree was constructed using phylogeny PhyML 3.0 using maximum likelihood bootstrap 100; substitution model HKY85 from the atgc-montpellier.fr website (www.atgc-montpellier.fr/phyml/) accessed on October 22, 2019. iTol interactive tree of Life V.5 (Letunic and Bork 2016) was used to generate the final tree.

date belong to the *Deltaproteobacteria* class. As shown in the phylogenetic tree of iron reducers (Figure 3.6), members of this class include a variety of *Geobacter* species as well as SRB species such as *Desulfobacter*, *Desulfuromusa*, *Desulfotalea*, and *Desulfolobus*.

Other than *Geobacter* and *Shewanella*, where the iron-reducing pathway has been studied, a number of bacteria (and archaea) have been observed to be capable of reducing iron. However, neither their pathways of iron reduction are known, nor whether they are representative of their phylogenetic groups in terms of iron reduction. For example, *Sinorhodobacter ferrireducens* (Yang et al. 2013), a non-phototrophic iron-reducing bacterium from the genus *Rhodobacter*, and *Desulfotalea arctica* (Knoblauch et al. 1999), a psychrophilic sulfate-reducing bacterium of the Class *Deltaproteobacteria*, have been isolated from marine sediments. They have been shown to reduce iron oxide in batch experiments but they have not yet been genetically characterized and their pathways of iron reduction remain unknown. We can nevertheless speculate whether representatives that are commonly found in shallow marine sediments use known metabolic pathways to reduce iron.

Other than members of the *Geobacteraceae*, members of the SRB have yet to be genetically investigated to determine which genes are necessary for their iron reduction abilities, including potential extracellular electron transport mechanisms (Reyes et al. 2016). Moreover, molecular models that track the electron transport from oxidation of substrate to the reduction of iron are also needed for these organisms. Recently, the proteome of an SRB belonging to the *Firmicutes*, *Desulfotomaculum reducens* strain MI-1, was studied in cultures with Fe-oxide as the only electron acceptor, and a multiheme cytochrome was shown to be highly expressed under these conditions, hinting at the possibility of its involvement in iron reduction (Otwell et al. 2016).

Overall, some over-representation of iron reducers within the *Deltaproterobacteria* may result from a cultivation bias. In particular, sulfate reducers have been intensively cultivated under diverse anaerobic conditions, including ferruginous conditions. Nevertheless, based on the wide distribution of iron reducers among many groups in the phylogenetic tree, we may conclude also that the electron-transporting machinery is widespread. While the pathways may not really be specific to iron as an electron acceptor, many microorganisms could modify their electron transport chains to take advantage of the electron potential difference and gain the maximum ATP. On the other hand, solid-phase electron acceptors pose a serious challenge, requiring complex strategies and extracellular electron transport pathways. It is known that certain genes involved in extracellular electron transport pathways result from horizontal and vertical gene transfer (Holmes et al., 2016; Baker et al. 2022). Such systems may already be established in many microorganisms using other metabolic pathways, and their application to reduce iron would require only minor modifications. The observation that iron reduction is a widespread ability in the microbial tree of life highlights that these systems require further investigation and perhaps hold the key to a better understanding of the ecological role microbial communities play in natural environments, such as marine sediments.

3.5. SUMMARY AND CONCLUSIONS

Microbial iron reduction may occur in sulfide-free zones in marine sediments to the extent where organisms outcompete other organisms using other electron acceptors to oxidize the same organic carbon substrate. Specificity towards iron reduction arises from the ability of the pathway to maximize its energy yield. Even though iron reduction per se may be non-specific, it requires an elaborate extracellular electron transport system or uptake system to access the solid phase iron. Strategies involve electron shuttles, siderophores, or direct contact via nanowires, pili, or connected cells forming a bacterial "cable". Redox zonation in natural sediments probably involves organization with respect to long-distance electron transport to access solid-phase electron acceptors, depending on the availability of other, dissolved electron acceptors. Recent phylogenetic studies have revealed that zones of ongoing iron reduction are preferentially inhabited by phylogenetic groups that have members known to reduce iron, although these may not necessarily be the known model organisms *Shewanella* and *Geobacter*. More often versatiles are present, correlating with the redox zonation, in particular suboxic zones or sulphide-free methanogenic zones. Future studies will not only require phylogenetic analysis but will also need to provide elucidation of the complex electron-transport mechanisms and how they interact among different microbial groups within the community.

REFERENCES

Afonso, M.D.S. and Stumm, W. (1992). Reductive dissolution of iron(III) (hydr)oxides by hydrogen sulfide. *Langmuir 8* (6): 1671–1675. https://doi:10.1021/la00042a030

Andrews, S.C., Robinson, A.K. and Quiñones, F.R. (2003). Bacterial iron homeostasis. *FEMS Microbiology Reviews 27* (2–3): 215–237. https://doi.org/10.1016/S0168-6445(03)00055-X

Baker, I.R., Conley, B.E., Gralnick, J.A., Girguis, P.R. (2022). Evidence for horizontal and vertical transmission of Mtr-mediated extracellular electron transfer among Bacteria. MBio 13 (1): e02904-21.https://doi.org/10.1128/mbio.02904-21

Bale, S., Goodman, K., Rochelle, P.A. et al. (1997). *Desulfovibrio profundas* sp. nov., a novel barophilic sulfate-reducing bacterium from deep sediment layers in the Jap an Sea. *International Journal of Systematic and Evolutionary Microbiology 47* (2): 515–521. https://doi.org/10.1099/00207713-47-2-515

Beal, E.J., House, C.H. and Orphan, V.J. (2009). Manganese-andiron-dependent marine methane oxidation. *Science 325* (5937): 184–187. doi:10.1126/science.1169984

Bennett, B.D., Redford, K.E. and Gralnick, J.A. (2018). MgtE homolog FicI acts as a secondary ferrous iron importer in *Shewanella oneidensis* strain MR-1. *Applied and Environmental Microbiology 84* (6): e01245-17. doi:10.1128/AEM.01245-17

Bond, D.R. and Lovley, D.R. (2003). Electricity production by *Geobacter sulfurreducens* attached to electrodes. *Applied and Environmental Microbiology 69* (3): 1548–1555.doi:10.1128/AEM.69.3.1548-1555.2003

Bretschger, O., Obraztsova, A., Sturm, C.A. et al. (2007). Current production and metal oxide reduction by *Shewanella oneidensis* MR-1 wild type and mutants. *Applied and Environmental Microbiology 73* (21): 7003–7012. doi:10.1128/AEM.01087–07

Breuer, M., Rosso, K.M., Blumberger, J. et al. (2015). Multi-haem cytochromes in *Shewanella oneidensis* MR-1: structures, functions and opportunities. *Journal of the Royal Society Interface 12* (102): 20141117. https://doi:10.1098/rsif.2014.1117

Bhushan, B., Halasz, A., Thiboutot, S. et al. (2004). Chemotaxis-mediated biodegradation of cyclic nitramine explosives RDX, HMX and CL-20 by *Clostridium* sp. EDB2. *Biochemical and Biophysical Research Communications 316* (3): 816–821.

Bhushan, B., Halasz, A. and Hawari, J. (2006). Effect of iron (III), humic acids and anthraquinone-2, 6-disulfonate on biodegradation of cyclic nitramines by *Clostridium* sp. EDB2. *Journal of Applied Microbiology 100* (3): 555–563. https://doi.org/10.1016/j.bbrc.2004.02.120

Bjerg, J.T., Boschker, H.T.S., Larsen, S. et al. (2018). Long-distance electron transport in individual, living cable bacteria. *Proceedings of the National Academy Of Sciences 115* (22): 5786–5791. https://doi.org/10.1073/pnas.1800367115

Bowman, J.P., McCammon, S.A., Nichols, D.S. et al. (1997). *Shewanella gelidimarina* sp. nov. and *Shewanella frigidimarina* sp. nov., novel Antarctic species with the ability to produce eicosapentaenoic acid (20: 5ω3) and grow anaerobically by dissimilatory Fe(III) reduction. *Microbiology 47* (4): 1040–1047. https://doi.org/10.1099/00207713-47-4-1040

Bozal, N., Montes, M.J., Tudela, E. et al. (2002). *Shewanella frigidimarina* and *Shewanella livingstonensis* sp. nov. isolated from Antarctic coastal areas. *International Journal of Systematic and Evolutionary Microbiology 52* (1): 195–205. https://doi.org/10.1099/00207713–52–1-195

Braud, A., Hannauer, M., Mislin, G.L.A. et al. (2009). The *Pseudomonas aeruginosa* pyochelin-iron uptake pathway and its metal specificity. *Journal of Bacteriology 191* (11): 3517–3525. doi:10.1128/JB.00010–09

Burns, J.L. (2010). Molecular mechanisms of microbial iron respiration by *Shewanella oneidensis* MR–1. PhD Thesis. Georgia Institute of Technology.

Caccavo, F., Jr., Lonergan, D.J., Lovley, D.R. et al. (1994). *Geobacter sulfurreducens* sp. nov., a hydrogen and acetate-oxidizing dissimilatory metal-reducing microorganism. *Applied and Environmental Microbiology 60* (10): 3752–3759.

Call, D.F. and Logan, B.E. (2011). Lactate oxidation coupled to iron or electrode reduction by *Geobacter sulfurreducens* PCA. *Applied and Environmental Microbiology 77* (24): 8791–8794. doi:10.1128/AEM.06434–11

Calvert, S.E. and Karlin, R.E. (1991).Relationships between sulphur, organic carbon and ironin the modern sediments of the Black Sea. *Geochimica et Cosmochimica Acta 55* (9): 2483–2490. https://doi.org/10.1016/0016–7037(91)90367-E

Canfield, D.E. (1989). Reactive iron in marine sediments. *Geochimica et Cosmochimica Acta 53*: 619–632. https://doi.org/10.1016/0016–7037(89)90005–7

Canfield, D.E., Thamdrup, B. and Hansen, J.W. (1993). The anaerobic degradation of organic matter in Danish coastal sediments: iron reduction, manganese reduction and sulfate reduction. *Geochimica et Cosmochimica Acta 57* (6): 3867–83. https://doi.org/10.1016/0016–7037(93)90340-3

Cartron, M., Maddocks, S., Gillingham, P. et al. (2006). Feo-transport of ferrous iron into bacteria. *Biometals 19* (2): 143–157. doi:10.1007/s10534/s10534–006–0003–2

Clarke, T.A., Edwards, M.J., Gates, A.J. et al. (2011). Structure of a bacterial cell surface decaheme electron conduit. *Proceedings of the National Academy of Sciences 108* (23): 9384–9389. doi:10.1073/pnas.1017200108

Coates, J.D., Lonergan, D.J. and Lovley, D.R. (1995). *Desulfuromonas palmitatis* sp. nov., a long-chain fatty acid oxidizing Fe(III) reducer from marine sediments. *Archives of Microbiology 164*: 406–413. https://doi.org/10.1007/BF02529738

Costa, N.L., Clarke, T.A., Philipp, L.A. et al. (2018). Electron transfer process in microbial electrochemical technologies: The role of cell-surface exposed conductive proteins. *Bioresource Technology 255*: 308–317. https://doi.org/10.1016/j.biortech.2018.01.133

Coursolle, D., Baron, D.B., Bond, D.R. et al. (2010). The Mtr respiratory pathway is essential for reducing flavins and electrodes in *Shewanella oneidensis*. *Journal of Bacteriology 192* (2): 467–474. doi:10.1128/JB.00925–09

de Castro, A.F. and Ehrlich, H.L. (1970). Reduction of iron oxide minerals by a marine Bacillus. *Antonie van Leeuwenhoek 36* (1): 317–327. https://doi.org/10.1007/BF02069033

de Haas, H., van Weering, C.E.T. and de Stigter, H. (2002). Organic carbon in shelf seas: sinks or sources, processes and products. *Continental Shelf Research 22* (5): 691–717. https://doi.org/10.1016/S0278–4343(01)00093–0

Dobbin, P.S., Carter, J.P., San Juan, C.G.S. et al. (1999). Dissimilatory Fe (III) reduction by *Clostridium beijerinckii* isolated from freshwater sediment using Fe (III) maltol enrichment. *FEMS Microbiology Letters 176* (1): 131–138. https://doi.org/10.1111/j.1574–6968.1999.tb13653.x

Dong, Y., Sanford, R., Boyanov, M.I. et al. (2016). *Orenia metallireducens* sp. nov. strain Z6, a novel metal-reducing member of the phylum Firmicutes from the deep subsurface. *Applied and Environmental Microbiology 82* (21): 6440–6453. doi:10.1128/AEM.02382–16

Edlund, A., Hårdeman, F., Jansson, J.K. et al. (2008). Active bacterial community structure along vertical redox gradients in Baltic Sea sediment. *Environmental Microbiology 10* (8): 2051–63. doi:10.1111/j.1462–2920.2008.01624.x

Edwards, M.J., Hall, A., Shi, L. et al. (2012). The crystal structure of the extracellular 11-heme cytochrome UndA reveals a conserved 10-hem motif and defined binding site for soluble iron chelates. *Structure 20* (7): 1275–1284. doi:10.106/j.str.2012.04.016

Egger, M., Rasigraf, O., Sapart, C.J. et al. (2015). Iron-mediated anaerobic oxidation of methane in brackish coastal sediments. *Environmental Science and Technology 49* (1): 227–283. https://doi:10.1021/es503663z

El-Naggar, M.Y., Wanger, G., Leung, K.M. et al. (2010). Electrical transport along bacterial nanowires from *Shewanella oneidensis* MR–1. *Proceedings of the National Academy of Sciences 107* (42): 18127–31. https://doi.org/10.1073/pnas.1004880107

Embree, M., Qiu, Y., Shieu, W. et al. (2014). The iron stimulon and Fur regulon of *Geobacter sulfurreducens* and their role in energy metabolism. *Applied and Environmental Microbiology 80* (9): 2918–2927. doi:10.1128/AEM.03916–13

Ettwig, K.F., Zhu, B., Speth, D. et al. (2016). Archaea catalyze iron-dependent anaerobic oxidation of methane. *Proceedings of the National Academy of Sciences 113* (45): 12792–12796. https://doi.org/10.1073/pnas.1609534113

Feliciano, C.T., da Silva, A.J.R., Reguera, G. et al. (2012). Molecular and electronic structure of the peptide subunit of *Geobacter sulfurreducens* conductive pili from first principles. *Journal of Physical Chemistry A 116* (30): 8023–8030. https://doi:10.1021/jp302232p

Fick, A. (1855). On liquid diffusion. *Philosophical Magazine 10* (63): 30–39. https://doi.org/10.1080/14786445508641925

Filman, D.J., Marino, S.F., Ward, J.E. et al. (2018). Structure of a cytochrome-based bacterial nanowire. bioRxiv. http://dx.doi.org/10.1101/492645

Finneran, K.T., Johnsen, C.V. and Lovley, D.R. (2003). *Rhodoferax ferrireducens* sp. nov., a psychrotolerant, facultatively anaerobic bacterium that oxidizes acetate with the reduction of Fe(III). *International Journal of Systematic and Evolutionary Microbiology 53* (3): 669–673. https://doi.org/10.1099/ijs.0.02298–0

Froelich, P., Klinkhammer, G., Bender, M. et al. (1979). Early oxidation of organic matter in pelagic sediments of the eastern equatorial Atlantic: suboxic diagenesis. *Geochimica et Cosmochimica Acta 43* (7): 1075–1090. https://doi.org/10.1016/0016–7037(79)90095–4

Gao, H., Obraztova, A., Stewart, N. et al. (2006). *Shewanella loihica sp.* nov., isolated from iron-rich microbial mats in the Pacific Ocean.*International Journal of Systematic and Evolutionary Microbiology 56*(8): 1911–1916. https://doi.org/10.1099/ijs.0.64354–0

Gorby, Y.A., Yanina, S., McLean, J.S. et al. (2006). Electrically conductive bacterial nanowires produced by *Shewanella oneidensis* strain MR-1 and other microorganisms. *Proceedings of the National Academy of Sciences 103* (30): 11358–11363. https://doi.org/10.1073/pnas.0604517103

Hamman, R. and Ottow, J.C.G. (1974). Reductive dissolution of Fe_2O_2 by saccharolytic *Clostrida* and *Bacillus polymyxa* under anaerobic conditions. *Zeitschrift für Pflanzenernährung und Bodenkunde 137* (2): 108–115. https://doi.org/10.1002/jpln.19741370205

Holmes, D.E., Bond, D.R. and Lovley, D.R. (2004a). Electron transfer by *Desulfobulbus propionicus* to Fe(III) and graphite electrodes. *Applied and Environmental Microbiology 70* (2): 1234–1237. doi:10.1128/AEM.70.2.1234–1237.2004

Holmes, D.E., Nicoll, J.S., Bond, D.R. et al. (2004b). Potential role of a novel psychrotolerant member of the family *Geobacteraceae*, *Geopsychrobacter electrodiphilus* gen. nov., sp., in electricity production by a marine sediment fuel cell. *Applied and Environmental MicrobiologyEnvironmental Microbiology 70* (10): 6023–6030. doi:10.1128/AEM.70.10.6023–6030.2004

Holmes,D.E., Dang, Y., Walker, D.J.F., Lovley, D.R. (2016). The electrically conductive pili of Geobacter species are a recently evolved feature for extracellular electron transfer. Microbial Genomics 2(8): e000072. doi: 10.1099/mgen.0.000072

Holmkvist, L., Ferdelman, T.G. and Jørgensen, B.B. (2011). A cryptic sulfur cycle driven by iron in the methane zone of marine sediment (Aarhus Bay, Denmark). *Geochimica et Cosmochimica Acta 75* (12): 3581–3599.https://doi.org/10.1016/j.gca.2011.03.033

Huang, J., Sun, B. and Zhang, X. (2010). *Shewanella xiamenensis* sp. nov., isolated from coastal sea sediments. *International Journal of Systematic and Evolutionary Microbiology 60* (7): 1585–1589. https://doi.org/10.1099/ijs.0.013300–0

Huber, H., Thomm, M., Konig, H. et al. (1982). *Methanococcus thermolithotrophicus*, a novel thermophilic lithotrophic methanogen. *Archives of Microbiology 132* (1): 47–50. https://doi.org/10.1007/BF00690816

Ivanova, E.P., Sawabe, T., Gorschkova, N.M. et al. (2001). Shewanella japonica sp. nov. *International Journal of Systematic and Evolutionary Microbiology 51* (3): 1027–1033. https://doi.org/10.1099/00207713–51–3–1027

Jensen, M.M., Thamdrup, B., Rysgaard, S. et al. (2003). Rates and regulation of microbial iron reduction in sediments of the Baltic-North transition. *Biogeochemistry 65* (3): 295–317. https://doi.org/10.1023/A:1026261303494

Jeong, Y.S., Song, S.K., Lee, S.J. et al. (2006). The growth and EPA synthesis of *Shewanella oneidensis* MR-1 and expectation of EPA biosynthetic pathway. *Biotechnology and Bioprocess Engineering 11* (2): 127–133. https://doi.org/10.1007/BF02931896

Kashefi, K., Tor, J.M., Holmes, D.E. et al. (2002). *Geoglobus ahangari* gen. nov., sp. nov., a novel hyperthermophilic archaeon capable of oxidizing organic acids and growing autotrophically on hydrogen with Fe(III) serving as the sole electron acceptor. *International Journal of Systematic and Evolutionary Microbiology 52* (3): 719–729. https://doi.org/10.1099/00207713–52–3-719

Kashefi, K., Holmes, D.E., Baross, J.A. et al. (2003). *Geothermobacter ehrlichii* gen. nov., sp. nov., a novel thermophilic member of the Geobacteraceae from "Bag City" hydrothermal vent. *Applied and Environmental Microbiology 69* (5): 2985–2993. doi: 10.1128/AEM.69.5.2985–2993.2003

Kim, B.H., Kim, H.J., Hyun, M.S. et al. (1999). Direct electrode reaction of Fe(III) reducing bacterium, *Shewanella putrefaciens*. *Journal of Microbiology and Biotechnology 9*: 127–31.

Kim, H.J., Park, H.S., Hyun, M.S. et al. (2002). A mediator-less microbial fuel cell using a metal reducing bacterium, *Shewanella putrefaciens. Enzyme and Microbial Technology 30* (2): 145–152. https://doi.org/10.1016/S0141–0229(01)00478–1

Kim, S.J., Park, S.J., Oh, Y.S. et al. (2012). *Shewanella arctica* sp. nov., an iron-reducing bacterium isolated from Arctic marine sediment. *International Journal of Systematic and Evolutionary Microbiology 62* (5): 1128–33.https://doi.org/10.1099/ijs.0.031401–0

Kjeldsen, K.U., Schreiber, L., Thorup, C.A. et al. (2019). On the evolution and physiology of cable bacteria. *Proceedings of the National Academy of Sciences 116* (38): 19116–19125. https://doi.org/10.1073/pnas.1903514116

Knoblauch, C., Sahm, K. and Jørgensen, B.B. (1999). Pschycrophilic sulfate-reducing bacteria isolated from permanently cold Arctic marine sediments: description of *Desulofrigus oceanense* gen. nov., *Desulfofrigus fragile* sp. nov., *Desulfofaba gelida* gen. nov., sp. nov., *Desulfotalea psychrophila* gen. nov., sp. nov. and *Desulfotalea arctica* sp. nov. *International Journal of Systematic and Evolutionary Microbiology 49* (4): 1631–1643. doi:10.1099/00207713–49–4-1631

König, I., Haeckel, M., Drodt, M. et al. (1999). Reactive Fe(II) layers in deep-sea sediments. *Geochimica et Cosmochimica Acta 63* (10): 1517–1526. https://doi.org/10.1016/S0016–7037(99)00104–0

Kraemer, S. (2004). The iron oxide dissolution and solubility in the presence of siderophores. *Aquatic Science 66* (1): 3–18. https://doi.org/10.1007/s00027–003–0690–5

Krumholz, L.R., Sharp, R., Fishbain, S.S. (1996). A freshwater anaerobe coupling acetate oxidation to tetrachloroethylene dehalogenation. *Applied Environmental Microbiology 62* (11): 4108–4113.

Kurr, M., Huber, R., König, H. et al. (1991). *Methanopyrus kandleri*, gen. and sp. nov. represents a novel group of hyperthermophilic methanogens growing at 110°C. *Archives of Microbiology 156*: 239–247. https://doi.org/10.1007/BF00262992

Lampa-Pastirk, S., Veazey, J.P., Walsh, K.A. et al. (2016). Thermally activated charge transport in microbial protein nanowires. *Scientific Reports 6* (1): 23517. https://doi.org/10.1038/srep23517

Laufer, K., Nordhoff, M., Røy, H. et al. (2016). Coexistence of microaerophilic, nitrate-reducing and phototrophic Fe(II) oxidizers and Fe(II) reducers in coastal marine sediment. *Applied and Environmental Microbiology Environmental Microbiology 82*(5): 1433–1447. doi:10.1128/AEM.03527–15

Lee, J.S., McBeth, J.M., Ray, R.I. et al. (2019). Iron cycling at corroding carbon steel surfaces. *Biofouling 29* (10): 1243–1252. https://doi.org/10.1080/08927014.2013.836184

Lemonie, F., Correia, D., Lefort, V. et al. (2019). NGPhylogeny.fr: new generation phylogenetic services for non-specialists. *Nucleic Acids Research* 47. doi:10.1093/nar/gkz303

Letunic, I. and Bork, P. (2016). Interactive tree of life (iTOL) v3: an online tool for the display and annotation of phylogenetic and other trees. *Nucleic Acids Research 44* (W1): W242–245. https://doi:10.1093/nar/gkw290

Levar, C.E., Chan, C.H., Mehta-Kolte, M.G. et al. (2014). An inner membrane cytochrome required only for reduction of high redox potential extracellular electron acceptors. *mBio 5* (6): e02034–14. doi:10.1128/mBio.02034–14

Li, J., Li, L., Bai, S. et al. (2019). New insight into the biogeochemical cycling of methane, S and Fe above the sulfate-methane transition zone in methane hydrate-bearing sediments. A case study in the Dongsha area, South China Sea. *Deep Sea Research Part I 145*: 97–108. https://doi.org/10.1016/j.dsr.2019.01.011

Li, M., Toner, B.M., Baker, B.J. et al. (2014). Microbial iron uptake as a mechanism for dispersing iron from deep-sea hydrothermal vents. *Nature Communications 5*: 3192. https://doi.org/10.1038/ncomms4192

Li, X., Zeng, X., Qiu, D. et al. (2019). Dissimilatory iron [Fe (III)] reduction by a novel fermentative, piezophilic bacterium *Anoxybacter fermentans* DY22613T isolated from East Pacific Rise hydrothermal sulfides. *Geomicrobiology Journal 36* (4): 291–302. https://doi.org/10.1080/01490451.2018.1526985

Liesack, W. and Finster, K. (1994). Phylogenetic analysis of five strains of gram-negative, obligately anaerobic, sulfur-reducing bacteria and description of *Desulfuromusa* gen. nov., including *Desulfuromusa kysingii* sp. nov., *Desulfuromusa bakii* sp.nov. and *Desulfuromusa succinoxidans* sp. nov. *International Journal of Systematic Bacteriology 44* (4): 753–758. https://doi.org/10.1099/00207713–44–4-753

Lin, B., Hyacinthe, C., Bonneville, S. et al. (2007). Phylogenetic and physiological diversity of dissimilatory ferric iron reducers in sediments of the polluted Scheldt estuary, Northwest Europe. *Environmental Microbiology 9* (8): 1956–1968. doi:10.1111/j.1462–2920.2007.01312.x

Lin, T., El Sebae, G., Jung, J.H. et al. (2016). *Pyrodictium delaneyi* sp. nov., a hyperthermophilic autotrophic archaeon that reduces Fe(III) oxide and nitrate. *International Journal of Systematic and Evolutionary Microbiology 66* (9): 3372. doi:10.1099/ijsem.0.001201

Liu, H. and Wang, H. (2016). Characterization of Fe(III)-reducing enrichment culture and isolation of Fe(III)-reducing bacterium *Enterobacter* sp. L6 from marine sediment. *Journal of Bioscience and Bioengineering122* (1): 92–96. https://doi.org/10.1016/j.jbiosc.2015.12.014

Liu, L., Li, S., Wang, S. et al. (2018). Complex iron uptake by the putrebactin-mediated and Feo systems in *Shewanella oneidensis*. *Applied and Environmental Microbiology 84* (20): e01752–18. doi:10.1128/AEM.01752–18

Lonergan, D.J., Jenter, H., Coates, J.D. et al. (1996). Phylogenetic analysis of dissimilatory Fe(III)-reducing bacteria. *Journal of Bacteriology 178* (8): 2404–2408. doi:10.1128/jb.178.8.2402-2408.1996

Lovley, D.R. (2004). Potential role of dissimilatory iron reduction in the early evolution of microbial respiration. In: *Origins. Cellular Origin, Life in Extreme Habitats and Astrobiology*, Vol 6, (ed. J. Seckback), 299–313. Dordrecht: Springer. https://doi.org/10.1007/1–4020–2522-X_19

Lovley, D.R. (2013). Dissimilatory Fe(III) and Mn(IV) reducing prokaryotes. In: *The Prokaryotes* (eds. E. Rosenberg, E.F. DeLong, S. Lory et al.), 287–308. Berlin, Heidelberg: Springer.

Lovley, D.R. (2017). Electrically conductive pili: biological function and potential applications in electronics. *Current Opinion in Electrochemistry 4* (1): 190–198. https://doi.org/10.1016/j.coelec.2017.08.015

Lovley, D.R. and Phillips, E.J.P. (1998). Novel mode of microbial energy metabolism: organic carbon oxidation coupled to dissimilatory reduction of iron or manganese. *Applied and Environmental MicrobiologyEnvironmental Microbiology 54* (6): 1472–1480.

Lovley, D.R., Derek, R., Roden, E.E. et al. (1993). Enzymatic iron and uranium reduction by sulfate-reducing bacteria. *Marine Geology 113* (1–2): 41–53. https://doi.org/10.1016/0025–3227(93)90148-O

Lovley, D.R., Phillips, E.J., Lonergan, D.J. et al. (1995). Fe(III) and S0 reduction by *Pelobacter carbinolicus*. *Applied Environmental Microbiology 61*(6): 2132–2138.

Lower, B.H., Shi, L., Yongsunthon, R. et al. (2007). Specific bonds between an iron oxide surface and outer membrane cytochromes MtrC and OmcA from *Shewanella oneidensis* MR–1. *Journal of Bacteriology 189* (13): 4944–52. doi:10.1128/JB.01518–06

Lower, B.H., Lins, R.D., Oestreicher, Z. et al. (2008). *In vitro* evolution of a peptide with a hematite binding motif that may constitute a natural metal-oxide binding archetype. *Environmental Science and Technology 42* (10): 3821–3827. https://doi:10.1021/es702688c

Lloyd, J.R., Leang, C., Hodges-Myerson, A.L. et al. (2003). Biochemical and genetic characterization of PpcA, a periplasmic c-type cytochrome in *Geobacter sulfurreducens*. *Biochemical Journal 369* (1): 159–161. doi:10.1042/BJ20020597

Malvankar, N.S., Vargas, M., Nevin, K.P. et al. (2011). Tunable metallic-like conductivity in microbial nanowire networks. *Nature Nanotechnology 6* (9): 573–579. https://doi.org/10.1038/nnano.2011.119

Malvankar, N.S., Vargas, M., Nevin, K. et al. (2015). Structural basis for metallic-like conductivity in microbial nanowires. *mBio 6* (2): e00084–15. doi:10.1128/mBio.00084–15

Marsili, E., Baron, D.B., Shikhare, I.D. et al.(2008). *Shewanella* secretes flavins that mediate extracellular electron transfer. *Proceedings of the National Academy of Sciences 105* (10): 3968–73. https://doi.org/10.1073/pnas.0710525105

Mehta, T., Coppi, M.V., Childers, S.E. et al. (2005). Outer membrane c-type cytochromes required for Fe(III) and Mn(IV) oxide reduction in *Geobacter sulfurreducens*. *Applied and Environmental Microbiology 71* (12): 8634–8641. doi:10.1128/AEM.71.12.8634–8641.2005

Meister, P., Bernasconi, S.M., Aiello, I.W., Vasconcelos, C., McKenzie, J.A. (2009). Depth and controls of Ca-rhodochrosite precipitation in bioturbated sediments of the Eastern Equatorial Pacific, ODP Leg 201, Site 1226 and DSDP Leg 68, Site 503. Sedimentology56: 1552–1568. https://doi.org/10.1111/j.1365-3091.2008.01046.x

Meister, P., Chapligin, B., Picard, A. et al. (2014). Early diagenetic quartz formation at a deep iron oxidation front in the Eastern Equatorial Pacific. *Geochimica et Cosmochimica Acta 137*: 188–207. https://doi.org/10.1016/j.gca.2014.03.035

Meister, P., Brunner, B., Picard, A. et al. (2019). Sulphur and carbon isotopes as tracers of past sub-seafloor microbial activity. *Nature Scientific Reports 9* (1): 604. https://doi.org/10.1038/s41598–018–36943–7.

Meitl, L.A., Eggleston, C.M., Colberg, P.J.S. et al. (2009). Electrochemical interaction of *Shewanella oneidensis* MR-1 and its outer membrane cytochromes OmcA and MtrC. *Geochimica et Cosmochimica Acta 73* (18): 5292–5307. https://doi:10.106/j.gca.2009.06.021

Mills, H.J., Reese, B.K., Shepard, A.K. et al. (2012). Characterization of metabolically active bacterial populations in subseafloor Nankai Trough sediments above, within and below the sulfate-methane transition zone. *Frontiers in Microbiology 3*: 113. https://doi:10.3389/fmicb.2012.00113

Nielsen, L.P., Peterson, N.R., Fossing, H. et al. (2010). Electric currents couple spatially separated biogeochemical processes in marine sediment. *Nature Letters 463* (7284): 1071–1074. https://doi.org/10.1038/nature08790

Oni, O., Miyatake, T., Kasten, S., Richter-Heitmann, T.et al. (2015). Distinct microbial populations are tightly linked to the profile of dissolved iron in the methanic sediments of the Helgoland mud area, North Sea. *Frontiers in Microbiology 6*: 365. https://doi.org/10.3389/fmicb.2015.00365

Otero, F.J., Chan, C.H. and Bond, D.R. (2018). Identification of different putative outer membrane electron conduits necessary for Fe(III) citrate, Fe(III) oxide, Mn(IV) oxide, or electrode reduction by *Geobacter sulfurreducens*. *Journal of Bacteriology 200* (19): e00347–18. doi:10.1128/JB.00347–18

Otte, S., Kuenen, J.G., Nielsen, L.P. et al. (1999). Nitrogen, carbon and sulfur metabolism in natural Thioploca samples. *Environmental Microbiology 65* (7): 3148–3157.

Otte, J.M., Harter, J., Laufer, K. et al. (2018) The distribution of active iron-cycling bacteria in marine and freshwater sediments is decoupled from geochemical gradients. *Environmental Microbiology 20*: 2483–2499. https://doi.org/10.1111/1462–2920.14260

Otwell, A.E., Callister, S.J., Zink, E.M. et al. (2016). Comparative proteomic analysis of *Desulfotomaculum reducens* MI-1: insights into the metabolic versatility of a gram-positive sulfate and metal-reducing bacterium. *Frontiers in Microbiology 7*: 191. https://doi.org/10.3389/fmicb.2016.00191

Park, H.S., Kim, B.H., Kim, H.S. et al. (2001). A novel electrochemically active and Fe(III)-reducing bacterium phylogenetically related to *Clostridium butyricum* isolated from a microbial fuel cell. *Anaerobe 7*(6): 297–306. https://doi.org/10.1006/anae.2001.0399

Parkhurst, D.L. and Appelo, C.A.J. (2013). Description of input and examples for PHREEQC version 3 – a computer program for speciation, batch-reaction, one-dimensional transport and inverse geochemical calculations. *US Geological Survey, Techniques and Methods, book 6, A43*, 497 p. Available at: https://pubs.usgs.gov/tm/06/a43/

Pfeffer, C., Larsen, S., Song, J.et al. (2012). Filamentous bacteria transport electrons over centimeter distances. *Proceedings of the National Academy of Sciences 491* (7423): 218–221. doi:10.1038/nature11586

Picard, A., Testemale, D., Wagenknecht, L. et al. (2015). Iron reduction by the deep-sea bacterium *Shewanella profunda* LT13a under subsurface pressure and temperature conditions. *Frontiers in Microbiology 5*: 796. https://doi.org/10.3389/fmicb.2014.00796

Picard, A., Gartman, A. and Girguis, P.R. (2016). What do we really know about the role of microorganisms in iron sulfide mineral formation? *Frontiers in Earth Sciences 4*: 68. https://doi.org/10.3389/feart.2016.00068.

Pirbadian, S., Barchinger, S.E., Leung, K.M.et al. (2014). *Shewanella oneidensis* MR-1 nanowires are outer membrane and periplasmic extensions of the extacellular electron transport components. *Proceedings of the National Academy of Sciences 111* (35): 12883–12888. https://doi.org/10.1073/pnas.1410551111

Pley, U., Schipka, J., Gambacorta, A. et al. (1991). *Pyrodictium abyssi* sp. nov. represents a novel heterotrophic marine archaeal hyperthermophile growing at 110°C. *Systematic and Applied Microbiology 14* (3): 245–253. https://doi.org/10.1016/S0723–2020(11)80376–0

Powell, S.M., Bowman, J.P., Snape, I. et al. (2003). Microbial community variation in pristine and polluted nearshore Antarctic sediments. *FEMS Microbiology Ecology 45* (2): 135–45. https://doi.org/10.1016/S0168–6496(03)00135–1

Ramette, A. (2007). Multivariate analyses in microbial ecology. *FEMS Microbiology Ecology 62* (2): 142–160. https://doi.org/10.1111/j.1574–6941.2007.00375.x

Reguera, G. (2018). Biological electron transport goes the extra mile. *Proceedings of the National Academy of Sciences 115* (22): 5632–5634. https://doi.org/10.1073/pnas.1806580115

Reguera, G., McCarthy, K.D., Mehta, T. et al. (2005). Extracellular electron transfer via microbial nanowires. *Nature 435* (7045): 1098–1101. doi:10.1038/nature03661

Reyes, C., Qian, F., Zhan, A. et al. (2012). Characterization of axial and proximal histidine mutations of the decaheme cytochrome MtrA from *Shewanella* sp. str. ANA-3 and implications for the electron transport system. *Journal of Bacteriology 194* (21): 5840–5847. doi:10.1128/JB.00890–12

Reyes, C., Dellwig, O., Dähnke, K. et al. (2016). Bacterial communities potentially involved in iron-cycling in Baltic Sea and North Sea sediments revealed by pyrosequencing. *FEMS Microbiology Ecology 92* (14): fiw054. https://doi:10.1093/femsec/fiw054.

Reyes, C., Schneider, D., Thürmer, A. et al. (2017). Potentially active iron, sulfur and sulfate reducing bacteria in Skagerrak and Bothnian Bay sediments. *Geomicrobiology Journal 34*(10): 840–850. https://doi.org/10.1080/01490451.2017.1281360

Rickard, D. and Luther, III, G.W. (2007). Chemistry of iron sulfides. *Chemical Reviews 107* (2): 514–562. https://doi.org/10.1021/cr0503658

Riedinger, N., Brunner, B., Krastel, S. et al. (2017). Sulfur cycling in an iron oxide-dominated, dynamic marine depositional system: the Argentine Continental Margin. *Frontiers in Earth Science 5*. https://doi:10.3389/feart.2017.00033.

Risgaard-Petersen, N., Revil, A., Meister, P. et al. (2012). Sulphur, iron and calcium cycling associated with natural electric currents in marine sediment. *Geochimica et Cosmochimica Acta. 92*: 1–13. https://doi.org/10.1016/j.gca.2012.05.036

Roden, E.E. and Lovley, D.R. 1993. Dissimilatory Fe(III) reduction by the marine microorganism, *Desulfuromonas acetoxidans Applied Environmental Microbiology 59* (3): 734–742.

Saha, R., Saha, N., Donofrio, R.S. et al. (2012). Microbial siderophores: a minireview. *Journal of Basic Microbiology 53* (4): 303–317. https://doi.org/10.1002/jobm.201100552

Saltikov, C.W., Cifuentes, A., Venkateswaran, K. et al. (2003). The ars detoxification system is advantageous but not required for As(V) respiration by the genetically tractable *Shewanella* species strain ANA-3. *Applied and Environmental Microbiology 69*(5): 2800–2809. doi:10.1128/AEM.69.5.2800–2809.2003

Sanjana, T.K., Ghangrekar, M.M. and Mitra, A. (2016). *In situ* bioremediation using sediment microbial fuel cell. *Journal of Hazardous, Toxic and Radioactive Waste 21* (2): 04016022. doi:10.1061/(asce)hz.2153-5515.0000339

Santos, T.C., Silva, M.A., Morgado, L. et al. (2015). Diving into the redox properties of *Geobacter sulfurreducens* cytochromes: a model for extracellular electron transfer. *Dalton Transactions 44* (20): 9335–9344. doi:10.1039/c5dt00556f

Satomi, M. 2014. The family *Shewanellaceae*. In: *The Prokaryotes* (eds. E. Rosenberg, E.F. DeLong, S. Lory et al.), 597–625. Berlin, Heidelberg: Springer. https://doi.org/10.1007/978-3-642-38922-1_226

Schauer, R., Petersen, N.R., Kjeldsen, K.U. et al. (2014). Succession of cable bacteria and electric currents in marine sediments. *ISME Journal 8* (6): 1314–1322. https://doi.org/10.1038/ismej.2013.239

Scheller, S., Yu, H., Chadwick, G.L., McGlynn, S.E., Orphan, V.J. (2016). Artificial electron acceptors decouple archaeal methane oxidation from sulfate reduction. Science351: 6274 703–707.

Schulz, H.D. and Zabel, M. ed. (2006). *Marine Geochemistry, 2e*. Berlin: Springer.

Shi, L., Squier, C., Zachara, J.M. et al. (2007). Respiration of metal (hydr)oxides by *Shewnaella* and *Geobacter*: a key role for multihaem c-type cytochromes. *Molecular Microbiology 65* (1): 12–20. https://doi.org/10.1111/j.1365–2958.2007.05783.x

Shi, L., Dong, H., Reguera, G. et al. (2016). Extracellular electron transfer mechanisms between microorganisms and minerals. *Nature Reviews Microbiology14 (10)*: 651–662. doi:10.1038/nrmicro.2016.93

Sinkko, H., Lukkari, K., Jama, A.S. et al. (2011). Phosphorus chemistry and bacterial community composition interact in brackish sediments receiving agricultural discharges. *PLoS One 6* (6): e21555. https://doi.org/10.1371/journal.pone.0021555

Sivan, O., Shusta, S.S., Valentine, D.L. (2016). Methanogens rapidly transition from methane production to iron reduction. *Geobiology 14* (2): 190–203. https://doi.org/10.1111/gbi.12172

Sleep, N.H. (2018). Geological and geochemical constraints on the origin and evolution of life. *Astrobiology18* (9): 1199–1219. https://doi.org/10.1089/ast.2017.1778

Slobodkina, G.B., Kolganova, T.V., Chernyh, N.A. et al. (2009). *Deferribacter autotrophicus* sp. nov., an iron (III)-reducing bacterium from a deep-sea hydrothermal vent. *International Journal of Systematic and Evolutionary Microbiology 59* (6): 1508–1512. https://doi.org/10.1099/ijs.0.006767–0

Smith, H., Abuyen, K., Tremblay, J. et al. (2018). Genome sequence of *Geothermobacter sp*. strain HR-1, an iron reducer from the Lō ʻihi Seamount Hawaiʻi. *Genome Announcements 6* (21): e00339–18. doi:10.1128/genomeA.00339–18

Stapleton, R., Sabree, Z., Palumbo, A.V. et al. (2005). Metal reduction at cold temperatures by *Shewanella* isolates from various marine environments. *Aquatic Microbial Ecology 38* (1): 81–91. doi:10.3354/amc038081

Subramanian, P., Pirbadian, S., El-Naggar, M.Y.et al. (2018). Ultrastructure of *Shewanella oneidensis* MR-1 nanowires revealed by electron cyrotomography. *Proceedings of the National Academy of Sciences 115* (14): E3246–E3255. https://doi:10.1073/pnas.1718810115

Tebo, B.M. and Obraztsova, A.Y. (1998). Sulfate-reducing bacterium grows with Cr(VI), U(VI), Mn(IV) and Fe(III) as electron acceptors. *FEMS Microbiology Letters 162* (1): 193–199. https://doi.org/10.1111/j.1574–6968.1998.tb12998.x

Timmers, P.H.A., Welte, C.U., Koehorst, J.J. et al. (2017). Reverse methanogenesis and respiration in methanotropic archaea. *Archaea*, *17*: 1–22. https://doi.org/10.1155/2017/1654237

Thang, N.M., Brüchert, V., Formolo, M. et al.(2013). The impact of sediment and carbon fluxes on the biogeochemistry of methane and sulfur in littoral Baltic Sea sediments (Himmerfjärden, Sweden). *Estuar. Coast*, *36* (1): 98–115. https://doi.org/10.1007/s12237–012–9557–0

Toulza, E., Tagliabue, A., Blain, S. et al. (2012). Analysis of the global ocean sampling (GOS) project trends in iron uptake by surface ocean microbes. *PLoS One*, *7*(2): e30931. https://doi.org/10.1371/journal.pone.0030931

Tully, B., Savalia, P., Abuyen, K. et al. (2017). Genome sequence of *Geothermobacter* sp. strain EPR-M, a deep-sea hydrothermal vent iron reducer. *Genome Announcements 5* (23): e00424–17. doi:10.1128/genomeA.00424–17

Turick, C.E., Louis, S.T. and Caccavo Jr, F. (2002). Melanin production and use as a soluble electron shuttle for Fe(III) oxide reduction and as a terminal electron acceptor by *Shewanella algae* BrY. *Applied and Environmental Microbiology 68* (5): 2436–2444. doi:10.1128/AEM.68.5.2436–2444.2002

Vargas, M., Kashefi, K., Blunt-Harris, E.L. et al. (1998). Microbiological evidence for Fe(III) reduction on early Earth. *Nature 395* (6697): 65.

Vargas, M., Malvankar, N.S., Tremblay, P.L. et al. (2013). Aromatic amino acids requires for pili conductivity and long-range extracellular electron transport in *Geobacter sulfurreducens*. *mBio 4*(2): e00105–13. doi:10.1128/mBio.00105–13

Venkateswaran, K., Dollhopf, M.E., Aller, R. et al. (1998). *Shewanella amazonensis* sp. nov., a novel metal-reducing facultative anaerobe from Amazonian shelf muds. *International Journal of Systematic and Evolutionary Microbiology 48* (3): 965–972. https://doi.org/10.1099/00207713–48–3-965

von Canstein, H., Ogawa, J., Shimizu, S. et al. (2008). Secretion of flavins by *Shewanella* species and their role in extracellular electron transfer. *Applied and Environmental Microbiology 74* (3): 615–623. doi:10.1128/AEM.01387–07

Walker, D.J.F., Martz, E., Holmes, D.E. et al. (2019). The archaellum of *Methanospirillum hungatei* is electrically conductive. *mBio 10* (2): e00579–19. doi:10.1128/mBio.00579–19.

Wang, F., Wang, J., Jian, H. et al. (2008). Environmental adaptation: genomic analysis of the piezotolerant and psychrotolerant deep-sea iron reducing bacterium *Shewanella piezotolerans* WP3. *PLoS One 3* (4): e1937. https://doi.org/10.1371/journal.pone.0001937

Wang, G., Qian, F., Saltikov, C.W. et al. (2011). Microbial reduction of graphene oxide by *Shewanella*. *Nano Research 4*(6): 563–570. https://doi.org/10.1107/s12274–011–0112–2

Wankel, S.D., Adams, M.M., Johnston, D.T. et al. (2012). Anaerobic methane oxidation in metalliferous hydrothermal sediments: influence on carbon flux and decoupling from sulfate reduction. *Environmental Microbiology 14* (1): 2726–2740. doi:10.1111/j.1462–2920.2012.02825.x

Watanabe, H.C., Yamashita, Y. and Ishikita, H. (2017). Electron transfer pathways in a multiheme cytochrome MtrF. *Proceedings of the National Academy of Sciences 114* (11): 2916–2921. https://doi.org/10.1073/pnas.1617615114

Waters, M.S., El-Naggar, M.Y., Hsu, L. et al. (2009). Simultaneous interferometric measurement of corrosive or demineralizing bacteria and their mineral interfaces. *Applied and Environmental Microbiology 75* (5): 1445–1449. doi:10.1128/AEM.02039–08

Weber, K.A., Achenback, L.A. and Coates, J.D. (2006). Microorganisms pumping iron: anaerobic microbial iron oxidation and reduction. *Nature Reviews Microbiology 4* (10): 752–764. doi:10.1038/nrmicro1490

Widdel, F. and Pfennig, N. (1981). Studies on dissimilatory sulfate-reducing bacteria that decompose fatty acids. *Archives of Microbiology 129* (5): 395–400. https://doi.org/10.1007/BF00406470

Wortmann, U.G., Chernyavsky, B., Bernasconi, S.M. et al. (2007). Oxygen isotope biogeochemistry of pore water sulfate in the deep biosphere: dominance of isotope exchange reactions with ambient water during microbial sulfate reduction (ODP Site 1130). *Geochimica et Cosmochimica Acta 71* (17): 4221–32. https://doi.org/10.1016/j.gca.2007.06.033

Yang, G., Chen, M., Zhou, S. et al. (2013). *Sinorhodobacter ferrireducens* gen. nov., sp. nov., a non-phototrophic iron-reducing bacterium closely related to phototrophic *Rhodobacter* species. *Antonie van Leeuwenhoek, International Journal of General and Molecular Microbiology 104* (5): 715–724. https://doi:10.1007/s10482–013–9979–0

Zacharoff, L., Chan, C.H. and Bond, D.R. (2015). Reduction of low potential electron acceptors requires the CbcL inner membrane cytochrome of *Geobacter sulfurreducens*. *Bioelectrochemistry 107*: 7–13. https://doi.org/10.1016/j.bioelechem.2015.08.003

Zeng, X., Zhang, Z., Li, X. et al. (2015). Caloranaerobacter ferrireducens sp. nov., an anaerobic, thermophilic, iron(III)-reducing bacterium isolated from deep-sea hydrothermal sulfide deposits. *International Journal of Systematic and Evolutionary Microbiology 65* (6): 1714–1718. https://doi.org/10.1099/ijs.0.000165

Zhao, JS., Manno, D., Leggiadro, C. et al. (2006). *Shewanella halifaxensis* sp. nov., a novel obligately respiratory and denitrifying psychrophile. *International Journal of Systematic and Evolutionary Microbiology 56* (1): 205–212. https://doi.org/10.1099/ijs.0.63829–0

4

Biogeochemistry of Nitrogen in the Marine System with Special Emphasis on the Arabian Sea and Bay of Bengal

Kalyani Sivan[§], Aninda Mazumdar*, Aditya Peketi, and Jittu Mathai

ABSTRACT

The marine nitrogen (N) cycle involves various N-transforming reactions that control the amount of N and its forms in the environment. The discovery of new biogeochemical processes and microorganisms responsible for driving the processes add to the complex picture of the N biogeochemical cycle. The N isotopic ratios of sinking particles indicate the water column redox conditions and the biogeochemical transformation processes. The $\delta^{15}N$ of sedimentary organic matter is a crucial tool for tracking past changes in the marine N cycle. This review summarizes the current understanding of the marine N cycle, its novel processes and biochemical pathways, and a brief account of the processes in the Northern Indian Ocean.

4.1. INTRODUCTION

Nitrogen (N) is a primary nutrient required for all living organisms for the biosynthesis of cellular components such as proteins and nucleic acids. In the marine realm, N biogeochemistry primarily depends on the redox transformations mediated by microorganisms (Falkowski, 1997; Berner, 2006; Canfield et al., 2010; Kuypers et al., 2018). Our understanding of the global N cycle has enhanced in recent years as a result of the identification of new processes and pathways performed by new groups of specialized organisms. This includes identifying denitrification and other dissimilatory N transformation pathways in eukaryotes, anammox, and anaerobic methane oxidation with NO_2^- (Thamdrup, 2012; Ramos and Pajares, 2018). In surface waters, N compounds belong to reactive (N_R) and non-reactive (N_{NR}) phases. Reactive N includes ammonia (NH_3), nitrate (NO_3^-), nitrite (NO_2^-), nitric oxide (NO), nitrous oxide (N_2O), and dissolved organic nitrogen (DON), which are biologically accessible to most species. Non-reactive includes dissolved N_2 gas (Galloway et al., 2003), which has a concentration exceeding 400 μmol l^{-1} in seawater. The main sources of N to the ocean include upwelling, winter convective mixing, atmospheric deposition, riverine and groundwater fluxes, eddies/gyres, and diffusion from a permanent thermocline, regeneration of organic particles, and diazotroph N_2 fixation (Figure 4.1; Galloway et al., 2004; Voss et al., 2013). Mass balance estimates show an increase in fluvial and atmospheric inputs of N to the coastal and the open ocean compared with preindustrial times (Duce et al., 2008) as a result of increased anthropogenic activities (Galloway et al., 2004, 2008; Gruber and Galloway 2008). The enhanced anthropogenic input of NH_3 and NO_x ($NO + NO_2$) is attributed to the increased use of fertilizers, fossil fuels, and sewage effluents (Rovira and Pardo, 2006).

Here we review the sources of N to the marine realm and biogeochemical cycle, with particular emphasis on processes in the Northern Indian Ocean.

CSIR-National Institute of Oceanography, Dona Paula, Goa, 403004

[§]Academy of Scientific and Innovative Research (AcSIR), Ghaziabad-201002, India

**Corresponding Author: Aninda Mazumdar (maninda@nio.org) ORCiD code: 0000-0002-7897-1646*

Systems Biogeochemistry of Major Marine Biomes, First Edition. Edited by Aninda Mazumdar and Wriddhiman Ghosh.

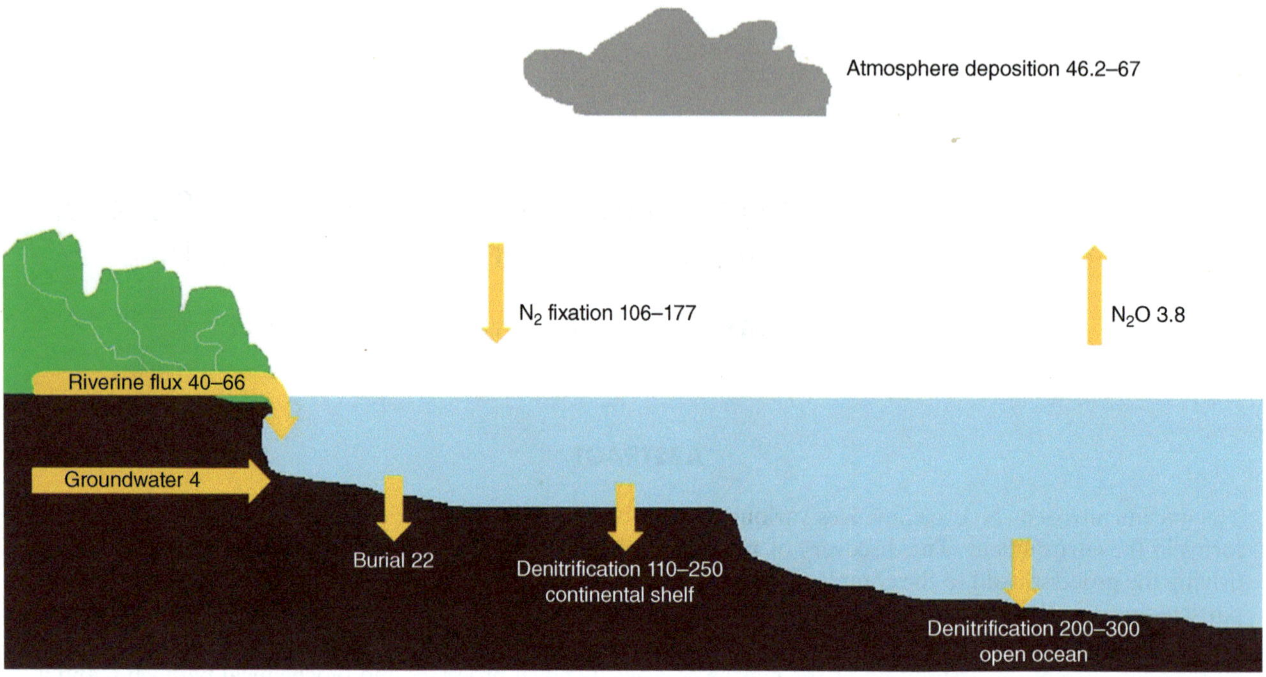

Figure 4.1 A simplified schematic diagram of nitrogen cycling in the global oceans. The fluxes are in Tg N year^{-1}. Modified from Voss et al. (2013).

4.2. SOURCES OF NITROGEN FOR THE OCEAN

4.2.1. Riverine and Groundwater Input

Rivers and groundwater supply most of the N_R to the coastal environment. Annually, rivers supply about 40–66 Tg N year^{-1} to the coastal ecosystem (Seitzinger et al., 2005; Boyer et al., (2006). According to Seitzinger et al. (2005), (25 Tg N year^{-1}) of riverine N is dissolved inorganic (DIN) while particulate nitrogen (PN) and dissolved organic nitrogen (DON) constitutes (30 Tg N year^{-1}) and (12 Tg N year^{-1}), respectively. Boyer et al. (2006) attributed contributions of ~11 Tg and 48 Tg N year^{-1} from dryland/internal watershed and coastal regions, respectively, to the total current efflux of riverine fixed N (59 Tg N year^{-1}). The riverine efflux to the global ocean has doubled compared with preindustrial times and is expected to increase further in the future (Seitzinger et al., 2005). Submarine groundwater discharges 4 Tg N year^{-1} to nearshore ecosystems (primarily as NO_3^-), of which the contribution from fertilizers is significant (Voss et al., 2011).

4.2.2. Atmospheric Deposition

Atmospheric N input falls into two main classes (Duce et al.,1991): (a) oxidized N compounds which include aerosol NO_3^- and gas-phase oxides of N (NO, NO_2, and related species), and (b) reduced N compounds, including aerosol NH_4^+, gaseous NH_3, and related species. Dentener et al. (2006) estimated a global aeolian input of 46.2 Tg N year^{-1}, while Duce et al. (2008) suggested a total N deposition of 67 Tg N year^{-1}. Apart from inorganic N, the atmosphere is also an important source of organic N which constitutes 20–30% of the total aerosol N (Cornell et al., 1995, 2003). Human activities have increased the aeolian fixed N input to the oceans from (3–8) × 10^{12} mol year^{-1} to (6.5–19) × 10^{12} mol year^{-1}, which has a potential impact on the productivity and carbon dynamics of the marine ecosystem (Cornell et al., 1995).

4.2.3. Volcanic Input

Undersea explosive volcanic eruptions and hydrothermal activity release N from the earth's mantle into the ambient ocean water and eventually to the atmosphere. The flux of N at mid-oceanic ridges is estimated to be 0.017–0.045 Tg N year^{-1} (Marty, 1995). Aerosols and metal salts adsorbed on tephra surfaces are highly soluble. Hence, they dissolve in contact with surface water and contribute to the nutrient reservoir (Jones and Gislason, 2008), thereby stimulating primary production.

4.2.4. Marine Upwelling

Upwelling is the slow upward movement of nutrient-rich deep water towards the surface. Coastal upwelling is strongly developed along the west coast of continents, resulting from offshore Ekman transport of surface water away from the coast due to winds along the coast. The upwelled water stimulates high primary production and remineralization, leading to O_2 deficiency in the subsurface water (Boje and Tomczak, 1978; Anderson and Lucas, 2009). Remineralization at depth delivers nutrients with a NO_3^- concentration of about 20–40 μmol l^{-1} (Voss et al., 2013). Nitrate is the dominant N form in coastal upwelling areas and is preferred by diatoms (Lomas and Glibert, 1999).

4.3. MARINE NITROGEN BIOGEOCHEMISTRY

Nitrogen in seawater functions as an electron acceptor and a donor owing to a wide range of oxidation states (–3 in NH_4^+ to +5 in NO_3^-) and therefore takes part in different respiratory metabolism (Pajares and Ramos, 2019). The marine N cycle (Figure 4.2) involves a complex set of redox transformations mediated by a specialized group of microorganisms (Falkowski, 1997; Canfield et al., 2010). Most microorganisms can assimilate both inorganic and organic N compounds in the marine ecosystem (Damashek and Francis, 2018). The balance between nitrogen fixation and N_2 production by dissimilatory processes regulates the fixed N availability

4.3.1. Biological Nitrogen Fixation

Biological nitrogen fixation (BNF) is responsible for the synthesis of fixed N which involves the conversion of N_2 to NH_3, followed by assimilation into amino acids by unicellular, heterocystous/nonheterocystous filamentous phototrophic/heterotrophic nitrogen-fixing bacteria and archaea (Capone et al., 1997; Canfield et al., 2010; Sohm et al., 2011). The BNF rates in coastal and benthic systems are ~15 Tg N $year^{-1}$ and 12 Tg N $year^{-1}$, respectively (Capone, 1988). The global N_2-fixation rate estimated by Gruber (2004) and Gruber and Galloway (2008) ranges from 106 to 120 Tg N $year^{-1}$. Based on incubation experiments, Deutsch et al. (2007) and Gro ßkopf et al. (2012) estimated a global N_2-fixation rate ranging from 140 to 177 Tg N $year^{-1}$.

Biological nitrogen fixation is an energy-requiring process and may be represented by the equation (1),

$$N_2 + 8H^+ + 16ATP + 8e^- \rightarrow 2NH_3 + H_2 + 16ADP + 16Pi \tag{1}$$

The fixation of N_2 by nitrogenase enzyme involves three major steps, including electron transport, ATP

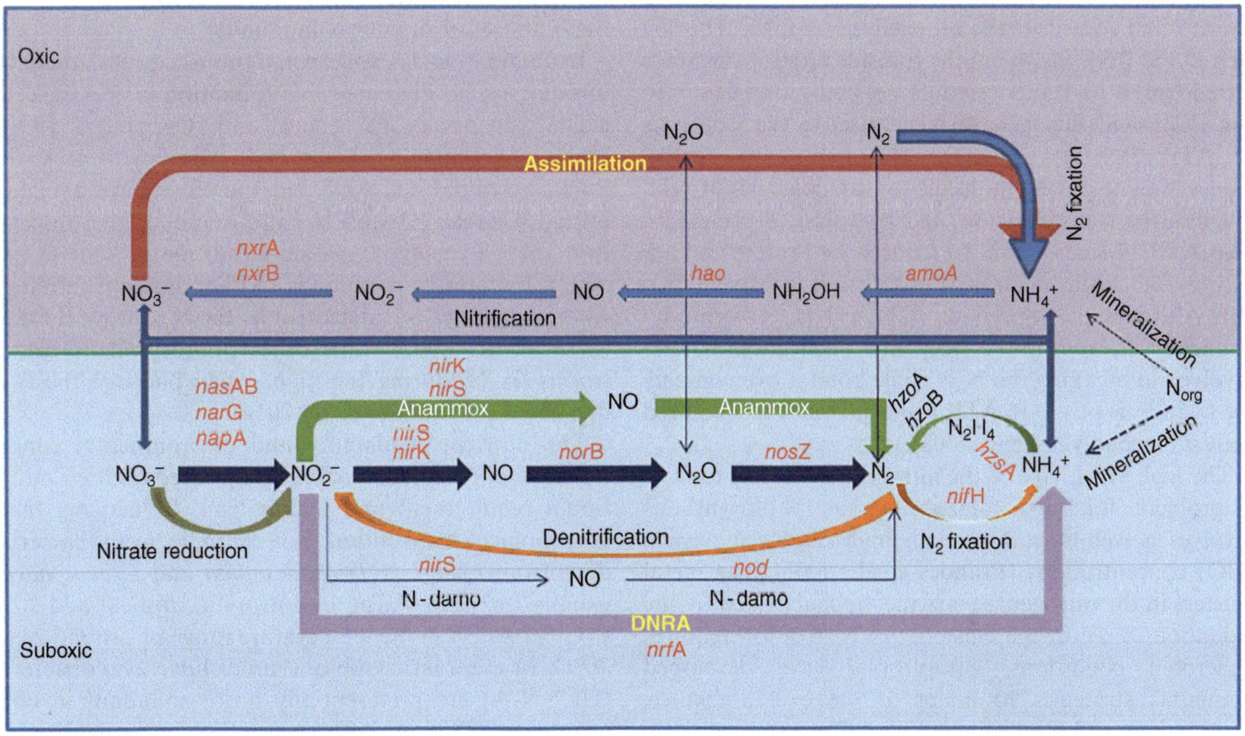

Figure 4.2 The major N cycling pathways and the marker genes associated with them. Modified from Canfield et al. (2010). The main marker genes involved in N transformations include: nitrogenase (*nifH*), nitrate reductases (*nasAB, narG, napA*), nitrite reductases (*nirK, nirS*), nitric oxide reductase (*norB*), nitrous oxide reductase (*nosZ*), anammox (*nirS, hzsA, hzoA, hzoB*), DNRA (*nrfA*), N-Damo (*nirS, nod*), ammonium monooxygenase (*amoA*), hydroxylamine oxidoreductase (*hao*), and nitrite oxidoreductase (*nxrA, nxrB*).

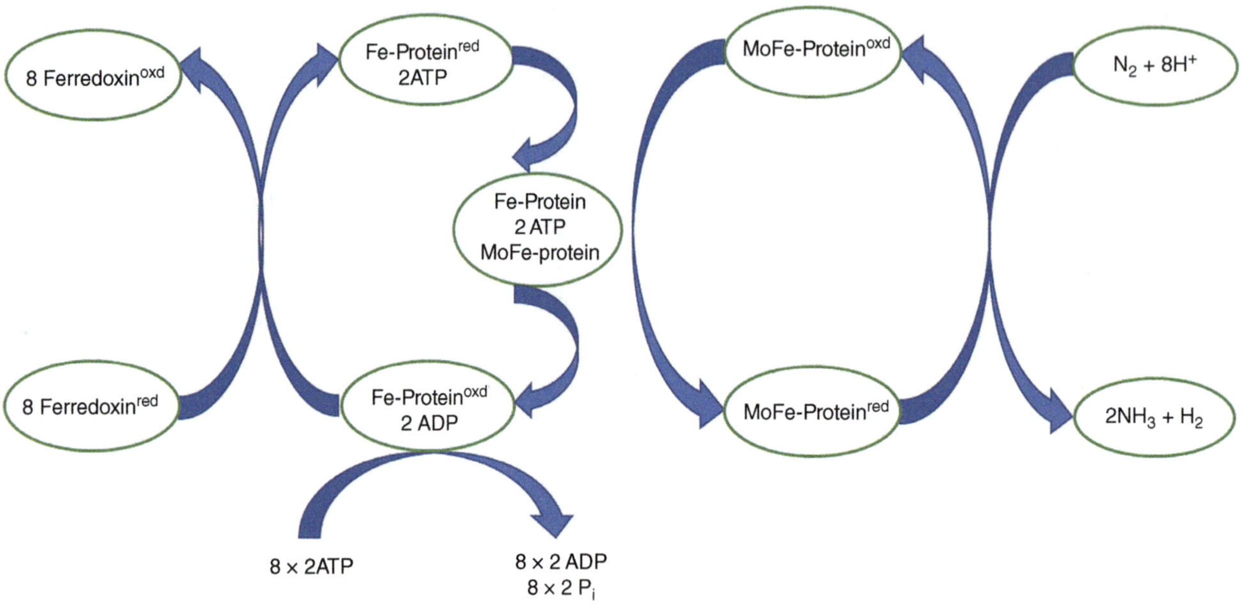

Figure 4.3 Mechanism of electron flow during biological N_2 fixation. Modified from Rees et al. (2005).

hydrolysis, and substrate reduction (Figure 4.3; Rees et al., 2005). Two specific proteins, dinitrogenase (Mo-Fe protein encoded in *nifDK* genes) and dinitrogenase reductase (Fe-protein encoded in the *nifH* gene: Zehr and Paerl, 2008) constitute the nitrogenase complex. The key step in the BNF involving the transfer of electrons from ferredoxin to Fe-Protein (reductase) and subsequently to Mo-Fe protein (nitrogenase) is energized by the hydrolysis of ATP (Rees et al., 2005; Hoffman et al., 2014). Conversion of one N_2 molecule to two molecules of NH_3 involves the transfer of $8e^-$ and hydrolysis of 16 ATPs. Two ATP molecules bind to reduced Fe-Protein and are hydrolyzed as the electron is transferred to Mo-Fe protein. Although the conversion of N_2 to NH_3 is thermodynamically favorable, the very high activation barrier involved in breaking the N-N triple bond is overcome by the hydrolysis of the 16 ATP, which provides the required activation energy (Cheng, 2008).

The high sensitivity of the nitrogenase enzyme to O_2 is a problem for most diazotrophs as photosynthesis involves a well-lit surface with high dissolved oxygen (DO) concentrations (Brandes et al., 2007). The metal clusters in the nitrogenase enzyme are inactivated by the oxidation of Fe in the clusters. As N_2 fixation requires anaerobic conditions, diazotrophs have developed numerous strategies to adapt to the oxic ambiance, including non-photosynthetic cells called heterocysts (in heterocystous cyanobacteria), which are N_2-fixing cells, and provide an anoxic environment (Staal et al., 2007). Heterocysts possess only photosystem I (they lack oxygenic photosystem II) and can maintain microaerophilic conditions with the help of their thick glycolipid cell walls (which act as an O_2 diffusion barrier) and by active respiration. In contrast, the other cells responsible for photosynthesis possess both photosystem I (used to generate ATP) and photosystem II (which produce O_2 during the splitting of water using light energy to generate H_2 for the synthesis of organic compounds).

In contrast, in *Trichodesmium* (non-heterocystous), O_2 interference is eliminated by N_2 fixation in specific cells called diazocytes (Bergman and Carpenter, 1991). However, relative to heterocysts, diazocytes lack the thick glycolipid cell wall and therefore have a higher influx of gasses (O_2 and N_2) and require higher respiration rates to achieve anoxic conditions (Staal et al., 2003). Temporal separation is another protective strategy where heterocystous diazotrophs fix N_2 during the day (Staal et al., 2007) while other photosynthetic diazotrophs fix N_2 during the night when photosynthesis is depressed (Moisander et al., 2014).

The nutrient availability and environmental conditions affect diazotrophs differently based on their nutritional requirement and metabolism, influencing their geographical distribution. For example, cyanobacterial diazotrophs such as *Trichodesmium* and *Crocosphaera* usually thrive in warm, oligotrophic tropical and subtropical waters in the temperature range of around 24 to 30°C. In contrast, symbiotic unicellular cyanobacteria (UCYN-A) are preferentially more abundant in cold waters (<25°C) in both eutrophic and oligotrophic waters (Sohm et al., 2011; Moreira-Coello et al., 2019). The availability of Fe affects nitrogenase activity because it is a constituent of the metalloproteins that makes up the subunits of the enzyme (Benavides et al., 2020). Phosphorus also limits N_2 fixation as it is required for the synthesis of lipids and nucleotides

(Dyhrman et al., 2007). However, *Trichodesmium* possesses the genetic capacity to compensate for phosphorous limitation by metabolizing dissolved DOP molecules such as phosphomonoesters and phosphonates (Dyhrman et al. 2006).

4.3.2. Nitrogen Assimilation

Nitrogen assimilation refers to the synthesis of cellular organic N compounds such as amino acids via assimilation of external inorganic and organic N (such as urea). In NH_4^+ assimilation, NH_3/NH_4^+ diffuses into the bacterial/phytoplankton cells and subsequently undergoes conversion to amino acids by the enzymes glutamine synthetase and glutamate synthase (Glass et al., 2009). Assimilation of NO_3^- is metabolically costly. NO_3^- needs to be reduced first to NH_3 (i.e., reducing N from an oxidation state of +5 to –3 with a transfer of eight electrons) by nitrate and nitrite reductases before assimilation into organic-N (Mulholland and Lomas, 2008). Phytoplanktons usually prefer NH_4^+ over NO_3^- as it is already in the redox state for the synthesis of amino acids, and thus requires less energy for assimilation.

Nitrate Assimilation

During NO_3^- assimilation, NO_3^- to NO_2^- reduction is catalyzed by the enzyme nitrate reductase (NarB) and involves transferring two electrons from ferredoxin through the [4Fe-4S] cluster to a Mo atom in molybdoprotein. As a result, Mo is reduced from the +6 to the +4 oxidation state. NO_3^- binds to reduced Mo and is subsequently reduced to NO_2^- while Mo oxidizes back to the +6 state (Flores and Herrero, 2005;Glass et al., 2009) following the overall reaction (equation 2).

$$NO_3^- + 2e^- + 2H^+ \rightarrow NO_2^- + H_2O \qquad (2)$$

During NO_2^- assimilation, six electrons (equation 3) are transferred from ferredoxin through the [4Fe-4S] cluster to the Fe in siroheme, where reduction of NO_2^- to NH_4^+ takes place, catalyzed by the enzyme nitrite reductase [Nir, Fe rich protein with one [4Fe-4S] cluster and one heme] (Kuznetsova et al., 2004).

$$NO_2^- + 6e^- + 8H^+ \rightarrow NH_4^+ + 2H_2O \qquad (3)$$

Ammonium Assimilation

This is the last step in the N assimilation pathway and involves incorporating NH_4^+ into alpha-ketoglutarate, forming glutamate catalyzed by glutamate dehydrogenase (Figure 4.4). The NH_4^+ ion is incorporated into glutamate to form glutamine catalyzed by the ATP-dependent enzyme glutamine synthetase (GS), represented by equation-4 (Muro-Pastor et al., 2005; Ohashi et al., 2011).

$$\text{Glutamate} + \text{ATP} + NH_4^+ \rightarrow \text{Glutamine} + \text{ADP} + P_i \qquad (4)$$

This is followed by the transfer of an amide group from glutamine to 2-oxoglutarate to yield two molecules of glutamate through a two-electron reduction (equation 5) catalyzed by the enzyme glutamate synthase (GltS or GOGAT) (Glass et al., 2009).

$$\text{Glutamine} + 2-\text{oxoglutarate} + 2e^- + 2H^+ \rightarrow 2\ \text{Glutamate} \qquad (5)$$

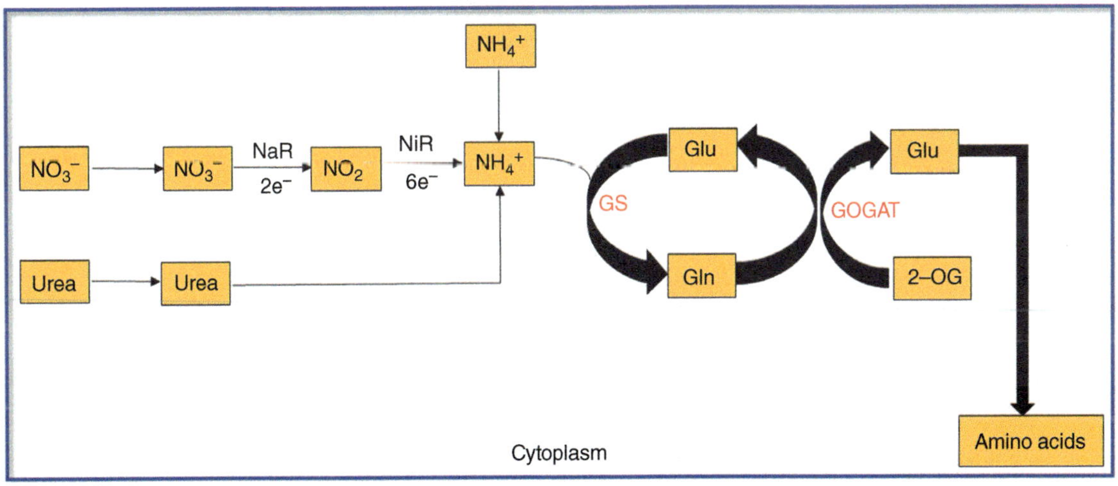

Figure 4.4 Nitrogen assimilation pathways in cyanobacteria. Glu, glutamate; Gln, glutamine; 2-OG, 2-oxoglutarate; NaR, nitrate reductase; NiR, nitrite reductase; GS, glutamine synthetase; GOGAT, glutamate synthase. Redrawn from Flores and Herrero (2005).

The two moles of glutamate can be used for the synthesis of amino acids.

The hydrolysis of urea (equation 6) also yields two molecules of NH_3 and one molecule of carbonic acid catalyzed by the enzyme urease (Glass et al., 2009).

$$H_2N-CO-NH_2+2H_2O \rightarrow 2NH_3+H_2CO_3 \qquad (6)$$

4.3.3. Nitrification

Nitrification is an aerobic process by which NH_3 is oxidized sequentially to hydroxylamine (NH_2OH), nitric oxide (NO), NO_2^-, and NO_3^- by a specialized group of bacteria and archaea (Kozlowski et al., 2016; Caranto and Lancaster, 2017). Nitrification is believed to be performed by three groups of microorganisms: Group I includes NH_3 oxidizers that oxidize NH_3 to NO_2^-, Group II includes NO_2^- oxidizers that oxidize NO_2^- to NO_3^- (nitratation), and Group III, which includes those that can oxidize NH_3 completely to NO_3^- (Comammox) (Ramos and Pajares, 2018). The NH_3 oxidation to NO_2^- is proposed to be carried out by both NH_3 oxidizing bacteria (AOB), including Betaproteobacteria (*Nitrosomonas* and *Nitrosospira*), Gammaproteobacteria including *Nitrosococcus* (Purkhold et al., 2000), and NH_3-oxidizing archaea (AOA) including the Thaumarchaeota phylum, such as *Nitrosopumilus maritimus* and *Cenarchaeum symbiosum* (Könneke et al., 2005). Fernandes et al. (2020) identified the bacterium *Nitrospina* and archaeon *Nitrosopumilus* from the Arabian Sea oxygen minimum zones (AS OMZ), thus suggesting the importance of nitrification.

While numerous studies have reported the relative abundance of AOB over AOA (Mosier and Francis, 2008), Smith et al. (2014) suggested that AOA possibly outnumbers AOB in both terrestrial and marine systems and are believed to play the primary role in the oxidation of NH_3. Microorganisms responsible for nitrification possess the enzyme ammonium monooxygenase (Amo, encoded in *amoA* gene), which converts NH_3 to NH_2OH. NH_2OH is then oxidized to NO_2^- using the enzyme hydroxylamine oxidoreductase (Hao, encoded in *haoA* gene). A recent study suggested that the conversion of NH_2OH to NO_2^- takes place through an intermediate step with the production of NO via a three-step process in AOB (Figure 4.5; Kozlowski et al., 2016; Caranto and Lancaster, 2017). NO is further oxidized to NO_2^- by an unknown enzyme hypothesized to be catalyzed by Cu-containing nitrite reductase (Cu-Nir). Incomplete oxidation of NH_2OH by Hao reductase to HNO or NO leads to the production of N_2O as a byproduct in AOB (Caranto and Lancaster, 2017). It is known that N_2O is also produced during nitrifier denitrification through NO_2^- reduction via an NO intermediate (Stein, 2011). The conversion of NO_2^- to NO_3^- (nitratation) by NO_2^- oxidizing bacteria (NOB, belonging to *Chloroflexi*, *Nitrospirae*, and *Nitrospinae*) is catalyzed by the enzyme nitrite oxidoreductase (Nxr, encoded in *nxrAB* gene) (Canfield et al., 2010; Kuypers et al., 2018). The overall reactions of nitrification may be represented by equations –(7–9). In addition, AOA produces N_2O by a distinct pathway through non-enzymatic reduction of intermediates NO or NH_2OH (Kozlowski et al., 2016). Electrons released during the nitrification pathway are used to fix inorganic carbon.

$$NH_3+O_2+2e^-+2H^+ \rightarrow NH_2OH+H_2O \qquad (7)$$

$$2NH_2OH+3/2O_2 \rightarrow 2NO+3H_2O \qquad (8)$$

$$NO_2^-+½O_2 \rightarrow NO_3^- \qquad (9)$$

Nitrification is controlled by concentrations of O_2, NH_4^+ and NO_2^- in addition to variations in light, pH, and salinity. The distribution and magnitude of nitrification rates in the oceans are also controlled by the composition and abundance of nitrifiers, particularly AOA (Smith et al., 2014). Being tolerant to low

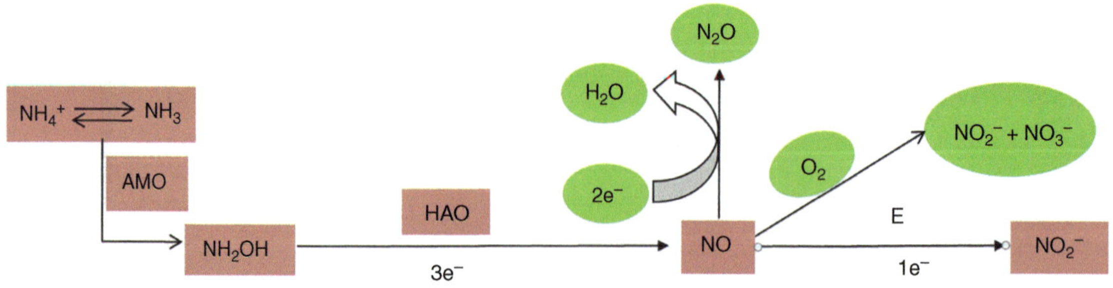

Figure 4.5 Model for NH_3 oxidation to NO_2^- in AOB (Caranto and Lancaster, 2017). Pathways in green are nonenzymatic. E is the proposed unknown enzyme required to oxidize NO to NO_2^-.

DO, AOA, and AOB (Park et al., 2010) may oxidize NH_3 to NO_2^-. Bristow et al. (2017) have shown that NOB have a high affinity for O_2, resulting in NH_4^+ and NO_2^- oxidation to NO_3^- even at nanomolar O_2 concentration in OMZs. Low nitrification rates in the sea surface may be to the result of substrate (NH_4^+ and NO_2^-) competition among phytoplankton and slow-growing nitrifiers (Smith et al., 2014).

Several studies of nitrification have reported that marine NH_4^+ and NO_2^- oxidation is photo-inhibited (Peng et al., 2016; Horak et al., 2018) while *amoA* genes have been identified from surface waters (Shiozaki et al., 2016), implying that light exhibits an indirect control on nitrification (Zakem et al., 2018). The highest rates of NH_3 oxidation occur below the sunlit surface (Yool et al., 2007) as the cytochromes in nitrifiers are sensitive to light (Ramos and Pajares, 2018). Olson (1981) proposed differential inhibition of light by nitrifiers such that NO_2^- oxidizers are more sensitive to light than NH_3oxidizers.

Temperature influences the composition and abundance of AOA in the San Francisco Bay estuary (Mosier and Francis, 2008). An increase in temperature may lead to uncoupling of NH_3^- and NO_2^- oxidation in coastal waters resulting in NO_2^- accumulation (Schaefer and Hollibaugh, 2017). However, Horak et al. (2018) suggested that AOA was least affected by temperature. The composition and abundance of NH_3 oxidizers seem to be strongly correlated with salinity in estuaries (Mosier and Francis, 2008). It also appears that NO_2^- oxidizers seem to be affected by high salinity. Nitrifying bacteria show high sensitivity to pH, such as *Nitrosomonas* and *Nitrobacter*, which are stable at a pH range of 7.0–8.0 and 7.5–8.0, respectively.

4.3.4. Denitrification

Denitrification is an anaerobic respiratory process found in both heterotrophic and autotrophic microbes that converts NO_3^- to N_2 gas through a series of reductive pathways ($NO_3^- \rightarrow NO_2^- \rightarrow NO \rightarrow N_2O \rightarrow N_2$) (Codispoti et al., 2001). Denitrification is restricted to environments where water column DO concentration is <5 μmol kg^{-1} (Codispoti et al., 2001). Denitrifiers include both prokaryotes (more than 60 genera of bacteria and archaea) and eukaryotes, including fungi, protozoa, and benthic foraminifera (Thamdrup, 2012). The overall reaction may be represented by equation (10):

$$\begin{aligned} &4/5NO_3^- + 1/2C_2H_3O_2^- + 3/10H^+ \\ &\rightarrow 2/5N_2 + HCO_3^- + 2/5H_2O \end{aligned} \tag{10}$$

In denitrifying microorganisms, NO_3^- reduction to NO_2^- is catalyzed by two enzymes: a membrane-bound nitrate reductase (Nar, encoded in *narG* genes) and periplasmic nitrate reductase (NAP, encoded in *napA* genes) (Zumft, 1997). The reduction of NO_2^- to NO is catalyzed by two periplasmic enzymes: a Cu-containing nitrite reductase (encoded by *nirK*) and a haem-containing cd1 nitrite reductase (encoded by *nirS*) (Kuypers et al., 2018). Two types of NO reductases (Nor, present in microorganisms such as *Pseudomonas denitrificans* and *Pseudomonas stutzeri* carry out NO reduction to N_2O: a cytochrome c-dependent complex (cNor) present almost exclusively in denitrifiers, and a quinol-dependent complex (qNor) present in microorganisms with incomplete denitrifying molecular characteristics encoded by two variables of the same gene *cnorB* and *qnorB* (Hino et al., 2010). The last step in the denitrification pathway (i.e., the reduction of N_2O to N_2) is catalyzed by the N_2O reductase (NOS) encoded in the *nosZ* gene (Zumft, 1997). Members of the Proteobacteria, Bacteroidetes, Chlorobi phyla, and archaea from Crenarchaeota and Halobacteria possess NOS (Cabello et al., 2004). Culture-dependent studies in OMZs of the Northern Indian Ocean have revealed the presence of NO_3^--reducing bacteria (e.g. belonging to Alphaproteobacteria and Gammaproteobacteria: see Mulla et al., 2018; Fernandes et al., 2019) thus suggesting a role of denitrification in the OMZ. Metagenomic studies have also identified the presence of denitrifiers (*Nitratireductor*) in the AS OMZ (Bandekar et al., 2018).

Denitrification rate depends on the concentrations of NO_3^- and NO_2^- (Thamdrup, 2012; Jayakumar et al., 2013), availability of labile organic matter, and availability of Cu since *nirK* and *nosZ* genes require Cu (Granger and Ward, 2003). The presence of low levels of sulfide can result in the accumulation of NO and N_2O by inhibiting the activity of nitrous oxide reductase as a result of the formation of chemically unreactive copper sulfide complexes, thus limiting the biological availability of Cu (Granger and Ward, 2003). Salinity also influences the abundance and composition of the denitrifier community in estuaries (Mosier and Francis, 2010).

4.3.5. Ammonium Production in the Marine Realm

Ammonium production in the marine realm occurs primarily by the breakdown and decomposition of N-containing organic compounds termed N-mineralization (Janssen, 1996). The organic N is degraded to NH_4^+ by microorganisms and other marine organisms as a metabolic byproduct. The first step in N mineralization (Figure 4.6) is the breakdown of complex macromolecules such as proteins, peptides, chitin, peptidoglycan, and nucleic acids to their constituent monomers by the activity of microbial enzymes proteinases, peptidases, chitinases, N-acetylglucosamidase, and nucleases, a process termed

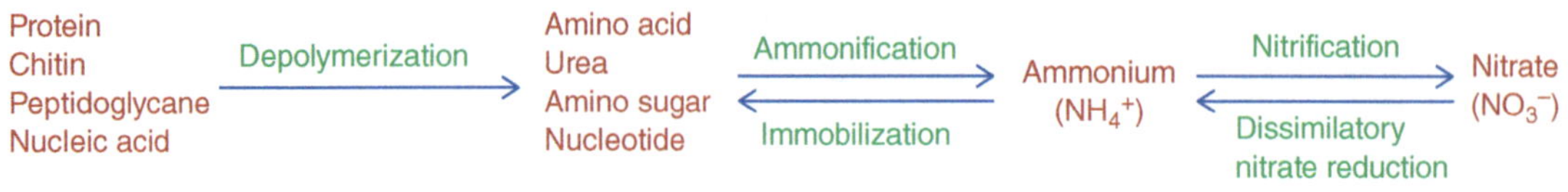

Figure 4.6 Mineralization in the N cycle. Adapted from Romillac (2019).

depolymerization (Jones et al., 2004; Romillac, 2019). Ammonification is the intermediate step between large organic molecule depolymerization and the nitrification step. Ammonification is the production of NH_4^+ from the mineralization of low molecular weight dissolved organic N molecules such as amino acids, urea, aminosugar by microorganisms via either intracellular or extracellular pathways.

4.3.6. Dissimilatory Nitrate Reduction to Ammonia

Dissimilatory nitrate reduction to ammonia (DNRA) is an anaerobic process in which NO_3^- is reduced to NO_2^- and subsequently to NH_4^+ (equation 11) (Kuypers et al., 2018). DNRA is considered primarily as a heterotrophic process where the electron donor utilized is an organic substrate (fermentative DNRA, carried out by bacteria such as clostridia, *Desulfovibrio*, and *Pseudomonas*) (Burgin and Hamilton, 2007).

$$NO_2^- + 3/4CH_3COO^- + 5/4H^+ + H_2O \rightarrow NH_4^+ + 3/2HCO_3^- \quad (11)$$

However, in some microorganisms such as *Thiomargarita* and *Thioploca*, DNRA is a chemolithoautotrophic process that uses inorganic reduced forms such as sulfide (H_2S or S^{2-}), elemental sulfur (S^0), or Fe^{2+} (equation 12 and 13) (Robertson et al., 2016; Slobodkina et al., 2017).

$$4S^0 + 3NO_3^- + 7H_2O \rightarrow 4SO_4^{2-} + 3NH_4^+ + 2H^+ \quad (12)$$

$$8Fe^{2+} + NO_3^- + 21H_2O \rightarrow NH_4^+ + 8Fe(OH)_3 + 14H^+ \quad (13)$$

The initial reduction of NO_3^- to NO_2^- in fermentative DNRA is catalyzed by membrane-bound nitrate reductase complex (NarGHI) or periplasmic nitrate reductase complex (NapAB) (Simon and Klotz, 2013). The NO_2^- reduction to NH_4^+ is catalyzed by the periplasmic cytochrome c nitrite reductase (ccNir) encoded by the *nrfA* gene (Kuypers et al., 2018). The NO_2^- reduction to NH_4^+ in a metal-reducing bacterium, *Shewanella oneidensis* MR-1 is catalyzed by octaheme tetrathionate reductase (Otr). In another chemolithoautotrophic sulfur oxidizing bacterium, *Thioalkalivibrio nitratireducens*, NO_2^- to NH_4^+ reduction is catalyzed by octaheme cytochrome c nitrite reductase (Onr) (Tikhonova et al., 2006).

4.3.7. Anammox

Anammox or anaerobic ammonium oxidation is a chemolithoautotrophic microbial process in which NH_4^+ oxidation is coupled to NO_2^- reduction (Strous et al., 1999). Anammox is an anaerobic exergonic process where N_2 gas is produced as a metabolic endproduct. Organisms belonging to the phylum Planctomycetes can perform anammox, and most of them belong to the genera Candidatus *Scalindua* (Lam and Kuypers, 2011). The presence of anammox bacteria Candidatus *Scalindula* has been identified from the AS OMZ (Bandekar et al., 2018; Fernandes et al., 2020). The overall anammox reaction may be represented by equation 14

$$NH_4^+ + 1.32NO_2^- + 0.066HCO_3^- + 0.13H^+ \rightarrow 0.26NO_3^- + 1.02N_2 + 0.066CH_2O_{0.5}N_{0.15} + 2.03H_2O \quad (14)$$

Anammox bacteria are slow-growing anaerobic chemolithoautotrophs that can fix CO_2 using NO_2^- as an electron donor. Approximately 0.07 mol carbon is fixed per mol NH_4^+ oxidized (Thamdrup, 2012). During anammox reaction, the reduction of NO_2^- to NH_2OH occurs and is catalyzed by a nitrite reducing enzyme. Then, NH_2OH reacts with NH_4^+ to form hydrazine using the enzyme hydrazine synthase (HZS, encoded by *hzsA* and *hzsB*). Hydrazine is subsequently oxidized by hydrazine dehydrogenase (HDH, encoded by *hzoAB* (Strous et al., 2006) or hydrazine oxidase (HZO) to N_2 (Simon and Klotz, 2013). This occurs in a membrane-bound intracytoplasmic compartment called the anammoxosome, composed of structurally rigid lipids, called ladderanes (Sinninghe-Damsté et al., 2002).

The availability of O_2, NO_3^-, NO_2^- and NH_4^+ regulates the anammox process in the marine environment (Risgaard-Petersen et al., 2004; Ramos and Pajares, 2018) and is identified to be coupled with nitrification, denitrification, and DNRA. Strous et al. (1997)

proposed that an O_2 concentration less than 1.1 μM is necessary for anammox to take place, while Bristow et al. (2017) reported anammox bacteria from the Bay of Bengal oxygen minimum zone (BoB OMZ) at O_2 concentrations as high as 9.5 μM. A higher abundance of anammox bacteria over denitrifiers is reported from the Arabian Sea OMZ (Bandekar et al., 2018). Canfield et al.(2010) linked anammox to the cryptic sulfur cycle in the OMZ waters of the Chilean coast. The NH_4^+ released during sulfate (SO_4^{2-}) reduction and DNRA contributes to the substrate requirements of the anammox bacteria. The hydrogen sulfide produced by SO_4^{2-} reduction is presumably oxidized by NO_3^- to the NO_2^- reduction process.

4.3.8. Nitrate/Nitrite-Dependent Anaerobic Methane Oxidation

Nitrate/nitrite-dependent anaerobic methane oxidation (N-Damo) involves anaerobic oxidation of CH_4 coupled to denitrification (Raghoebarsing et al., 2006). This process is carried out by a microbial consortium of NC10 bacteria (Candidatus *Methylomirabilis oxyfera*) and ANME archaea (Candidatus *Methanoperedens nitroreducens*) where they oxidize CH_4 coupled to denitrification in the absence of O_2 (Ettwig et al., 2008; Ettwig et al., 2010; Haroon et al., 2013).

Methanoperedens nitroreducens couples anaerobic CH_4 oxidation with NO_3^- reduction to NO_2^- by reverse methanogenesis using the enzyme methyl-coenzyme M (methyl-CoM) reductase for CH_4 activation (equation 15) (Haroon et al., 2013). The endproduct NO_2^- can be used by NO_2^--dependent methanotrophs. Species belonging to the phylum NC10, *M. oxyfera*, oxidizes CH_4 to CO_2 with simultaneous reduction of NO_2^- to N_2 through an intra-aerobic CH_4 oxidation pathway (equation 16) (Ettwig et al., 2010).

$$CH_4 + 4NO_3 \rightarrow CO_2 + 4NO_2^- + 2H_2O \, \Delta G^{0'} = -523\,kJ\,mol^{-1}\,CH_4 \quad (15)$$

$$3CH_4 + 8NO_2^- + 8H^+ \rightarrow 3CO_2 + 4N_2 + 10H_2O \, \Delta G^{0'} = -928\,kJ\,mol^{-1}\,CH_4 \quad (16)$$

In an experimental culture, it was observed that *M. oxyfera* can produce O_2 through dismutation of NO (equation 17) catalyzed by the enzyme nitric oxide dismutase (Nod):

$$2NO \rightarrow O_2 + N_2 \quad (17)$$

Methylomirabilis oxyfera prefers NO_2^- over NO_3^- and oxidizes CH_4 using methane monooxygenase (encoded by *pmoA* gene) found in aerobic methanotrophs (Ettwig et al., 2010). In coculture experiments with anammox and CH_4 oxidizers, it is observed that if not NH_4^+ limited, anammox bacteria outcompete CH_4 oxidizers for NO_2^- (Luesken et al. 2011).

The NC10 genes have been identified from anoxic freshwater systems (Shen et al., 2015) and NO_2^--rich sediments of the South China Sea (Chen et al., 2015). Members of the phylum NC10 have been reported from OMZs, for example in the ETNP OMZ associated with high NO_2^- and CH_4 concentrations (Padilla et al., 2016). Unfortunately, N-Damo has only been studied recently, and very little is known about ANME archaea as most of the studies are focused on NC10 bacteria. The *mcrA* gene specific to archaea such as *M. nitroreducens* has been recovered from ETNP OMZ (Thamdrup et al., 2019), opening the pathway for detailed research of archaea ANME-2d in marine environments.

4.4. N-CYCLE IN MARINE SEDIMENTS

Sediments play a fundamental role in marine N cycling through microbial remineralization processes occurring across the oxic/anoxic interface. The burrowing and irrigation activities of macrofauna and meiofauna (Widdicombe et al., 2000) can significantly affect these microbial processes by changing the physicochemical properties of the sediment, including depth (aerobic–anaerobic interface) of O_2 penetration (Kristensen et al., 1991; Nielsen et al., 2004). Reworking of the sediment during burrowing activity results in higher meiofaunal and microbial abundance and biodiversity (Laverock et al., 2010) in the burrow compared with the surface and the subsurface sediment.

4.4.1. Nitrogen Fixation in Sediments

Burrowing activity can lead to the development of reducing micro-niches (due to an increase in organic matter accumulation), resulting in high SO_4^{2-} reduction rates in the sediment and subsequent high N_2 fixation (Goldhaber et al., 1977; Bertics et al., 2010). Bertics et al. (2013) reported the presence of *nifH* genes (nitrogenase) responsible for N_2 fixation in the sulfate-reducing bacteria (*Desulfovibrio vulgaris*, and *Desulfonema limicola*) from seasonally hypoxic Eckernforde Bay, Baltic Sea. Whereas, Dekas et al. (2009) and Hu et al. (2020) showed that AOM consortia can mediate N_2 fixation or NH_4^+ assimilation. Hu et al. (2020) attributed depleted N-isotopic signature ($\delta^{15}N_{org}$ up to -5‰) of sediment organic matter to ANME-mediated N uptake. The N_2 fixation in marine sediments is also positively correlated to bottom water temperature, bioturbation, and bioirrigation. In contrast, bottom water O_2 concentrations inhibit N_2 uptake because O_2 prevents nitrogenase enzyme activity (Bertics et al., 2013).

4.4.2. Nitrification in Marine Sediments

Nitrification can be an important process in shallow and deep-water sediments (Henriksen et al., 1993). Oxygen infusion into the bioturbated/bioirrigated sediments and the presence of MnO_2 have been found to enhance nitrification rates in marine sediments. Bioirrigation extends the depth of O_2 penetration below the sediment–water interface (Kristensen et al., 1991; Nielsen et al., 2004), thereby supporting higher rates of coupled nitrification–denitrification reactions. The Mn oxide may drive anoxic nitrification where NH_4^+ generated during organic matter remineralization is oxidized to NO_3^- by autotrophic bacteria (Mortimer et al., 2004), which are subsequently reduced to N_2 by denitrification (Luther et al., 1997). Several pieces of evidence suggest the production of NO_3^- and NO_2^- anaerobically in the marine sediments supported by the presence of both *Nitrosomonas* and *Nitrosospira*-like 16S rRNA gene sequences below the O_2 penetration depth in organic-rich sediments undergoing MnO_2-driven nitrification (Anschutz et al., 2000; Mortimer et al., 2002, 2004). Anoxic NH_4^+ incubations by Hulth et al. (1999), Thamdrup and Dalsgaard (2002), Bartlett et al. (2008), and Javanaud et al. (2011) demonstrated the simultaneous accumulation of NO_3^-, NO_2^- and Mn^{2+} in marine sediments. The anoxic nitrification by Mn oxides in marine sediments is also confirmed by the production of dissolved Mn^{2+} and a NO_3^- maximum in pore water profiles (Anschutz et al., 2000). However, porewater NO_3^- and NO_2^- may be consumed by several processes such as denitrification, anammox, and DNRA (Fernandes et al., 2015). The close microbial coupling of N and Mn cycles in continental margin sediments may be represented (Mortimer et al., 2004) by equations (18) and (19):

$$4MnO_2 + NH_4^+ + 6H^+ \rightarrow 4Mn^{2+} + NO_3^- + 5H_2O \quad (18)$$

$$8MnOOH + NH_4^+ + 14H^+ \rightarrow 8Mn^{2+} + NO_3^- + 13H_2O \quad (19)$$

4.4.3. Denitrification in Marine Sediments

Denitrification is an important reductive N transformation pathway in coastal marine sediments and is estimated to remove ~24 Tg of fixed N per year globally (Galloway et al., 2004). Sediments are favorable sites for denitrification as it is often associated with low oxygen concentrations, sharp oxic/anoxic interfaces and a high flux of both NO_3^- and organic matter (Seitzinger et al., 2006). Both heterotrophic and autotrophic denitrifiers are found in marine sediments depending on the availability of NO_3^- and electron donors, such as organic matter, sulfide, etc. (Canfield et al., 2005).

In marine sediments, several species of foraminifera were found to concentrate NO_3^- intracellularly and perform denitrification under anoxic conditions (Risgaard-Petersen et al., 2006; Bernhard et al., 2012). Risgaard-Petersen et al. (2006) proposed a complete denitrification pathway in a benthic foraminifera *Globobulimina turgida* and suggested that the denitrification rates measured in the species cannot be attributed to bacterial denitrification because of the low abundance of endobiotic bacteria in their cells. In contrast, an endobiotic bacteria belonging to the genus *Pseudomonas* capable of performing denitrification is associated with allogromiid foraminifera (NO_3^- accumulating) (Bernhard et al., 2012). It is suggested that these endobiotic bacteria survive on the fermentation products formed during metabolic activity in allogromiids, which in return help the foraminifera to live under anoxic conditions. Recently, a novel eukaryotic denitrification pathway (of prokaryotic origin) was found in foraminifera species of the genus *Globobulimina* and was attributed to the identification of the enzymes nitrite reductase (NirK), nitric oxide reductase (Nor), and nitrate transporters (Nrt) (Woehle et al., 2018).

In anaerobic marine sediments, the pore water NO_3^- concentrations may become zero within few centimeters below the sediment surface. The ability of foraminifera to store NO_3^- from the sediment surface in their vacuoles and bury deeper into the sediment well below the depth of penetration of O_2 and NO_3^- helps them to perform denitrification in deeper sediments (Risgaard-Peterson et al., 2006). The release of this intracellular NO_3^- to the surrounding NO_3^--deficient porewater after cell lysis may contribute to the initiation of N biogeochemical processes such as denitrification, anammox, and DNRA deeper in the sediment.

The generation of N_2 gas from NO_3^- via oxidation of H_2S, termed chemodenitrification is also an important process in marine sediments, e.g. by *Thiobacillus denitrificans* and *Thiomicrospira denitrificans* (Burgin and Hamilton, 2007). The particle reworking and irrigational activities by macrofauna (Karlson et al., 2007; Bonaglia et al., 2013) and meiofauna (Bonaglia et al., 2014) are known to enhance both denitrification and DNRA.

4.4.4. Anammox in Marine Sediments

Anammox in marine sediments may contribute <1% to up to 80% of benthic N_2 production, depending on the locations (Dalsgaard et al., 2005). The contribution of anammox in shallow marine sediments is <10% as a result of direct competition for substrates with denitrifiers (Trimmer and Engström 2011). Daalsgard et al. (2005) observed a relative increase in the significance of sedimentary anammox activity with water depth as a result of a decrease in denitrification activity in deeper waters. Less sedimentation of organic detritus with increasing water depth leads to low rates of benthic respiration, and the increasing importance of anammox with depth is linked to the fact that denitrification is attenuated with depth. The low rates of denitrification relative to anammox in deeper water locations rich in manganese oxide may be due to competition with them for electron donors to fuel dissimilatory reduction of manganese dioxide rather than denitrification (Dalsgaard and Thamdrup, 2002). The presence of sulfide in the low micromolar range also appears to inhibit anammox in shallow marine sediments (Jensen et al., 2008). The contribution of anammox to total N loss depends on the organic carbon content and nature, bottom water temperature, and NO_3^- concentration. Dalsgaard and Thamdrup (2002) suggested that anammox would be more in agreement with lower temperatures.

Anammox linked to SO_4^{2-} reduction (Sulfammox) and Fe reduction (Feammox) promoting significant N loss has recently been identified in coastal sediments (Rios-Del Toro et al., 2018). The mechanism of sulfammox and Feammox process is still unclear.

Sulfate and ferrihydrite reduction can fuel anammox in marine sediments in the absence of NO_2^- according to the following reactions represented by equations (20–22) (Rios-Del Toro et al., 2018):

$$8NH_4^+ + 3SO_4^{2-} \rightarrow 4N_2 + 3HS^- + 12H_2O + 5H^+ \, \Delta G^{0'} = -17.6\,kJ\,mol^{-1} \quad (20)$$

$$2NH_4^+ + SO_4^{2-} \rightarrow N_2 + S^0 + 4H_2O \, \Delta G^{0'} = -45.3\,kJ\,mol^{-1} \quad (21)$$

$$NH_4^+ + 3Fe(OH)_3 + 5H^+ \rightarrow 3Fe^{2+} + 0.5N_2 + 9H_2O \, \Delta G^{0'} = -245\,kJ\,mol^{-1} \quad (22)$$

During this process, NH_4^+ is partly oxidized to NO_2 using SO_4^{2} as an electron acceptor followed by the oxidation of NH_4^+ to N_2 by NO_2^- (Liu et al., 2008, 2020; Rios-Del Toro et al., 2018;). Another mechanism has also been proposed where NH_4^+ is partly oxidized to NO_2^- and some of the NO_2^- is reduced to N_2 by sulfide and subsequently NO_2^- and NH_4^+ is converted to N_2 (Fdz-Polanco et al., 2001). Schrum et al. (2009) suggested that the sulfammox could account for the NH_4^+ disappearance in the SO_4^{2-}–NH_4^+ interface in the marine sediments. The elemental sulfur or sulfide produced during sulfammox can be reoxidized to SO_4^{2-} by autotrophic denitrification using NO_2^- (the intermediate of sulfammox) (Liu et al., 2020).

4.4.5. Dissimilatory Nitrate Reduction to Ammonia In Marine Sediments

Chemolithoautotrophic DNRA is particularly important in anoxic sediments and sediments with sufficient sulfide concentrations where nitrification and denitrification are inhibited (Gardner et al., 2006). Even though DNRA has a lower energy yield than denitrification, it accepts more electrons ($8e^-$) per mole of NO_3^- reduced while denitrification accepts only five electrons (Tiedje, 1988). Therefore, DNRA may be favored over denitrification in NO_3^--limiting anoxic sediments rich in electron donors. Other than high S^{2-} concentration and $C:NO_3^-$ ratios, other factors such as higher temperature and salinity also favor DNRA over denitrification (Song et al., 2014). In coastal ecosystems, DNRA shows a large seasonal variation with higher rates during summer. This may be attributed to the reduced O_2 solubility at a higher temperature and enhanced metabolism (O_2 consumption in the sediment), leading to more reducing conditions in the sediment (Ferrón et al., 2009). The exact relationship between salinity and DNRA is uncertain. The DNRA rates have been found to increase with salinity in some estuaries (Gardner et al., 2006), while Dong et al. 2009 proposed a negative correlation.

The giant sulfur-oxidizing bacteria *Beggiatoa* performs DNRA using S^{2-} as an electron donor (Burgin and Hamilton, 2007). In contrast, both *Thiomargarita* and *Thioploca* exhibit the capacity to concentrate NO_3^- in their vacuoles and perform NO_3^- fermentation by coupling NO_3^- reduction and sulfide oxidation deeper in the sediment (Thamdrup, 2012).

4.5. NITROGEN CYCLING IN THE NORTHERN INDIAN OCEAN

The northern Indian Ocean constitutes about 21% of the total volume (15×10^6 km^3) of hypoxic waters in the global ocean (Acharya and Panigrahi, 2016). A significant portion of this lies in the AS (2.5×10^6 km^3), with the BoB covering a much smaller volume (0.6×10^6 km^3). The northern Indian Ocean OMZ is more intense in the AS than in the BoB. The high productivity and poor ventilation in the Northern Indian Ocean leads to suboxic conditions in intermediate waters with intense denitrification in the AS (Naqvi et al., 2006, 2010). Other regions where significant denitrification occurs in the water column are the eastern tropical North Pacific (ETNP) and eastern Tropical South Pacific (ETSP) (Codispoti et al., 2001).

4.5.1. Pelagic Nitrogen Cycle in Oxygen Minimum Zones

The AS hosts one of the earth's most prominent and thickest OMZ (O_2 concentration <5 μM) with a vertical span of 150–1200 m (Bange et al., 2005). The amount of N lost by pelagic denitrification in the AS is proposed to range from 21 to 60 Tg N year^{-1} (Bange et al., 2000, Codispoti et al., 2001; Devol et al., 2006) compared with the global estimate of 147 Tg N year^{-1} (Rixen et al., 2020). The AS OMZ favors denitrification as the most dominant N-loss pathway (Ward et al., 2009). Denitrification is detected at a depth (150–700 m: Banse et al., 2014) of secondary nitrite maxima within the core of OMZ (Ward et al., 2009). Isotope pairing experiments (Nicholls et al., 2007; Ward et al., 2009; Bulow et al., 2010) and relative microbial abundances (Devol et al., 2006; Ward et al., 2009) suggest active denitrification (>90% of overall N_2 production) in the AS. The dominance of denitrification over anammox may be attributed to a high labile organic matter supply (Ward et al., 2009) and the relatively slow-growing nature of anammox bacteria compared with the dynamic fast-growing denitrifiers (Strous et al., 1998). Interestingly, based on anoxic incubations of ^{15}N-labeled NO_2^- and NH_4^+ and functional gene analysis, Jensen et al. (2011) showed a close coupling between anammox and DNRA, indicating anammox as the key player in the Omani shelf, while the role of denitrification was hardly detectable. The dominant role of anammox has also been shown in the OMZs off Peru (Lam et al., 2009), Chile (Thamdrup et al., 2006), and Benguela (Kuypers et al., 2003).

Denitrifying conditions develop along the western continental shelf of India seasonally during late summer and autumn (Naqvi et al., 2000). Coastal upwelling during southwest monsoon brings O_2-depleted waters onto the shelf: as the upwelled water moves over the shelf, it loses O_2 due to organic matter (high productivity) remineralization. Moreover, increased precipitation and river discharge lead to strong stratification, thereby enhancing O_2 deficiency below the pycnocline. Coastal eutrophication has enhanced O_2 deficiency in recent years (Naqvi et al., 2000) and may increase further. Denitrification in the seasonal OMZ (5 Tg N year^{-1}) is smaller than that of perennial OMZ (Bange et al., 2005). Based on the consumption of initial NO_3^- concentration in upwelled water over a month, Naik (2003) found the annual denitrification rate from the western continental shelf of India during seasonal anoxia as 1.3–3.8 Tg N, which is smaller than the estimate (3.7–11 Tg N year^{-1}) by Sarkar et al. (2020). The fixed N removal as a result of denitrification was estimated to be 20–60% of the total N loss from perennial OMZ. Based on ^{15}N-labeled NH_4^+ incubations, anammox rates were estimated within 0.09–0.27 Tg N year^{-1}. In addition, Sarkar et al. (2020) indicated occurrences of DNRA, although these were sporadic along the west coast shelf region, and estimated a DNRA rate ~109.25 ± 22.1 nmol N_2 l^{-1} day^{-1}.

Biological nitrogen fixation is considered an active process in the AS and is favored by NO_3^-:PO_4^{3-} <15. The low NO_3^-:PO_4^{3} ratios are attributed to denitrification (Bange et al., 2005). Bange et al. (2000) estimated the N_2 fixation rate in the AS as 3.3 Tg N year^{-1}. Ahmed et al. (2017) reported an N_2 fixation rate between 0.8 and 225 μmol N m^{-3}day^{-1} from the Eastern AS (EAS) with the supporting evidence of low δ^{15}N (<2%) in particulate organic matter from regions with large surface blooms of *Trichodesmium*. The bloom is further supported by the iron supply from atmospheric deposition, upwelling, and transport from the shelf and slope sediments. Large surface blooms of *Trichodesmium* are found in the AS during spring intermonsoon, particularly in NO_3^- depleted surface water (Bange et al., 2005).

In the BoB, there has been no evidence for fixed N loss even though O_2 levels were below the detection limit. Based on the identification of denitrifiers and anammox bacteria and incubation experiments with ^{15}N-labeled N compounds, Bristow et al. (2017) suggested N_2 production as a persistent phenomenon in the BoB. Bristow et al. (2017) also showed that the low O_2 concentration (10–200 nM) within the BoB OMZ waters enables NO_2^- oxidation to dominate denitrification or anammox, resulting in reduced N loss due to limitation of NO_2^-. A pelagic N_2 production rate of 1.7 Tg N year^{-1} was estimated from the BoB (Bristow et al., 2017), which constitutes about 12% of the N_2 production in the AS. Also, δ^{15}N of NO_3^- in the OMZ waters exhibited a small enrichment of 1.5‰ compared with deep water, indicating NO_3^- reduction.

4.5.2. Sedimentary Denitrification and Anammox Rates in the Arabian Sea

Schwartz et al. (2009) reported sedimentary denitrification rates during intermonsoon (0.40–3.17 mmol N m^{-2} day^{-1}), postmonsoon periods (0.66–3.78 mmol N m^{-2} day^{-1}) periods, and a net annual denitrification rate of 1.1–10.5 Tg N year^{-1} off the Pakistan margin (northern AS) over a water depth of 140–1200 m. This constitutes 6.5% of the mean global sedimentary denitrification rate of 160 Tg N year^{-1}. Sokoll et al. (2012) reported sedimentary denitrification (1.46 to 2.73 mmol N m^{-2} day^{-1}) and anammox rates (0.21–0.89 mmol N m^{-2} day^{-1}) from water depths of 360–1430 m off the Pakistan margin. The benthic denitrification rates measured from slurry incubations decreased with water depth while anammox rates increased. The increase in anammox rates with water depth may be attributed to the reactivity of organic matter, where the availability of

Table 4.1 Sedimentary denitrification rates in the Arabian Sea.

Site	Denitrification rate (mmol N m^{-2} day^{-1})	Anammox rate (mmol N m^{-2} day^{-1})	Annual N loss rate (Tg N $year^{-1}$)	Source
Indian Arabian Sea continental shelf	0.17–1.33	–	0.4–3.5	Naik and Naqvi (2002)
Arabian Sea shelf sediments	–	–	6.8	Bange et al. (2000)
Pakistan margin Arabian Sea OMZ	0.07–3.68	–	1.1–10.5	Schwartz et al. (2009)
Pakistan margin Arabian Sea OMZ	1.46–2.73	0.21–0.89	14.7	Sokoll et al. (2012)

refractory carbon favors anammox. This trend was also supported by the relative increase in anammox *nirS* genes with water depth. The mean benthic N loss rate estimated from the AS shelf sediments is as high as 14.7 Tg N $year^{-1}$ and ranged between 12.3 and 17.0 Tg N $year^{-1}$ (Sokoll et al., 2012). This study reflects the importance of the benthic anammox N loss pathway in the Pakistan margin, particularly in deeper water depths. Sedimentary denitrification rates in the Arabian Sea are summarized in Table 4.1.

Using acetylene block-based measurements, Naik and Naqvi (2002) estimated sedimentary denitrification rate of 0.4–3.5 Tg N $year^{-1}$ from the central west coast of India during the upwelling period. Using the empirical relationship between sedimentary denitrification and primary production, Bange et al. (2000) calculated a seasonally weighted annual N loss of 6.8 Tg N at a water depth <200 m from the AS shelf sediments.

4.6. NITROGEN ISOTOPIC VALUES IN SINKING PARTICULATES

The N isotope composition of sinking particles may provide an insight into various biological processes in the water column such as denitrification, N_2 fixation, and extent of NO_3^- assimilation by phytoplankton (Montoya and Voss, 2006). Gaye-Haake et al. (2005) reported $\delta^{15}N$ values (5.8–7.8‰) of sinking particulates from five sedimentation traps deployed in the AS over a depth range of 500–3200 m (Figure 4.7). However, Montoya and Voss (2006) reported significantly higher $\delta^{15}N$ values (7.1–11.2‰) of sinking particulates from the AS as part of the JGOFS program from N21° to 0° latitudes. The ^{15}N enrichment of the particulates ($\delta^{15}N_{PN}$) is attributed to the assimilation of ^{15}N-rich residual fixed N by the primary producers. The low $\delta^{15}N$ values recorded at N10° station are attributed to BNF which produces ^{14}N-rich organic particulate. The upwelled water from the deeper denitrification zone injects ^{15}N-rich fixed N (Altabet et al., 1999) into the shallow/surface waters leading to the production of ^{15}N-rich particulate organic nitrogen (PON). The $\delta^{15}N_{NO3}$ values range from 7 to 9‰ within the euphotic zone, while from the northeastern AS OMZ (300 mbsl) and central/western AS (300 mbsl) a $\delta^{15}N_{NO3}$ range of 15 to 18‰ and 12‰ (Naqvi et al., 1998; Altabet et al., 1999), respectively, were reported. Gandhi and Ramesh (2016) reported an increasing trend in $\delta^{15}N_{PN}$ from the Omani shelf (6.8 to 11.8‰) to the central AS (6.6‰ to 12.3‰) and south-central AS (8.7 to 25.4‰) for the top 50 m of water during late SW monsoon. They attributed the change in $\delta^{15}N$ to the advection of the upwelled and unutilized NO_3^- with lower $\delta^{15}N$ surface values from the Oman coast to the Central AS imparting its signature to the PON.

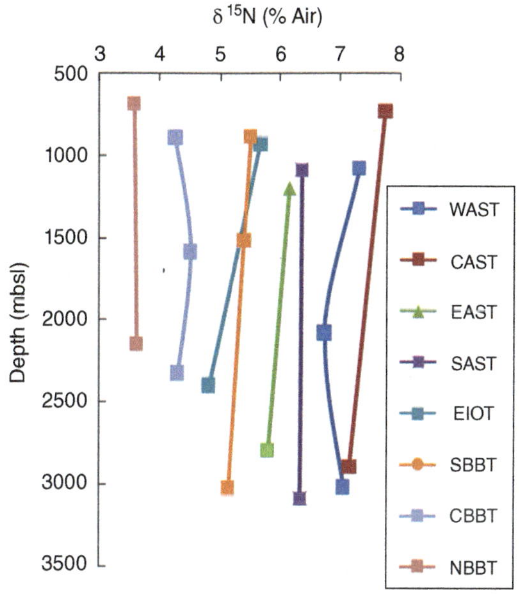

Figure 4.7 Vertical profile of nitrogen isotope of sinking particulates from the northern Indian Ocean (*Source*: Gaye-Haake et al., 2005). NBBT, Northern Bay of Bengal;SBBT, Southern Bay of Bengal; CBBT, Central Bay of Bengal; CAST, Central Arabian Sea; WAST, Western Bay of Bengal; EAST: Eastern Arabian Sea; SAST, Southern Arabian Sea; EIOT, Equatorial Indian Ocean.

Table 4.2 The $\delta^{15}N$ of sinking particles from northern Indian Ocean.

Site	$\delta^{15}N$ (‰)	References
Bay of Bengal[a]	2–7.6	Kumar et al. (2004)
Arabian Sea[b]	5.8–7.8	Gaye-Haake et al. (2005)
Bay of Bengal[b]	3.43–.51	
Bay of Bengal[b]	2–8	Unger et al. (2006)
Central Arabian Sea[a]	6.6–12.3	Gandhi and Ramesh (2016)
Oman Shelf[a]	8.2–12.8	
Offshore Oman[a]	6.0–13	
South Central Arabian Sea[a]	8.7–25.4	

[a]Represents $\delta^{15}N_{PN}$ from samples collected in CTD rosette.
*[b]Represents $\delta^{15}N$ of sinking particles collected in a sediment trap.

In the BoB, the $\delta^{15}N$ values decreased from southern to northern BoB. The average $\delta^{15}N$ (3.6–4.5‰) of sinking particles in the central and northern BoB is much lower than that of the AS (Table 4.2) because of the absence of denitrification and the high influx of terrestrial particulate nitrogen in the BoB (Gaye-Haake et al., 2005). A multi-annual time-series study by Unger et al. (2006) reported a range of $\delta^{15}N$ values (2–8‰) of sinking particles from the southern BoB. Data were collected at shallow, intermediate, and deep waters over a depth range of 886–3024 m. The range of $\delta^{15}N$ values observed is attributed to a contribution from terrestrially derived organic particulates with low $\delta^{15}N$ values and the flux of ^{15}N-rich particulates from the AS during the southwest monsoon. The mixing of continental and marine-derived particulate N is also evident from Kumar et al. (2004), who reported $\delta^{15}N_{PN}$ varying from 2 to 7.6‰ from coastal to open ocean, indicating mixing of continental and marine sources (Kumar et al., 2004). High $\delta^{15}N$ values (~7.6‰) from the open Ocean may be attributed to the uptake of regenerated NH_4^+ (Wada and Hattori, 1976).

4.6.1. $\delta^{15}N$ in Marine Sediments from the Arabian Sea and Bay of Bengal

The N isotopc ratios of sedimentary organic matter are critical in the evaluation of the historical marine N cycle. In the AS, the prevalence of suboxic conditions in intermediate waters results in denitrification and accumulation of isotopically heavier NO_3^- in the water column. Denitrification is associated with an isotopic effect of 20–30‰, leaving the residual NO_3^- with $\delta^{15}N$ values above 20‰ (Naqvi et al., 1998; Voss et al., 2001). When monsoon-driven upwelling and winter convective mixing brings isotopically heavy NO_3^- to the euphotic zone, organic matter produced is also expected to become enriched in ^{15}N due to NO_3 assimilation. The net $\delta^{15}N$ of organic particulate would depend on the N isotopic fractionation during NO_3 assimilation and BNF (Altabet et al., 1999). Under nutrient limiting conditions, phytoplankton utilizes available reactive N completely, resulting in similar Nisotopic ratios of the phytoplankton with the isotopic composition of the source (Altabet and McCarthy, 1985). Thus, the sedimentary $\delta^{15}N$ values may provide information on the past fluctuations in the OMZ intensity and water column denitrification. Gaye-Haake et al. (2005) reported $\delta^{15}N$ values of >11‰ from surface sediments of the central AS which is significantly greater than that recorded from the southeastern AS ($\delta^{15}N$ 5–7‰).

Gaye-Haake et al. (2005), Agnihotri et al. (2003) and Cowie et al. (2009) reported north to south variation in $\delta^{15}N$ values in sinking particles and surface sediments of the AS. Along the west coast of India, Kurian et al. (2013) reported enriched sedimentary $\delta^{15}N$ values in the northern transects (>7‰) north of Ratnagiri relative to the southern transects (4.8–7.1‰). This is attributed to the intense denitrification between 15 and 21°N (Naqvi, 1987; Naqvi and Noronha, 1991). Table 4.3 summarizes the $\delta^{15}N$ values in sediments reported from the northern Indian Ocean.

Several studies have reported the prevalence of intense denitrification in the AS during interglacial periods and weakening during the glacials (Ganeshram et al., 2000; Gaye et al., 2018; Tripathi et al., 2020). Denitrification in the EAS is mainly governed by intermediate water ventilation, while in the western Arabian Sea (WAS), it is primarily influenced by variability in southwest monsoon-driven productivity (Pichevin et al., 2007; Tripathi et al., 2020).

Kessarkar et al. (2018) reported $\delta^{15}N$ values ranging between 6‰ and 8.4‰ from a sediment core (water depth· 500 mbsl) from the EAS. The average sediment $\delta^{15}N$ value of 7.8‰ indicated the prevalence of an intense denitrifying condition in the EAS OMZ during the early Holocene, reflecting less ventilation of OMZ during that time. However, by mid-Holocene, the oxygenation of the OMZ led to the suppression of denitrification, indicated by the slightly lower sedimentary $\delta^{15}N$ of 6.2‰ (mean value). The OMZ condition intensified again since around 5.5 ka BP, accompanied by intense denitrification (Kessarkar et al., 2018). In contrast, Kao et al. (2015) reported relatively lower sedimentary $\delta^{15}N$ values (~5–6.5‰) from the EAS since 6 ka BP due to N_2 fixation and was attributed as a local effect.

During the glacial periods, the reduced sea level restricts the inflow of highly saline

Red Sea Water (RSW) and Persian Gulf Water (PGW) to the Arabian Sea which in turn favors the northward influx of O_2-rich Antarctic Intermediate

Table 4.3 The $\delta^{15}N$ values in sediments from the northern Indian Ocean.

Source	Site	$\delta^{15}N$ (‰)
Agnihotri et al. (2003)[b]	Western continental margin of India	5.9 – 9.6
Gaye-Haake et al. (2005)[b]	Central Arabian Sea	>11 (maximum)
	Southeastern Arabian Sea	5 – 7
	Central Indian Ocean	7 – 8
	Northern Bay of Bengal	<4 (minimum)
Kurian et al. (2013[b])	West coast of India	4.6 – 9.1
Kao et al. (2015[a])	South west coast of India	4.7 – 7.1
Kessarkar et al. (2018[a])	Eastern Arabian Sea	6 – 8.4
Contreras-Rosales et al. (2016)	Northern Bay of Bengal	4.4 – 5.0
Kim et al., 2018[a]	Eastern Arabian Sea	5.3 – 10.1
Tripathi et al. (2020[a])	Northeastern Arabian Sea	3.4 – 9.1
Nagoji and Tiwari (2021[a])	Southeastern Arabian Sea	4.91– 8.74

[a]The $\delta^{15}N$ range recorded in the whole core.
[b]The $\delta^{15}N$ in surface sediments.

Water (AAIW) in contrast to interglacial periods (Pichevin et al., 2007; Tripathi et al., 2020). During the Last Glacial Maxima (LGM) and the mid/late Holocene, denitrification in the WAS declined due to enhanced ventilation by AAIW and reduced monsoon productivity respectively. In contrast, denitrification intensified in the EAS during Holocene (Gaye et al., 2018; Kim et al., 2018; Tripathi et al., 2020) and the LGM despite less monsoon productivity (Nagoji and Tiwari, 2021), which may be attributed to reduced thermohaline circulation. During the Holocene, in contrast to the WAS, the enhanced influx of oxygen-poor RSW and PGW into the EAS owing to a sea level rise reduced the inflow of oxygen-rich AAIW, resulting in reduced ventilation of the EAS followed by enhanced denitrification (Tripathi et al., 2020; Nagoji and Tiwari, 2021). The $\delta^{15}N$ data of bulk sediments (>6 during the last 800 ka – mid-Pleistocene) indicated the prevalence of a persistent OMZ and strong denitrification in the EAS while there was an intermittent OMZ breakdown during glacial periods in the WAS (Kim et al., 2018). Gaye et al., (2018) also reported weak denitrification during LGM and stadials ($\delta^{15}N$ = 4–6‰) of bulk sediments from the northern Oman margin. Tripathi et al (2017) showed that the modern strength of the OMZ where denitrification is a permanent feature was attained at ~1.0 Ma (early Pleistocene).

In contrast to the AS, BoB sedimentary records do not show convincing evidence of past OMZ denitrification (Pattan et al. 2013; Contreras-Rosales et al., 2016). In the Northern BoB, Contreras-Rosales et al. (2016) reported $\delta^{15}N$ between 4.4 and 5.0‰ during the Holocene and slightly lower values during the LGM. The lack of denitrification in the BoB compared with the AS is also evident from the difference in the distribution and diversity of living benthic foraminifera assemblages from both the basins (Suokhrie et al., 2020). Benthic foraminiferal species capable of denitrification, such as *Bolivina seminude* and *Buliminella tenuata* have been reported from the AS OMZ (Mazumder and Nigam, 2014; Verma et al., 2018). The identification of benthic foraminifera performing denitrification from the AS is an implication of the prevalence of intense denitrification in the basin. In contrast, none of the benthic foraminifera reported so far from the BoB OMZ possess the capacity for denitrification (Suokhrie et al., 2020),

In contrast to the water column particulates, the sedimentary organic matter may undergo additional alteration due to diagenetic processes, altering the $\delta^{15}N_{PN}$ (Bordovskiy, 1965; Meyers, 1997; Altabet et al., 1999; Ganeshram et al., 2000; Gaye-Haake et al., 2005). The extent of alteration appears to be a function of water depth, sedimentation rate, and the microbiological processes explained earlier.

In summary, denitrification has prevailed in the AS during interstadials of the Pleistocene and entire Holocene, as evident from the recorded $\delta^{15}N$ values (>6‰) (Agnihotri et al., 2003; Kessarkar et al., 2018).

4.7. SUMMARY

This review summarizes the marine N biogeochemical processes mediated by microorganisms. The processes include BNF, N assimilation, denitrification, nitrification, anammox, DNRA in the water column and sediments. In the past few years, the discovery of new metabolic processes such as N-Damo, Feammox, sulfammox, comammox has added new pathways to the marine N biogeochemical cycle. However, better knowledge of the physiology and ecology of the microorganisms involved is required to understand the factors that mediate N transformation pathways. Future research should focus on detecting novel processes and the metabolic pathways, ecology, and distribution of the least understood processes such as N-Damo, sulfammox, Feammox, and Comammox in order to gain a better picture of the marine N cycle. The present review has emphasized the significance of the N-cycle in the Northern Indian Ocean (AS and BoB). The prevalence of hypoxic conditions in the AS and BoB is attributed to the landlocked northern boundary, poor ventilation of intermediate water, and respiration of organic matter (marine productivity). Compared with the intense denitrification observed in the OMZ of AS, the BoB shows a diminished signal of denitrification owing to the NO_2^- oxidation to NO_3^-. The intense denitrification signal in the AS is also imprinted on sinking particulates as high $\delta^{15}N_{PN}$ values, whereas in BoB, particulates primarily bear the signature of terrestrial organic matter. Studies of sediment cores from AS show a notable shift in denitrification intensity during the glacial and interglacial periods through Pleistocene to Holocene: more than 90% of N_2 loss from AS is attributed to denitrification.

ACKNOWLEDGEMENT

We thank the editors for inviting us to contribute to this book. We acknowledge Director, CSIR-NIO, for funding the research program. We also thank the CSIR-NIO's publication committee for carrying out the internal review.

REFERENCES

Acharya, S.S. and Panigrahi, M.K. (2016). Eastward shift and maintenance of Arabian Sea oxygen minimum zone: understanding the paradox. *Deep Sea Research Part I: Oceanographic Research Papers 115*: 240–252. https://doi.org/10.1016/j.dsr.2016.07.004

Agnihotri, R., Bhattacharya, S.K., Sarin, M.M. et al. (2003). Changes in surface productivity and subsurface denitrification during the Holocene: a multiproxy study from the eastern Arabian Sea. *The Holocene 13* (5): 701–713. https://doi.org/10.1191/0959683603hl656rp

Ahmed, A., Gauns, M., Kurian, S. et al. (2017). Nitrogen fixation rates in the eastern Arabian Sea. *Estuarine, Coastal and Shelf Science 191*: 74–83. http://doi.org/10.1016/j.ecss.2017.04.005

Altabet, M.A. and McCarthy, J.J. (1985). Temporal and spatial variations in the natural abundance of 15N in PON from a warm-core ring. *Deep Sea Research Part A. Oceanographic Research Papers 32* (7): 755–772. https://doi.org/10.1016/0198-0149 (85)90113-X

Altabet, M.A., Murray, D.W. and Prell, W.L. (1999). Climatically linked oscillations in Arabian Sea denitrification over the past 1 m.y.: Implications for the marine N cycle. *Paleoceanography 14*: 732–743. https://doi.org/10.1029/1999PA900035

Anderson, T.R. and Lucas, M.I. (2009). Upwelling ecosystems. In: *Ecosystem Ecology* (ed. S.E. Jorgensen). Amsterdam: Elsevier

Anschutz, P., Sundby, B., Lefrancois, L. et al. (2000). Interactions between metal oxides and species of nitrogen and iodine in bioturbated marine sediments. *Geochimica et Cosmochimica Acta 64*: 2751–2763. https://doi.org/10.1016/S0016-7037(00)00400-2

Bandekar, M., Ramaiah, N., Jain, A. et al. (2018). Seasonal and depth-wise variations in bacterial and archaeal groups in the Arabian Sea oxygen minimum zone. *Deep Sea Research Part II: Topical Studies in Oceanography 156*: 4–18. https://doi.org/10.1016/j.dsr2.2017.12.015

Bange, H.W., Naqvi, S.W.A. and Codispoti, L.A. (2005). The nitrogen cycle in the Arabian Sea. *Progress in Oceanography 65* (2–4): 145–158. https://doi.org/10.1016/j.pocean.2005.03.002

Bange, H.W., Rixen, T., Johansen, A.M. et al. (2000). A revised nitrogen budget for the Arabian Sea. *Global Biogeochemical Cycles,14*: 1283–1297. https://doi.org/10.1029/1999GB001228

Banse, K., Naqvi, S.W.A., Narvekar, P.V. et al. (2014). Oxygen minimum zone of the open Arabian Sea: variability of oxygen and nitrite from daily to decadal timescales. *Biogeosciences 11* (8): 2237–2261. https://doi.org/10.5194/bg-11-2237-2014

Bartlett, R., Mortimer, R.J. and Morris, K. (2008). Anoxic nitrification: evidence from Humber Estuary sediments (UK). *Chemical Geology 250* (1–4): 29–39. https://doi.org/10.1016/j.chemgeo.2008.02.001

Benavides, M., Duhamel, S., Van Wambeke, F. et al. (2020). Dissolved organic matter stimulates N_2 fixation and nifH gene expression in Trichodesmium. *FEMS Microbiology Letters 367* (4): fnaa034. https://doi.org/10.1093/femsle/fnaa034

Bergman, B. and Carpenter, E.J. (1991). Nitrogenase confined randomly to the marine cyanobacterium *Trichodesmium thiebautii. Journal of Phycolology 27*: 158–165. https://doi.org/10.1111/j.0022-3646.1991.0015

Berner, R.A. (2006). Geological nitrogen cycle and atmospheric N_2 over Phanerozoic time. *Geology 34* (5): 413–415. https://doi.org/10.1130/G22470.1

Bernhard, J.M., Edgcomb, V.P., Casciotti, K.L. et al. (2012). Denitrification likely catalyzed by endobionts in an allogromiid foraminifer. *The ISME Journal 6* (5): 951–960. https://doi.org/10.1038/ismej.2011.171

Bertics, V.J., Löscher, C.R., Salonen, I. et al. (2013). Occurrence of benthic microbial nitrogen fixation coupled to sulfate reduction in the seasonally hypoxic Eckernförde Bay, Baltic Sea. *Biogeosciences, 10* (3): 1243–1258. https://doi.org/10.5194/bg-10-1243-2013

Bertics, V.J., Sohm, J.A., Treude, T. et al. (2010). Burrowing deeper into benthic nitrogen cycling: the impact of bioturbation on nitrogen fixation coupled to sulfate reduction. *Marine Ecology Progress Series, 409*: 1–15. https://doi.org/10.3354/meps08639

Boje, R. and Tomczak, M. (1978). Ecosystem analysis and the definition of boundaries in upwelling regions. In: *Upwelling Ecosystems* (eds. R. Boje and M. Tomczak), 3–11. Heidelberg, Berlin: Springer. https://doi.org/10.1007/978-3-642-66985-9_1

Bonaglia, S., Bartoli, M., Gunnarsson, J.S. et al. (2013). Effect of reoxygenation and *Marenzelleria* spp. bioturbation on Baltic Sea sediment metabolism. *Marine Ecology Progress Series 482*: 43–55. https://doi.org/10.3354/meps10232

Bonaglia, S., Nascimento, F.A., Bartoli, M. et al. (2014). Meiofauna increases bacterial denitrification in marine sediments. *Nature Communications 5* (1):1–9. https://doi.org/10.1038/ncomms6133

Bordovskiy, O.K. (1965). Accumulation of organic matter in bottom sediments. *Marine Geology 3* (1–2): 33–82. https://doi.org/10.1016/0025-3227(65)90004-6

Boyer, E.W., Howarth, R.W., Galloway, J.N. et al. (2006). Riverine nitrogen export from the continents to the coasts. *Global Biogeochemical Cycles 20*: GB1S91. https://doi.org/10.1029/2005GB002537

Brandes, J.A., Devol, A.H., Jayakumar et al. (1998). Isotopic composition of nitrate in the central Arabian Sea and eastern tropical North Pacific: a tracer for mixing and nitrogen cycles. *Limnology and Oceanography 43*: 1680–1689. https://doi.org/10.4319/lo.1998.43.7.1680

Brandes, J.A., Devol, A.H. and Deutsch, C. (2007). New developments in the marine nitrogen cycle. *Chemical Reviews,107* (2): 577–589. https://doi.org/10.1021/cr050377t

Bristow, L.A., Callbeck, C.M., Larsen, M. et al. (2017). N_2 production rates limited by nitrite availability in the Bay of Bengal oxygen minimum zone. *Nature Geoscience 10* (1): 24–29. http://doi.org/10.1038/ngeo2847

Bulow, S.E., Rich, J.J., Naik, H.S. et al. (2010). Denitrification exceeds anammox as a nitrogen loss pathway in the Arabian Sea oxygen minimum zone. *Deep Sea Research Part I: Oceanographic Research Papers 57* (3): 384–393. https://doi.org/10.1016/j.dsr.2009.10.014

Burgin, A.J. and Hamilton, S.K. (2007). Have we overemphasized the role of denitrification in aquatic ecosystems? A review of nitrate removal pathways. *Frontiers in Ecology and the Environment 5*. 89–96. https://doi.org/10.1890/1540-929520075

Cabello, P., Roldan, M.D. and Moreno-Vivian, C. (2004). Nitrate reduction and the nitrogen cycle in archaea. *Microbiology 150* (11): 3527–3546. https://doi.org/10.1099/mic.0.27303-0

Canfield, D.E., Kristensen, E. and Thamdrup, B. (2005). The sulfur cycle. *Advances in Marine Biology 48*: 313–381. https://doi.org/10.1016/S0065-2881(05)48009-8

Canfield, D.E., Glazer, A.N. and Falkowski, P.G. (2010). The evolution and future of Earth's nitrogen cycle. *Science 330*: 192–96. https://doi.org/10.1126/science.1186120

Capone, D.G. (1988). Benthic nitrogen fixation. In: *Nitrogen Cycling in Coastal Marine Environments* (ed.T.H. Blackburn), 85–123. NewYork: John Wiley and Sons.

Capone, D.G., Zehr, J.P., Paerl, H.W. et al. (1997). *Trichodesmium*, a globally significant marine cyanobacterium. *Science 276*: 1221–1229. https://doi.org/10.1126/science.276.5316.1221

Capone, D., Subramaniam, A., Montoya, J. et al. (1998). An extensive bloom of the N_2-fixing cyanobacterium *Trichodesmium erythraeum* in the central Arabian Sea. *Marine Ecology Progress Series 172*,: 281–292. https://doi.org/10.3354/meps172281

Caranto, J.D. and Lancaster, K.M. (2017). Nitric oxide is an obligate bacterial nitrification intermediate produced by hydroxylamine oxidoreductase. *Proceedings of the National Academy of Sciences 114* (31): 8217–8222. https://doi.org/10.1073/pnas.1704504114

Chen, J., Jiang, X.W. and Gu, J.D. (2015). Existence of novel phylotypes of nitrite-dependent anaerobic methane-oxidizing bacteria in surface and subsurface sediments of the South China Sea. *Geomicrobiology Journal 32* (1): 1–10. https://doi.org/10.1080/01490451.2014.917742

Cheng, Q. (2008). Perspectives in biological nitrogen fixation research. *Journal of Integrative Plant Biology 50* (7): 786–798. https://doi.org/10.1111/j.1744-7909.2008.0070

Codispoti, L.A., Brandes, J.A., Christensen, J.P. et al. (2001). The oceanic fixed nitrogen and nitrous oxide budgets: moving targets as we enter the anthropocene? *Scientia Marina 65* (Suppl. 2): 85–105. https://doi.org/10.3989/scimar.2001.65s285

Contreras-Rosales, L.A., Schefuß, E., Meyer, V. et al. (2016). Origin and fate of sedimentary organic matter in the northern Bay of Bengal during the last 18 ka. *Global and Planetary Change, 146*: 53–66. https://doi.org/10.1016/j.gloplacha.2016.09.008

Cornell, S., Randell, A. and Jickells, T. (1995). Atmospheric inputs of dissolved organic nitrogen to the oceans. *Nature 376* (6537): 243–246. https://doi.org/10.1038/376243a0

Cornell, S.E., Jickells, T.D., Cape, J.N. et al. (2003). Organic nitrogen deposition on land and coastal environments: a review of methods and data. *Atmospheric Environment 37*: 2173–2191. https://doi.org/10.1016/S1352-2310(03)00133-X

Cowie, G.L., Mowbray, S., Lewis, M., et al. (2009). Carbon and nitrogen elemental and stable isotopic compositions of surficial sediments from the Pakistan margin of the Arabian Sea. *Deep Sea Research Part II: Topical Studies in Oceanography 56* (6–7): 271–282. https://doi.org/10.1016/j.dsr2.2008.05.031

Dalsgaard, T. and Thamdrup, B. (2002). Factors controlling anaerobic ammonium oxidation with nitrite in marine sediments. *Applied and Environmental Microbiology 68* (8): 3802–3808. https://doi.org/10.1128/AEM.68.8.3802-3808.2002

Dalsgaard, T., Canfield, D.E., Petersen, J. et al. (2003). N_2 production by the anammox reaction in the anoxic water column of Golfo Dulce, *Costa Rica. Nature 422* (6932): 606–608. https://doi.org/10.1038/nature01526

Dalsgaard, T., Thamdrup, B. and Canfield, D.E. (2005). Anaerobic ammonium oxidation (anammox) in the marine environment. *Research in Microbiology 156*: 457–464. https://doi.org/10.1016/j.resmic.2005.01.011

Damashek, J. and Francis, C.A. (2018). Microbial nitrogen cycling in estuaries: from genes to ecosystem processes. *Estuaries and Coasts 41* (3): 626–66 https://doi.org/10.1007/s12237-017-0306-2

Sinninghe Damsté, J., Strous, M., Rijpstra, W.I.C. et al. (2002). Linearly concatenated cyclobutane lipids form a dense bacterial membrane. *Nature 419* (6908): 708–712. https://doi.org/10.1038/nature01128

Dekas, A.E., Poretsky, R.S. and Orphan, V.J. (2009). Deep-sea archaea fix and share nitrogen in methane-consuming microbial consortia. *Science 326* (5951): 422–426. https://doi.org/10.1126/science.1178223

Dentener, F., Drevet, J., Lamarque, J.F. et al. (2006). Nitrogen and sulfur deposition on regional and global scales: a multi-model evaluation. *Global Biogeochemical Cycles 20* (4): 1–21. https://doi.org/10.1029/2005GB002672

Deutsch, C., Sarmiento, J.L., Sigman, D.M. et al. (2007). Spatial coupling of nitrogen inputs and losses in the ocean. *Nature 445* (7124): 163–167. https://doi.org/10.1038/nature05392

Devol, A.H., Uhlenhopp, A.G., Naqvi, S.W.A. et al. (2006). Denitrification rates and excess nitrogen gas concentrations in the Arabian Sea oxygen deficient zone. *Deep Sea Research Part I: Oceanographic Research Papers 53* (9): 1533–1547. https://doi.org/10.1016/j.dsr.2006.07.005

Dong, L.F., Smith, C.J., Papaspyrou, S. et al. (2009). Changes in benthic denitrification, nitrate ammonification and anammox process rates and nitrate and nitrite reductase gene abundances along an estuarine nutrient gradient (the Colne Estuary, United Kingdom). *Applied and Environmental Microbiology 75* (10): 3171–3179. https://doi.org/10.1128/AEM.02511–08

Duce, R.A., Liss, P.S., Merrill, J.T. et al. (1991). The atmospheric input of trace species to the world ocean. *Global Biogeochemical Cycles 5*: 193–259. https://doi.org/10.1029/91GB01778

Duce, R.A., LaRoche, J., Altieri, K. et al. (2008). Impacts of atmospheric anthropogenic nitrogen on the open ocean. *Science 320*: 893–897. https://doi.org/10.1126/science.1150369

Dyhrman, S.T., Chappell, P.D., Haley, S.T. et al. (2006). Phosphonate utilization by the globally important marine diazotroph *Trichodesmium*. *Nature 439* (7072): 68–71. https://doi.org/10.1038/nature04203

Dyhrman, S.T., Ammerman, J.W. and Van Mooy, B.A. (2007). Microbes and the marine phosphorus cycle. *Oceanography 20* (2): 110–116. https://doi.org/10.5670/oceanog.2007.54

Ettwig, K.F., Shima, S., Van De Pas-Schoonen, K.T. et al. (2008). Denitrifying bacteria anaerobically oxidize methane in the absence of Archaea. *Environmental Microbiology 10* (11): 3164–3173. https://doi.org/10.1111/j.1462–2920.2008.0172

Ettwig, K.F., Butler, M.K., Le Paslier, D. et al. (2010). Nitrite-driven anaerobic methane oxidation by oxygenic bacteria. *Nature 464* (7288): 543–548. https://doi.org/10.1038/nature08883

Falkowski, P. (1997). Evolution of the nitrogen cycle and its influence on the biological sequestration of CO_2 in the ocean. *Nature 387*, 272–275. https://doi.org/10.1038/387272a0

Fdz-Polanco, F., Fdz-Polanco, M., Fernandez, N. et al. (2001). New process for simultaneous removal of nitrogen and sulphur under anaerobic conditions. *Water Research 35*: 1111–1114. https://doi.org/10.1016/s0043–1354(00)00474–7

Fernandes, G.L., Shenoy, B.D., Menezes, L.D. et al. (2019). Prokaryotic diversity in oxygen depleted waters of the Bay of Bengal inferred using culture-dependent and -independent methods. *Indian Journal of Microbiology 59* (2): 193–199. https://doi.org/10.1007/s12088–019–00786–1

Fernandes, G.L., Shenoy, B.D. and Damare, S.R. (2020). Diversity of bacterial community in the oxygen minimum zones of Arabian Sea and Bay of Bengal as deduced by Illumina sequencing. *Frontiers in Microbiology 10*: 3153. https://doi.org/10.3389/fmicb.2019.03153

Fernandes, S.O., Javanaud, C., Aigle, A. et al. (2015). Anaerobic nitrification–denitrification mediated by Mn-oxides in meso-tidal sediments: Implications for N_2 and N_2O production. *Journal of Marine Systems 144*: 1–8. https://doi.org/10.1016/j.jmarsys.2014.11.011

Ferrón, S., T. Ferron, S., Ortega, T. et al. (2009). Benthic fluxes in a tidal salt marsh creek affected by fish farm activities: Río San Pedro (Bay of Cádiz, SW Spain). *Marine Chemistry 113* (1–2): 50–62. https://doi.org/10.1016/j.marchem.2008.12.002

Flores, E. and Herrero, A. (2005). Nitrogen assimilation and nitrogen control in cyanobacteria. *Biochemical Society Transactions 33* (1): 164–167. https://doi.org/10.1042/BST0330164

Galloway, J.N., Aber, J.D., Erisman, J.W. et al. (2003). The nitrogen cascade. *Bioscience 53* (4): 341–356. https://doi.org/10.1641/0006–3568 (2003)053[0341:TNC]2.O;2

Galloway, J.N., Dentener, F.J., Capone, D.G. et al. (2004). Nitrogen cycles: past, present and future. *Biogeochemistry 70* (2): 153–226. https://doi.org/10.1007/s10533–004–0370–0

Galloway, J.N., Townsend, A.R., Erisman, J.W. et al. (2008). Transformation of the nitrogen cycle: recent trends, questions and potential solutions. *Science 320* (5878): 889–892. https://doi.org/10.1126/science.1136674

Gandhi, N. and Ramesh, R. (2016). Natural isotopic composition of particulate organic nitrogen in the eastern and central Arabian sea. *Proceedings of the National Academy of Sciences, India 82* (4): 1283–1291. https://doi.org/10.16943/ptinsa/2016/48574

Ganeshram, R.S., Pedersen, T.F., Calvert, S.E. et al. (2000). Glacial–interglacial variability in denitrification in the world's oceans: causes and consequences. *Paleoceanography 15* (4): 361–376. https://doi.org/10.1029/1999PA000422

Gardner, W.S., McCarthy, M.J., An, S. et al. (2006). Nitrogen fixation and dissimilatory nitrate reduction to ammonium (DNRA) support nitrogen dynamics in Texas estuaries. *Limnology and Oceanography 51* (1 Pt 2): 558–568. https://doi.org/10.4319/lo.2006.51.1_part_2.0558

Gaye, B., Böll, A., Segschneider, J. et al. (2018). Glacial–interglacial changes and Holocene variations in Arabian Sea denitrification. *Biogeosciences 15* (2): 507–527. https://doi.org/10.5194/bg-15–507–2018

Gaye-Haake, B., Lahajnar, N., Emeis, K.C. et al. (2005). Stable nitrogen isotopic ratios of sinking particles and sediments from the northern Indian Ocean. *Marine Chemistry 96* (3–4): 243–255. https://doi.org/10.1016/j.marchem.2005.02.001

Glass, J.B., Wolfe-Simon, F. and Anbar, A.D. (2009). Coevolution of metal availability and nitrogen assimilation in cyanobacteria and algae. *Geobiology 7* (2): 100–123. https://doi.org/10.1111/j.1472–4669.2009.0019

Goldhaber, M.B., Aller, R.C., Cochran, J.K. et al. (1977). Sulfate reduction, diffusion and bioturbation in Long Island Sound sediments; report of the FOAM Group. *American Journal of Science 277* (3): 193–237. https://doi.org/10.2475/ajs.277.3.193

Granger, J. and Ward, B.B. (2003). Accumulation of nitrogen oxides in copper-limited cultures of denitrifying bacteria. *Limnology and Oceanography 48* (1): 313–318. https://doi.org/10.4319/lo.2003.48.1.0313

Großkopf, T., Mohr, W., Baustian, T. et al. (2012). Doubling of marine dinitrogen-fixation rates based on direct measurements. *Nature 488* (7411): 361–364. https://doi.org/10.1038/nature11338

Gruber, N. (2004). The dynamics of the marine nitrogen cycle and its influence on atmospheric CO2 variations. In: *The Ocean Carbon Cycle and Climate* (eds. M. Follows and T. Oguz), 97–148. Dordrecht: Springer. https://doi.org/10.1007/978-1-4020-2087-2_4

Gruber, N. and Galloway, J.N. (2008). An Earth-system perspective of the global nitrogen cycle. *Nature 451* (7176): 293–296. https://doi.org/10.1038/nature06592

Haroon, M.F., Hu, S., Shi, Y. et al. (2013). Anaerobic oxidation of methane coupled to nitrate reduction in a novel archaeal lineage. *Nature 500*: 567–570. https://doi.org/10.1038/nature12375

Henriksen, K., Blackburn, T.H., Lomstein, B.A. et al.(1993). Rates of nitrification, distribution of nitrifying bacteria and inorganic N fluxes in northern Bering–Chukchi shelf sediments. *Continental Shelf Research 13* (5–6): 629–651.https://doi.org/10.1016/0278-4343 (93)90097-H

Hino, T., Matsumoto, Y., Nagano, S. et al. (2010). Structural basis of biological N_2O generation by bacterial nitric oxide reductase. *Science 330* (6011): 1666–1670. https://doi.org/10.1126/science.1195591

Hoffman, B.M., Lukoyanov, D., Yang, Z.Y. et al. (2014). Mechanism of nitrogen fixation by nitrogenase: the next stage. *Chemical Reviews 114* (8): 4041–4062. https://doi.org/10.1021/cr400641x

Horak, R.E., Qin, W., Bertagnolli, A.D. et al. (2018). Relative impacts of light, temperature and reactive oxygen on thaumarchaeal ammonia oxidation in the North Pacific Ocean. *Limnology and Oceanography 63* (2): 741–757. https://doi.org/10.1002/lno.10665

Hu, Y., Feng, D., Peng, Y. et al. (2020). A prominent isotopic fingerprint of nitrogen uptake by anaerobic methanotrophic archaea. *Chemical Geology 558*; 119972. https://doi.org/10.1016/j.chemgeo.2020.119972

Hulth, S., Aller, R.C. and Gilbert, F. (1999). Coupled anoxic nitrification/manganese reduction in marine sediments. *Geochimica et Cosmochimica Acta 63* (1): 49–66. https://doi.org/10.1016/S0016-7037(98)00285-3

Janssen, B.H. (1996). Nitrogen mineralization in relation to C:N ratio and decomposability of organic materials. In: *Progress in Nitrogen Cycling Studies*, Vol. *68* (eds. O. Van Cleeput, G. Hofman and A. Vermoesen), 69–75). Dordrecht: Springer. https://doi.org/10.1007/978-94-011-5450-5_13

Javanaud, C., Michotey, V., Guasco, S. et al. (2011). Anaerobic ammonium oxidation mediated by Mn-oxides: from sediment to strain level. *Research in Microbiology 162* (9): 848–857. https://doi.org/10.1016/j.resmic.2011.01.011

Jayakumar, A., Peng, X. and Ward, B.B. (2013). Community composition of bacteria involved in fixed nitrogen loss in the water column of two major oxygen minimum zones in the ocean. *Aquatic Microbial Ecology 70* (3): 245–259. https://doi.org/10.3354/ame01654

Jensen, M.M., Kuypers, M.M., Gaute, L. et al. (2008). Rates and regulation of anaerobic ammonium oxidation and denitrification in the Black Sea. *Limnology and Oceanography 53* (1): 23–3 https://doi.org/10.4319/lo.2008.53.1.0023

Jensen, M.M., Lam, P., Revsbech, N.P. et al. (2011). Intensive nitrogen loss over the Omani Shelf due to anammox coupled with dissimilatory nitrite reduction to ammonium. *ISME Journal 5* (10): 1660–1670. https://doi.org/10.1038/ismej.2011.44

Jones, D.L., Shannon, D., Murphy, D.V. et al. (2004). Role of dissolved organic nitrogen (DON) in soil N cycling in grassland soils. *Soil Biology and Biochemistry 36* (5): 749–756. https://doi.org/10.1016/j.soilbio.2004.01.003

Jones, M.T. and Gislason, S.R. (2008) Rapid releases of metal salts and nutrients following the deposition of volcanic ash into aqueous environments. *Geochimica et Cosmochimica Acta 72*: 3661–3680. https://doi.org/10.1016/j.gca.2008.05.030

Kao, S.J., Wang, B.Y., Zheng, L.W. et al. (2015). Spatiotemporal variations of nitrogen isotopic records in the Arabian Sea. *Biogeosciences 12* (1): 1–14. https://doi.org/10.5194/bg-12-1-2015

Karlson, K., Bonsdorff, E. and Rosenberg, R. (2007). The impact of benthic macrofauna for nutrient fluxes from Baltic Sea sediments. *AMBIO: A Journal of the Human Environment 36* (2): 161–167. https://doi.org/10.1579/0044-7447(2007)36[161:TIOBMF]2.O;2

Kessarkar, P.M., Naqvi, S.W.A., Thamban, M. et al. (2018). Variations in denitrification and ventilation within the Arabian Sea oxygen minimum zone during the Holocene. *Geochemistry, Geophysics, Geosystems 19* (7): 2179–2193. https://doi.org/10.1029/2017GC007286

Kim, J.E., Khim, B.K., Ikehara, M. et al. (2018). Orbital-scale denitrification changes in the Eastern Arabian Sea during the last 800 kyrs. *Scientific Reports 8* (1): 1–8. https://doi.org/10.1038/s41598-018-25415-7

Könneke, M., Bernhard, A.E., José, R. et al. (2005). Isolation of an autotrophic ammonia-oxidizing marine archaeon. *Nature 437* (7058): 543–546.

Kozlowski, J.A., Stieglmeier, M., Schleper, C. et al. (2016). Pathways and key intermediates required for obligate aerobic ammonia dependent chemolithotrophy in bacteria and Thaumarchaeota. *The ISME Journal 10*: 1836–1845. https://doi.org/10.1038/ismej.2016.2

Kristensen, E., Hjorth Jensen, M. and Aller, R.C. (1991). Direct measurement of dissolved inorganic nitrogen exchange and denitrification in individual polychaete (*Nereis virens*) burrows. *Journal of Marine Research 49* (2): 355–377. https://doi.org/10.1357/002224091784995855

Kumar, S., Ramesh, R., Bhosle, N.B. et al. (2004). Natural isotopic composition of nitrogen in suspended particulate matter in the Bay of Bengal. *Biogeosciences 1* (1): 63–70. https://doi.org/10.5194/bg-1-63-2004

Kurian, S., Nath, B.N., Kumar, N.C. et al. (2013). Geochemical and isotopic signatures of surficial sediments from the western continental shelf of India: inferring provenance, weathering and the nature of organic matter. *Journal of Sedimentary Research 83* (6): 427–442. https://doi.org/10.2110/jsr.2013.36

Kuypers, M.M., Sliekers, A.O., Lavik, G. et al. (2003). Anaerobic ammonium oxidation by anammox bacteria in the Black Sea. *Nature 422* (6932): 608–61 https://doi.org/10.1038/nature01472

Kuypers, M.M., Marchant, H.K. and Kartal, B. (2018). The microbial nitrogen-cycling network. *Nature Reviews Microbiology*, *16* (5): 26. https://doi.org/10.1038/nrmicro.2018.9

Kuznetsova, S., Knaff, D.B., Hirasawa, M., et al (2004). Mechanism of spinach chloroplast ferredoxin-dependent nitrite reductase: spectroscopic evidence for intermediate states. *Biochemistry 43*, 510–51 https://doi.org/10.1021/bi035662q

Lam, P. and Kuypers, M.M. (2011). Microbial nitrogen cycling processes in oxygen minimum zones. *Annual Review of Marine Science 3*: 317–345. https://doi.org/10.1146/annurev-marine-120709–142814

Lam, P., Lavik, G., Jensen, M.M. et al. (2009). Revising the nitrogen cycle in the Peruvian oxygen minimum zone. *Proceedings of the National Academy of Sciences 106* (12): 4752–4757. https://doi.org/10.1073/pnas.0812444106

Laverock B., Smith C., Tait K. et al. (2010). Bioturbating shrimp alter the structure and diversity of bacterial communities in coastal marine sediments. *ISME Journal 4*: 1531–1544. https://doi.org/10.1038/ismej.2010.86

Liu, L.Y., Xie, G.J., Xing, D.F. et al. (2020). Sulfate dependent ammonium oxidation: a microbial process linked nitrogen with sulfur cycle and potential application. *Environmental Research 192*: 11028. https://doi.org/10.1016/j.envres.2020.110282

Liu, S., Yang, F., Gong, Z. et al. (2008). Application of anaerobic ammonium-oxidizing consortium to achieve completely autotrophic ammonium and sulfate removal. *Bioresource Technology 99* (15): 6817–6825.https://doi.org/10.1016/j.biortech.2008.01.054

Lomas, M.W. and Glibert, P.M. (1999). Interactions between $NH4^+$ and $NO3^-$ uptake and assimilation: comparison of diatoms and dinoflagellates at several growth temperatures. *Marine Biology 133* (3): 541–55 https://doi.org/10.1007/s002270050494

Luesken, F.A., Sánchez, J., van Alen, T.A. et al. (2011). Simultaneous nitrite-dependent anaerobic methane and ammonium oxidation processes. *Applied and Environmental Microbiology 77* (19): 6802–6807. https://doi.org/10.1128/AEM.05539–11

Luther, III, G.W., Sundby, B., Lewis, B.L. et al. (1997). Interactions of manganese with the nitrogen cycle: alternative pathways to dinitrogen. *Geochimica et Cosmochimica Acta 61* (19): 4043–4052. https://doi.org/10.1016/S0016–7037 (97)00239–1

Marty, B. (1995). Nitrogen content of the mantle inferred from N–Ar correlation in oceanic basalts. *Nature 377*; 326–329. https://doi.org/10.1038/377326a0

Mazumder, A. and Nigam, R. (2014). Bathymetric preference of four major genera of rectilinear benthic foraminifera within oxygen minimum zone in Arabian Sea off central west coast of India. *Journal of Earth System Science 123* (3): 633–639. https://doi.org/10.1007/s12040–014–0419-y

Meyers, P.A. (1997). Organic geochemical proxies of paleoceanographic, paleolimnologic and paleoclimatic processes. *Organic Geochemistry 27* (5–6): 213–250.

Moisander, P.H., Serros, T., Paerl, R.W. et al. (2014). Gammaproteobacterialdiazotrophs and nifH gene expression in surface waters of the South Pacific Ocean. *ISME Journal 8*: 1962–1973. https://doi.org/10.1038/ismej.2014.49

Montoya, J.P. and Voss, M. (2006). Nitrogen cycling in suboxic waters: isotopic signatures of nitrogen transformation in the Arabian Sea oxygen minimum zone. In: *Past and Present Water column Anoxia, Nato Science Series: IV: Earth and Environmental Sciences*, Vol. *64* (ed. L. Neretin), 257–281. Dordrecht: Springer. https://doi.org/10.1007/1-4020-4297-3_11

Moreira-Coello, V., Mouriño-Carballido, B., Marañón, E. et al. (2019). Temporal variability of diazotroph community composition in the upwelling region off NW Iberia. *Scientific Reports 9* (1): 1–13. https://doi.org/10.1038/s41598–019–39586–4

Mortimer, R.J., Harris, S.J., Krom, M.D. et al. (2004). Anoxic nitrification in marine sediments. *Marine Ecology Progress Series 276*: 37–52. https://doi.org/10.3354/meps276037

Mortimer, R.J., Krom, M.D., Harris, S.J. et al. (2002). Evidence for suboxic nitrification in recent marine sediments. *Marine Ecology Progress Series 236*; 31–35. https://doi.org/10.3354/meps236031

Mosier, A.C. and Francis, C.A. (2008). Relative abundance and diversity of ammonia-oxidizing archaea and bacteria in the San Francisco Bay estuary. *Environmental Microbiology 10* (11): 3002–301 https://doi.org/10.1111/j.1462–2920.2008.0176

Mosier, A.C. and Francis, C.A. (2010). Denitrifier abundance and activity across the San Francisco Bay estuary. *Environmental Microbiology Reports 2* (5): 667–676. https://doi.org/10.1111/j.1758–2229.2010.0015

Mulholland, M.R. and Lomas, M.W. (2008). Nitrogen uptake and assimilation. In: *Nitrogen in the Marine Environment* (eds D.G. Capone, D.A Bronk, M.R Mullholand et al.), 303–384. Amsterdam: Elsevier Inc.

Mulla, A., Fernandes, G., Menezes, L. et al. (2018). Diversity of culturable nitrate-reducing bacteria from the Arabian Sea oxygen minimum zone. *Deep Sea Research Part II: Topical Studies in Oceanography 156*: 27–33. https://doi.org/10.1016/j.dsr2.2017.12.014

Muro-Pastor, M.I., Reyes, J.C. and Florencio, F.J. (2005). Ammonium assimilation in cyanobacteria. *Photosynthesis Research 83* (2): 135–150. https://doi.org/10.1007/s11120–004–2082–7

Nagoji, S. and Tiwari, M. (2021). Causes and climatic influence of centennial-scale denitrification variability in the south-eastern Arabian Sea since the last glacial period. *Quaternary Research 101*: 156–168. https://doi.org/10.1017/qua.2020.118

Naik, H.S. (2003). *Benthic nitrogen cycling with special reference to nitrous oxide in the coastal and continental shelf environments of the eastern Arabian Sea*. Doctoral dissertation, Goa University.

Naik, H. and Naqvi, S.W.A. (2002). Sedimentary nitrogen cycling over the western continental shelf of India. *EOS – Transactions of the American Geophysical Union 83* (4): Abstract OS12I-05.

Naqvi, S.W.A. (1987). Some aspects of the oxygen-deficient conditions and denitrification in the Arabian Sea. *Journal of Marine Research 45* (4): 1049–1072. https://doi.org/10.1357/002224087788327118

Naqvi, S.W.A. and Noronha, R.J. (1991). Nitrous oxide in the Arabian Sea. *Deep Sea Research Part A. Oceanographic Research Papers 38* (7): 871–890. https://doi.org/10.1016/0198–0149(91)90023–9.

Naqvi, S.W.A., Yoshinari, T., Jayakumar, D.A. et al. (1998). Budgetary and biogeochemical implications of N_2O isotope signatures in the Arabian Sea. *Nature 394* (6692): 462–464. https://doi.org/10.1038/28828

Naqvi, S.W.A., Jayakumar, D.A., Narvekar, P.V. et al. (2000). Increased marine production of N_2O due to intensifying anoxia on the Indian continental shelf. *Nature 408*: 346–349. https://doi.org/10.1038/35042551

Naqvi, S.W.A., Naik, H., Pratihary, A. et al. (2006). Coastal versus open-ocean denitrification in the Arabian Sea, *Biogeosciences 3*: 621–633. https://doi.org/10.5194/bg-3-621-200

Naqvi, S.W.A., Naik, H., D'Souza, W. et al. (2010). Carbon and nitrogen fluxes in the North Indian Ocean. In: *Carbon and Nutrient Fluxes in Continental Margins: a Global Synthesis* (eds. K.K. Liu, L. Atkinson, R. Quinones and M.L. Talaue), 180–192. Berlin: Springer.

Nicholls, J.C., Davies, C.A. and Trimmer, M. (2007). High-resolution profiles and nitrogen isotope tracing reveal a dominant source of nitrous oxide and multiple pathways of nitrogen gas formation in the central Arabian Sea. *Limnology and oceanography 52* (1): 156–168. https://doi.org/10.4319/lo.2007.52.1.0156

Nielsen, O.I., Gribsholt, B., Kristensen, E. et al. (2004). Microscale distribution of oxygen and nitrate in sediment inhabited by *Nereis diversicolor*: spatial patterns and estimated reaction rates. *Aquatic Microbial Ecology 34* (1): 23–32. https://doi.org/10.3354/ame034023

Ohashi, Y., Shi, W., Takatani, N. et al. (2011). Regulation of nitrate assimilation in cyanobacteria. *Journal of Experimental Botany 62* (4): 1411–1424. https://doi.org/10.1093/jxb/erq427

Olson, R. (1981). Differential photoinhibition of marine nitrifying bacteria: a possible mechanism for the formation of the primary nitrite maximum. *Journal of Marine Research 39*: 227–238.

Padilla, C.C., Bristow, L.A., Sarode, N. et al. (2016). NC10 bacteria in marine oxygen minimum zones. *The ISME Journal 10* (8): 2067–2071. https://doi.org/10.1038/ismej.2015.262

Pajares, S. and Ramos, R. (2019). Processes and microorganisms involved in the marine nitrogen cycle: knowledge and gaps. *Frontiers in Marine Science 6*: 739. https://doi.org/10.3389/fmars.2019.00739

Park, B.J., Park, S.J., Yoon, D.N. et al. (2010). Cultivation of autotrophic ammonia-oxidizing archaea from marine sediments in coculture with sulfur-oxidizing bacteria. *Applied and Environmental Microbiology 76* (22): 7575–7587. https://doi.org/10.1128/AEM.01478-10.

Pattan, J.N., Mir, I.A., Parthiban, G. et al. (2013). Coupling between suboxic condition in sediments of the western Bay of Bengal and southwest monsoon intensification: a geochemical study. *Chemical Geology 343*: 55–66 https://doi.org/10.1016/j.chemgeo.2013.02.011.

Peng, X., Fuchsman, C.A., Jayakumar, A. et al. (2016). Revisiting nitrification in the Eastern Tropical South Pacific: a focus on controls. *Journal of Geophysical Research: Oceans 121* (3): 1667–1684. https://doi.org/10.1002/2015JC011455

Purkhold, U., Pommerening-Röser, A., Juretschko, S. et al. (2000). Phylogeny of all recognized species of ammonia oxidizers based on comparative 16S rRNA and amoA sequence analysis: implications for molecular diversity surveys. *Applied and Environmental Microbiology 66* (12): 5368–5382. https://doi.org/10.1128/AEM.66.12.5368–5382.2000

Raghoebarsing, A.A., Pol, A., Van de Pas-Schoonen, K.T. et al. (2006). A microbial consortium couples anaerobic methane oxidation to denitrification. *Nature 440* (7086): 918–921. https://doi.org/10.1038/nature04617

Ramos, R. and Pajares, S. (2018). Reviews and syntheses: processes and functional genes involved in nitrogen cycling in marine environments. *Biogeosciences Discussions, 1–50.* https://doi.org/10.5194/bg-2018–2279

Rees, D.C., Akif Tezcan, F., Haynes, C.A. et al. (2005). Structural basis of biological nitrogen fixation. *Philosophical Transactions of the Royal Society A: Mathematical, Physical and Engineering Sciences 363* (1829): 971–984. https://doi.org/10.1098/rsta.2004.1539

Rios-Del Toro, E.E., Valenzuela, E.I., López-Lozano, N.E. et al. (2018). Anaerobic ammonium oxidation linked to sulfate and ferric iron reduction fuels nitrogen loss in marine sediments. *Biodegradation 29*: 429–442. https://doi.org/10.1007/s10532–018–9839–9838

Risgaard-Petersen, N., Meyer, R.L., Schmid, M. et al. (2004). Anaerobic ammonium oxidation in an estuarine sediment. *Aquatic Microbial Ecology 36* (3): 293–30 https://doi.org/10.3354/ame036293

Risgaard-Petersen, N., Langezaal, A.M., Ingvardsen, S. et al. (2006). Evidence for complete denitrification in a benthic foraminifer. *Nature 443* (7107): 93–96. https://doi.org/10.1038/nature05070

Rixen, T., Cowie, G., Gaye, B. et al. (2020). Reviews and syntheses: present, past, and future of the oxygen minimum zone in the northern Indian Ocean. *Biogeosciences 17* (23): 6051–6080. https://doi.org/10.5194/bg-17-6051-2020

Robertson, E.K., Roberts, K.L., Burdorf, L.D. et al. (2016). Dissimilatory nitrate reduction to ammonium coupled to Fe (II) oxidation in sediments of a periodically hypoxic estuary. *Limnology and Oceanography 61* (1): 365–381. https://doi.org/10.1002/lno.10220

Romillac, N. (2019). Ammonification. In: *Encyclopedia of Ecology*, Vol. 2 (ed. B.D. Fath), 256–263. Amsterdam, The Netherlands: Elsevier. https://doi.org/10.1016/B978–0-12–409548–9.10889–9

Rovira, J.L. and Pardo, P. (2006). Nutrient pollution of waters: eutrophication trends in European marine and coastal environments. *Contributions to Science 3* (2): 181–186. https://doi.org/10.2436/20.7010.01.4

Sarkar, A., Naqvi, S.W.A., Lavik, G. et al. (2020). Massive nitrogen loss over the western indian continental shelf during seasonal anoxia: evidence from isotope pairing technique. *Frontiers in Marine Science 7*: 678. https://doi.org/10.3389/fmars.2020.00678

Schaefer, S.C. and Hollibaugh, J.T. (2017). Temperature decouples ammonium and nitrite oxidation in coastal waters. *Environmental Science and Technology 51* (6): 3157–3164. https://doi.org/10.1021/acs. est.6b03483

Schrum, H.N., Spivack, A.J., Kastner, M. et al. (2009). Sulfate-reducing ammonium oxidation: a thermodynamically feasible metabolic pathway in subseafloor sediment. *Geology 37* (10): 939–942. https://doi.org/10.1130/G30238A.1

Schwartz, M.C., Woulds, C. and Cowie, G.L. (2009). Sedimentary denitrification rates across the Arabian Sea oxygen minimum zone. *Deep Sea Research Part II: Topical Studies in Oceanography 56* (6–7): 324–332. https://doi.org/10.1016/j.dsr2.2008.05.028

Seitzinger, S.P., Harrison, J.A., Dumont, E. et al. (2005). Sources and delivery of carbon, nitrogen and phosphorous to the coastal zone: an overview of global nutrient export from watersheds (NEWS) models and their application. *Global Biogeochemical Cycles 19* (4): GB4S01. https://doi.org/10.1029/2005GB002606

Seitzinger, S., Harrison, J.A., Böhlke, J.K. et al. (2006). Denitrification across landscapes and waterscapes: a synthesis. *Ecological Applications 16*: 2064 –2090. https://doi.org/10.1890/1051–0761 (2006)016[2064:dalawa]2.o;2

Shen, L.D., Liu, S., He, Z.F. et al. (2015). Depth-specific distribution and importance of nitrite-dependent anaerobic ammonium and methane-oxidising bacteria in an urban wetland. *Soil Biology and Biochemistry 83*: 43–51. https://doi.org/10.1016/j.soilbio.2015.01.010

Shiozaki, T., Ijichi, M., Isobe, K. et al. (2016). Nitrification and its influence on biogeochemical cycles from the equatorial Pacific to the Arctic Ocean. *The ISME Journal 10* (9): 2184–2197. https://doi.org/10.1038/ismej.2016.18

Simon, J. and Klotz, M.G. (2013). Diversity and evolution of bioenergetic systems involved in microbial nitrogen compound transformations. *Biochimica et Biophysica Acta. Bioenergetics, 1827*: 114–135. https://doi.org/10.1016/j.bbabio.2012.07.005

Slobodkina, G.B., Mardanov, A.V., Ravin et al. (2017). Respiratory ammonification of nitrate coupled to anaerobic oxidation of elemental sulfur in deep-sea autotrophic thermophilic bacteria. *Frontiers in Microbiology 8*: 87. https://doi.org/10.3389/fmicb.2017.00087

Smith, J.M., Chavez, F.P. and Francis, C.A. (2014). Ammonium uptake by phytoplankton regulates nitrification in the sunlit ocean. *PLoS One 9*: e108173. https://doi.org/10.1371/journal.pone.0108173

Sohm, J.A., Webb, E.A. and Capone, D.G. (2011). Emerging patterns of marine nitrogen fixation. *Nature Reviews Microbiology 9* (7): 499–508. https://doi.org/10.1038/nrmicro2594

Sokoll, S., Holtappels, M., Lam, P. et al. (2012). Benthic nitrogen loss in the Arabian Sea off Pakistan. *Frontiers in Microbiology 3*: 395. https://doi.org/10.3389/fmicb.2012.00395

Song, B., Lisa, J.A. and Tobias, C.R. (2014). Linking DNRA community structure and activity in a shallow lagoonal estuarine system. *Frontiers in Microbiology 5*: 460. https://doi.org/10.3389/fmicb.2014.00460

Staal, M., Meysman, F.J. and Stal, L.J. (2003). Temperature excludes N_2-fixing heterocystous cyanobacteria in the tropical oceans. *Nature 425* (6957): 504–507. https://doi.org/10.1038/nature01999

Staal, M., Rabouille, S. and Stal, L.J. (2007). On the role of oxygen for nitrogen fixation in the marine cyanobacterium *Trichodesmium* sp. *Environmental Microbiology 9*: 727–736. https://doi.org/10.1111/j.1462–2920.2006.01195.x

Stein, L.Y. (2011). Heterotrophic nitrification and nitrifier denitrification. *Nitrification 95–114*. https://doi.org/10.1128/978155581714h5

Strous, M., Van Gerven, E., Kuenen, J.G.,et al. (1997). Effects of aerobic and microaerobic conditions on anaerobic ammonium-oxidizing (anammox) sludge. *Applied and Environmental Microbiology, 63* (6): 2446–2448. https://doi.org/10.1128/AEM.63.6.2446–2448.1997

Strous, M., Heijnen, J.J., Kuenen, J.G. and Jetten, M.S.M. (1998). The sequencing batch reactor as a powerful tool for the study of slowly growing anaerobic ammonium-oxidizing microorganisms. *Applied Microbiology and Biotechnology 50* (5): 589–596. https://doi.org/10.1007/s002530051340

Strous, M., Fuerst, J.A., Kramer, E.H. et al. (1999). Missing lithotroph identified as new planctomycete. *Nature 400* (6743): 446–449. https://doi.org/10.1038/22749

Strous, M., Pelletier, E., Mangenot, S., et al. (2006). Deciphering the evolution and metabolism of an anammox bacterium from a community genome. *Nature 440* (7085): 790–794. https://doi.org/10.1038/nature04647

Suokhrie, T., Saraswat, R. and Nigam, R. (2020). Lack of denitrification causes a difference in benthic foraminifera living in the oxygen deficient zones of the Bay of Bengal and the Arabian Sea. *Marine Pollution Bulletin 153*: 110992. https://doi.org/10.1016/j.marpolbul.2020.110992

Thamdrup, B. (2012). New pathways and processes in the global nitrogen cycle. *Annual Review of Ecology Evolution and Systematics43*:407–428.https://doi.org/10.1146/annurev-ecolsys-102710–145048

Thamdrup, B. and Dalsgaard, T. (2002). Production of N_2 through anaerobic ammonium oxidation coupled to nitrate reduction in marine sediments. *Applied and Environmental Microbiology 68*: 1312–1318. https://doi.org/10.1128/AEM.68.3.1312–1318. 2002

Thamdrup, B., Dalsgaard, T., Jensen, M.M. et al. (2006). Anaerobic ammonium oxidation in the oxygen-deficient waters off northern Chile. *Limnology and Oceanography 51* (5): 2145–2156. https://doi.org/10.4319/lo.2006.51.5.2145

Thamdrup, B., Steinsdóttir, H.G., Bertagnolli, A.D. et al. (2019). Anaerobic methane oxidation is an important sink for methane in the ocean's largest oxygen minimum zone. *Limnology and Oceanography 64* (6): 2569–2585. https://doi.org/10.1002/lno.11235

Tiedje, J.M. (1988). Ecology of denitrification and dissimilatory nitrate reduction to ammonium. In: *Biology of Anaerobic Microorganisms, Environmental Microbiology of Anaerobes* (ed. A.J.B. Zehnder), 179–244. New York, N.Y: John Wiley and Sons, Inc.

Tikhonova, T.V., Slutsky, A., Antipov, A.N. et al. (2006). Molecular and catalytic properties of a novel cytochrome c nitrite reductase from nitrate-reducing haloalkaliphilic sulfur-oxidizing bacterium Thioalkalivibrio nitratireducens. *Biochimica et Biophysica Acta – Proteins and Proteomics 1764*(4):715–723. https://doi.org/10.1016/j.bbapap.2005.12.021

Trimmer, M. and Engström, P. (2011). Distribution, activity and ecology of anammox bacteria in aquatic environments. *Nitrification201–235*.https://doi.org/10.1128/978155581714h9

Tripathi, S., Tiwari, M., Lee, J. et al. (2017). First evidence of denitrification vis-à-vis monsoon in the Arabian Sea since Late Miocene. *Scientific Reports 7* (1): 1–7. https://doi.org/10.1038/srep43056

Tripathi, S., Behera, P. and Tiwari, M. (2020). Evolution and dynamics of the denitrification in the Arabian Sea on millennial to million-year timescale. *Current Science 119* (2): 282. https://doi.org/:d10.18520/cs/v119/i2/282–290

Unger, D., Schäfer, P., Ittekkot, V. et al. Nitrogen isotopic composition of sinking particles from the southern Bay of Bengal: Evidence for variable nitrogen sources. *Deep Sea Research Part I: Oceanographic Research Papers 53* (10): 1658–1676. https://doi.org/10.1016/j.dsr.2006.07.004

Verma, K., Bharti, S.K. and Singh, A.D. (2018). Late Glacial–Holocene record of benthic foraminiferal morphogroups from the eastern Arabian Sea OMZ: paleoenvironmental implications. *Journal of Earth System Science 127* (2): 1–15. https://doi.org/10.1007/s12040–018–0920–9

Voss, M., Dippner, J.W. and Montoya, J.P. (2001). Nitrogen isotope patterns in the oxygen-deficient waters of the Eastern Tropical North Pacific Ocean. *Deep Sea Research Part I: Oceanographic Research Papers 48* (8): 1905–1921. https://doi.org/10.1016/S0967–0637(00)00110–2

Voss, M., Baker, A., Bange, H.W. et al. (2011). *Nitrogen Processes in Coastal and Marine Ecosystems, 147–176.* Cambridge: Cambridge University Press. https://doi.org/10.1017/CBO9780511976988.011

Voss, M., Bange, H.W., Dippner, J.W. et al. (2013). The marine nitrogen cycle: recent discoveries, uncertainties and the potential relevance of climate change. *Philosophical Transactions of the Royal Society B: Biological Sciences 368* (1621): 20130121. https://doi.org/10.1098/rstb.2013.0121

Wada, E. and Hattori, A. (1976). Natural abundance of ^{15}N in particulate organic matter in the North Pacific Ocean. *Geochimica et Cosmochimica Acta 40* (2): 249–251.https://doi.org/10.1016/0016–7037 (76)90183–6

Ward, B.B., Devol, A.H., Rich, J.J. et al. (2009). Denitrification as the dominant nitrogen loss process in the Arabian Sea. *Nature 461* (7260): 78–81. https://doi.org/10.1038/nature08276

Widdicombe, S., Austen, M.C., Kendall, M.A. et al. (2000). Bioturbation as a mechanism for setting and maintaining levels of diversity in subtidal macrobenthic communities. In: *Island, Ocean and Deep-Sea Biology* (eds. M.B. Jones, J.M.N. Azevedo, A.I. Neto et al.), 369–377). Dordrecht: Springer. http://dx.doi.org/10.1023/A:1004120621842

Woehle, C., Roy, A.S., Glock, N. et al. (2018). A novel eukaryotic denitrification pathway in foraminifera. *Current Biology 28* (16): 2536–2543. https://doi.org/10.1016/j.cub.2018.06.027

Yool, A., Martin, A.P., Fernández, C. and Clark, D.R. (2007). The significance of nitrification for oceanic new production. *Nature 447* (7147): 999–1002. https://doi.org/10.1038/nature05885

Zakem, E.J., Al-Haj, A., Church, M.J. et al. (2018). Ecological control of nitrite in the upper ocean. *Nature Communications 9* (1): 1–13. https://doi.org/10.1038/s41467–018–03553-w

Zehr, J.P. and Paerl, H.W. (2008). Molecular ecological aspects of nitrogen fixation in the marine environment. In: *Microbial Ecology of the Oceans* (ed. D.L. Kirchman), 481–525). Hoboken, NJ: John Wiley and Sons, Inc. https://doi.org/10.1002/9780470281841h13

Zumft, W.G. (1997). Cell biology and molecular basis of denitrification. *Microbiology and Molecular Biology Reviews 61* (4): 533–616. https://doi.org/10.1128/.61.4.533–616.1997

5

Organic Carbon in Sediments of the Western Indian Margin

Pratima M. Kessarkar[1], Venigalla P. Rao[1,2], Lina L. Fernandes[1], and Siby Kurian[1]

ABSTRACT

The Arabian Sea (AS) is known to be one of the most productive regions of the world ocean with very high organic carbon (OC). The very high OC in the sediments of the eastern AS (western margin of India) are well debated with hypothesis revolving around high primary productivity and preservation due to the oxygen minimum zone (OMZ). Studies on sediment OC from the AS date back to the 1930s. The papers published from 1960 to 1970 were mostly focused on fisheries and carbon productions. From 1970 to 1980, papers published were on the data collected during International Indian Ocean Expedition (IIOE), with less focus on the western margin of India. The researchers observed high OC in the sediments covering large areas and suggested productivity as the major reason. It was only after the 1980s that OC preservation was related to the OMZ. After 1990 there came two strong schools of thought, where one supported productivity and the other preservation. Subsequently, factors such as slope width and downslope transport were introduced. Studies on OC content in sediment cores started only after the 1990s aiming to look into changes in the OC content and to infer past productivity. The major observation on OC distribution in sediment cores was that the OC content during Glacial and interglacial times is not uniform throughout the Arabian Sea. The richness of OC in sediment cores affected the extent and intensity of early diagenesis is in sediments within the OMZ and those in the oxygenated waters. High OC was observed to have an effect on magnetic minerals in sediment cores. Here, we compile the studies on OC distribution from the AS, especially the western margin of India, and put together the processes responsible for high OC in the sediments. We suggest that the Arabian Sea is complex in physiology, hydrodynamics, and circulation (atmospheric and ocean) patterns that lead to multifaceted conditions involving differential productivity and accumulation and, preservation and productivity are insufficient to explain the OC distribution.

5.1. INTRODUCTION

Oceans store 60 times more carbon compared with the atmosphere (Berger et al., 1989), with physical and chemical processes in the ocean considered important for organic carbon (OC) distribution and global carbon balancing (Premuzic et al., 1982). Understanding OC distribution is important as CO_2 from the ocean/atmosphere gets buried into the marine sediments through OC and drives atmospheric oxygenation at glacial–interglacial timescale (Cartapanis et al., 2016). There has been a change in OC content in the ocean during past wherein accumulation of OC was 50% higher during the glacial maximum compared with the interglacials (Cartapanis et al., 2016). Changes in OC from the oceans have been related to atmospheric CO_2 during glacial times that was lower (~200 ppm) compared with the Holocene (300 ppm) (Berger et al., 1989). Productivity is considered to be an important factor that helps CO_2 content of the ocean, as

[1] *CSIR-National Institute of Oceanography, Dona Paula, 403004, Goa, India*

[2] *Department of Civil Engineering, Vignan's University, Vadlamudi, 522213, Guntur, India*

Corresponding Author: Pratima M. Kessarkar (pratimak@nio.org)

ORCiD code: 0000-0002-9974-2682

Systems Biogeochemistry of Major Marine Biomes, First Edition. Edited by Aninda Mazumdar and Wriddhiman Ghosh.

OC is exported from the photic zone via photosynthesis, and CO_2 in the surface water is lowered due to dissolved organic carbon that acts as biological pump (Berger et al., 1989). During high primary productivity, CO_2 is removed from the atmosphere and is fixed into the oceans (Premuzic et al., 1982) and is accumulated into organic matter (OM) on the continental slope. Organic carbon can be marine or terrestrial and can be in dissolved or particulate form in the water column. The western margin of India is one such place where there is an accumulation of high OC in the sediments. This Chapter reviews the studies on organic carbon distribution in surface sediments and sediment cores of the AS and reports on the factors controlling its distribution in the present day and during the late Quaternary. The influence of OC content on sediment components during the early diagenesis and in establishing the sedimentary environment is also discussed.

5.2. ORGANIC MATTER AND ORGANIC CARBON

Organic matter consists of organic compounds such as amino acids, sugars, phenols, polysaccharides, lipids, pigments, and constituents of living organisms (Premuzic et al., 1982), and occurs in dissolved and particulate form. Geologists usually analyze total OC and OM is calculated by multiplying total OC by a factor of 1.724 (OM = OC × 1.724). Out of the total productivity of OM in the surface water, only a small quantity reaches the sediment (Berger et al., 1989). The quantity of OM that finally gets into the sediments depends on several parameters, such as chemical stability, oxygen concentrations, interaction with clays, and biochemical usefulness (Premuzic et al., 1982). Organic compounds are adsorbed onto carbonate particles and clays (with the larger surface area) and settle through the water column, ultimately reaching the sediments, and become enriched in the OM (Premuzic et al., 1982). Fine clays are accumulators of OM, and their ability to adsorb depends on their physical and chemical properties (Premuzic et al., 1982 and references therein). A high concentration of clay minerals has been observed on the continental slope areas in some parts of the world (Emery and Uchupi, 1972) and this can have a role in creating high OC contents. Organic carbon from the sediments is used to infer primary productivity (Figure 5.1a). Based on fluxes of the OC in coastal waters about 120 g C m^{-2} $year^{-1}$ is produced in the top 200 m but only 1 g C m^{-2} $year^{-1}$ is deposited on the ocean floor. While in the open ocean, it is much less, where, from a primary productivity of 30 g C m^{-2} $year^{-1}$, only 0.01 g C m^{-2} $year^{-1}$ (Figure 5.1a) reaches the sediments (Berger et al., 1989). Primary production (PP) in the AS is not iron (Fe)-limiting as the concentrations of Fe are much higher. It is believed that control of PP by

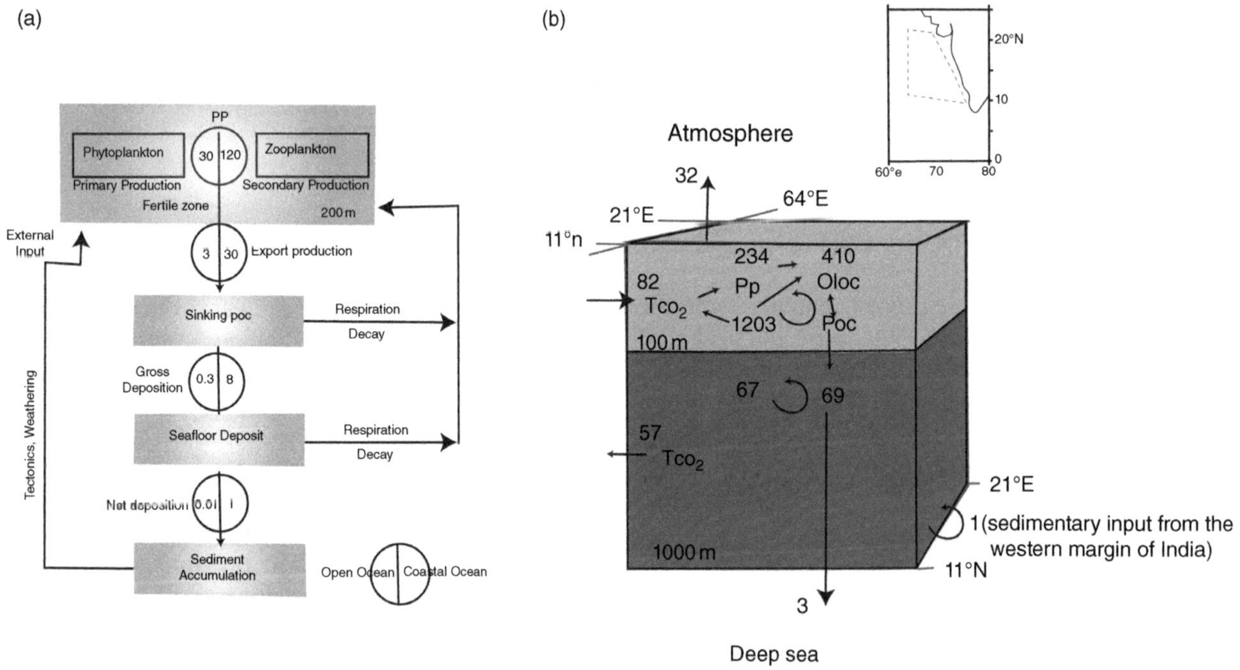

Figure 5.1 (a) Schematic diagram showing transfer of organic carbon from surface to the deep ocean (Berger et al 1989), the carbon flux is shown (g C m^{-2} $year^{-1}$); PP-primary productivity. (b) Carbon budget in the Arabian Sea (Tg C $year^{-1}$) in the area shown in the insight, living carbon production rates (OLOC) in (microzoooplankton, mesozooplankton, and bacteria; Sharma et al., 2003).

macronutrients mostly depends on the availability of nitrate (Sen Gupta and Naqvi, 1984; Measures and Vink, 1999; Naqvi et al., 2003). A low rate of PP, despite nutrient-rich upwelled water within the euphotic zone is related to the unsuitability of a low oxygen environment which hampers the growth of oxygenic photosynthesizers (Naqvi et al., 2010).

5.3. ORGANIC CARBON DISTRIBUTION IN THE WORLD OCEAN AND THE ARABIAN SEA

Primary production in the world ocean is estimated to be 1.5 to 1.8 × 10^{10} tons C fixed per year (Ryther, 1969). For roughly 90% of the ocean (326 × 10^6 km^2), annual primary production in the open sea is 25–75 g C fixed m^{-2} and averages to ~50 gC m^{-2}. Marine sediments receive only 1% of the carbon fixed per year through primary production (Dunne et al., 2007; Middelburg, 2011). On the basis of new and old carbon, Westberry and Behrenfeld (2014) estimated that the annual global oceanic primary productivity of C was between 70 and 145 Gt. The coastal zone and offshore high productivity regions together represent 10% of the total area of the oceans. Of this, 7.5% of the total world ocean primary production occurs within the 180 m depth range of coastal waters, with ~100 g C m^{-2} $year^{-1}$ (Ryther, 1969). Upwelling regions cover not more than one-tenth of 1% of the ocean surface (Ryther, 1969), but PP is >1 g C m^{-2} day^{-1} (in some areas may exceed 10.01 g C m^{-2} day^{-1}). The AS is one such upwelling region (upwelling of subsurface water is a seasonal phenomenon related to the monsoon winds) with a carbon flux of 6.4 g C m^{-2} day^{-1} (Ryther and Menzel, 1965). Apart from autochthonous production, Kandasamy and Nath (2016) suggested that the terrestrial OM on the continental shelves brought down by the rivers was an important component of the global carbon cycle. A large amount of OC received from the land and oceans is stored on continental margins and this is therefore an important area of the global carbon cycle (Bianchi et al., 2018). High OC content was reported in river deltas (114 Tg C $year^{-1}$) and non-deltaic shelf (70 Tg C $year^{-1}$) compared with the open ocean (6 Tg C $year^{-1}$; Burdige, 2005). Bianchi et al. (2018) reviewed the fate of OC on the continental margin sediments in different depositional environments at the land–ocean interface.

Several studies reported OC fluxes, fate of OC, and CO_2 transport in the AS (Sarma, 2002; Sarma et al., 2003, 2007). It is estimated that there is a net supply of 25 Tg C $year^{-1}$ to the upper 1000 m of the water column (JGOFS India study region) in the AS by a physical pump. Total carbon demand (1203 Tg C $year^{-1}$) by mesozooplankton, microzooplankton, and bacteria in the surface layers is not supported by photosynthetic activity (234 Tg C $year^{-1}$) alone and is supplied through circulation. Carbon demand is nearly double that of the total living biomass production rate (644 Tg C $year^{-1}$), suggesting that most of the demand might be met from internal cycling involving grazing/excretion activities of microzooplankton and mesozooplankton to sustain a perennial supersaturation of CO_2 in surface waters of the AS and emission of 32 Tg C $year^{-1}$ to the atmosphere (Figure 5.1b). Surface box becomes richer by 82 Tg C $year^{-1}$, while 57 Tg C is removed annually from the subsurface one (Figure 5.1b). A total community demand of 1203 ± 662Tg C $year^{-1}$ is required to sustain heterotrophic living biomass production rates. Sarma et al. (2007) reported that export of carbon through a biological pump from the surface to the deep ocean has a direct influence on the removal of CO_2 from the atmosphere. The carbon is sequestered for only a few days to months on the surface, while the carbon removed from the surface to deep waters takes hundreds of years to re-enter the atmosphere. The deep water dissolved inorganic carbon (DIC) in the northern Indian Ocean is greater than elsewhere in the World Oceans. An efficient biological pump appears to be operating in the northern Indian Ocean and transports surface-derived organic/inorganic carbon to deeper layers, where it is converted and stored for longer periods in dissolved inorganic form. Accumulation of DIC in deep waters is found to be the greatest in the North Indian Ocean compared with elsewhere in the world oceans. The deep waters of the AS receive greater amounts of both organic and inorganic carbon through sinking from the euphotic zone. Their subsequent decomposition and dissolution has resulted in the accumulation of DIC (6–39.5 μmol kg^{-1}) in deep waters. Although this region acts as a source for atmospheric CO_2, the efficient biological pump here facilitates the storage of a significant amount of carbon for longer periods in dissolved inorganic form. Model results have revealed that the physical processes associated with the biogeochemical cycling of carbon and oxygen regulate the OMZ in the AS. The sinking of low OC fluxes and oxygen supply by physical pumping during the non-monsoon period maintains OMZ.

5.4. METHODS ADOPTED TO DETERMINE OC IN THE ARABIAN SEA

The production of OM (primary productivity in surface waters) was first measured in the equatorial part of the Indian Ocean during the Danish *Galathea* Expedition using radioactive carbon (C^{14};Steemann, 1952; Steemann and Jensen, 1957). However, the most popular method used to estimate the OC content is that by El-Wakeel and Riley (1957), and is represented as a percentage. The other method used is one where $CaCO_3$ is removed from the sediments by treating with mild acid and then the sample is combusted in two cycles in a Perkin-Elmer carbon-hydrogen analyzer at temperatures of 750–800°C in the presence of cupric oxide, which acts as a catalyst, (Kolla et al., 1981). Paropkari et al. (1992) determined OC content, following the method of El Wakeel and Riley (1957) along most parts of the western margin of India (Figure 5.2). However, in the margin off Mangalore–Cochin, organic carbon was determined using a Perkin-Elmer elemental analyzer after removing inorganic carbon by acidulation (Paropkari, 1990). The chemical data were subjected to correlation analysis using a program developed at the CSIR-National Institute of Oceanography (Paropkari, 1990). In contrast, Reemtsma et al. (1990) determined OC with a Carlo Erba Elemental Analyser after removal of carbonate, whereas Naidu et al. (1992) determined OC using wet oxidation method (El Wakeel and Riley, 1957). Haake et al. (1993) measured nitrogen and carbon using Carlo-Erba Nitrogen Analyzer, and the OC was calculated by subtracting carbonate carbon from total carbon: total OM was assumed to be OC multiplied by 1.8. Calvert et al. (1995) determined total carbon and nitrogen by combustion/gas chromatography (CNS analyzer). Here, carbonate carbon was determined by acid liberation of CO_2 followed by coulometry and OC values were derived from the difference between the total and carbonate carbon values. PrakashBabu et al. (1999) revisited the study Paropkari et al. (1992) and followed a different method of estimating OC. The method used was sample combustion at high temperatures (1200°C) in an oxygenated environment and detection of carbon dioxide by coulometry. Total carbon (TC) content was cross-checked by a second method, based on combustion coupled with infrared detection (Sirocko et al., 2000); both TC and total nitrogen (TN) of the sediment were measured using Carlo Erba CN-analyser. Bhushan et al. (2001) determined TC with an elemental analyzer, and IC by using Ca with a

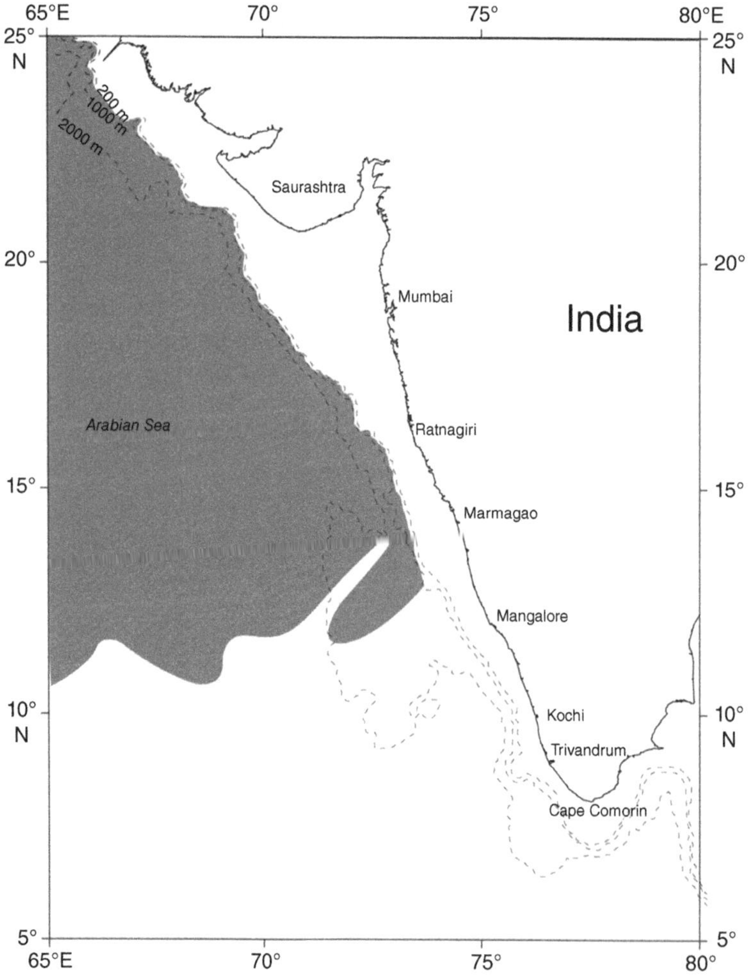

Figure 5.2 Bathymetric contours from the western margin of India, gray area is the denitrification zone (modified after Naqvi, 1991).

Perkin Elmer Atomic Absorption Spectrometer (AAS), OC = TC–IC. Gauns et al. (2005) determined particulate organic carbon (POC) suspended in the water column using a CHN analyzer. They filtered 3 l of water sample onto a precombusted 47 mm diameter GF/F filter (0.7 μm pore size). The filter papers were dried at 60C and later exposed to HCl to remove $CaCO_3$ and analyzed on a Perkin Elmer 2400 CHN analyzer. Kurian et al. (2020) measured OC together with N content, using a Delta V Plus isotope ratio mass spectrometer (IRMS) coupled with an elemental analyzer in continuous-flow mode. It is notable that earlier investigators mostly used the El Walkeel and Riley (1957) method and more recent investigators have used a CN analyzer.

5.5. PROCESSES RESPONSIBLE FOR PRIMARY PRODUCTIVITY AND TRANSPORT OF ORGANIC CARBON

High productivity on the western margin of India is related to (i) upwelling induced productivity (Shankar and Shetye, 1997), (ii) deep convective mixing (Madhupratap et al., 1996), (iii) eddy diffusion (e.g. Rejomon et al., 2013), (iv) formation of fronts (Sarma et al., 2018), and (5) ballast effect (Rixen et al., 2019). The AS is considered to be one of the most highly productive regions of the world. The high productivity in the AS is related to its physiography, monsoon forcing, and other oceanic processes. The AS undergoes seasonal variations in primary production under the influence of atmospheric monsoonal reversals of winds (Tomczak and Godfrey, 1994; Smith et al., 1991). During the northeast monsoon, the AS experiences winter convective mixing that leads to the upwelling of nutrient-rich subsurface waters resulting in high productivity (Madhupratap et al., 1996). During the southwest monsoon, upwelling takes place along the Arabian coast and southwest margin of India (Morrison et al., 1998; Shetye et al., 1990), resulting in increased productivity. Primary production ranges from 190 to 770 mg C m^{-2} day^{-1} in the open ocean and 200 to 1760 mg C m^{-2} day^{-1} in the coastal regions of the AS (Bhattathiri et al., 1996). Higher primary production during the southwest monsoon is related to the advection of nutrient-rich waters offshore (Young and Kindle, 1994). Changes in gross primary productivity also vary with changing intensity of the southwest monsoon (Rajagopalan et al., 1992). The sinking of the carbon to deeper depths is observed during the monsoons (Lee et al., 1998). The surface water pCO_2 of ~40–170 μatm gets drawdown in the AS as result of biological processes (Takahashi et al., 2002) and, AS is a perennial source of CO_2 to the atmosphere (Sarma et al., 1998). The AS is considered an important environment for the biogeochemical cycling of carbon under the influence of monsoon forcing (Scientific Committee on Ocean Research, 1996).

Increased productivity in the AS resulting from eddy diffusion (Rejomon et al., 2013) has impacts on the annual fluxes of OC to the deep sea. In the eastern AS, cold-core eddies result in rapid biological activity and increased production. There are also regional to basin-wide scales of enhanced biological production mediated by eddies. The increased export flux associated with grazing results in increased biogenic particle sedimentation. Fronts during the monsoon are associated with high productivity (Sarma et al., 2018). Ballast effect: large surface areas for OM (~800–1000 m^2 g^{-1}) and clay-smectite (~600–800 m^2 g^{-1}) compared with the carbonates, sand, and silt (<1 m^2 g^{-1}) help in transporting cations from the surface to depth (Sirocko et al., 2000). Aggregates pellets and lithogenic matter enhance OC flux rates on an average by 45% and at river-influenced areas by 62% in the Indian Ocean (Rixen et al., 2019).

5.6. OXYGEN MINIMUM ZONE AND ORGANIC CARBON

Several investigators reported the importance of OMZ in the OC preservation, and have varied opinions on the vertical depth extent of OMZ. The biologists view is that the OMZ is a low oxygen layer in the ocean where oxygen gets used by the primary productivity and is observed all over the world. Chemists use terms such as suboxic, anoxic, euxinic and their definitions are based on the concentrations of oxygen, nitrite, and nitrate in the water column (Naqvi et al., 2000). Geologists consider zones in the water column with dissolved oxygen (DO) of <0.5 ml l^{-1} as OMZ. Helly and Levin (2004) estimated that the OMZ (DO <0.5 ml l^{-1}) covers an area of 1 148 000 km^2 of seafloor. Of the total OMZ area, approximately 31% occurs in the eastern Pacific Ocean, 59% in the Indian Ocean (AS and Bay of Bengal), and 10% in the southeastern Atlantic. The vertical extent of the OMZ is thicker and more homogeneous in the Northern Indian Ocean and is notably thicker (and deeper) in the AS than in the Bay of Bengal. The presence of an OMZ helps preserve OC in the sediments (Kolla et al., 1981; Paropkari et al., 1992).

5.7. STUDIES CARRIED OUT ON ORGANIC CARBON FROM THE WESTERN MARGIN OF INDIA

5.7.1. Particulate/Dissolved Organic Carbon from the Water Column

From the 1980s sediment trap experiments have been useful in understanding the type of sediment flux, including OM, settling at traps located at different depths in the same mooring or trap moorings spatially located at different places in the oceans. It was found that remineralization takes place in POC in the sediment traps, depending on the time it remains in the trap. To answer some of the unsolved questions regarding marine carbon cycling, Ittikkot et al. (1992) suggested using remote sensing techniques for monitoring sea surface and the use of moored sediment traps to monitor the increased rate of the biological pump. Results of long-term sediment trap studies (see Figure 5.6) in three locations from the AS (western, central, and eastern) were reported by Haake et al.(1993). These authors showed seasonal and interannual variations in particle fluxes settling in different traps. On the basis of measurements of bulk component fluxes at the central Ttrap station for 3 years and western and eastern stations for 4 years, they suggested control of the particulate fluxes to the deep sea by monsoon, with generally higher fluxes (biogenic and lithogenic) during both the southwestern and northeastern monsoons and lower fluxes during the inter-monsoon period. The OC content in the traps ranged from 4% to 9%, and TN varied from 0.4% to 1.4%, with lower values during periods of high fluxes. The eastern trap, close to the western margin of India, showed an interannual variability of particle fluxes and showed maximum fluxes (44%) during August to October. This trap did not show much relationship between particulate fluxes and monsoon-related processes in the surface waters. Comparison of average annual accumulation rates in traps with sediment accumulation rates in surface sediments indicated slower biogenic component accumulation rates in the sediments than in sediment traps.

Gauns et al. (2005) studied the biological productivity and carbon fluxes to the deep Bay of Bengal and the AS and reported higher POC fluxes in the AS (224–284 mg C m^{-2} day^{-1}, average 250 mg C m^{-2} day^{-1}) during the southwest monsoon than in northeast monsoon (135–260 mg C m^{-3} day^{-1}, average 192 mg C m^{-2} day^{-1}) and the spring inter-monsoon (161–202 mg C m^{-2} day^{-1}, average 182 mg C m^{-2} day^{-1}). Further, the amount of carbon reaching a depth of 1000 m shows seasonal differences: 0.09 mg C m^{-3} day^{-1} during the spring inter-monsoon, 0.24 mg C m^{-3} day^{-1} during northeast monsoon, and 0.37 mg C m^{-3} day^{-1} during southwest monsoon. On the basis of the characteristics of OM, Bange et al. (2005) summarized important processes, such as denitrification, N_2 fixation, and nitrous oxide emissions, and reported that the AS is a major source of N_2O and sink for fixed N, and attributed this to enhanced rates of denitrification that take place in the suboxic part of the water column. They also suggested that water column denitrification is much higher compared with sedimentary denitrification, and additions of fixed N via N_2 fixation are also small compared with pelagic denitrification. Monsoon forcing results in nutrient input owing to upwelling in summer, and convective mixing in winter that sustain high rates of CO_2 fixation in the surface layer and supports the export of particulate OM to the deep sea (Rixen et al., 2005). The northern Indian Ocean has a continental boundary to the north that prevents the deep convective regime; as a result, the O_2 demand for OM decomposition cannot be met because of less oxygenated intermediate waters.

The sediment community oxygen consumption (SCOC) *in situ* measurements were conducted during 1995–1998 at five abyssal sites with a depth range of 3190–4450 m to understand the variations of benthic carbon remineralization (Witte and Pfannkuche, 2000). The studies indicated that the SCOC ranged from 0.9 to 6.3 mmol O_2 m^{-2} day^{-1} and was considerably higher in the western, northern, and eastern AS sediment trap stations than in the global ocean abyssal areas. The POC flux (vertical) of benthic carbon remineralization (BCR) is only 20–50%. High rates of BCR are related to the faster rate of deposition of very labile OM resulting from dust particles (aeolian), enhanced rain efficiencies, and lateral advection. Major temporal variations in SCOC were noted in stations of the western and eastern AS. There was a strong seasonality with lower particle fluxes during the inter-monsoon than in monsoon periods in the northern Indian Ocean (Ramaswamy and Nair 1994). Results of POC flux point to only 4.3–8.2% of the carbon fixed by primary producers being exported to a depth of 100 m, as revealed in the 1995 US JGOFS AS process study, which is similar to that in the equatorial Pacific (Lee et al., 1998). Ramaswamy et al. (2005) collected suspended particulate matter (SPM) using drifting sediment traps in the northern AS and reported export flux derived from ^{234}Th removal rates in the water column and POC/^{234}Th ratios measured in settling particles. They measured PP at eight depths between the surface and 120 m in February 1997 using ^{14}C uptake in bottles that were incubated *in situ* during the daytime. The average export flux of carbon was ~332 dpm m^{-2} day^{-1} which represents 36% of the daily PP. The average ^{234}Th flux out of the photic zone was 2300 dpm m^{-2} day^{-1} and the average POC/^{234}Th ratio in trap-derived particles was 0.14 mg dpm^{-1}. Export of about one-third of the daily PP at the end of the winter monsoon is related to the episodic nature of Salp swarms. Salp swarms in the AS are considered to be an important pathway for the quick export of carbon from the euphotic zone. Off Goa, at intra-annual variations in δ^{13}C and δ^{15}N in the water column, particulate organic matter (POM) showed very low δ^{15}N values (up to –4.17‰) during the pre-monsoon season (Maya et al., 2011). The δ^{15}N of POM appeared to be lower than the mean value (7.38‰) for surficial sediments, most probably because of diagenetic enrichment. Based on results, they suggested that the sedimentary

$\delta^{15}N$ may not always be related to the denitrification intensity signature of the overlying intermediate waters. During the southwestern monsoon, when a large input of terrestrial OM is expected through runoff from land, the OC/TN ratio remains low, but a significant difference was observed between $\delta^{13}C$ data in 2007 and 2008. Inputs of soil OM may have elemental and isotopic signatures that differ from those of the conventional (C3 plant-derived) OM and could explain the constancy of the OC/TN ratio.

Using sediment trap experiments along with satellite observations, Rixen et al. (2019) illustrated the influence of PP and ballast effect on OC flux. They suggested that most of the sites in the Indian Ocean PP are the main control on the spatial variability of OC fluxes except at sites influenced by river discharge: here, the lithogenic matter content influences the spatial variability of the OC flux. Flux calculations suggest that lithogenic matter ballast enhances the sinking speeds and reduces the time of OM respiration in the water column with increasing OC fluxes in the Indian Ocean. The increased input of lithogenic material as a result of changes in land use may significantly affect the CO_2 uptake of the OC pump in receiving areas of the ocean. Lithogenic matter can enhance OC flux rates by 45% and in the river-influenced locations by up to 62% at trap locations. The lithogenic matter ballast may increase the mean sinking speeds by 52% and the mean OC flux by 45%.

5.7.2. Organic Carbon in Surface Sediments

Studies on Organic Carbon Before 1980

Trask (1932) was the first to report OC content of 1–1.3% in surface sediments of the AS. These sediments are presently kept in the British Museum. London, UK. The high OC content in the AS sediments was believed to be the result of high productivity in the water column. The highest OC content (2.2%) was reported in the sediments off Mumbai (Trask, 1932, Wiseman and Bennett, 1940). Subsequent studies on OC were on the sediments collected during or after the International Indian Ocean Expedition (IIOE: 1961–1965). In 1961, 52 bottom samples were collected on a grid pattern (at one-degree square) as a part of the IIOE from the northern AS using the US Naval Oceanographic vessel, *USS Requisite.* Menzel (1964) reported dissolved OC concentrations at depths between 0 and 2000 m along a section from 20°S to 20°N in the western part of the AS and found no correlation between primary production and dissolved carbon in the euphotic zone. Carbon concentration along with temperature/salinity was used to characterize deep water masses. They reported a period of a minimum 260 years to produce the OC present in the AS, allowing for 90% conversion of total particulate OM to carbon dioxide by grazing organisms. Menzel (1964) concluded that carbon in solution and in particulate form in the ocean is stable and subject to limited change by biological action. On the basis of 52 sediment samples, Stewart et al. (1965) prepared a contour map of the northern AS with OC ranges between 1% and 6% (Figure 5.3). Later, more detailed studies were carried out on primary productivity off Kerala. Productivity in the water column off the Alleppey shelf during the post-monsoon period in 1967 was 806.6 mg C m^{-2} day^{-1} (Radhakrishna, 1969) and diatoms constituted 88% of the phytoplankton during upwelling. Murthy et al. (1969) reported a range of OC values (0.5–6.45%) in surface sediments at depths between 16.5 and 676.7 m off Mumbai to Alleppey, with the highest values off Mumbai. Here, high OC content is related to high productivity in overlying waters. The coarser-grained sediments on continental slope associated with *Globigerina* and *Globorotalia* were interpreted as favorable conditions for the deposition of high OC. Further, the glauconite presence within foraminiferal tests was related to the reducing environment. Texture of the sediments in this region was found to favor oxidation of OM by infusing of water through sediments. There is a mention of low oxygen intermediate water by Murthy et al. (1969), but its relationship with OM preservation is not explained in detail. RaghuPrasad (1967) reported the relationship between organic and/or plankton production and their relation to large-scale fishing along the Indian coast. They observed that the organic production peak periods correspond to low periods of fishery and vice versa, with high fishery after a production period of high OM (time lag

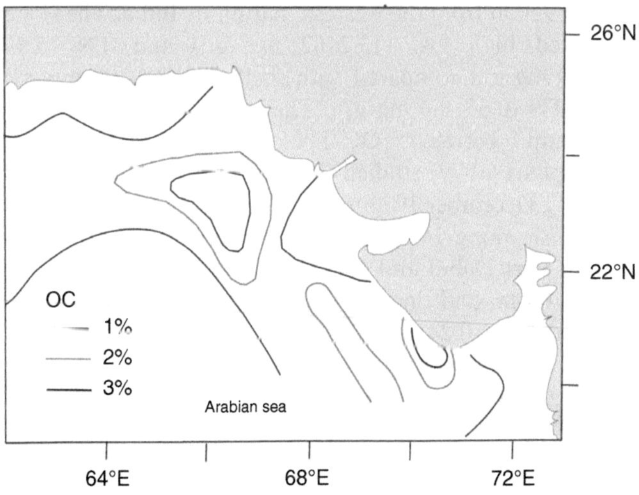

Figure 5.3 One of the oldest contour maps showing organic carbon distribution in the surface sediments modified after Stewart et al. (1965). OC, organic carbon.

taken for conversion of OM to fisheries). In that study, it was revealed for the first time that there is a region with a very high rate of production off the southeast coast of India that is comparable to most productive regions of the world (Raghu Prasad, 1967).

Nair (1970) reviewed and compiled the available information on productivity and related factors in the Indian seas. The continental shelf of India covers an area of ~57 million hectares and contains 61 × 10^6 tonnes of carbon; of this, three-quarters is from the west coast of India. Yield as a percentage of carbon on the west coast of India is 0.225% C, which is much higher than the world ocean (average 0.030% C; Nair, 1970). Qasim and Sankaranarayanan (1970) studied the concentration of POC in the vicinity of the Kavaratti Atoll and found that it is much higher (342 C mg m^{-3}) than that of the surrounding sea (average 225 C mg m^{-3}). About 20% of the total gross production of POC is largely produced by the coral reef. Szekielda (1970) calculated the liberated energy from the oxidation of organic material using stoichiometrical models. Organic matter in the oceans is produced during photosynthesis by the uptake of solar energy. This energy is later liberated during the oxidation of the organic substances. The greatest liberated energy is found in the region with the greatest oxygen consumption and denitrification. Rajamanickam and Setty (1973) reported on the OC content in surface sediments off Goa and showed low OC (0.85%) at 5–10 m depths, high OC (>5%) at 10–20 m depths, and adifferent range of values (high and low) between 20 m and 50 m depths (Figure 5.4). They found a proportional relationship of OC with clay size sediments. Phosphorus content in sediments during the post-monsoon periods was related to the supply of apatite transported by rivers. Bhosle et al. (1978) reported OC and TN in 34 surface sediments at depths between 17 and 1950 m from the western margin of India. The slope showed high OC (13.2–62 mg g^{-1}) and TN (5.4–17.99 mg g^{-1}) compared with shelf OC (3.6–4.8 mg g^{-1}) and TN (0.32–4.4 mg g^{-1}). They observed a direct relationship between OC-TN and OC-carbohydrate. Paropkari (1979) studied 90 sediment samples collected during December 1973 to May 1974 from cruises of *INS Darshak* along 14 sections of the continental shelf between Port Dabol and Indus Canyon at depths of 20 m and 100 m, and one sample at 250 m. The OC content varied from 0.24 to 2.24 %, with an average value of 0.95%, which corresponds to 1.64% of OM, and is lower than the world average (2.5%) for nearshore sediments. They reported much lower OC content in the northern Indian shelf compared with the other upwelling areas such as southwest coast of India, the continental shelf of West Africa, and Peru. The low OC in the northern Indian shelf is related to the oxidation of OM resulting from a high DO content or grazing by zooplankton and

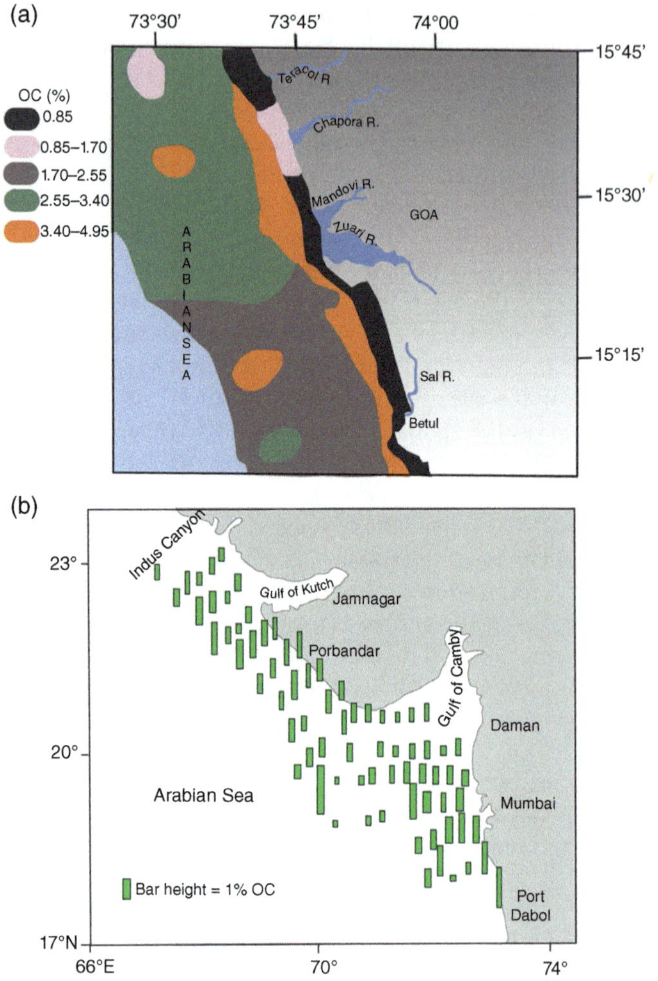

Figure 5.4 (a) Distribution of organic carbon (OC) in the surface sediments off Goa, modified after Rajamanickam and Setty (1973). (b) The OC content in the surface sediments modified after Paropkari, (1979).

higher organisms. The grazing is suggested to be the most likely reason as high zooplankton biomass was recorded in this region.

Studies on Organic Carbon from 1980 to 2000

Kolla et al. (1981) prepared the first detailed map (Figure 5.5a) of OC distribution in sediments of the AS. The important findings of this study are: (i) the overall high OC in the sediments close to the land, and (ii) the highest OC values (5–10%) at depths between 200 and 1500 m on the western continental margin of India. The findings are primarily related to preservation, which results from the impingement of low-oxygenated waters on the seafloor and high sedimentation rates. This is the first mention of OM preservation related to low-oxygenated waters. Calculations suggest that there is 15–80 times greater preservation of OM along the Indian

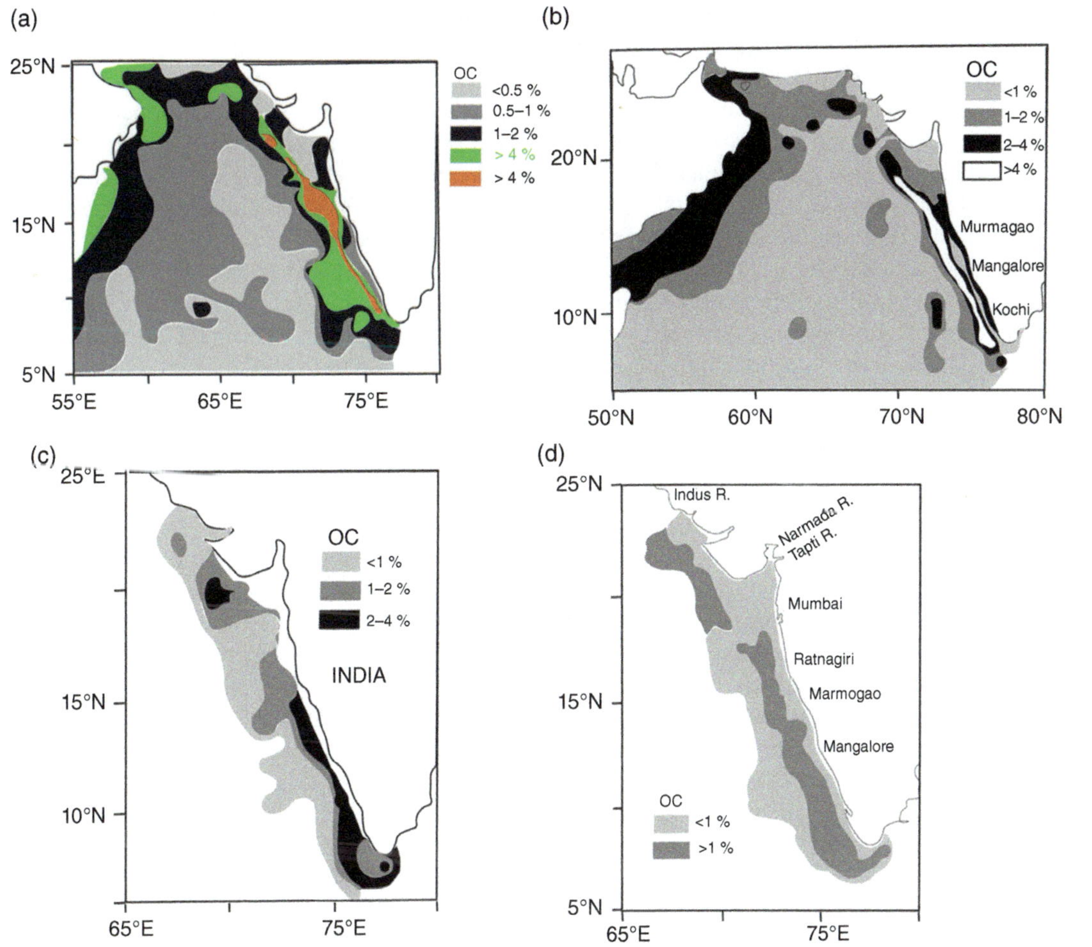

Figure 5.5 Maps showing the distribution of organic carbon in the Arabian Sea: (a) Kolla et al., 1981; (b) Paropkari et al. (1992); (c) Naidu et al. (1992); (d) PrakashBabu et al. (1999).

margin compared with the region off the Arabian Peninsula. Sediments from the continental slopes of India showed high OC (5–10%) compared with that off Arabian Peninsula and Somalia (<4.0%). Later, Qasim (1982) compiled data based on 13 cruises of IIOE from 1960 to 1965 (samples from all months except October), four cruises of *R. V. Gaveshani* (samples collected during February, March, October, and November 1976) and five cruises of *INS Darshak* (samples from December 1973, January, February, March, April, and May 1974) and reported two oxygen minima, one at a depth of 100–400 m and second at 800–1500 m. The formation of OMZ at shallower depths (100–400 m) is related to high organic production and sinking of a huge amount of OM. The land-locked nature of the sea lacks horizontal advection and the upper water layers have high salinity. In contrast, the OMZ at deeper depths (800–1500 m), is related to physical processes in the AS. In the northern AS, primary production is ~835 mg C m^{-2} day^{-1} or 530 or10^6 t C $year^{-1}$, with about 25% of the total photosynthetic production taking place in 8% of the area. In this area, zooplankton biomass is much higher than the rest of the AS and mostly limited to the upper 200 m. The OM on the western shelf of India is mostly detrital in the inner shelf and marine in the upper slope (Paropkari, 1979, 1987).

Paropkari (1990) studied 136 surface sediments from the shelf and upper slope off Mangalore–Cochin and reported varying OC contents (0.04–0%). Phosphate in these sediments is associated with OM. Trace elements (Ni, Cu, and Zn) were correlated with OC suggesting a biogenous source. The high OC (up to 10%) in the inner shelf (1.86%) and upper slope (4.11 %) sediments are associated with clays and related to the high productivity in overlying waters (0.50–0.75 g C m^{-2} day^{-1}; Qasim et al., 1978). They suggested better preservation of OM in fine-grained sediments and higher sedimentation rates, resulting in rapid burial. In addition, the impingement on the OMZ (150–1500 m water depth) results in a strongly reducing environment and subsequent preservation of OM. In contrast, low OC content (average 0.68%) in outer shelf sediments is mainly attributed to a high dissolved oxygen content of bottom waters (4–5 mll^{-1}), coarse-grained sediments, which permit easy diffusion of oxidizing waters and, to some extent, washing out of the fine-grained organic material by strong currents. Rao and Murty (1990) studied 226 sediment samples from the central west coast of India between 17° and 13°N. Major and trace elements of these sediments were compared with the OC data of Paropkari et al. (1987). The study revealed that the clays, Fe, Mn oxyhydroxides, OC, and/or sulfides act as carriers for most of the elements. Organic carbon and/or sulfides played a major role in fixing excess metal content from the slope sediments. Rao and Murty (1990) suggested the use of high concentrations of OC, PO_4, Ni, and Zn contents and low quantities of Mn to identify upwelling environments in slope sediments. Naidu (1990) reported OC and $CaCO_3$ contents, planktonic foraminifers, and radiolarians in 21 surficial sediments at depths between 275 m and 2050 m off Marmugao–Karwar in order to report the upwelling signature. The OC content ranged between 0.26% and 5 %, with high OC (>3%) in the sediments of the upper slope and low values (<1.5%) in the lower slope sediments (below 1000 m depth). The relative abundance of planktonic foraminifers and radiolarians was correlated with high $CaCO_3$ and OC in the upper slope (<1000 m) inferring that the intensity of upwelling is greater on the upper slope. Reemtsma et al. (1990) reported fatty acid and OC content in particulate matter from the Central AS for the summer season and found that OC content and fatty acid decrease with increasing particle fluxes. Particulate organic matter during high productivity and high flux periods show signs of a lesser amount of degradation within the surface layers. The biological alteration of sinking particulate matter appears to occur mainly on particles. The differences in fatty acid composition in the particle flux collected at 732 m and 2914 m indicate that organisms, especially bacteria, are involved in the degradation of POM. Somasundar et al. (1990) used available data from the AS (0–25°N, 50–80°E) to estimate the C budget, taking into consideration the possible sources of fluxes. Annual fluxes of carbon into and out of the AS are estimated to be 446 and 530 Tg, and the C budget is negatively balanced by 84 Tg year^{-1}. The northward movement of Antarctic bottom water enriched in CO_2 may compensate for this deficit. The amount of carbon escaping into the atmosphere (as CO_2) from the AS is higher (~74 Tg year^{-1}) than the tropical ocean's global average fluxes. The loss of carbon to the atmosphere is 14.0%. Fluxes of C from rivers and burial of C through the sedimentation process are not significant to the total flux of C. The AS was found to be a C source for the Persian Gulf and the Red Sea and the residence time of C in the AS is around 944 years based on the standing crop and net outfluxes.

Naidu et al. (1992) studied 63 surface sediments from the entire west coast of India collected during various cruises of *O. R. V. Sagar Kanya* and *R. V. Gaveshani*. They reported high OC (2–4%) in the upper slope off southern India between the southern tip of India and Marmugao (Goa) and relatively low OC (1–2%) in the northern slope, except off Saurashtra (Gujarat), where the OC values were 2–4% (Figure 5.5c). Sediments in the lower slope had a low OC content (<2%). The abundance of foraminiferal species and OC is related to productivity and upwelling. Paropkari et al. (1992) studied OC distribution in the more than 600 surficial sediments from the AS (Figure 5.5b). They observed that the OC content somewhat reflects the surface productivity in the regions and is associated with low oxygen in the water column. The OC content on the slope of the Indian western margin is two- to four-fold higher (maximum 16.71%) than the slope of the Arabian Peninsula (maximum 7.54%), but the productivity in the latter region is three to four times greater than in the former. They suggested that the control of productivity on OC enrichment is of minor importance. This was the first report where high OC (> 4%) exactly coinciding with the OMZ was reported, and clearly reflects the significance of the anoxic bottom waters in the preservation of OC. It is of note that the organic-rich band on the western Indian slope is wider, longer, and more concentrated compared with that of the Arabian Peninsula slope. Although equally intense oxygen minima impinge on the surrounding continental slope of the AS, unlike the Indian margin no organic enrichment is found on other continental slopes. The enrichment of OC on the slope is attributed to factors such as shelf width, sediment grain size, sedimentation rates, bottom currents, the gradient of the slope, and the adsorption capacity of clays. Further, Paropkari et al. (1992) suggested that productivity is not a major control factor in OC enrichment of bottom sediments and it is the bottom water anoxia in conjunctions with various depositional parameters that determine the degree of preservation. In the same year, Pedersen et al. (1992) reported OC measurements at 0–1 cm intervals in 14 box cores in the top sediments from the continental shelf-upper slope area off Oman. Based on carbon and nitrogen, $\delta^{13}C$ and Rock-Eval measurements, Cr/Al, Zr/Al, and I/OC, ratios, they suggested that the OM is of marine origin. Lack of significant correlation was found among the abundance of sedimentary OC, OC:TN, and hydrogen index (HI) with that of the bottom water O_2 concentration. They argued that the OMZ on the Oman Margin does not help in the preservation of organic material in the sediments. Instead, they suggested that the post-depositional fate of OM in the Oman margin (highly productive area) could be controlled largely by winnowing processes and reworking. They further suggested that active bottom currents could erode the clay and silt and influence the OM from the continental slope off Oman. Later, Paropkari et al. (1993a) disagreed with Pedersen et al. (1992) and commented that the productivity in the Indian margin is almost three-fold lower and OC three- to five-fold higher, compared with the Oman margin. By referring to Kolla et al. (1981), who reported that the degree of preservation of OM along the Indian margin is 15–80 times higher than that of the Oman margin, these authors suggested that it is the result of impingement of low oxygen waters on the slope and high sedimentation rates. They also proposed that the interaction of various factors, such as the texture of sediments, sedimentation rate, overlying water productivity, shelf width, bottom currents, the gradient of the slope, and the nature of clay minerals in an exclusively anoxic setting play a decisive role in the preservation and spatially variable concentration of OC on the slopes of the AS. Further, Paropkari et al. (1993b) tested the validity of the hypothesis that OC preservation is favored by the anoxic waters of the OMZ by studying 25 surficial sediments in the depth range of 85–2297 m, covering the OMZ and the oxic layers above and below the it from the western margin of India: the OC content was found to vary between 0.14% and 6.18%. They used rock-evalpyrolysis, which indicated that OM on the slope is hydrogen-rich, Type II kerogen, and mostly marine. They also suggested that the overlying waters had low productivity and could not be the possible cause for the enrichment of OC. The high HI values (average 306) associated with high OC content (average 3.28%) in the upper slope sediments in contact with the OMZ indicates enhanced preservation under reducing conditions in the anoxic waters of the OMZ. It is also suggested that the sediments of the upper slope in contact with the OMZ have excellent hydrocarbon generation potential, and that the slope beds should be studied for hydrocarbon exploration. Calvert et al. (1995) disagreed with Paropkari et al. (1992), where the sedimentary OC maximum is said to result from preferential preservation of OM, where OMZ intersects with sediments on the seafloor. Calvert et al. (1995), by analyzing sediments with more extensive geographical coverage, reported that the lowest oxygen levels occurred at depths between 200 m and 800 m, whereas maximum OC and nitrogen contents occurred at depths between 200 m and 1600 m. The OM was found to be marine based on OC/TN ratios, and because the $\delta^{13}C$ and HI values for the sediments within and outside the OMZ are similar. Taking into account the fact the OC maximum extends over a greater depth range than the OMZ, Calvert et al. (1995) suggested that there is no evidence for preferential preservation of the OM within the OMZ. They emphasized that the distribution of OM on the western Indian continental margin is controlled by (1) the supply of OC that decreases away from centers of coastal upwelling and decreases with increasing water depths, (2) dilution by other sediments, and (3) texture of the sediments. Further, they suggested that the oxygen minima cannot be used to explain the preservation of OC at intermediate palaeo depths in the geological record. Later, Rao and Wagle (1997) also reviewed the sedimentary processes associated with the sediments of the continental shelf and slope off western India. PrakashBabu et al.(1999) studied OC in 112 surface sediments at depths of 28 m and 3700 m along the eastern AS. The OC content varied from 0.70 to 5.86% (Figure 5.5d), with the highest values (5.12–5.86%) on the marginal highs, with intermediate values (3.53–4.22%) on shelf margin, and lower values (1.80-3.84%) on the continental slope and Arabian Basin (0.98–1.46%). The OC distribution in the surface sediments matches well with the productivity variations in the region. They suggested that productivity to be important for OC accumulation. They did not rule out the possibility of the influence of the OMZ and other controlling factors such as clay minerals, sediment texture, dilution, sedimentation rate, and topography for higher OC in sediments. They added that the diverse topographic features and associated hydrodynamic processes play important roles in the enrichment of OC.

Studies on the Distribution of Organic Carbon from 2000 to 2020

By the year 2000, researchers had linked high OC content in sediments to other geochemical and isotopic parameters. Farrenkopf and Luther III (2002) measured dissolved iodine species and total iodine concentrations during the spring inter-monsoon of 1995 with the objective of assessing iodine as a proxy in the AS carbon and nitrogen processes. The study highlighted two separate regimes of iodine chemistry: (i) iodide concentrations were in the range 58–558 nM in the well-oxygenated surface layer (WOSL) and (ii) iodide concentrations varied from ~200 to 950 nM in the OMZ. Estimates showed that an excess total iodine of only 3.6% can result from remineralization of sinking particulate organic iodine (POI) from the surface waters to the OMZ. About ~96% of the total excess iodine to the OMZ could be advected from the margin sediments of India. Sirocko et al. (2000) studied 37 deep-sea sediment cores from the AS at four age-intervals during the last glacial and Holocene, although they had few samples covering the north-western margin of India. However, this study is important as the distribution patterns of OC concentration in three ages were reported for the first time. The surface sediments in the deep AS revealed that the lowest OC values were associated with low sedimentation rates and low biogenic productivity. High OC values were found to be associated with high surface productivity and high bulk accumulation rates in the western AS. Reddy (2003) reported OM content from the silt-clay fraction of 48 sediments from the continental shelf off Kochi. Organic matter content was higher in the recent, near-shore silt-clay-dominated sediments towards Alleppey (average 3.8%) than in those towards Azhikode (average 1.97%). The OM content in the near shore sediments (up to 30 m water depth) varied from a minimum of 1.03% to a maximum of 3.99%. The OM content in the offshore (water depth greater than 30 m) sediments is very low, ranging from 0.8 to 1.74%, with an average of 0.71%. The high OC in the inner shelf sediments is mainly controlled by the texture of the recent sediments in conjunction with other parameters such as detrital organic supply and productivity in the overlying waters, while in relict sediments the low OC content is related to high-energy conditions, high DO, along-shore currents, and sandy grain size of the sediment. It is known that OM can also be used as a tool to identify the recent and relict sediment zones. The sand-dominated sediments at a depth of 30 m are from the mid-Holocene strandline deposit that extends continuously all along the study area parallel to the coast from Azhikode to Alleppey. Cowie (2005) reviewed the literature on the biogeochemistry of surficial sediments from the AS up to the year 1990 and focused on benthic, biological, and geochemical studies. The results highlighted that the AS benthic system is highly dynamic, with strong benthic–pelagic coupling that seen as a cross-basin trophic gradient and in benthic response to seasonal productivity variations and carbon flux. Benthic biogeochemical processes such as phosphogenesis, denitrification on the upper slope within the OMZ, and fluxes of nutrients, dissolved OM and trace metals may be of global significance but remain poorly quantified. Oxygen availability becomes a primary environmental control on the benthos in the laminated sediments of the OMZ on the continental margin off Pakistan and India only when the oxygen levels fall below the threshold. One important suggestion is that the carbon cycling mechanisms and differences in the efficiency of carbon burial, in contrast to redox settings, could be an answer to what the primary controls are on OC preservation and distribution. Organic carbon, TN, proteins (PRT), and total hydrolysable carbohydrates (TCHO) from the shelf sediments showed seasonal changes (Jacob et al., 2009). During late summer, the average OC content in the surface sediments was 12.78 ± 9.77 mg g^{-1}. Organic carbon along the outer shelf was a maximum of 36.70 mg g^{-1} at 15°N and 28.38 mg g^{-1} at 17°N.During the pre-monsoon period, the average TOC was 10.84 ± 4.63 mg g^{-1} with an exceptionally high value at the outer shelf (35.34 mg g^{-1}) at 15°N. During late summer, the persistence of upwelling resulted in low oxygenated (DO <180 μM), cold (SST <28°C), nutrient-rich (NO_3>2 μM) surface waters, and the subsurface suboxia, whereas, during the pre-monsoon period, oligotrophic and oxygen saturated (DO >200 μM; NO_3 <2 μM) conditions prevailed. The OC/TN ratios indicated the presence of marine OM during both seasons. The OM was aged as indicated by TCHO + PRT:OC and PRT:TCHO ratios and did not change between the seasons. Coupling between OM and surface ocean productivity during the late summer was indicated by high OC and reactive OM (TCHO and PRT) concentrations in surface sediments along the active upwelling regions (13–17°N).

Based on macrofaunal distribution on the western continental margin of India, Ingole et al. (2010) reported the highest amounts of chlorophyll-a (Chl-a) and OC in the mid-slope areas. Faunal abundance and biomass decreased in the slope region as a result of the reduced oxygen in the bottom-water. Sajan et al. (2010) studied OM in the sediments at depths of 30, 50, 100, and 200 m along 17 transects along the western margin of India (between Dwaraka to Cape Comorin). The OM content varied from 0.2 to 6.7% in 30–50 m depths, 0.4–5.5% in 51–100 m depths, 0.9–5.1% in 101–150 m depths, and 0.8–6.2% in >150 m depths. Epibenthic megafaunal zonation studies across the OMZ of the Indian continental margin indicated changes in the megafaunal assemblage and were correlated to both oxygen availability and sediment OM (Hunter et al., 2011). Kurian et al. (2013) studied OC, $CaCO_3$, major elements, trace and rare earth elements (REE) in surface sediments from 13 transects from the western continental margin of India, between Dwaraka in the north to Cape Comorin in the south. The studies indicated that, based on OC/TN ratios and $\delta^{13}C$ values of OC, the OC is of marine origin. Ramya et al.

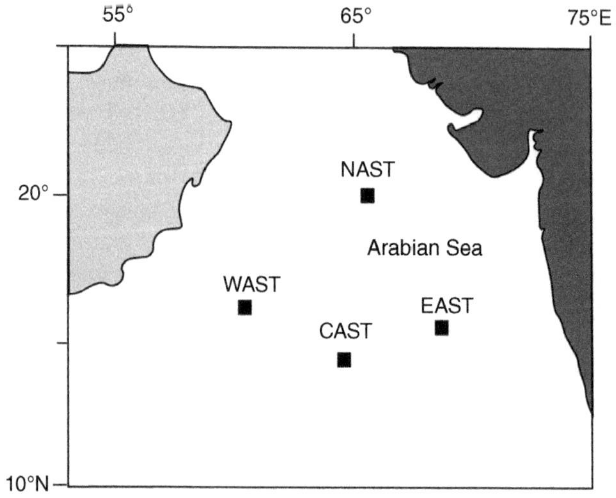

Figure 5.6 Location of sediment traps in the Arabian Sea: NAST, northern trap; WAST, western trap; CAST, central trap; EAST, eastern trap.

(2013) reported the composition and variability in heterotrophic bacteria from the shelf sediments of southwest of India and its relationship with the sediment biogeochemistry. The OC content in the estuarine sediments of the west coast of India (Pradhan et al., 2014) indicated abundant terrigenous OM contribution (80%). The OC content was high (0.5 to 5%), and nitrogen content was low (with OC/TN ratio >15–20) in the estuarine sediments. Johnson et al. (2014) examined the inventory and sources of OC in the sediments of the Kerala–Konkan Basin and found dominant marine OC. Robin et al. (2012) reported on OC content in the nearshore (2 km, 5 km, and 10 km) sediments across six transects along the southern Kerala coast. The OC content ranged from 14.65 ± 0.37 mg g^{-1} to 1.79 ± 0.87 mg g^{-1}, with the highest values in the transect off Cochin and the lowest values off Paravur. Sreekanth et al. (2015) studied 14 sediments along five transects off Mangalore–Cochin at depths of 30, 50, and 100 m in each transect to study trace metal enrichment and OM sources. The OC content was found to range between 0.14 and 4.67%. Cheriyan et al. (2015) reported on the OC along four transects, covering the inner and outer shelf off Cochin–Trivandrum at depths between 30 m and 200m. The OC content ranged between 0.12 and 6.81%, with the lowest values found in the nearshore sediments off Trivandrum and the highest values off Kollam. A high OC content is related to minor dilution with the terrigenous fraction and high upwelling induced phytoplankton productivity, and an additional supply of OM to the shelf. High OC content associated with fine-grained sediment is related to the capability of cohesive sediments with higher retention. Low OC content in the near coastal regions is related to coarse (sand-sized) sediments. Acharya and Panigrahi (2016) put together the published and unpublished data on OC, using geostatistical analysis in the eastern AS, and suggested that primary productivity may not be responsible for a high OM content. The inverse correlation of bottom water DO with OC, and the Ti-normalized fraction of OC is related to the importance of the role of anoxia in the preservation of OM. Organic carbon accumulation rates are strongly dependent on the sedimentation rate, and mass accumulation rates of OC that show a strong dependency on sedimentation rates and are important controlling factors in the EAS OM accumulation. Chakraborty et al. (2016) investigated surface sediments off Gujarat at depths between 80 and 1544 m. They reported that the OC content (1.06–4.38 %) increased with increasing fine-grained sediments and decreasing DO levels. With increasing OC content (under suboxic overlying water) there is increased accumulation of Hg in the sediments. In these sediments, with increasing OC and total Hg concentrations, there is a decrease in the concentration of Hg-methylation (MeHg). Overall, Hg-methylation in the surface sediments is controlled by the nature of the OC in the OMZ of the AS. Enge et al. (2014) studied benthic foraminifers that are exposed to particularly severe levels of oxygen depletion within the OMZ sediments on the western margin of India. For the benthic community, high food supply is in distinct pulses during water mixing (winter monsoon) and upwelling (summer monsoon). An *in situ* isotope labeling experiment ($\delta^{13}C$,$\delta^{15}N$) was performed in the OMZ core region with <4.5 μmol l^{-1} DO to study the response to the pulses of food on the benthic foraminifera. Benthic foraminifers of the Indian margin were found to be highly adapted to low oxygen and can utilize OM at low oxygen concentrations (<0.1 ml l^{-1}). Bao et al. (2018) observed that the age of ^{14}C in OM systematically increased with a fine grain size of <20 μm, and decreased in coarsest grain size fraction of >500 μm in deep-sea sediments from the AS (4000 m). Bulk sediment ^{14}C of OM in specific grain sizes and protein thermal OC decomposition products suggest that grain size is a significant factor controlling ^{14}C age in OM in deep-sea sediments of the northwest AS. In marine surface sediments, ^{14}C signatures of OC are highly variable, but the causes for this heterogeneity are unclear.

Lengger et al.(2019) studied dark carbon fixation from the OMZ of the AS, which contributes to sedimentary OC. Biogeochemical models are used for estimation of carbon drawdown; however, carbon fixation in the mid- and lower water column are ignored. Chemoautotrophic carbon fixation manifests in a ^{13}C-depleted OC signature in the AS OMZ sediments. Dark carbon fixation rates are determined experimentally in the AS by Jensen et al. (2011). Combining the $\delta^{13}C$ values of bacteriohopanetetrol stereoisomer (BHT), a biomarker (anammox bacteria), and $\delta^{13}C$ of sedimentary OM are used to calculate the contribution of anammox to sedimentary organic matter (SOM) in the AS and it comes up to 30% of OM deposited within the OMZ. These bacterial contributions result in a shift in the $\delta^{13}C$ values of OM by 1–2‰ in the sediments, but under what conditions and how well these signatures are preserved remain unclear. From results, the burial of carbon in OMZ by the chemoautotrophs (e.g. anammox bacteria) is much higher than previously believed, as the higher remineralization rates are higher than inferred from OM decreases. Therefore, while evaluating past occurrences of OMZs based on $\delta^{13}C$ values of SOM, there is a need to consider the ^{13}C-depletion contributed by bacteria. The chemoautotrophic process represents a mechanism of CO_2 removal by carbon fixation from the pelagic water column which contributes to sinking OC fluxes. Chemoautotrophs should be included in biogeochemical models for predicting feedbacks to a warming world as it explains the carbon budget mismatches observed between experimental and modeling estimates.

5.7.3. Variations in Organic Carbon in the Sediment Cores

Studies on variations of OC in sediment cores are helpful to better understand past changes in the OC content and use it as proxy for paleo productivity. Murthy et al. (1969) were the first to record OC distribution in sediment cores; OC decreased with depth in some cases but in the majority of the cases they observed unsystematic fluctuations. Unfortunately, these cores did not have any age control. Sarkar et al.(1993) reported past changes in the OC content in a sediment core from the southeastern AS. Organic carbon in the core is marine, with a $\delta^{13}C$ level >–20‰. Anoxic depositional environment was inferred based on OC enrichment and authigenic U (as there was removal of U from the sea water owing to anoxic conditions and precipitation antigenic U in the sediments) during the glacial time. Caratini et al. (1994) studied two sediment cores covering 3500 years off Karwar (Karnataka). The OC values were in the range 1–2.5%, with slightly higher values at the top of the core. Rostek et al. (1997) studied sea surface temperature (SST) and past productivity for the last 240 000 years in a core from the Gulf of Aden (AS). They found agreement among different palaeo-productivity proxies. In another core from the southeastern AS, glacial stages were associated with relatively high productivity and warm interstadials were associated with low productivity. Thamban et al. (1997) reported OC in six gravity cores recovered from different physiographic domains from the continental slope off Goa–Cochin at depths between 280 and 350 m. Organic carbon content varied from 0.94–8.7%, with higher values in Holocene sections than in Pleistocene sections in all the cores. The OC found was marine or a mixture of marine and terrestrial in the Holocene sediments. Productivity is the main controlling factor for OC distribution but reworking and texture of sediments also influenced the OC content. Schulte et al.(1999) studied two sediment cores: the core recovered off Pakistan from an OMZ with high productivity area and another core off the Maldives, located in moderate primary productivity and well-oxygenated bottom water. Palaeo-productivity changes were reconstructed in both the cores using biomarkers (alkenones, dinosterol, and brassicasterol). Results suggest that PP increased during glacial stages and decreased during interstadials. Variations in the OC in the core off Pakistan were mainly linked to changing bottom water oxygenation. In contrast, the accumulation of OC in a core off the Maldives is mainly controlled by PP. A proxy C_{35}/C_{31}-*n*-alkane ratio is used to infer the past redox conditions in the AS. The results showed that the intensity of the OMZ in the northeastern AS has varied considerably over the last 65, 000 years. Further, there were times where OMZ even vanished completely when oceanic circulation reduced during Heinrich and Younger Dryas events, suggesting the importance of oxygen-depleted intermediate water to sustain OMZ in addition to higher PP.

Bhushan et al. (2001) studied OC and nitrogen content from 11 sediment cores from the eastern AS. The OC content ranged from ~1% to 6% and was mostly marine, with low OC/TN ratios (6–15; average 9.3). Better preservation of OC is related to a faster deposition rate and higher detrital (clay) content. Burial fluxes of OC were higher (13.8 g m^{-2} $year^{-1}$) in the northern locations compared with southern locations (2.14 g m^{-2}/ $year^{-1}$). Thamban et al. (2001) studied sediment core from the southwestern continental margin of India. The OC content varied from 1.3 to 3.8% in the sediments for the past 18 000 years. The high OC content during the glacial period (~18 000–15 000 years BP) was attributed to high primary productivity during the winter monsoon. Reduced surface productivity during the early Holocene (13 000–6000 years BP) was related to the water circulation system along the southwestern margin of India. Pailler et al. (2002) studied sediment core from the AS (eastern equatorial), covering age of 350 000 years. They used metals (U, Cd, Mo), OC, and alkenone concentrations and observed a cyclicity of 23000 years in PP resulting in coupled variations in inorganic and organic constituents. They also suggested that the high productivity was in phase with low northern hemisphere summer insolation and is dominated by precessional forcing.

Agnihotri et al. (2002) used OC and TN concentrations to deduce surface productivity variations and monsoon intensity for the last millennium based on cores from northeastern AS. They observed coupling between productivity, monsoon intensity and total solar irradiance (TSI). Agnihotri et al. (2003a) also studied three sediment cores, at different water depths from the eastern AS, using $CaCO_3$, OC, TN, elemental concentrations of Sr, Ba, Al, Fe, Mg, Ti, Mn, Cr, and V in sediments, to assess the paleoproductivity changes and factors influencing OC accumulation due to glacial–interglacial oscillations. The OC content in the cores ranges from 0.11 to 2.25%. Significant enrichments of OC and TN during the Last Glacial Maximum (LGM) were related to the input of organic-rich sediments from the shelf during lowered sea levels and preservation as a result of increased sedimentation rates. Agnihotri et al. (2003b) studied a sediment gravity core recovered off Mangalore at a water depth of 600 m. The OC and TN contents in the core ranged from 1.3 to 5.7% and 0.2 to 0.7 %, respectively. The $\delta^{13}C$ content of OC (–21.8 to – 19.5‰) indicates marine OC. Changes in surface productivity during Holocene are related to an increase in the intensity of monsoon and hence the intensity of the upwelling.

Pattan et al. (2003) reported low marine OC (<0.8%) with mass accumulation rates of <0.02 $g^{-1}cm^2$ per 1000 years from the southeast AS, with low values during isotope stages 1, 2, 3, 4, and 6. Agnihotri et al. (2003a), however, reported that OC distribution patterns neither coincided with productivity proxies nor exhibited any trend with interglacial–glacial periods. Organic carbon is low (0.3%) and of marine origin during interstadials of oxygen isotope stage 5. In contrast, the OC content is high (1.8%) and of terrigenous origin (fluvially derived) during stadials (5.2 and 5.4). Reichart et al. (2004) reported pronounced OMZ during interstadials and well oxygenated waters during stadials. They further reported good correlations between AS OC and Greenland Ice Sheet Project (GISP2) $\delta^{18}O$ records. Prabhu and Shankar (2005) reconstructed palaeo-productivity for the past 200 000 years, using proxies such as coccolithophore, OC and $CaCO_3$ contents in two sediment cores from the eastern AS. The OC content varied from 0.37 to 2.64%, with high OC content during glacial periods (marine isotope stage, MIS 2, 4, and 6). Palaeoproductivity proxies such as the percentage of *Florisphaera profunda* (coccolithophore) and OC indicate high productivity during the glacial periods. They disagreed with the finding of Agnihotri et al. (2003a) and suggested that high OC content is not an artifact caused by enhanced preservation of OM and high rate of sedimentation, but that PP is the main source for OC.

Guptha et al.(2005) studied carbonate and OC content in three sediment cores covering the past 140 000 years to infer productivity changes. Organic carbon in the northern core (0.015–0.97%) and central core (0.54–1.3%) showed no systematic trends with glacials and interglacials, while the OC content in the southern core ranged from 0.3–1.8%, with the highest values for MIS 5. Differential variation in $CaCO_3$ and OC content at the northeastern and southeastern AS during interglacials and glacial periods were related to regional differences in sedimentation rates, dilution and preservation. Tiwari et al. (2006) reported variations in OC in a core collected from the east of Maldive Islands (equatorial Indian Ocean). These variations represented preservation rather than surface productivity. The burial flux of OC closely follows the OC content, indicating the importance of a burial flux. Kessarkar and Rao (2007) studied OC content from the southeastern AS and found that it was greater in the core off Thiruvananthpuram (1.4–4.4%) than in the core off Mangalore (0.8–2.4%) (Figure 5.7). They suggested that the variations in productivity and downslope movement of sediment controlled OC enrichment. Singh (2007) studied a sediment core off Goa that had OC content of 1.2–5.6%, with the highest value at the top of the core. Thamban et al. (2007) studied laminated sediment cores collected within the OMZ off Maharashtra.

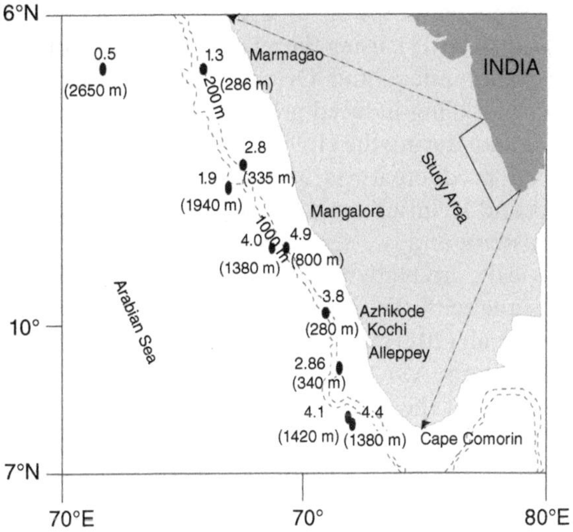

Figure 5.7 Organic carbon from the surface sediments (core tops). The organic carbon content (%) is given in italics and the numbers in brackets are the depths. Modified after Kessarkar and Rao (2007).

The accumulation rates of OC ranged between 0.54 and 2.12 g cm^{-2} per 1000 years. Agnihotri et al. (2008) studied gravity cores from the inner shelf region off Goa at a depth of ~45 m with seasonal OMZ, covering 700 years. Both the OC (2.32–3.4%) and TN (0.24 to 0.40%) were predominantly marine. Surface productivity varied with external solar irradiance forcing during the Little Ice Age (LIA: 650–1750 AD) with low productivity related to weakened summer monsoon. Kurian et al.(2009) studied 1-m long cores from the mid-shelf (depth 45 m) off Goa using biomarkers (dinosterol, phytol, stigmasterol, and β-sitosterol), OC, and $CaCO_3$. The OC content varied between 2.3% and 3.4%, whereas the $CaCO_3$ content varied from 4% to 13.6%. They attributed the increase in biomarkers during recent years to enhanced productivity, high solar irradiance, and anthropogenic inputs of nutrients: they suggested that solar forcing can control coastal upwelling strength and biological productivity in the eastern AS on timescales of decadal to centennial.

Kessarkar et al. (2010) reported productivity changes for the past 70 000 years in a sediment core recovered off Mangalore, using paleo-productivity proxies such as OC, TN, and $CaCO_3$. They reported high productivity at the core top, during the LGM and MIS 4, and low productivity during mid MIS1 and mid-MIS 3. Reduced primary productivity during warmer stages (mid MIS1 and MIS3) was related to intense SW monsoon associated with near-surface stratification. Rao et al. (2010) reported high OC (1.5 to 4.4%) in a sediment core from the southeast AS collected at a depth of 1420 m in oxygenated waters (below the OMZ) covering the past 35 000 years. Paleoproductivity records have been inferred using sediment cores off Goa for the past 80 000 years (Singh et al. 2011), using the relative abundance of planktonic foraminifera and OC content. They reported a higher upwelling-induced productivity during the glacial period than during the Holocene (it should be noted that reduced productivity is associated with Atlantic cold events and Heinrich events). Combining the results from the eastern and western AS, they reported that the biological productivity diminished during Heinrich events and with the overall year-round low productivity. Gupta et al. (2011) reported data from Hole ABP-25, 02 (northeastern AS) from 20 000 to 10 000 years, and observed that the deep northeast AS was better oxygenated with low a OC supply, except during the period 14 000–12 500 years, suggesting a weaker summer monsoon and a weak OMZ. Kessarkar et al. (2013) studied sediment cores off Mangalore covering an age of 16.7000 years found an OC content ranging from 0.6 to 5%. The lowest OC was observed in the early Holocene and increased gradually thereafter towards the late Holocene and was linked to ISM-induced productivity. On the basis of aragonite contents and abundance of pteropods in a core recovered from the OMZ, Naidu et al. (2014) suggested preservation of aragonite during Heinrich Events associated with the decline in monsoon-induced productivity and more northward movement of Antarctica Intermediate Water (AAIW). Rasiq et al. (2016) studied sedimentary pigments, OC, and TN contents, and their isotopes in three cores from the OMZ of the Eastern AS and reported that the OC content might not have much control over the preservation of sedimentary pigments. They also suggested that the bottom water DO and fauna controlled the accumulation and preservation of OC in OMZ sediments of the Indian margin. The OC was mostly marine, as seen from OC/TN ratios and the $\delta^{13}C$ contentof OC. Dissolved oxygen, sediment texture (grain size), and the effect of macrofauna controlled the preservation of sedimentary pigments and OM from the OMZ sediments.

Nagoji and Tiwari (2017) studied sediment cores covering age of 47 000 years from the Southeastern AS. Based on OC/TN ratio and $\delta^{13}C$ of OC plot, they reported a mixed productivity signal that limits OC as a proxy for productivity. The high OC content is related to better preservation in response to high sedimentation rates in suboxic bottom waters, and is indicative of intensified OMZs after mid-Holocene. Naik et al. (2017a) suggested that the land runoff, coastal upwelling, inputs (benthic and atmospheric) make the western Indian shelf waters biologically productive during the monsoon, resulting in high OM mineralization in the sediments. They suggested that dissimilatory sulphate reduction (SR) is the pathway of OM mineralization as a result of the depletion of other electron acceptors (O_2, NO_3, Mn, and Fe-oxides) just below the sediment–water interface. The top 10 cm had the highest SR rates, and depth integrated SR rate (0.066–0.46 mol m^{-2} $year^{-1}$) decreased with increasing water depth. The OC remineralized via sulphate reduction was ~0.52 mol m^{-2} $year^{-1}$ with an estimated average OC accumulation rate of ~5.6 (±0.5) mol m^{-2} $year^{-1}$. They suggested that the bulk of the OM gets preserved in sediments. Naik et al. (2017b) reconstructed paleoproductivity and bottom water oxygenation changes in the last 32 000 years, using the absolute abundance of planktic foraminifera, OC, $CaCO_3$, and *Globigerinoides ruber* $\delta^{18}O$ and $\delta^{13}C$ from the SE AS. The OC was 1.5–5.4%, with elevated OC values during most of the last glacial interval, indicating higher productivity in the eastern AS. The overall resultant carbon sequestration was confined only to a restricted zone and was not large enough to substantially alter atmospheric CO_2. Kurian et al. (2020) studied sediment cores recovered from the upper (500 m), middle (812 m), and lower (1168 m) OMZs from the continental slope off Ratnagiri. The OC was high (3.6–4.9%) in the lower OMZ core. The results suggest that the bottom water DO concentration did not play an important role in the preservation of OC, whereas other factors such as productivity, sedimentation rates, winnowing and type of OC (reworked) played major roles in OC distribution in sediments. Recent studies (Tripathi et al., 2017; Khim et al., 2018) based on expedition 355 of the International Ocean Discovery Program (IODP) from the AS have reported long term variations of OC to infer past monsoon conditions and carbon isotopes to infer terrestrial weathering conditions since the late Miocene.

5.7.4. Early Diagenesis of Organic Matter and the Sedimentary Environment

Organic matter content along with sedimentation rate, bottom water oxygenation, and degree of mixing in the upper sediment layers of sediment cores are helpful to understand the intensity of diagenesis or post-depositional evolution of sediments. Rock magnetic properties are sensitive in detecting authigenic magnetite and ultrafine magnetite formed by bacteria (Kirschvink and Chang, 1984; Karlin et al., 1987; Stolz et al., 1990) and can also be used to elucidate the sedimentary environment as magnetic minerals are diagenetically altered in the presence of suboxic to anoxic bottom/porewater conditions and organic matter. These features of sediment will ultimately determine whether or not suboxic/anoxic conditions become established in the sediment (Bloemendal et al., 1988, 1993). Therefore, understanding the impact of early diagenesis on magnetic minerals has been important for interpreting the sedimentary environment (Snowball and Thompson, 1990; Verosub and Roberts, 1995).

Anilkumar et al. (2005) were the first to report rock magnetic properties along with grain size, OC, $CaCO_3$, and $\delta^{18}O$ of planktonic foraminifera in the sediments of 15 gravity cores retrieved along the western margin of India. They reported an abundant continental supply of sediments on the northwest margin of India during glacial times compared with the southwest margin of India. Conversely, the early Holocene sediments of all cores were characterized by high magnetic susceptibility, low OC and $CaCO_3$. The late Holocene sediments of the SW margin of India were subjected to early diagenesis at different intervals in the cores. Rao et al. (2010) reported the geochemical and rock magnetic properties of sediments in a core recovered at a depth of 1420 m from the southwest margin of India. The magnetic properties and OC of sediments and $\delta^{18}O$ of planktonic foraminifera were used to identify the signatures of Bølling-Allerød and Younger Dryas events in this core. Glacial sediments show higher OC, $CaCO_3$, Ba, Mo, U, and Cd, while the early-to-late Holocene sediments show increasing concentrations of OC, $CaCO_3$, Ba, Cu, Ni, and Zn and decreasing concentrations of Mo, U, and Cd. High productivity-induced low-oxygenated bottom waters associated with reducing sedimentary conditions during glaciation, and high productivity in oxygenated bottom waters in the Holocene were inferred based on variations in sediment composition. The cores exhibited different stages of diagenesis at different sediment intervals. The occurrence of fine-grained, low-coercivity, ferrimagnetic mineral during glacial periods is indicative of its formation in organic-rich, anoxic sediments, which may be analogous to the diagenetic magnetic enhancement known in sapropels of the Mediterranean Sea (Roberts et al., 1999) and Japan Sea (Hayashida et al., 2007). Similarly, magnetic properties together with organic carbon, calcium carbonate, redox-sensitive elements, $\delta^{18}O$ of *Globigerinoides ruber*, and radiocarbon dating were carried out on two sediment cores off Mangalore – one core collected at a depth of 1380 m in oxygenated waters below the OMZ and another at 800 m within the OMZ. The sediments in the core from the 1380 m depth, which were deposited during the last 14 000 years, were subjected to diagenetic changes at several intervals, resulting in the dissolution of fine-grained magnetites and enrichment of redox-sensitive trace elements (Cu, Ni, Zn, V, Mo, and U). The sediments in the same core that were deposited at 18–27 000 years BP were characterized by a distinct decrease in magnetic parameters, high OC, low concentrations of redox-sensitive trace elements, and abundant pyritized tubules (Rao et al., 2008). The reductive diagenetic conditions indicated by the properties of magnetic rock contrast with the weak suboxic conditions revealed by low concentrations of trace elements in the sediments. This is because of intermittent bioturbation that allowed oxidants to penetrate into the sediments, remobilize redox-sensitive trace elements into the water column, and modify the primary geochemical signal of the sedimentary environment. The sediments deposited in a core recovered from the OMZ AAS62/1 showed a distinct diagenetic zone extending from 3000 to 6000 years BP (Figure 5.8), that was much larger and much more intense than found another core (AAS62/2) at a depth of 1380 m (Rao et al 2008). This zone is also enriched with redox sensitive trace metals (Cu, Ni, Zn, and U). Around 3000 to 6000 years BP, the drastic decrease in magnetic concentration, the increase in magnetic grain size, and an increasing contribution of antiferromagnetic mineral suggest a selective dissolution of more detrital magnetite minerals by the process of reductive diagenesis. The oxidative decomposition of organic matter in sediments might have started a series of reduction processes during sulfate reduction, where detrital/biogenic magnetites and other iron oxides progressively reduced or transformed into pyrites or gregites. Bloemendal et al. (1992) suggested reductive diagenesis is a considerable determinant of sediment magnetic properties in high-productivity areas, where the fine-grained ferrimagnetic fraction is preferentially removed and an increase in coarse-grained ferrimagnetic grain size leads to an increase in the antiferromagnetic to ferrimagnetic ratio. Mohamed et al. (2011) noted that the depletion of magnetic minerals associated with a progressive reduction of detrital iron oxides is mainly controlled by the diagenesis of organic matter in the coastal marine sediments of Ria de Vigo. Reductive diagenesis in sediments has been reported from different continental margins, the Japan Sea (Vigliotti, 1997; Yamazaki et al., 2003; Hayashida et al., 2007), the continental shelf of Korea (Liu et al., 2004), the Okhotsk Sea (Kawamura et al., 2007), the east coast of New Zealand (Rowan and Roberts, 2006), and the Chukchi-Alaska margin (Brachfeld et al., 2009).The extent and degree of early diagenesis are much greater in the sediments associated with OMZs than in the sediments associated with oxygenated waters. Similarly, the concentrations of redox-sensitive elements are higher in the early diagenetic zone but lower in the oxygenated waters.

5.8. SUMMARY AND FUTURE WORK

Considering the OC data from the western margin of India and the AS, it is seen that the OC content varies across and along the margin. The high OC content in surface sediments still needs a suitable explanation as it revolves around the productivity, preservation, ballast effect, shelf width, downslope transfer, and texture of the sediments. This has to be seen in regard to the type of productivity associated with this region and physical

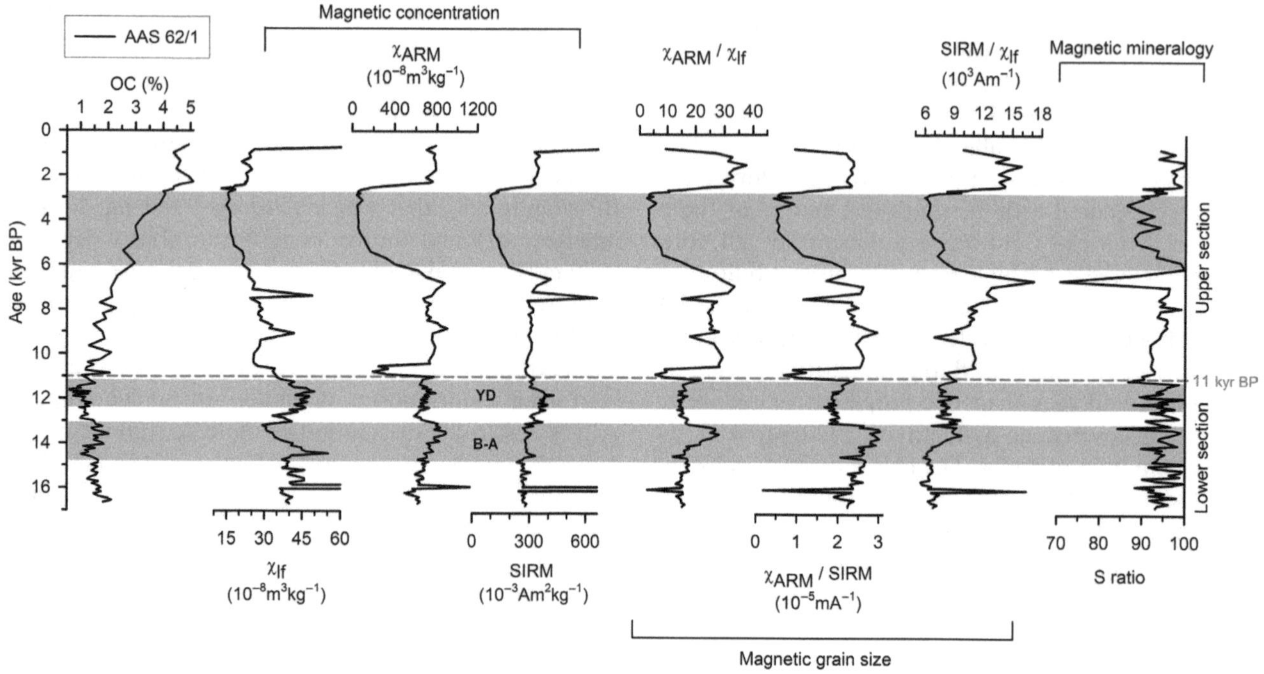

Figure 5.8 Down-core variations of organic carbon, magnetic concentration, grain size and mineralogy parameters in core AAS65/1 for the last 17 000 years BP collected off Mangalore. This core is collected from Oxygen Minimum Zone depths. One can see a large trough that shows early diagenetic dissolution of fine-grained magnetites between 3000 and 6000 years BP and a peak in magnetic grain size parameters showing formation of fine-grained magnetite at about 14 000 years BP.

forcing. The AS is one of the few regions where there is a high OC content during glacial times. There are also different interpretations for the changes in OC content at glacial–interglacial levels. The variation in OC content together with magnetic rock and geochemical properties in sediment cores are helpful in order to understand the extent and intensity of diagenesis and zones of reductive diagenesis and diagenetic magnetic enhancement. Multiproxy studies are to be encouraged in order to understand the diagenetic sedimentary environment, zones for the formation of new magnetic minerals, and the increasing concentration of reductive elements. Therefore further systematic studies are required and a workshop should be organized to bring together the researchers working on OC and different aspects of sedimentary OM.

ACKNOWLEDGEMENT

The authors thank the Director, CSIR-National Institute of Oceanography, Goa, for the facilities and encouragement. P.M.K. would like to thank Drs A.L. Paropkari and C.K. Haridevi for discussions and suggestions on OC from the AS.

REFERENCES

Acharya, A.A. and Panigrahi, M.K. (2016). Evaluation of factors controlling the distribution of organic matter and phosphorus in the Eastern Arabian Shelf: A geostatistical reappraisal. *Continental Shelf Research 126*: 79–88.

Agnihotri, R., Dutta, K., Bhushan, R. et al. (2002). Evidence for solar forcing on the Indian monsoon during the last millennium. *Earth and Planetary Science Letters 198*: 521–527.

Agnihotri, R., Bhattacharya, S.K., Sarin, M.M. et al. (2003a). Changes in surface productivity and subsurface denitrification during the Holocene: a multiproxy study from the eastern Arabian Sea. *The Holocene 13*: 701–713. doi:10.1191/0959683603hl656rp.

Agnihotri, R., Sarin, M.M., Somayajulu, B.L.K. et al. (2003b). Late-Quaternary biogenic productivity and organic carbon deposition in the Eastern Arabian Sea. *Palaeogeography, Palaeoclimatology, Palaeoecology 197*: 43–60.

Agnihotri, R. Kurian, S., Fernandes, M. et al. (2008). Variability of subsurface denitrification and surface productivity in the coastal eastern Arabian Sea over the past seven centuries. *The Holocene 18* (5): 755–764

Anilkumar, A., Rao, V.P., Patil, S.K. et al. (2005). Rock magnetic records of the sediments of the eastern Arabian Sea: Evidence for late Quaternary climatic change. *Marine Geology 220* (1–4): 59–82.

Bange, H.W., Naqvi, S.W.A. and Codispoti, L.A. (2005). The nitrogen cycle in the Arabian Sea. *Progress in Oceanography 65*: 145–158.

Bao, R., McNichol, A.P., McIntyre, C.P. et al. (2018). Dimensions of radiocarbon variability within sedimentary organic matter. *Radiocarbon 60* (3): 775–790.

Berger, W.H., Smetacek, V.S. and Wefer, G. (1989). Ocean productivity and Paleoproductivity- An overview. In: *Productivity of the Oceans: Present and Past* (eds. W.H. Berger, V.S. Smetacek and G. Wefer), 1–34. John Wiley and Sons Limited.

Bhattathiri, P.M.A., Pant, A., Sawant, S. et al. (1996). Phytoplankton production and chlorophyll distribution in the eastern and central Arabian Sea in 1994–1995. *Current Science 71*: 857–862.

Bhosle, N.B., Dhargalkar, V.K. and Braganca, A.M. (1978). Distribution of biochemical compounds in sediments of the shelf and slope regions of the west coast of India. *Indian Journal Of Marine Sciences 7*: 155–158.

Bhushan, R., Dutta, K. and Somayajulu, B.L.K. (2001). Concentrations and burial fluxes of organic and inorganic carbon on the eastern margins of the Arabian Sea. *Marine Geology178*: 95–112.

Bianchi, T.S., Cui, X., Blair, N.E. et al. (2018). Centers of organic carbon burial and oxidation at the land–ocean interface. *Org. Geochem. 115*: 138–155. doi: 10.1016/j.orggeochem.2017.09.008

Bloemendal, J., Lamb, B. and King, J.W. (1988). Palaeoenvironmental implications of rockmagneticproperties of late Quaternary sediment cores from the eastern equatorial Atlantic. *Palaeocenography 3*: 61–87.

Bloemendal, J., King, J.W., Hall, F.R. et al. (1992). Rock-magnetism of late Neogene and Pleistocene deep-sea sediments: relationship of sediment source, diagenetic processes and sediment lithology. *The Journal of Geophysical Research 97*: 4361–4375.

Bloemendal, J., King, J.W., Hunt, A., et al., (1993). Origin of sedimentary magnetic record atOcean Drilling Program sites on the Owen Ridge, Western Arabian Sea. *Journal of Geophysical Research 98*: 4199–4219.

Brachfeld, S., Barletta, F., St-Onge, G. et al. (2009). Impact of diagenesis on the environmental magnetic record from a Holocene sedimentary sequence from the Chukchi-Alaskan margin, Arctic Ocean. *Global and Planetary Change 68*: 100–114.

Burdige, D.J. (2005). Burial of terrestrial organic matter in marine sediments: a re-assessment. *Global Biogeochemical Cycles 19*: GB4011.

Calvert, S.E., Pederson, T.F., Naidu, P. D. et al. (1995). On the organic carbon maximum on the continental slope of the eastern Arabian Sea. *Journal of Marine Research 53*: 269–296.

Caratini, C., Bentaleb, I., Fontugne, M. et al. (1994). A less humid climate since ca. 3500 yr BP from marine cores off Karwar, western India. *Palaeogeography, Palaeoclimatology, Palaeoecology 109*: 371–384.

Cartapanis, O., Bianchi, D., Jaccard, S.L. et al. (2016). Global pulses of organic carbon burial in deep-sea sediments during glacial maxima. *Nature Communications 7* (1), 1–7.

Chakraborty, P., Chakraborty, S., Jayachandran, S. et al. (2016). Effects of bottom water dissolved oxygen variability on copper and lead fractionation in the sediments across the oxygen minimum zone, Western Continental margin of India. *Science of the Total Environment 566*: 1052–1061.

Cheriyan, E., Sreekanth, A., Mrudulrag, S.K. et al. (2015). Evaluation of metal enrichment and trophic status on the basis of biogeochemical analysis of shelf sediments of the south-eastern Arabian Sea, India. *Cont. Shelf Res. 108*: 1–11.

Cowie, G. L. (2005). The biogeochemistry of Arabian Sea surficial sediments: a review of recent studies. *Progress in Oceanography 65*: 260–289.

Dunne, J.P., Sarmiento, J.L. and Gnanadesikan, A. (2007). A synthesis of global particle export from the surface ocean and cycling through the ocean interior and on the seafloor. *Global Biogeochemical Cycles 21*: GB4006.doi:10.1029/2006GB002907

El-Wakeel, S.K. and Riley, J.P.(1957). The determination of organic carbon in marine muds. *Conseil Permanent International pour l'Exploration de la Mer 22*: 180–183.

Emery, K.O., and Uchupi, E. (1972). Western North Atlantic Ocean: Topography, rocks, structure water, life and sediments. *American Association of Petroleum Geologists Memoir 17*.

Enge, A.J., Witte, U., Kucera, M. et al. (2014). Uptake of phytodetritus by benthic foraminifera under oxygen depletion at the Indian margin (Arabian Sea). *Biogeosciences 11* (7), 2017–2026.

Farrenkopf, A. and Luther III, G.W. (2002). Iodine chemistry reflects productivity and denitrification in the Arabian Sea: evidence for flux of dissolved species from sediments of western India into the OMZ. *Deep-Sea Research II 49*: 2303–2318.

Gauns, M., Madhupratap, M., Ramaiah, N. et al. (2005). Comparative accounts of biological productivity characteristics and estimates of carbon fluxes in the Arabian Sea and the Bay of Bengal. *Deep-Sea Research Part II: Topical Studies in Oceanography 52*: 2003–2017.

Gupta, A.K., Mohan, K., Sarkar, S. et al. (2011). East–West similarities and differences in the surface and deep northern Arabian Sea records during the past 21 kyr. *Palaeogeography, Palaeoclimatology, Palaeoecology 301*: 75–85.

Guptha, M.V.S., Naidu, P.D., Haake, B.G. et al. (2005). Carbonate and carbon fluctuations in the Eastern Arabian Sea over 140 ka: implications on productivity changes? *Deep Sea Research Part II: Topical Studies in Oceanography 52* (14–15): 1981–1993.

Haake, B., Ittekkot, V., Rixen, T. et al. (1993). Seasonality and interannual variability of particle fluxes to the deep Arabian Sea. *Deep Sea Research I 40*: 1323–1344.

Hayashida, A., Hattori S. and H. Oda (2007): Diagenetic modifications of magnetic properties observed in a piston core(MD01–2407) from the Oki Ridge, Japan Sea. *Palaeogeography, Palaeoclimatology, Palaeoecology 247*: 65–73.

Helly, J.J. and Levin, L.A. (2004). Global distribution of naturally occurring marine hypoxia on continental margins. *Deep-Sea Research I 51*: 1159–116.

Hunter, W.R., Oguri, K., Kitazato, H. et al. (2011). Epi-benthic megafaunal zonation across an oxygen minimum zone at the Indian continental margin, *Deep-Sea Research, 58*: 699–710.

Ingole, B.S., Sautya, S., Sanitha, S. et al. (2010). Macrofaunal community structure in the western Indian continental margin including the oxygen minimum zone. *Marine Ecology 31*. doi:10.1111/j.1439–0485.2009.00356.x.

Ittekkot, V., Haake, B., Bartsch, M. et al. (1992). Organic carbon removal in the sea: The continental connection. In: *Upwelling Systems: Evolution Since the Early Miocene. Publication No. 64*. (eds. C.P. Summerhayes, W.L. Prell and K.C. Emeis), 167–176. London: Geological Society.

Jacob, J., Jayaraj, K.A., Rehman, H.H. et al. (2009). Biogeochemical characteristics of the surface sediments along the western continental shelf of India. *Chemistry and Ecology 25* (2): 135–149.

Jensen, M.M., Lam, P., Revsbech, N.P. et al. (2011). Intensive nitrogen loss over the Omani Shelf due to anammox coupled with dissimilatorynitrite reduction to ammonium. *The ISME Journal 5* (10): 1660–1670. https://doi.org/10.1038/ismej.2011.44

Johnson, J.E., Phillips, S.C., Torres, M. et al (2014). Influence of total organic carbon deposition on the inventory of gas hydrate in the Indian continental margins. *Marine and Petroleum Geology 58*: 406–424.

Kandasamy, S. and Nath, B.N. (2016). Perspectives on the terrestrial organic matter transport and burial along the land-deep sea continuum: caveats in our understanding of biogeochemical processes and future needs. *Frontiers in Marine Science 3*: 1–18. doi:10.3389/fmars.2016.00259

Karlin, R., Lyle, M. and Heath, G.R. (1987). Authigenic magnetite formation in suboxic marine sediments. *Nature 326*: 490–493.

Kawamura, N., Oda, H., Ikehara, K. et al. 2007. Diagenetic effect on magnetic properties of marine core sediments from the southern Okhotsk Sea. *Earth, Planets and Space 59*: 83–93.

Kessarkar, P.M. and Rao, V.P. (2007). Organic carbon in sediments of the southwestern margin of India: Influence of productivity and monsoon variability during the Late Quaternary. *Journal of the Geological Society of India 69*: 42–52.

Kessarkar, P.M., Rao, V.P., Naqvi, S.W.A. et al. (2010). Fluctuations in productivity and denitrification in the southeastern Arabian Sea during the Late Quaternary. *Current Science 99*: 485–491.

Kessarkar, P.M., Rao, V.P., Naqvi, S.W.A. et al. (2013). Variation in the Indian summer monsoon intensity during the Bølling-Allerød and Holocene. *Paleoceanography 28*: 413–425.

Khim B.-K., Lee J., Ha S. et al. (2018). Variations in $\delta^{13}C$ values of sedimentary organic matter since late Miocene time in the Indus Fan (IODP Site 1457) of the eastern Arabian Sea. *Geological Magazine* https://doi.org/10.1017/S0016756818000870

Kirschvink, J.L. and Chang, S.B.R. (1984). Ultrafine-grained magnetite in deep sea sediments. Possible bacterial magnetofossils. *Geology 12*: 559–562.

Kolla, V., Ray, P.K. and Kostecki, J.A. (1981). Surficial sediments of the Arabian Sea. *Marine Geology 41*: 183–204.

Kurian, S., Agnihotri, R., Borole, D.V. et al. (2009). Possible solar control on primary production along the Indian west coast on decadal to centennial timescale. *Journal of Quaternary Science 24*:109–116.

Kurian, S., Nath, B.N., Kumar, N.C.et al. (2013). Geochemical and isotopic signatures of surficial sediments from the western continental shelf of India: inferring provenance, weathering, and the nature of organic matter. *Journal of Sedimentary Research 83* (6): 427–442. 10.2110/jsr.2013.36

Kurian, S., Kessarkar, P.M., Rao, V.P., Reshma, K., Sarkar, A. (2020). Controls on organic matter distribution in oxygen minimum zone sediments from the continental slope off western India. *JourNal of Marine of Systems 207*: 103118.

Lee, C., Murray, D.W., Barber, R.T. et al. (1998). Particulate organic carbon fluxes: Compilation of results from the 1995 US JGOFS Arabian Sea Process Study. *Deep Sea Research Part II 45*: 2489–2501.

Lengger, S.K., Rush, D., Mayser, J.P. et al. (2019). Dark carbon fixation in the Arabian Sea oxygen minimum zone contributes to sedimentary organic carbon (SOM). *Global Biogeochemical Cycles 33*(12), 1715–1732.

Liu, J., Zhu, R., Roberts, A.P. et al. (2004). High-resolution analysis of early diagenetic effects on magnetic minerals in post-middle-Holocene continental shelf sediments from the Korea Strait. *Journal of Geophysical Research 109* (B03103): 1–15.

Madhupratap, M., Kumar, S.P., Bhattathiri, P.M.A. et al. (1996).Mechanism of the biological response to winter cooling in the northeastern Arabian Sea. *Nature 384*: 549–552.

Maya, M.V., Soares, M., Agnihotri, R. et al. (2011). Variations in some environmental characteristics including C and N stable isotopic composition of suspended organic matter in the Mandovi estuary. *Environmental Monitoring and Assessment 175* (1–4): 501–517.

Measures, C.I. and Vink, S. (1999). Seasonal variations in the distribution of Fe and Al in the surface waters of the Arabian Sea. *Deep Sea Research II 46*: 1597–1622.

Menzel, D.W. (1964). The distribution of dissolved organic carbon in the Western Indian Ocean. *Deep-Sea Research 11* (5): 757–766.

Middelburg, J.J. (2011). Chemoautotrophy in the ocean. *Geophysical Research Letters 38*: L24604.

Mohamed, K.J., Rey, D., Rubio, B., et al. (2011). Onshore-offshore gradient in reductive early diagenesis in coastal marine sediments of the Ria de Vigo, Northwest Iberian Peninsula. *Continental Shelf Resources 31*: 433–447.

Morrison, J.M., Codispoti, L., Gaurin, S. et al. (1998). Seasonal variation of hydrographic and nutrient fields during the US JGOFS Arabian Sea Process Study. *Deep Sea Research II 45*: 2053–2101.

Muller, G. (1967). *Methods in Sedimentary Petrology*. New York, NY: Hafner.

Murthy, P., Reddy, S.N., Gand, C.V. et al (1969). Distribution of organic matter in the marine sediments off the west coast of India. *Proceedings of the National Institute of Sciences of India 35*: 377–384.

Nagoji, S.S. and Tiwari, M. (2017). Organic carbon preservation inSoutheastern Arabian Sea sedimentssince mid-Holocene: implications to South Asian Summer Monsoon variability. *Geochemistry, Geophysics, Geosystems 18*: doi:10.1002/2017GC006804.

Naidu, P.D. (1990). Distribution of upwelling index planktonic foraminifera in the sediments. *Oceanologica Acta 13*: 327–334.

Naidu, P.D., Prakash Babu, C. and Rao, C.M. (1992). The upwelling record in the sediments of the western continental margin of India. *Deep-Sea Research 39*: 715–723.

Naidu, P.D., Singh, A.D., Ganeshram, R. et al. (2014). Abrupt climate-induced changes in carbonate burial in the Arabian Sea: causes and consequences. *Geochemistry, Geophysics, Geosystems 15*: 1398–1406.

Naik, R., Naqvi, S.W.A., Araujo, J. (2017a). Anaerobic carbon mineralisation through sulphate reduction in the inner shelf sediments of Eastern Arabian Sea. *Estuaries and Coasts 40*: 134–144.

Naik, D.K., Saraswat, R., Lea, D.W. et al (2017b). Last glacial–interglacial productivity and associated changes in the eastern Arabian Sea. *Palaeogeography, Palaeoclimatology, Palaeoecology 483*: 147–56.

Nair, P.V.R. (1970). *Primary productivity in the Indian seas. Bulletin of the Central Marine Fisheries Research Institute 22*. Kochi: Central Marine Fisheries Research Institute.

Naqvi, W.A. (1991). Geographical extent of denitrification in the Arabian Sea in relation to some physical processes. *Oceanologica Acta 14* (3): 281–290.

Naqvi, S.W.A., Jayakumar, D.A., Narvekar, P.V. et al. (2000). Increased marine production of N_2O due to intensifying anoxia onthe Indian continental shelf. *Nature 408*: 346–349.

Naqvi, S.W.A., Naik, H. and Narvekar, P.V. (2003). The Arabian Sea. Sea. In: *Biogeochemistry of Marine Systems* (eds. K. Black and G. Shimmield), 156–206. Oxford: Blackwell.

Naqvi, S.W.A., Moffett, J.W., Gauns, M.U. et al. (2010). The Arabian Sea as a high-nutrient, low-chlorophyll region during the late Southwest monsoon. *Biogeoscience 7*: 2091–2100.

Pailler, D. Bard, E., Rostek, F. et al. (2002). Burial of redox-sensitive metals and organic matter in the equatorial Indian Ocean linked to precession. *Geochimica et Cosmochimica Acta 66*: 849–865.

Paropkari, A.L. (1979). Distribution of organic carbon in sediments of the NorthWestern Continental shelf of India. Distribution of organic carbon in sediments from the Arabian Sea. *Indian Ocean Journal of Marine Science 8*: 127–129.

Paropkari, A.L. (1990). Geochemistry of sediments from the Mangalore-Cochin shelf and upper slope off Southwest India: geological and environmental factors controlling dispersal of elements. *Chemical Geology 81*: 99–119.

Paropkari, A.L., Rao, C.M. and Murty, P.S.N. (1987). Environmental controls on the distribution of organic matter in recent sediments of the Western Continental Margin of India. In: *Petroleum Geochemistry and Exploration in the Afro-Asian Region* (eds. R.K. Kumar, P. Dwivedi and V. Banerjee), 347–361. Routledge.,.

Paropkari, A.L., PrakashBabu, C. and Mascarenhas, A. (1992). A critical evaluation of depositional parameters controlling the variability of organic carbon in Arabian Sea sediments. *Marine Geology 107*: 213–226.

Paropkari, A.L., Mascarenhas, A. and PrakashBabu, C. (1993a). Comment on "Lack of enhanced preservation of organic matter in sediments under the oxygen minimum on the Oman Margin" by T. F. Pedersen, G. B. Shimmield, and N.B. Price, *Geochimica et Cosmochimica Acta 57*: 2399–2401

Paropkari, A.L., PrakashBabu, C. and Mascarenhas, A. (1993b). New evidence for enhanced preservation of organic carbon in contact with oxygen minimum zone on the western continental slope of India. *Marine Geology 111*: 7–13.

Pattan, J.N., Masuzawab, T., Divakar Naidu, P. et al. (2003). Productivity fluctuations in the southeastern Arabian Sea during the last 140 ka. *Palaeogeography, Palaeoclimatology, Palaeoecology 193*: 575–590.

Pedersen, T.F., Shimmield, G.B., and Price, N.B. (1992). Lack of enhanced preservation of organic matter in sediments under the oxygen minimum on the Oman Margin. *Geochimica et Cosmochimica Acta 56* (1): 545–551.

Prabhu, C.N. and Shankar, R. (2005). Palaeoproductivity of the eastern Arabian Sea during the past 200 ka: a multi-proxy investigation. *Deep-Sea Research II 52*: 1994–2002.

Pradhan, U., Wu, Y.,Shirodkar, P. et al. (2014). Sources and distribution of organic matter in thirty five tropical estuaries along the west coast of India-a preliminary assessment. *Estuarine, Coastal and Shelf Science 151*: 21–33, 10.1016/j.ecss.2014.09.010

PrakashBabu, C., Brumsack, H.J., Schnetger, B. (1999). Distribution of organic carbon in surface sediments along the eastern Arabian Sea: a revisit. *Marine Geology 162*: 91–103.

Premuzic, E.T., Benkovitz, C.M., Gaffney, J.F. et al. (1982). The nature and distribution of organic matter in the surface sediments of world oceans and seas. *Organic Geochemistry 4*: 63–77.

Qasim, S.Z. (1982) Oceanography of the northern Arabian Sea. *Deep-Sea Research 29*: 1041–68.

Qasim, S.Z. and Sankaranarayanan, V.N. (1970). Production of particulate organic matter by the reef on Kavaratti atoll (Laccadives). *Limnology and Oceanography 15*: 574–578.

Qasim, S.Z., Wafer, M.V.M., Vijayaraghavan, S. et al. (1978). Biological productivity of the coastal waters of India from Dhabol to Tuticorin. *Indian Ocean Journal of Marine Science 7*: 84–93.

Radhakrishna, K. (1969). Primary productivity studies in the shelf waters off Alleppey, south-west India, during the post-monsoon, 1967. *Marine Biology 4*: 174–181.

RaghuPrasad, R. (1967). Fishes and their environment. *Seafood Exporter 2* (1): 1–5.

Rajagopalan M.S., Thomas, P.A., Mathew, K.J. et al. (1992). Productivity of the Arabian Sea along the southwest coast of India. *CMFRI Bulletin 45*: 9–37.

Rajamanickam, G.V. and Setty, M.G.A.P. (1973). Distribution of phosphorus and organic carbon in the near shore sediments of Goa. *Indian Ocean Journal of Marine Science 2*: 84–89.

Ramaswamy, V. and Nair, R.R. (1994). Fluxes of material to the Arabian Sea and Bay of Bengal – sediment trap studies. *Proceedings of the Indian Academy of Science (Earth and Planetary Sciences) 103*: 189–210.

Ramaswamy, V., Sarin, M.M. and Rengarajan, R. (2005). Enhanced export of carbon by salps during the northeast monsoon period in the northern Arabian Sea. *Deep-Sea Research II 52*: 1922–1929.

Ramya, K.D., Jimly, C.J., Neil, S.C. et al. (2013). Biogeochemistry of the shelf sediments of south eastern Arabian Sea: effect on benthic bacterial heterotrophs. *Advances in Applied Science Research* 4315–328.

Rao, Ch.M.,Murty, P.S.N. (1990). Geochemistry of the Continental Margin Sediments, of the Central West Coast of India. *Journal of the Geological Society of India 35*: 19–37.

Rao, V.P. and Wagle, W.G. (1997). Geomorphology and surficial geology of the western continental shelf and slope of India: a review. *Current Science 73*: 330– 350.

Rao, V.P., Kessarkar, P.M., Patil S.K. et al. (2008). Rock magnetic and geochemical record in a core from the eastern Arabian Sea: diagenetic and environmental implications during the late Quaternary. *Palaeogeography, Palaeoclimatology, Palaeoecology 270*: 46–52.

Rao, V.P., Kessarkar, P.M., Thamban, M. et al. (2010). Paleoclimatic and diagenetic history of the Late Quaternary sediments in a core from the Southestern Arabian sea: Geochemical and magnetic signals. *Journal of Oceanography 66*: 133–146.

Rasiq, K.T., Kurian, S., Karapurkar, S.G. et al. (2016). Sedimentary pigments and nature of organic matter within the oxygen minimum zone (OMZ) of the Eastern Arabian Sea (Indian margin). *Estuarine, Coastal and Shelf Science 176*: 91–101.

Reddy, N.P.C. (2003). Organic matter distribution in the continental shelf sediments, off Kochi, west coast of India. *Indian Journal of Petroleum Geology 12* (2): 41–47.

Reemtsma, T., Haake, B., Ittekkot, V. et al. (1990). Downward flux of particulate fatty acids in the Central Arabian Sea. *Marine Chemistry 29*: 183–202.

Reichart G.-J., Brinkhuis H., Huiskamp F. et al. (2004). Hyper-stratification following glacial overturning events in the northern Arabian Sea. *Paleoceanography 19*: PA2013. doi: 10.1029/2003PA000900

Rejomon, G., Kumar, D.P.K. and Nair, M. (2013). Biogeochemical significance of eddies of the eastern Arabian Sea. *Applied Ecology and Environmental Research 11*: 237–248.

Rixen, T., Guptha, M.V.S. and Ittekkot, V. (2005). Deep ocean fluxes and their link to surface ocean processes and the biological pump. *Progress in Oceanography 65*: 240–259. doi:10.1016/j.pocean.2005.03.006.

Rixen, T., Gaye, B., Emeis, K. et al. (2019). The ballast effect of lithogenic matter and its influences on the carbon fluxes in the Indian Ocean. *Biogeosci. 16*: 485–503.

Roberts, A. P., Stoner J.S. and C. Richter (1999): Diagenetic magnetic enhancement of sapropels from the eastern Mediterranean Sea. *Marine Geology 153*: 103–116.

Robin, R.S., Muduli, P.R., Vardhan, K.V. et al. (2012). Heavy metal contamination and risk assessment in the marine environment of Arabian Sea, along the southwest coast of India. *American Journal of Chemistry 2* (4): 191–208.

Rostek, F., Bard, E., Beaufort, L. et al. (1997). Sea surface temperature and productivity records for the past 240 kyr in the Arabian Sea. *Deep Sea Research II 44* (6–7): 1461–1480.

Rowan, C.J. and Roberts, A.P. (2006). Magnetite dissolution, diachronous greigite formation, and secondary magnetizations from pyrite oxidation: unravelling complex magnetizations in Neogene marine sediments from New Zealand. *Earth and Planetary Science Letters 241*: 119–137.

Ryther, J.H. (1969). Photosynthesis and fish production in the sea. *Science 166*: 72–76.

Ryther, J.H. and Menzel. D. W. (1965). On the production, composition, and distribution of organic matter in the Western Arabian Sea. *Deep-Sea Research 12*: 199–209.

Sajan, S., Joydas, T.V. and Damodaran R. (2010). Meiofauna of the western continental shelf of India, Arabian Sea. *Estuarine, Coastal and Shelf Science 86*: 665–674.

Sarkar, A., Bhattacharya, S. K. and Sarin, M. M. (1993). Geochemical evidence for anoxic deep water in the Arabian Sea during the last glaciations. *Geochimica et Cosmochimica Acta 51*: 1009–1016

Sarma, V.V.S.S. (2002). An evaluation of physical and biogeochemical processes regulating perennial suboxic conditions in the water column of the Arabian Sea. *Global Biogeochemical Cycles 16*: 1082. https://doi.org/10.1029/2001GB001461.

Sarma, V.V.S.S., Kumar, M.D. and George, M.D. (1998). The eastern and central Arabian Sea as a perennial source for atmospheric carbon dioxide, *Tellus. Series B 50*: 179–184.

Sarma, V.V.S.S., Swathi, P.S., Dileep Kumar, M. et al. (2003). Carbon budget in the eastern and central Arabian sea: an Indian JGOFS synthesis. *Global Biogeochemical Cycles 17*(4): 1102. doi:10.1029/2002GB001978.

Sarma, V.V.S.S., Dileep Kumar, M. and Saino, T. (2007). Impact of sinking carbon flux on accumulation of deep-ocean carbon in the Northern Indian Ocean. *Biogeochemistry 82*: 89–100

Sarma, V.V.S.S., Desai, D.V., Patil, J.S. et al. (2018). Ecosystem response in temperature fronts in the northeastern Arabian Sea. *Progress in Oceanography 165*: 317–331.

Schulte, S., Rostek, F., Bard, E. et al. (1999). Variations of oxygen-minimum and primary productivity recorded in sediments of the Arabian Sea. *Earth and Planetary Science Review 173* (3): 205–221.

Scientific Committee on Ocean Research (1997). *Proceedings of the Scientific Committee on Oceanic Research*. https://scor-int.org/Publications/1996Proceedings.PDF

Sen Gupta, R. and Naqvi, S.W.A. (1984). Chemical oceanography of the Indian Ocean, north of the equator. *Deep Sea Research 31*: 671–706.

Shankar, D. and Shetye, S.R. (1997). On the dynamics of the Lakshadweep high and low in the southeastern Arabian Sea. *Journal of Geophysical Research 102* (12): 551–12 562.

Shetye, S.R., Gouveia, A.D., Shenoi, S.S.C. et al. (1990). Hydrography and circulation off the west coast of India during the southwest monsoon 1987. *Journal of Marine Research 48*: 359–378.

Singh, A.D. (2007). Episodic preservation of pteropods in the eastern Arabian Sea: Monsoonal change, oxygen minimum zone intensity and aragonite compensation depth. *Indian Journal Of Marine Sciences 36*: 378–383.

Singh, A.D., Jung, S.J.A., Darling, K. et al. (2011). Productivity collapses in the Arabian Sea during glacial cold phases. *Paleoceanography 26*: PA3210. doi:10.1029/2009PA001923.

Sirocko, F., Garbe-Schönberg, D. and Devey, C. (2000). Processes controlling trace element geochemistry of Arabian Sea sediments during the last 25,000 years. *Global and Planetary Change 26* (1–3): 217–303.

Smith, S.L., Banse, K., Cochran, J.K. et al. (1991). *US JGOFS: Arabian Sea Process Study. US JGOFS Planning Report No. 13*. Woods Hole: Woods Hole Oceanographic Institution.

Snowball, I.F. and Thompson, R., 1990. A stable chemical remanence in Holocene sediments. *Journal of Geophysical Research 95*: 4471–4479.

Somasundar, K., Rajendran, A. and Kumar, M.D. et al. (1990). Carbon and nitrogen budgets of the Arabian Sea. *Marine Chemistry 30*: 363–377.

Sreekanth, A., Mrudulrag S.K., Cheriyan E. et al. (2015). Trace metal enrichment and organic matter sources in the surface sediments of Arabian Sea along southwest India (Kerala coast). *Marine Pollution Bulletin 101*: 938–946.

Steemann, N. (1952). The use of radio-active carbon (C**) for measuring organic production in the sea. *Journal du Conseil 18*: 117–140.

Steemann, N. and Jensen, E.A. (1957). Primary oceanic production. The autotrophic production of organic matter in the oceans. *Galathea Reports 1*: 49–136.

Stewart, R.A., Pilkey, O.H. and Nelson, B.W. (1965). Sediments of the northern Arabian Sea. *Marine Geology 3*: 411–427.

Stolz, J.F., Lovley, D.R., Haggerty, S.E., 1990. Biogenic magnetite and the magnetization of sediments. *Journal of Geophysical Research 95*: 4355–4361.

Szekielda, K. H. (1970). The liberated energy potentially available from oxidation processes in the Arabian Sea. *Deep Sea Research and Oceanographic Abstracts 17* (3): 641–646.

Takahashi, T., Sutherland, S.C., Sweeney, C. et al. (2002). Global sea–air CO_2 flux based on climatological sur-face ocean pCO_2, and seasonal biological and temperature effects. *Deep Sea Research II 49*: 1601–1612.

Thamban, M., Rao, V.P. and Raju, S.V. (1997). Controls on organic carbon distribution in the sediment cores from the eastern Arabian Sea. *Geo-Marine Letters 17* (3): 20–27.

Thamban, M., Rao, V.P., Schneider, R.R. et al. (2001).Glacial to Holocene fluctuations in hydrography and productivity along the southwestern continental margin of India. *Palaeogeography, Palaeoclimatology, Palaeoecology 165*: 113–127. doi:10.1016/S0031–0182 (00)00156.

Thamban, M., Kawahata, H. and Rao, V.P. (2007). Indian summer monsoon variability during the Holocene as recorded in sediments of the Arabian Sea: timing and implications. *Journal of Oceanography 63*: 1009–1020. 10.1007/s10872–007–0084–8

Tiwari, M., Ramesh, R., Somayajulu, B.L.K., et al. S. (2006). Paleomonsoon precipitation deduced from a sediment core from the equatorial Indian Ocean. *Geo-Marine Letters 26* (1): 23–30

Tomczak, M. and Godfrey, J.S. (1994). *Regional Oceanography: An Introduction*. Oxford: Pergamon.

Trask, P.D. (1932). Hydrological investigative work by petroleum companies. *Transactions. American Geophysical Union 13*. doi:10.1029/TR013i001p00306.

Tripathi S., Tiwari M., Lee J. et al. (2017). First evidence of denitrification vis-à-vis monsoon in the Arabian Sea since Late Miocene. *Scientific Reports 7*: 43056. doi: 10.1038/srep43056.

Verosub, K.L. and Roberts, A.P. (1995). Environmental magnetism: past, present and future. *Journal of Geophysical Research 100*: 2175–2192.

Vigliotti, L. (1997). Magnetic properties of light and dark sediment layers from the Japan Sea: diagenetic and paleoclimatic implications. *Quaternary Science Reviews 16*: 1093–1114.

Westberry, T.K. and Behrenfeld, M.J. (2014). Oceanic net primary production. In: *Biophysical Applications of Satellite Remote Sensing* (ed. J.M. Hanes), 205–230. Berlin: Springer

Wiseman, J.D.H and Bennett, H. (1940). The distribution of organic carbon and nitrogen in the sediments of the Arabian Sea. John Murray Expedition 1933–34. *Scientific Reports 3*: 193–221.

Witte, U. and Pfannkuche, O. (2000). High rates of benthic carbon remineralization in the abyssal Arabian Sea. *Deep Sea Research II 47*(14): 2785–2804.

Yamazaki, T., Abdeldayem, A.L. and Ikehara, K. (2003). Rock-magnetic changes with reduction diagenesis in Japan Sea sediments and preservation of geomagnetic secular variation in inclination during the last 30,000 years. *Earth Planets and Space 55*: 327–340.

Young, D.K. and Kindle, J.C. (1994). Physical processes affecting the availability of dissolved silicate for diatom production in the Arabian Sea. *Journal of Geophysical Research 99* (C11): 22619–22632.

Biome II
Ocean Depths

6

Deep Subsurface Microbiomes of the Marine Realm

Jagannath Sarkar, Nibendu Mondal, Subhrangshu Mandal, Sumit Chatterjee, and Wriddhiman Ghosh*

ABSTRACT

The oceans cover ~71% of the Earth's surface, so their sediments constitute one of the largest biomes of the planet. Moreover, the marine deep subsurface is not only a colossal microbial habitat but also one of the most enigmatic and difficult to explore microbiomes on Earth. Albeit microbial cell density in the subsurface is manifold lower than in surficial ecosystems, the sheer vastness of the deep and dark extremities of the biosphere makes them the largest reservoir of organic carbon. The progressively organic carbon impoverished and energy-limited marine deep subsurface poses a unique set of physicochemical challenges (including high temperature and mechanical pressure) to metabolism, notwithstanding which the habitat is replete with microbial life and activity, often up to several hundreds of meters below the seafloor. The last few decades have witnessed quantum leaps in our understanding of the microbiology of the deep subsurface of the marine realm, and in tune with that aspects pertaining to physiology, adaptation, evolution, and astrobiology have been reviewed extensively. Here, we compile all the information available for in situ microbial processes and their geochemical manifestations (including typical geological features such as gas hydrate deposits, mud volcanoes and carbonate mounds) and aim to visualize a comprehensive biogeochemical picture of ecosystem functioning in the marine deep subsurface.

6.1. INTRODUCTION

The marine deep subsurface is an enormous geomicrobiome, perhaps second only to the ocean waters, and so has emerged as a hotspot of cross-disciplinary researches on the dynamics of geosphere–biosphere coevolution. The marine deep subsurface biome is considered to begin at 1 mbsf (meter below seafloor) and continue for more than 1 km into the sediment-depths, or for all practical purposes until where microbial cells and DNA remain detectable (Jørgensen and Boetius, 2007; Biddle et al., 2012; Teske et al., 2013). Alternative views, however, consider the marine subsurface horizon below 5 mbsf as "deep" (Valentine, 2011). In any case, the marine deep subsurface biosphere, comprising the microbiomes within sediments, pore-fluids, and upper basaltic crusts (including the fluids circulating through the crusts), extends from the continental shelves and slopes up to the abyssal plains, and is one of the most mysterious ecosystems on Earth (D'Hondt et al., 2004, Jørgensen and D'Hondt, 2006; Orcutt et al., 2013; Schrenk et al., 2013; Morono et al., 2020). Hydrothermal vents situated on the mid-oceanic ridges, and the aging as well as newly-formed ocean crusts, are also considered to be parts of the marine deep biosphere (Cowen et al., 2003; Huber et al., 2006; Santelli et al., 2008; Orcutt et al., 2011a; Edwards et al., 2012), but as these are distinctive physiographies with different biogeochemical dynamics they are reviewed separately in Chapter 14 of this book (Lankalapalli et al., 2022).

Department of Microbiology, Bose Institute, P-1/12 CIT Scheme VII M Kolkata, 700054 India,

**Corresponding Author: Wriddhiman Ghosh (wriman@jcbose.ac.in). ORCiD code: 0000-0002-4922-7404*

Systems Biogeochemistry of Major Marine Biomes, First Edition. Edited by Aninda Mazumdar and Wriddhiman Ghosh.

With increasing sediment-depth microbial life gets progressively constrained by the high *in situ* heat and severe limitation of bioenergetic resources such as utilizable organic carbon, electron sources, and respiratory electron acceptors (Arndt et al., 2013; Lever et al., 2015). These pressures of natural selection, together with the absence/nominal presence of many of the standard forces in ecology and evolution (for example, dispersal of organisms in and out of the system, connectivity with other ecosystems and communities, high metabolic activity and flexibility, and constant thermodynamic fluctuations) not only shape the peculiar structures and functions of the deep subsurface biosphere but also make them unique sites for revealing hitherto unseen facets of geobiological evolution (Biddle et al., 2012). All the ecological peculiarities and physicochemical constraints to habitability notwithstanding, rich microbiomes have been revealed by the international ocean discovery program (IODP; http://www.iodp.org/expeditions), over the past two decades, in tens of 100 m deep sub-seafloor sediments (Inagaki et al., 2006; Biddle et al., 2008, 2011; Teske and Sørensen, 2008). What is more, the sprawling deep sub-seafloor microbial communities, or at least parts of them, are thought to be metabolically active, and therefore playing potent roles in the biogeochemistry of the benthic realm (D'Hondt et al., 2004, 2009; Jørgensen and D'Hondt, 2006; Schrenk et al., 2010; Orsi et al., 2013a). Consequently, marine deep subsurface microbiomes hold strong implications for long-term carbon sequestration and climate control (Farmer et al., 2019), subsurface elemental cycling (Wasmund et al., 2017), and geobiological evolution of the seabed (Biddle et al., 2005; Picard and Ferdelman, 2011).

While research publications on this topic are burgeoning by the day, a number of structural and functional aspects of this gargantuan geomicrobiome remain unknown or unexplained: for example, a clear resolution is needed with regard to what proportions of the microbial diversities reported from various deep subsurface samples are actually indigenous to those habitats. Owing to the very low abundance of microbial cells in all deep subsurface habitats, possibilities are high that even trace amounts of contaminants introduced from the fluid chemicals and/or equipment used in the collection of samples and subsequent experimental procedure (including DNA sequencing) would generate significant false-positive signals in the total diversity determined (Sheik et al., 2018). Furthermore, microorganisms are said to remain viable for millions of years in the extremely nutrient-depleted oxic/anoxic deep sub-seafloor sediments by minimizing energy and nutrient requirements and metabolizing at ultraslow rates (Morono et al., 2011; Lomstein et al., 2012; Røy et al., 2012; D'Hondt et al., 2015). This hypothesis awaits corroboration by physiological and molecular microbiological experiments *in vitro*. The body of knowledge accrued on this mega-microbiome of the marine realm is as multifaceted as it is voluminous, so this chapter provides only an overview of its biogeochemistry and highlights the constraints and opportunities that the system affords for life through geological times. Implications are discussed for global cycling and sequestration of carbon, geobiological evolution of Earth, and certain pertinent aspects of astrobiology.

6.2. ECOSYSTEM CONSTRAINTS IN THE MARINE DEEP SUBSURFACE

Like all other subsurface environments of this planet, deep sub-seafloor ecosystems are constrained by the absence of light and, therefore, lack of phototrophic primary productivity. While organic carbon and organic/inorganic oxidants of photolithoautotrophic and chemoorganoheterotrophic origins were buried in these systems during their sedimentary formation, further trickling of nutrients from benthic territories gets limited as porosity and permeability decrease with the deepening and aging of the sediment (Jones et al., 2018). In this way, over geological time, deep subsurface sediments become increasingly deficient in bioavailable sources of energy, carbon, electron, and respiratory substrates or electron acceptors (D'Hondt et al., 2009, 2015; Hoehler and Jørgensen, 2013; Lever et al., 2015). However, wide areas of deeply-buried sediments at open-ocean sites in the eastern Pacific feature less-negative redox potentials, and therefore have electron acceptors such as nitrate and O_2 supplied to the deepest sedimentary communities through the underlying basaltic aquifer (D'Hondt et al. 2004; Batzke et al., 2007). Furthermore, because of low rates of sedimentation and microbial respiration *in situ*, the entire subsurface in South Pacific Gyre remains oxic, from the seafloor until basement (D'Hondt et al., 2009, 2015). On the other hand, whether O_2 is present or not, acute nutrient scarcity is a common characteristic unifying all deep subsurface ecosystems of the marine realm. Oligotrophic stress in these extreme habitats, which are subjected to much higher pressures compared with habitats on the Earth's surface, and occasionally experience elevated temperature and extreme pH (see Jones et al., 2018, and references therein), is accentuated by the fact that microorganisms, whether living[1] as active or

[1] *All three of the following states of metabolism are considered to be reflective of life: (i) dormant, which involves minimal but definite metabolic activities to support basic cellular maintenance even though organisms under such a condition cannot influence the in situ biogeochemistry; (ii) not growing or dividing but physiologically functional enough to have biogeochemical impact; and (iii) physiologically active, and growing or dividing (Johnson et al., 2007; Lomstein et al., 2012; Blazewicz et al., 2013; Kempes et al., 2017).*

dormant cells, must eventually rely on exogenous (i.e. external substrate derived) rather than endogenous (i.e. biomass derived) catabolism to persist in these habitats over long (geological) timescales (Bradley et al., 2018).

Notwithstanding the extreme bioenergetic and/or biophysical constrains associated with marine deep subsurfaces, diverse lines of evidence indicate that considerable proportions of the microorganisms present *in situ* are metabolically active. Such corroborative data come from across disciplinary approaches, which include microbial cell enumeration (Cragg et al., 1990, 1992; Cragg and Kemp, 1995; Mauclaire et al., 2004; Schippers et al. 2005), enrichments and growth of live cultures (Cragg et al., 1990; D'Hondt et al., 2004; Batzke et al., 2007; Smith et al., 2011), study of preserved geochemical records (D'Hondt et al., 2002; Røy et al., 2012), metatranscriptome analyses (Orsi et al., 2013a, 2013b), extraction and characterization of intact polar lipids (Lipp et al., 2008), isotopic data (Morono et al., 2011), and substrate turnover measurements based on *in situ* conditions as well as *ex situ* incubation experiments (Schippers et al., 2005; also see Orcutt et al., 2013 and references therein). Unanimously, however, culture-independent as well as culture-dependent studies point out that metabolism in deep subsurface microbial communities is rendered at extremely slow paces (Lever et al., 2015; Rédou et al., 2015; Pachiadaki et al., 2016; Braun et al., 2017).

6.3. FACTORS CONSTRAINING THE STUDY OF MARINE DEEP SUBSURFACES

Sample contamination, and consequent introduction of spurious microbial entities into the results of culture-independent as well as culture-dependent investigations, has been a major problem for researches in deep subsurface microbiology ever since its inception. There is, therefore, a continuous quest for finer methods to assess and minimize contamination so as to achieve maximum accuracy in revealing the microbial communities that are actually native to these habitats (Sheik et al., 2018). While deep subsurface explorations rely essentially on drilling, long core recovery requires addition of drilling fluids that inevitably infiltrate into the cores (Kallmeyer et al., 2006; Lever et al., 2006; Kallmeyer, 2011). As the drilling fluids carry foreign microbes, sources and causes of contamination for the deep subsurface microbiome starts right at the sample collection stage. As drilling practices are indispensable for most subsurface studies, and because infiltration of drilling fluids into the cores cannot be avoided, several methodologies to minimize core contamination and at the same time assess the degree of contamination to decipher the actual native microflora has been proposed and practiced. Drilling methods that can detect contaminations and authenticate the cored samples have been developed over time and used successfully in deep microbiome explorations (Colwell et al., 1992; Griffin et al., 1997; Smith et al., 2000; Biddle et al., 2006; Kieft et al., 2007; Orsi et al., 2013a; Inagaki et al., 2015; Morono and Inagaki, 2016). Major techniques used to assess microbial contamination of cores from drill fluids include fluorescent dyes and pigments (Friese et al., 2017), tracer techniques based on microspheres, and perfluorocarbon (Senum and Dietz, 1990; Smith et al., 2000; House et al., 2003; Kallmeyer et al., 2006; Lever et al., 2006; Yanagawa et al., 2013).

While substantial progress has been made in the direction of more sanitized drilling strategies and formulation of methods for detecting and filtering out drilling-derived contaminations from microbiome analyses (Wilkins et al., 2014; Friese et al., 2017), other crucial sources of contamination for deep sub-seafloor samples include the molecular biological reagents and water purification systems that are used in the laboratories, and potentially the researchers' bodies (Laurence et al., 2014; Salter et al., 2014). Albeit all metagenome-based community analyses run the risk of contaminations coming in from these sources the problem becomes especially accentuated in case of the extremely-low biomass ecosystems, such as those of the deep sub-seafloor, where the introduction of even minute levels of contaminants can skew the apparent architecture of the communities being revealed (Yanagawa et al., 2013). The latest emphasis, therefore, is on scrutinizing all sequence datasets emanating from deep sub-seafloor sediment samples by testing "no DNA template" controls in every polymerase chain reaction and sequencing-based community analysis, so as to discriminate true representatives of the deep biosphere from spurious inclusions (Sheik et al., 2018).

6.4. BIOGEOCHEMISTRY OF MARINE DEEP SUBSURFACES

Although metabolic rates in deep subsurface microbial communities are in general very low (and keep decreasing with increasing depth below the seafloor), they still render active biogeochemical cycling, especially that of carbon and sulfur, up to several meters below the seafloor in distinct sites of the global ocean (Jørgensen et al., 2019). Microbiome architecture, and therefore the biogeochemistry, of marine deep subsurface ecosystems changes significantly with distance from the coastline (water-depth and *in situ* sediment deposition dynamics) and depth of the sediment (Bhattacharya et al., 2021), both of which are important determinants of whether molecular oxygen (O_2) is present in the system to support aerobic microbial processes (D'Hondt et al., 2009, 2015; Røy et al., 2012). Oxic deep subsurfaces are located in the mid-oceanic gyres whereas their anoxic counterparts are encountered mostly across the continental margins (D'Hondt et al., 2009; Kallmeyer et al., 2012). Both types of systems are discussed in this section: first as an introduction to the major study sites, then in a holistic manner to rationalize the discrete geomicrobiological information available for the distinct study sites. In the second approach, we compile and compare the biogeochemical data under two sediment-depth ranges, namely 1–5 mbsf and >5 mbsf, and then under each depth range category we classify the sediment systems on the basis of the absence/presence of dissolved O_2. In most marine subsurface systems, the geomicrobial regime changes drastically after around 5 mbsf (Parkes et al., 1994; Estes et al., 2019). Moreover, according to some researchers 5 mbsf should be considered as the starting depth of the marine deep subsurface biome (Valentine, 2011). The biogeochemistry is, therefore, discussed separately for the subsurface systems situated on the two sides of the 5 mbsf depth boundary. In the process of this systemic consideration, we try to envisage potentially comprehensive biogeochemical frameworks for the two individual depth tiers, which in turn are further categorized based on their anoxic or oxic sedimentary context.

6.4.1. Major Sites of Exploration and Their Geological Contexts

Discrete marine sediment horizons feature distinct physiographic and geochemical characteristics. For instance, continental shelf/slope sediments are rich in organic carbon; those featuring mud volcanoes are replete with reduced inorganic chemicals; on the other hand, sediments underlying open oceans far away from the continental margins are poor in both organic and inorganic nutrients (Lalli and Parsons, 1997; D'Hondt et al., 2009; Røy et al., 2012). Owing to a decrease in sedimentation rate in the coast to open ocean trajectory, a few meters of deep-sea sediment encompass the diagenetic history of several tens of thousands of years, while equivalent depths of shallow coastal sediments record only a few thousand years (Bhattacharya et al., 2021). Concurrently, vertical spans of the different biogeochemical zones in the sediment column are longer in open ocean sites compared with shallow coastal locations. The marine deep subsurface has been explored for *in situ* geochemistry and/or microbial life to a variety of depths, at a large numbers of physiographically distinct oceanic sites. Of these, some typically well-studied ones are discussed below.

Eastern Equatorial Pacific Ocean

During the ODP Leg 201 expedition, the entire 319.6 m thick and organic matter rich sediment package (having an age of 11 million years at the base) was explored at Open Ocean Site 1225 (indicated as Site 1a in Figures 6.1 and 6.2a), located at a water-depth of 3760 m in the eastern equatorial Pacific Ocean, near the boundary of the South Equatorial Current and the North Equatorial Countercurrent (Baker et al., 1991; Oyun et al., 1995). Here, seawater flows through the basaltic crust and diffuses into the sediment bottom. Methane, ammonium, dissolved inorganic carbon (DIC), and alkalinity are individually highest in the middle of the sediment package and decline towards the seawater–sediment as well as sediment–basement interfaces. Sulfate concentration is lowest in the middle of the core (only 7% of the seawater value), while nitrate and dissolved O_2 are highest near the two interfaces and decrease gradually downward from the seafloor and upward from the basement (D'Hondt et al., 2003).

In the same ODP Leg 201 expedition in the eastern equatorial Pacific Ocean, a 420 m thick surface-to-basement sediment package was investigated at a water-depth of 3297 m (Open Ocean Site 1226, indicated as site 1b in Figures 6.1 and 6.2b). This exploration site was located at the boundary of South Equatorial Current and Peru Current (300 km south of the Galapagos Islands), where advection water of the latter current renders the water-column highly productive (Chavez and Barber, 1987). In this sediment horizon too (age of the sample, just above the base: 16.5 million years), seawater flows through the basaltic crust, so the seawater–sediment and the sediment–basement interfaces have similar but oppositely oriented geochemical profiles. While sulfate (present consistently at levels >68% of the seawater sulfate concentration) is the predominant electron acceptor throughout the sediment package, the broad range of DIC and ammonium concentrations recorded along the sediment horizon are indicative of organic matter mineralization throughout the core (water that flows through basement is the major sink of the DIC). Methane concentration increases to highs of 2–3 μM in the 100–250 mbsf zone (D'Hondt et al., 2003).

Furthermore, in ODP Leg 201, an organic matter and methane-rich Pleistocene–Holocene sediment package was investigated up to 194.4 mbsf, at a shallower (150.5 m) water-depth on the Peru shelf (Open Ocean Site 1229) (indicated as Site 1c in Figures 6.1 and 6.2b). This sediment system features brine ingression at its bottom, so sulfate is present in the upper strata (due to downward diffusion from the overlying seawater) and then in the deepest layers (due to upward diffusion from the brine, which has a considerably higher sulfate concentration than the sea water), but not in the middle layer spanning 35–88 mbsf. While both the sulfate depletion depths (35 and 88 mbsf) coincide with local maxima of dissolved sulfide concentration (a broad peak featuring between ~20 and 40 mbsf and a sharper one at ~90 mbsf), depth-trends of methane mirror the sulfate profiles in both the territories (D'Hondt et al., 2003).

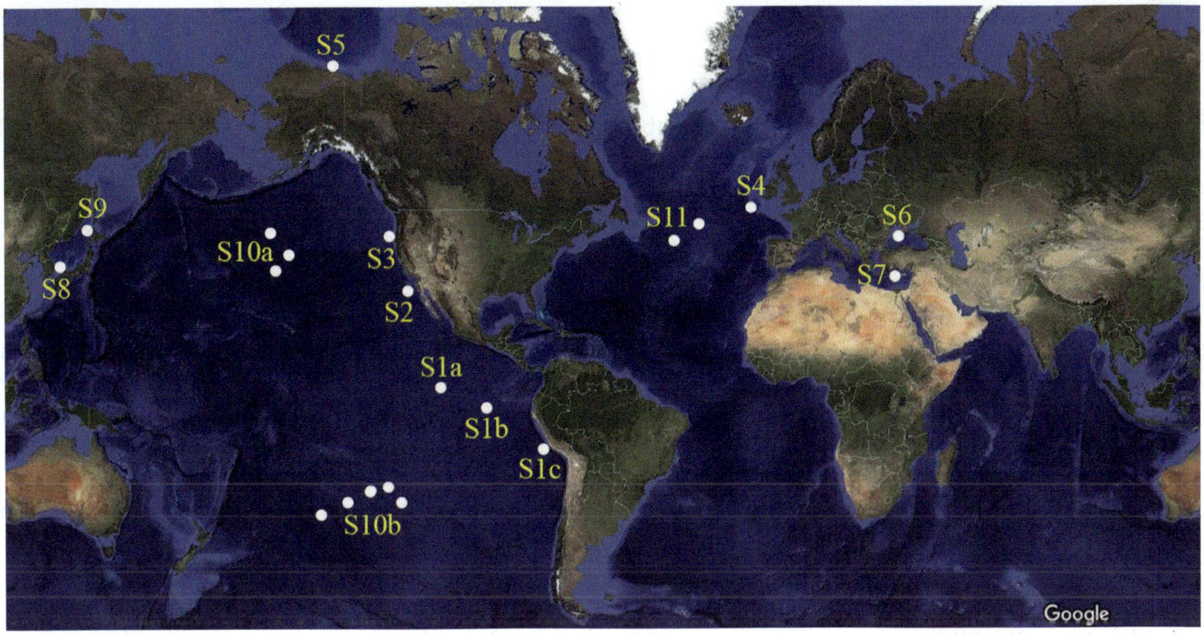

Figure 6.1 Approximate localization of the major sites (numbered S1 –S11) of marine deep subsurface exploration (indicated site positions are not in scale with actual latitudes and longitudes). The world map shown has been sourced from Google Maps. S1, Eastern Equatorial Pacific Ocean (S1a, Open Ocean Site 1225; S1b, Open Ocean Site 1226; S1c, Open Ocean Site 1229); S2, Santa Barbara Basin; S3, Cascadia Margin; S4, North-eastern Atlantic Ocean; S5, Alaskan Beaufort Sea; S6, North-western Black Sea; S7, Eastern Mediterranean Sea; S8, Ulleung/Tsushima Basin; S9, Eastern Japan Sea; S10, Pacific Gyres (S10a, North Pacific Gyre; S10b, South Pacific Gyre); S11, North Atlantic Gyre.

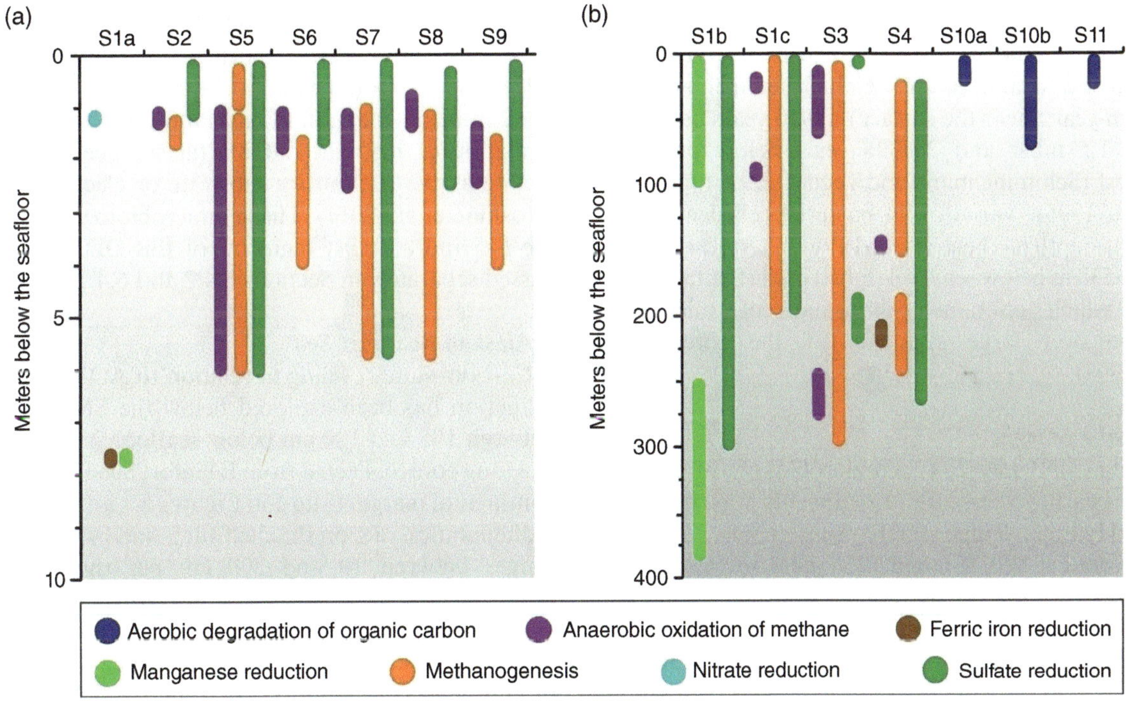

Figure 6.2 Major metabolic processes discovered via microbiological experiments in the various depth zones below the seafloor (in meters), at the major sites (numbered S1 –S11) of marine deep subsurface exploration. (a) Sites where <10 mbsf sediment horizons have been explored; (b) sites explored for sediments > 10 mbsf. The number codes used to indicate the different sites are same those used in Figure 6.1. Lengths of the vertical bars shown under each site indicate the depths (in meters) below the seafloor over which the individual metabolic processes have been detected. Information pertaining to the in situ metabolic processes have been taken from Lauer et al. (2016) for Site S1a; Parkes et al. (2005) for Sites S1b and S1c; Harrison et al. (2009) for Site S2; Cragg et al. (1996) and Colwell et al. (2008) for Site S3; Webster et al. (2009) for Site S4; Treude et al. (2014) for Site S5; Leloup et al. (2007) for Site S6; Lazar et al. (2012) for Site S7; Cho et al. (2017) for Site S8; Yanagawa et al. (2014) for Site S9; Røy et al. (2012) for Site S10a; D'Hondt et al. (2009, 2015) for Site S10b; Estes et al. (2019) for Site S11.

The geomicrobiological characteristics of the 1–5 mbsf and >5 mbsf segments of the three deeply explored Open Ocean Sites 1225, 1226, and 1229 are discussed separately in Sections 6.4.2 and 6.4.3, respectively.

Santa Barbara Basin

In the western American continental margin, microbiome structures and functions have been studied in the shallow sediment horizon (just below 1 mbsf) of the Santa Barbara Basin (indicated as Site 2 in Figures 6.1 and 6.2a), off the southern Californian coast (water-depth 587 m; bottom-water O_2 0.2 μM), focusing on the microbiota native to the sulfate–methane transition zone (SMTZ); here sulfate depletes completely at 1.4 mbsf while methane reaches 40 mM at 1.2 mbsf; Harrison et al., 2009). This submarine basin is bound by the Channel Islands in the south and the California coast in the north. To the west and east, narrow sills constrain the basin's ventilation, making the waters hypoxic (and scarce of benthic fauna) and sediments unbioturbated. Deeper parts of the basin, however, periodically encounter influx of cool, oxygenated water across the sills situated on the western flank; this critically impacts the benthic biology and sedimentary biogeochemistry of Santa Barbara Basin (Bograd et al., 2002). The age–depth relationship for the organic carbon-rich sediments (total organic carbon, TOC, content 2–6 wt%), which underlie a highly productive coastal ocean having high sedimentation rate (3–5 mm year^{-1} near the surface), is 950 years before present at 1.5 mbsf and 28 896 years before present at 43.3 mbsf (Schimmelmann and Tegner, 1991; Ingram and Kennett, 1995). Variation in bacterial cell density with sediment-depth has been studied along a 68 mbsf horizon, at a site 538 m below sea level (mbsl) in the Santa Barbara Basin, which contained thousands of micromolar methane over large segments of the core (Cragg et al., 1995).

Cascadia Margin

During ODP Leg 204, geomicrobiology of methane hydrate-bearing sub-seafloor sediments was studied at South Hydrate Ridge (ODP sites 1244, 1245, and 1251 located at 890, 870 and 1216 mbsl, respectively) on the Cascadia Margin, off the Oregon coast (Site 3 in Figures 6.1 and 6.2b; Tréhu et al., 2003; Nunoura et al., 2008). Here, methane is stored within the accretionary wedge of a major subduction zone (https://www.pmel.noaa.gov/eoi/Cascadia-margin.html), and detected throughout the 0–130 mbsf and 0–140 mbsf gas hydrate stability zones of sites 1244 and 1245, respectively; they are also present in the 110–120 mbsf and 190–200 mbsf layers of Site 1251 (Nunoura et al., 2008). Geomicrobiology of gas hydrate-bearing sediments has also been studied at ODP sites 889, 890, and 892 within this region (Cragg et al., 1996). The geomicrobiological characteristics of the 1–5 mbsf and >5 mbsf segments of this deeply explored ODP territory are discussed separately in Sections 6.4.2 and 6.4.3, respectively.

North-eastern Atlantic Ocean

Geomicrobiology of deeply-buried, cold-water coral carbonate deposits has been explored along the 155 m tall Challenger Mound (ODP Site U1317 Hole A) that is partially buried within the sediments and coral rubbles of Porcupine Seabight (indicated as Site 4 in Figures 6.1 and 6.2b), a deep-sea basin located on the Irish continental margin in north-eastern Atlantic Ocean (Webster et al., 2009; Hoshino et al., 2011). In this territory, the seafloor, almost devoid of live corals, is situated at 815 mbsl, from where the mound rises 34 m (i.e. up to 781 mbsl) while the remaining 121 m of the mound is buried in the sediment and coral debris. Along the U1317 core, metagenome-based characterization of the microbiota has been carried out at 4, 20, and 105 mbsf; sulfate concentration was found to decrease from 27.3 mM at 1.4 m from the mound top to 2.15 mM beneath the base of the mound (at 168.42 m from the top); simultaneous to this drop in sulfate, methane concentration rose to >3 mM and remained consistently high for the rest of the sediment-depth explored. Furthermore, detection of sulfide in the SMTZ (a peak of 0.2 mM sulfide was recorded at ~2 m below the mound base) indicated that anaerobic oxidation of methane (AOM) takes place *in situ* using sulfate as the respiratory substrate or electron acceptor (Hoshino et al., 2011). The geomicrobiological details of the 1–5 and >5 mbsf segments of this ODP site are discussed separately in Sections 6.4.2 and 6.4.3 respectively.

Alaskan Beaufort Sea

Carbon–sulfur cycling in relation to AOM and sulfate reduction has been explored below the SMTZs located between 105 and 120 cm below seafloor (cmbsf), within 6 m long cores collected from Beaufort Sea in the Alaskan continental margin (Site 5 in Figures 6.1 and 6.2a), where sedimentation rate on the shelf (at a water-depth ~300 m) ranges between 10 and 300 cm per thousand years (Drenzek et al., 2007; Treude et al., 2014). Trends of variation in sulfate and methane concentrations are similar across the sediment cores explored in Beaufort Sea, even though depth-wise values differ from core to core. Overall, sulfate concentration decreases linearly from 27 mM at the sediment surface to zero at the SMTZ; in contrast, methane concentration increases linearly below the SMTZ and reaches the core-wise maxima (ranging between 10 and 14.2 mM) at different sediment-depths, i.e. between 130 and 560 cmbsf (Treude et al., 2014).

North-western Black Sea

During the Holocene epoch, Black Sea (indicated as Site 6 in Figures 6.1 and 6.2a) which is the largest euxinic basin in the marine realm, converted to a brackish water system from a limnic one (Leloup et al., 2007). Pleistocene deposits having limnic history are therefore found preserved across the >100 cmbsf sediment horizons of the present-day Black Sea (Holmkvist et al., 2011). Within the 500 cmbsf horizon, upper sediment layers encompass laminated coccolith ooze and have high TOC (4 wt%) while the underlying sapropels contain even higher (15 wt%) TOC (Jørgensen et al., 2001). In a 4.6 m sediment horizon explored at 1024 mbsl, sulfate concentration has been recorded as ~15 mM and almost zero at the core top and 180 cmbsf, respectively; methane has been found to intrude the sulfate zone (SMTZ lies within 150–180 cmbsf), albeit increasing steeply below 150 cmbsf and attaining a partial pressure of 100 kPa at 175 cmbsf (with a further increase in sediment-depth methane supersaturates); sulfate reduction-derived H_2S increases from 0.7 mM at the core top to 2.8 mM in the SMTZ, after which it depletes completely at 400 cmbsf (Leloup et al., 2007). In another study, the sedimentary microbiology of Black Sea was investigated in 10 m sediment cores collected from 2000 mbsl (Schippers et al., 2012).

Eastern Mediterranean Sea

Deep subsurface biogeochemistry has been studied in the Eastern Mediterranean Sea in the context of the Amsterdam Mud Volcano (Site 7 in Figures 6.1 and 6.2a) located at a water-depth of 2028 m (Lazar et al., 2012). Within the 6 m sediment horizon explored, temperature was found to increase with depth, indicating potential methane seepage. Sulfate decreased with sediment-depth along the first ~100 cmbsf; methane concentration was highest at the deepest part of the core, i.e at 600 cmbsf, and lowest at 50 cmbsf, thereby indicating a broad SMTZ between 50 and 100 cmbsf (Lazar et al., 2012). Remarkably, acetate concentration in the pore-water was also reported to be very high in this sediment system (maximum 2120 µM at 450 cm).

Ulleung/Tsushima Basin

Gas hydrate-bearing sediments of the Ulleung Basin (or Tsushima Basin, indicated as Site 8 in Figures 6.1 and 6.2a) located at the meeting point of Korea Strait and Sea of Japan (East Sea) have been explored for their geomicrobiology with reference to the SMTZs, along 4–7 m cores collected at 1450 mbsl (Park et al., 2006; Cho et al., 2017). The high TOC content (2–4 wt%) of this sediment system is comparable to that of the Santa Barbara Basin; its age is 9300 years at 154 mbsf; and the average sedimentation rate of the area varies between 17.61 and 19.96 cm per thousand years (Park et al., 2006; Cho et al., 2017).

Eastern Japan Sea

Microbial communities thriving in gas hydrate associated sediments at >900 mbsl of the eastern Japan Sea (Site 9 in Figures 6.1 and 6.2a) have been studied in 40 m long cores (Yanagawa et al., 2014). In these terrigenous and diatomaceous sediments, the SMTZ lies at depths of <5 m, with sulfate concentration at the seafloor ranging between 20 and 30 mM and decreasing to near zero at 5 mbsf; methane increases from 0 mM at the seafloor to 15–20 mM at 5 mbsf, and remains at ~10 mM up to 30 mbsf (Yanagawa et al., 2014). Bacterial and geochemical profiling of deep sediments (up to 518 mbsf) was carried out in another study based on an exploration site located at a water-depth of 900 m of the eastern Japan Sea (Cragg et al., 1992).

Pacific Gyres

The non-fossiliferous red clay sediments of the North Pacific Gyre (NPG, Site 10a in Figures 6.1 and 6.2b; Yamazaki and Shimono, 2013) have been explored for their geomicrobiology and the oxic deep subsurface was found to harbor aerobic microbial communities characterized by extremely low metabolic activities (Kallmeyer et al., 2012). Here, aerobic microbial respiration has been detected in sediments 86 million years old in the process of investigating up to 28 m long cores from different locations in the equatorial and northern Pacific Ocean, where water-column productivity decreases by 50%, and O_2 penetration depth increases from 90 cmbsf to >30 mbsf, in the northward trajectory (total sediment thickness at the NPG coring sites, however, is 88–100 m; Straume et al., 2019).

The South Pacific Gyre (SPG, indicated as Site 10b in Figures 6.1 and 6.2b) is the largest oligotrophic region within the marine realm and covers ~10% of the Earth's surface. It features astonishingly low primary productivity and sedimentation rates (0.1–1 m per million years) compared with the rest of the global ocean, so the SPG sediments experience oxygen flux throughout the package, plausibly until the basaltic ocean crust (Fischer et al., 2009). The center of the SPG is far away from continent margins, and therefore almost disconnected from the shelf/slope biogeochemical dynamics, yet live aerobic microbiomes characterized by extremely low areal respiration rates are present throughout the oxic sediment horizons of the SPG (D'Hondt et al., 2009, 2015). Whereas sonar and seismic reflection data estimated the total thickness of the sediment package above the oceanic basement to vary between 1 and 130 m, geomicrobial explorations (D'Hondt et al., 2009) have thus far covered sediments up to 74.5 mbsf (located at a water-depth of 5697 m). Within this territory, aerobic microbial life has been found to persist in O_2-containing, zeolitic and/or metalliferous pelagic clays up to 101.5 million years old (Morono et al., 2020).

The geomicrobiological characteristics of the 1–5 mbsf and >5 mbsf zones of the deep subsurface sites explored within the Pacific Gyres are discussed separately in Sections 6.4.2 and 6.4.3, respectively.

Atlantic Gyres

The geomicrobiology of the sediments of the North Atlantic Gyre (NAG, indicated as Site 11 in Figures 6.1 and 6.2b) has been explored across 14 different coring stations in the North Pond area (Picard and Ferdelman, 2011). These sediments underlying a water-depth of 4580 m and overlying an active 7–8 million year old basaltic crust (this potentially harbors a vigorous flow of cold seawater), reach a thickness up to 300 m (Langseth et al., 1984), of which up to 9 m was investigated by Picard and Ferdelman (2011). Aerobic chemoorganoheterotrophic activity of microorganisms was detected throughout this sediment horizon, albeit at very low rates, consistent with the low flux of organic matter in the system. More recent study of organic geochemistry in the NAG sediment system involved approximately 25–27 m cores collected from >5350 mbsl, where total thickness of the 70–98 million year old sediment package ranges between 98 and 100 m (Müller et al., 2008; Whittaker et al., 2013). The geomicrobiological characteristics of the 1–5 mbsf and >5 mbsf zones of this deep subsurface site are discussed separately in Sections 6.4.2 and 6.4.3, respectively.

Deep sub-seafloor sediments of the South Atlantic Gyre (SAG, which also has acutely oligotrophic waters; Morel et al., 2010) are yet to be explored, although a few microbiological studies have been conducted in the suboxic surface sediments of the abyssal plains off the western coast of Africa (Cape, Angola and Guinea basins; Schauer et al., 2010), and in an assortment of biogeochemically distinct sediment systems starting from the rim of the SAG into the Benguela upwelling system on the inner continental shelf off Namibia (Lagostina et al., 2015).

Indian Ocean Gyres

The gyre located in the southern Indian Ocean is formed by the interactions of South Equatorial Current and West Australian Current; it is also a major oligotrophic zone of the marine realm which expands at an average annual rate of 4.46% per year (Jena et al., 2013). Whereas the deep sedimentary geomicrobiology of the Indian Oceanic gyre is yet to be investigated (so the territory is not included in Figures 6.1 and 6.2), a number of sites have been explored at the level of the sediment surface and/or within 1 mbsf (Talbot and Bianchi, 1997; Siegert et al., 2011). Notably, however, the methane seep at the fore-arc sediment basin of Sumatra has been studied by means of nine different sediment cores, of which two located at water-depths >1000 mbsl covered >1 m sediment-depth. Here, free methane was detected in the water-column, and sulfate concentration decreased along the sediment-depth while sulfide concentration increased; the SMTZ, however, remained uncharacterized (Siegert et al., 2011). Furthermore, in the Bengal fan region, IODP expeditions 353 and 354 conducted geological and geochemical, but no direct microbiological, investigations (France-Lanord et al., 2016; Peketi et al.,2020), so their findings remain outside the purview of this chapter.

Northern Indian Ocean

Several >1 m sediment cores have been explored for biogeochemistry, microbiology, and paleo-ecology, across the Arabian Sea and Bay of Bengal (Mazumdar et al.,2009, 2012, 2014; Orsi et al., 2017; Fernandes et al., 2018; Bhattacharya et al., 2020, 2021; Mandal et al., 2020). These sediment horizons are mostly located within the oxygen minimum zones (OMZs) of the Northern Indian Ocean, and as this book includes a separate chapter (Chapter 1) dealing with the geomicrobiology of marine OMZs and upwelling regions (Mazumdar et al., 2022), the present review does not include the information available for the deep subsurface biogeochemistry of these sites (the territory is also not included in Figures 6.1 and 6.2).

6.4.2. Geomicrobiology of Marine Deep Subsurfaces

Albeit microbial cell density decreases with increasing sediment-depth irrespective of whether O_2 is present *in situ*, copious prokaryotic life is harbored within deep sub-seafloor ecosystems. The density of microbial biomass and/or microbial diversity may not be high in a given cubic centimeter or meter of deep subsurface sediment, but owing to the sheer enormity of the overall territory that comes under this megabiome, the total amount of microbial abundance and diversity potentially harbored in this marine physiographic regime is expected to be very high. Early research estimated that microbial biomass in the global marine deep subsurface could be equivalent to 10% of the total biosphere carbon (Parkes et al., 1994). Subsequently, 50–80%, and 10–30%, respectively, of the total microbial biomass and total biomass of all life-forms on Earth was estimated to be present in these extremities of the biosphere (Whitman et al., 1998; Parkes et al., 2000). Here we aim to compile and rationalize the existing knowledge on the microbial ecology of marine deep subsurfaces by classifying the systems into two broad depth-categories: 1–5 mbsf and >5 mbsf, and each category is further considered under two subcategories based on the presence or absence of O_2 *in situ*.

Geomicrobiology of 1–5 mbsf Sediment Horizons Having an Anoxic Milieu

Microbial ecology of these sediment horizons has mostly been explored across continental margins and upwelling zones, and typically with reference to the spatial dynamics of sulfate- and methane-metabolizing organisms (Bowles et al., 2014; Bhattacharya et al., 2021). Dominant segments of these shelf/slope subsurfaces are anoxic owing to high productivity in the water-columns and consequential depletion of O_2 within shallow sediment-depths, and due to relatively higher sedimentation rates than the open oceans, which in turn result in low oxygen exposure time for the breakdown and remineralization of the organic matter that is delivered to the seafloor (Seiter et al., 2005; Burdige, 2007). Thus direct relations between sedimentation rate, organic matter degradation, methane production/oxidation, and depths of SMTZs have been envisaged for these ecosystems harboring rich taxonomic and metabolic diversities of anaerobic microorganisms (Canfield, 1994; Komada et al., 2016; Bhattacharya et al., 2021).

At ODP Site 1225 located in the eastern region of the equatorial Pacific Ocean (Site 1a in Figure 6.2a), the sedimentary microbiota has been explored by environmental 16S rRNA gene sequencing and analysis to reveal archaeal communities at 1.1 and 7.8 mbsf (subgroups of Marine Group I microorganisms were predominant in the nitrate-containing 1.05–1.10 mbsf and manganese-/iron-containing 7.7–7.8 mbsf zones; Lauer et al., 2016). Geochemical data from this site further envisaged aerobic metabolism and nitrate reduction within 2 mbsf, below which manganese and iron reduction prevailed (D'Hondt et al., 2004).

In the Santa Barbara Basin (Site 2 in Figure 6.2a), the microbiota has been explored with special reference to the SMTZ which is located between 1.25 and 1.35 mbsf; here, nominal sulfate reduction has been found to take place at 0.5 mbsf while methanogenesis predominates the community metabolism at 1.63 mbsf (Harrison et al., 2009). Metagenome analyses further indicated that archaea belonging to the group 1 anaerobic methanotrophs (ANME-1), and *Desulfobacterales*- and *Desulfosarcina–Desulfococcus*-related Deltaproteobacteria, could be the main methane oxidizers and sulfate-reducers in the SMTZ, respectively (Harrison et al., 2009). In the Santa Barbara sediments, the overall abundance of bacterial cells was found to correlate with TOC concentration and decrease from 1.27×10^9 cm^{-3} at the seafloor to 4.77×10^6 cm^{-3} at 12 mbsf, following which it decreased very slowly (Cragg et al., 1995).

In the Cascadia Margin, off Oregon (Site 3 in Figure 6.2b), sedimentary microbial communities, as revealed by 16S rRNA gene cloning and sequencing (Tréhu et al., 2003), encompass highly diversified groups of bacteria alongside members of the Deep Sea Archaeal Group (DSAG) throughout the cores explored (only a few >140 mbsf sample-sites lacked DSAG). While the total copy number of prokaryotic rRNA genes, a measure of overall bacterial and archaeal abundance, decreased with increasing sediment-depth, SMTZs in the various cores investigated occurred between 4.5 and 8 mbsf; below the SMTZs methanogenesis prevailed (Tréhu et al., 2003; Nunoura et al., 2008).

The microbiology of Beaufort Sea sediments in the north Alaskan margin, (Site 5 in Figure 6.2a) has also been explored with reference to the SMTZ, which in this territory lies between 105 and 120 cmbsf (Treude et al., 2014). Peculiarly, in this sediment system, methane is present both above and below the SMTZ but not within it (the highest methane concentrations were detected between 260 and 551 cmbsf, indicating that these sediment-depths were the main sites of methanogenesis). Rate of AOM, which again was apparently rendered by novel and unique microbial groups, is also low between the sediment surface and SMTZ, but increases with sediment-depth below the SMTZ. While sulfate reduction was detectable throughout the 6 m sediment horizon explored, sulfide was detected from just above the SMTZ to a little more than 4 mbsf (Treude et al., 2014).

In the north-western Black Sea sediments (Site 6 in Figure 6.2a) abundance of sulfate-reducing microorganisms increases from the core-top until the SMTZ, and then decreases subsequently, even though they persist through the methane zone. Sulfate reduction rate is also highest in the SMTZ, apparently owing to AOM-dependent activities of the sulfate-reducers (Leloup et al., 2007). From seafloor to 1 m beneath, microbial cell count decreases ten to a hundred times, while it increases again in the organic-carbon-rich 8–9 mbsf layer; in the entire sediment system bacterial candidate division JS-1 and classes *Anaerolineae* and *Caldilineae* of the phylum Chloroflexi predominate alongside *Crenarchaeota* (Schippers et al., 2012).

The subsurface microbiome of Amsterdam Mud Volcano (Site 7 in Figure 6.2a) has been explored with special reference to archaeal diversity, which in turn was reported to be dominated by methanogens and ANME (Lazar et al., 2012).

In the Ulleung Basin (Site 8 in Figure 6.2a), sulfate reduction rate and dsrA (dissimilatory sulfate reductase A) gene abundance have been found to decline overall with increasing sediment-depth; in the process, however, both parameters exhibit multiple phases of synchronized rise and fall in the sediment surface to core bottom trajectory (Cho et al., 2017). Similar trends of population dynamics have also been reported for sulfate-reducing microorganisms in other 3 m sediment horizons of the global ocean such those underlying the Arabian Sea OMZ (Fernandes et al., 2018; Bhattacharya et al., 2021). In the Ulleung Basin, SMTZs have been identified within 0.6–1.0 and 3.5–5.0 mbsf, and marked by an abundance of methanogens, as indicated by copious mcrA (methyl coenzyme M reductase) genes in the metagenomes. Analysis of 16S rRNA gene libraries further revealed Marine Benthic Group B as the predominant archaea above and within the SMTZ, while ANME-1 prevailed in the methane-containing layers below the SMTZ, alongside the bacterial candidate division JS1 (Cho et al., 2017).

In the eastern Japan Sea (Site 9 in Figure 6.2a), the microbial community structure has been delineated (on the basis of 16S rRNA gene libraries) for the gas hydrate associated sub-SMTZ sediments (<5 mbsf) and found to be quite simple and uniform (largely dominated by uncultivated bacterial lineages related to JS1 and NT-B2; Yanagawa et al., 2014). Here, bacterial cell density decreases over a 6 mbsf sediment horizon; sulfate reduction rate remains high within 3 mbsf alongside high diversities of heterotrophic microorganisms and other microbial metabolic types such as nitrate reducers, aerobic/anaerobic ammonifiers and acetogens; methanogenesis rate increases below the sulfate reduction peak at ~3 mbsf (Cragg et al., 1992).

Geomicrobiology of 1–5 mbsf Sediments Having Oxic Milieu

Albeit open ocean sediments contain O_2, the abundance of microbial cells *in situ* is several orders of magnitude lower compared with shelf/slope sediments; moreover, information is scarce on what types of microorganisms inhabit deep subsurfaces underlying open oceans (D'Hondt et al., 2009; Orcutt et al., 2011b). For example, no pure culture-based taxonomic information of microbial diversity is available for the NPG sediments; only genera such as *Alteromonas*, *Marinobacter*, *Moritella*, *Pseudoalteromonas*, and *Vibrio* are known to be present in the water-column in association with sinking particulate matter (Fontanez et al., 2015). Furthermore, microbial community dynamics along sediment-depths have not been resolved for any O_2-containing deep subsurface site of the marine realm. However, 16S rRNA gene sequence analysis has shown that a 1.2–2.6 mbsf sediment horizon located just south-west of the South Pacific Gyre encompasses considerable microbial diversity, with Bacteroidetes, Chloroflexi, Deferribacteres, Deltaproteobacteria, Firmicutes and Verrucomicrobia, and Marine Benthic Group B (related to the crenarchaea) and Marine Group I (related to the thaumarchaea) and *Methanomicrobia* and *Thermococci* (belonging to the euryarchaea) being the major bacterial and archaeal groups present *in situ* (Durbin and Teske, 2011; Steele et al., 2012). Microbial community structure along the sediment-depth is also unexplored in North Atlantic Gyre. A major limitation of microbiome exploration in oligotrophic marine sediments is the difficulty in retrieving undisturbed sediment packages for the upper layers as these generally get contaminated with seawater during ODP borehole initiation (Lever et al., 2006; Durbin and Teske, 2011). That said, active aerobic degradation of organic carbon takes place across these sediment horizons of the global ocean. Although organic matter in these sediments get prolonged O_2 exposure because of the low sedimentation rate, it becomes more recalcitrant with increasing sediment-depth and escapes remineralization by microorganisms; density of microbial cells also decreases sharply with increasing sediment-depth.

In the 100 m deep oxic sediment package of NPG (Site 10a in Figure 6.2b), 90% of the organic carbon delivered to the seafloor is mineralized within the first few centimeters, and below 1 m only ~1% of the organic carbon remains as a catabolic substrate (Røy et al., 2012). Here the rate of sedimentation is approximately 0.2 mm per 1000 years, while the O_2 consumption rate (theoretically estimated based on the variation of O_2 concentration with sediment-depth) decreases from 10 mmol O_2 l^{-1} $year^{-1}$at 0.5 mbsf to 0.001 mmol O_2 l^{-1} $year^{-1}$ at 20–30 mbsf (at the latter depths the sediments are older than 66 million years). Whereas 10^8 microbial cells cm^{-3} are present in the surface sediment of NPG, 10^3 cells cm^{-3} are detected at 20 mbsf; further below, the density of cells, despite being extracted from the sediments, remain lower than the limit of enumeration by fluorescence microscopy. In this territory, per cell O_2 consumption rate also decreases with increasing sediment-depth and reaches ~10^{-3} fmol $cell^{-1}$ day^{-1} at ~30 mbsf; low respiration rates such as these are comparable to those of cultures in prolonged stationary state (Røy et al., 2012).

In South Pacific Gyre (Site 10b in Figure 6.2b), Redfield stoichiometry of porewater O_2 and nitrate is indicative of aerobic respiration coupled marine organic matter oxidation in the sediments, especially the shallow upper layers (D'Hondt et al., 2015). Below a few 10s of cmbsf, TOC content declines gradually with increasing sediment-depth, and 1.5 mbsf downward, the annual rate of TOC burial ranges between 4.4×10^{-10} and 8.2×10^{-9} mol C cm^{-2} (D'Hondt et al., 2009). Although the rate of burial of organic carbon is extremely low, aerobic microorganisms occur throughout the ~75 mbsf (up to 100 million years old) sediment horizon of SPG explored thus far (D'Hondt et al., 2009, 2015; Morono et al., 2020). A large fraction of the sedimentary organic carbon is thought to be aggregated or adsorbed to minerals, thereby giving some selective advantage to the native microbial types which can grow on minerals (Durbin and Teske, 2011; Steele et al., 2012). The SPG sediment system has extremely low abundance of microbial cells (~10^6 cells cm^{-3} at 1.5 mbsf to <10^3 cells cm^{-3} at ~25 mbsf) – here, every depth below the seafloor has a three to four orders of magnitude lower cell count compared with the global average at that subseafloor depth (Steele et al., 2012). In the greater part of the SPG sediment package, cell density is below the detection limit of standard counting methods (D'Hondt et al., 2009). Corroborating the potential adherence of organic carbon to iron minerals, magnetic separation of microbial cells followed by 16s rRNA gene sequencing and analysis has identified the 1.2–2.6 mbsf sediment layer as the most biodiverse zone of the SPG sediments (Durbin and Teske, 2011; Steele et al., 2012).

In NAG (Site 11a in Figure 6.2b) also, the sediment system is extremely poor in organic carbon content – TOC decreases from <0.3% to <0.1% within just 1.5 mbsf (which corresponds to an age of 1 million years); subsequently, the decline is slow and linear, reaching ≤0.05% at the core bottom (25–27 mbsf, up to 24 million years). Whereas cell density at the sediment surface across coring sites is 1.4×10^2 to 1.5×10^5 cells cm^{-3}, it declines below the quantification limit for fluorescence microscopy (13.7 cells cm^{-3}) at the core bottom (Estes et al., 2019). Sedimentary organic matter at the NAG is characterized by low C/N ratio; 40–60% of it encompasses proteinaceous amide and carboxylic compounds and, notably, its overall chemical composition does not vary with sediment-depth or core location. Furthermore, in NAG sediments, aromatic carbons are present in greater amounts than the Pacific gyre sediments; apparently, mineral surface adsorption makes the organic matter inaccessible to many heterotrophic microorganisms present (Estes et al., 2019).

6.4.3. Geomicrobiology of Marine Subsurfaces >5 mbsf

The >5 mbsf horizons are geobiologically less explored compared with the sediments deposited above 5 mbsf; the overall pace of exploration here is also slow because coring the marine subsurface at these depths entails a huge financial cost and highly specialized technological expertise (D'Hondt et al., 2009, 2015). That said, a number of studies have shown that all marine subsurface ecosystems >5 mbsf deep are characterized by lower cell concentration than their overlying layers, irrespective of whether the sediment system is anoxic or oxic. For example, at 518 mbsf in the Pacific Ocean, the microbial population is 120 times lower than that at the sediment surface (Parkes et al., 1994). Globally, a rapid decrease of microbial abundance takes place within 6 mbsf, and below that level populations decrease more slowly. At these sediment-depths, buried organic carbon becomes increasingly recalcitrant and nondegradable by microorganisms. Fermentation, sulfate reduction, methanogenesis (in anoxic sediment systems), and aerobic organic carbon degradation (in oxic sediment systems) are the major metabolisms in these horizons (D'Hondt et al., 2009). It has been found that SMTZs occur at >5 mbsf sediment-depths only in open ocean territories, whereas continental margin sediments generally have SMTZs above 5 mbsf. However, there are a number of examples where study sites located in shallow water territories close to the continents were found to have SMTZs at sediment-depths greater than 5 mbsf; these features were attributed to potential lithological properties or basement composition and not sedimentation rate (Chen et al., 2017; Akam et al., 2020). Furthermore, at deeply explored sites where seawater flows inside the basaltic crust, basements of the sediment packages harbor diversified microbiomes characterized by a reverse pattern of metabolic stratification with respect to the sediment layers adjacent to the seafloor (D'Hondt et al., 2004).

Geomicrobiology of >5 mbsf Subsurfaces Having an Anoxic Milieu

In the eastern equatorial Pacific Ocean, Site 1229 (Site 1c in Figure 6.2b) has consistently high TOC (2–8 wt%; Whelan et al., 1990), microbial cell density, and anaerobic sulfate reduction throughout the sediment package, with sulfate concentration reaching zero at approximately 35 mbsf. Although biogenic methane builds up to its highest concentrations (~2 mM) between 65 and 75 mbsf, methanogenesis also takes place alongside sulfate reduction within the 0–35 mbsf zone (methane concentration, however, is lower here because of AOM activity); at 30 mbsf both sulfate reduction and hydrogenotrophic methanogenesis intensify (Parkes et al., 2005). In the deep subsurface of Site 1229, 16S rRNA gene sequence analysis identified *Crenarchaeota*, DSAG, South African Gold Mine lineages affiliated to *Euryarchaeota*, Methanobacteriales and Methanosarcinales as the major archaeal groups, while *Alphaproteobacteria*, *Bacteroidetes*, *Betaproteobacteria*, *Chloroflexi*, *Deltaproteobacteria*, *Gammaproteobacteria*, Candidate Divisions JS1 and OP8, *Planctomycetes*, and Novel Group NT-B2 were found to be the major bacterial groups (Parkes et al., 2005). In contrast, only Marine Group I members were detected in the 7.75–7.80 mbsf zone, below the penetration depth of porewater nitrate at Site 1225 (Site 1a in Figure 6.2a; Lauer et al., 2016). In these sediment horizons, remarkably, microbial activity and cell density is again rejuvenated below the SMTZ at ~90 mbsf, on account of a mirror image microbiome profile shaped from the core bottom upwards by nitrate and oxygen supplied from the underlying basaltic aquifer through which oxic water circulates (D'Hondt et al., 2004). Sulfate reduction, thus, peaks again at ~95 mbsf, alongside methanogenesis from acetate (Parkes et al., 2005). Remarkably, in some of the deepest sediment layers of this territory, facultative aerobes (these have also been isolated as pure cultures) are sustained by the small but definite amounts of NO^{3-}, Mn^{4+}, Fe^{3+} and O_2 present (Batzke et al., 2007). Repeated rejuvenation of the microbiota at greater sediment-depths is also encountered in the ODP site 1226 (Site 1b in Figure 6.2b), where such phenomena remarkably coincide with high diatom content of the sediments (three diatom-rich layers characterized by enhanced sulfate reduction and cell density have been identified between the surface and 400 mbsf) due to reasons that are not clear (Parkes et al., 2005). Notably, strains isolated from the deepest (420 mbsf of Site 1226, where the sediment is ~16.5 million years old) and oldest (101 mbsf of Site 1231, where the sediment is ~40 million years old) samples of this region belonged to *Firmicutes* and *Actinobacteria* (D'Hondt et al., 2003; Aiello et al., 2006; Batzke et al., 2007).

The sedimentary microbiome along the first 200 mbsf of the Cascadia Margin (Site 3 in Figure 6.2b) follows the general trend of decreasing microbial populations, but in the 222–234 mbsf horizon the total bacterial population again increases in association with the existence of methane hydrates (Cragg et al., 1996; Colwell et al., 2008). Anaerobic microbial processes associated with the latter sediment-depths are still incompletely known, although sulfate-reduction and methane oxidation are apparently active *in situ* (Cragg et al., 1996). Copious bacterial populations (1.37×10^9 cells cm^{-3} at the sediment surface to 1.11×10^7 cells cm^{-3} at the base of sediment core) are also known to be present throughout the 518 m sediment horizon explored thus far in the east Japan Sea (Cragg et al., 1992).

Low rates of methanogenesis, both hydrogenotrophic and acetoclastic, have been recorded above the base of the carbonate mound in Porcupine Seabight (Site 4 in Figure 6.2b), where microbial communities are largely bacteria-dominated (Webster et al., 2009). Below the mound base, i.e. at >121 mbsf, only acetoclastic methanogenesis has been detected alongside an increase in methane concentration and the processes of sulfate reduction and acetate oxidation. The 16S rRNA gene sequence-based microbial diversity is highest at ~110.1 meters above the mound base (mamb), after which it decreases until 53.1 mamb (Webster et al., 2009). Within 17.1–24.1 mamb, again, diversity increases by a few orders of magnitude and then remains unchanged. The bacterial taxa present above the mound base are Candidate Division JS1, *Chloroflexi*, *Deltaproteobacteria* and *Gammaproteobacteria*, while *Betaproteobacteria* and *Gammaproteobacteria* prevail below the base; the archaeal population, in turn, is dominated by *Euryarchaeota* SAGMEG, although they occur only above the mound base together with *Crenarchaeota* Group C3 (Webster et al., 2009).

Geomicrobiology of >5 mbsf Subsurfaces Having an Oxic Milieu

Owing to low sedimentation rate, the sub-seafloor territories of all the major oceanic gyres remain oxic throughout their depth (D'Hondt et al., 2009; Røy et al., 2012). Not only is the seafloor organic carbon flux extremely low in these regions, most of the buried organic matter, in turn, is mineralized within shallow depths of the sediment; with further increase in sediment-depth, the organic matter becomes recalcitrant and un-utilizable for microorganisms. Although sediment-depth-wise fine resolution analysis of microbial community structure is still lacking for the deep subsurfaces of the oligotrophic oceans, aerobic heterotrophs, having extremely low respiration rates, are found throughout these sediment regimes (D'Hondt et al., 2015).

Aerobic microorganisms, including aerobic non-spore formers, characterized by extremely low respiration rates, are present throughout the deep subsurface (up to 101.5 million years old) of the SPG (Site 10b in Figure 6.2b), even though cell numbers are too low for counting by standard methods in slurry culture (up to seven orders of magnitude less than the average count for the same sediment-depths in costal and upwelling regions; D'Hondt et al., 2009; Morono et al., 2020). Although sulfate and nitrate are present in these sediment systems microbes cannot use them as terminal electron acceptors because O_2 is also abundant; in the process, anaerobes are potentially eliminated from these sedimentary microbiomes over geological times (Morono et al., 2020). Although pure culture based microbiological reports from these deep subsurface ecosystems of the marine realm are few, strains of *Achromobacter* and *Luteococcus* have been isolated from 18.1–18.2 mbsf and 123.2–123.3 mbsf of the SPG, respectively (Fan et al., 2014; Zhang et al., 2014). Furthermore, the presence of bacterial magnetites (magnetofossils) throughout the SPG sediments has indicated past activity of magnetotactic bacteria (Yamazaki and Shimono, 2013).

The deep subsurface territories of NPG (Site 10a in Figure 6.2b) also remain oxic throughout because of extremely low sedimentation rates, while the microbiomes too have very low respiration rates; density of microbial cells in this region decreases from 10^8 cm^{-3} at the sediment surface to 10^3 cm^{-3} at 20 mbsf (Røy et al., 2012). Although the sediment horizon has been empirically investigated up to 28 mbsf, extrapolation of O_2 concentration data has suggested that the ~100 m sediment package is entirely oxygenated.

Deep persistence of organic matter despite the presence of O_2 and heterotrophic microbial activities also characterizes the NAG sediments. Microbial cell abundance, here, decreases from 1.45×10^5 cm^{-3} at the sediment surface to 13.7 cm^{-3} at the core bottom (Estes et al., 2019).

In this context, it is noteworthy that the microbiology of the deep sediments of SAG is unexplored; the same is also true for the sediments of Indian Ocean Gyre. For the SPG, NPG and NAG, depth-wise community metabolisms and population dynamics, geological consequences and manifestations, and evolutionary implications of their deep subsurface life are still far from being clear. This said, from the point of view of physiology, it is extraordinary that territorially contained microbial populations persist through millions of years of diagenetic maturation of the sediment that they have been buried into. Aging sediments undergo continued nutrient impoverishment, so considerable efforts have gone into explaining how microbes remain alive, turn over cellular elements, and then grow, if at all, under acute nutrient scarcity (Jørgensen, 2011; Langerhuus et al., 2012; Lomstein et al., 2012; Braun et al., 2017). In the context of the O_2-containing SPG sediments (D'Hondt et al., 2009, 2015) it is remarkable that aerobic bacterial populations have remained viable *in situ* for hundreds of millions of years under extreme oligotrophy and ultraslow rates of metabolism (facultative anaerobes, although present, have apparently remained less viable; Morono et al., 2020). Hitherto unknown principles of bioenergetics underlying such life in ultraslow motion not only have the potentials of redefining life's adaptations and sustainability on a planetary scale but also change the paradigm of our understanding aging and death. Furthermore, from the perspective of the resilience of aerobic life in particular, it is noteworthy that whereas in the million years old SPG sediments it is constrained by low carbon and energy (but not O_2), in another deep subsurface ecosystem of the marine realm, namely the thousands of years old, highly-sulfidic sediments of Arabian Sea OMZ, its main constraint is O_2 itself (Bhattacharya et al., 2019, 2020). In the biogeochemical sense, the two sediment systems are mirror images of each other. While extremely low rates of sedimentation and organic carbon delivery to the SPG sediment surface constrain aerobic catabolic pathways *in situ* (and facilitate oxygenation of entire sediment horizons), OMZ sediments have very shallow depth of O_2 infiltration because copious influx of labile organic matter across the seabed leads to rapid O_2 consumption within a few centimeters below the seafloor (Breuer et al., 2009; Cavan et al., 2017; Jessen et al., 2017); in addition, high concentrations of porewater H_2S often preclude the presence/influx of O_2 in the OMZ sediments (Bhattacharya et al., 2019, 2020). Thus, the two ecosystems in tandem elucidate the extraordinary resilience of aerobic bacteria in the face of resource scarcity, be it for reduced metabolites (carbon and energy source) as in the SPG sediments, or oxidants (O_2 in the form of a terminal electron acceptor) as in the OMZ sediments. Overall, it is evident from these reports that aerobic bacteria have very strong metabolic aptitudes for frugal lifestyle in terms of resource requirement and utilization. This said, we know little about how this is achieved mechanistically. For example, we still do not know the genetic regulations, cell signaling pathways, and metabolomic controls for life in ultraslow motion. It is also still unclear why organic matter depletes so slowly along SPG sediment-depths despite live aerobic bacteria and O_2 both being there. Similarly, we also do not know what cryptic source(s) of O_2 sustains obligate aerobes in the highly sulfidic sediments of Arabian Sea OMZ (Bhattacharya et al., 2020). Nevertheless, microbial ecological data available are sound enough to evoke the hypothesis that aerobic life could have already been well established on Earth long before organic matter became copious across the chemosphere or the atmosphere became overwhelmingly oxic.

6.5. SUMMING UP THE GEOMICROBIOLOGY OF THE MARINE DEEP SUBSURFACE

Marine deep subsurfaces are peculiar ecosystems because, with time, they get progressively disconnected from the rest of the biosphere in terms of linear flow of energy and cyclic flow of nutrients (redox cycling of elements). Organic matter precipitation from the overlying water-column is the main source of carbon for the microorganisms which live here under extreme energy and nutrient limitation. Sediments of continental margins and upwelling regions receive steady supply of organic matter by virtue of higher productivity in the water-columns and sedimentation rate. In contrast, both these parameters are low in the case of open ocean sediments, which consequently receive typically scant organic carbon supply over time. Microbial metabolisms in the different strata of a marine sediment system are eventually controlled by the sedimentation rate of that region: with increasing distance from the continental landmass sedimentation rate decreases and the vertical expanse of different metabolic zones/layers in the sediment increase. On top of this, in the oceanic gyres, entire sediment packages are characterized by O_2 penetration because of ultralow sedimentation rates and very low respiration rates of the sedimentary microbiota.

In the continental shelf and slope sediments anoxic/euxinic zones appear often within a few centimeters below the seafloor. Sulfate and methane, in conjunction with their microbial transformers, shape the overall geomicrobiology of these sediment systems. Sulfate–methane transition zones, where concentrations of both sulfate and methane approach zero, often feature enhanced microbial density, diversity and/or activity. Apart from the typical geomicrobiological profiles of shelf/slope and open ocean sediments, special physiographic features such as gas hydrate deposits, crustal seawater intruded basement sediments, mud volcanoes, and carbonate mounds have their individual geomicrobiological characteristics. Over the decades, as progressive improvement of coring techniques has made investigation of surface to basement sediments feasible, viable microorganisms have been discovered in O_2-containing ocean gyre sediments that are 100 million years old. In such diagenetically highly-matured, carbon-deprived, ancient habitats, cell system maintenance and evasion of death seem to be primarily dependent on the microbes ability to slow down their metabolic rates in sync with the extremely low energy/carbon flux in the physicochemical milieu.

6.6. ULTRASLOW METABOLISM AND SUSTAINABILITY OF DEEP LIFE: IMPLICATIONS FOR EVOLUTION AND ASTROBIOLOGY

Apparent metabolic rates of microbial communities in these nutrient-impoverished extremities of the biosphere (calculated based on the turnover of major substrates in bulk sediments) are said to be four to five orders of magnitude less than the microbial metabolic rates typically encountered in ecosystems on the surface of the Earth (D'Hondt et al., 2002; Price and Sowers, 2004; Jørgensen and D'Hondt, 2006; Lin et al., 2006). Furthermore, in deep subsurface microbiomes where energy and nutrient flux is extremely low, significant fractions of bacterial, archaeal and fungal populations have been found to retain potentials for rapid *in vitro* metabolism and growth (if supplied with excess substrates), even after wintering millions of years of 'starvation' and ultralow metabolic activities (Inagaki et al., 2005; Lin et al., 2006; Morono et al., 2011; Lomstein et al., 2012; Colwell and D'Hondt, 2013; Orsi et al., 2013a, 2013b; Inagaki et al., 2015; Rédou et al., 2015; Pachiadaki et al., 2016; Rajala and Bomberg, 2017). Remarkably, rates of active metabolism recorded during *in vitro* 'revival' of deep subsurface populations were more than thousand times faster than the rates of metabolic activities[2] estimated for the natural populations of deep biospheres (Morono et al., 2011).

The metabolic state of deep subsurface microbial communities – i.e. the physiological, biochemical and molecular mechanisms that microbial cells utilize to remain alive *in situ* – is one of the most remarkable intrigues in contemporary microbiology. The above-mentioned research works collectively envisage that deep sediments, with aging, become increasingly impoverished in terms of nutrients because of a progressively decreasing influx, and concomitant utilization by residential microbial populations; the latter also causes the pool of organic carbon to become more and more refractory to biodegradation (a process known as aging of the organic matter). In such scenarios, community metabolism, and therefore community size, is thought to be balanced against the ambient energy/carbon flux[3]. Microorganisms are said to recalibrate their metabolic rates down to the bare minimum levels required for sheer maintenance of cellular machineries (prevent the slow, spontaneous chemical/radiolytic degradation of vital cell components, including DNA); cell populations apparently achieve millions of years of longevity in this way by metabolizing such ultralow amounts of carbon that are far below their species-specific molar growth yields (i.e. the amount needed, relative to the cells' own biomass, to double their cell size and divide) (Jørgensen, 2011). Theoretical calculations based on organic geochemistry data from deep, energy-limited marine sediments (Jørgensen, 2011; Langerhuus et al., 2012; Lomstein et al., 2012) further estimate:

- *in situ* generation time of anaerobic microbial cells to be in the order of 1000 years;
- turnover times[4] of vegetative cells to be much lower (anywhere between tens to hundreds of years) than the microbial necromass turnover times and cellular generation times (Langerhuus et al., 2012; Braun et al., 2017);
- turnover time of microbial necromass to be in the range of 0.5–1.0 × 10^5 years.

---

[2] *Microbial activities in deep biospheres are quantified based on (i) numerical transport-reaction modeling of porewater data (Kimura et al., 1997; Morris et al., 2003), (ii) experiments involving radioactive tracers, or (iii) the extent of racemization of D:L-aspartic acid in the necromass of the microbiota (Braun et al., 2017).*

[3] *In an ecosystem, flux of carbon through the biomass is defined as "carbon content of the biomass (expressed in kg C vol^{-1}) ÷ biomass turnover time (expressed in year)".*

[4] *Biomass/necromass turnover time – with respect to any element (let's say, for example, carbon or nitrogen) – is defined as "the standing C or N content in the biomass/necromass (expressed in kg vol^{-1}) ÷ annual input of C or N into the biomass/necromass (expressed in kg vol^{-1} year^{-1})". Measurements of turnover of biomass carbon are based on measurements of inputs and declines of 14C in sediments; similarly, turnover of biomass nitrogen is measured by adding 15N-labeled nitrate or other salts to the sediments; notably, turnover rates of different nutrients within the biomass may be quite different.*

Whereas such exorbitantly high generation time would, on the ground, mean that the cells only turn over their biomass extremely slowly and in reality never divide, the above findings also imply that many microorganisms of the deep subsurface exist in the form of dormant entities (Kminek et al., 2003; Langerhuus et al., 2012; Lomstein et al., 2012), in a metabolic state comparable to that reported for Siberian permafrost microflora and which involves extremely low level of metabolism that is sufficient to achieve continuous DNA repair (Johnson et al., 2007). The extremely high values estimated for *in situ* generation time may have been reached because analyses were carried out by integrating bulk data. It may well be the case that there actually are a few cells that have generation times of days/weeks and a vast majority of the biomass is actually dead (or at least never dividing). As bulk analyses cannot discern these two extremes, one is prone to derive exorbitantly low metabolic rates that probably are not exemplified anywhere else in nature. In order to elucidate the microbiology of such eternal lag phases[5] in natural populations, and, as a corollary, to understand the biology of aging, such ingenious culture-based experiments are needed which simulate progressively acute energy and nutrient scarcity, and reveal the comprehensive metabolomics of the microorganisms that are subjected to those conditions.

Deep subsurface ecosystems of the marine realm are totally deprived of phototrophic primary productivity, and so depend completely on chemotrophic nutrition and get progressively constrained (with depth and age) for electron donors, electron acceptors and carbon. These features make the microbial communities of these low-energy habitats ideal models for understanding the biology of extreme oligotrophy (limits of low energy requirement for life). Furthermore, it is now a well-known hypothesis in evolution that when impacts of asteroids and comets endangered the existence of early life on Earth, the last common ancestor of all extant life on Earth survived in deep water or deep subsurface where it was insulated from the thermal radiation (Sleep and Zahnle, 1998; Michalski et al., 2018). While such a refugium role has also been envisaged in relation to the Martian subsurface and potential early life in that planet, from an astrobiological perspective, the deep marine subsurface of Earth serves as an analog for studying the thermodynamic/bioenergetic feasibility of subsurface life in a number of extraterrestrial bodies such as Mars, Europa, Enceladus and Titan, where potentially habitable subsurfaces exist beneath biochemically/biophysically inhospitable surface conditions (Jones et al., 2018). All the above points in unison guarantee that in the days to come the marine deep subsurface shall remain one of the hottest spots of research in geobiology, exobiology, and evolution.

[5] *Lag phase is regarded as that period of microbial growth in laboratory cultures when bacterial cells do not divide but only adapt themselves to the culture conditions. In this phase of the growth cycle, which lasts for varied time-frames (ranging from 1 h to several days) depending on the microorganism concerned and the conditions provided, cells synthesize RNA, enzymes, and other biomolecules that are necessary for subsequent division (Tortora et al., 2018).*

REFERENCES

Aiello, I.W., Kellett, K. and Jørgensen, B.B. (2006). Sedimentology of open ocean biogenic sediments from ODP Leg 201, Eastern Equatorial Pacific (Sites 1225 and 1226). Proceedings of the Ocean Drilling Program, Scientific results (Vol. 101).

Akam, S.A., Coffin, R.B., Abdulla, H.A. et al. (2020). Dissolved Inorganic Carbon Pump in Methane-Charged Shallow Marine Sediments: State of the Art and New Model Perspectives. *Frontiers in Marine Science 7*: 206. https://doi.org/10.3389/fmars.2020.00206

Arndt, S., Jørgensen, B.B., LaRowe, D.E. et al. (2013). Quantifying the degradation of organic matter in marine sediments: a review and synthesis. *Earth-Science Reviews 123*: 53–86. https://doi.org/10.1016/j.earscirev.2013.02.008

Baker, P.A., Stout, P.M., Kastner, M. et al. (1991). Large-scale lateral advection of seawater through oceanic crust in the central equatorial Pacific. *Earth and Planetary Science Letters,105*: 522–533. https://doi.org/10.1016/0012–821X(91)90189-O

Batzke, A., Engelen, B., Sass, H. et al. (2007). Phylogenetic and physiological diversity of cultured deep-biosphere bacteria from equatorial Pacific Ocean and Peru Margin sediments. *Geomicrobiology Journal 24*: 261–273. https://doi.org/10.1080/01490450701456453

Bhattacharya, S., Roy, C., Mandal, S. et al. (2019). Metabolically-active obligate aerobes in anoxic (sulfidic) marine sediments. *bioRxiv*, 728287. https://doi.org/10.1101/728287

Bhattacharya, S., Roy, C., Mandal, S. et al. (2020). Aerobic microbial communities in the sediments of a marine oxygen minimum zone. *FEMS Microbiology Letters 367*: fnaa157. https://doi.org/10.1093/femsle/fnaa157

Bhattacharya, S., Mapder, T., Fernandes, S. et al. (2021). Sedimentation rate and organic matter dynamics shape microbiomes across a continental margin. *Biogeosciences 18*: 5203–5222. https://doi.org/10.5194/bg-18-5203-2021

Biddle, J.F., Fitz-Gibbon, S., Schuster, S.C. et al. (2008). Metagenomic signatures of the Peru Margin subseafloor biosphere show a genetically distinct environment. *Proceedings of the National Academy of Sciences USA 105*: 10583–10588. https://doi.org/10.1073/pnas.0709942105

Biddle, J.F., House, C.H. and Brenchley, J.E. (2005). Microbial stratification in deeply buried marine sediment reflects changes in sulfate/methane profiles. *Geobiology 3*: 287–295. https://doi.org/10.1111/j.1472–4669.2006.0006

Biddle, J.F., Lipp, J.S., Lever, M.A. et al. (2006). Heterotrophic Archaea dominate sedimentary subsurface ecosystems off Peru. *Proceedings of the National Academy of Sciences USA 103*: 3846–3851. https://doi.org/10.1073/pnas.0600035103

Biddle, J.F., Sylvan, J.B., Brazelton, W.J. et al. (2012). Prospects for the study of evolution in the deep biosphere. *Frontiers in Microbiology 2*: 285. https://doi.org/10.3389/fmicb.2011.00285

Biddle, J.F., White, J.R., Teske, A.P. et al. (2011). Metagenomics of the subsurface Brazos-Trinity Basin (IODP site 1320): comparison with other sediment and pyrosequenced metagenomes. *The ISME Journal 5*: 1038. https://doi.org/10.1038/ismej.2010.199

Blazewicz, S.J., Barnard, R.L., Daly, R.A. et al. (2013). Evaluating rRNA as an indicator of microbial activity in environmental communities: limitations and uses. *The ISME Journal 7*: 2061. https://doi.org/10.1038/ismej.2013.102

Bograd, S.J., Schwing, F.B., Castro, C.G. et al. (2002). Bottom water renewal in the Santa Barbara Basin. *Journal of Geophysical Research Oceans 107*: 3216. https://doi.org/10.1029/2001JC001291

Bowles, M.W., Mogollón, J.M., Kasten, S. et al. (2014). Global rates of marine sulfate reduction and implications for sub-sea-floor metabolic activities. *Science 344*: 889–891. https://doi.org/10.1126/science.1249213

Bradley, J.A., Amend, J.P. and LaRowe, D.E. (2018). Survival of the fewest: microbial dormancy and maintenance in marine sediments through deep time. *Geobiology 17*: 43–59. https://doi.org/10.1111/gbi.12313

Braun, S., Mhatre, S.S., Jaussi, M. et al. (2017). Microbial turnover times in the deep seabed studied by amino acid racemization modelling. *Scientific Reports 7*: 5680. https://doi.org/10.1038/s41598–017–05972-z

Breuer, E.R., Law, G.T., Woulds, C. et al. (2009). Sedimentary oxygen consumption and microdistribution at sites across the Arabian Sea oxygen minimum zone (Pakistan margin.). *Deep Sea Research Part II: Topical Studies in Oceanography 56*: 296–304. https://doi.org/10.1016/j.dsr2.2008.06.010

Burdige, D.J. (2007). Preservation of organic matter in marine sediments: controls, mechanisms and an imbalance in sediment organic carbon budgets? *Chemical Reviews 107*: 467–485. https://doi.org/10.1021/cr050347q

Canfield, D.E. (1994). Factors influencing organic carbon preservation in marine sediments. *Chemical Geology 114*: 315–329. https://doi.org/10.1016/0009–2541(94)90061–2

Cavan, E.L., Trimmer, M., Shelley, F. et al. (2017). Remineralization of particulate organic carbon in an ocean oxygen minimum zone. *Nature Communications 8*: 14847. https://doi.org/10.1038/ncomms14847

Chavez, F.P. and Barber, R.T. (1987). An estimate of new production in the equatorial Pacific. *Deep Sea Research Part A. Oceanographic Research Papers 34*: 1229–1243. https://doi.org/10.1016/0198–0149(87)90073–2

Chen, M., Kim, J.H., Choi, J. et al. (2017). Biological early diagenesis and insolation-paced paleoproductivity signified in deep core sediment organic matter. *Scientific Reports 7*: 1581. https://doi.org/10.1038/s41598–017–01759–4

Cho, H., Hyun, J.H., You, O.R. et al. (2017). Microbial community structure associated with biogeochemical processes in the sulfate–methane transition zone (SMTZ) of gas-hydrate-bearing sediment of the Ulleung Basin, East Sea. *Geomicrobiology Journal 34*: 207–219. https://doi.org/10.1080/01490451.2016.1159767

Colwell, F.S., Boyd, S., Delwiche, M.E. et al. (2008). Estimates of biogenic methane production rates in deep marine sediments at Hydrate Ridge, Cascadia margin. *Applied and Environmental Microbiology 74*,: 3444–3452. https://doi.org/10.1128/AEM.02114–07

Colwell, F.S. and D'Hondt, S. (2013). Nature and extent of the deep biosphere. *Reviews in Mineralogy and Geochemistry 75*: 547–574. https://doi.org/10.2138/rmg.2013.75.17

Colwell, F.S., Stormberg, G.J., Phelps, T.J. et al. (1992). Innovative techniques for collection of saturated and unsaturated subsurface basalts and sediments for microbiological characterization. *Journal of Microbiological Methods 15*: 279–292. https://doi.org/10.1016/0167-7012(92)90047-8

Cowen, J.P., Giovannoni, S.J., Kenig, F. et al. (2003). Fluids from aging ocean crust that support microbial life. *Science 299*: 120–123. https://doi.org/10.1126/science.1075653

Cragg, B.A. and Kemp, A.E.S. (1995). 26. Bacterial profiles in deep sediment layers from the eastern Equatorial Pacific Ocean, Site 851. *Proceedings of the Ocean Drilling Program, Scientific Results 138*: 599–604. https://doi.org/10.2973/odp.proc.sr.138.130.1995

Cragg, B.A., Parkes, R.J., Fry, J.C. et al. (1990). Bacterial biomass and activity profiles within deep sediment layers. *Proceedings of the Ocean Drilling Program Scientific Results 112*: 607–619. https://doi.org/10.2973/odp.proc.sr.112.161.1990.

Cragg, B.A., Harvey, S.M., Fry, J.C. et al. (1992). 46. Bacterial biomass and activity in the deep sediment layers of the Japan Sea, Hole 798B. *Proceedings of the Ocean Drilling Program, Scientific Results 127*/128: 761–776. https://doi.org/10.2973/odp.proc.sr.127128-1.184.1992

Cragg, B.A., Parkes, R.J., Fry, J.C. et al. (1995). 10. Bacterial profiles in deep sediments of the Santa Barbara basin, SITE 893. *Proceedings of the Ocean Drilling Program, Scientific Results 146*: 1–25. https://doi.org/10.2973/odp.proc.sr.146-2.302.1995

Cragg, B.A., Parkes, R.J., Fry, J.C. et al. (1996). Bacterial populations and processes in sediments containing gas hydrates (ODP Leg 146: Cascadia Margin). *Earth and Planetary Science Letters 139*: 497–507. https://doi.org/10.1016/0012–821X(95)00246–9

D'Hondt, S., Rutherford, S. and Spivack, A.J. (2002). Metabolic activity of subsurface life in deep-sea sediments. *Science 295*: 2067–2070. https://doi.org/10.1126/science.1064878

D'Hondt, S.L., Jørgensen, B.B., Miller, D.J. et al. (2003). *Proceedings of the Ocean Drilling Program, Initial Reports 201*. https://doi.org/10.2973/odp.proc.ir.201.2003

D'Hondt, S., Jørgensen, B.B., Miller, D.J. et al. (2004). Distributions of microbial activities in deep subseafloor sediments. *Science 306*: 2216–2221. https://doi.org/10.1126/science.1101155

D'Hondt, S., Spivack, A.J., Pockalny, R. et al. (2009). Subseafloor sedimentary life in the South Pacific Gyre. *Proceedings of the National Academy of Sciences 106*: 11651–11656. https://doi.org/10.1073/pnas.0811793106

D'Hondt, S., Inagaki, F., Zarikian, C.A. et al. (2015). Presence of oxygen and aerobic communities from sea floor to basement in deep-sea sediments. *Nature Geoscience 8*: 299–304. https://doi.org/10.1038/ngeo2387

Drenzek, N.J., Montluçon, D.B., Yunker, M.B. et al. (2007). Constraints on the origin of sedimentary organic carbon in the Beaufort Sea from coupled molecular 13C and 14C measurements. *Marine Chemistry 103*: 146–162. https://doi.org/10.1016/j.marchem.2006.06

Durbin, A.M. and Teske, A. (2011). Microbial diversity and stratification of South Pacific abyssal marine sediments. *Environmental Microbiology 13*: 3219–3234. https://doi.org/10.1111/j.1462-2920.2011.0254

Edwards, K.J., Fisher, A.T. and Wheat, C.G. (2012). The deep subsurface biosphere in igneous ocean crust: frontier habitats for microbiological exploration. *Frontiers in Microbiology 3*: 8. doi:10.3389/fmicb.2012.00008

Estes, E.R., Pockalny, R., D'Hondt, S. et al. (2019). Persistent organic matter in oxic subseafloor sediment. *Nature Geoscience 12*: 126–131. https://doi.org/10.1038/s41561-018-0291-5

Fan, X., Zhang, Z., Li, Z. et al. (2014). *Luteococcussediminum* sp. nov., isolated from deep subseafloor sediment of the South Pacific Gyre. *International Journal of Systematic and Evolutionary Microbiology 64*: 2522–2527. https://doi.org/10.1099/ijs.0.058529-0

Farmer, J.R., Hönisch, B., Haynes, L.L. et al. (2019). Deep Atlantic Ocean carbon storage and the rise of 100,000-year glacial cycles. *Nature Geoscience 12*: 355. https://doi.org/10.1016/j.earscirev.2013.02.008

Fernandes, S., Mazumdar, A., Bhattacharya, S. et al. (2018). Enhanced carbon-sulfur cycling in the sediments of Arabian Sea oxygen minimum zone center. *Scientific Reports 8*: 1–15. https://doi.org/10.1038/s41598-018-27002-2

Fischer, J.P., Ferdelman, T.G., D'Hondt, S. et al. (2009). Oxygen penetration deep into the sediment of the South Pacific gyre. *Biogeosciences 6*: 1467–1478. https://doi.org/10.5194/bg-6-1467-2009

Fontanez, K.M., Eppley, J.M., Samo, T.J. et al. (2015). Microbial community structure and function on sinking particles in the North Pacific Subtropical Gyre. *Frontiers in Microbiology 6*: 469. https://doi.org/10.3389/fmicb.2015.00469

France-Lanord, C., Spiess, V., Klaus, A. et al. (2016). Site U1451. *Proceedings of the International Ocean Discovery Program 354*. https://doi.org/10.14379/iodp. proc.354.105.2016

Friese, A., Kallmeyer, J., Kitte, J.A. et al. (2017). A simple and inexpensive technique for assessing contamination during drilling operations. *Limnology and Oceanography Methods 15*: 200–211. https://doi.org/10.1002/lom3.10159

Griffin, W.T., Phelps, T.J., Colwell, F.S. et al. (1997). Methods for obtaining deep subsurface microbiological samples by drilling. In: *The Microbiology of the Terrestrial Deep Subsurface* (eds. S.A. Penny and L.H. Dana), 23–44. Boca Raton, FL: CRC Press.

Harrison, B.K., Zhang, H., Berelson, W. et al. (2009). Variations in archaeal and bacterial diversity associated with the sulfate-methane transition zone in continental margin sediments (Santa Barbara Basin, California). *Environmental MicrobiologyApplied and environmental Microbiology 75*: 1487–1499. https://doi.org/10.1128/AEM.01812-08

Hoehler, T.M. and Jørgensen, B.B. (2013). Microbial life under extreme energy limitation. *Nature Reviews Microbiology 11*: 83. https://doi.org/10.1038/nrmicro2939

Holmkvist, L., KamyshnyJr, A., Vogt, C. et al. (2011). Sulfate reduction below the sulfate–methane transition in Black Sea sediments. *Deep Sea Research Part I: Oceanographic Research Papers 58*: 493–504. https://doi.org/10.1016/j.dsr.2011.02.009

Hoshino, T., Morono, Y., Terada, T. et al. (2011). Comparative study of subseafloor microbial community structures in deeply buried coral fossils and sediment matrices from the challenger mound in the porcupine seabight. *Frontiers in Microbiology2*: 231. https://doi.org/10.3389/fmicb.2011.00231

House, C.H., Cragg, B.A., Teske, A. et al. (2003). Drilling contamination tests during ODP Leg 201 using chemical and particulate tracers. *Proceedings of the Ocean Drilling Program, Initial Reports 201*: 1–19. https://doi.org/10.2973/odp. proc. ir.201.102.2003

Huber, J.A., Johnson, H.P., Butterfield, D.A. et al. (2006). Microbial life in ridge flank crustal fluids. *Environmental Microbiology 8*: 88–99. https://doi.org/10.1111/j.1462-2920.2005.0087

Inagaki, F., Okada, H., Tsapin, A.I. et al. (2005). Microbial survival: the paleome: a sedimentary genetic record of past microbial communities. *Astrobiology 5*: 141–153. https://doi.org/10.1089/ast.2005.5.141

Inagaki, F., Nunoura, T., Nakagawa, S. et al. (2006). Biogeographical distribution and diversity of microbes in methane hydrate-bearing deep marine sediments on the Pacific Ocean Margin. *Proceedings of the National Academy of Sciences USA 103*: 2815–2820. https://doi.org/10.1073/pnas.0511033103

Inagaki, F., Hinrichs, K.U., Kubo, Y. et al. (2015). Exploring deep microbial life in coal-bearing sediment down to ~2.5 km below the ocean floor. *Science 349*: 420–424. https://doi.org/10.1126/science.aaa6882

Ingram, B.L. and Kennett, J.P. (1995). Radiocarbon chronology and planktonic-benthic foraminiferal ^{14}C age differences in Santa Barbara Basin sediments, Hole 893A. *Proceedings of the Ocean Drilling Program, Scientific Results 146*: 19–27. https://doi.org/10.2973/odp.proc.sr.146-2.278.1995

Jena, B., Sahu, S., Avinash, K. et al. (2013). Observation of oligotrophic gyre variability in the south Indian Ocean: Environmental forcing and biological response. *Deep Sea Research Part I: Oceanographic Research Papers 80*: 1–10. https://doi.org/10.1016/j.dsr.2013.06.002

Jessen, G.L., Lichtschlag, A., Ramette, A. et al. (2017). Hypoxia causes preservation of labile organic matter and changes seafloor microbial community composition (Black Sea). *Science Advances 3*: e1601897. https://doi.org/10.1126/sciadv.1601897

Johnson, S.S., Hebsgaard, M.B., Christensen, T.R. et al. (2007). Ancient bacteria show evidence of DNA repair. *Proceedings of the National Academy of Sciences USA 104*: 14401–14405. https://doi.org/10.1073/pnas.0706787104

Jones, R.M., Goordial, J.M. and Orcutt, B.N. (2018). Low energy subsurface environments as extraterrestrial analogs. *Frontiers in Microbiology 9*: 1605. https://doi.org/10.3389/fmicb.2018.01605

Jørgensen, B.B., Findlay, A.J. and Pellerin, A. (2019). The biogeochemical sulfur cycle of marine sediments. *Frontiers in Microbiology 10*: 849. https://doi.org/10.3389/fmicb.2019.00849

Jørgensen, B.B., Weber, A. and Zopfi, J. (2001). Sulfate reduction and anaerobic methane oxidation in Black Sea sediments. *Deep Sea Research Part I: Oceanographic Research Papers 48*: 2097–2120. https://doi.org/10.1016/S0967-0637(01)00007-3

Jørgensen, B.B. (2011). Deep subseafloor microbial cells on physiological standby. *Proceedings of the National Academy of Sciences USA 108*: 18193–18194. https://doi.org/10.1073/pnas.1115421108

Jørgensen, B.B. and Boetius, A. (2007). Feast and famine-microbial life in the deep-sea bed. *Nature Reviews Microbiology 5*: 770. https://doi.org/10.1038/nrmicro1745

Jørgensen, B.B. and D'Hondt, S. (2006). A starving majority deep beneath the seafloor. *Science 314*: 932–934. https://doi.org/10.1126/science.1133796

Kallmeyer, J. (2011). Detection and quantification of microbial cells in subsurface sediments. In *Advances in Applied Microbiology*, Vol. *76* (eds. I. Allen, S.S. Laskin and M.G. Geoffrey), 79–103. Academic Press.

Kallmeyer, J., Mangelsdorf, K., Cragg, B. et al. (2006). Techniques for contamination assessment during drilling for terrestrial subsurface sediments. *Geomicrobiology Journal 23*: 227–239. https://doi.org/10.1080/01490450600724258

Kallmeyer, J., Pockalny, R., Adhikari, R.R. et al. (2012). Global distribution of microbial abundance and biomass in subseafloor sediment. *Proceedings of the National Academy of Sciences 109*: 16213–16216. https://doi.org/10.1073/pnas.1203849109

Kempes, C.P., van Bodegom, P.M., Wolpert, D. et al. (2017). Drivers of bacterial maintenance and minimal energy requirements. *Frontiers in Microbiology 8*: 31. https://doi.org/10.3389/fmicb.2017.00031

Kieft, T.L., Phelps, T.J. and Fredrickson, J.K. (2007). Drilling, coring and sampling subsurface environments. In *Manual of Environmental Microbiology 3e* (eds. J. Hurst, R.L. Crawford, G.R. Knudsen et al.), 799–817. Washington, DC: American Society of Microbiology.

Kimura, G., Silver, E., Blum, P. et al. (1997). Costa Rica Accretionary Wedge Sites 1039–1043. *Proceedings of the Ocean Drilling Program, Initial Reports 170*. https://doi.org/10.2973/odp.proc.ir.170.1997

Kminek, G., Bada, J.L., Pogliano, K. et al. (2003). Radiation-dependent limit for the viability of bacterial spores in halite fluid inclusions and on Mars. *Radiation Research 159*: 722–729. https://doi.org/10.1667/0033-7587(2003)159[0722:rlftvo]2.o;2

Komada, T., Burdige, D.J., Li, H.L. et al. (2016). Organic matter cycling across the sulfate-methane transition zone of the Santa Barbara Basin, California Borderland. *Geochimica et Cosmochimica Acta 176*: 259–278. https://doi.org/10.1016/j.gca.2015.12.022

Lagostina, L., Goldhammer, T., Røy, H. et al. (2015). Ammonia-oxidizing Bacteria of the *Nitrosospira* cluster 1 dominate over ammonia-oxidizing Archaea in oligotrophic surface sediments near the South Atlantic Gyre. *Environmental Microbiology Reports 7*: 404–413. https://doi.org/10.1111/1758-2229.12264

Lalli, C. and Parsons, T.R. (1997). *Biological Oceanography: An Introduction. Elsevier Butterworth-Heinemann.* https://doi.org/10.1016/B978-0-7506-3384-5056-7

Langerhuus, A.T., Røy, H., Lever, M.A. et al. (2012). Endospore abundance and D:L-amino acid modeling of bacterial turnover in holocene marine sediment (Aarhus Bay). *Geochimica et Cosmochimica Acta 99*: 87–99. https://doi.org/10.1016/j.gca.2012.09.023

Langseth, M.G., Hyndman, R.D., Becker, K. (1984). The hydrogeological regime of isolated sediment ponds in mid-oceanic ridges. In: *Initial Reports DSDP, Leg 78B, San Juan to Las Palmas*, 825–837. US Govt. Printing Office (UK distributors, IPOD Committee, NERC, Swindon).

Lauer, A., Sørensen, K.B. and Teske, A. (2016). Phylogenetic characterization of marine benthic archaea in organic-poor sediments of the eastern equatorial pacific ocean (ODP site 1225). *Microorganisms 4*: 32. https://doi.org/10.3390/microorganisms4030032

Laurence, M., Hatzis, C. and Brash, D.E. (2014). Common contaminants in next-generation sequencing that hinder discovery of low-abundance microbes. *PLoS One 9*: e97876. https://doi.org/10.1371/journal.pone.0097876

Lazar, C.S., John Parkes, R., Cragg, B.A. et al. (2012). Methanogenic activity and diversity in the centre of the Amsterdam Mud Volcano, Eastern Mediterranean Sea. *FEMS Microbiology Ecology 81*: 243–254. https://doi.org/10.1111/j.1574-6941.2012.0137

Leloup, J., Loy, A., Knab, N.J. et al. (2007). Diversity and abundance of sulfate-reducing microorganisms in the sulfate and methane zones of a marine sediment, Black Sea. *Environmental Microbiology 9*: 131–142. https://doi.org/10.1111/j.1462-2920.2006.0112

Lever, M.A., Alperin, M., Engelen, B. et al. (2006). Trends in basalt and sediment core contamination during IODP Expedition 301. *Geomicrobiology Journal 23*: 517–530. https://doi.org/10.1080/01490450600897245

Lever, M.A., Rogers, K.L., Lloyd, K.G. et al. (2015). Life under extreme energy limitation: a synthesis of laboratory-and field-based investigations. *FEMS Microbiology Reviews 39*: 688–728. https://doi.org/10.1093/femsre/fuv020

Lin, L.H., Wang, P.L., Rumble, D. et al. (2006). Long-term sustainability of a high-energy, low-diversity crustal biome. *Science 314*: 479–482. https://doi.org/10.1126/science.1127376

Lipp, J.S., Morono, Y., Inagaki, F. et al. (2008). Significant contribution of Archaea to extant biomass in marine subsurface sediments. *Nature 454*: 991–994. https://doi.org/10.1038/nature07174

Lomstein, B.A., Langerhuus, A.T., D'Hondt, S. et al. (2012). Endospore abundance, microbial growth and necromass turnover in deep sub-seafloor sediment. *Nature 484*: 101. https://doi.org/10.1038/nature10905

Mandal, S., Bhattacharya, S., Roy, C. et al. (2020). Cryptic roles of tetrathionate in the sulfur cycle of marine sediments: microbial drivers and indicators. *Biogeosciences 17*: 4611–4631. https://doi.org/10.5194/bg-17-4611-2020

Mauclaire, L., Zepp, K., Meister, P. et al. (2004). Direct *in situ* detection of cells in deep-sea sediment cores from the Peru Margin (ODP Leg 201, Site 1229). *Geobiology 2*: 217–223. https://doi.org/10.1111/j.1472-4677.2004.0003

Mazumdar, A., Dewangan, P., Joäo, H.M. et al. (2009). Evidence of paleo-cold seep activity from the Bay of Bengal, offshore India. *Geochemistry Geophysics Geosystems 10*: Q0600. https://doi.org/10.1029/2008GC002337

Mazumdar, A., Joäo, H.M., Peketi, A. et al. (2012). Geochemical and geological constraints on the composition of marine sediment pore fluid: possible link to gas hydrate deposits. *Marine and Petroleum Geology 38*: 35–52. https://doi.org/10.1016/j.marpetgeo.2012.07.004

Mazumdar, A., Peketi, A., Joäo, H.M. et al. (2014). Pore-water chemistry of sediment cores off Mahanadi Basin, Bay of Bengal: Possible link to deep seated methane hydrate deposit. *Marine and Petroleum Geology 49*: 162–175. https://doi.org/10.1016/j.marpetgeo.2013.10.011

Michalski, J.R., Onstott, T.C., Mojzsis, S.J. et al. (2018). The Martian subsurface as a potential window into the origin of life. *Nature Geoscience 11*: 21–26. https://doi.org/10.1038/s41561-017-0015-2

Morel, A., Claustre, H. and Gentili, B. (2010). The most oligotrophic subtropical zones of the global ocean: similarities and differences in terms of chlorophyll and yellow substance. *Biogeosciences 7*: 3139. https://doi.org/10.5194/bg-7-3139-2010

Morono, Y. and Inagaki, F. (2016). Chapter three – analysis of low-biomass microbial communities in the deep biosphere. *Advances in Applied Microbiology 95*: 149–178. https://doi.org/10.1016/bs.aambs.2016.04.001

Morono, Y., Terada, T., Nishizawa, M. et al. (2011). Carbon and nitrogen assimilation in deep subseafloor microbial cells. *Proceedings of the National Academy of Sciences USA 108*: 18295–18300. https://doi.org/10.1073/pnas.1107763108

Morono, Y., Ito, M., Hoshino, T. et al. (2020). Aerobic microbial life persists in oxic marine sediment as old as 101.5 million years. *Nature Communications 11*: 1–9. https://doi.org/10.1038/s41467-020-17330-1

Morris, J.D., Villinger, H., Klaus, A. et al. (2003): Proceedings of the Ocean Drilling Program, 205 Initial Reports. *Proceedings of the Ocean Drilling Program, Ocean Drilling Program, 205*. https://doi.org/10.2973/odp. proc.ir.205.2003

Müller, R.D., Sdrolias, M., Gaina, C. et al. (2008). Age spreading rates and spreading asymmetry of the world's ocean crust in *Geochemistry, Geophysics, Geosystems 9*: Q04006.

Nunoura, T., Inagaki, F., Delwiche, M.E. et al. (2008). Subseafloor microbial communities in methane hydrate-bearing sediment at two distinct locations (ODP Leg 204) in the Cascadia Margin. *Microbes and Environments 23*: 317–325. https://doi.org/10.1264/jsmeE08514

Orcutt, B.N., Bach, W., Becker, K. et al. (2011a). Colonization of subsurface microbial observatories deployed in young ocean crust. *The ISME Journal 5*: 692–703.

Orcutt, B.N., Sylvan, J.B., Knab, N.J. et al. (2011b). Microbial ecology of the dark ocean above, at and below the seafloor. *Microbiology and Molecular Biology Reviews 75*: 361–422. https://doi.org/10.1128/MMBR.00039-10

Orcutt, B.N., LaRowe, D.E., Biddle, J.F. et al. (2013). Microbial activity in the marine deep biosphere: progress and prospects. *Frontiers in Microbiology 4*: 189. https://doi.org/10.3389/fmicb.2013.00189

Orsi, W.D., Edgcomb, V.P., Christman, G.D. et al. (2013a). Gene expression in the deep biosphere. *Nature 499*: 205. https://doi.org/10.1038/nature12230

Orsi, W., Biddle, J.F. and Edgcomb, V. (2013b). Deep sequencing of subseafloor eukaryotic rRNA reveals active fungi across marine subsurface provinces. *PLoS One 8*: e56335. https://doi.org/10.1371/journal.pone.0056335

Orsi, W.D., Coolen, M.J., Wuchter, C. et al. (2017). Climate oscillations reflected within the microbiome of Arabian Sea sediments. *Scientific Reports 7*: 1–12. https://doi.org/10.1038/s41598-017-05590-9

Oyun, S., Elderfield, H. and Klinkhammer, G.P. (1995). 41. Strontium isotopes in pore waters of east equatorial pacific sediments: indicators of seawater advection through oceanic crust and sediments. *Proceedings of the Ocean Drilling Program, Scientific Results 138*: 813–819. https://doi.org/10.2973/odp. proc. sr.138.156.1995

Pachiadaki, M.G., Rédou, V., Beaudoin, D.J. et al. (2016). Fungal and prokaryotic activities in the marine subsurface biosphere at Peru Margin and Canterbury Basin inferred from RNA-based analyses and microscopy. *Frontiers in Microbiology 7*: 846. https://doi.org/10.3389/fmicb.2016.00846

Park, M.H., Kim, J.H., Ryu, B.J. et al. (2006). AMS radiocarbon dating of the marine late Pleistocene Holocene sediment cores from the western Ulleung Basin, East/Japan Sea. *Nuclear Instruments and Methods in Physics Research Section B: Beam Interactions with Materials and Atoms 243*: 211–215. https://doi.org/10.1016/j.nimb.2005.08.120

Parkes, R.J., Cragg, B.A. and Wellsbury, P. (2000). Recent studies on bacterial populations and processes in subseafloor sediments: a review. *Hydrogeology Journal 8*: 11–28. https://doi.org/10.1007/PL00010971

Parkes, R.J., Cragg, B.A., Bale, S.J. et al. (1994). Deep bacterial biosphere in Pacific Ocean sediments. *Nature 371*, 410–413.

Parkes, R.J., Webster, G., Cragg, B.A. et al. (2005). Deep sub-seafloor prokaryotes stimulated at interfaces over geological time. *Nature 436*: 390–394. https://doi.org/10.1038/nature03796

Peketi, A., Mazumdar, A., Pillutla, S.P.K. et al. (2020). Influence of dual sulfate reduction pathways on pore-fluid chemistry and occurrences of methane hydrate in sediment cores (IODP-353) off Mahanadi basin, Bay of Bengal. *Geochemical Journal 54*: 1–11. https://doi.org/10.2343/geochemj.2.0576

Picard, A. and Ferdelman, T.G. (2011). Linking microbial heterotrophic activity and sediment lithology in oxic, oligotrophic subseafloor sediments of the North Atlantic Ocean. *Frontiers in Microbiology 2*: 263. https://doi.org/10.3389/fmicb.2011.00263

Price, P.B. and Sowers, T. (2004). Temperature dependence of metabolic rates for microbial growth, maintenance and survival. *Proceedings of the National Academy of Sciences USA 101*: 4631–4636. https://doi.org/10.1073/pnas.0400522101

Rajala, P. and Bomberg, M. (2017). Reactivation of deep subsurface microbial community in response to methane or methanol amendment. *Frontiers in Microbiology 8*: 431. https://doi.org/10.3389/fmicb.2017.00431

Rédou, V., Navarri, M., Meslet-Cladière, L. et al. (2015). Species richness and adaptation of marine fungi from deep-subseafloor sediments. *Applied and Environmental Microbiology 81*: 3571–3583. https://doi.org/10.1128/AEM.04064-14

Røy, H., Kallmeyer, J., Adhikari, R.R. et al. (2012). Aerobic microbial respiration in 86-million-year-old deep-sea red clay. *Science 336*: 922–925. https://doi.org/10.1126/science.1219424

Salter, S.J., Cox, M.J., Turek, E.M. et al. (2014). Reagent and laboratory contamination can critically impact sequence-based microbiome analyses. *BMC Biology 12*: 87. https://doi.org/10.1186/s12915-014-0087-z

Santelli, C.M., Orcutt, B.N., Banning, E. et al. (2008). Abundance and diversity of microbial life in ocean crust. *Nature 453*: 653. https://doi.org/10.1038/nature06899

Schauer, R., Bienhold, C., Ramette, A., (2010). Bacterial diversity and biogeography in deep-sea surface sediments of the South Atlantic Ocean. *The ISME Journal 4*: 159–170. https://doi.org/10.1038/ismej.2009.106

Schimmelmann, A. and Tegner, M.J. (1991). Historical oceanographic events reflected in13C/12C ratio of total organic carbon in laminated Santa Barbara Basin Sediment. *Global Biogeochemical Cycles 5*: 173–188. https://doi.org/10.1029/91GB00382

Schippers, A., Neretin, L.N., Kallmeyer, J. et al. (2005). Prokaryotic cells of the deep sub-seafloor biosphere identified as living bacteria. *Nature 433*: 861. https://doi.org/10.1038/nature03302

Schippers, A., Kock, D., Höft, C. et al. (2012). Quantification of microbial communities in subsurface marine sediments of the Black Sea and off Namibia. *Frontiers in Microbiology 3*: 16. https://doi.org/10.3389/fmicb.2012.00016

Schrenk, M.O., Huber, J.A. and Edwards, K.J. (2010). Microbial provinces in the subseafloor. *Annual Review of Marine Science 2*: 279–304. https://doi.org/10.1146/annurev-marine-120308-081000

Schrenk, M.O., Brazelton, W.J. and Lang, S.Q. (2013). Serpentinization, carbon and deep life. *Reviews in Mineralogy and Geochemistry 75*: 575–606. https://doi.org/10.2138/rmg.2013.75.18

Seiter, K., Hensen, C. and Zabel, M. (2005). Benthic carbon mineralization on a global scale. *Global Biogeochemical Cycles 19*: 1–26. https://doi.org/10.1029/2004GB002225

Senum, G.I. and Dietz, R.N., (1990). Perfluorocarbon tracer tagging of drilling muds for the assessment of sample contamination (No. BNL-43145; CONF-900171-1). Brookhaven National Lab. (BNL), Upton, NY, USA

Sheik, C.S., Reese, B.K., Twing, K.I. et al. (2018). Identification and removal of contaminant sequences from ribosomal gene databases: lessons from the census of deep life. *Frontiers in Microbiology 9*: 840. https://doi.org/10.3389/fmicb.2018.00840

Siegert, M., Krüger, M., Teichert, B. et al. (2011). Anaerobic oxidation of methane at a marine methane seep in a forearc sediment basin off Sumatra, Indian Ocean. *Frontiers in Microbiology 2*: 249. https://doi.org/10.3389/fmicb.2011.00249

Sleep, N.H. and Zahnle, K. (1998). Refugia from asteroid impacts on early Mars and the early Earth. *Journal of Geophysical Research: Planets 103*: 28529–28544. https://doi.org/10.1029/98JE01809

Smith, A., Popa, R., Fisk, M. et al. (2011). *In situ* enrichment of ocean crust microbes on igneous minerals and glasses using an osmotic flow-through device. *Geochemistry, Geophysics, Geosystems 12*. https://doi.org/10.1029/2010GC003424

Smith, D.C., Spivack, A.J., Fisk, M.R. et al. (2000). Tracer-based estimates of drilling-induced microbial contamination of deep sea crust. *Geomicrobiology Journal 17*: 207–219. https://doi.org/10.1080/01490450050121170

Steele, J.A., Dekas, A.E., Harrison, B.K. et al. (2012). *Mineral-Association and Activity of Bacteria and Archaea in the Deep Subsurface South Pacific Gyre Sediment*. AGUFM, 2012, B43G-0492.

Straume, E.O., Gaina, C., Medvedev, S. et al. (2019). GlobSed: Updated total sediment thickness in the world's oceans. *Geochemistry, Geophysics, Geosystems 20*. https://doi.org/10.1029/2018GC008115

Talbot, V. and Bianchi, M. (1997). Bacterial proteolytic activity in sediments of the Subantarctic Indian Ocean sector. *Deep Sea Research Part II: Topical Studies in Oceanography 44*: 1069–1084. https://doi.org/10.1016/S0967-0645(96)00107-5

Teske, A. and Sørensen, K.B. (2008). Uncultured archaea in deep marine subsurface sediments: have we caught them all? *The ISME Journal 2*: 3. https://doi.org/10.1038/ismej.2007.90

Teske, A., Biddle, J.F., Edgcomb, V.P. et al. (2013). Deep subsurface microbiology: a guide to the research topic papers. *Frontiers in Microbiology 4*: 122. https://doi.org/10.3389/fmicb.2013.00122

Tortora, G.J., Funke, B.R., Case, C.L. et al. (2018). *Microbiology: An Introduction*. London, UK: Pearson.

Tréhu, A.M., Bohrmann, G., Rack, F.R. et al. (2003). Leg 204 summary. *Proceedings of the Ocean Drilling Program, Initial Reports 204*: 1–75. https://doi.org/10.2973/odp. proc.ir.206.2003

Treude, T., Krause, S., Maltby, J. et al. (2014). Sulfate reduction and methane oxidation activity below the sulfate-methane transition zone in Alaskan Beaufort Sea continental margin sediments: Implications for deep sulfur cycling. *Geochimica et Cosmochimica Acta 144*: 217–237. https://doi.org/10.1016/j.gca.2014.08.018

Valentine, D.L. (2011). Emerging topics in marine methane biogeochemistry. *Annual Review of Marine Science 3*: 147–171. https://doi.org/10.1146/annurev-marine-120709-142734

Wasmund, K., Mußmann, M. and Loy, A. (2017). The life sulfuric: microbial ecology of sulfur cycling in marine sediments. *Environmental Microbiology Reports 9*: 323–344. https://doi.org/10.1111/1758-2229.12538

Webster, G., Blazejak, A., Cragg, B.A. et al. (2009). Subsurface microbiology and biogeochemistry of a deep, cold-water carbonate mound from the Porcupine Seabight (IODP Expedition 307). *Environmental Microbiology 11*: 239–257. https://doi.org/10.1111/j.1462-2920.2008.0175

Whelan, J.K., Kanyo, Z., Tarafa, M. et al. (1990). 38. Organic matter in Peru upwelling sediments – analysis by pyrolysis, pyrolysis-gas chromatography and pyrolysis-gas chromatography mass spectrometry. *Proceedings of the Ocean Drilling Program, Scientific Results 112*: 573–590. https://doi.org/10.2973/odp.proc.sr.112.158.1990

Whitman, W.B., Coleman, D.C. and Wiebe, W.J. (1998). Prokaryotes: the unseen majority. *Proceedings of the National Academy of Sciences 95*: 6578–6583. https://doi.org/10.1073/pnas.95.12.6578

Whittaker, J.M., Goncharov, A., Williams, S.E. et al. (2013). Global sediment thickness data set updated for the Australian-Antarctic Southern Ocean. *Geochemistry, Geophysics, Geosystems 14*: 3297–3305. https://doi.org/10.1002/ggge.20181

Wilkins, M.J., Daly, R., Mouser, P.J. et al. (2014). Trends and future challenges in sampling the deep terrestrial biosphere. *Frontiers in Microbiology 5*: 481. https://doi.org/10.3389/fmicb.2014.00481

Yamazaki, T. and Shimono, T. (2013). Abundant bacterial magnetite occurrence in oxic red clay. *Geology 41*: 1191–1194. https://doi.org/10.1130/G34782.1

Yanagawa, K., Nunoura, T., McAllister, S. et al. (2013). The first microbiological contamination assessment by deep-sea drilling and coring by the D/V Chikyu at the Iheya North hydrothermal field in the Mid-Okinawa Trough (IODP Expedition 331). *Frontiers in Microbiology 4*: 327. https://doi.org/10.3389/fmicb.2013.00327

Yanagawa, K., Kouduka, M., Nakamura, Y. et al. (2014). Distinct microbial communities thriving in gas hydrate-associated sediments from the eastern Japan Sea. *Journal of Asian Earth Sciences 90*: 243–249. https://doi.org/10.1016/j.jseaes.2013.10.019

Zhang, Z., Fan, X., Gao, X. et al. (2014). *Achromobacter sediminum* sp. nov., isolated from deep subseafloor sediment of South Pacific Gyre. *International Journal of Systematic and Evolutionary Microbiology 64*: 2244–2249. https://doi.org/10.1099/ijs.0.062265-0

7

Biogeochemistry of Marine Petroleum Systems

Subhrangshu Mandal[1,*], Nibendu Mondal[1], Sabyasachi Bhattacharya[1], Wriddhiman Ghosh[1,*], and Bhaskar Bhadra[2,*]

ABSTRACT

Within marine (as well as terrestrial) petroleum systems, source rocks, where petrified organic matter transforms into oil, are essentially devoid of microbial life, whereas the reservoir rocks to which the oil migrates and gets stored harbor biogeochemically active microbiotas. Although the oil reservoir microbiome was discovered as early as the 1930s, long before the International Ocean Discovery Program started making sensational discoveries regarding microbial life in the marine deep subsurface, overall knowledge on the oil reservoir ecosystem did not increase proportionately owing to a number of systemic problems. While methanogenesis, sulfate reduction, sulfide oxidation, acetogenesis, fermentation, and plausibly also methanotrophy, are the main microbial processes detected in oil reservoirs, there is still no convincing conclusion regarding what fraction of the communities reported are actually indigenous. This knowledge gap concerning the geomicrobiology of oil reservoirs is accentuated by the conspicuous absence of corroborative information from preserved geochemical records. Cross-disciplinary research initiated at the exploration and prospecting stage seems to be the way forward in building translational knowledge on biosphere–geosphere interactions in these geophysically dynamic and economically crucial deep subsurface ecosystems of the marine realm.

7.1. INTRODUCTION

Petroleum basins and reservoirs are unique deep subsurface environments in the terrestrial as well as marine realms where organic matter[1] is buried undegraded and subsequently transformed into oil and natural gas with increases in the basins' depth, temperature, and pressure (Hyne, 2001; Peters et al., 2012). When sediments accumulate in subsiding areas or deep depressions created in the Earth's crust by plate tectonics (depositional basins and troughs that are often delimited by faults), hydrocarbon-rich rocks form and potential petroleum systems develop, depending on the depth/temperature and duration of burial, plus the location of the basin with respect to the position of the tectonic plates (for terms see Schlumberger Ltd, 2021). Characterized by the dominance of multiple fluid phases (oil, gas, and water), these environments are hosted in multiple oil-bearing strata of low porosity/permeability sedimentary rocks, and represent nutrient-constrained (oligotrophic) yet complex microbial ecosystems (Gao et al., 2016). The three basic components of a petroleum generation and accumulation system (Magoon and

[1] *While terrestrial organic matter such as woody plant parts transform into coal and methane under similar circumstances, remains of algae and soft-bodied lower plant groups transform into oil and then natural gas at yet higher temperatures.*

[1]*Department of Microbiology, Bose Institute, P-1/12 CIT Scheme VIIM, Kolkata, 700054, West Bengal, India*

[2]*Reliance Technology Group, Reliance Industries Limited, Navi Mumbai, 400701, India*

**Corresponding Authors: Bhaskar Bhadra (bhaskar.bhadra@ril.com), ORCiD code: 0000-0003-4656-885X; Wriddhiman Ghosh (wriman@jcbose.ac.in), ORCiD code: 0000-0002-4922-7404; Subhrangshu Mandal (s.mandalbolpur@gmail.com)*

Systems Biogeochemistry of Major Marine Biomes, First Edition. Edited by Aninda Mazumdar and Wriddhiman Ghosh.

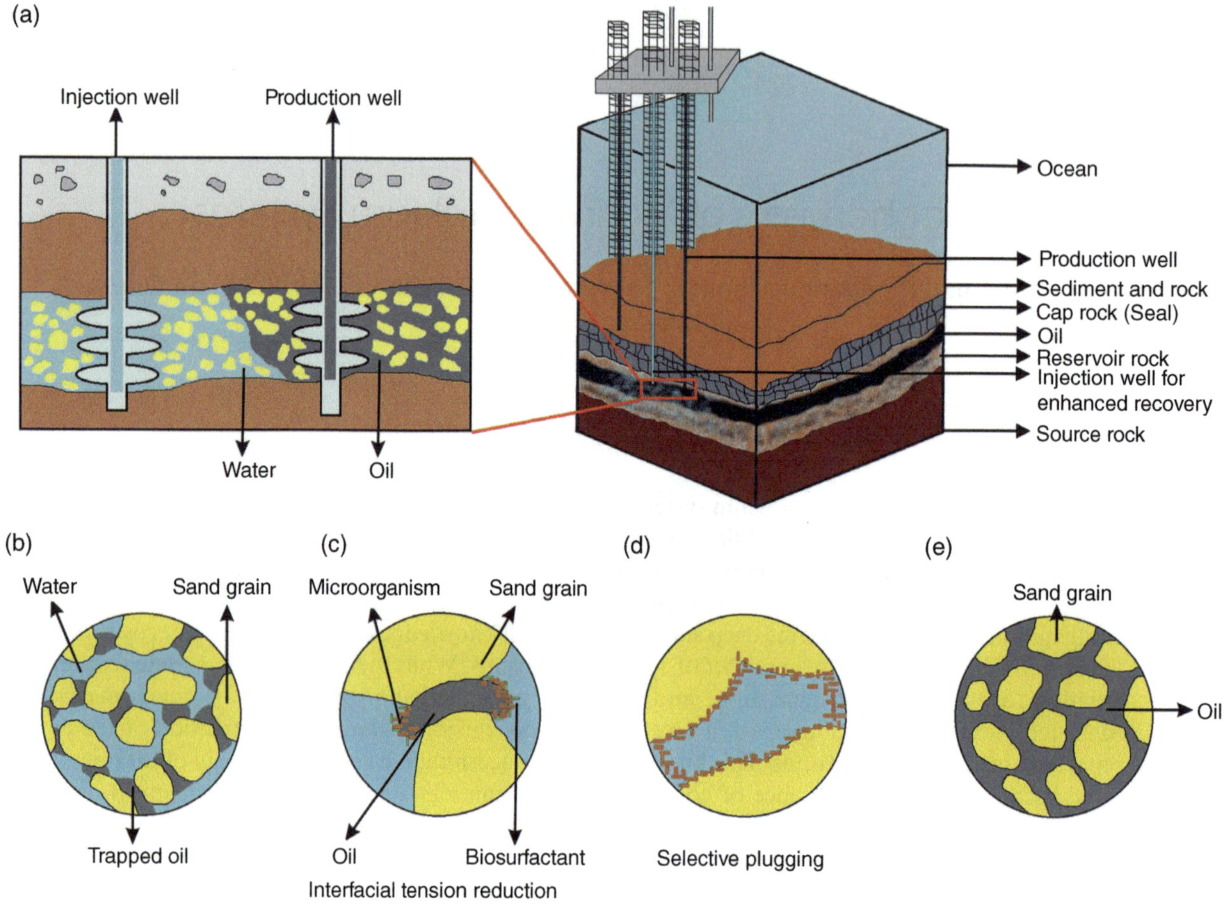

Figure 7.1 Schematic diagram showing the typical physical features of a marine petroleum system: (a) stratigraphic view, with details of the oil recovery scenario shown in the inset; (b–e) physical circumstances constraining oil production and key strategies involved in the enhanced oil recovery procedure.

Dow, 1994; Hyne, 2001; Levorsen, 2001; Peters et al., 2017; also see Figure 7.1a) are:

• source rock (typically, these are composed of fine-grained, organic-carbon-rich, shale or limestone; whereas the former rock-type is clayey, the latter is composed of calcite, the chief constituent of sea shells);

• reservoir rock (these high-porosity high-permeability rocks, mostly sandstone and carbonates consisting calcite and dolomite, store the oil and also transmit the same); and

• cap rock (seal) that can form a trap-like structure to accumulate the oil and gas (these shale rocks, or limestones and sandstones immersed in shale, prevent further transmission, ascension and surficial seepage of the oil; a trap can be an anticline or dome-like formation over the reservoir rock).

With continued deposition, organic matter petrified in the source rock transforms into oil at ~2–5 km basin depths where *in situ* temperature reaches 65–150°C; deeper down, at further elevated temperatures, oil gets converted to natural gas and graphite (Hyne, 2001; Dembicki Jr, 2009). In addition to temperature, *in situ* pressure and burial time are also crucial factors governing oil formation (Peters et al., 2012). For example, in basins that have low sedimentation rates, the source rocks remain at shallow depths (consequently, at lower temperatures) for prolonged times, thereby taking millions of years to form oil; under such conditions conventional oil may not form at all and the rocks may become sources of shale oil (Hyne, 2001; Jiang et al., 2016).

Most of the large oil reservoirs of the world occur within deformed physiographies in between the Earth's mobile tectonic plates (i.e. within orogenic regions). Tectonic activities not only create the basin structures needed to host deep sedimentary rocks where oil is formed, they also deform the sedimentary rocks to create the traps required for the accumulation of the oil. Accordingly, the most oil- and gas-rich territories of the marine as well as terrestrial realms – for example those in the east and west coasts of America, Caspian sea, Persian

Gulf, Siberia, Ural mountains, and north of Africa – are all located within basins and orogens (Buryakovsky et al., 2001; Konyuhov and Maleki, 2006; Arzate et al., 2009; Deng, 2019). Furthermore, ~80% of the world's known oil-bearing basins encompass marine source rocks, while the source rocks of the remaining 20% are lacustrine in origin (Deng, 2019).

Offshore petroleum reservoirs (oil fields and natural gas deposits that lie under the ocean) feature high temperature and pressure, absence of dissolved oxygen, and often highly saline formation water (Brakstad et al., 2008). These hostile environmental conditions notwithstanding, offshore petroleum reservoirs are long known to harbor active microbial communities rich in taxonomic and metabolic diversities (Bastin, 1926; Krejci-Graf et al., 1932; Brakstad et al., 2008). Microbiome analyses in offshore petroleum reservoirs over the years have addressed and evaluated (i) the role of microbes in the genesis of petroleum over geological time (Magot et al., 2000), (ii) maintenance and biodegradation of petroleum reservoirs (Head et al., 2003), (iii) long-standing effects of petroleum pollution (Lindstrom et al., 1999), (iv) petro-waste remediation (Ka et al., 2001), and (v) control of harmful microbial behavior during petroleum production (Eaton and Chapman, 1992). Thus, in the applied frontier, in-depth understanding of the microbiology of offshore oil reservoirs has become central to the development of advanced and economical drilling and recovery strategies. For example, contrivances for the avoidance of souring (microbial production of hydrogen sulfide *in situ*, which causes significant damage to the equipment involved in oil/gas recovery; Youssef et al., 2009; Kadhim and Ali, 2017) and enhancement of oil recovery efficiency (Kakati et al., 2020) both depend on our knowledge of *in situ* geomicrobiology. However, our knowledge of the structures and functions of petroleum reservoir microbiomes is small compared with other marine deep subsurface ecosystems such as sediments several meters deep, hydrothermal vents on mid-oceanic ridges and ocean crusts, gas hydrate deposits, and cold methane seeps. This chapter, therefore, presents an overview of the geophysical and biogeochemical mechanisms that lead to the formation of marine oil fields. It discusses their physiochemical features, microbiome architectures, and ecosystems constraints, and highlights the roles of *in situ* microbial processes (biodegradation of hydrocarbons, souring, etc.) in the physicochemical dynamics (formation, transformation and depletion) of oil reservoirs. Finally, the entire gamut of knowledge available on the past and present geomicrobiology of marine petroleum systems is considered in the context of sustainable utilization and management of these economically resourceful and ecologically sensitive extremities of the Earth's biosphere.

7.2. FORMATION OF OIL RESERVOIRS

Petroleum (crude oil in both liquid and gaseous states), across the marine sedimentary realm, originates from the deposition and burial of organic matter along continental shelves, under anoxic conditions, over hundreds of millions of years. During this protracted geological time-period, layers of depositional materials such as sand and silt settle over the organic matter before extensive decomposition can take place; increasing temperature and pressure gradually transform the organic matter (which, in the course of the burial process, gets hosted in fine-grained sedimentary rocks known as the "source rocks" of the oil fields) into hydrocarbons such as kerogen, oil, and gas (Schenk et al., 1997; McCarthy et al., 2011; Al-Areeq, 2018). In the process, organic material becomes progressively enriched in carbon and hydrogen and depleted in oxygen and nitrogen. Oil formation is a continuous geodynamic process, as reflected in its occurrence in sedimentary rocks of variable geological age, i.e. those ranging from Precambrian (>500 million years old) to Pleistocene (approximately one million year old) origin (Hyne, 2001).

In all the conventional[2] petroleum systems, oil is generated in thermally mature source rocks buried under hundreds of meters of sedimentary overburden (Magoon and Dow, 1994; Peters et al., 2017). After its formation in a source rock, petroleum generally gets transmitted to a permeable and porous carrier bed, eventually migrating to and accumulating in a reservoir rock, where it gets structurally/stratigraphically trapped under an impermeable seal (Al-Hajeri et al., 2009). Notably, however, this entire pathway of oil generation, expulsion, migration, and trapping is not necessarily completed in all petroleum-forming basins.

Upon being deposited in an anoxic basin, part of the sedimentary organic matter is microbially converted to methane. However, with increasing burial depth, high *in situ* heat transforms the organic matter into kerogen, a unique solvent-insoluble, hydrophobic, particulate, complex organic macromolecule that is regarded as comprising "the largest single mass of organic carbon on Earth" (Petsch, 2014). Although far less reactive (in low-temperature aqueous chemical milieu) than commonplace organic matter, kerogen is not totally inert to geochemical transformation: as heat increases with basin-depth,

[2] *In unconventional petroleum systems such as those producing shale oil and gas (this involves extraneous application of heat during exploitation) the source rock itself serves as the reservoir and the seal; oil generated in such self-contained systems does not migrate, but remains trapped within the porous/fractured source rock having low permeability (Allix et al., 2010; McCarthy et al., 2011).*

kerogen converts to bitumen and petroleum (Peters et al., 2005). As kerogen[3] keeps converting to petroleum it becomes increasingly depleted of hydrogen. With maturity of the system the petroleum compounds also become structurally simplified: from oil to wet gas that has substantial amounts of gaseous alkanes other than methane, and then dry gas, which is composed almost exclusively of methane (Selley, 2003; McCarthy et al., 2011). During petroleum genesis, where *in situ* temperature reaches >130°C, the crude oil starts converting to gas (this process is also dependent on the pressure existing in the depositional basin). The wet gas and condensates which form initially contain copious C4–C10 compounds, whereas further rise in temperature leads to the formation of dry gas that is rich in lighter (C1–C3) hydrocarbon compounds (for more details, see Offshore Engineering, 2021). Whereas the temperature under which maturation takes place determines the chemical nature of the resultant hydrocarbon, the depositional environment and age of the depositional basin also act as key drivers of thermal maturation. After maturation, migration of the hydrocarbon fluids takes place from the low-permeability source rocks towards the porous and permeable reservoir rocks that eventually harbor the petroleum (Magoon and Dow, 1994; Levorsen, 2001; Peters and Fowler, 2002; Peters et al., 2017). Depositional basin-specific geological parameters such as sedimentation and burial dynamics, the basins' geothermal gradients, tectonic activities, etc., bring about subtle characteristic variations in this universal mechanism of oil and gas formation (Magoon and Dow, 1994; Hallett, 2002; McCarthy et al., 2011; Peters et al., 2017; Carvajal-Arenas et al., 2020). While the factors controlling oil reservoir formation vary from basin to basin, the biogeochemical background of the oil source is apparently the most important and universal control factor (Deng, 2019). Based on the conditions under which the source rocks were formed, it has been concluded that major oil-bearing basins (such as those in the Atlantic Ocean, East Africa coast, Gulf of Mexico, North Sea, Persian Gulf, Sichuan region, Tarim region, Western Siberia, and also the Paleo-Tethys Ocean) were all partially land-locked gulf physiographies (having only one or two or three water channels connecting them with the global ocean) when their main source rocks were formed (Guzman-Vega et al., 1991; Guzman-Vega and Mello, 1999; Galloway et al., 2000; Isaksen and Ledje, 2001; Pinous et al., 2001; Abadi et al., 2008; Arzate et al., 2009; Beglinger et al., 2012). The gulfs hosted large riverine estuaries, so were rich in weathered and denuded mineral nutrients for phytoplankton growth (high primary productivity), even as they were deprived of nutrients via oceanic upwelling; poor ventilation of the gulf waters ensured high nutrient accumulation, and thereby long-term productivity and organic matter preservation, which in turn resulted in oil-rich source rocks (Deng, 2019).

7.3. ECOSYSTEM CONSTRAINTS AND HABITABILITY OF PETROLEUM BASINS

In oil and natural gas-forming basins, lack of O_2 (anoxic condition) is the key predisposing factor that prevents the decomposition of organic matter before petrifaction, while high *in situ* sedimentation rates cause increases in pressure and temperature that transform organic matter into hydrocarbons by a process commonly called maturation (Connan, 1984). Where oil formation takes place in sediments that are fine- to coarse-grained, and deposited under oxic conditions at the surface, O_2 is depleted within shallow sediment-depths owing to rapid organic matter remineralization by aerobic microorganisms, and anaerobes are able to occupy only the deeper horizons of the sediment package (Wellsbury et al., 1997).

The first stage of crude oil genesis involves the formation of kerogen from buried organic matter. Subsequently, with prolonged deposition, as the temperature and pressure of the source sedimentary rock increase, kerogen gets transformed into petroleum (in general there is an average 3°C rise in temperature with every 100 m increase in sediment-depth; Head et al., 2003; see also, Offshore Engineering, 2021). Conversion of organic matter to hydrocarbon compounds takes place over an "oil window" of 60–120°C (2–4 km basin depth) and a "gas window" of 120–180°C (4–6 km basin depth). However, most oil formation takes place at >80°C (Head et al., 2003; Bastianoni et al., 2005; McCarthy et al., 2011). As this is a critical temperature barrier for the growth of most microbial groups (Roy et al., 2020) freshly formed oil is largely sterile in nature (Wilhelms et al., 2001; Head et al., 2003).

Once the oils settle in the reservoir rocks, their chemistry is impacted primarily by *in situ* biodegradation (Brooks and Welte, 1984). Microbial degradation of oil

[3] *Type II kerogens (that have H:C and O:C atomic ratios neither too high nor too low) are mostly deposited in marine environments as they are derived mainly from sea algae/planktons; they have the potentials for the generation of oil as well as gas. In contrast, Type I (having high H:C and low O:C atomic ratios) and Type III (having low H:C and high O:C) kerogens are deposited in lacustrine/marine and terrestrial environments respectively. While Type I kerogens have the potential for generating oil, Type III can produce only gas (Moldowan and Seifert, 1980; Prauss and Riegel, 1989; Kandiyoti et al., 2006). The seminal van Krevelen diagram can be referred to for the elemental dynamics of kerogen maturation; it plots O:C versus H:C ratios of source rock organic matter, whereas modified van Krevelen diagrams use the oxygen versus hydrogen indices from Rock-Eval pyrolysis (Peters et al., 2016).*

takes place mostly in reservoirs having temperatures ≤80°C (Larter et al., 2003; Grabowski et al., 2005). Accordingly, reservoirs with ~50°C temperature often have 70% of their oils in heavily degraded conditions (Pepper and Santiago, 2001). Correspondingly, maximum biodegradation of stored oil is encountered in reservoirs situated ~2.5 km from the sediment surface, and seldom beyond a depth of ~4 km below the seafloor (Brooks and Welte 1984; Head et al., 2003; Grabowski et al., 2005). Up to such depths where reservoir temperatures remain lower than 80°C, the probability of encountering biodegraded oil apparently decreases with increasing temperature, and such trends are apparently independent of the chemical composition of the oil present (Evans et al., 1971; Brooks and Welte, 1984; Pepper and Santiago, 2001). Notably, however, not all reservoirs that have low temperature necessarily contain degraded petroleum; this is possibly because they have either been recently infused with freshly formed oil or uplifted from deep-seated (>80°C) territories. These instances are reflective of negligible introduction of surficial fluids and/or microorganisms into such "paleopasteurized" deep subsurface petroleum reservoirs; furthermore, they indicate that biodegradation in <80°C petroleum reservoir basins is rendered essentially by microorganisms that were present in the reservoir since its genesis (Wilhelms et al., 2001). The latter hypothesis may be explained as follows. Down the sediment depths in marine depositional basins, depleting O_2, increasing temperature, and increasing predominance of anaerobic microbial communities, are associated with increasing consumption of labile organic matter and generation of residual recalcitrant organic carbon (Arndt et al., 2013). Although this scenario is generally accompanied by dwindling microbial diversity and activity down the sediment depths (Lever et al., 2015), some versatile heterotrophic microbial communities can still establish themselves in these virtually nutrient-limited (albeit replete with recalcitrant organic matter) extremities of the biosphere (Wellsbury et al., 1997; Bradley et al., 2018; Jones et al., 2018). Subsequently, as and when such sedimentary depositional regimes give rise to potential petroleum reservoir rocks (and upon arrival of oil in the reservoir), these indigenous heterotrophic anaerobic microbial communities can play significant roles in oil biodegradation (Wellsbury et al., 1997).

Other dynamic interplays between geological and biological processes also impact the physicochemical constitution of oil and natural gas reservoirs (Head et al., 2003). For example, during the millions of years of residence of the hydrocarbon fluids in the reservoir rock, plate tectonics (drifting of continent over the underlying mantle) and other geophysical forces rearrange the oceanic and continental crusts (Thompson, 1976; Shuichang et al., 2011), and in doing so influence the physiographies of offshore (as well as onshore) oil basins and their built-in microbiomes. These phenomena have enormous potentials to alter comprehensively the biogeochemical structure of these systems, including the *in situ* microbial community architectures as well as the dynamics of formations, availabilities and distributions of organic carbon and redox inorganic nutrients that are central to the *in situ* community metabolisms (Head et al., 2003).

7.4. MICROBIOME FRAMEWORK OF PETROLEUM RESERVOIRS

As described above, *in situ* physicochemical conditions, especially the elevated temperatures of the subsurface, critically influence the habitability of petroleum ecosystems and shape the indigenous microbiome structure and functions. Accordingly, microbial degradation of oil becomes largely constrained at basin depths having temperatures ≥80°C; deeper oil reservoirs having temperatures up to 150°C have no microbial activity at all (Philippi, 1977; Fisher, 1987; Barth, 1991; Bernard et al., 1992). Although a number of hyperthermophiles have been isolated from >103°C reservoirs, in many of the cases the authors have suggested that the microorganisms could be exogenous in origin, potentially coming from the seawaters injected during drilling, or secondary recovery as a part of the enhanced oil recovery procedure (Stetter et al., 1993; Grassia et al., 1996). Furthermore, *in situ* pH of the high-pressure reservoirs differs considerably from the pH measured at atmospheric pressure; this also adds selection pressure to the microbial communities of the oil reservoirs (Magot et al., 2000). For example, pH values in oil reservoirs can range between 3 and 7 owing to dissolution of gasses under high *in situ* pressure (~50000 kPa), even though the measured pH of fresh to salt-saturated water ranges between 5 and 8 (Magot et al., 2000). In this context it is also noteworthy that the physiology and metabolism of the microorganisms inhabiting offshore oil reservoirs are essentially adapted to this high *in situ* pressure (Röling et al., 2003). For example, their cell membranes contain high proportions of unsaturated fatty acids which confer fluidity and stability under high pressure: such microorganisms also have high levels of expression for pressure-regulated genes (Kato and Bartlett, 1997; Yano et al., 1998). In addition to temperature, pressure, and pH, scarcity of electron donors and acceptors also constrain bacterial metabolism in oil field environments, where nitrogen and phosphorus are especially in short supply (Magot et al., 2000).

Compared with other well-studied marine microbiomes, there is less information about the geomicrobiology and biogeochemical cycles of petroleum reservoir habitats. Nevertheless, a number of chemical characteristics of these secluded environments have been correlated with their native microbiota. Oil reservoirs have low redox potentials; while electron donors such as hydrogen gas and diverse organic molecules such as simple fatty acids, sugars, etc. are present, electron acceptors such as oxygen, nitrate and ferric iron are absent (Youssef et al., 2009). Sulfate and carbonate, present at various concentrations in the stratal waters[4] are known to support sulfate reducers, methanogenes, acetogenes and fermentative microbes *in situ*. While acetate, benzoate, butyrate, formate, propionate, and members of the complex group of naphtenic acids, are often present at high concentrations in marine oil reservoirs (Magot et al., 2000), degradation of *n*-alkanes, homocyclic and polycyclic aromatic compounds, and heterocyclic compounds containing nitrogen or sulfur, are the common forms of catabolism in oil reservoir ecosystems (Grbic-Galic, 1990; Aeckersberg et al., 1991; Rueter et al., 1994; Krurmholz et al., 2005). Furthermore, the complex hydrocarbons, which are found in the resin and asphaltene fractions of crude oil, can become potential electron donors for anaerobic metabolism, and in being so these nitrogen/sulfur/oxygen hetero atom-containing hydrocarbon compounds can assist the alkanes and light aromatic components of oil to remain unaffected by microbial degradation (Magot et al., 2000).

7.5. MICROBIAL COMMUNITY STRUCTURES AND FUNCTIONS IN MARINE OIL RESERVOIRS

One of the earliest investigations of deep subsurface microbiology was based on the produced-waters[5] of an oil well system, and incidentally, that investigation involved the isolation of sulfate-reducing bacteria (Bastin, 1926). Recent advances in meta-omics techniques have led to a more comprehensive revelation of microbial community architecture and ecosystem functions in oil reservoir environments. Acetogenesis, fermentation, iron and manganese reduction, methanogenesis, nitrate reduction, sulfate reduction, sulfide oxidation, and also potentially methanotrophy, have thus been identified as the key microbial processes in oil reservoirs. Presence of molecular indicators such as genes for chlorite-disproportionating enzymes has suggested the occurrence of (per)chlorate reducers in oil reservoirs and adjoining ecological niches (Liebensteiner et al., 2014), even though microorganisms possessing such a phenotype have seldom been isolated from these environments. (Per) chlorate disproportionation, which can nonphototrophically produce O_2 *in situ*, has also been envisaged for potential applications in oil recovery, as the O_2 generated can be used by associated aerobic microorganisms (Coates et al., 1998) to degrade carbon compounds that are otherwise undegradable under the anaerobic conditions of the subsurface (Liebensteiner et al., 2014).

Overall, in the context of microbiome architecture, it is noteworthy that oil reservoirs are some of the few ecosystems where culture-independent analyses of community composition have largely corroborated the findings of culture-dependent investigations (Youssef et al., 2009). This, in turn, may be a potential reason why indigenous oil-degrading microorganisms have played significant roles in reducing the overall environmental impact of marine oil spills and runoffs from onshore oil refineries and petrochemical industries (Pritchard and Costa, 1991; Bragg et al., 1992, 1994; Readman et al., 1992; Coates et al., 1997; Harayama et al., 1999; Atlas and Hazen, 2011). In contrast, genetically engineered microorganisms that routinely receive high attention in laboratory researches aimed at improving the degradation of hazardous environmental pollutants have never been employed for *in situ* bioremediation (Das and Chandran, 2011).

Members belonging to the genera *Thermococcus*, *Thermotoga*, *Methanococcus*, *Methanobacterium*, and *Thermodesulfobacterium* that essentially encompass thermophiles and hyperthermophiles, have been reported from geographically distinct marine petroleum reservoirs via culture-dependent and/or culture-independent investigations (Slobodkin et al., 1999; Orphan et al., 2000; Stetter and Huber, 2000; Takahata et al., 2000; Gao et al., 2016). Culture-independent investigations have also identified members of the phylum *Bacteroidetes* classified under the genus *Anaerophaga* from offshore high-temperature reservoirs (Dahle et al., 2008). At the phylum level, *Crenarchaeota*, *Euryarchaeota*, *Frimicutes*, *Nitrospira*, *Proteobacteria*, and *Thermodesulfobacteria* are the predominant groups encountered across geographically distinct marine oil reservoirs (Nilsen et al., 1996; Greene et al., 1997; Tardy-Jacquenod et al., 1998; Slobodkin et al., 1999; Myhr and Torsvik, 2000; Miroshnichenko et al., 2001; Takahata et al., 2001; Miranda-Tello et al., 2003; Jeanthon et al., 2005; Belyakova et al., 2006; Cheng et al., 2007; Dahle et al., 2008; Youssef et al., 2009). However, ascertaining which components of the microflora isolated/identified are truly indigenous to oil reservoirs explored still remains a problem (Magot, 2005) owing to a number of factors. For example, contamination with exogenous microorganisms (alongside sulfate or oxygen that alters the

---

[4] In geology, stratal water refers to the water layer(s) that lie below the standing water level. In the context of oil fields, stratal water, with reference to the oil-bearing stratum/strata, can occur as upper, bottom, intermediate as well as edge waters.

[5] This term refers to the water that is produced alongside oil and gas as a byproduct. Such waters, which are more copious in oil reservoirs compared with reservoirs rich in gas, either reside in the same zone that harbors the hydrocarbons or in a zone just underneath.

in situ biogeochemistry) can occur when seawater is reinjected, or seawater/brine infiltrates into oil reservoirs after exposure to surface conditions and disequilibrium of pressure on account of drilling (Youssef et al., 2009); systematic shifts can also occur in microbial community compositions because of anthropogenic activities around oil reservoirs (Vigneron et al., 2017).

7.5.1. Anaerobic Breakdown of Hydrocarbons and Fermentation

Anaerobic oxidation is now considered to be a major process influencing the physical and chemical characteristics of hydrocarbon reservoirs (Gilbert et al., 2019). Thermal decomposition of organic matter within the depositional basins also yields non-methane hydrocarbons such as ethane, propane, etc. (Etiope and Ciccioli, 2009; Dalsøren et al., 2018). In transit through sediments and/or water columns (often via seepage), these hydrocarbons are utilized by *in situ* microorganisms as the sources of carbon and energy under nitrate /sulfate /ferric iron respiring or methanogenic conditions (Widdel and Grundmann, 2010). The prospect of the oxidation of hydrocarbon (to CO_2) in anoxic environments opened up in the early 1990s when growth on C12 to C20 alkanes, 1-hexadecene, 1-hexadecanol, 2-hexadecanol, palmitate, or stearate as sole carbon source was demonstrated in a sulfate-reducing deltaproteobacterial strain isolated from the resin/asphaltene precipitates of an oil–water separator in an oil field near Hamburg (Germany). This study further demonstrated that *n*-hexadecane, when added (as an energy, electron, and carbon source) to anoxic/sulfidic marine sediment samples from distinct geographical locations, could support the dissimilatory reduction of sulfate to sulfide (Aeckersberg et al., 1991). Subsequently, a sulfate-reducing *Desulfobacteraceae* strain capable of using propane or *n*-butane as sole carbon source under anoxic conditions was isolated from a hydrocarbon seep area in the Guayamas Basin (Kniemeyer et al., 2007). However, despite significant culture-based evidence in favor of anaerobic oxidation of hydrocarbons, *in situ* identification and quantification of such processes remain a significant challenge except in rare cases of large-scale degradation/consumption of the hydrocarbon pool (Quistad and Valentine, 2011; Meng et al., 2017; Gilbert et al., 2019). Some headway in establishing methods for ascertaining *in situ* hydrocarbon oxidation (and assessing its impact on the reservoirs) seems to have been made recently in the form of specific stable isotope ratios revealed for the position-specific as well as bulk hydrogen and carbon atoms of hydrocarbons that have been degraded by microorganisms or have remained untouched since their formation/preservation (Quistad and Valentine, 2011; Meng et al., 2017; Gilbert et al., 2019).

With regard to fermentative microorganisms, several such thermophiles have thus far been isolated from offshore oil reservoirs (Grassia et al., 1996). These belonged to either the phylum *Thermotogae* or the order *Thermoanaerobacteriales* of the phylum *Firmicutes* (Cayol et al., 1995). All oil reservoir isolates of *Thermotogae*, in turn, were found to be affiliated to the genera *Geotoga*, *Petrotoga*, *Thermosipho*, and *Thermotoga*. Of these genera, again, *Geotoga* and *Petrotoga* appear to be endemic to oil reservoir environments (Davey et al., 1993; Miranda-Tello et al., 2007; Youssef et al., 2009). Halophilic fermentative bacteria belonging to the genus *Halanaerobium* (order *Halanaerobiales* within the class *Clostridia* of *Firmicutes*) have been isolated from the Emeraude oil field, off the Congo coast (Ravot et al., 1997).

7.5.2. Methanogenesis

Mesophilic and hydrogenotrophic methanogens such as members of *Methanobacterium* (order *Methanobacteriales*) and *Methanoplanus* (order *Methanomicrobiales*) have been isolated from low-salinity oil reservoirs such as those located in Tataria, southern Kara Sea of West Siberia, and the offshore oil fields of Gulf of Guinea, West Africa (Davydova-Charakhch'yan et al., 1992; Ollivier et al., 1997). On the other hand, thermophilic and hydrogenotrophic strains of methanogens belonging to *Methanobacterium*, *Methanoculleus*, *Methanothermobacter* (all three classified under the order *Methanomicrobiales*), and *Methanococcus* and *Methanothermococcus* (both classified under the order *Methanococcales*) have been retrieved from the South Elwood and Santa Clara oil fields, off California (Davydova-Charakhch'yan et al., 1992; Ollivier et al., 1997; Orphan et al., 2000; Jeanthon et al., 2005). Mesophilic and methylotrophic methanogens such as *Methanhalophilus euhalobius* and *Methanosarcina siciliae*, and thermophilic and methylotrophic methanogens such as *Methermicoccus shengliensis* have been isolated alongside aceticlastic methanogens such as *Methanosarcina mazei* (all these genera belong to the order *Methanosarcinales*) from oil fields located in the Gulf of Mexico and the Yellow River delta in the Bohai Sea (Obraztsova et al., 1987; Ni and Boone, 1991; Cheng et al., 2007).

7.5.3. Sulfate Reduction

Strains of sulfate-reducing microorganisms belonging to the Delta class of *Proteobacteria*, and the bacterial phyla *Firmicutes*, *Nitrospira*, and *Thermodesulfobacterium* have been retrieved from oil reservoirs located in distinct territories of the marine realm. Sulfate-reducing members of the archaeal phyla *Euryarchaeota* and *Crenarchaeota* have been isolated from the Hutton oil field, East Shetland Basin, North Sea (Stetter et al., 1993; Youssef et al., 2009). Within *Deltaproteobacteria*, members of the orders *Desulfobacterales*, *Desulfovibrionales* and *Syntrophobacterales*, and particularly strains belonging to the genera *Desulfobacter*, *Desulfobacterium*, *Dsulfobulbus*, *Desulfomicrobium*, *Desulfotignum*, *Desulfovibrio*, and *Thermosulforhabdus* have been isolated from the Emeraude shallow water oil field (off the Congo shore), a North Sea oil field (off the Norwegian coast), and the White Tiger oil field (Cuu Long basin, East Sea, offshore east of the Mekong Delta, Vietnam), amongst others (Galushko and Rosanova, 1991; Beeder et al., 1995; Lien and Beeder, 1997; Lien et al., 1998; Rozanova et al., 2001; Magot et al., 2004; Birkeland, 2005; Ommedal and Torsvik, 2007; dos Santos et al., 2020). Members of the order *Clostridiales* of *Firmicutes*, for example, *Desulfotomaculum*, *Desulfurispora*, *Desulfovirgula*, *Thermodesulfobium*, and *Thermodesulfobacter* have been reported from various regions of the Norwegian sector of the North Sea, including the Gullfaks oil field (Christensen et al., 1992; Nilsen et al., 1996; Euzeby, 2008).

7.6. SUMMARY VIEW OF THE POTENTIAL NETWORK OF BIOGEOCHEMICAL PROCESSES IN OFFSHORE OIL RESERVOIRS

It is difficult to delineate the biogeochemical processes actually operating in offshore oil reservoirs because of problems in deciphering the origin of the microorganisms identified from these environments and the knowledge gap in corroborative *in situ* geochemical data. However, collating all the microbiological data that are available from geographically distinct petroleum reservoirs, we attempt here to envisage a potential network of biogeochemical processes that may operationalize carbon–sulfur–nitrogen cycling in these ecosystems (Figure 7.2). New organic carbon can be formed *in situ* by autotrophic CO_2 fixation, while organic matter breakdown can eventually result in the formation of acetate, lactate, and other low carbon number organic acids, and CO_2, by the action of various anaerobic/fermentative microbial processes. Subsequently, sulfate reduction using the electrons released from catabolism of the already simplified organic matter such as lactate, acetate, etc. (to CO_2) can couple the *in situ* carbon and sulfur cycles. On the other hand, anaerobic oxidation of sulfide using nitrate as the electron acceptor potentially completes the recycling of sulfur, in addition to coupling the carbon-sulfur cycle with the nitrogen cycle. Acetate, in addition to serving as a carbon source for sulfate-reducers, can be transformed into methane and CO_2 by the activity of the *in situ* methanogen populations; furthermore, hydrogenotrophic methanogenesis (using CO_2 and H_2) also appears to be operational in these ecosystems, according to *in situ* microbial community analyses. Whereas the majority of methane generated is potentially trapped in the reservoir rocks, acetate can be supplemented in the system (from CO_2) by the autotrophic acetogens. Thus, autotrophy and organic matter remineralization, and methanogenesis and acetogenesis may well complement each other to run a sustainable and self-contained carbon cycle in offshore oil reservoir ecosystems. There is, however,

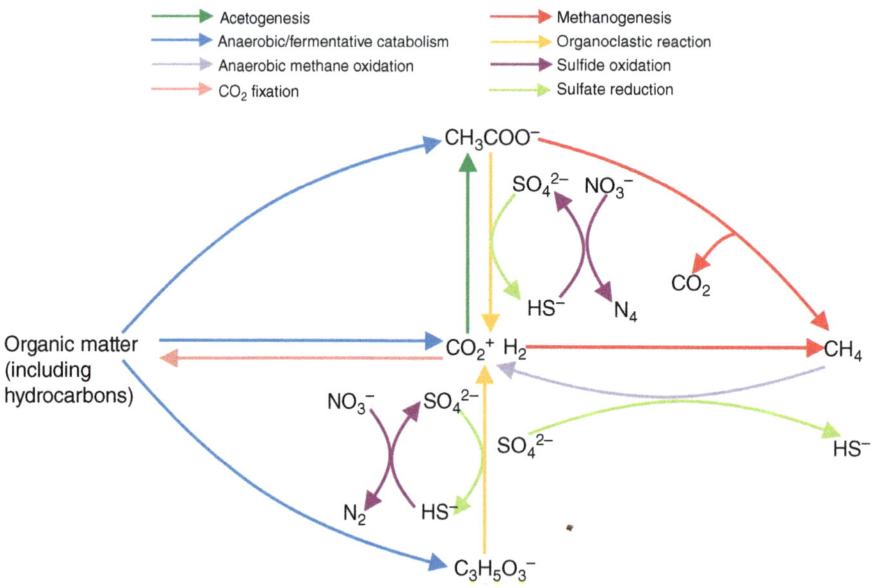

Figure 7.2 Schematic diagram showing the potential network of biogeochemical processes (as suggested by microbial diversity studies) in offshore oil reservoirs (notably, all microbial metabolic types have not been identified from a single site).

little evidence for anaerobic methane oxidation in these environments, as only a few studies of microbial ecology have suggested that such a symbiotic existence of methanogenic archaea and sulfate reducers may possibly exist in these ecosystems.

7.7. EFFECTS OF BIODEGRADATION ON PETROLEUM PROPERTIES

Crude oil is a heterogeneous mixture of paraffinic (straight chain *n*-paraffins or branched isoparaffins), cycloparaffinic (naphthenic), and aromatic (multiple-ringed) hydrocarbons; it also contains small but definite amounts of sulfur, nitrogen and/or oxygen-bearing compounds along with minute quantities of metals such as cadmium, cobalt, lead, mercury, nickel, and vanadium. The main fractionation products include hydrocarbon gases such as ethane, propane and butane; naphtha such, as asphalt, fuel oils, gas oil, gasoline, kerosene, lubricating oils and paraffin wax (Burcik, 1957; McCain, 1990; Gad, 2014; Marigómez, 2014). Albeit the overall chemical constitution of crude oil is largely universal, relative abundance of the aforementioned chemical constituents and the physical properties of the crude oil vary with its source or geological background (Stauffer et al., 2008; Marigómez, 2014; Speight, 2014). For example, proportion of paraffinic compounds is greater in the volatile fractions than in the heavier ones; less-branched isoparaffins, overall, are more abundant than the highly-branched ones; the most abundant cycloalkanes are those which are based on cyclopentane and cyclohexane rings; polycyclic compounds are more abundant in fractions having high molecular weights; aromatic species are also copious, with >30% of any crude oil being made up of polycyclic aromatic hydrocarbons (Stauffer et al., 2008; Marigómez, 2014, Speight, 2014). Oils, having high specific gravity (heavy and viscous oils) are rich in hetero-compounds and aromatic compounds, and yield asphalt, pitch, or tar. The light oils (gaseous petroleum that condenses into liquid oil at the surface) are rich in alkanes and naphthenes, and yield methane, ethane, propane, and butane, and sometimes traces of carbon dioxide, hydrogen sulfide, and nitrogen (Selley, 2003). Odor, expansion coefficient (compressibility factor), aqueous solubility, surface tension, bubble point, dew point, pour point, and flash point are some of the most important properties of crude oil. Furthermore, physicochemical properties such as wax content, sulfur content, and the nature of the predominant hydrocarbon groups determine or affect the economic value of crude oil (National Academies of Sciences, Engineering, and Medicine, 2016).

Biodegradation of petroleum is the return of carbon to inorganic form or shortening of the long chains to smaller ones via electron transfer. Microbial degradation critically affects the composition as well as physical properties of crude oil and natural gases. Oxidation of the C6+ components of oil results in the lowering of the saturated hydrocarbon content, and therefore the economic value, of oil. Increase in the acidity, density, metal content, sulfur content, or viscosity of oil (Evans et al., 1971; Volkman et al., 1984; Rowland et al., 1986; Meredith et al., 2000; Wenger et al., 2002) lowers its value, in addition to which these parameters negatively affect oil production (by reducing the rates of oil flow in the production wells) and downstream refining operations. During biodegradation of oil, hydrocarbons are destroyed preferentially (Fedorak and Westlake, 1984; Huang et al., 2003), acyclic and cyclic compounds, aromatic and saturated carboxylic acids, and phenols are generated (Meredith et al., 2000), while various acidic non-hydrocarbons emanate from the aromatic heterocyclic compounds present (Thorn and Aitkin, 1998). Besides the long chains, light (C2–C5) hydrocarbons and gases present in crude oil can also act as substrates of *in situ* microbial degradation (James and Burns, 1984; Larter, 1997; Borcham et al., 2001). Although there is little evidence of methane oxidation during crude oil biodegradation within reservoirs, the discovery of archaeal 16S rRNA gene sequences belonging to anaerobic methane oxidizers indicates that anaerobic oxidation of methane potentially occurs at some stage of petroleum reservoir genesis and development (Head et al., 2003; Suflita and Mcinerney, 2008).

7.8. DELETERIOUS MICROBIAL ACTIVITIES: HYDROGEN SULFIDE PRODUCTION (SOURING) AND ITS REMEDIATION WITH NITRATE

Seawater is injected into offshore oil reservoirs (Figure 7.1a) to maintain fluid pressure in the systems, and at the same time drive residual oils towards the production wells (this operation is commonly termed as water-flooding; see the inset in Figure 7.1a). At the later stages of reservoir flooding, the amount of produced water increases in the production wells. Consequent to this, the produced waters are separated from the oil fractions, mixed with "make-up" water (which is often drawn from a nearby water plant) to compensate for the oil volume, and reinjected into the reservoirs (Figure 7.3; also see Voordouw, 2011). This entire procedure introduces ~28 mM sulfate into the system, thereby ushering in the production of large amount of sulfide by sulfate-reducing microorganisms, a process commonly termed as souring. Even in high temperature oil reservoirs, sulfate-reducing hyperthermophilic archaea such as *Archaeoglobus fulgidus*, which can grow on diverse organic carbon compounds (including complex polymers) as energy, electron, and carbon sources, act as souring agents (Kaster et al., 2009; Yamane et al., 2011). Enhanced

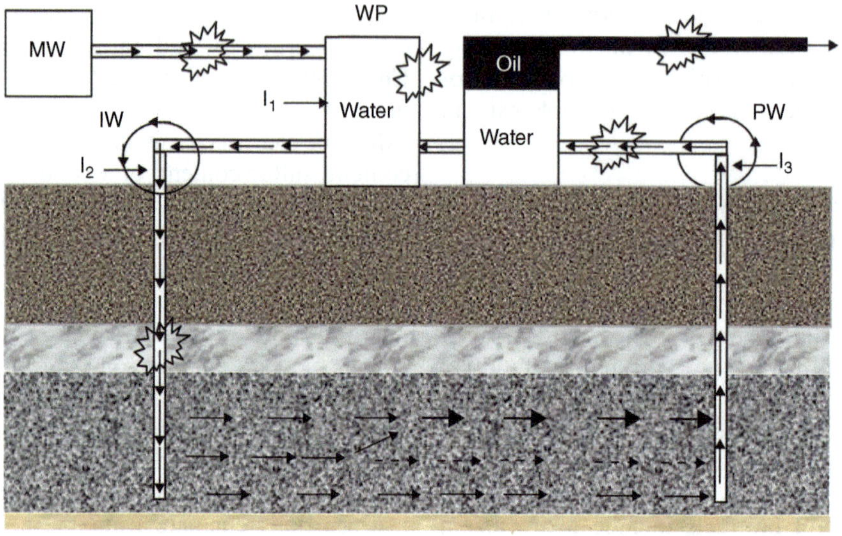

Figure 7.3 Geomicrobiological contour of a petroleum reservoir in relation to production of crude oil (Figure reproduced from Voordouw, 2011, with kind permission from Elsevier). Water is introduced through the injection wells (IW) to drive the oil towards the production wells (PW). After getting piped out from the subsurface, the oil:water mixture is separated into produced oil and produced water; the latter is then supplemented with make-up water (MW) in a water plant (WP) and reinjected into the reservoir. Souring is controlled by injecting nitrate into the WP (at location I_1) and/or at specific injection wells (location I_2). Biocides, inhibitors of corrosion, scavengers of sulfide, etc. are introduced at I_1 and/or I_3. Within the pipelines, corrosion-prone sites are indicated by multicornered 'star' shapes.

build-up of H_2S not only corrodes pipelines, platform structures, and other drilling equipment (thereby increasing the cost of production), but also increases crude oil refining cost, plugs the reservoir channels with sulfidic mineral deposits, and increases health hazards owing to H_2S toxicity (Nemati et al., 2001; Myhr et al., 2002; Vigneron et al., 2018). While hydrogen embrittlement (hydrogen assisted/induced cracking) of metallic equipment is caused indirectly by hydrogen sulfide-producing microorganisms, in many cases members of *Enterobacteriaceae* have been reported to be directly involved in biocorrosion of the metallic equipment installed in marine oil fields (Bermont-Bouis et al., 2007).

In the initial stages of understanding, souring was thought to be a consequence of some abiotic process native to petroleum reservoirs (Herbert, 1987), but its subsequent attribution to *in situ* sulfate-reducing bacteria (Ligthelm et al., 1991) led to the formulation of remedies (for example, injection of nitrate or biocide into reservoir) that in turn were also microbe-based (Sunde et al., 1993). Although physicochemical removal of sulfate is also an effective strategy, the prohibitive cost of implementation limits its use in practice. The effectiveness of biocide application is also constrained by its limited depth reach within reservoir formations. In contrast, injection of nitrate is routinely employed for the remediation or prevention of souring in offshore oil fields; such techniques are also reportedly successful in reservoirs having *in situ* temperatures ≥60°C (Sunde and Torsvik, 2005) and homogeneous permeability (nitrate injection, nevertheless, is also considerably effective in reservoirs having heterogeneous distribution of permeability; Liebensteiner et al., 2014). Mechanistically, nitrate supports the growth of anaerobic (nitrate reducing) sulfide oxidizers (Hubert and Voordouw, 2007) leading to the recycling of sulfide back to sulfate (thereby averting souring of the system); at the same time, the sulfate-reducing microorganisms present *in situ* are inhibited by the nitrite produced from nitrate reduction as dissimilatory sulfite reductase, the central enzyme of sulfate reduction, is inhibited by nitrite (Wolfe et al., 1994; He et al., 2010; Voordouw, 2011). Furthermore, nitrate introduction enables the more energy-efficient nitrate reducers to outcompete sulfate reducers in the race for the same electron donors such as fatty acids and hydrocarbons (Liebensteiner et al., 2014). This said, nitrate-mediated mitigation of souring is purportedly not that successful in low-temperature oil reservoirs, and may well not alleviate souring in deep nitrate-depleted zones where injected waters often usher in far greater amounts of sulfate than the amount of nitrate that eventually reaches there (Voordouw et al., 2009; Callbeck et al., 2011, 2013). In view of these limitations, it has been suggested that (per) chlorate, as a reservoir-temperature-independent souring control agent, could be a potent alternative or force multiplier to nitrate (Liebensteiner et al., 2014) because these oxyanions of chlorine are potentially inhibitory/toxic to sulfate-reducers, including the thermophilic ones (Liebensteiner et al., 2013; Engelbrektson et al., 2014), and can serve as respiratory substrates for the oxidation of sulfide (Gregoire et al., 2014).

7.9. *IN SITU* MICROBIAL PROCESSES BENEFICIAL TO OIL RECOVERY

Recovery of oil from reservoir rocks is constrained by the latter's geological features (for example, permeability; Figure 7.1b) and the physicochemical properties of the oil (for example, the difference in viscosity between water and the oil). Microbe-mediated enhancement of oil recovery (MEOR) was propounded early in the history of industrial oil production (Beckman, 1926). Of the many mechanisms of MEOR known, *in situ* biosurfactant generation (Figure 7.1c) is thought to help mobilize residual oil, while microbe-mediated plugging (Figure 7.1d) of high-permeability zones (Gray et al., 2008) is also widely used. Stimulation of indigenous microbes that can utilize a fraction of the hydrocarbons as energy, electron and/or carbon sources constitute a major aspect of MEOR, and in that direction (per)chlorate reduction has been proposed as a versatile enabler as the chlorite and O_2 generated from this metabolic process can chemically/microbially oxidize HS^-, Fe^{2+} as well as hydrocarbons *in situ* (Liebensteiner et al., 2014). Several other *in situ* microbial activities (for example, pathways of hydrocarbon metabolism) and their resultant products (for example, acids, biopolymers, biosurfactants, emulsifiers, gases, and solvents) are also reportedly helpful in oil recovery procedures. Key microbial process-based technologies involved in oil recovery include paraffin removal, well stimulation[6], and enhanced water-flooding (Knapp et al., 1990). Viscous and immobilized substances such as paraffin are required to be removed from oil-producing wells so as to improve drainage and enhance oil flow, and in doing so make the systems more cost effective (Sousa et al., 2019). While buttressing the metabolism of hydrocarbons by native microbial communities is a common strategy for removing *in situ* paraffin deposits, many hydrocarbon-degrading aerobes (Van Hamme et al., 2003) and anaerobes (Spormann and Widdel, 2000) have been isolated from offshore oil reservoirs. *Pseudomonas* strains have been reported to be capable of reducing paraffin/wax precipitation from crude oil by emulsifying immiscible hydrocarbons such as kerosene, toluene, xylene, and crude oil, in addition to lowering the concentration of long chain (C22+) hydrocarbons (Sifour et al. 2007; Etoumi et al., 2008). Field tests have also shown that strains of *Bacillus* and *Pseudomonas* species can effectively remove paraffin and enhance oil production in test wells (He et al., 2003; Xiao et al., 2012). Acids, gases and solvents produced by microorganisms *in situ* help mobilize the oil entrapped in the course of water flooding (Figure 7.1b), so after successful initiation in Statoil's Norne field (located in the Norwegian Sea), metabolites resulting from microbial treatment are routinely utilized to shore up production at the level of individual wells (Rassenfoss, 2011). In relation to MEOR, strains of *Bacillus* and *Clostridium* are generally the acid/gas/solvent producers of choice, whereas a wider diversity of microorganisms is utilized as producers of surface-active agents, which facilitate recovery by lowering surficial or interfacial tension (Bryant, 1990; Youssef et al., 2004).

[6] *Oil well stimulation involves large-scale microbial production of acids, biosurfactants, emulsifiers, gases and/or solvents inside, or in the close vicinity of, production wells to enhance production rates.*

7.10. CONCLUDING REMARKS

It is peculiar that despite an early start, biogeochemical knowledge on marine petroleum reservoirs is still thought to be in its infancy. The main issue holding back a robust elucidation of the geomicrobiological dynamics of oil reservoirs surrounds the basic question regarding the origin (indigenous or non-indigenous nature) of the microbial communities reported from these deep subsurface environments. Although a few studies have envisaged microbial community succession during water-flooding of oil reservoirs (Vigneron et al., 2017), the overall paucity of consensus on the actual compositions of the *in situ* communities precludes any comprehensive estimation of how microbiome structures and/or functions change in oil reservoirs over time in relation to the different exploitation practices. Tandem investigation and interpretation of preserved geochemical records from the same samples used for reporting *in situ* microflora is needed to unambiguously elucidate petroleum reservoir biogeochemistry. Accordingly, new lines of cross-disciplinary research involving all technical and academic partners of the petroleum exploration and production industry need to be initiated at the exploration and prospecting stages so that a translational knowledge base can be created for the geomicrobiology of marine petroleum systems.

ACKNOWLEDGEMENTS

The authors thank Aaheli Bhattacharjee, Senior Geologist (subsurface), Cairn India, for reading the manuscript and offering a valuable review.

REFERENCES

Abadi, A.M., Van Wees, J.D., Van Dijk, P.M. et al. (2008). Tectonics and subsidence evolution of the Sirt Basin, Libya. *AAPG Bulletin 92*: 993–1027. https://doi.org/10.1306/03310806070

Aeckersberg, F., Bak, F. and Widdel, F. (1991). Anaerobic oxidation of saturated hydrocarbons to CO_2 by a new type of sulfate-reducing bacterium. *Archives of Microbiology 156*: 5–14. https://doi.org/10.1007/BF00418180

Al-Areeq N.M. (2018). Petroleum source rocks characterization and hydrocarbon generation. In *Recent Insights in Petroleum Science and Engineering* (ed. M. Zoveidavianpoor), 1–29. Rijeka, Croatia: InTech Publisher. https://doi.org/10.5772/intechopen.70092

Al-Hajeri, M.M., Al Saeed, M., Derks, J. et al. (2009). Basin and petroleum system modeling. *Oilfield Review 21*: 14–29.

Allix, P., Burnham, A., Fowler, T. et al. (2010). Coaxing oil from shale. *Oilfield Review 22*: 4–15.

Arndt, S., Jørgensen, B.B., LaRowe, D.E. et al. (2013). Quantifying the degradation of organic matter in marine sediments: a review and synthesis. *Earth-Science Reviews 123*: 53–86. https://doi.org/10.1016/j.earscirev.2013.02.008

Arzate, S.O., Moguel, C.R., Lopez, M.M. et al. (2009). Basin modeling of the Upper Jurassic petroleum systems (Tithonian and Oxfordian) in the Akalan-Chilam area of the Campeche Sound in the southern Gulf of Mexico. In: *Petroleum Systems in the Southern Gulf of Mexico* (eds. C. Bartolini and J.R.R. Ramos), 285–313. American Association of Petroleum Geologists. https://doi.org/10.1306/13191089M903333

Atlas, R.M. and Hazen, T.C. (2011). Oil biodegradation and bioremediation: a tale of the two worst spills in US history. *Environmental Science and Technology 45*: 6709–6715. https://doi.org/10.1021/es2013227

Barth, T. (1991). Organic acids and inorganic ions in waters from petroleum reservoirs, Norwegian continental shelf: a multivariate statistical analysis and comparison with American reservoir formation waters. *Applied Geochemistry 6*, 1–15. https://doi.org/10.1016/08832927(91)90059-X

Bastianoni, S., Campbell, D., Susani, L. et al. (2005). The solar transformity of oil and petroleum natural gas. *Ecological Modelling 186*: 212–220. https://doi.org/10.1016/j.ecolmodel.2005.01.015

Bastin, E. (1926). Microorganisms in oilfields. *Science 63*: 21–24.

Beckman, J.W. (1926). The action of bacteria on mineral oil. *Industrial and Engineering Chemistry, News Edition 4*: 23–26.

Beeder, J., Torsvik, T. and Lien, T. (1995). *Thermodesulforhabdus norvegicus* gen. nov., sp. nov., a novel thermophilic sulfate-reducing bacterium from oil field water. *Archives of Microbiology 164*: 331–336. https://doi.org/10.1007/BF02529979

Beglinger, S.E., Doust, H. and Cloetingh, S. (2012). Relating petroleum system and play development to basin evolution: West African South Atlantic basins. *Marine and Petroleum Geology 30*, 1–25. https://doi.org/10.1016/j.marpetgeo.2011.08.008

Belyakova, E.V., Rozanova, E.P., Borzenkov, I.A. et al. (2006). The new facultatively chemolithoautotrophic, moderately halophilic, sulfate-reducing bacterium *Desulfovermiculus halophilus* gen. nov., sp. nov., isolated from an oil field. *Mikrobiologiia 75*: 161–171. https://doi.org/10.1134/S0026261706020093

Bermont-Bouis, D., Janvier, M., Grimont, P.A.D. et al. (2007). Both sulfate-reducing bacteria and Enterobacteriaceae take part in marine biocorrosion of carbon steel. *Journal of Applied Microbiology 102*: 161–168. https://doi.org/10.1111/j.1365–2672.2006.0305

Bernard, F.P., Connan, J. and Magot, M. (1992). Indigenous microorganisms in connate water of many oil fields: a new tool in exploration and production techniques. In: *SPE Annual Technical Conference and Exhibition, Society of Petroleum Engineers*. https://doi.org/10.2118/24811-MS

Birkeland, N.K. (2005). Sulfate-reducing bacteria and archaea. In: *Petroleum Microbiology* (eds. B. Ollivier and M. Magot), 35–54. Washington, DC: ASM Press. https://doi.org/10.1128/978155581758h3

Boreham, C.J., Hope, J.M. and Hartung-Kagi, B. (2001). Understanding source, distribution and preservation of Australian natural gas: a geochemical perspective. *The APPEA Journal 41*: 523–547. https://doi.org/10.1071/AJ00026

Bradley, J.A., Amend, J.P. and LaRowe, D.E. (2018). Survival of the fewest: microbial dormancy and maintenance in marine sediments through deep time. *Geobiology 17*: 43–59. https://doi.org/10.1111/gbi.12313

Bragg, J.R., Prince, R.C., Wilkinson, J.B. et al. (1992). *Bioremediation for shoreline cleanup following the 1989 Alaskan oil spill*. Exxon Company.

Bragg, J.R., Prince, R.C., Harner, E.J. et al. (1994). Effectiveness of bioremediation for the Exxon Valdez oil spill. *Nature 368*: 413–418. https://doi.org/10.1038/368413a0

Brakstad, O.G., Kotlar, H.K. and Markussen, S. (2008). Microbial communities of a complex high-temperature offshore petroleum reservoir. *International Journal of Oil, Gas and Coal Technology 1*: 211–228. https://doi.org/10.1504/IJOGCT.2008.019843

Brooks, J. and Welte, D.H. (1984). *Advances in Petroleum Geochemistry*. Academic Press. https://doi.org/10.1016/C2009–0-21875–7

Bryant, R.S. (1990). Microbial enhanced oil recovery and compositions therefor. IIT Research Institute, US Patent 4,905,761.

Burcik, E.J. (1957). *Properties of Petroleum Reservoir Fluids*. John Wiley and Sons.

Buryakovsky, L., Aminzadeh, F. and Chilingarian, G.V. (2001). *Petroleum geology of the South Caspian Basin*. Elsevier. https://doi.org/10.1016/B978-0-88415-342-5000–0

Callbeck, C.M., Agrawal, A. and Voordouw, G. (2013). Acetate production from oil under sulfate-reducing conditions in bioreactors injected with sulfate and nitrate. *Applied and Environmental Microbiology 79*, 5059–5068. https://doi.org/10.1128/aem.01251–13

Callbeck, C.M., Dong, X., Chatterjee, I. et al. (2011). Microbial community succession in a bioreactor modeling a souring low-temperature oil reservoir subjected to nitrate injection. *Applied Microbiology and Biotechnology 91*: 799–810. https://doi.org/10.1007/s00253–011–3287–2

Carvajal-Arenas, L.C., Torrado, L., Mann, P. et al. (2020). Basin modeling of Late Cretaceous/Mio-Pliocene petroleum system of the deep-water eastern Colombian Basin and South Caribbean Deformed Belt. *Marine and Petroleum*

Geology 121: 104511. https://doi.org/10.1016/j.marpetgeo.2020.104511

Cayol, J.L., Ollivier, B., Patel, B.K.C. et al. (1995). Description of *Thermoanaerobacter brockii* subsp. *lactiethylicus* subsp. nov., isolated from a deep subsurface French oil well, a proposal to reclassify *Thermoanaerobacter finnii* as *Thermoanaerobacter brockii* subsp. *finnii* comb. nov., and an emended description of *Thermoanaerobacter brockii. International Journal of Systematic Bacteriology 45*: 783–789. https://doi.org/10.1099/00207713-45-4-783

Cheng, L., Qiu, T.L., Yin, X.B. et al. (2007). *Methermicoccus shengliensis* gen. nov., sp. nov., a thermophilic, methylotrophic methanogen isolated from oil-production water, and proposal of *Methermicoccaceae* fam. nov. *International Journal of Systematic and Evolutionary Microbiology 57*: 2964–2969. https://doi.org/10.1099/ijs.0.65049-0

Christensen, B., Torsvik, T. and Lien, T. (1992). Immunomagnetically captured thermophilic sulfate-reducing bacteria from North Sea oil field waters. *Applied and Environmental Microbiology 58*: 1244–1248.

Coates, J.D., Bruce, R.A. and Haddock, J.D. (1998). Anoxic bioremediation of hydrocarbons. *Nature 396*: 730. https://doi.org/10.1038/25470

Coates, J.D., Woodward, J., Allen, J. et al. (1997). Anaerobic degradation of polycyclic aromatic hydrocarbons and alkanes in petroleum-contaminated marine harbor sediments. *Applied and Environmental Microbiology 63*: 3589–3593.

Connan, J. (1984). Biodegradation of crude oils in reservoirs. *Advances in Petroleum Geochemistry 1*: 299–335.

Dahle, H., Garshol, F., Madsen, M. et al. (2008). Microbial community structure analysis of produced water from a high-temperature North Sea oil-field. *Antonie Van Leeuwenhoek 93*: 37–49. https://doi.org/10.1007/s10482-007-9177-z

Dalsøren, S.B., Myhre, G., Hodnebrog, Ø. et al. (2018). Discrepancy between simulated and observed ethane and propane levels explained by underestimated fossil emissions. *Nature Geoscience 11*: 178–184. https://doi.org/10.1038/s41561-018-0073-0

Das, N. and Chandran, P. (2011). Microbial degradation of petroleum hydrocarbon contaminants: an overview. *Biotechnology Research International 2011*: 941810. http://dx.doi.org/10.4061/2011/941810

Davey, M.E., Wood, W.A., Key, R. et al. (1993). Isolation of three species of *Geotoga* and *Petrotoga*: two new genera, representing a new lineage in the bacterial line of descent distantly related to the "Thermotogales". *Systematic and Applied Microbiology 16*: 191–200. https://doi.org/10.1016/S0723-2020(11)80467-4

Davydova-Charakhch'yan, I.A., Mileeva, A.N., Mityushina, L.L. et al. (1992). Acetogenic bacteria from oil fields of Tataria and wester Siberia. *Microbiology New York 61*: 208–216.

DembickiJr, H. (2009). Three common source rock evaluation errors made by geologists during prospect or play appraisals. *AAPG Bulletin 93*: 341–356.

Deng, Y. 2019. Control effect of the gulf on marine petroleum. *Petroleum Research 4*: 283–294. https://doi.org/10.1016/j.ptlrs.2019.09.003

dos Santos, J.C., Lopes, D.R.G., Da Silva, J.D. et al. (2020). Diversity of sulfate-reducing prokaryotes in petroleum production water and oil samples. *International Biodeterioration and Biodegradation 151*: 104966. https://doi.org/10.1016/j.ibiod.2020.104966

Eaton, R.W. and Chapman, P.J. (1992). Bacterial metabolism of naphthalene: construction and use of recombinant bacteria to study ring cleavage of 1,2-dihydroxynaphthalene and subsequent reactions. *Journal of Bacteriology 174*: 7542–7554. https://doi.org/10.1128/jb.174.23.7542-7554.1992

Engelbrektson, A., Hubbard, C.G., Tom, L.M. et al. (2014). Inhibition of microbial sulfate reduction in a flow-through column system by (per) chlorate treatment. *Frontiers in Microbiology 5*: 315. https://doi.org/10.3389/fmicb.2014.00315

Etiope, G. and Ciccioli, P. (2009). Earth's degassing: a missing ethane and propane source. *Science 323*: 478–478. https://doi.org/10.1126/science.1165904

Etoumi, A., El Musrati, I., El Gammoudi, B. and El Behlil, M. (2008). The reduction of wax precipitation in waxy crude oils by *Pseudomonas* species. *Journal of Industrial Microbiology and Biotechnology 35*: 1241–1245. https://doi.org/10.1007/s10295-008-0420-z

Euzeby, J.P. (2008). *List Of Prokaryotic Names with Standing in Nomenclature*. http://www.bacterio.net.

Evans, C.R., Rogers, M.A. and Bailey, N.J.L. (1971). Evolution and alteration of petroleum in western Canada. *Chemical Geology 8*: 147–170. https://doi.org/10.1016/0009-2541(71)90002-7

Fedorak, P.M. and Westlake, D.W. (1984). Microbial degradation of alkyl carbazoles in Norman Wells crude oil. *Applied and Environmental Microbiology 47*: 858–862.

Fisher, J.B. (1987). Distribution and occurrence of aliphatic acid anions in deep subsurface waters. *Geochimica et Cosmochimica Acta 51*: 2459–2468. https://doi.org/10.1016/0016-7037(87)90297-3

Gad, S.C. (2014). Petroleum Hydrocarbons. In: *Encyclopedia of Toxicology* (ed. P. Wexler), 838–840. Academic Press. https://doi.org/10.1016/B978-0-12-386454-3.00899-X

Galloway, W.E., Ganey-Curry, P.E., Li, X. et al. (2000). Cenozoic depositional history of the Gulf of Mexico basin. *AAPG Bulletin 84*: 1743–1774. https://doi.org/10.1306/8626C37F-173B-11D7-8645000102C1865D

Galushko, A.S. and Rosanova, E.P. (1991). *Desulfobacterium cetonicum* spec. nov., a sulfate reducing bacterium oxidizing fatty acids and ketones. *Mikrobiologiia 60*: 102–107.

Gao, P., Tian, H., Wang, Y. et al. (2016). Spatial isolation and environmental factors drive distinct bacterial and archaeal communities in different types of petroleum reservoirs in China. *Scientific Reports 6*: 20174. https://doi.org/10.1038/srep20174

Gilbert, A., Lollar, B.S., Musat, F. et al. (2019). Intramolecular isotopic evidence for bacterial oxidation of propane in subsurface natural gas reservoirs. *Proceedings of the National Academy of Sciences USA 116*: 6653–6658. https://doi.org/10.1073/pnas.1817784116

Grabowski, A., Nercessian, O., Fayolle, F. et al. (2005). Microbial diversity in production waters of a low-temperature biodegraded oil reservoir. *FEMS Microbiology Ecology 54*: 427–443. https://doi.org/10.1016/j.resmic.2005.03.009

Grassia, G.S., McLean, K.M., Glénat, P. et al. (1996). A systematic survey for thermophilic fermentative bacteria and archaea in high temperature petroleum reservoirs. *FEMS Microbiology Ecology 21*: 47–58. https://doi.org/10.1111/j.1574-6941.199b0033

Gray, M.R., Yeung, A., Foght, J.M. et al. (2008). Potential microbial enhanced oil recovery processes: a critical analysis. In: *Proceedings of the SPE Annual Technical Conference and Exhibition, Denver, Colorado*. Society of Petroleum Engineers.

Grbic-Galic, D. (1990). Anaerobic microbial transformation of nonoxygenated aromatic and alicyclic compounds in soil, subsurface and freshwater sediments. In: *Soil biochemistry*, Vol. *6* (eds. J.M. Bollag and G. Stotzky), 117–186. New York: Marcel Dekker.

Greene, A.C., Patel, B.K.C. and Sheehy, A. (1997). *Deferribacter thermophilus* gen. nov., sp. nov., a novel thermophilic manganese-and iron-reducing bacterium isolated from a petroleum reservoir. *International Journal of Systematic and Evolutionary Microbiology 47*: 505–509.

Gregoire, P., Engelbrektson, A., Hubbard, C.G. et al. (2014). Control of sulfidogenesis through bio-oxidation of H_2S coupled to (per) chlorate reduction. *Environmental Microbiology Reports 6*: 558–564. https://doi.org/10.1111/1758-2229.12156

Guzman-Vega, M.A. and Mello, M.R. (1999). Origin of oil in the Sureste Basin, Mexico. *AAPG Bulletin 83*: 1068–1095. https://doi.org/10.1306/E4FD2E79-1732-11D7-8645000102C1865D

Guzman-Vega, M.A., Espitalié, J. and Delfaud, J. (1991). Source rocks characterization of Upper Jurassic of Tampico-Tuxpan 2 basin, east of Mexico. In: *Advances in Exploration Technology, Source-Rock Geology. Proceedings of the Thirteenth World Petroleum Congress, Buenos Aires, Argentina* (eds. L. Montadert and A. Ortiz), 137–138.

Hallett, D. (2002). Petroleum systems. In: *Petroleum Geology of Libya* (ed. D. Hallett), 355–416. Elsevier. https://doi.org/10.1016/B978-044450525-5/50010-X

Harayama, S., Kishira, H., Kasai, Y. et al. (1999). Petroleum biodegradation in marine environments. *Journal of Molecular Microbiology and Biotechnology 1*: 63–70.

He, Q., He, Z., Joyner, D.C. et al. (2010). Impact of elevated nitrate on sulfate-reducing bacteria: a comparative study of *Desulfovibrio vulgaris*. *The ISME Journal 4*: 1386–1397. https://doi.org/10.1038/ismej.2010.59

He, Z., Mei, B., Wang, W. et al. (2003). A pilot test using microbial paraffin-removal technology in Liaohe oilfield. *Petroleum Science and Technology 21*: 201–210. https://doi.org/10.1081/LFT-120016942

Head, I.M., Jones, D.M. and Larter, S.R. (2003). Biological activity in the deep subsurface and the origin of heavy oil. *Nature 426*: 344–352. https://doi.org/10.1038/nature02134

Herbert, B.N. (1987). Reservoir Souring. In: *Microbial Problems in the Offshore Oil Industry* (eds. E.C. Hill, J.L. Shennan and A.J. Watkinson), 63–71. London: John Wiley and Sons.

Huang, H., Bowler, B.F., Zhang, Z. et al. (2003). Influence of biodegradation on carbazole and benzocarbazole distributions in oil columns from the Liaohe basin, NE China. *Organic Geochemistry 34*: 951–969. https://doi.org/10.1016/S0146-6380(03)00033-0

Hubert, C. and Voordouw, G. (2007). Oil field souring control by nitrate-reducing *Sulfurospirillum* spp. that outcompete sulfate-reducing bacteria for organic electron donors. *Applied and Environmental Microbiology 73*: 2644–2652. https://doi.org/10.1128/aem.02332-06

Hyne, N.J. (2001). *Nontechnical Guide to Petroleum Geology, Exploration, Drilling and Production*. Oklahoma: Pennwell Books.

Isaksen, G.H. and Ledje, K.H.I. (2001). Source rock quality and hydrocarbon migration pathways within the greater Utsira High area, Viking Graben, Norwegian North Sea. *AAPG Bulletin 85*: 861–883. https://doi.org/10.1306/8626CA23-173B-11D7-8645000102C1865D

James, A.T. and Burns, B.J. (1984). Microbial alteration of subsurface natural gas accumulations. *AAPG Bulletin 68*: 957–960.

Jeanthon, C., Nercessian, O., Corre, E. et al. (2005). Hyperthermophilic and methanogenic archaea in oil fields. In: *Petroleum Microbiology* (eds. B. Ollivier and M. Magot), 55–69. Washington, DC: ASM Press. https://doi.org/10.1128/9781555817589.ch4

Jiang, Z., Zhang, W., Liang, C. et al. (2016). Basic characteristics and evaluation of shale oil reservoirs. *Petroleum Research 1*: 149–163. https://doi.org/10.1016/S2096-2495(17)30039-X

Jones, R.M., Goordial, J.M. and Orcutt, B.N. (2018). Low energy subsurface environments as extraterrestrial analogs. *Frontiers in Microbiology 9*: 1605. https://doi.org/10.3389/fmicb.2018.01605

Ka, J.O., Yu, Z. and Mohn, W.W. (2001). Monitoring the size and metabolic activity of the bacterial community during biostimulation of fuel-contaminated soil using competitive PCR and RT-PCR. *Microbial Ecology 42*: 267–273. https://doi.org/10.1007/s00248-001-0003-2

Kadhim, M.G. and Ali, M.T. (2017). A critical review on corrosion and its prevention in the oilfield equipment. *Journal of Petroleum Research and Studies 14*: 162–189.

Kakati, A., Kumar, G. and Sangwai, J.S. (2020). Oil recovery efficiency and mechanism of low salinity-enhanced oil recovery for light crude oil with a low acid number. *ACS Omega 5*: 1506–1518. https://doi.org/10.1021/acsomega.9b03229

Kandiyoti, R., Herod, A.A., Bartle, K.D. et al. (2006). Pyrolysis: thermal breakdown of solid fuels in a gaseous environment. In: *Solid Fuels and Heavy Hydrocarbons Liquids* (eds. R. Kandiyoti, A.A. Herod, K.D. Bartle et al.), 36–90. Elsevier Science. https://doi.org/10.1016/B978-008044486-4/50003-9

Kaster, K.M., Bonaunet, K., Berland, H. et al. (2009). Characterisation of culture-independent and-dependent microbial communities in a high-temperature offshore chalk petroleum reservoir. *Antonie Van Leeuwenhoek 96*: 423–439. https://doi.org/10.1007/s10482-009-9356-1

Kato, C. and Bartlett, D.H. (1997). The molecular biology of barophilic bacteria. *Extremophiles 1*: 111–116. https://doi.org/10.1007/s007920050023

Knapp, R.M., Chisholm, J.L. and McInerney, M.J. (1990). Microbially enhanced oil recovery. In: *Proceedings of the SPE/UH Emerging Technologies Conference, Houston*, 229–234. Institute for Improved Oil Recovery, University of Houston.

Kniemeyer, O., Musat, F., Sievert, S.M. et al. (2007). Anaerobic oxidation of short-chain hydrocarbons by marine sulphate-reducing bacteria. *Nature 449*: 898–901. https://doi.org/10.1038/nature06200

Konyuhov, A.I. and Maleki, B. (2006). The Persian Gulf Basin: geological history, sedimentary formations and petroleum potential. *Lithology and Mineral Resources 41*: 344–361. https://doi.org/10.1134/S0024490206040055

Krejci-Graf, K. (1932). Rule of density of oils. *AAPG Bulletin 16*: 1038.

Krurmholz, L.R., Caldwell, M.E. and Suflita, J.M. (2005). Biodegradation of 'BTEX'hydrocarbons under anaerobic conditions. In: *Bioremediation: principles and applications*, Vol. *6* (eds. R.L. Crawford and D.L. Crawford), 61–99. New York: Cambridge University Press.

Larter, I.H.S. (1997). Petroleum migration, alteration and remigration within Troll field, Norwegian North Sea. *AAPG Bulletin 81*: 222–248.

Larter, S., Wilhelms, A., Head, I. et al. (2003). The controls on the composition of biodegraded oils in the deep subsurface-part 1: biodegradation rates in petroleum reservoirs. *Organic Geochemistry 34*: 601–613. https://doi.org/10.1016/S0146-6380(02)00240-1

Lever, M.A., Rogers, K.L., Lloyd, K.G. et al. (2015). Life under extreme energy limitation: a synthesis of laboratory-and field-based investigations. *FEMS Microbiology Reviews 39*: 688–728. https://doi.org/10.1093/femsre/fuv020

Levorsen, A.I. (2001). *Geology of Petroleum*. Oklahoma: American Association of Petroleum Geologists. https://doi.org/10.1306/SP812

Liebensteiner, M.G., Pinkse, M.W., Schaap, P.J. et al. (2013). Archaeal (per) chlorate reduction at high temperature: an interplay of biotic and abiotic reactions. *Science 340*: 85–87. https://doi.org/10.1126/science.1233957

Liebensteiner, M.G., Tsesmetzis, N., Stams, A.J. et al. (2014). Microbial redox processes in deep subsurface environments and the potential application of (per) chlorate in oil reservoirs. *Frontiers in Microbiology 5*: 428. https://doi.org/10.3389/fmicb.2014.00428

Lien, T. and Beeder, J. (1997). *Desulfobacter vibrioformis* sp. nov., a sulfate reducer from a water-oil separation system. *International Journal of Systematic Bacteriology 47*: 1124–1128. https://doi.org/10.1099/00207713-47-4-1124

Lien, T., Madsen, M., Steen, I.H. (1998). *Desulfobulbus rhabdoformis* sp. nov., a sulfate reducer from a water-oil separation system. *International Journal of Systematic Bacteriology 48*: 469–474. https://doi.org/10.1099/00207713-48-2-469

Ligthelm, D.J., de Boer, R.B., Brint, J.F. et al. (1991). Reservoir souring: an analytical model for H2S generation and transportation in an oil reservoir owing to bacterial activity. In: *Offshore Europe*. Aberdeen: Society of Petroleum Engineers. https://doi.org/10.2118/23141-MS

Lindstrom, J.E., Barry, R.P. and Braddock, J.F. (1999). Long-term effects on microbial communities after a subarctic oil spill. *Soil Biology and Biochemistry 31*: 1677–1689. https://doi.org/10.1016/S0038-0717(99)00081-4

Magoon, L.B. and Dow, W.G. (1994). *The petroleum system. In: The Petroleum System – From Source to Trap* (eds. L.B. Magoon and W.G. Dow). American Association of Petroleum Geologists. https://doi.org/10.1306/M60585C1

Magot, M. (2005). Indigenous microbial communities in oil Fields. In: *Petroleum Microbiology* (eds. B. Ollivier and M. Magot), 21–33. Washington, DC: ASM Press. https://doi.org/10.1128/978155581758h2

Magot, M., Ollivier, B. and Patel, B.K. (2000). Microbiology of petroleum reservoirs. *Antonie Van Leeuwenhoek 77*: 103–116. https://doi.org/10.1023/A:1002434330514

Magot, M., Basso, O., Tardy-Jacquenod, C. et al. (2004). *Desulfovibrio bastinii* sp. nov. and *Desulfovibrio gracilis* sp. nov., moderately halophilic, sulfate-reducing bacteria isolated from deep subsurface oilfield water. *International Journal of Systematic and Evolutionary Microbiology 54*: 1693–1697. https://doi.org/10.1099/ijs.0.02977–0

Marigómez, I. (2014). Oil, Crude. In: *Encyclopedia of Toxicology* (ed. P. Wexler), 663–669. Academic Press. https://doi.org/10.1016/B978-0-12-386454-3.00524-8

McCainJr, W.D. (1990). *The Properties of Petroleum Fluids*. PennWell Publishing Company.

McCarthy, K., Rojas, K., Niemann, M. et al. (2011). Basic petroleum geochemistry for source rock evaluation. *Oilfield Review 23*: 32–43.

Meng, Q., Wang, X., Wang, X. et al. (2017). Gas geochemical evidences for biodegradation of shale gases in the Upper Triassic Yanchang Formation, Ordos Basin, China. *International Journal of Coal Geology 179*: 139–152. https://doi.org/10.1016/j.coal.2017.05.018

Meredith, W., Kelland, S.J. and Jones, D.M. (2000). Influence of biodegradation on crude oil acidity and carboxylic acid composition. *Organic Geochemistry 31*: 1059–1073. https://doi.org/10.1016/S0146–6380(00)00136–4

Miranda-Tello, E., Fardeau, M.L., Fernández, L. et al. (2003). *Desulfovibrio capillatus* sp. nov., a novel sulfate-reducing bacterium isolated from an oil field separator located in the Gulf of Mexico. *Anaerobe 9*: 97–103. https://doi.org/10.1016/S1075–9964(03)00064–7

Miranda-Tello, E., Fardeau, M.L., Joulian, C. et al. (2007). *Petrotoga halophila* sp. nov., a thermophilic, moderately halophilic, fermentative bacterium isolated from an offshore oil well in Congo. *International Journal of Systematic and Evolutionary Microbiology 57*: 40–44. https://doi:10.1099/ijs.0.64516–0

Miroshnichenko, M.L., Hippe, H., Stackebrandt, E. et al. (2001). Isolation and characterization of *Thermococcus sibiricus* sp. nov. from a Western Siberia high-temperature oil reservoir. *Extremophiles 5*: 85–91. https://doi.org/10.1007/s007920100175

Moldowan, J.M. and Seifert, W.K. (1980). First discovery of botryococcane in petroleum. *Journal of the Chemical Society, Chemical Communications* 912–914. https://doi.org/10.1039/C39800000912

Myhr, S. and Torsvik, T. (2000). *Denitrovibrio acetiphilus*, a novel genus and species of dissimilatory nitrate-reducing bacterium isolated from an oil reservoir model column. *International Journal of Systematic and Evolutionary Microbiology 50*: 1611–1619. https://doi.org/10.1099/00207713-50-4-1611

Myhr, S., Lillebø, B.L., Sunde, E. et al. (2002). Inhibition of microbial H_2S production in an oil reservoir model column by nitrate injection. *Applied Microbiology and Biotechnology 58*: 400–408. https://doi.org/10.1007/s00253-001-0881-8

National Academies of Sciences, Engineering and Medicine. (2016). Chemical and Physical Properties of Crude Oils. In: *Spills of Diluted Bitumen from Pipelines: a Comparative Study of Environmental Fate, Effects and Response*, 21–34.

Washington, DC: National Academies Press. https://doi.org/10.17226/21834

Nemati, M., Jenneman, G.E. and Voordouw, G. (2001). Mechanistic study of microbial control of hydrogen sulfide production in oil reservoirs. *Biotechnology and Bioengineering 74*: 424–434. https://doi.org/10.1002/bit.1133

Ni, S. and Boone, D.R. (1991). Isolation and characterization of a dimethyl sulfide-degrading methanogen, *Methanolobus siciliae* HI350, from an oil well, characterization of M. siciliae T4/MT and emendation of *M. siciliae*. *International Journal of Systematic Bacteriology 41*: 410–416. https://doi.org/10.1099/00207713-41-3-410

Nilsen, R.K., Beeder, J., Thorstenson, T. et al. (1996). Distribution of thermophilic marine sulfate reducers in north sea oil field waters and oil reservoirs. *Applied and Environmental Microbiology 62:* 1793–1798.

Obraztsova, A.Y., Laurinavinchus, K.S., Bezrukova, L.V. et al. (1987). Biological properties of *Methanosarcina* not utilizing carbonic-acid and hydrogen. *Mikrobiologiia 56*: 807–812.

Offshore Engineering. (2021). Hydrocarbon Formation. https://www.offshoreengineering.com/oil-and-gas/petroleum-geology/1-hydrocarbon-formation

Ollivier, B., Cayol, J.L., Patel, B.K.C. et al. (1997). *Methanoplanus petrolearius* sp. nov., a novel methanogenic bacterium from an oil-producing well. *FEMS Microbiology Letters 147*: 51–56. https://doi.org/10.1111/j.1574-6968.199b1021

Ommedal, H. and Torsvik, T. (2007). *Desulfotignum toluenicum* sp. nov., a novel toluene-degrading, sulphate-reducing bacterium isolated from an oil-reservoir model column. *International Journal of Systematic and Evolutionary Microbiology 57*: 2865–2869. https://doi.org/10.1099/ijs.0.65067-0

Orphan, V.J., Taylor, L.T., Hafenbradl, D. et al. (2000). Culture-dependent and culture-independent characterization of microbial assemblages associated with high-temperature petroleum reservoirs. *Applied and Environmental Microbiology66*:700–711.https://doi.org/10.1128/AEM.66.2.700-711.2000

Pepper, A. and Santiago, C. (2001). Impact of biodegradation on petroleum exploration and production: observations and outstanding problems. *Abstracts of Earth Systems Processes. Edinburgh* 24–28. Geological Society of London and Geological Society of America.

Peters, K.E. and Fowler, M.G. (2002). Applications of petroleum geochemistry to exploration and reservoir management. *Organic Geochemistry 33*: 5–36. https://doi.org/10.1016/S0146-6380(01)00125-5

Peters, K.E., Walters, C.C. and Moldowan, J.M. (2005). *The Biomarker Guide*. Cambridge: Cambridge University Press.

Peters, K.E., Curry, D.J. and Kacewicz, M. (2012). An overview of basin and petroleum system modeling: Definitions and concepts. In (eds. K.E. Peters, J.C. David and K. Marek), 1–16. *Basin modeling: new horizons in research and applications*. American Association of Petroleum Geologists. https://doi.org/10.1306/13311426H4139

Peters, K.E., Xia, X., Pomerantz, A.E. et al. (2016). Geochemistry applied to evaluation of unconventional resources. In: *Unconventional Oil and Gas Resources Handbook*, 71–126. Gulf Professional Publishing. https://doi.org/10.1016/B978-0-12-802238-2.00003-1

Peters K.E., Schenk O., Scheirer AH. et al. (2017). Basin and petroleum system modeling. In: *Springer Handbook of Petroleum Technology* (eds. C.S. Hsu and P.R. Robinson), 381–417. Springer Handbooks. https://doi.org/10.1007/978-3-319-49347-3_11

Petsch, S.T. (2014). Weathering of Organic Carbon. In: *Treatise on Geochemistry* (eds. H.D. Holland and K.K. Turekian), 217–238. Elsevier. https://doi.org/10.1016/B978-0-08-095975-7.01013-5

Philippi, G.T. (1977). On the depth, time and mechanism of origin of the heavy to medium-gravity naphthenic crude oils. *Geochimica et Cosmochimica Acta 41*: 33–52. https://doi.org/10.1016/0016-7037(77)90185-5

Pinous, O.V., Levchuk, M.A. and Sahagian, D.L. (2001). Regional synthesis of the productive Neocomian complex of West Siberia: sequence stratigraphic framework. *AAPG Bulletin 85*: 1713–1730. https://doi.org/10.1306/8626D04F-173B-11D7-8645000102C1865D

Prauss, M. and Riegel, W. (1989). Evidence from phytoplankton associations for causes of black shale formation in epicontinental seas. *Neues Jahrbuch für Geologie und Paläontologie-Monatshefte 671–682*. https://doi.org/10.1127/njgpm/1989/1989/671

Pritchard, P.H. and Costa, C.F. (1991). EPA's Alaska oil spill bioremediation project. Part 5. *Environmental Science and Technology 25*: 372–379. https://doi.org/10.1021/es00015a002

Quistad, S.D. and Valentine, D.L. (2011). Anaerobic propane oxidation in marine hydrocarbon seep sediments. *Geochimica et Cosmochimica Acta 75*: 2159–2169. https://doi.org/10.1016/j.gca.2011.02.001

Rassenfoss, S. (2011). From bacteria to barrels: microbiology having an impact on oil fields. *Journal of Petroleum Technology 63*: 32–38. https://doi.org/10.2118/1111-0032-JPT

Ravot, G., Magot, M., Ollivier, B. et al. (1997). *Haloanaerobium congolense* sp. nov., an anaerobic, moderately halophilic, thiosulfate-and sulfur-reducing bacterium from an African oil field. *FEMS Microbiology Letters 147*: 81–88. https://doi.org/10.1111/j.1574-6968.199b1022

Readman, J.W., Fowler, S.W., Villeneuve, J.P. et al. (1992). Oil and combustion-product contamination of the Gulf marine environment following the war. *Nature 358*: 662–665. https://doi.org/10.1038/358662a0

Röling, W.F., Head, I.M. and Larter, S.R. (2003). The microbiology of hydrocarbon degradation in subsurface petroleum reservoirs: perspectives and prospects. *Research in Microbiology 154*: 321–328. https://doi.org/10.1016/S0923-2508(03)00086-X

Rowland, S.J., Alexander, R., Kagi, R.I. et al. (1986). Microbial degradation of aromatic components of crude oils: a comparison of laboratory and field observations. *Organic Geochemistry 9*: 153–161. https://doi.org/10.1016/0146-6380(86)90065-3

Roy, C., Mondal, N., Peketi, A. et al. (2020). Geomicrobial dynamics of Trans-Himalayan sulfur-borax spring system reveals mesophilic bacteria's resilience to high heat. *Journal of Earth System Science 12*:, 157. https://doi.org/10.1007/s12040-020-01423-y

Rozanova, E.P., Borzenkov, I.A., Tarasov, A.L. et al. (2001). Microbiological processes in a high-temperature oil field. *Mikrobiologiia 70*: 102–110. https://doi.org/10.1023/A:1004809308305

Rueter, P., Rabus, R., Wilkest, H. et al. (1994). Anaerobic oxidation of hydrocarbons in crude oil by new types of sulphate-reducing bacteria. *Nature 372*: 455–458. https://doi.org/10.1038/372455a0

Schenk, H.J., Di Primio, R. and Horsfield, B. (1997). The conversion of oil into gas in petroleum reservoirs. Part 1: Comparative kinetic investigation of gas generation from crude oils of lacustrine, marine and fluviodeltaic origin by programmed-temperature closed-system pyrolysis. *Organic Geochemistry 26*: 467–481. https://doi.org/10.1016/S0146–6380(97)00024–7

Schlumberger Ltd. (2021). *The Oilfield Glossary*. https://www.glossary.oilfield.slb.com/en/Terms/b/basin.aspx

Selley, R.C. (2003). Petroleum Geology. In: *Encyclopedia of Physical Science and Technology* (ed. R.A. Meyers), 729–740. Academic Press. https://doi.org/10.1016/B0–12–227410–5/00555-X

Shuichang, Z., Baomin, Z., Benliang, L. et al. (2011). History of hydrocarbon accumulations spanning important tectonic phases in marine sedimentary basins of China: taking the Tarim Basin as an example. *Petroleum Exploration and Development 38*: 1–15. https://doi.org/10.1016/S1876–3804(11)60010–4

Sifour, M., Al-Jilawi, M.H. and Aziz, G.M. (2007). Emulsification properties of biosurfactant produced from *Pseudomonas aeruginosa* RB 28. *Pakistan Journal of Biological Sciences: PJBS 10*: 1331–1335. https://doi.org/10.3923/pjbs.2007.1331.1335

Slobodkin, A.I., Jeanthon, C., L'haridon, S. et al. (1999). Dissimilatory reduction of Fe (III) by thermophilic bacteria and archaea in deep subsurface petroleum reservoirs of Western Siberia. *Current Microbiology 39*: 99–102. https://doi.org/10.1007/s002849900426

Sousa, A.L., Matos, H.A. and Guerreiro, L.P. (2019). Preventing and removing wax deposition inside vertical wells: a review. *Journal of Petroleum Exploration and Production Technology 9*: 2091–2107. https://doi.org/10.1007/s13202–019–0609-x

Speight, J.G. (2014). *The Chemistry and Technology of Petroleum*. CRC Press.

Spormann, A.M. and Widdel, F. (2000). Metabolism of alkylbenzenes, alkanes and other hydrocarbons in anaerobic bacteria. *Biodegradation 11*: 85–105. https://doi.org/10.1023/A:1011122631799

Stauffer, E., Dolan, J.A. and Newman, R. (2008). Flammable and Combustible Liquids. In: *Fire Debris Analysis* (eds. E. Stauffer, J.A. Dolan and R. Newman), 199–233. Academic Press. https://doi.org/10.1016/B978–012663971–1.50011–7

Stetter, K.O. and Huber, R. (2000). The role of hyperthermophilic prokaryotes in oil fields. In: *Microbial Biosystems: New Frontiers: Proceedings of the 8th International Symposium on Microbial Ecology* 369–375. Halifax: Atlantic Canada Society for Microbial Ecology.

Stetter, K.O., Hoffmann, A. and Huber, R. (1993). Microorganisms adapted to high temperature environments. *Trends in Microbial Ecology*, 25–28.

Suflita, J. and Mcinerney, M. (2008). Microbial approaches for the enhanced recovery of methane and oil from mature reservoirs. In: *Bioenergy* (eds. J. Wall, C. Harwood and A. Demain), 389–403. Washington, DC: ASM Press. https://doi.org/10.1128/9781555815547.ch30

Sunde, E. and Torsvik, T. (2005). Microbial control of hydrogen sulfide production in oil reservoirs. In: *Petroleum Microbiology* (eds. B. Ollivier and M. Magot), 201–213. Washington, DC: ASM Press. https://doi.org/10.1128/9781555817589.ch10

Sunde, E., Thorstenson, T., Torsvik, T. et al. (1993). Field-related mathematical model to predict and reduce reservoir souring. In: *Proceedings of the SPE International Symposium on Oilfield Chemistry*. New Orleans: Society of Petroleum Engineers. https://doi.org/10.2118/25197-MS

Takahata, Y., Nishijima, M., Hoaki, T. et al. (2000). Distribution and physiological characteristics of hyperthermophiles in the Kubiki oil reservoir in Niigata, Japan. *Applied and Environmental Microbiology 66*: 73–79. https://doi.org/10.1128/AEM.66.1.73-79.2000

Takahata, Y., Hoaki, T. and Maruyama, T. (2001). Starvation survivability of *Thermococcus* strains isolated from Japanese oil reservoirs. *Archives of Microbiology 176*: 264–270. https://doi.org/10.1007/s002030100318

Tardy-Jacquenod, C., Magot, M., Patel, B.K.C., et al. (1998). *Desulfotomaculum halophilum* sp. nov., a halophilic sulfate-reducing bacterium isolated from oil production facilities. *International Journal of Systematic Bacteriology 48*: 333–338. https://doi.org/10.1099/00207713-48-2-333

Thompson, T.L. (1976). Plate tectonics in oil and gas exploration of continental margins. *AAPG Bulletin 60*: 1463–1501.

Thorn, K.A. and Aiken, G.R. (1998). Biodegradation of crude oil into nonvolatile organic acids in a contaminated aquifer near Bemidji, Minnesota. *Organic Geochemistry 29*: 909–931. https://doi.org/10.1016/S0146–6380(98)00167–3

Van Hamme, J.D., Singh, A. and Ward, O.P. (2003). Recent advances in petroleum microbiology. *Microbiology and Molecular Biology Reviews 67*: 503–549. https://doi.org/10.1128/MMBR.67.4.503–549.2003

Vigneron, A., Alsop, E.B., Lomans, B.P. et al. (2017). Succession in the petroleum reservoir microbiome through an oil field production lifecycle. *The ISME Journal 11*: 2141–2154. https://doi.org/10.1038/ismej.2017.78

Vigneron, A., Head, I.M. and Tsesmetzis, N. (2018). Damage to offshore production facilities by corrosive microbial biofilms. *Applied Microbiology and Biotechnology 102*: 2525–2533. https://doi.org/10.1007/s00253-018–8808–9

Volkman, J.K., Alexander, R., Kagi, R.I. et al. (1984). Biodegradation of aromatic hydrocarbons in crude oils from the Barrow Sub-basin of Western Australia. *Organic Geochemistry 6*: 619–632. https://doi.org/10.1016/0146–6380(84)90084–6

Voordouw, G. (2011). Production-related petroleum microbiology: progress and prospects. *Current Opinion in Biotechnology 22*: 401–405, https://doi.org/10.1016/j.copbio.2010.12.005

Voordouw, G., Grigoryan, A.A., Lambo, A. et al. (2009). Sulfide remediation by pulsed injection of nitrate into a low temperature Canadian heavy oil reservoir. *Environmental Science and Technology 43*: 9512–9518. https://doi.org/10.1021/es902211j

Wellsbury, P., Goodman, K., Barth, T. et al. (1997). Deep marine biosphere fuelled by increasing organic matter availability during burial and heating. *Nature 388*: 573–576. https://doi.org/10.1038/41544

Wenger, L.M., Davis, C.L. and Isaksen, G.H. (2002). Multiple controls on petroleum biodegradation and impact on oil quality. *SPE Reservoir Evaluation and Engineering 5*: 375–383. https://doi.org/10.2118/80168-PA

Widdel, F. and Grundmann, O. (2010) Biochemistry of the Anaerobic Degradation of Non-Methane Alkanes. In: *Handbook of Hydrocarbon and Lipid Microbiology* (ed. K.N. Timmis). Berlin, Heidelberg: Springer. https://doi.org/10.1007/978-3-540-77587-4_64

Wilhelms, A., Larter, S.R., Head, I. et al. (2001). Biodegradation of oil in uplifted basins prevented by deep-burial sterilization. *Nature 411*: 1034–1037. https://doi.org/10.1038/35082535

Wolfe, B.M., Lui, S.M. and Cowan, J.A. (1994). Desulfoviridin, a multimeric-dissimilatory sulfite reductase from *Desulfovibrio vulgaris* (Hildenborough) Purification, characterization, kinetics and EPR studies. *European Journal of Biochemistry 223*: 79–89. https://doi.org/10.1111/j.1432-1033.199b1896

Xiao, M., Li, W.H., Lu, M. et al. (2012). Effect of microbial treatment on the prevention and removal of paraffin deposits on stainless steel surfaces. *Bioresource Technology 124*: 227–232. https://doi.org/10.1016/j.biortech.2012.07.063

Yamane, K., Hattori, Y., Ohtagaki, H. et al. (2011). Microbial diversity with dominance of 16S rRNA gene sequences with high GC contents at 74 and 98 C subsurface crude oil deposits in Japan. *FEMS Microbiology Ecology 76*: 220–235. https://doi.org/10.1111/j.1574-6941.2011.0104

Yano, Y., Nakayama, A., Ishihara, K. et al. (1998). Adaptive changes in membrane lipids of barophilic bacteria in response to changes in growth pressure. *Applied and Environmental Microbiology 64*: 479–485. https://doi.org/10.1128/AEM.64.2.479-485.1998

Youssef, N.H., Duncan, K.E., Nagle, D.P. et al. (2004). Comparison of methods to detect biosurfactant production by diverse microorganisms. *Journal of Microbiological Methods 56*: 339–347. https://doi.org/10.1016/j.mimet.2003.11.001

Youssef, N., Elshahed, M.S. and Mcinerney, M.J. (2009). Microbial processes in oil fields: culprits, problems and opportunities. In: *Advances in Applied Microbiology*, Vol. *66* (Eds. A. Laskin, S. Sariaslani, G. Gadd), 141–251. San Diego: Elsevier. https://doi.org/10.1016/S0065-2164(08)00806-X

Biome III
Polar Oceans

8

Biogeochemical Processes in the Arctic Ocean

Rupesh K. Sinha and K.P. Krishnan*

ABSTRACT

The strongest indicators of climate change and acidification of the ocean are expected to be found in the Arctic Ocean (AO). The AO biogeochemical system is expected to experience feedback with respect to climate change, which in turn can influence the energy flow throughout the Arctic food web. Climate change is also responsible for increasing freshwater discharge from glaciers across the Arctic, which raises several questions regarding potential downstream effects in the marine environment. Arctic glacier fjords are often found to exhibit distinct discharge–productivity relationships, and several studies need to be considered in order to understand the overall effects of glacier discharge on Arctic marine ecosystems. High inflow of sediments through glacial meltwater has an impact on light penetration to a large extent in the surface layer, thereby affecting primary production. Similarly, sea ice and the snow cover on surface of sea ice reduces light penetration in the water column. Consequently, a reduction in sea-ice cover with increasing global temperature could significantly enhance the light reaching to the water column, and a progressively increasing sea ice-free summer will possibly increase annual primary production in the AO close to a glacier. The Arctic terrestrial freshwater system is likely to experience a transition from a surface-water dominated system to a ground-water dominated system because of a progressive thaw of the permafrost. As a result of this transition, mineral-rich groundwater may become an important contributor to streamflow in addition to the currently dominant contribution from mineral-poor surface water. This chapter presents a review of recent studies investigating the biogeochemistry of Arctic Ocean, fjords and permafrost including consideration of likely impacts of warming-induced changes on these ecosystems.

8.1. INTRODUCTION

Polar ecosystems have been identified to be particularly susceptible to the ongoing environmental changes and both the Arctic and Antarctic have been recognized as key regions in the regulation of the global climate. Consequently, in recent years, international efforts in polar research have been enhanced considerably, and a wealth of new findings is being produced at a growing rate by the international community of polar researchers (Hop and Wiencke, 2019). The Arctic region is defined as the region north of the Arctic Circle (66.5°N). This straight-forward differentiation does not, however, consider the terrestrial geography and the oceanic currents as well as the associated gradients in climate. Therefore, an alternative classification is widely accepted where Arctic is defined as a region in which the mean air temperature of the warmest period (summer) is below 10°C. The Arctic is covered largely by the Arctic Ocean (AO), which is a relatively small, comprising around 14.06×10^6 km^2 or ~4% of the total ocean surface and is almost completely landlocked. Geographically, at the North American side, the AO is connected with the Pacific Ocean through the

ESSO-National Center for Polar and Ocean Research, Ministry of Earth Sciences, Headland Sada, Vasco da Gama, 403804, Goa, India
**Corresponding Author: K.P. Krishnan (krishnan@ncpor.res.in) ORCiD code: 0000-0003-1101-657X*

Systems Biogeochemistry of Major Marine Biomes, First Edition. Edited by Aninda Mazumdar and Wriddhiman Ghosh.

Bering Strait, and with the Atlantic Ocean through the Baffin Strait; and at the European side it is connected with Hudson Strait and with the Fram Strait and the Norwegian Seagate. The continental shelf of Arctic Ocean is relatively broad and shallow (<200 m deep) constituting ~53% of the total area that surround a deeper central basin (>2000 m deep) (Bates and Mathis, 2009). The AO is characterized by a large freshwater input receiving ∼11% of global river discharge. There are six major rivers that flow into the Arctic: the Ob, Yenisey, Kolyma, and Lena Rivers, which are the four largest in Asia, and the Mackenzie and Yukon in North America (McClelland et al., 2012). As per reports of IPCC (IPCC, 2014), a global increase in temperature has been observed in the last decades that has been linked with increasing concentrations of greenhouse gases in the atmosphere. As a consequence of this build-up of greenhouse gasses, global radiative forcing and consequently the global temperature has been drastically impacted (Forster et al., 2007). Moreover, these changes have been the most significant at higher latitudes and the AO is currently experiencing rapid change owing to warming, sea ice loss and increased freshwater fluxes (Overland et al., 2008). Thus, the aim of this review chapter is to bring out the key advancements made in the Arctic particularly in the field of biogeochemical studies of the AO, fjords and permafrost, which are experiencing progressive warming.

8.2. THE ARCTIC OCEAN AND ITS BIOGEOCHEMISTRY

The AO comprises 4% of the world ocean by area, and is best analysed via a panArctic approach (Carmack and Wassmann, 2006) that takes into account the surrounding areas with which the AO interacts (Figure 8.1). The AO is influenced by the inflow of warm and saline waters from the Atlantic and the Pacific Ocean, as well as by the inflow of freshwater rivers from surrounding land.

The surface of the AO is covered with ice, some of which is multi-year ice, and some of which melts in summer and freezes again in winter (Silyakova, 2013). It is known that CO_2 has high solubility in the cold and relatively fresh waters of the AO. The present AO reservoir of inorganic carbon is consequently largely in the form of bicarbonate ions and the concentration of carbonate ions is low. The higher pCO_2 values over winter until April and May could be respiration driven when the surface snow cover has thinned because of melting, allowing the phytoplankton to bloom beneath the ice. During June and July when the sea ice has melted away the pCO_2 can be seen to be elevated as a result of re-equilibration of the surface layer with the atmosphere (Figure 8.2). This causes a movement towards more CO_2 being dissolved in seawater (sw) and pCO_2 increases. Sea ice in the AO greatly influences carbonate dynamics. When sea ice forms, salts and gases (including CO_2) are released with the brine rejection and further CO_2 is

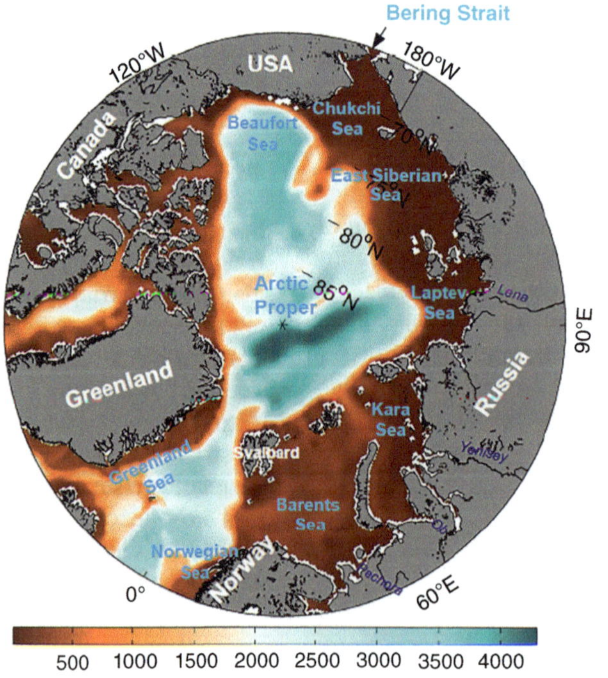

Figure 8.1 Schematic of the Arctic Ocean. Colors show depth in meters (the topography was adopted from a coupled physical–biogeochemical ecosystem model, SINMOD; see Slagstad and McClimans, 2005).

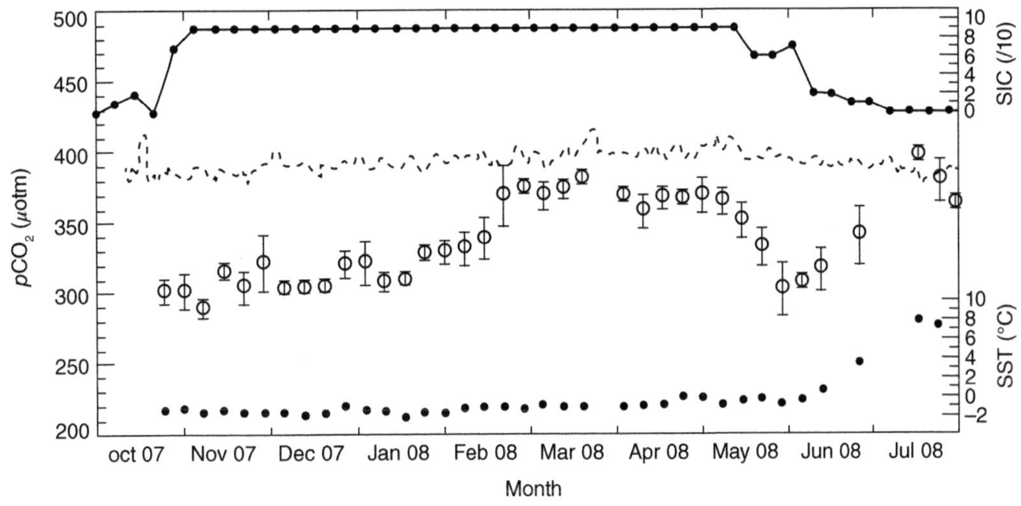

Figure 8.2 Weekly average seawater (sw) pCO_2 measurements in Amundsen Gulf (Canadian Arctic) (open circles); error bars are standard deviation. Sea surface temperatures from the equilibration system (solid grey circles); atmospheric (atm) pCO_2 from the meteorological tower (gray dashed line); and sea ice concentration from weekly Canadian Ice Service charts (solid line with gray circles) (from Else et al., 2012, and Silyakova, 2013). SST, sea surface temperature; SIC, sea-ice concentration

released with the formation of calcium carbonate particles, known as ikaite crystals (Dieckmann et al., 2008, 2010; Rysgaard et al., 2012). The melting of sea ice leads a reverse process, which reduces pCO_2(sw). These processes include dilution of surface saline waters with fresh melt water, which contain almost no gases and dissolution of ikaite crystals, which are a source of excess total alkalinity (Rysgaard et al., 2012). The carbonate system of the AO is further influenced by sea ice dynamics and properties. For example, the CO_2 flux could be enhanced through the sea ice, which has high porosity (Loose et al., 2011), and the direction of this flux is defined by the difference between pCO_2(sw) and atmospheric (atm) pCO_2 (Silyakova, 2013). Owing to increased thinning of sea ice there is increment in the air–sea fluxes with more open-ocean areas, which leads to over undersaturated waters. Similarly an increase in CO_2-rich brine export could lead to an increase in the Arctic Ocean CO_2 sink. This additional sink would be balanced by increased stratification and outgassing in some regions because of enhanced vertical mixing with deep CO_2-rich waters. Thus, there are projections that the Arctic Ocean will transfer from a carbon export system to a carbon retention system (Lannuzele et al. 2020).

In coastal regions of the AO, the carbonate system is greatly influenced by the inflow of freshwater rivers, in two ways. First, the rivers have very low concentrations of calcium so the saturation state of calcium carbonate in these regions is very low. Second, the rivers that flow into the AO contain the highest reported concentrations of organic matter in the world (Dittmar and Kattner, 2003). This organic matter respires in the water column, leading to an increase in pCO_2(sw), fall in pH, and saturation states of calcium carbonate (Figure 8.3). An example of river discharge that leads to an increase in pCO_2(sw) is given in the study of Anderson et al. (2009), who investigated surface pCO_2(sw) in the Laptev and East Siberian sea. This region is known for the inflow of large Siberian rivers (Figure 8.3a). It was found that areas where rivers enter the ocean are always high in pCO_2(sw), which can reach levels of 60.8–141 Pa (600–1400 μatm) (Figure 8.3b–d) (Silyakova, 2013). The intensive remineralization of organic material in the bottom water could be significant in increasing the bottom water pCO_2 value when undersaturation enforced by stable stratification of the water column during the productive seasons diminishes the entrainment of remineralized material from below the thermocline into the surface water (Brasse et al., 2002).

8.3. RESPONSE OF THE ARCTIC OCEAN AND ARCTIC FJORDS TO CLIMATE CHANGE

The AO sea ice is reducing drastically in response to climate change and a new record of minimum sea ice extent was observed in September 2012. This exceptional reduction in the extent of sea ice makes it plausible that the AO surface in the summer season will be ice free in a few years, not decades as projected previously (Kerr, 2012). Less sea ice in the AO means a decrease in albedo, penetration of more solar radiation into the surface waters, and a rise in the sea surface temperature. Mainly because of the effect of positive albedo feedback (IPCC AR4, 2007; http://www.ipccdata.org/), the Arctic is warming two to three times faster than other areas of the

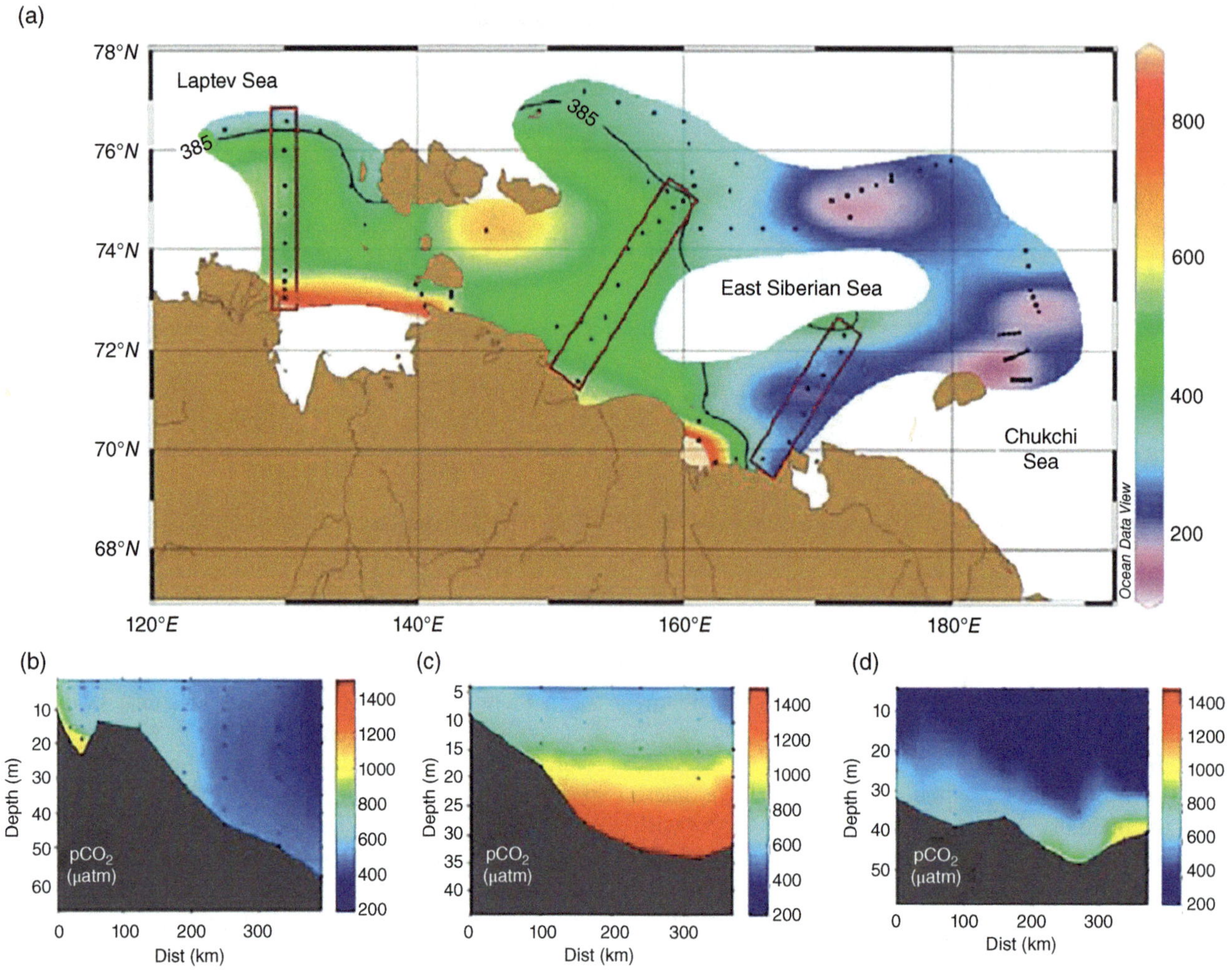

Figure 8.3 (a) The top panel shows the distribution of surface pCO_2 in summer 2008. The lower panel shows sections of pCO_2 in the Laptev Sea (b), western East Siberian Sea (c) and eastern East Siberian Sea (d) (from Anderson et al., 2009, and Silyakova, 2013).

planet; this phenomenon is known as the Arctic amplification (Loeng, 2005; Trenberth and Josey, 2007; Serreze and Barry, 2011).

The thinning of the Arctic ice has accelerated in the last four decades (Kwok, 2018) in response to warming. As a consequence, there is rapid shrinking of older ice that has survived multiple summers, which is being replaced by first-year ice that melts completely during the spring and summer each year (Meier et al., 2014). It has been noted that the freezing of the sea ice is delayed in the annual freeze–thaw cycle while melt onset is earlier than in the recent past, leading to a longer ice-free period (Stroeve et al., 2016). Owing to thinning of the snow cover there is a considerable rise in the biologically productive marginal ice zones in summer, which mostly advance poleward towards regions where sea ice is increasingly younger and thinner (Strong and Rigor, 2013). With the current scenario of increasing carbon emissions this trend is projected to continue. There are several model-based studies that predict a nearly ice-free Arctic Ocean in the summer by the end of the century or even earlier under the Representative Concentration Pathway (RCP) 8.5 termed as "worst-case" emission scenario (Lannuzel et al., 2020). Reduction in the Arctic sea ice, being one of the most prominent manifestations of global climate change, has strong implications for the planetary albedo, ocean stratification, acceleration of global warming, and possible effects global ocean circulation and weather patterns, particularly in the northern hemisphere. In this review, by providing a comprehensive understanding of Arctic biogeochemistry in the changing climate we intend to motivate future scientific efforts, raise public awareness, and facilitate policy making.

Annual freshwater discharge from glaciers has increased globally in recent decades (reviewed in Hopwood et al. 2020) and under a RCP 4.5 climate scenario it will continue to do so across most Arctic regions until the middle of this century (Bliss et al., 2014; Huss and Hock, 2018). This increased surface runoff and subsurface discharge into the ocean raises questions about the downstream effects in marine ecosystems dynamics, particularly with respect to carbon sequestration (Meire et al., 2015, 2017; Hopwood et al. 2020). In order to understand the effect of such enhanced glacial discharge on the marine environment, studies addressing the physical and chemical perturbations occurring in the water column as a result of this discharge and the structure, function, and resilience of ecosystems within these regions should be made. Hopwood et al. (2020) have stated that quantifying the magnitude of environmental perturbations from glacial discharge is complicated by the multiple concurrent and occasionally counteracting effects that glacial discharge has in the marine environment. For example during glacial discharge the ice–rock abrasion that occurs results in glacially fed rivers carrying higher sediment loads than temperate rivers (Overeem et al., 2017; Hopwood et al., 2020). The high load of sediment particles in such glacier discharge plumes can limit penetration of light into the water column (Halbach et al., 2019), and ingestion of glacial particles was found to be hazardous, or even fatal, to zooplankton, krill and benthic fauna (Hopwood et al., 2020). However, these plumes also contribute to high concentrations of inorganic components including calcium carbonate which affects seawater alkalinity, dissolved silicic acid (hereafter Si) and iron (Fe) which can potentially support increased marine primary production (Statham et al., 2008; Fransson et al., 2015; Meire et al., 2016).

The impacts of glacier discharge also depend upon the spatial and temporal scales investigated (van de Poll et al., 2018; Hopwood et al., 2020). The summertime discharge in the semi-enclosed Arctic coastal regions and fjord systems can typically produce strong, near-surface stratification. This can result in a shallow, nutrient-poor water layer with reduced primary production, which drives phytoplankton biomass deeper in the water column (Juul-Pedersen et al., 2015). At the broader scales across continental shelves, such glacial-fed freshening can reduce vertical nutrient supply throughout summer but at the same time may also delay the breakdown of stratification in autumn, therefore extending the phytoplankton growing season (Coupel et al., 2015; Oliver et al., 2018). Hopwood et al. (2020) have synthesized field studies from Svalbard and Greenland glacier catchments with different characteristics (Figure 8.4), and provided insights on the spatial and temporal timescales at which these different effects interact to increase or reduce marine primary production.

In the Arctic and sub-Antarctic, most glacial discharge enters the ocean through fjord systems (Straneo and Cenedese, 2015). The glacial discharge is characterized by strong lateral gradients and seasonal changes in environmental conditions that differentiate these ecosystems from offshore systems (Krawczyk et al., 2018). Fjords are critical zones for deep mixing that determine how glacially modified waters are exchanged with the coastal ocean (Beaird et al., 2018). They are also efficient sinks for organic carbon and CO_2 (Fransson et al., 2015; Smith et al., 2015). Study of fjord-scale processes could therefore answer questions related to the effect of glacial discharge on Arctic coastal primary production (Arimitsu et al., 2012; Renner et al., 2012; Meire et al., 2017).

Fjords play an important role in providing a pathway for the exchange of heat, salt, and nutrients between near-glacier waters and adjacent coastal regions acting as highly stratified estuaries (Hopwood et al., 2020). In deeper fjords, for example those around periphery of Greenland, stratification can be observed during summer with warm and saline water found at depths below 200 m overlaid by cold and fresher water with a thin layer (~50 m) of relatively warm water near the surface (Straneo et al., 2012). This freshwater injection into fjords from subglacial discharge and melting of terminus and iceberg play substantial role in buoyancy-driven flows in the fjord (Carroll et al., 2017). Although these process largely govern circulation and exchange between glacier fjords and shelf waters, and have been studied primarily in terms of their effects on ocean physics and glacier melting, they also have profound impacts on marine productivity which also need to be addressed (Torsvik et al., 2019).

8.4. BIOCHEMICAL EFFECTS OF GLACIAL DISCHARGE ON MARINE RESOURCES

One of the most direct mechanisms through which glacial discharge affects downstream marine primary production is by altering the availability of light, macronutrients (such as nitrate, NO_3; phosphate, PO_4; and silicic acid, Si) and/or micronutrients (such as iron and manganese) in the ocean. The chemical composition of glacial discharge is relatively well constrained, especially around Greenland, Alaska, and Svalbard (Schroth et al., 2011; Hodson et al., 2016; Meire et al., 2016). Although high sediment particle loads and Si are generally associated with glacially fed waters around the Arctic (Figure 8.5a), the concentrations of macronutrients in glacial discharge are relatively low (Figure 8.5a–c) and similar to those of coastal seawater (Meire et al., 2016; Overeem et al., 2017), while the level of macronutrients in Arctic rivers can be higher than in glacial discharge (Holmes et al., 2012) (Figure 8.5d–f).

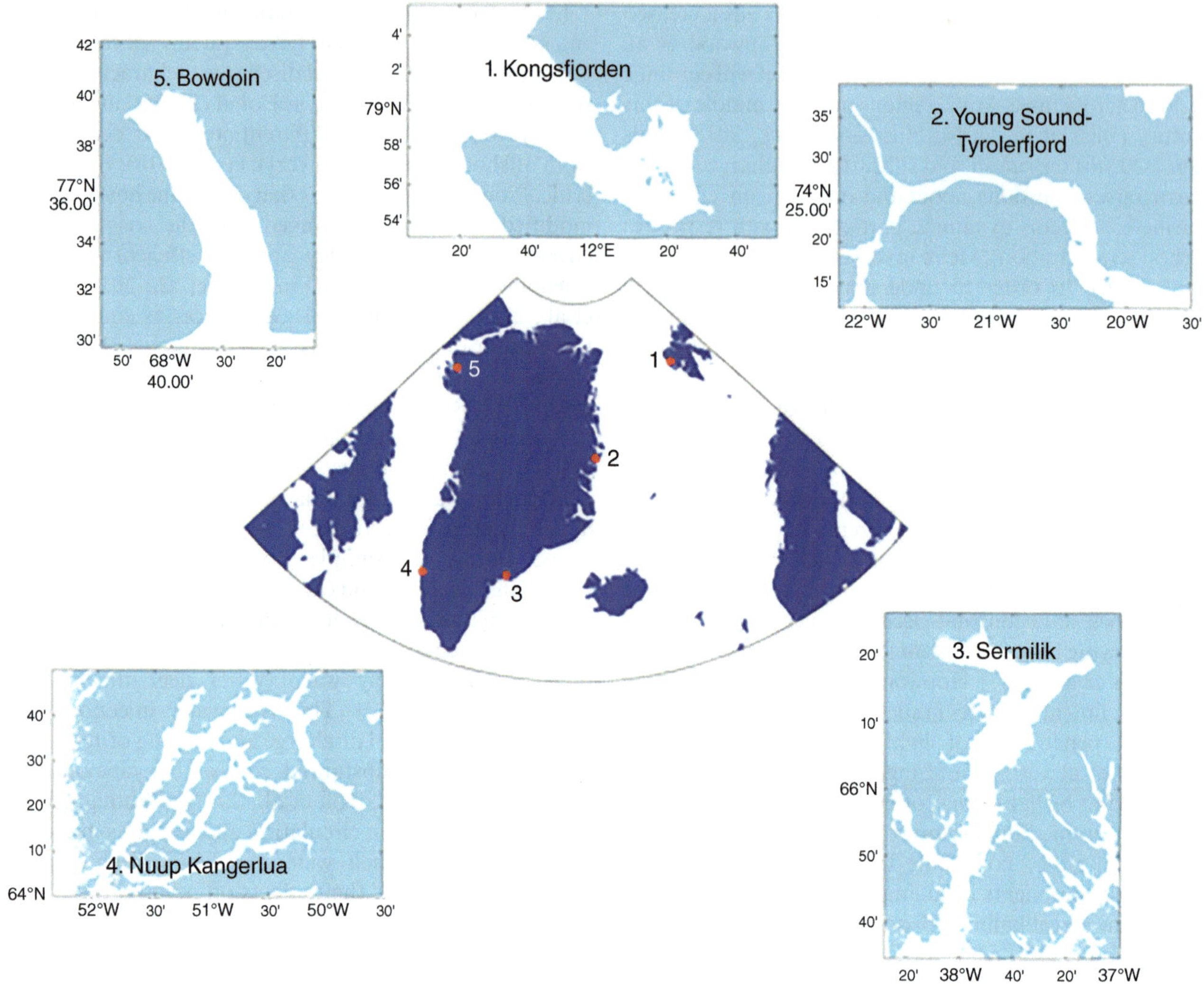

Figure 8.4 Locations of five key Arctic field sites, where extensive work bridging the glacier and marine domains has been conducted. 1, Kongsfjorden (Svalbard); 2, Young Sound (East Greenland); 3, Sermilik (Southeast Greenland); 4, Nuup Kangerlua/Godthåbsfjord (Southwest Greenland); 5, Bowdoin Fjord/Kangerluarsuup Sermia (Northwest Greenland) (from Hopwood et al. 2020).

Both river and glacier meltwater discharge do not significantly increase the PO_4 concentration in Arctic coastal waters (Figure 8.5c,f). As suggested by several studies, the river water is comparatively a much more important source of NO_3 (Cauwet and Sidorov, 1996; Emmerton et al., 2008; Hessen et al., 2010) and river estuaries show a sharp decline in this nutrient with increasing salinity owing to biological uptake and mixing (Figure 8.5e) while concentrations of Si are more variable (Cauwet and Sidorov, 1996; Emmerton et al., 2008; Hessen et al., 2010). The concentration of dissolved Si (DSi) is higher at low salinity in rivers than in glacier discharge (Figure 8.5a,d), although a variety of estuarine behaviors are evident across the Arctic. High concentrations of DSi can be observed at varying salinity because of the opposing effects of Si release from particles and uptake of DSi by diatoms (Figure 8.5d).

Arctic glacier fjords are characterized by highly turbid fresh water that outflows into the ocean. As discussed earlier, the high turbidity in surface waters of the glacier fjords primarily results from the high transport of sediment particles into these drainage systems from melting of iceberg together with the resuspension of fine sediments (Chu et al., 2012; Stevens et al., 2016). The high sediment load is evident in the glacially derived freshwater around Greenland, which is reported to contribute ~1% of the annual freshwater discharge into the ocean and 7–9% of the annual fluvial sediment load (Overeem et al., 2017; Hopwood et al. 2020). Spatial and temporal variation was observed in the sediment load, which leads to pronounced inter- and intra-catchment differences

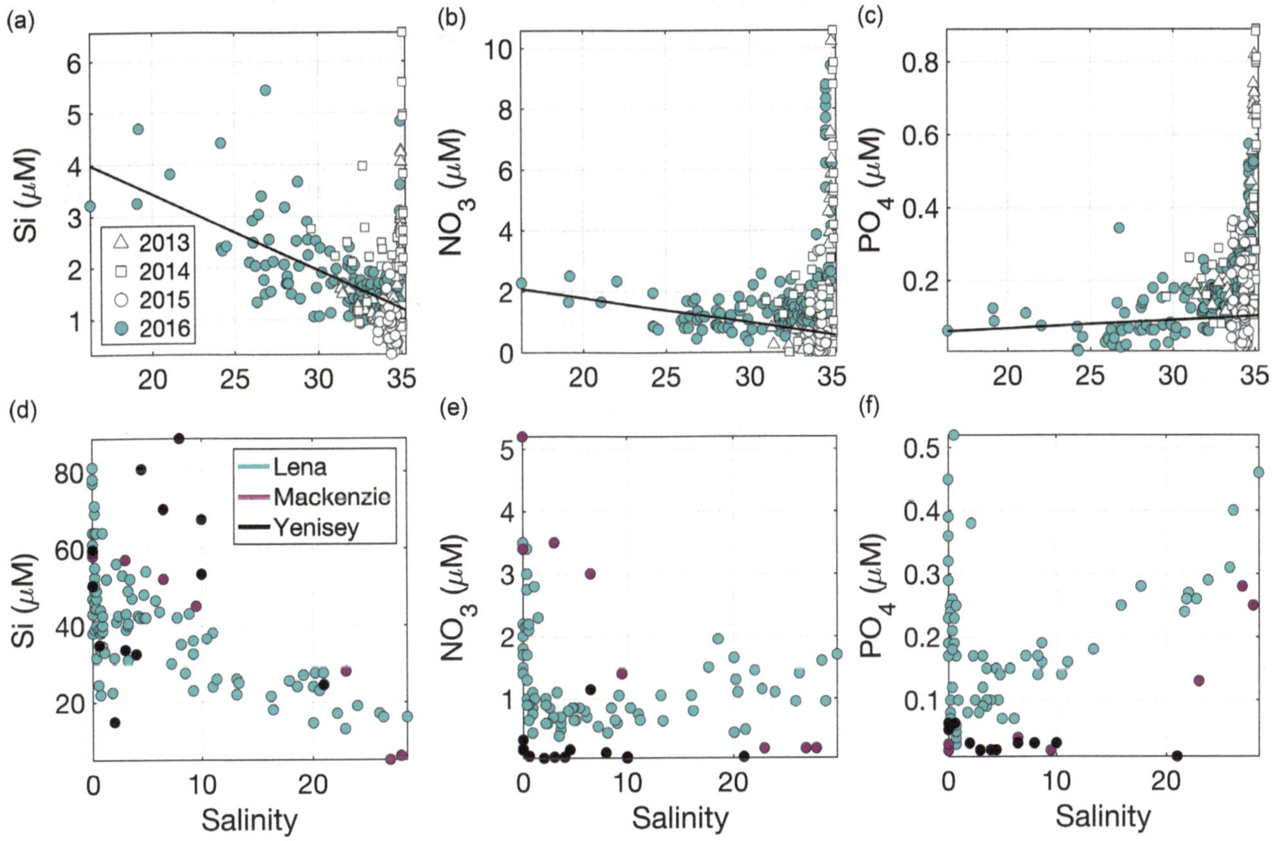

Figure 8.5 Distributions of (a) Si, (b) NO_3 and (c) PO_4 across the measured salinity gradient in Kongsfjorden in summer 2013, 2014, 2015 and 2016 (Fransson et al., 2016; van de Poll et al., 2018; Cantoni et al., 2019). For summer 2016, a full depth data are shown, with a linear regression (black line) for glacially modified waters (S <34.2). The location of stations varies between the datasets, with the 2016 data providing the broadest coverage of the inner fjord. Distributions of (d) Si, (e) NO_3 andand (f) PO_4 in surface waters of three major Arctic river estuaries: the Lena, Mackenzie, and Yenisey (Cauwet and Sidorov, 1996; Emmerton et al., 2008; Hessen et al., 2010) (from Hopwood et al. 2020).

(Murray et al., 2015). For example, the satellite-derived measurements of sediment load for 160 Greenlandic glacier outflows indicate a median sediment load of 992 mg l^{-1}, however, some catchments exhibited >3000 mg l^{-1} (Overeem et al., 2017). Moreover, it is also suggested that >25% of the total annual sediment load is released in a single outflow from the Sermeq glacier (Overeem et al., 2017).

Therefore, the extent to which high turbidity in glacier outflows limits the availability of light in downstream marine environments is highly variable between catchments and varies with distance from glacier outflows (Murray et al., 2015; Mascarenhas and Zielinski, 2019; Hopwood et al., 2020). There is lack of studies on the occurrence and effects of subsurface turbidity close to glaciers and it has been suggested that the subsurface turbidity may be even more spatially and temporally variable than their surface counterparts (Stevens et al., 2016; Moskalik et al., 2018). Generally, an expansion in spatial coverage is expected with increasing glacier discharge for near-surface turbid plumes, but this tendency is not always evident at the catchment scale (Chu et al., 2012; Hudson et al., 2014). Furthermore, in case of long-term glacier retreat the proglacial lakes act as efficient sediment traps resulting in decline in the sediment load discharge at the coastline (Normandeau et al., 2019).

8.5. EFFECT OF BIOCHEMICAL CHANGES ON PRIMARY AND SECONDARY PRODUCTION

Phytoplankton are the main drivers of primary production in high-latitude waters while in sea ice covered regions the total primary production is significantly contributed by the sea-ice algae (<1–66%; Meire, 2016). Primary production is largely determined by light and nutrient availability, which in turn are the functions of stratification of surface waters (Wassmann and Reigstad, 2011). In high-latitude marine ecosystems, the strong seasonal cycle of primary productivity is largely determined by the strong seasonality in incoming radiation and light penetration (Sakshaug, 2004). The attenuation of light in the water column results in the formation of photic or euphotic zone, till where the light can penetrate. The underwater light is largely attenuated by dissolved and particulate substances (both organic and inorganic) present in the water column (Kirk, 1994). The inflow of sediments suspended particles through glacial meltwater can largely weaken the light penetration in the surface layer restricting the photic zone to 5–10 m, instead of a normal photic zone depth of around 30–40 m in absence of such sediment input (Murray et al., 2015). In addition, sea ice and the snow cover on top of sea ice could strongly reduce the light reaching the water column (Nicolaus et al., 2013). Hence, a reduction in sea-ice cover with increasing global temperature could enhance light penetration to the water column and consequently a longer sea ice-free summer period will potentially increase annual primary production in the AO (Arrigo et al., 2008) (Figure 8.6). Nutrients are another crucial factor that can limit the primary production. Phosphorous and nitrogen are important drivers of primary production, with nitrogen being reported as the crucial limiting nutrient in the Arctic (Tremblay and Gagnon, 2009). Silicate is a vital nutrient for diatoms that use silica to build their cell frustule. Limitation of silica could therefore limit diatom population and cause a major change in the species composition of phytoplankton (Meire, 2016).

The Arctic is primarily influenced by two ocean end members: Pacific water with high nutrients (silica, ~30 μM; phosphate, ~2 μM; nitrate, ~15 μM) and an Atlantic water with reduced nutrients (silica, ~6 μM; phosphate, ~1 μM; nitrate, ~8 μM) (McClelland et al. (2012). As per the Redfield ratio (Si:N:P, 15:16:1) of diatom stoichiometry, silica is present in excess quantity in Pacific water, while Si:N ratios in Atlantic water indicate that silica is limiting compared to nitrate. A study at Godthåbsfjord (SW Greenland) bottom water indicated a similar low Si:N (0.7) ratio. Silica concentrations were also found to be lower than nitrate in the deep water at west coast of Greenland (Jensen et al., 1999). In the photic zone, the new production is often limited by a rapid consumption of inorganic nutrients and low rates of nutrient recycling (Wassmann and Reigstad, 2011). Therefore, the supply of nutrients to the photic zone following the spring bloom is a crucial factor that determine the productivity (Carmack and Wassmann, 2006; Wassmann and Reigstad, 2011). Experimentally, the difference in the new and regenerated production can be made based on the source of nitrogen (Eppley and Peterson, 1979). New nitrate can be supplied to the photic zone from deeper water layers through upwelling or diffusion of nitrate from deeper waters or though riverine or glacial discharge, driving the new production (Wassmann and Reigstad, 2011). The released nitrogen through remineralization of organic material in the photic zone can fuel regenerated production. The start of the productive season in spring which is characterized by an intense phytoplankton bloom ensues with the increasing solar radiation (Hodal et al., 2012; Sakshaug, 2004). In many sub-Arctic and Arctic marine ecosystems (50–65%), this spring bloom plays an important role in annual primary production and thus is crucial for sustaining the secondary production in these high-latitude systems (Sakshaug, 2004). As the sea ice limits solar radiation to the surface water layers, the timing of the blooms is significantly affected by the presence of sea ice coverage (Rysgaard and Glud, 2007). A significant link between the ice-free period and annual primary production in the Arctic and specifically in Greenland has been demonstrated in earlier observations (Arrigo et al., 2008; Rysgaard and Glud, 2007) (Figure 8.6). Therefore, break-up of sea ice is often associated with the sudden exposure of increased light to the underlying nutrient-rich water layers and subsequent development of a strong spring bloom. The sea ice melt also increases stability of the water column and creates favorable conditions for the development of bloom (Wu et al., 2007). In addition, physical factors play a crucial

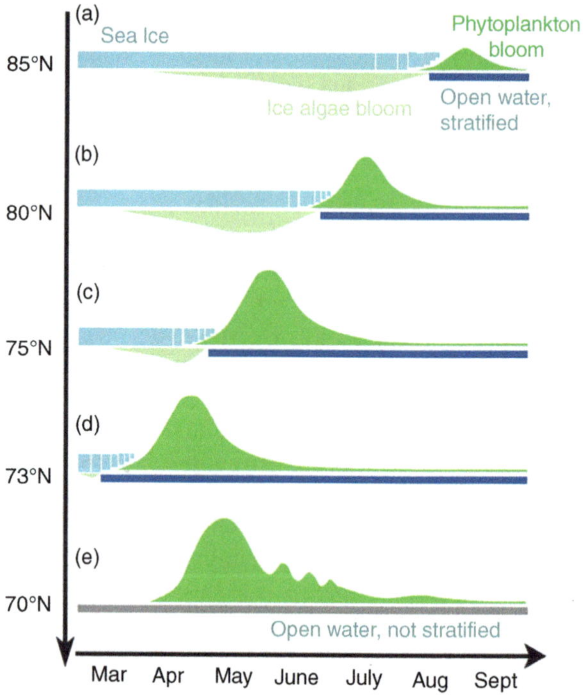

Figure 8.6 Conceptual model of the timing of ice algae and phytoplankton bloom development along a latitudinal axis of the open water–seasonal ice zone region (ranging from 75 to 85°N) with a long to short productive season (Wassmann and Reigstad, 2011) (from Meire, 2016).

role in the timing and development of the spring bloom. During the winter months, a high level of turbulent mixing and strong light limitation prevents growth of phytoplankton (Eilertsen and Frantzen, 2007), while a rapid increase in incoming solar radiation with high nutrient concentrations in surface waters, and the development of a more stratified water column instigate the development of an intense phytoplankton bloom during spring (Sakshaug, 2004). Stratification of the surface waters helps to prevent the cells from being transported out of the euphotic zone below the compensation depth (Sverdrup, 1953). Temperature and salinity are the main factors responsible for this increased stratification for e.g. temperature driven increased insolation, and elevated air–sea heat exchange, and/or change in salinity during sea ice melt or during river runoff by snow melt can create stratification. Spring bloom often can cause rapid depletion of nutrients above the pycnocline, and consequently the nutrient flux to the upper water layer determines post spring bloom production (Tremblay and Gagnon, 2009). A higher and long-term primary production can be observed in areas where hydrodynamics conditions continuously supply nutrients, for example through tidal mixing, turbulence in current junctions, or though upwelling in straits and along shelf-breaks (Sakshaug, 2004). With the changing climate the increase in ice melt and freshwater runoff could cause strengthening of stratification and further isolate the photic layer from turbulent vertical nutrient supply and this is expected to significantly reduce Arctic primary productivity (Wassmann and Reigstad, 2011). There is still a knowledge gap regarding factors governing biogeochemistry of glacial meltwater and therefore the productivity in coastal areas. Greenland's fjords are specifically characterized by high input of glacial meltwater during summer. Seasonal biological studies in Godthåbsfjord and research cruises along the west coast of Greenland indicate that the dominant microplankton groups are diatoms such as *Chaetoceros* spp. and *Thalassiosira* spp., dinoflagellates, silicoflagellates, and ciliates.

Primary producers within the ice (sympagic) and in the underlying ocean (pelagic) depend on light and nutrients to grow. Under optimal conditions, sea ice can harbors dense communities of algae with chlorophyll *a* concentrations among the highest recorded for any aquatic environment (Arrigo, 2017). Ice algae and phytoplankton that form the base of the food web in the AO supports key foraging species which sustain further higher trophic species such as ringed seals and beluga whales (Steiner et al. 2019). Benthic secondary production is predicted to decline in the AO as a result of reduced sympago-benthic coupling and a lack of ice-algae downfalls (Lannuzel et al. 2020). While increased light and ice-scouring in the shallow region due to sea-ice retreat might positively impact growth of macroalgal population such as kelp (Filbee-Dexter et al. 2019), studies on the higher trophic levels and marine living resources suggests that the biomass of polar cod and other cold-adapted fish resident of the AO (Steiner et al. 2019) will continue to decline in many of the Arctic shelf regions as the sub-polar and Atlantic fish expand their ranges towards north (Haug et al. 2017). There are reports on the decline of the sympagic predators such as ringed seals, beluga whales, and polar bears with the continued declines in sympagic habitat and key prey fish such as polar cod (Steiner et al. 2019). Consequently, these fauna may face local- to regional-scale extinctions in the Arctic shelf domains. In contrast, it is also expected that the presence of generalist predators such as baleen whales, orcas, and certain seabird species could increasingly expand into Arctic shelf seas (Descamps et al. 2017).

8.6. ARCTIC PERMAFROST

8.6.1 Biogeochemistry and Significance of Arctic Permafrost

Permafrost covers ~25% of the Northern Hemisphere land surface and ~80% of the Arctic drainage basin, and is defined as any subsurface material that remains below 0°C for at least two consecutive years. Permafrost underlies approximately 22.79×10^6 km^2 or 23.9% of the exposed land area of the Northern Hemisphere (Jones, 2019 and references therein). This is ~15% of the global exposed land area (Zhang et al., 2008). In the continuous permafrost zone, permafrost exists beneath the entire exposed land surface, except for widely scattered sites (Figure 8.7; van Everdingen, 2005). In the extensive discontinuous permafrost zone, permafrost underlies between 65% and 90% of the exposed land surface, and in the sporadic permafrost zone, permafrost underlies between 10% and 35% of the exposed land surface (van Everdingen, 2005). In the International Polar Year (2007–2009), improvements were made to the permafrost monitoring network in the Northern Hemisphere polar regions by drilling new boreholes for long-term ground temperature monitoring, resuming ground temperature measurements at previously abandoned sites, and continuing to measure ground temperatures at existing permafrost observatory sites (Romanovsky et al., 2010; Figure 8.7). The overall trend in the mean annual ground temperature (MAGT) is to decrease northward. This trend is modified by factors such as elevation, ocean currents, snow thickness, vegetation type, ground composition, and landforms (Romanovsky et al., 2010). For example, thicker snow cover is associated with a higher MAGT (Smith et al., 2010). In northwestern Russia, northern Scandinavia and Svalbard, warm ocean currents cause

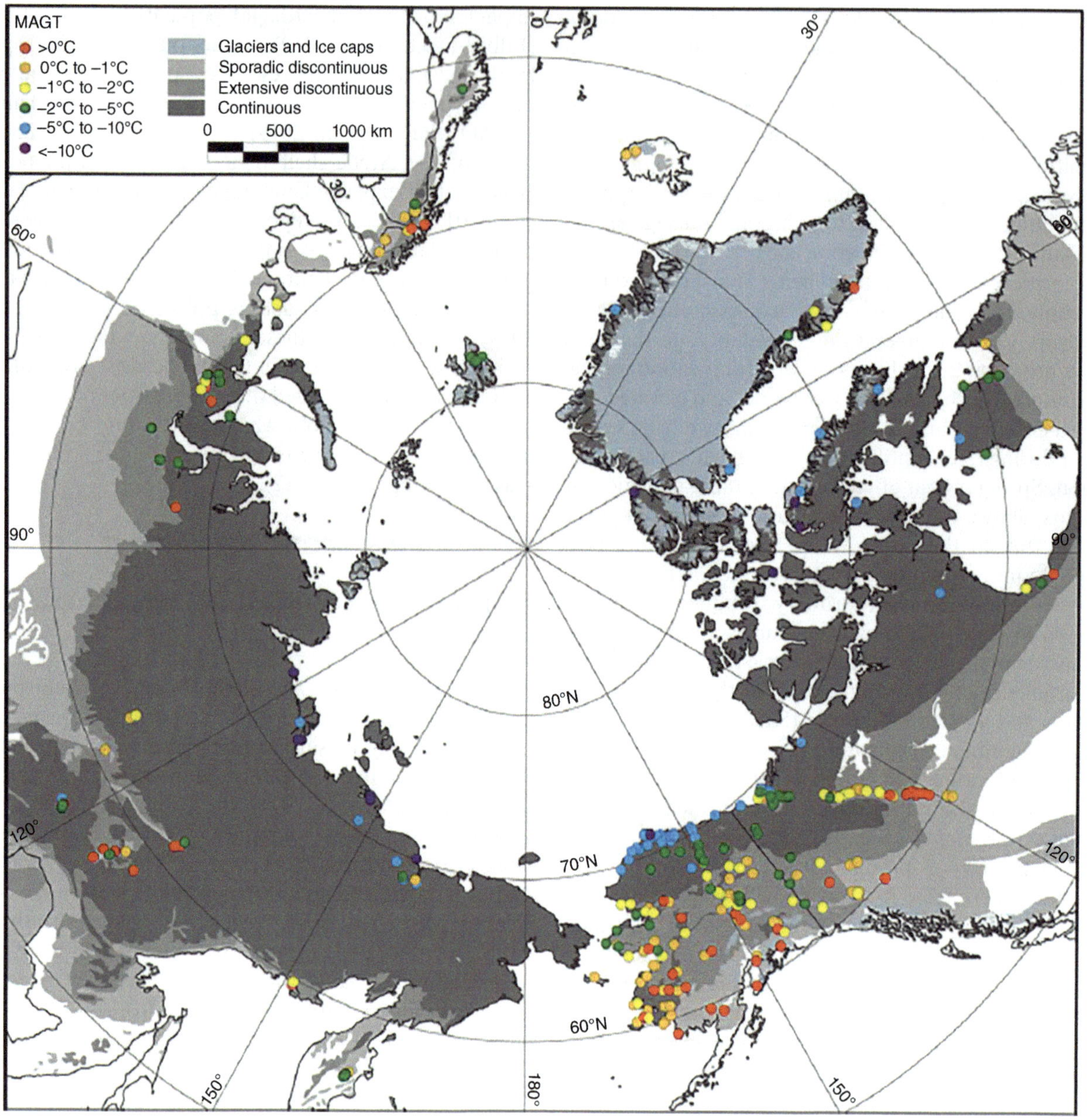

Figure 8.7 The mean annual ground temperature (MAGT) during the International Polar Year (2007–2009). The MAGT, at the depth of zero annual amplitude, is represented by colored dots. The greyscale permafrost zones are after Brown et al. (1997) (from Jones, 2019).

the MAGT to be higher than in other locations with a similar latitude in the high Arctic (Romanovsky et al., 2010). Overall, Northern Hemisphere permafrost has warmed since the late 1970s (Romanovsky et al., 2010). The rates of warming are higher for colder permafrost (<−2°C), such as in Svalbard, than for warmer permafrost because of latent heat effects (close to 0°C) (Jones, 2019).

As glaciers retreat, fjord valleys have rapid sedimentation and organic carbon burial rates, forming valley infills that are ideal for methanogenesis (Gilbert et al., 2017; Gilbert et al., 2018). Over the past 10 000 years, almost one-quarter of global marine sedimentation has occurred in fjords (Syvitski et al., 1987). In the lower elevation, coastal sites of Svalbard, glacial retreat, and postglacial isostatic rebound have played a cardinal role in landscape evolution (Milne and Shennan, 2007). In western Svalbard (Figure 8.8), for example, there were three glacial advances during the late Pleistocene (Mangerud et al., 1998). In the most recent glaciation, the Last Glacial Maximum (LGM; 20 000 years ago), the entire archipelago of Svalbard was covered by an ice sheet

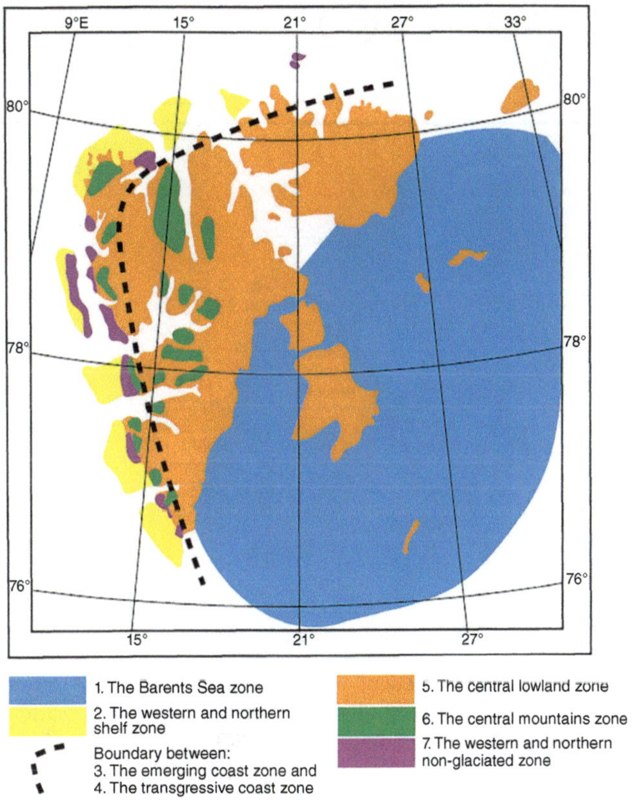

Figure 8.8 A schematic representation of permafrost zones in Svalbard. The Barents Sea was glaciated during the Last Glacial Maximum (LGM) and permafrost is absent, owing to a lack of subaerial exposure (Zone 1). Limited permafrost aggraded in areas that were not glaciated during the LGM, but were inundated by the sea in the Holocene (Zone 2).

(Landvik et al., 2005). Warm-based glaciers have basal ice at the pressure melting point. During the LGM, the valleys of western Spitsbergen were filled with warm-based, erosive ice streams, as indicated by the glacial striae up to 200 m above sea level in central Spitsbergen (Landvik et al., 2005; Humlum, 2005). Warm-based ice is a key factor in regulating the age of permafrost in Svalbard.

In the east, permafrost is younger and thinner owing to its recent subaerial exposure (Zone 3). In the north and west, coastal erosion and transgressions are occurring, but there is permafrost beneath the beaches (Zone 4). Zone 5 deglaciated after the LGM (except for areas covered by present-day glaciers), since which time permafrost has developed. Nunataks (LGM unglaciated mountains) contain permafrost up to 100 000 years old (Zone 6). Some areas were not glaciated during the LGM, but are above the present sea level (Zone 7) (from Jones, 2019).

The Arctic cryosphere is a far more dynamic system than deep marine sediments and does not exist in steady state: it is changing in response to past and present climate perturbations (Jakobsson et al., 2014), as well as in response to seasonal rainfall and temperature variations. Cycling (of electrons) occurs between different elements of the cascade and is represented in Figure 8.9. For example, even in marine sediments in a steady state, the re-oxidation of 70–90% of sulfide (produced by sulfate reduction) occurs, with the amount of re-oxidation increasing with an increased input of Fe(III) (Thamdrup et al., 1994). Experiments on salt marsh sediments indicate that this recycling can occur with sulfide acting as an electron donor for Fe(III) reduction (Mortimer et al., 2011). Hence, the definition of biogeochemical zones in Figure 8.9 is only a proxy for process, as the concentrations of reactants and products relate to the difference between the rates of consumption and production, rather than the absolute rates of consumption and production. In addition, sediment microheterogeneity facilitates the existence of bacterial microniches, where isolated reducing conditions exist in confined areas of high organic matter. Moreover, the biogeochemical zonation (Figure 8.9) is entangled with physical processes, such as hydrological input to and outflows from the active layer, and annual freezing and thawing of the active layer (Throckmorton et al., 2016). Finally, the speciation and

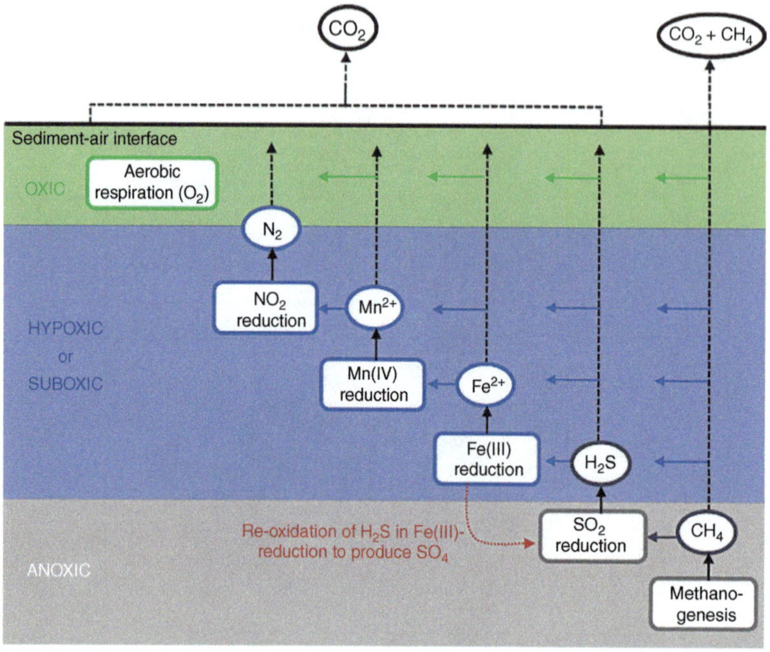

Figure 8.9 Conceptual diagram of the theoretical vertical sequence in water-saturated sediments of organic matter remineralization (vertical arrows from rounded rectangles to ovals) and reoxidation pathways (horizontal arrows). Reoxidation of H_2S during Fe(III) reduction to produce SO_4 is shown by red dotted line and arrow. Above the sediment–air interface are arrows indicating the flux of greenhouse gases (CH_4 and CO_2) to the atmosphere (from Jones, 2019).

availability of electron acceptors (dependent partly on substrate material and sedimentation) and organic carbon result in a geochemical complexity that is not fully represented by the conventional redox ladder illustrated in Figure 8.9 (Bethke et al., 2011). For example, the reduction of ferric iron was found to increase carbon dioxide production relative to methane production in an Arctic peat soil in Barrow, which is situated on the coastal plain of Alaska. Large quantities of siderite, which contains reduced iron and carbon, were found in these sediments. The pathway of decomposition of organic carbon was also found to be heavily influenced by the mineralogy of iron oxides (Herndon et al., 2017). For example, ferrihydrite minerals suppressed methanogenesis, whereas the crystalline iron oxides did not decrease methane production (Herndon et al., 2015). In summary, a range of factors influence methane and carbon dioxide production in the complex permafrost and active layer sediments of the Arctic cryosphere (Jones, 2019).

8.6.2 Impact of Thawing Permafrost on Arctic Environment

Arctic is warming at a rate that is twice the global average (Overland et al., 2015) with an increase in the air temperatures of almost 3°C since the beginning of the 20th century. Nutrient cycling and productivity in Arctic ecosystems could be drastically altered with the associated increase in permafrost temperatures (Romanovsky et al., 2013) and the depth of the summer active layer/top layer of soil that freezes and thaws seasonally (Keller et al., 2010). Substantial quantities of carbon (C), nitrogen (N), and phosphorus (P) are contained within permafrost layers (Belshe et al., 2013) and are an important source of nutrients. Dissolved organic carbon (DOC) leached from newly thawed permafrost soils can provide a labile C source for microbial and photochemical processing to CO_2 and CH_4 (Vonk et al., 2015). As temperatures rise, the simulated DOC increases and eventually the active layer becomes too deep to completely refreeze during winter, forming a talik (Schaefer et al., 2014). The southern margins of northern hemisphere permafrost regions have the warmest permafrost and will result in the greatest talik formation. In due course, the permafrost will become patchy and then disappear, and the boundaries of continuous and discontinuous permafrost will move north (Schaefer et al., 2014). Deeper permafrost may persist for many years or even centuries, although near-surface permafrost in the top few meters of soil may disappear. Over time the remaining permafrost will contract around the coldest regions such as in the Northern hemisphere, Northern Siberia, and the islands of Northeast Canada, where the permafrost is most resistant to thaw. The wider range of projected permafrost degradation indicates that ALT will increase and the areal extent of near-surface permafrost will decrease. Much of the spread in estimated permafrost degradation has resulted from assumptions of different emissions scenarios and associated warming (Koven et al., 2013). Models with no organic soil layer simulated deeper active layers and less permafrost than currently observed and were more sensitive to permafrost degradation in response to future climate change (Koven et al., 2013). It is suggested that improvements in projections of permafrost degradation should focus on improving the representation of the soil organic layer, snow processes, and excess ground ice. Although the models vary widely, they all indicate that permafrost degradation will be drastic in the future, resulting in substantial changes to the landscape caused by abrupt changes in soil physical properties and hydrology. Expensive and extensive damage to buildings, roads and other key infrastructure can occur quickly once permafrost begins to thaw, impacting on national and regional budget planning and public services. However, there are very few studies and reports that quantify the risks, costs, and mitigation associated with property and infrastructure damage caused by permafrost degradation.

Another important consequence across the Arctic region is the impact of permafrost degradation on the biogeochemistry of streams and rivers. One of the most profound changes to occur with future Arctic warming may be the transition of the Arctic system from a surface water dominated system to a groundwater dominated system, with resulting cascading effects on hydrology, ecosystems, and biogeochemical cycling (Nelson, 2003; Smith et al., 2005). Although this transition is difficult to assess over large areas, the relatively steady and straightforward relationship between concentrations of major ions and groundwater contribution to rivers may allow the geochemistry of river waters to be utilized as an indicator of changes in watershed. In general, concentrations of major ions (e.g. Ca^{2+}, Mg^{2+}, K^{+}, and Na^{+}) are predicted to increase with permafrost degradation and lowering of water tables, as a result of less interaction with organic shallow soils and greater interaction with deep mineral horizons (Frey and McClelland, 2009). Similarly, concentrations of phosphate and silicate in rivers are expected to increase. In contrast, considerable ambiguity exists about how dissolved inorganic nitrogen and organic matter in arctic rivers will respond to the degradation of permafrost. Deeper flow paths under lower permafrost conditions seem to ease nitrate export for sites in Alaska, but the presence or absence of permafrost shows little effect on nitrate export in watersheds in West Siberia. The impacts of permafrost degradation on organic matter in rivers are of particular interest because concentrations of organic matter in Arctic rivers are some of the highest among rivers globally, owing to large stores of organic-rich soils in the pan-Arctic watershed. While organic matter export is expected to decrease in Alaskan watersheds with permafrost degradation and increased active layers depths, the presence of deep, anoxic, organic-rich peat soils in West Siberia may be the reason for projections of increased organic matter export in watersheds in this region. The lability and age of organic matter found in rivers are also expected to show clear changes with permafrost degradation, and may be utilized as integrative measures of permafrost variability. Regardless of the uncertainties in future predictions, however, it is clear that significant impacts on biogeochemical cycling in rivers will transpire with continued warming and degradation of permafrost. All of these changes in the delivery of organic matter, inorganic nutrients, and major ions to Arctic rivers and ultimately to coastal waters may have important consequences for primary production and carbon cycling on Arctic shelves and in the Arctic Ocean basin interior (Frey and McClelland, 2009).

8.7. SUMMARY AND FUTURE PERSPECTIVES

The biogeochemical system in the AO is largely defined by whether the region is influenced by waters from the Atlantic Ocean, the Pacific Ocean, and/or by freshwater runoff from ice melt and terrestrial systems. The AO must be studied together with surrounding areas (pan-Arctic), as significant signals of climate change and ocean acidification enter the AO via their effects on surrounding areas. Spatial variability in the AO biogeochemistry is also specified by whether the region has high or low biological activity. The biological activity of low trophic level organisms plays a large role in the annual and long-term dynamics of the carbon biogeochemical system. The effects of climate change and ocean acidification on the carbon biogeochemical system might therefore be greatly mitigated by future biological activity.

In the Arctic, marine-terminating glaciers are linked with the enhanced vertical fluxes of macronutrients, which is crucial in the development of summertime phytoplankton blooms throughout the meltwater season. In contrast, land-terminating glaciers are generally associated with the local suppression of primary production, owing to limitation of light and stratification obstructing vertical mixing of nutrient supply. Compared with other coastal environments, the primary production in Arctic glacier fjords without marine-terminating glaciers is generally low. With the continued decrease in the extent of sea ice, benthic secondary production is predicted to decline in the AO owing to reduced sympago-benthic coupling and lack of ice algae downfalls. As a result of loss of sympagic habitat the abundance of higher trophic species endemic or common to the Arctic such as beluga whales, polar bears, and polar cod is projected to decline while subpolar species will become increasingly abundant in Arctic waters. Thus, with the continued projections of sea ice loss, the Arctic fauna face the risk of local to regional extinction.

Thawing of near-surface permafrost and ground ice has been observed in continuous and discontinuous permafrost zones across the pan-Arctic region over the past several decades. One of the most important changes to the Arctic system caused by ongoing and future warming will be the shift from a surface water-dominated system to a groundwater dominated system. As the degradation of permafrost thaws previously frozen soils and significantly changes the flow paths of waters through the system, this in turn may have profound consequences for biogeochemical cycling in Arctic watersheds. These consequences can be highly complex and will certainly exhibit spatial and temporal variabilities associated with current permafrost conditions such as sensitivity to permafrost thaw, mode of permafrost degradation (extensive permafrost thaw, active layer deepening, thermokarst processes), and environmental characteristics of watersheds (e.g. land cover, soil type, and topography).

This review, therefore, emphasizes the importance of investigating Arctic biogeochemistry using different methods, including *in situ* observations, large-scale experiments on the response of the marine pelagic ecosystem to future scenarios, as well as three-dimensional ecosystem modeling because understanding of the biogeochemical system under future scenarios involves looking at the combined effects of physical, chemical, and biological processes. A combination of the results from different methodological approaches is more likely to provide a comprehensive understanding of how the Arctic ecosystem works and its responses to climatic changes.

ACKNOWLEDGEMENTS

The authors express their gratitude to the Director, National Centre for Polar and Ocean Research, Ministry of Earth Sciences, Government of India for his interest and support for this work. This is NCPOR contribution number B-6/2020-21.

REFERENCES

Anderson, L.G., Jutterström, S., Hjalmarsson, S. et al. (2009). Out-gassing of CO_2 from Siberian Shelf seas by terrestrial organic matter decomposition. *Geophysical Research Letters 36*: L20601.

Arimitsu, M.L., Piatt, J.F., Madison, E.N. et al. (2012). Oceanographic gradients and seabird prey community dynamics in glacial fjords. *Fisheries Oceanography 21*: 148–169.

Arrigo, K.R. (2017). Sea ice as a habitat for primary producers. In: *Sea Ice*. (ed. D.N. Thomas), 352–369.Chichester: John Wiley & Sons.

Arrigo, K.R., van Dijken, G. and Pabi, S. (2008). Impact of a shrinking Arctic ice cover on marine primary production. *Geophysical Research Letters 35* (19): 1–6.

Bates, N. and Mathis, J. (2009). The Arctic Ocean marine carbon cycle: evaluation of air–sea CO_2 exchanges, ocean acidification impacts and potential feedbacks. *Biogeosciences 6*: 2433–2459.

Beaird, N.L., Straneo, F. and Jenkins, W. (2018). Export of strongly diluted Greenland meltwater from a major glacial fjord. *Geophysical Research Letters 43*: 4163–4170.

Belshe, E.F., Schuur, E.A.G. and Bolker, B.M. (2013). Tundra ecosystems observed to be CO_2 sources due to differential amplification of the carbon cycle. *Ecology Letters 16*: 1307–1315.

Bethke, C.M., Sanford, R.A., Kirk, M.F. et al. (2011). The thermodynamic ladder in geomicrobiology. *American Journal of Science 311* (3): 183–210.

Bliss, A., Hock, R. and Radić, V. (2014). Global response of glacier runoff to twenty-first century climate change. *Journal of Geophysical Research-Earth Surface 119*: 717–730.

Brasse, S., Nellen, M., Seifert, R. et al. (2002). The Carbon dioxide system in the Elbe estuary. *Biogeochemistry 59*: 25–40.

Brown, J., Ferrians, Jr., O.J., Heginbottom J.A. et al. (1997). *Circum-Arctic Map of Permafrost and Ground-ice Conditions. US Geological Survey, Map CP-45*. US Department of the Interior.

Cantoni, C., Hopwood, M., Clarke, J. et al. (2019). Hydrological, biogeochemical and carbonate system data in coastal waters and in glacier drainage systems in Kongsfjorden (Svalbard), during July–August 2016, Data set, *PANGAEA*. doi:10.1594/PANGAEA.904171

Carmack, E,. and Wassmann, P. (2006). Food webs and physical–biological coupling on pan-Arctic shelves: unifying concepts and comprehensive perspectives. *Progress in Oceanography 71*: 446–477.

Carroll, D., Sutherland, D.A., Shroyer, E.L. et al. (2017). Subglacial discharge-driven renewal of tidewater glacier fjords. *Journal of Geophysical Research: Oceans 122*: 6611–6629.

Cauwet, G. and Sidorov, I. (1996). The biogeochemistry of Lena River: organic carbon and nutrients distribution. *Marine Chemistry 53*: 211–227.

Chu, V.W., Smith, L.C., Rennermalm, A.K. et al. (2012). Hydrologic controls on coastal suspended sediment plumes around the Greenland Ice Sheet. *The Cryosphere 6*: 1–19.

Coupel, P., Ruiz-Pino, D., Sicre, M.A. et al. (2015). The impact of freshening on phytoplankton production in the Pacific Arctic Ocean. *Progress in Oceanography 131*: 113–125.

Descamps, S., Aars, J., Fuglei E. et al. (2017). Climate change impacts on wildlife in a High Arctic archipelago – Svalbard, Norway. *Global Change Biology 23*: 490–502.

Dieckmann, G.S., Nehrke, G., Papadimitriou, S. et al. (2008). Calcium carbonate as ikaite crystals in Antarctic sea ice. *Geophysical Research Letters 35*: L08501.

Dieckmann, G.S., Nehrke, G., Uhlig, C. et al. (2010). Brief Communication: Ikaite ($CaCO_3$*$6H_2O$) discovered in Arctic sea ice. *The Cryosphere 4*: 227–230.

Dittmar, T. and Kattner, G. (2003). The biogeochemistry of the river and shelf ecosystem of the Arctic Ocean: a review. *Marine Chemistry 83*: 103–120.

Eilertsen, H.C. and Frantzen, S. (2007). Phytoplankton from two sub-Arctic fjords in northern Norway 2002–2004: I. Seasonal variations in chlorophyll a and bloom dynamics. *Marine Biology Research 3*(5): 319–332.

Else, B.G.T., Papakyriakou, T.N., Galley, R.J. et al. (2012). Annual cycles of pCO_2sw in the southeastern Beaufort Sea: New understandings of air-sea CO_2 exchange in arctic polynya regions. *Journal of Geophysical Research 117*: C00G13.

Emmerton, C.A., Lesack, L.F.W. and Vincent, W.F. (2008). Nutrient and organic matter patterns across the Mackenzie River, estuary and shelf during the seasonal recession of sea-ice. *Journal of Marine System 74*: 741–755.

Eppley, R.W. and Peterson, B.J. (1979). Particulate organic matter flux and planktonic new production in the deep ocean. *Nature 282*: 677–680.

Filbee-Dexter, K., Wernberg, T., Fredriksen, S. et al. (2019). Arctic kelp forests: diversity, resilience and future. *Global and Planetary Change 172*: 1–14.

Forster, P., Ramaswamy, V., Artaxo, P. et al. (2007). Changes in atmospheric constituents and in radiative forcing. In: *Climate Change 2007. The Physical Science Basis. Contribution of Working Group I to the Fourth Assessment Report of the Intergovernmental Panel on Climate Change* (ede. by S. Solomon, D. Qin, M. Manning et al.). Cambridge University Press

Fransson, A., Chierici, M., Hop, H. et al. (2016). Late winter-to-summer change in ocean acidification state in Kongsfjorden, with implications for calcifying organisms. *Polar Biology 39*: 1841–1857.

Fransson, A., Chierici, M., Nomura, D. et al. (2015). Effect of glacial drainage water on the CO_2 system and ocean acidification state in an Arctic tidewater-glacier fjord during two contrasting years. *Journal of Geophysical Research: Oceans 120*: 2413–2429.

Frey, K.E. and McClelland, J.W. (2009). Impacts of permafrost degradation on arctic river biogeochemistry. *Hydrological Processes 23*: 169–182.

Gilbert, G.L., Cable, S., Thiel, C. et al. (2017). Cryostratigraphy, sedimentology and the late Quaternary evolution of the Zackenberg River delta, northeast Greenland. *The Cryosphere 11*(3): 1265–1282.

Gilbert, G.L., O'Neill, H.B., Nemec, W., (2018). Late Quaternary sedimentation and permafrost development in a Svalbard fjord-valley, Norwegian high Arctic. *Sedimentology 65*(7): 2531–2558.

Halbach, L., Vihtakari, M., Duarte, P. et al. (2019). Tidewater glaciers and bedrock characteristics control the phytoplankton growth environment in a fjord in the Arctic. *Frontiers in Marine Science 6*: 254.

Haug, T. Bogstad, B., Chierici, M. et al. (2017). Future harvest of living resources in the Arctic Ocean north of the Nordic and Barents Seas: a review of possibilities and constraints. *Fisheries Research 188*: 38–57.

Herndon, E., AlBashaireh, A., Singer, D. et al. (2017). Influence of iron redox cycling on organo-mineral associations in Arctic tundra soil. *Geochimica et Cosmochimica Acta 207*: 210–231.

Herndon, E.M., Yang, Z., Bargar, J. et al. (2015). Geochemical drivers of organic matter decomposition in arctic tundra soils. *Biogeochemistry 126*(3): 397–414.

Hessen, D.O., Carroll, J., Kjeldstad, B. et al. (2010). Input of organic carbon as determinant of nutrient fluxes, light climate and productivity in the Ob and Yenisey estuaries. *Estuarine and Coastal Shelf Science 88*: 53–62.

Hodal, H., Falk-Petersen, S., Hop, H. et al. (2012). Spring bloom dynamics in Kongsfjorden, Svalbard: Nutrients, phytoplankton, protozoans and primary production. *Polar Biology 35*: 191–203.

Hodson, A., Nowak, A. and Christiansen, H. (2016). Glacial and periglacial floodplain sediments regulate hydrologic transfer of reactive iron to a high arctic fjord. *Hydrological Processes 30*: 1219–1229.

Holmes, R.M., McClelland, J.W., Peterson, B.J. et al. (2012). Seasonal and annual fluxes of nutrients and organic matter from large rivers to the Arctic Ocean and surrounding seas. *Estuaries and Coasts 35*: 369–382.

Hop, H. and Wiencke, C. (2019). *The Ecosystem of Kongsfjorden, Svalbard*. Cham: Springer.

Hopwood, M.J., Carroll, D., Dunse, T., Hodson, A. et al. (2020). Review article: How does glacier discharge affect marine biogeochemistry and primary production in the Arctic? *The Cryosphere 14*: 1347–1383.

Hudson, B., Overeem, I., McGrath, D. et al. (2014) MODIS observed increase in duration and spatial extent of sediment plumes in Greenland fjords. *The Cryosphere 8*: 1161–1176.

Humlum, O. (2005). Holocene permafrost aggradation in Svalbard. *Geological Society of London, Special Publications 242* (1): 119–129

Huss, M. and Hock, R. (2018). Global-scale hydrological response to future glacier mass loss. *Nature Climate Change 8*: 135–140.

IPCC AR4 (2007) *Climate Change: Synthesis Report. Contribution of Working Groups I, II and III to the Fourth Assessment Report of the Intergovernmental Panel on Climate Change*. Geneva: IPCC.

IPCC (2014). *Climate Change: Synthesis Report. Contribution of Working Groups I, II and III to the Fifth Assessment Report of the Intergovernmental Panel on Climate Change*. Geneva: IPCC.

Jakobsson, M., Ingolfsson, O., Long, A.J. et al. (2014). The dynamic Arctic. *Quaternary Science Reviews 92*: 1–8.

Jensen, H.M., Pedersen, L., Burmeister, A. et al. (1999). Pelagic primary production during summer along 65–72°N off West Greenland. *Polar Biology 21*(5): 269–278.

Jones, E.L. (2019). Ph.D. thesis: *The biogeochemistry of the active layer and shallow permafrost in a high Arctic fjord valley, Svalbard.* Department of Geography, University of Sheffield, Sheffield, UK.

Juul-Pedersen, T., Arendt, K.E., Mortensen, J. et al. (2015). Seasonal and interannual phytoplankton production in a sub-Arctic tidewater outlet glacier fjord, SW Greenland. *Marine Ecology Progress Series 524*: 27–38.

Keller, K., Blum, J.D. and Kling, G.W. (2010). Stream geochemistry as an indicator of increasing permafrost thaw depth in an Arctic watershed. *Chemical Geology 273*: 76–81.

Kerr, R.A. (2012). Ice-free Arctic sea may be years, not decades, away. *Science 337*: 1591.

Kirk, J.T.O. (1994). *Light and Photosynthesis in Aquatic Ecosystems*. Cambridge: Cambridge University Press.

Koven, C.D., Riley, W.J. and Stern, A. (2013). Analysis of permafrost thermal dynamics and response to climate change in the CMIP5 earth system models. *Journal of Climate 26*: 1877–1900.

Krawczyk, D.W., Meire, L., Lopes, C. et al. (2018). Seasonal succession, distribution and diversity of planktonic protists in relation to hydrography of the Godthåbsfjord system (SW Greenland). *Polar Biology 41*: 2033–2052.

Kwok, R. (2018). Arctic sea ice thickness, volume and multiyear ice coverage: losses and coupled variability (1958–2018). *Environmental Research Letters 13*: 105005.

Landvik, J.Y., Ingolfsson, O., Mienert, J. et al. (2005). Rethinking Late Weichselian ice-sheet dynamics in coastal NW Svalbard. *Boreas 34* (1): 7–24.

Lannuzel, D., Tedesco, L., van Leeuwe, M. et al. (2020). The future of Arctic sea-ice biogeochemistry and ice-associated ecosystems. *Nature Climate Change 10*: 983–992.

Loeng, H. (2005). Marine systems. In: *Arctic Climate Impact Assessment (ACIA)* (eds. C. Symon, L. Arris and B. Heal), 453–538. Cambridge: Cambridge University Press. Loose, B., Schlosser, P., Perovich, D. et al. (2011). Gas diffusion through columnar laboratory sea ice: implications for mixed-layer ventilation of CO_2 in the seasonal ice zone. *Tellus B* 63: 23–39.

Mangerud, J., Dokken, T., Hebbeln, D. et al. (1998). Fluctuations of the Svalbard–Barents sea ice sheet during the last 150 000 years. *Quaternary Science Reviews 17*: 11–42.

Mascarenhas, V.J. and Zielinski, O. (2019). Hydrography-driven optical domains in the Vaigat-Disko Bay and Godthabsfjord: effects of glacial meltwater discharge. *Frontiers in Marine Science 6*: 335.

McClelland, J.W., Holmes, R.M., Dunton, K.H. et al. (2012). The Arctic Ocean Estuary. *Estuaries and Coasts 35* (2): 353–368.

Meire, L. (2016). PhD thesis. *Biogeochemical cycling in a subarctic fjord adjacent to the Greenland Ice Sheet. Marine Biology*, Ghent University, Belgium.

Meier, W.N., Hovelsrud G.K., van Oort B.E.H. et al. (2014). Arctic sea ice in transformation: a review of recent observed changes and impacts on biology and human activity. *Reviews of Geophysics 52*: 185–217.

Meire, L., Søgaard, D.H., Mortensen, J. et al. (2015). Glacial meltwater and primary production are drivers of strong CO_2 uptake in fjord and coastal waters adjacent to the Greenland Ice Sheet. *Biogeosciences 12*: 2347–2363.

Meire, L., Meire, P., Struyf, E. et al. (2016). High export of dissolved silica from the Greenland Ice Sheet. *Geophysical Research Letters 43*: 9173–9182.

Meire, L., Mortensen, J., Meire, P. et al. (2017). Marine-terminating glaciers sustain high productivity in Greenland fjords. *Global Change Biology 23*: 5344–5357.

Milne, G. and Shennan, I. (2007). Sea level studies: Isostasy. In: *Encyclopedia of Quaternary Science* (ed. S. Elias), 3043–3051. Oxford, Elsevier.

Mortensen, J., Rysgaard, S., Arendt, K.E. et al. (2018). Local coastal water masses control heat levels in a West Greenland tidewater outlet glacier fjord. *Journal of Geophysical Research: Oceans 123*: 8068–8083.

Mortimer, R.J.G., Galsworthy, A.M.J., Bottrell, S.H. et al. (2011). Experimental evidence for rapid biotic and abiotic reduction of Fe (III) at low temperatures in salt marsh sediments: a possible mechanism for formation of modern sedimentary siderite concretions. *Sedimentology 58* (6): 1514–1529.

Moskalik, M., Ćwiąkała, J., Szczuciński, W. et al. (2018). Spatiotemporal changes in the concentration and composition of suspended particulate matter in front of Hansbreen, a tidewater glacier in Svalbard. *Oceanologia 60*: 446–463.

Murray, C., Markager, S., Stedmon, C.A. et al. (2015). The influence of glacial melt water on bio-optical properties in two contrasting Greenland fjords. *Estuarine and Coastal Shelf Science 163*: 72–83.

Nelson, F.E. (2003). (Un)frozen in time. *Science 299*: 1673–1675.

Nicolaus, M., Petrich, C., Hudson, S.R. et al. (2013). Variability of light transmission through Arctic land–fast sea ice during spring. *Cryosphere 7*: 977–986.

Normandeau, A., Dietrich, P., Hughes Clarke, J. et al. (2019). Retreat pattern of glaciers controls the occurrence of turbidity currents on high-latitude fjord deltas (Eastern Baffin Island). *Journal of Geophysical Research – Earth Surface 124*: 1559–1571.

Oliver, H., Luo, H., Castelao, R.M. et al. (2018). Exploring the potential impact of Greenland meltwater on stratification, photosynthetically active radiation and primary production in the Labrador Sea. *Journal of Geophysical Research: Oceans 123*: 2570–2591.

Overeem, I., Hudson, B.D., Syvitski, J.P.M. et al. (2017). Substantial export of suspended sediment to the global oceans from glacial erosion in Greenland. *Nature Geoscience 10*: 859–863.

Overland, J.E., Wang, M. and Salo, S. (2008). The recent Arctic warm period. *Tellus, Series A: Dynamic Meteorology and Oceanography 60A* (4): 589–597.

Overland, J., Hanna, E., Hanssen-Bauer, I. et al. (2015). Surface air temperature. In: *Arctic Report Card 2015*. (eds. M. Jeffries, J. Richter-Menge and J. Overland), 10–16. Washington, DC: NOAA.

Renner, M., Arimitsu, M.L., Piatt, J.F. et al. (2012). Structure of marine predator and prey communities along environmental gradients in a glaciated fjord. *Canadian Journal of Fisheries and Aquatic Sciences 69*: 2029–2045.

Romanovsky, V.E., Smith, S.L. and Christiansen, H.H. (2010). Permafrost thermal state in the polar northern hemisphere during the international polar year 2007–2009: a synthesis. *Permafrost and Periglacial Processes 21* (2): 106–116.

Romanovsky, V. E, Smith, S.L., Christiansen, H.H. et al. (2013). *Permafrost, Arctic Report Card*. Washington, DC: NOAA.

Rysgaard, S. and Glud, R.N. (2007). Carbon cycling and climate change: Predictions for a High Arctic marine ecosystem (Young Sound, NE Greenland). *Bioscience 58*: 206–214.

Rysgaard, S., Glud, R.N., Lennert, K. et al. (2012). Ikaite crystals in melting sea ice – implications for pCO_2 and pH levels in Arctic surface waters. *The Cryosphere 6*: 901–908.

Sakshaug, E. (2004). Primary and secondary production in the Arctic Seas. In: *The Organic Carbon Cycle in the Arctic Ocean*. (eds. R. Stein and R.W. MacDonald), 57–81 Berlin, Heidelberg: Springer.

Schaefer, K., Lantuit, R.V.E., Schuur, E.A.G. et al. (2014). The impact of the permafrost carbon feedback on global climate. *Environmental Research Letters 9*: 085003.

Schroth, A.W., Crusius, J., Chever, F. et al. (2011). Glacial influence on the geochemistry of riverine iron fluxes to the Gulf of Alaska and effects of deglaciation. *Geophysical Research Letters 38*: L16605.

Serreze, M.C. and Barry, R.G. (2011). Processes and impacts of Arctic amplification: a research synthesis. *Global and Planetary Change 77*: 85–96.

Silyakova A. (2013). PhD thesis. *Arctic Ocean carbon biogeochemistry under climate change and ocean acidification*. University of Bergen, Bergen, Norway.

Slagstad, D., McClimans, T.A. (2005). Modeling the ecosystem dynamics of the Barents sea including the marginal ice zone: I. Physical and chemical oceanography. *Journal of Marine Systems 58*: 1–18.

Smith, L.C., Sheng, Y., MacDonald et al. (2005). Disappearing Arctic lakes. *Science 308*: 1429.

Smith, S.L., Romanovsky, V.E., Lewkowicz, A.G. et al. (2010). Thermal state of permafrost in North America: a contribution to the international polar year. *Permafrost and Periglacial Processes 21*(2): 117–135.

Smith, R.W., Bianchi, T.S., Allison, M. et al. (2015). High rates of organic carbon burial in fjord sediments globally. *Nature Geoscience 8*: 450–453.

Statham, P.J., Skidmore, M. and Tranter, M. (2008). Inputs of glacially derived dissolved and colloidal iron to the coastal ocean and implications for primary productivity. *Global Biogeochemical Cycles 22*: Gb3013.

Steiner, N.S. Cheung, W.W.L., Cisneros-Montemayor, A.M. et al. (2019). Impacts of the changing ocean–sea ice system on the key forage fish Arctic cod (*Boreogadus saida*) and subsistence fisheries in the western Canadian Arctic – evaluating linked climate, ecosystem and economic (CEE) models. *Frontiers in Marine Science 6*: 179.

Stevens, L.A., Straneo, F., Das, S.B. et al. (2016). Linking glacially modified waters to catchment-scale subglacial discharge using autonomous underwater vehicle observations. *The Cryosphere 10*: 417–432.

Straneo, F. and Cenedese, C. (2015). The dynamics of Greenland's glacial fjords and their role in climate. *Annual Reviews of Marine Science 7*: 89–112.

Straneo, F., Sutherland, D.A., Holland, D. et al. (2012). Characteristics of ocean waters reaching Greenland's glaciers. *Annals of Glaciology 53*: 202–210.

Stroeve, J.C., Crawford, A.D. and Stammerjohn, S. (2016). Using timing of ice retreat to predict timing of fall freeze-up in the Arctic. *Geophysical Research Letters 43*: 6332–6340.

Strong, C. and Rigor, I.G. (2013). Arctic marginal ice zone trending wider in summer and narrower in winter. *Geophysical Research Letters, 40*: 4864–4868.

Sverdrup, H. (1953). On conditions for the vernal blooming of phytoplankton. *ICES Journal of Marine Science 18*(3): 287–295.

Syvitski, J.P.M., Burrell, D.C. and Skei, J.M. (1987). *Fjords: Processes and Products*. New York: Springer-Verlag Inc.

Thamdrup, B., Fossing, H. and Barker Jorgensen, B. (1994). Manganese, iron and sulfur cycling in a coastal marine sediment, Aarhus Bay, Denmark. *Geochimica et Cosmochimica Acta 58* (23): 5515–5129.

Throckmorton, H.M., Newman, B.D., Heikoop, J.M. et al. (2016). Active layer hydrology in an arctic tundra ecosystem: quantifying water sources and cycling using water stable isotopes. *Hydrological Processes 30* (26): 4972–4986.

Torsvik, T., Albretsen, J., Sundfjord, A. et al. (2019). Impact of tidewater glacier retreat on the fjord system: modeling present and future circulation in Kongsfjorden, Svalbard. *Estuarine and Coastal Shelf Science 220*: 152–165.

Tremblay, J.É. and Gagnon, J. (2009). The effects of irradiance and nutrient supply on the productivity of Arctic waters: a perspective on climate change. In: *Influence of Climate Change on the Changing Arctic and Sub-Arctic Conditions. NATO Science for Peace and Security Series C: Environmental Security* (eds. J.C.J. Kostianoy and A.G. Nihoul), 73–93 and,. Dordrecht: Springer.

Trenberth, K.E. and Josey, S.A. (2007). Observations: surface and atmospheric climate change. In: *Climate Change 2007: the Physical Science Basis: Contribution of Working Group I to the Fourth Assessment Report of the Intergovernmental Panel on Climate Change*. (eds. S. Solomon, D. Qin, M. Manning, et al.), 235–336. Cambridge: Cambridge University Press.

van de Poll, W.H., Kulk, G., Rozema, P.D. et al. (2018). Contrasting glacial meltwater effects on post-bloom phytoplankton on temporal and spatial scales in Kongsfjorden, Spitsbergen. *Elementa: Science of the Anthropocene 6*: 50.

van Everdingen, R.O. (2005). *Multi-Language Glossary of Permafrost and Related Ground Ice Terms*. Calgary: University of Calgary, International Permafrost Association, The Arctic Institute of North America.

Vonk, J.E., Tank, S.E., Bowden, W.B. et al. (2015). Reviews and syntheses: Effects of permafrost thaw on Arctic aquatic ecosystems. *Biogeosciences 12*: 1–97.

Wassmann, P. and Reigstad, M. (2011) Future Arctic Ocean seasonal ice zones and implications for pelagic-benthic coupling. *Oceanography 24* (3): 220–231.

Wu, Y., Peterson, I.K., Tang, C.C.L. et al. (2007). The impact of sea ice on the initiation of the spring bloom on the Newfoundland and Labrador Shelves. *Journal of Plankton Research 29* (6): 509–514.

Zhang, T., Barry, R.G., Knowles, K. et al. (2008). Statistics and characteristics of permafrost and ground-ice distribution in the Northern Hemisphere. *Polar Geography 31* (1–2): 47–68.

9

Biogeochemistry and Ecology of the Indian Sector of the Southern Ocean

Rahul Mohan[1]*, Suhas S. Shetye[2], Shramik Patil[1], Sabu Prabhakaran[1], Vidya P. Jayapalan[1], and Melena Soares[1]

ABSTRACT

The Southern Ocean (SO) is a crucial contributor in determining the Earth's climate as it stores and exchanges a large amount of carbon dioxide (CO_2) and heat between the oceans and the atmosphere. The Antarctic Circumpolar Current (ACC) is a deep-reaching, highly variable eastward flowing current in the SO, driven by the westerly winds; therefore, the SO plays a key role in regulating the global atmospheric temperatures. Since 2006, it is estimated that 60–90% of global ocean heat content change associated with global warming is based in the SO; however, the warming of its water mass is heterogeneous. In the northern part of the ACC, the upper 1000 m water column is warming at a rate of 0.1–0.2°C per decade, while the regions to the south of ACC are not warming or are slightly cooling. In this scenario, changes in the atmospheric circulation, eddy activity, interaction of the ocean ice and atmosphere, and frontal migration become an essential aspect in assessing the biogeochemical balance of the SO. In the coming decades, the oceanic frontal zones will be affected by a decrease in primary production, and subsequent CO_2 uptake, altering the biogeographic boundaries of calcifying and silicifying protists. In a future warming scenario, rapidly melting glaciers, sea ice and icebergs in the coastal Antarctic region may release macronutrients and micronutrients which will boost protist production and support the higher trophic levels. However, under the thick Antarctic sea ice, production of few protist species will decline because of glacier/sea ice melting and an increase in freshwater stratification. This chapter, provides a firm grounding with new insights into the biogeochemistry of the SO based on recent investigations in the Indian sector of the SO.

9.1. INTRODUCTION

Human-induced global warming has already caused many observed changes in the climate system (IPCC, 2019). World oceans store more than 90% of the excess heat that has accumulated and this has caused a rise in the ocean temperature (Stocker et al., 2013; Abram et al., 2019). The increase in oceanic temperatures promotes thermal expansion of seawater, and influences the rise in the sea water level (Nerem et al., 2018). Among the world oceans, the Southern Ocean (SO), the region referred to south of 40°S (for scientific studies), and the Pacific Ocean have experienced significant warming compared with the Atlantic and Indian Oceans (Shi et al., 2018; Swart et al., 2018). It has been suggested that warming of the Indian Ocean is linked to anthropogenic forcing as well as the Pacific Decadal Oscillation (Li et al., 2018). More importantly, the observed warming

[1]*ESSO-National Centre for Polar and Ocean Research, Ministry of Earth Sciences, Vasco-da-Gama, Goa, 403804, India*

[2]*CSIR-National Institute of Oceanography, Dona Paula, Goa, 403004, India*

**Corresponding Author: Rahul Mohan (rahulmohan@ncpor.res.in) ORCiD code: 000-0001-8758-2215*

Systems Biogeochemistry of Major Marine Biomes, First Edition. Edited by Aninda Mazumdar and Wriddhiman Ghosh.

and freshening of the SO since the year 1950 (Gille, 2002) are attributable to an increase in the human-induced greenhouse gas emission (Swart et al., 2018). The SO is known as a primary global sink for CO_2 as well as heat, and plays a vital role in the global ocean circulation and, consequently, global climate (Sarmiento and Le Quere, 1996; Caldeira and Duffy, 2000; Sigman and Boyle, 2000; Mikaloff Fletcher et al., 2006; Gruber et al., 2009; Gent and Danabasoglu, 2011; Swart and Fyfe 2012; Lenton et al., 2013). At present, the SO CO_2 net uptake is –0.8 Pg C year^{-1} (Takahashi et al., 2009) and –1.0 Pg C year^{-1} (Landschützer et al., 2015), accounting for about half of the global oceanic anthropogenic CO_2 uptake over the past 125 years (Frölicher et al., 2015).

The ongoing decrease in oceanic pH caused by the rise in oceanic CO_2 has been termed ocean acidification (OA) (Houghton 2007). The solubility of CO_2 in the seawater depends on the sea surface temperature (SST). The solubility of CO_2 increases with the decreasing SST. Thus, the effect of OA is more prominent in the colder oceans (such as the SO and the Arctic Ocean) than in the warmer tropical oceans (Feely et al., 2009). In the past century, average ocean pH has dipped from 8.21 to 8.10 and the average open oceanic pH in the past 15–30 years has decreased at the rate of ~0.0018 year^{-1} (Bates et al., 2014). A model study predicts that, the global average pH without mitigation is expected to fall to between 7.67 and 7.81 by 2100, accounting for about five times the amount of acidification that has already occurred (Bernie et al., 2010). Increasing OA is expected to affect mineralizing as well as nonmineralizing marine organisms and ecosystem, including those that support fisheries and higher trophic levels (Fabry et al., 2008; Doney et al., 2009; Kroeker et al., 2013). Ongoing OA makes it difficult for calcifying marine organisms such as corals, coccolithophores, pteropods, shellfishes, and foraminifera among others), to calcify (Orr et al., 2005; Bednaršek et al., 2012; Kroeker, et al., 2013).

The SO is getting warmer and the carbonate chemistry is changing very rapidly (Sallée 2018; Gille, 2002). Future predictions show that the SO will become warmer, fresher, and more acidic, and it will have more exposure to photosynthetically active radiation and ultraviolet radiation (Orr et al., 2005; Marinov et al., 2010; Boyd et al., 2016). Such changes will affect the composition of the SO plankton, their biogeographic boundaries, and the food web structure. In recent studies, southward shifts of SO plankton community structure have been observed, and these shifts are likely to continue in the coming years (Siegert et al., 2019; Patil et al., 2014, 2017, 2020). The increase in solar radiation and subsequent warming will also make the surface of the oceans more stratified. Vertical mixing will be reduced in the water column and nutrient pumping from the deeper depths to the surface is expected to decrease. This will affect large-sized phytoplankton (such as microplankton (>20 μm) and will lower the downward flux of biogenic carbon (Matear and Hirst, 1999; Petrou et al., 2016). However, small-sized flagellates (Matear and Hirst 1999; Petrou et al., 2016) will be favored in such a scenario, and flagellate abundance will likely increase, resulting in high grazing pressure (Evans et al., 2011).

A better understanding of ongoing changes in the SO will be useful to predict the future changes in the SO oceanography and ecosystem, changes in the marine biological and carbonate counter pumps, and the evolution of the global climate system (Swart et al., 2018). However, despite the role of SO in the oceanic carbon cycle, it has remained as one of the least studied oceanic regions, especially with regard to biogeochemical variables (Gray et al., 2018), plankton ecology and DNA studies. Hence, it warrants rigorous field, probe-based, and satellite-based studies to understand the feedback mechanisms of a changing SO. This chapter provides a brief assessment on the biogeochemistry of the Indian Sector of the SO (ISSO) based on investigations carried out during the last decade. The ISSO has a quasi-zonal oceanic frontal system (Anilkumar et al., 2015a), and the fronts join near the Crozet Plateau to form one of the world's strongest oceanic frontal systems (Orsi et al., 1995; Belkin and Gordon 1996; Sparrow et al. 1996; Holliday and Read 1998; Kostianoy et al. 2004).

9.2. ROLE OF CURRENTS AND OCEANIC FRONTS IN THE SOUTHERN OCEAN

The ACC is the strongest, continuous eastward-flowing barotropic current, which encircles the Antarctic continent (Dong et al., 2006). This inter-basin connectivity of the ACC permits distribution of heat, and water masses in the Pacific, Atlantic, and Indian Ocean (Rintoul et al., 2001; Sallée, 2018). The strong westerly winds are the primary mechanisms for driving the ACC (Rintoul, 2009). The ACC seems to exist as multiple narrow jets imbedded within, or associated with, density fronts (the Subantarctic and Polar fronts) which appear to be circumpolar in extent. The strength and latitudinal locations of these fronts are influenced by the bathymetry of the region (Pollard et al., 2002). The ACC is also characterized by a high degree of eddy activity generally as a result of the dynamic instabilities of the fronts (Hughes, 2005).

Compared with the other sectors of the SO, the ISSO consists of notable regional dissimilarities controlled by its topography (Kostianoy et al., 2004), and variability of the Agulhas Return Current (Belkin and Gordon, 1996). The Agulhas Return Front (ARF), associated with the

Agulhas retroflection current, merges with the Southern Subtropical Front (SSTF) in some parts of the ISSO (Lutjeharms and Ansorge, 2001). The region where these fronts merge is a highly dynamic zone as a result of eddy activity. The eddies occuring in this region are known to transport Subtropical Surface Water (STSW) southward utilizing the enhanced meridional currents along its periphery (Chacko et al., 2014, Sabu et al., 2015). Further south, the major fronts and zones identified in the ISSO are: the Subantarctic Zone (SAZ), the Subantarctic Front (SAF), the Polar Frontal Zone (PFZ), the Polar Front (PF), the Antarctic Zone (AZ), the Southern ACC Front (SACCF), and the southern boundary of the ACC (SB) (Holliday and Read, 1998). There is no surface expression for SB and it lies very close to the SACCF (Orsi et al., 1995). The STF, SAF, and PF also consist of northern and southern branches (Whitworth and Nowlin, 1987). The SST characteristics are used as an indicator to demarcate various fronts in the ISSO fronts (Figure 9.1). The STF separates the STSW from the Subantarctic Surface Waters, and the ARF appears between the northern and southern branch of the STF. Each front embedded within the ACC is associated with an intense gradient in the biogeochemical properties, temperature, and salinity (Orsi et al., 1995; Rintoul, 2009).

The AZ is an important region in the ISSO, characterized by winter water in the upper 150 m and overlain by warmer and fresher Antarctic Surface water during the summer (Orsi et al., 1995). The large-scale meridional and zonal circulation patterns (Bindoff et al., 2000), seasonal surface water mass transformations (Williams et al., 2008), and the seasonal formation and dissipation of sea ice (Gloersen et al., 1993) makes the AZ a physically dynamic region. The Antarctic Slope Current (ASC) is one of the important circulation features in this region which controls the flow of water towards the coastal region of Antarctica. The structure and variability of the ASC, influence key processes such as glacier/sea ice melting and freshening of the bottom water masses in the ISSO (Thompson et al., 2018).

Studies have shown that the SO has experienced a complex set of changes over the past several decades. There is also evidence that the strong westerly winds in the SO are intensifying and shifting towards the south (Fyfe and Saenko, 2006) owing to the positive trend in the Southern Annular Mode (SAM). The SAM is the dominant mode of variability in the SO, defining the pressure difference between the high latitudes (Antarctica) and mid-latitudes. This trend in SAM has been directly linked to the changes in the ozone layer and global climate change (Arblaster

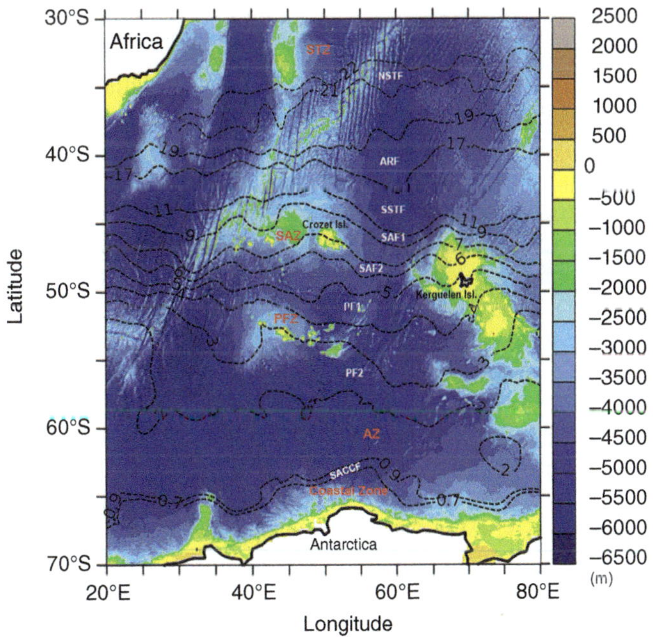

Figure 9.1 Schematic representation of various fronts in the Indian Sector of the Southern Ocean (ISSO). The position of the fronts is marked based on their southern subtropical characteristics using the WOA18 (Locarnini et al., 2018) climatology data. Background colour represents the bathymetry of the ISSO from ETOPO data (location of fronts modified from Holliday and Read, 1998, and Anilkumar et al., 2014). AZ, Antarctic Zone; NSTF, Northern Subtropical Front; PF, Polar Front; PFZ, Polar Frontal Zone; SACCF, Southern Antarctic Circumpolar Current Front; SAF, Subantarctic Front; SAZ, Subantarctic Zone; SSTF, Southern Subtropical Front; STZ, Subtropical Zone.

and Meehl, 2006). The southward shift of the PF in the ISSO is attributed to a positive SAM (Anilkumar et al., 2014).

The uppermost 1000 m of the northern part of the ACC, is warming rapidly at a rate of 0.1–0.2°C per decade (Sallée, 2018), while the surface subpolar seas south of this region are not warming or are slightly cooling (Armour et al., 2016). The mid-depth (700–1100 m) temperatures of the SO within the ACC have risen by 0.17°C between 1950 and 1980 (Gille, 2002). A study by Swart et al. (2018) also showed warming and salinification of the Upper Circumpolar Deep Water south of the 45°S and cooling and freshening of the waters north of this latitude. A recent study by Auger et al. (2021) has also reported the warming and shoaling of the Upper Circumpolar Deep Water in the subpolar SO. Maheshwari et al. (2013) reported a higher rate (0.02 ± 0.01°C $year^{-1}$) of warming in the ISSO; similarly, Aoki (1997) also reported a significant warming trend at depths between 0 and 400 m during the period 1983–1995. Using 32 years' of temperature data, Aoki et al. (2003) reported a significant warming trend in the ISSO at almost all depths from 200 m to 900 m. The Antarctic Bottom Water (AABW), which is coldest and densest water mass has warmed (Purkey and Johnson, 2010), freshened (Swift and Orsi, 2012), and decreased in volume (Purkey and Johnson, 2013) in recent decades. All these changes show a global scale contraction of AABW, suggesting a possibility of a slowdown of the bottom, southern limb of the meridional overturning circulation (Morrison, 2015).

The study by Anilkumar et al. (2015b) has reported an increased rate of freshening of the AABW in the ISSO. Furthermore, Jena et al. (2018), using satellite-derived Antarctic sea ice extent from 1979 to 2015, reported an increasing trend in sea ice in the ISSO region resulting from the strengthening of westerly winds. Therefore, the enhanced sea ice formation and its subsequent melting may increase the freshwater input, resulting in changes in the properties of the surface as well as the deeper water masses of the ISSO. This enhanced meltwater may increase the surface stratification, which has important implications on the biogeochemical cycle of the ISSO.

9.3. NUTRIENTS IN THE SOUTHERN OCEAN

The biogeochemical cycling of nutrients and its supply to the photic zone are critical for phytoplankton production, ecosystem functioning and CO_2 uptake (Deutsch and Weber, 2012). The SO plays a significant role in the global circulation of nutrients (Pollard et al., 2006). The nutrient supply from the SO is the major source of nutrients in the northern region (north of 30°S) and supports up to three-quarters of total biological production (Sarmiento et al., 2004). An increase in the biological production and nutrient drawdown in the SO tends to decrease the supply of nutrient and hence the biological production in the rest of the oceans (Sarmiento and Orr, 1991; Sarmiento et al., 2004; Marinov et al., 2006; Oschlies et al., 2010). In the Antarctic continental shelf region, CO_2 and nutrients are supplied to the upper ocean layer through upwelling and mixing of the Circumpolar Deep Water (Prézelin et al., 2000; Carrillo et al., 2004; Serebrennikova and Fanning, 2004; Martinson et al., 2008).

Although the SO is characterized by elevated macronutrient concentrations, the primary productivity in the SO is low because of low availability of micronutrients (such as Fe) in the surface waters (Martin, 1990), which makes the SO one of the high nutrient–low chlorophyll regions in the world (Sarmiento et al., 2004; Martin et al., 2013). Generally, biological production is low in the SO, although high production is often documented around the mid-oceanic islands (such as South Georgia, Crozet, and Kerguelen) (Pollard et al., 2007) and in coastal Antarctic regions. In the coastal Antarctic regions, high biological productivity is attributed to the release of nutrients from melting of the sea ice and runoff from the land. During the austral summer, melting of the winter sea ice and shelf ice releases essential macro nutrients and micronutrients, including Fe, into the surface waters (Wang et al., 2014) which boost phytoplankton production. Dissolved Fe concentrations up to 36.8 $nmol^{-1}$ have been reported in the sea ice of the Weddell Sea (Lannuzel et al., 2010). This accumulation of Fe in the sea ice depends on suspended sediments which contribute about 4.1×10^6 mol Fe $month^{-1}$ in the SO during summer (Wang et al., 2014).

A few studies have reported N:P ratios <16 in the western Antarctic Peninsula region (Henley et al., 2017). Such uptake ratios, which are lower than the Redfield ratio of 16N:1P, result from the dominance of diatoms in the coastal Antarctic areas (Annett et al., 2010; Clarke et al., 2008). Similarly, SiO_4^{4-} depletion along with the NO_3^- excess indicates the dominance of diatoms (Goeyens et al., 2000; Goffart et al., 2000).

The availability of macronutrients in the SO is strongly influenced by eddies, seamounts, islands, meltwater intrusion, and remineralization (Dufour et al., 2015; Read and Pollard, 2017). A recent study by George et al. (2018) suggested that eddy shedding and associated vertical and zonal advection supply nutrients to the Agulhas Retroflection Frontal Zone (ARFZ) and Southern Subtropical Frontal Zone (STFZ) surface waters. In the ISSO, macronutrient concentrations increase from the subtropical region towards the Antarctic coast (Figure 9.2) (Rajakumar et al., 2008; Shetye et al., 2015; Soares et al., 2015; Patil et al., 2017). Although the concentrations of nutrient increase in the Antarctic coastal region,

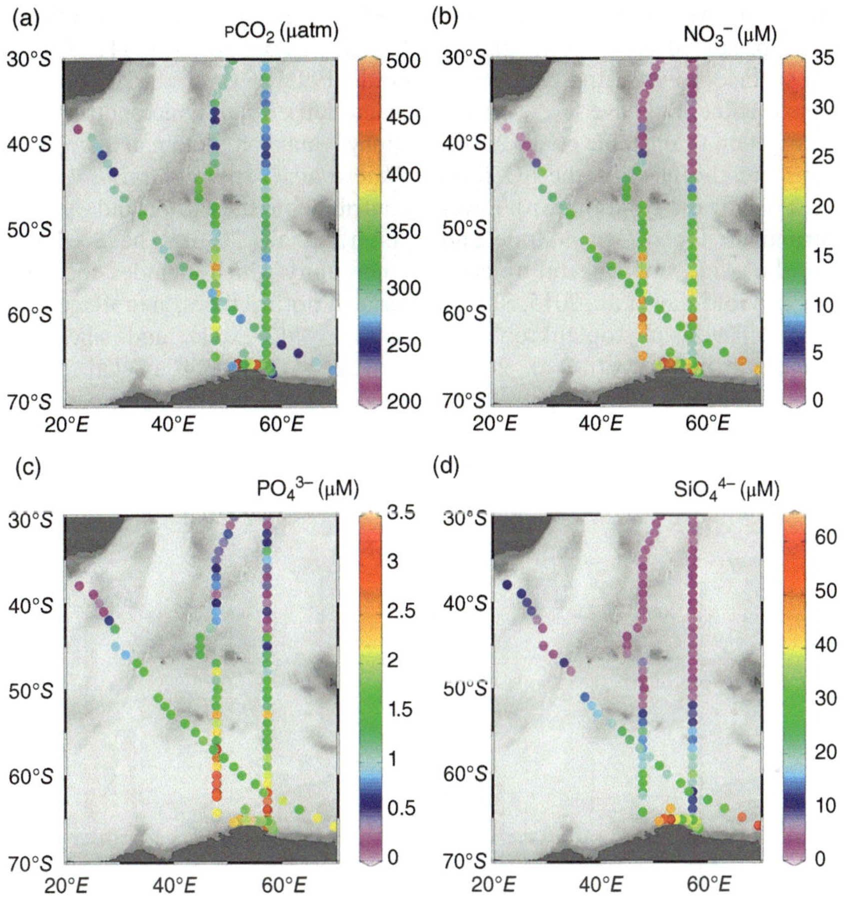

Figure 9.2 Sea surface distribution of (a) pCO_2, (b) NO^{3-} (c) PO_4^{3-}, and (d) SiO_4^{4-} in the Indian Sector of the Southern Ocean. The data was collected as part of Indian scientific expedition to the Southern Ocean and Antarctica during austral summers (2009 and 2012) onboard *ORV Akademik Boris Petrov and MV Ivan Papanin,* respectively (modified from Shetye et al., 2015). Data was plotted using Ocean Data View software (Schlitzer, 2015).

there is rapid SiO_4^{4-} utilization and drawdown (<5 μM) under the Antarctic land–fast ice (Shetye et al., 2014; Shetye et al., 2019).

9.4. SOUTHERN OCEAN PLANKTON ECOLOGY

Investigations on Southern Ocean plankton ecology have provided information on their standing stock, biogeographic boundaries and response to changing climatic conditions. Several studies were carried out on the governing factors for plankton ecology and biogeographic distribution across the oceanic frontal zones, which include coastal Antarctic regions, sea ice related regions, and sea ice covered coastal Antarctic regions (Patil et al., 2013, 2017, 2020; Naik et al., 2015, 2020; Shetye et al., 2019). The SO holds one of the most extensive microscopic ecosystems in the world and comprises more than 500 protist species of which ~350 are phytoplanktonic and ~150 are microheterotrophic (Scott and Marchant, 2005). These protists are known to co-occur with heterotrophic prokaryotes (archaea and bacteria) and viruses. They collectively embrace the microbial food web that sequester, respire, and/or mineralize much of the carbon (Fenchel, 2008; Kirchman, 2008). The seasonal variation in the physicochemical parameters, sea ice extent and frontal migrations may have a pronounced effect on plankton community structure in the SO. During the early austral summer, melting of sea ice results in large phytoplankton blooms, which support a profusion of the Antarctic marine life (Riaux-Gobin et al., 2011). In contrast, during winter, the SO experiences a reduction in phytoplankton which is attributed to a low light intensity, thick ice cover, and low nutrient supply as a result of strong upper water stratification. An increase in the global average temperatures and the absorption of the excess anthropogenic carbon emissions in high latitude oceans is expected to alter the marine biogeochemical cycle in coming decades.

Southward migration of westerlies and oceanic fronts, increased SST, ocean stratification, glacier melting, decreased sea surface salinity, and changes in the sea ice cover are also predicted to have severe effects on the polar marine ecosystem (Constable et al., 2014). Marine plankton forms are the base of the SO food web; however, it shows a patchy distribution, with lower production in the Permanent Open Ocean Zone and higher production in the Antarctic coastal regions (Smith and Gordon, 1997; Goldman et al., 2015; Shetye et al., 2019). In the ISSO, siliceous phytoplankton such as diatoms dominate south of the PF (Patil et al., 2013), whereas, carbonate producers (coccolithophores, foraminifera) control the subtropical and subantarctic regions (Goldman, 1988; Patil et al., 2017, 2020) (Figures 9.3 and 9.4).

More than 50% of the SO consist of three major oceanic frontal regions/zones (STF, SAF, and the PF) (Orsi et al., 1995; Park et al., 2019). The PFZ is rich in both macronutrients and micronutrients, which form deep chlorophyll maxima dominated by large-sized diatoms (e.g. *Thalassiothrix* spp. and *Rhizosolenia* spp.), and significantly contributes to the carbon and silica flux (Kemp et al., 2006; Assmy et al., 2013). The SAZ hosts a diverse protist community which includes coccolithophores, cyanobacteria, dinoflagellates, nanoflagellates, small diatoms (de Salas et al., 2011), and silicifying haptophytes (Patil et al., 2014, 2015; Jordan et al., 2016). This circum-global region has been recently recognized as the Great Calcite Belt based on remote sensing of particulate inorganic carbon at estimated concentrations of ~1 μmol l^{-1} compared with the global average of 0.2 μmol l^{-1} (Balch et al., 2011).

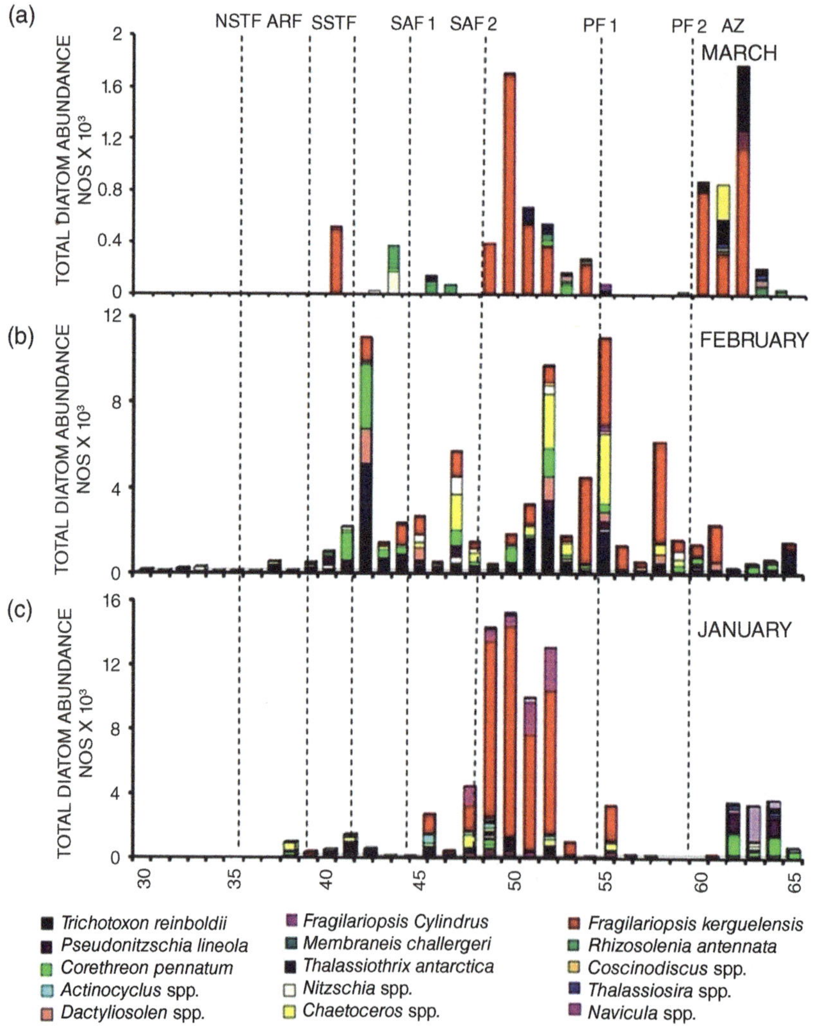

Figure 9.3 Latitudinal variations in the diatom abundances (n × 10^3 cells l^{-1}) during (a) March 2009, (b) February 2010, and (c) January 2012 in the Indian Sector of the Southern Ocean (Shetye et al., 2015). (Note: different scale was used for diatom abundances). ARF, Agulhas Return Front; AZ, Antarctic Zone; NSTF, Northern Subtropical Front; PF, Polar Front; SAF, Subantarctic Front; SSTF, Subtropical Front.

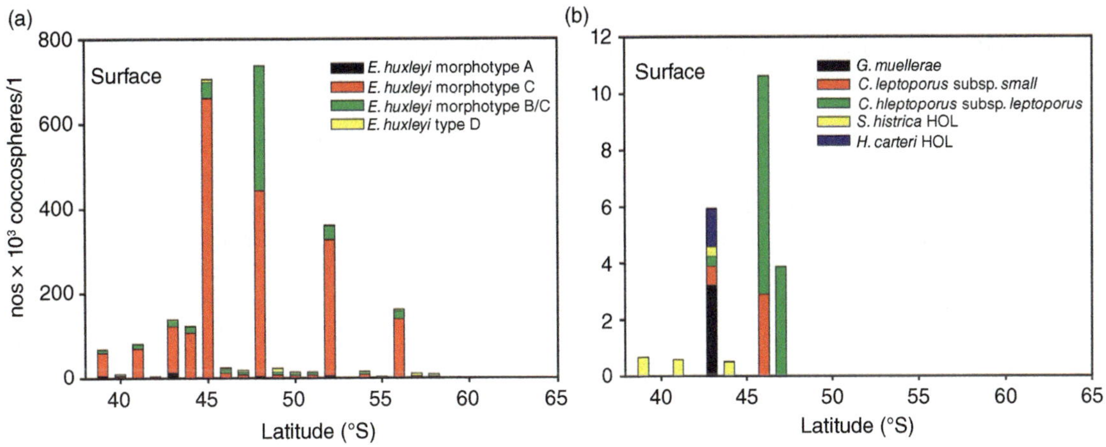

Figure 9.4 (a) *Emiliania huxleyi* morphotypes distribution, (b) other dominant coccolithophores recorded during February 2010 In the Indian Sector of the Southern Ocean (Patil et al., 2017). (*E. huxleyi* type D represents malformed *E. huxleyi*: e.g. Mohan et al., 2008; Patil et al., 2017).

The main primary and secondary consumers in the SAZ are copepods, mesopelagic fish, myctophids which serve as alternative food for penguins, squids, and mesopelagic fish population (Murphy et al., 2016). All these organisms together consume the majority of the primary producers, making phytoplankton abundance relatively low (Banse, 1996). In the ISSO, there is a high abundance of diatoms in the PFZ and AZ after the breakup of sea ice, which is attributed to nutrient and micronutrient availability. However, a few diatoms (such as *Fragilariopsis kerguelensis*) take advantage of this over other diatoms and dominate the total assemblage. Diatoms are more widespread in the ISSO during the mid-austral summer (e.g. February) with an increasing dominance from SAZ to AZ. However, during the late austral summer, diatom abundance reduces because of macronutrient and micronutrient depletion and increased grazing pressure (Patil et al., 2013) (Figure 9.3). Similarly, higher coccolithophore abundance in the ISSO was reported during the early austral summer (e.g. December) in the subtropical–subantarctic zone; the greatest diversity was reported during the mid-austral summer and reduced diversity was found during the late austral summer (Patil et al., 2020). The studies by Patil et al. (2017, 2020) show that the abundance of coccolithophores during the austral summer in the subtropical and subantarctic regions reaches up to 3000 × 10^3 cells l^{-1}. The PF is observed as the southernmost boundary of coccolithophore extent and south of this, the abundance of diatoms increases.

9.5. OCEAN CARBONATE CHEMISTRY

Globally the oceans take up 48% of the CO_2 emitted via fossil-fuel burning and cement manufacture (Sabine et al., 2004). Takahashi et al. (2009) (based on ~3 million measurements of surface water pCO_2 obtained from 1970 to 2007) reported an increase in the surface ocean water pCO_2 at a mean rate of 1.5 μatm $year^{-1}$ with basin specific rates varying between 1.27 ± 0.5 and 2.17 ± 0.4 μatm $year^{-1}$. The Antarctic continental shelf regions are particularly important as they are responsible for high primary production and CO_2 uptake (Arrigo et al., 2008a), and affect large-scale biogeochemical cycles and global climate (Arrigo et al., 2008b). In the SO, outgassing of CO_2 driven by the upwelling of carbon-rich deep water (Hoppema 2004) is counteracted by CO_2 drawdown by biological production (Metzl et al., 2006; Gruber et al., 2009). Several studies have suggested stagnation and reduction in the SO carbon sink from the 1980s to the early 2000s (Wetzel et al., 2005; Le Quéré et al., 2007; Lovenduski et al., 2008; Lenton et al., 2013), and this was attributed to the increase in the speed of westerlies in response to the positive SAM (Thompson et al. 2011; Marshall, 2003). However, this decrease in the carbon uptake capacity stopped around 2002, and by 2012 the SO regained its expected strength (Landschützer et al., 2015) because of the weakening of the upper ocean overturning circulation (DeVries et al., 2017).

The power of the SO as a carbon sink has been shown to vary seasonally across the different SO zones (Metzl et al., 1991; Metzl, 2009). The ISSO acts as a sink for atmospheric pCO_2 during the austral summer, but its capacity decreases from the early summer to the early winter (Shetye et al., 2015, 2017). Padin et al. (2010) reported that the subtropical waters of the Atlantic sector of the SO act as a sink for atmospheric CO_2 during spring, and they serve as a source during the autumn season. In the ISSO, high pCO_2 concentrations are reported in the Polar front when compared with the SAF, STF, and the Antarctic zone (Figure 9.2) (Shetye et al., 2015). Metzl (2009) reported a fugacity (fCO_2) growth rate of 2.39 per year for the region south of the PF in the ISSO during 1991–2007, which was faster than that of the Pacific sector of the SO. Metzl (2009) attributed this high growth rate of fCO_2 to a change in the rise in sea level. Since 1993, a decrease in the rise in sea level has been observed in the southwestern Indian Antarctic sector and an increase has been reported in the southwestern Pacific Antarctic sector (Morrow et al., 2008). Furthermore, studies by Prasanna et al. (2015) identified regions in the ISSO favorable for the venting of CO_2 from the ocean owing to the prevalence of strong winds and wind-driven divergence causing an upwelling of dissolved inorganic carbon rich deeper waters. Such upwelling regions and the low production/high CO_2 regions in the PF are also regions of low pH conditions (Shetye et al., 2015).

The ongoing rise in anthropogenic CO_2 emissions and its oceanic absorption has already decreased the SO surface pH levels by ~0.1 units (Key et al., 2004; McNeil and Matear 2007). Modeling studies project a decrease in the SO pH levels up to ~0.3 units by the end of the century (Caldeira and Wickett 2003; Orr et al., 2005; McNeil and Matear 2007). Such changes in the SO pH may influence the carbon and silicon cycles in the future (Petrou et al., 2019) and pose a significant threat to the marine organisms (Mostofa et al. 2016). Reduction in the pH and Ω (calcium carbonate saturation) may lower biogenic calcification in many marine taxa (Kroeker et al., 2013), increase the $CaCO_3$ dissolution (Caldeira and Wickett, 2003; Eyre et al., 2018), and hence reduce habitat suitability for calcifying organisms (Schulz et al., 2019). Changes in the oceanic pH have been shown to affect marine calcifying organisms in previous studies (Orr et al., 2005; Riebesell et al., 2000; Fabry et al., 2008; Doney et al., 2009) and it is expected that these changes will be detrimental in the coming decades.

High latitude ecosystems will experience OA to a greater extent because of seasonal amplification, colder temperatures enhancing CO_2 solubility, and freshening of seawater (Doney et al. 2009; Fabry et al. 2009). In the SO, OA may affect calcifying plankton by lowering the carbonate ion concentration, $[CO_3^{2-}]$, to levels where $CaCO_3$ shells begin to dissolve (McNeil and Matear, 2008). McNeil and Matear (2008) reconstructed the SO seasonal cycle of pH and found that it varies by 0.06 units between summer and winter. Intense wintertime minimum $[CO_3^{2-}]$ in the south of the PF when combined with the anthropogenic CO_2 uptake is likely to induce aragonite undersaturation when atmospheric CO_2 levels reach 450 ppm (McNeil and Matear 2008). Model studies project that with the current CO_2 emission rates, the SO surface waters will become undersaturated with respect to aragonite by the end of 21st century (Orr et al., 2005; McNeil and Matear, 2008; Fabry et al., 2009; Hauri et al., 2016). In reality, the SO aragonite undersaturation events have already occurred and are predicted to be prolonged (Bednaršek et al., 2012; Hauri et al., 2016; Gardner et al., 2018). Negrete-García et al. (2019) reported the emergence of a shallow aragonite saturation horizon in the SO and predicted earlier undersaturation events during the winter. More importantly, this shallowing will be seen even under emission-stabilizing scenarios, indicating habitat destruction for aragonite mineralizing organisms. Gardner et al. (2018) reported 39% larval mortality in the pteropod *Limacina helicina antarctica* exposed to low pH levels (pH 7.6) and extensive shell dissolution was induced, thus increasing shell fragility. Bednaršek et al. (2014) reported that *L. helicina antarctica* specimens showed dissolution immediately after exposure to undersaturated conditions (saturation state of aragonite (Ω_{ar}) ~0.8). It was observed that *L. helicina antarctica* lost 1.4% of shell mass per day. A few studies have reported that most of the SO phytoplankton are unaffected by OA in terms of growth and/or carbon fixation (Boelen et al. 2011; Hoogstraten et al. 2012; Trimborn et al. 2013, 2017; Gao and Campbell 2014; Xu et al. 2014; Boyd et al. 2015; Heiden et al. 2016; Petrou et al. 2016). Studies on a pennate diatom *Nitzschia lecointei* showed a slight stimulation by OA, which was followed by a decline in its growth (Torstensson et al. 2013, 2015). Beszteri et al., (2018) reported that haptophytes such as *Phaeocystis antarctica* are better adapted to OA than diatoms such as *F. kerguelensis* and *Chaetoceros debilis*. Hoppe et al. (2013) reported that OA is expected to enhance primary productivity and the abundance of heavily silicified diatoms in Fe-enriched waters. Contrary to Hoppe et al. (2013), a recent work by Petrou et al. (2019) showed that OA sharply diminishes diatom silicification in a natural Antarctic community. Donahue et al. (2019) concluded from studies in the SO that, there are taxon-specific and locality specific differences in natural phytoplankton community responses to increased pCO_2. However, despite all the above-mentioned studies, there is a need for *in situ* free-ocean CO_2 enrichment experiments such as the one described in Stark et al. (2018).

The response of coccolithophores to high pCO_2 levels has provided different results in culture and field-based investigations. Various laboratory studies on the coccolithophore species *Emiliania huxleyi* have observed decreased (Riebesell et al., 2000; Zondervan et al., 2002), increased (Fiorini et al., 2011), or unchanged (Jones et al., 2013) calcification patterns when subjected to high pCO_2 levels. In field-based investigations several authors reported a decrease in coccolith calcification in *E. huxleyi* in high CO_2 environments, using water column and sediments (e.g. Beaufort et al., 2011); however, several studies also show the opposite, i.e. an increase in calcification at high pCO_2 levels (e.g. Smith et al., 2012). Several extant coccolithophores species may be affected by high pCO_2 levels but few coccolithophore species may find adaptation strategies to cope with more acidic conditions (Irie et al., 2010).

9.6. SEA ICE AND IMPLICATIONS FOR FUTURE GLOBAL CHANGE

Sea ice covers as much as ~40% of the SO in September/October and acts as a key indicator and a thermal barrier for climate change (Thomas and Dieckmann, 2002). Primary production in these regions is regulated by the annual sea ice cycle, as it modulates the timing and distribution of the phytoplankton blooms (Smith and Comiso, 2008; Vernet et al., 2008; Venables et al., 2013). As ice cover retreats during the early summer, large phytoplankton blooms develop because of increased light availability, stratified conditions, nutrient availability, and possible seeding by sea ice algae (Clarke et al., 2008; Smith and Nelson, 1985). The Antarctic winters are characterized by higher sea ice, deeper mixed layers and very low chlorophyll (Henley et al., 2017).

Climate model projections suggest that Antarctic Peninsular air temperatures will increase more than the global average in coming decades (Hoegh-Guldberg et al., 2018). The rise in Antarctic shelf water (upper 300 m) temperatures by 0.6°C (Meredith and King, 2005; Martinson and McKee, 2012), have already resulted in the Antarctic glacier retreat, a decline in the sea ice season, and absence of the perennial ice cover in west Antarctica (Cook et al., 2005; Martinson et al., 2008; Stammerjohn et al., 2008; Wouters et al., 2015; Shepherd et al., 2018; Rignot et al., 2019). The length of the sea ice season on the west of the Antarctic Peninsula has been reduced by ~4 days (Massom et al., 2018), while sea ice coverage to the east has increased during the austral summer and decreased during the winter (Siegert et al., 2019). The extent of summer sea ice for the Antarctic as a whole reached historic lows in 2016/2017 (Schlosser et al., 2018).

A model study has predicted that the sea ice will decline in the 21st century owing to increased heating (ocean and atmosphere) and rain (Liu and Curry 2010). However, a recent study by Kumar et al. (2020) showed that the extent of sea ice has increased (~14.5 km compared with 2001) in the Amery shelf, east Antarctica and is expected to further increase. A few studies have suggested that the Antarctic coastal regions will become warmer and fresher, will be more acidic, and will have greater exposure to photosynthetically active radiation and ultraviolet light (Sarmiento et al., 2004; Orr et al., 2005; Boyd et al., 2008). Sea ice melting is expected to decrease surface salinity in the Antarctic coastal region, and this will affect most phytoplankton species as they may experience osmotic stress upon exposure to varying salinity levels (Bisson and Kirst, 1995; Lionard et al., 2005). Schofield et al. (2017) and Mendes et al. (2013) have already reported a phytoplankton community shift from diatom to cryptophytes in the low saline conditions from the Bransfield Strait (Antarctic Peninsula).

A recent study by Shetye et al. (2019) showed the presence of mucilaginous aggregates of sea ice algae attached to the under flanks of thick sea ice (1.7 m) near the Larsemann Hills, east Antarctica. Under the sea ice melting scenario, ice-related algae, which benefit from the UV protection of ice may decline. Some studies have reported an 80% decline in the regional krill density between 1976 and 2004. Any decline in krill population, which feeds on the algae living underneath the Antarctic sea ice, will have a negative impact on higher trophic level organisms (such as penguins, seals, and whales) (Lucas et al., 2014). Figure 9.5 explains the effects of melting on sea ice algae, krill, and whales. Thinner sea ice may therefore drive low algal biomass, which will affect zooplankton and thus the whole Antarctic food web (Murphy et al., 2006; Ballerini et al., 2014).

A reduction in sea ice-related algae will also affect the silica budget in the coastal Antarctic region. Shetye et al. (2019) observed a rapid utilization of SiO_4^{4-} at the sea ice–water interface. In the rapid sea ice melting scenario it is expected that more SiO_4^{4-} will be available in the shallow Antarctic coastal regions (<300 m). The SiO_4^{4-} concentrations would not only increase in the surface waters but might also increase at deeper depths, allowing the benthic organisms to utilize it without competition.

9.7. SUMMARY

The ISSO comprises a diverse array of environments from the STF to the sea ice zone, which affects phytoplankton production and higher trophic levels. The climate-induced oceanic and atmospheric changes, ocean warming, eddy activity, ocean stratification, sea ice/glacier melting, frontal shifts, and lowering of oceanic pH will have severe effects on plankton standing stock, diversity, physiology, and the marine food web. An increase in ocean warming and stratification will affect the biogeographic boundaries of calcifying and silicifying protists in the ISSO. Calcifying phytoplankton are projected to expand their distribution in current sea ice zone waters. In the coastal Antarctic region, melting of sea ice and glaciers will result in an excess supply of macronutrients and micronutrients supporting the production of some protist species, but increased freshening and stratification below the coastal sea ice will reduce large phytoplankton blooms. In addition, lowering pH levels will affect calcite–aragonite shelled organisms, which may suffer from

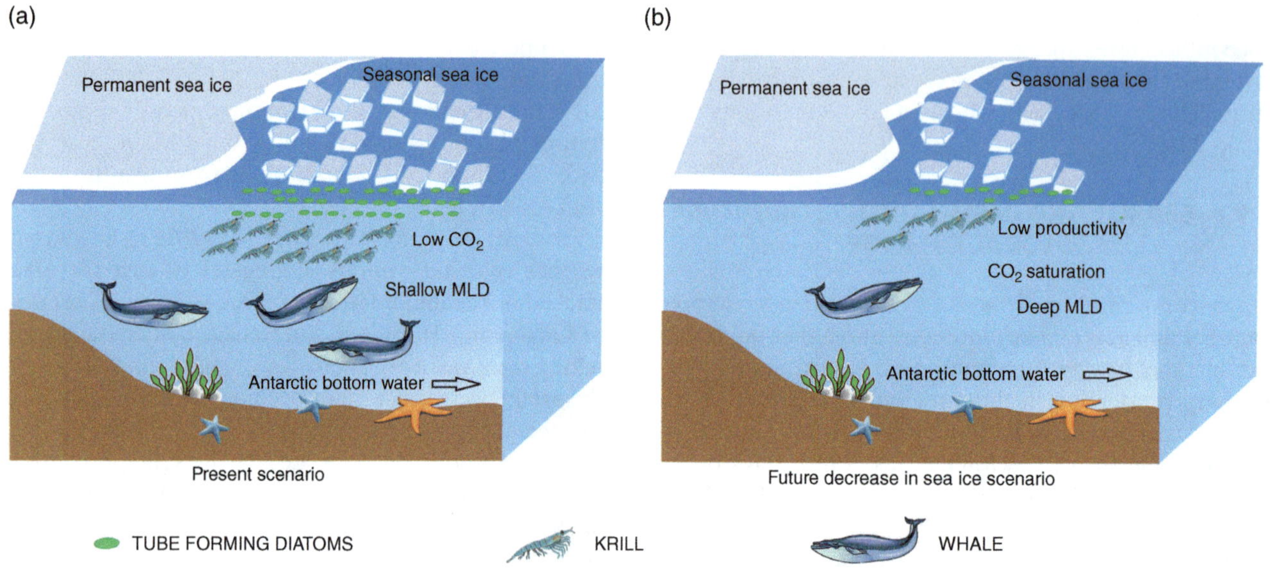

Figure 9.5 Schematic diagram showing the impact of sea ice melting on marine biota. MLD, mixed layer depth.

dissolution and impaired calcite–aragonite production. Continuous multidisciplinary research on the ISSO biogeochemical cycles and its response to global climate is required in order to better understanding the effects of warming, acidification, and coastal processes.

ACKNOWLEDGEMENTS

The authors are grateful to the Editors of the Book for inviting us to write a chapter. We are grateful to the Secretary, Ministry of Earth Sciences, Government of India, New Delhi and the Director, NCPOR, Goa, for their constant encouragement and support under the project 'Polar Micropaleontology and Past Climate'. S.P. would thanks the DST-INSPIRE, Government of India for providing funds for the project DST/INSPIRE/04/2015/001969. This is NCPOR contribution No J 55/2020–21.

REFERENCES

Abram, N., Gattuso, J.P., Prakash, A. et al. (2019). Framing and context of the report. *IPCC Special Report on the Ocean and Cryosphere in a Changing Climate* 73–129.

Anilkumar, N., Chacko, R., Sabu, P. et al. (2014). Biological response to physical processes in the Indian Ocean sector of the Southern Ocean, a case study in the coastal and oceanic waters. *Environmental Monitoring and Assessment 186*: 8109–8124. https://doi.org/10.1007/s10661–014–3990–4.

Anilkumar, N., George, J.V., Chacko, R. et al. (2015a). Variability of fronts, fresh water input and chlorophyll in the Indian Ocean sector of the Southern Ocean, New Zealand *Journal of Marine and Freshwater Research 49* (1): 20–40. https://doi.org/10.1080/00288330.2014.924972.

Anilkumar, N., Chacko, R., Sabu P. et al. (2015b), Freshening of Antarctic Bottom Water in the Indian Ocean Sector of Southern Ocean. *Deep Sea Research II 118* (B): 162–169. https://doi.org/10.1016j.dsr2.2015.03.009.

Annett, A.L., Carson, D.S., Crosta, X. et al. (2010). Seasonal progression of diatoms assemblages in surface waters of Ryder Bay, Antarctica. *Polar Biology 33*: 13–29. https://doi.org/10.1007/s00300–099–0681–7.

Aoki, S. (1997). Trends and interannual variability of surface layer temperature in the Indian Sector of the Southern Ocean observed by Japanese Antarctic Research Expeditions. *Journal of Oceanography 53*: 623–631.

Aoki, S., Yoritaka, M. and Masuyama, A. (2003). Multidecadal warming of subsurface temperature in the Indian Sector of the Southern Ocean. *Journal of Geophysical Research 108* (C4): 8081. https://doi.org/10.1029/2000JC000307.

Arblaster, J.M. and Meehl, G.A. (2006). Contributions of External Forcings to Southern Annular Mode Trends. *Journal of Climate 19*: 2896–2905. https://doi.org/10.1175/JCLI3774.1.

Armour, K.C., Marshall, J., Schott, J.R. et al. (2016). Southern Ocean warming delayed by circumpolar upwelling and equatorward transport. *Nature Geoscience 9*: 549–554. https://doi.org/10.1038/ngeo2731.

Arrigo, K.R., van Dijken, G. and Long, M. (2008a). Coastal Southern Ocean: A strong anthropogenic CO2 sink. *Geophysical Research Letters 35*: L21602. https://doi.org/10.1029/2008GL035624.

Arrigo, K.R., van Dijken, G.L. and Bushinsky, S., 2008b. Primary production in the Southern Ocean, 1997–2006. *Journal of Geophysical Research 113* (C8): C08004. https://doi.org/10.1029/2007JC004551.

Assmy, P., Smetacek, V., Montresor, M. et al. (2013). Thick-shelled, grazer-protected diatoms decouple ocean carbon and silica cycles in the iron-limited Antarctic Circumpolar Current. *PNAS 110* (51): 20633–20638. https://doi.org/10.1073/pnas.1309345110.

Auger, M., Morrow, R., Kestenare, E. et al. (2021). Southern Ocean *in-situ* temperature trends over 25 years emerge from interannual variability. *Nature Communications 12*: 514. https://doi.org/10.1038/s41467–020–20781–1.

Balch, W.M., Drapeau, D.T., Bowler, B.C. et al. (2011). The contribution of coccolithophores to the optical and inorganic carbon budgets during the Southern Ocean Gas Exchange Experiment: new evidence in support of the "Great Calcite Belt" hypothesis. *Journal of Geophysical Research 116*: C00F06. https://doi.org/10.1029.2011JC006941.

Ballerini, T., Hofmann, E.E., Ainley, D.G. et al. (2014). Productivity and linkages of the food web of the southern region of the western Antarctic Peninsula continental shelf. *Progress in Oceanography 122*: 10–29. https://dx.doi.org/10.1016/j.pocean.2013.11.007.

Banse, K. (1996). Low seasonality of low concentrations of surface chlorophyll in the Subantarctic water ring: underwater irradiance, iron, or grazing? *Progress In Oceanography 37* (3–4): 241–291. https://doi.org/10.1016/S0079–6611(96)00006–7.

Bates, N.R., Astor, Y.M., Church, M.J. et al. (2014). A time-series view of changing surface ocean chemistry due to ocean uptake of Anthropogenic CO_2 and ocean acidification. *Oceanography 27*: 126–141. https://doi.org/10.5670/oceanog.2014.16.

Beaufort, L., Probert, I., de Garidel-Thoron, T. et al. (2011). Sensitivity of coccolithophores to carbonate chemistry and ocean acidification. *Nature 476*, 80–83. https://doi.org/10.1038/nature10295.

Bednaršek, N., Tarling, G.A., Bakker, D.C.E. et al. (2012). Description and quantification of pteropod shell dissolution: a sensitive bioindicator of ocean acidification. *Global Change Biology 18*(7):2378–2388. https://doi.org/10.1111/j.1365–2486.2012.02668.x.

Bednaršek, N., Tarling, G.A., Bakker, D.C.E. et al. (2014). Dissolution Dominating Calcification Process in Polar Pteropods Close to the Point of Aragonite Undersaturation. *PLoS ONE 9* (10): e109183. https://doi.org/10.1371/journal.pone.0109183.

Belkin, I.M. and Gordon, A.L. (1996). Southern Ocean fronts from the Greenwich meridian to Tasmania, *Journal of Geophysical Research 101* (C2): 3675–3696. https://doi.org/10.1029/95JC02750.

Bernie, D., Lowe, J., Tyrrell, T. et al. (2010). Influence of mitigation policy on ocean acidification, *Geophysical Research Letters 37*: L15704. https://doi.org/10.1029/2010GL043181.

Beszteri, S., Thoms, S., Benes, V. et al. (2018). The response of three Southern Ocean phytoplankton species to ocean acidification and light availability: a transcriptomic study. *Protist 169* (6): 958–975. https://doi.org/10.1016/j.protis.2018.08.003.

Bindoff, N.L., Rosenberg, M.A. and Warner, M.J. (2000). On the circulation and water masses over the Antarctic continental slope and rise between 80 and 150oE. *Deep Sea Research II 47* (12–13): 2299–2326. https://doi.org/10.1016/S0967–0645 (00) 00038–2.

Bisson, M.A. and Kirst, G.O. (1995). Osmotic acclimation and turgor pressure regulation in algae. *Naturwissenschaften 82*: 461–471. https://doi.org/10.1007/BF01131597.

Boelen, P., van de Poll, W.H., van der Strate, H.J.et al. (2011). Neither elevated nor reduced CO_2 affects the photophysiological performance of the marine Antarctic diatom *Chaetoceros brevis*. *Journal of Experimental Marine Biology and Ecology 406*(2011):38–45. https://doi.org/10.1016/j.jembe.2011.06.012.

Boyd, P.W., Dillingham, P.W., McGraw, C.M. et al. (2016). Physiological responses of a Southern Ocean diatom to complex future ocean conditions. *Nature Climate Change 6*: 207–213. https://doi.org/10.1038/nclimate2811.

Boyd, P.W., Doney, S.C., Strzepek, R. et al. (2008). Climate-mediated changes to mixed-layer properties in the Southern Ocean: assessing the phytoplankton response. *Biogeosciences 5*: 847–864. https://doi.org/10.5194/bg-5–847–2008.

Boyd, P.W., Lennartz, S.T., Glover, D.M. et al. (2015). Biological ramifications of climate-change-mediated oceanic multi-stressors. *Nature Climate Change 5*: 71–79. https://doi.org/10.1038/nclimate2441.

Caldeira, K. and Duffy, P.B. (2000). The role of the Southern Ocean in uptake and storage of anthropogenic carbon dioxide. *Science 287*: 620–622. https://doi.org/10.1126/science.287.5453.620.

Caldeira, K. and Wickett, M.E. (2003). Anthropogenic carbon and ocean pH. *Nature 425*, 365. https://doi.org/10.1038/425365a.

Carrillo, C.J., Smith, R.C. and Karl, D.M. (2004). Processes regulating oxygen and carbon dioxide in surface waters west of the Antarctic Peninsula. *Marine Chemistry 84*: 161–179. https://doi.org/10.1016/j.marchem.2003.07.004.

Chacko, R., Nuncio, M., George, J.V. et al. (2014). Observational evidence of the southward transport of watermasses by eddies in the Indian sector of the Southern Ocean. *Current Science 107*: 1573–1581.

Clarke, A., Meredith, M.P., Wallace, M.I., Brandon, M.A., Thomas, D.N. (2008). Seasonal and interannual variability in temperature, chlorophyll and macronutrients in northern Marguerite Bay, Antarctica. *Deep Sea Research II 55* (18–19): 1988–2006. https://doi.org/10.1016/j.dsr2.2008.04.035.

Constable, A.J., Melbourne-Thomas, J., Corney, S.P. et al. (2014). Climate change and Southern Ocean ecosystems I: how changes in physical habitats directly affect marine biota, *Global Change Biology 20* (10): 3004–3025. https://doi.org/10.1111/gcb.12623.

Cook., A.J., Fox, A.J., Vaughan, D.G. et al. (2005). Retreating glacier fronts on the Antarctic Peninsula over the past half-century. *Science 308* (5721): 541–544. https://doi.org/10.1126/science.110423

de Salas, M.F., Eriksen, R., Davidson, A.T. et al. (2011). Protistan communities in the Australian sector of the Sub-Antarctic Zone during SAZ-Sense. *Deep Sea Research II 58* (21–22): 2135–2149. https://doi.org/10.1016/j.dsr2.2011.05.032.

Deutsch, C. and Weber, T. (2012). Nutrient ratios as a tracer and driver of Ocean Biogeochemistry, *Annual Review of Marine Science 4*: 113–141. https://doi.org/10.1146/annurev-marine-120709–142821.

DeVries, T., Holzer, M. and Primeau, F. (2017). Recent increase in oceanic carbon uptake driven by weaker upper-ocean overturning, *Nature 542*: 215–218. https://doi.org/10.1038/nature21068.

Donahue, K., Klaas, C., Dillingham, P.W. et al. (2019). Combined effects of ocean acidification and increased light intensity on natural phytoplankton communities from two

Southern Ocean water masses. *Journal of Plankton Research 41* (1): 30–45. https://doi.org/10.1093/plankt/fby048.

Doney, S.C., Fabry, V.J., Feely, R.A. et al. (2009). Ocean acidification: the other CO_2 problem. *Annual Review of Marine ScienceScience 1* (1): 169–192. https://doi.org/10.146/annurev.marine.010908.163834.

Dong, S., Sprintall, J. and Gille, S.T. (2006). Location of the polar front from AMSR-E satellite sea surface temperature measurements. *Journal of Physical Oceanography 36*: 2075–2089. https://doi.org/10.1175/JPO2973.1.

Dufour, C.O., Griffies, S.M., de Souza, G.F. et al. (2015). Role of mesoscale eddies in cross-frontal transport of heat and biogeochemical tracers in the Southern Ocean. *Journal of Physical Oceanography 45*: 3057–3081. https://doi.org/10.1175/JPO-D-14-0240.1.

Evans, C., Thomson, P.G., Davidson, A.T. et al. (2011). Potential climate change impacts on microbial distribution and carbon cycling in the Australian Southern Ocean. *Deep Sea Research Part II 58* (21–22): 2150–2161. https://doi.org/10.1016/j.dsr2.2011.05.019.

Eyre, B.D., Cyronak, T., Drupp, P. et al. (2018). Coral reef will transition to net dissolving before end of century. *Science 359* (6378): 908–911. https://doi.org/10.1126/science.aao1118.

Fabry, V.J., McClintock, J.B., Mathis, J.T. et al. (2009). Ocean acidification at high latitudes: the bellweather. *Oceanography 22* (4): 160–171. https://doi.org/10.5670/oceanog.2009.105.

Fabry, V.J., Seibel, B.A., Feely, R.A. et al. (2008). Impacts of ocean acidification on marine fauna and ecosystem processes. *ICES Journal of Marine Science 65* (3): 414–432. https://doi.org/10.1093/icesjms/fsn048.

Feely, R.A., Orr, J., Fabry, V.J. et al. (2009). Present and future changes in seawater chemistry due to ocean acidification. Section 3. In: *Carbon Sequestration and its Role in the Global Carbon Cycle. Geophysical Monograph Series* Vol. *83* (eds. B.J. McPherson and E.T. Sundquist), 175–188. Washington, DC: American Geophysical Union.

Fenchel, T. (2008). The microbial loop-25 years later. *Journal of Experimental Marine Biology and Ecology 366*: 99–103. https://doi.org/10.1016/j.jembe.2008.07.013.

Fiorini, S., Middelburg, J.J. and Gattuso, J.-P. (2011). Testing the effects of elevated PCO_2 on coccolithophores (Prymnesiophyceae): comparison between haploid and diploid life stages (1). *Journal of Phycology 47* (6): 1281–1291. https://doi.org/10.1111/j.1529-8817.2011.01080.x.

Frölicher, T.L., Sarmiento, J.L., Paynter, D.J. et al. (2015). Dominance of the Southern Ocean in Anthropogenic carbon and heat uptake in CMIP5 models. *Journal of Climate 28*: 862–886. https://doi.org/10.1175/JCLI-D-14-00117.1.

Fyfe, J.C. and Saenko, O.A. (2006). Simulated changes in the extratropical Southern Hemisphere winds and currents. *Geophysical Research Letters 33* (6): L06701. https://doi.org/10.1029/2005GL025332.

Gao, K. and Campbell, D.A. (2014). Photophysiological response of marine diatom to elevated CO_2 and decreased pH: a review, *Plant Biology 41* (5): 449–459. https://doi.org/10.1071/FP13247.

Gardner, J., Manno, C., Bakker, D.C.E. et al. (2018). Southern Ocean pteropods at risk from ocean warming and acidification. *Marine Biology 165*: 8. https://doi.org/10.1007/s00227-017-3261-3.

Gent, P. and Danabasoglu, G. (2011). Response to increasing Southern hemisphere winds in CCSM4, *Journal of Climate 24*: 4992–4998. https://doi.org/10.1175/JCLI-D-10-05011.1.

George, J.V., Anilkumar, N., Nuncio, M. et al. (2018). Upper layer diapycnal mixing and nutrient flux in the subtropical frontal region of the Indian sector of the Southern Ocean. *Journal of Marine Systems 187*: 197–205. https://doi.org/10.1016/j.jmarsys.2018.07.007.

Gille, S.T. (2002). Warming of the Southern Ocean since the 1950s. *Science 295* (5558): 1275–1277. https://doi.org/10.1126/science.1065863.

Gloersen, P., Campbell, W.J., Cavalieri, D.J. et al. (1993). Satellite passive microwave observations and analysis of Arctic and Antarctic sea ice. 1978–1987. *Annals of Glaciology 17*: 149–154. https://doi.org/10.3189/S0260305500012751.

Goeyens, L., Elskens, M., Catalano, G. et al. (2000). Nutrient depletions in the Ross Sea and their relation with pigment stocks. *Journal of Marine Systems 27* (1–3): 195–208. https://doi.org/10.1016/S0924-7963 (00) 00067-1.

Goffart, A., Catalano, G. and Hecq, J.H. (2000). Factors controlling the distribution of diatoms and Phaeocystis in the Ross Sea. *Journal of Marine Systems 27* (1–3): 161–175. https://doi.org/10.1016/S0924-7963 (00) 00065-8.

Goldman, J.A.L., Kranz, S.A., Young, J.N. et al. (2015). Gross and net production during the spring bloom along the West Antarctic Peninsula. *New Phytologist 205* (1): 182–191. https://doi.org/10.1111/nph.13125.

Goldman, J.C. (1988). Spatial and temporal discontinuities of biological processes in pelagic surface waters. In: *Towards a Theory On Biological–Physical Interactions in the World Oceans* (ed. B.J. Rothschild), 273–296. Dordrecht: Kluwer Academic.

Gray, A.R., Johnson, K.S., Bushinsky, S.M. et al. (2018). Autonomous biogeochemical floats detect significant carbon dioxide outgassing in the high-latitude Southern Ocean. *Geophysical Research Letters 45* (17): 9049–9057. https://doi.org/10.1029/2018GL078013.

Gruber, N., Gloor, M., Mikaloff Fletcher, S.E. et al. (2009). Oceanic sources, sinks, and transport of atmospheric CO_2. *Global Biogeochemical Cycles 23* (1): GB1005. https://doi.org/10.1029/2008G003349.

Hauri, C., Friedrich, T. and Timmermann, A. (2016). Abrupt onset and prolongation of aragonite undersaturation events in the Southern Ocean. *Nature Climate Change 6*: 172–176. https://doi.org/10.1038/nclimate2844.

Heiden, J.P., Bischof, K. and Trimborn, S. (2016). Light intensity modulates the response of two antarctic diatom species to ocean acidification. *Frontiers in Marine Science 3*: 260. https://doi.org/10.3389/fmars.2016.00260.

Henley, S.F., Tuerena,R.E., Annett, A.L. et al. (2017). Macronutrient supply, uptake and recycling in the coastal ocean of the west Antarctic Peninsula. *Deep Sea Research II 139*: 58–76. https://doi.org/10/1016/j.dsr2.2016.10.003.

Hoegh-Guldberg, O., Poloczanska, E.S., Skirving, W. et al. (2018). Coral reef ecosystem under climate change and ocean acidification. *Frontiers in Marine Science 4*: 158. https://doi.org/10.3389/fmars.2017.00158.

Holliday, N.P. and Read, J.F. (1998). Surface oceanic fronts between Africa and Antarctica. *Deep Sea Research I 45* (2–3): 217–238. https://doi.org/10.1016/S0967-0637 (97) 00081-2.

Hoogstraten, A., Timmermans, K.R. and deBar, H.J.W. (2012). Morphological and Physiological effects in *Proboscia Alata* (Bacillariophyceae) grown under different light and CO_2 conditions of the modern Southern Ocean. *Journal of Phycology 48* (3): 559–568. https://doi.org/10.1111/j.1529-8817.2012.01148.x.

Hoppe, C.J.M., Hassler, C.S., Payne, C.D. et al. (2013). Iron limitation modulates ocean acidification effects on Southern Ocean phytoplankton Communities. *PLOS ONE 8* (11): e70890. https://doi.org/10.1371/journal.pone.0079890.

Hoppema, M. (2004). Weddell Sea turned from source to sink for atmospheric CO_2 between pre-industrial time and present. *Global and Planetary Change 40* (3–4): 219–231. https://doi.org/10.1016/j.gloplacha.2003.08.001.

Houghton, R.A. (2007). Balancing the global carbon budget. *Annual Review of Earth and Planetary Science 35*: 313–347. http://doi.org/10.1146/annurev.earth.35.031306.140057.

Hughes, C.W. (2005). Nonlinear vorticity balance of the Antarctic Circumpolar Current. *Journal of Geophysical Research 110*: C11008.

IPCC (2019). Summary for Policymakers. In: *IPCC Special Report on the Ocean and Cryosphere in a Changing Climate* (eds. H.-O. Pörtner, D.C. Roberts, V. Masson-Delmotte et al.). IPCC.

Irie, T., Bessho, K., Findlay, H.S. *et al.* (2010). Increasing costs due to ocean acidification drives phytoplankton to be more heavily calcified: optimal growth strategy of coccolithophores. *PLoS ONE 5* (10): e13436. https://doi.org/10.1371/journal.pone.0013436.

Jena, B., Kumar, A., Ravichandran, M. et al. (2018). Mechanism of sea-ice expansion in the Indian Ocean sector of Antarctica: Insights from satellite observation and model reanalysis, *Journal of Physical Research 13* (10): e0203222. https://doi.org/10.1371/journal.pone.0203222.

Jones, B.M., Iglesias-Rodriguez, M.D., Skipp, P.J. et al. (2013). Response of the *Emiliania huxleyi* proteome to ocean acidification. *PLOS One 8* (4): e61868. https://doi.org/10.1371/journal.pone.0061868.

Jordan, R.W., Abe, K., Cruz, J. et al. (2016). Observations on the morphological diversity and distribution of two siliceous nanoplankton genera, *Hyalolithus* and *Petasaria. Micropaleontology 61* (6): 439–455.

Kemp, A.E.S., Pearce, R.B., Grigorov, I. et al. (2006). Production of giant marine diatoms and their export at oceanic frontal zones: Implications for Si and C flux from stratified oceans. *Global Biogeochemical Cycles 20* (4): GB4S04. https://doi.org/10.1029/2006GB002698.

Key, R.M., Kozyr, A., Sabine, C.L. et al. (2004). A global ocean carbon climatology: results from Global Data Analysis Project (GLODAP). *Global Biogeochemical Cycles 18* (4): GB4031. https://doi.org/10.1029/2004GB002247.

Kirchman, D.L. (2008). *Microbial Ecology of the Oceans*, 2e. Wiley-Blackwell. https://doi.org/10.1002/9780470281840.

Kostianoy, A.G., Ginzburg, A.I., Frankignoulle, M. et al. (2004). Fronts in the Southern Indian Ocean as inferred from satellite sea surface temperature data. *Journal of Marine Systems 45* (1–2): 55–73. https://doi.org/10.1016/j.jmarsys.2003.09.004.

Kroeker, K.J., Kordas, R.L., Crim, R. et al. (2013). Impacts of ocean acidification on marine organisms: quantifying sensitivities and interaction with warming, *Global Change Biology 19* (6): 1884–1896. https://doi.org/10.1111/gcb.12179.

Kumar, A., Srivastava, A., Yadav, J. et al. (2020). Spatio-temporal changes and prediction of Amery ice shelf, east Antarctica: A remote sensing and statistics-based approach. *Journal of Environmental Management 267*: 110648. https://doi.org/10.1016/j.jenvman.2020.110648.

Landschützer, P., Gruer, N., Haumann, F.A. et al. (2015). The reinvigoration of the Southern Ocean carbon sink. *Science 349* (6253): 1221–1224. https://doi.org/10.1126/science.aab2620.

Lannuzel, D., Schoemann, V., de Jong, J. et al. (2010). Distribution of dissolved iron in Antarctic sea ice: Spatial, seasonal, and inter-annual variability. *Journal of Geophysical Research 115* (G3): G03022. https://doi.org/10.1029/2009JG001031.

Le Quéré, C., Rödenbeck, C., Buitenhuis, E.T. et al. (2007). Saturation of the Southern Ocean CO_2 sink due to recent climate change. *Science 316* (5832): 1735–1738. https://doi.org/10.1126/science.1136188.

Lenton, A., Tilbrook, B., Law, R.M. et al. (2013). Sea-air CO2 fluxes in the Southern Ocean for the period 1990–2009. *Biogeosciences 10* (6): 4037–4054. https://doi.org/10.5194/bg-10-4037-2013.

Li, Y.L., Han, W.Q., Hu, A.X. et al. (2018). Multidecadal changes of the upper Indian Ocean heat content during 1965–2016. *Journal of Climate 31*: 7863–7884. https://doi.org/10.1175/JCLI-D-18-0116.1.

Lionard, M., Muylaert, K., Gansbeke, D.V. et al. (2005). Influence of changes in salinity and light intensity on growth of phytoplankton communities from the Schelde River and estuary (Belgium/The Netherlands). *Hydrobiologia 540*: 105–115. https://doi.org/10.1007/s10750-004-7123.

Liu, J. and Curry, J.A. (2010). Accelerated warming of the Southern Ocean and its impacts on the hydrological cycle and sea ice. *PNAS 107* (34): 14987–14992. https://doi.org/10.1073/pnas.100336107.

Locarnini, R.A., Mishonov, A.V., Baranova, O.K. et al. (2019). *World Ocean Atlas 2018*, Vol. 1: *Temperature*. NOAA Atlas. NESDIS 81.

Lovenduski, N.S., Gruber, N., Doney, S.C. (2008). Towards a mechanistic understanding of the decadal trends in the Southern Ocean carbon sink. *Global Biogeochemical Cycles 22* (3): GB3016. https://doi.org/10.1029/2007GB003139.

Lucas, C.H., Jones, D.O.B., Hollyhead, C.J. et al. (2014). Gelatinous zooplankton biomass in the global oceans: geographic variation and environmental drives. *Global Ecology and Biology 23* (7): 701–714. https://doi.org/10.1111/geb.12169.

Lutjeharms, J.R.E. and Ansorge, I.J. (2001). The Agulhas Return Current, *Journal of Marine Systems 30* (1–2): 115–138. https://doi.org/10.1016/S0924-7963 (01) 00041-0.

Maheshwari, M., Singh, R.K., Oza, S.R. et al. (2013). An investigation of the southern ocean surface temperature variability using long-term optimum interpolation SST data. *Oceanography 392632*: 1–9. https://doi.org/10.5402/2013/392632.

Marinov, I., Doney, S.C. and Lima, I.D. (2010). Response of ocean phytoplankton community structure to climate change over the 21st century: partitioning the effects of nutrients, temperature and light. *Biogeosciences 7*: 3941–3959. https://doi.org/10.5194/bg-7–3941–2010.

Marinov, I., Gnanadesikan, A., Toggweiler, J.R. et al. (2006). The Southern Ocean biogeochemical divide, *Nature 441*: 964–967. https://doi.org/10.1038/nature04883.

Marshall, G.J. (2003). Trends in the Southern Annular Mode from observations and reanalyses. *Journal of Climate 16*: 4134–4143. https://doi.org/10.1175/1520–0442(2003) 016<4134:TITSAM >2.0.CO;2.

Martin, J.H. (1990). Glacial-Interglacial CO_2 change: the Iron Hypothesis. *Paleoceanography and Paleoclimatology 5* (1): 1–13. https://doi.org/10.1029/PA005i001p00001.

Martin, P., van der Loeff, M.R., Cassar, N. et al. (2013). Iron fertilization enhanced net community production but not downward particle flux during the Southern Ocean iron fertilization experiment LOHAFEX. *Global Biogeochemical Cycles 27* (3): 871–881. https://doi.org/10.1002/gbc.200077.

Martinson, D.G. and McKee, D.C. (2012). Transport of warm Upper Circumpolar Deep Water onto the western Antarctic Peninsula continental shelf. *Ocean Science 8*: 433–442. https://doi.org/10.5194/os-8–433–2012.

Martinson, D.G., Stammerjohn, S.E., Lannuzzi, R.A. et al. (2008). Western Antarctic Peninsula physical oceanography and spatio-temporal variability. *Deep Sea Research II 55* (18–19): 1964–1987. https://doi.org/10.1016/j.dsr2.2008.04.038.

Massom, R.A., Scambos, T.A., Bennetts, L.G. et al. (2018). Antarctic ice shelf disintegration triggered by sea ice loss and ocean swell. *Nature 558*: 383–389. https://doi.org/10.1038/s41586–018–0212–1.

Matear, R.J. and Hirst, A.C. (1999). Climate change feedback on the future oceanic CO_2 uptake. *Tellus B 51*: 722–733. https://doi.org/10.3402/tellusb.v51i3.16472.

McNeil, B.I. and Matear, R.J. (2007). Climate change feedbacks on future oceanic acidification. *Tellus B 59*: 191–198. https://doi.org/10.1111/j.1600–0889.2006.00241.x.

McNeil, B.I. and Matear, R.J. (2008). Southern Ocean acidification: a tipping point at 450 p.p.m. atmospheric CO_2. *PNAS 105* (48): 18860–18864. https://doi.org/10.1073/pnas.0806318105.

Mendes C.R.B., Tavano V.M., Leal M.C. et al. (2013). Shifts in the dominance between diatoms and cryptophytes during three late summers in the Bransfield Strait (Antarctic Peninsula). *Polar Biology 36*: 537–547.

Meredith, M.P. and King, J.C. (2005). Rapid climate change in the ocean west of the Antarctic Peninsula during the second half of the 20th century. *Geophysical Research Letters 32* (19): L19604. https://doi.org/10.1029/2005GL024042.

Metzl, N. (2009). Decadal increase of oceanic carbon dioxide in Southern Indian Ocean surface waters (1991–2007). *Deep Sea Research II 56* (8–10): 607–619. https://doi.org/10.1016/j.dsr2.2008.12.007.

Metzl, N., Beauverger, C., Brunet, C. et al. (1991). Surface water carbon dioxide in the southwest Indian sector of the Southern Ocean: a highly variable CO_2 source/sink region in summer. *Marine Chemistry 35* (1–4): 85–95. https://doi.org/10.1016/S0304–4203 (09) 90010-X.

Metzl, N., Brunet, C., Jabaud-Jan, A. et al. (2006). Summer and winter air–sea CO_2 fluxes in the Southern Ocean, *Deep Sea Research I 53* (9): 1548–1563. https://doi.org/10.1016/j.dsr.2006.07.006.

Mikaloff Fletcher, S.E., Gruber, N., Jacobson, A.R. et al. (2006). Inverse estimates of anthropogenic CO_2 uptake, transport, and storage by the ocean. *Global Biogeochemical Cycles20*(2):GB2002.https://doi.org.10.1029/2005GB002530.

Mohan, R., Mergulhao, L.P., Guptha, M.V.S. et al. 2008. Ecology of coccolithophores in the Indian sector of the Southern Ocean. *Marine Micropaleontology 67*(1–2): 30–45.

Morrison, A.K., England, M.H. and Hogg, A.M. (2015). Response of Southern Ocean convection and abyssal overturning to surface buoyancy perturbations. *Journal of Climate 28*: 4263–4278.

Morrow, R., Valladeau, G. and Sallée, J.B. (2008). Observed subsurface signature of Southern Ocean sea level rise. *Progress in Oceanography 77* (4): 351–366. https://doi.org/10.1016/j.pocean.2007.03.002.

Mostofa, K.M.G., Liu, C.-Q., Zhai, W. et al. (2016). Reviews and synthesis: ocean acidification and its potential impacts on marine ecosystem. *Biogeosciences 13*: 1767–1786. https://doi.org/10.5194/bg-13–1767–2016.

Murphy, E.J., Watkins, J.L., Trathan, P.N. et al. (2006). Spatial and temporal operation of the Scotia sea ecosystem: a review of large-scale links in a krill centred food web. *Philosophical Transactions of the Royal Society B 362*:113–148. https://doi.org/10.1098/rstb.2006.1957.

Murphy, E.J., Cavanagh, R.D., Drinkwater, K.F. et al. (2016). Understanding the structure and functioning of polar pelagic ecosystems to predict the impacts of change. *Proceedings of the Royal Society B 283* (1844): 2161646. https://doi.org/10.1098/rspb.2016.1646.

Naik, R.K., George, J., Soares, M. et al. (2020). Observations of surface water phytoplankton community in the Indian Ocean: a transect from tropics to polar latitudes. *Deep Sea Research II 178*: 104848. doi:10.1016/j.dsr2.2020.104848.

Naik, R.K., George, J.V., Soares, M.A. et al. (2015). Phytoplankton community structure at the juncture of the Agulhas Return Front and Subtropical Front in the Indian Ocean sector of the Southern Ocean: bottom-up and top-down control. *Deep Sea Research II 118* (B): 233–239. https://doi.org/10.1016/j.dsr2.2015.01.002.

Negrete-García, G., Lovenduski, N.S., Hauri, C. et al. (2019). Sudden emergence of a shallow aragonite saturation horizon in the Southern Ocean. *Nature Climate Change 9*: 313–317. https://doi.org/10.1038/s41558–019–0418–8.

Nerem, R.S., Beckley, B.D., Fasullo, J.T. et al. (2018). Climate-change-driven accelerated sea-level rise detected in the altimeter era. *PNAS 115* (9): 2022–2025. https://doi.org/10.1073/pnas.1717312115.

Orr, J.C., Fabry, V.J., Aumont, O. et al. (2005). Anthropogenic ocean acidication over the twenty-first century and its impact on calcifying organisms. *Nature 437*: 681–686. https://doi.org/10.1038/nature04095.

Orsi, A.H., Whitworth III, T. and Nowlin, Jr., W.D. (1995). On the meridional extent and fronts of the Antarctic Circumpolar Current. *Deep-Sea Research I 42* (5): 641–673. https://doi.org/10.1016/0967 (95) 00021-W.

Oschlies, A., Pahlow, M., Yool, A. et al. (2010). Climate engineering by artificial ocean upwelling: channelling the sorcerer's apprentice. *Geophysical Research Letters 37* (4): L04701. https://doi.org/10.1029/2009GL041961.

Padin, X.A., Vázquez-Rodríguez, M., Castaño, M., et al. (2010). Air–sea CO2 fluxes in the Atlantic as measured during boreal spring and autumn. *Biogeosciences 7*: 1587–1606. https://doi.org/10.5194/bg-7–1587–2010.

Park, Y.-H., Park, T., Kim, T.W. et al. (2019). Observations of the Antarctic Circumpolar Current Over the Udintsev Fracture Zone, the narrowest choke point in the Southern Ocean. *Journal of Geophysical Research: Oceans 124* (7): 4511–4528. https://doi.org/10.1029/2019JC015024.

Patil, S., Mohan, R., Shetye, S. et al. (2013). Phytoplankton abundance and community structure in the Antarctic polar frontal region during austral summer of 2009. *Chinese Journal of Oceanology and Limnology 31*: 21–30. https://doi.org.10.1007/s00343–013–1309-x.

Patil, S., Mohan, R., Shetye, S. et al. (2014). *Prymnesium neolepis* (Prymnesiaceae): a siliceous Haptophyte from the Southern Indian Ocean. *Micropaleontology 60* (5): 475–481.

Patil, S., Mohan, R., Gazi, S., Shetye, S., Jafar, S.A. (2015). *Petasaria heterolepis* (Prymnesiaceae) from the Southern Indian Ocean. *Micropaleontology 61* (3): 171–176.

Patil, S.M., Mohan, R., Shetye, S.S. et al. (2017). Biogeographic distribution of extant Coccolithophores in the Indian sector of the Southern Ocean. *Marine Micropaleontology 137*: 16–30. https://doi.org/10.1016/j.marmicro.2017.08.002.

Patil, S.M., Mohan, R., Shetye, S.S. et al. (2020). Interannual changes of austral summer coccolithophore assemblages and southward expanse in the Southern Indian Ocean. *Deep Sea Research II 178*: 104765. https://doi.org/10.1016/j.dsr2.2020.104765.

Petrou, K., Baker, K.G., Nielsen, D.A. et al. (2019). Acidification diminishes diatom silica production in the Southern Ocean. *Nature Climate Change 9*: 781–786. https://doi.org/10.1038/s41558–019–0557-y.

Petrou, K., Kranz, S.A., Trimborn, S. et al. (2016). Southern Ocean phytoplankton physiology in a changing climate. *Journal of Plant Physiology 203*: 135–150. https://doi.org/10.1016/j.jplph.2016.05.004.

Pollard, R., Tréguer, P. and Read, J. (2006). Quantifying nutrient supply to the Southern Ocean. *Journal of Geophysical Research 111* (C5): C05011. https://doi.org/10.1029/2005JC003076.

Pollard, R.T., Lucas, M.I. and Read, J.F. (2002). Physical controls on biogeochemical zonation in the Southern Ocean. *Deep Sea Research 49*: 3289–3305. https://doi.org/10.1016/S0967–0645 (02) 00084-X.

Pollard, R.T., Venables, H.J., Read, J.F. et al. (2007). Large scale circulation around the Crozet Plateau controls an annual phytoplankton bloom in the Crozet Basin. *Deep Sea Research II 54* (18–20): 1915–1929. https://doi.org/10.1016/j.dsr2.2007.06.012.

Prasanna, K., Ghosh, P. and Anil Kumar, N. (2015). Stable isotopic signature of Southern Ocean deep water CO_2 ventilation. *Deep Sea Research II 118* (B): 177–185. https://doi.org.10.1016/j.dsr2.2015.04.009.

Prézelin, B.B., Hofmann, E.E., Mengelt, C. et al. (2000). The linkage between Upper Circumpolar Deep Water (UCDW) and phytoplankton assemblages on the west Antarctic Peninsula continental shelf. *Journal of Marine Research 58*: 165–202. https://doi.org/10.1357/002224000321511133.

Purkey, S.G. and Johnson, G.C. (2010).Warming of global abyssal and deep Southern Ocean waters between the 1990s and 2000s: Contributions to global heat and sea level rise budgets. *Journal of Climate 23*: 6336–6351.

Purkey, S.G., and Johnson, G.C. (2013). Antarctic Bottom Water warming and freshening: Contributions to sea level rise, ocean freshwater budgets, and global heat gain. *Journal of Climate 26*: 6105–6122.

Rajakumar, A., Alagarsamy, R., Khare, N., Saraswat, R., Subramaniam, M.M. (2008). Studies on the nutrient distribution in the Southern Ocean waters along the 45oE transect. Indian Journal of Marine*Science 37* (4): 424–429.

Read, J. and Pollard, R. (2017). An introduction to the physical oceanography of six seamounts in the southwest Indian Ocean, *Deep Sea Research II 136*: 44–58. https://doi.org/10.1016/j.dsr2.2015.06.022.

Riaux-Gobin, C., Poulin, M., Dieckmann, G. et al. (2011). Spring phytoplankton onset after the ice break-up and sea-ice signature (Adélie Land, East Antarctica). *Polar Research 30* (1): 5910. https://doi.org/10.3402/polar.v30i0.5910.

Riebesell, U., Zondervan, I., Rost, B. et al. (2000). Reduced calcification of marine plankton in response to increased atmospheric CO_2. *Nature 407*: 364–367. https://doi.org/10.1038/35030078.

Rignot, E., Mouginot, J., Scheuchl, B. et al. (2019). Four decades of Antarctic Ice sheet mass balance from 1979 to 2017. *PNAS 116*. 1095–1103. https://doi.org/10.1073/pnas.1812883116.

Rintoul, R.S (2009). Antarctic circumpolar current. In: *Encyclopedia of Ocean Sciences* (eds. J. Steel, S. Thorpe and K.K. Turekian, K.K.), 178–190. Oxford: Oxford Academic Press.

Rintoul, S.R., Hughes, C. and Olbers, D. (2001). The Antarctic Circumpolar System. In: *Ocean Circulation and Climate* (eds. G. Siedler, J. Church and J. Gould), 271–302. Academic Press.

Sabine, C.L., Feely, R.A., Gruber, N. et al. (2004). The Oceanic Sink for Anthropogenic CO_2. *Science 305* (5682): 367–371. https://doi.org/10.1126/science.1097403.

Sabu, P., George, J.V, Anilkumar, N. et al. (2015). Observations of watermass modification by mesoscale eddies in the southern Subtropical Indian Ocean. *Deep Sea Research II 118* (B): 152–161.

Sallée, J.-B. (2018). Southern Ocean warming. *Oceanography 31* (2). https://doi.org/10.5670/oceanog.2018.215.

Sarmiento, J.L. and Le Quere, C. (1996). Oceanic carbon dioxide uptake in a model of century scale global warming. *Science 274* (5291): 1346–1350. https://doi.org/10.1126/science.274.5291.1346.

Sarmiento, J.L. and Orr, J.C. (1991). Three-dimensional simulations of the impact of Southern Ocean nutrient depletion on atmospheric CO_2 and ocean chemistry. *Limnology and Oceanography 36* (8): 1928–1950. https://doi.org/10.4319/lo.1991.36.8.1928.

Sarmiento, J.L., Slater, R., Barber, R. et al. (2004). Response of Ocean Ecosystems to *Climate Warming. 18* (3): GB3003. https://doi.org.10.1029/2003GB002134.

Schlitzer, R. (2015). Data analysis and visualization with Ocean Data View. *CMOS Bulletin SCMO 43* (1): 9–13.

Schlosser, E., Haumann, F.A. and Raphael, M.N. (2018). Atmospheric influences on the anomalous 2016 Antarctic sea ice decay. *The Cryosphere 12*: 1103–1119. https://doi.org/10.5194/tc-12–1103–2018.

Schofield, O., Saba, G., Coleman, K. et al. (2017). Decadal variability in coastal phytoplankton community composition in a changing West Antarctic Peninsula. *Deep Sea Research Part I 124*, 42–54. https://doi.org/10.1016/j.dsr.2017.04.014.

Schulz, K.G., Hartley, S. and Eyre, B. (2019). Upwelling amplifies ocean acidification on the east Australian Shelf: implications for marine ecosystem. *Frontiers in Marine Science 6*: 636. https://doi.org/10.3389/fmars.2019.00636.

Scott, F.J. and Marchant, H.J. (2005). *Antarctic Marine Protists*. Canberra: Australian Biological Resources Study.

Serebrennikova, Y.M. and Fanning, K.A. (2004). Nutrients in the Southern Ocean GLOBEC region: variations, water circulation, and cycling, *Deep Sea Research II 51* (17–19): 1981–2002. https://doi.org/10.1016/j.dsr2.2004.07.023.

Shepherd, A., Ivins, E., Rignot, E. et al. (2018). Mass balance of the Antarctic Ice Sheet from 1992 to 2017. *Nature 558*: 219–212. https://doi.org/10.1038/s41586-018-0179-y.

Shetye, S., Mohan, R., Jafar, S.A. et al. (2014). Diatom burst-driven silica depletion under the Antarctic sea ice: evidence from sponge spicules. *Current Science 107* (2): 273–277.

Shetye, S.S., Mohan, R., Patil, S. et al. (2015). Oceanic pCO_2 in the Indian sector of the Southern Ocean during the austral summer-winter transition phase. *Deep Sea Research II 118* (B): 250–260. https://doi.org/10.1016/j.dsr2.2015.05.017.

Shetye, S.S., Mohan, R., Patil, S. et al. (2019). Hidden biogeochemical anonymities under Antarctic fast ice. *Regional Studies in Marine Science 31*: 100789. https://doi.org/10.1016/j.rsma.2019.100789.

Shi, J.R., Xie, S.P. and Talley, L.D. (2018). Evolving relative importance of the Southern Ocean and north Atlantic in anthropogenic ocean heat uptake, *Journal of Climate 31*: 7459–7479. https://doi.org/10.1175/JCLI-D-18-0170.1.

Siegert, M., Atkinson, A., Banwell, A. et al. (2019). The Antarctic Peninsula Under a 1.5°C Global Warming Scenario. *Frontiers in Environmental Science 7*: 102. https://doi.org/10.3389/fenvs.2019.00102.

Sigman, D.M. and Boyle, E.A. (2000). Glacial/interglacial variations in atmospheric carbon dioxide. *Nature 407*: 859–869. https://doi.org/10.1038/35038000.

Smith, Jr, W.O. and Comiso, J.C. (2008). Influence of sea ice on primary production in the Southern Ocean: a satellite perspective. *Journal of Geophysical Research 113* (C5): C05S93. https://doi.org/10.1029/2007JC004251.

Smith, H.E.K., Tyrrell, T., Charalampopoulou, A. et al. (2012). Predominance of heavily calcified coccolithophores at low $CaCO_3$ saturation during winter in the Bay of Biscay. *PNAS 109* (23): 8845–8849. https://doi.org/10.1073/pnas.1117508109.

Smith, W.O. and Nelson, D.M. (1985). Phytoplankton bloom produced by a receding ice edge in the Ross Sea: spatial coherence with the density field. *Science 227* (4683): 163–166. https://doi.org/10.1126/science.227.4683.163.

Smith, W.O.J. and Gordon, L.I. (1997). Hyperproductivity of the Ross Sea (Antarctica) polynya during austral spring. *Geophysical Research Letters 24*: 233–236. https://doi.org/10.1029/96GL03926.

Soares, M.A., Bhaskar, P.V., Naik, R.K. et al. (2015). Latitudinal δ13C and δ15N variations in particulate organic matter (POM) in surface waters from the Indian Ocean sector of Southern Ocean and the Tropical Indian Ocean in 2012. *Deep Sea Research II 118* (B): 186–196. https://doi.org/10.1016/j.dsr2.2015.06.009.

Sparrow, M.D., Heywood, K.J., Brown, J. et al. (1996). Current structure of the south Indian Ocean, *Journal of Geophysical Research 101* (C3): 6377–6391. https://doi.org/10.1029/95JC03750.

Stammerjohn, S.E., Martinson, D.G., Smith, R.C. et al. (2008). Trends in Antarctic annual sea ice retreat and advance and their relation to El Niño-Southern Oscillation and Southern Annular Mode variability. *Journal of Geophysical Research 113* (C3): C03S90. https://doi.org/10.1029/2007JC004269.

Stark, J.S., Roden, N.P., Johnstone, G.J. et al. (2018). Carbonate chemistry of an *in-situ* free-ocean CO_2 enrichment experiment (antFOCE) in comparison to short term variation in Antarctic coastal waters. *Scientific Reports 8*: 2816. https://doi.org/10.1038/s41598-018-21029-1.

Stocker, T.F., Qin, D. and Plattner, G.K. (2013). *Observations: Ocean. Climate Change 2013: the Physical Science Basis. Contribution of Working Group I to the 5th Assessment Report of the Intergovernmental Panel on Climate Change*. Cambridge: Cambridge University Press.

Swart, N.C. and Fyfe, J.C. (2012). Observed and simulated changes in the Southern Hemisphere surface westerly wind-stress. *Geophysical Research Letters 39* (16): L16711. https://doi.org/10.1029/2012GL052810.

Swart, N.C., Gille, S.T., Fyfe, J.C. et al. (2018). Recent Southern Ocean warming and freshening driven by greenhouse gas emissions and ozone depletion. *Science Nature Geoscience 11*: 836–841. https://doi.org/10.1038/s41561-018-0226-1.

Swift, J. H. and Orsi, A.H., 2012, Sixty-four days of hydrography and storms: *RVIB Nathaniel B. Palmer*'s 2011 S04P cruise. *Oceanography 25*: 54–55. https://doi.org/10.5670/oceanog.2012.74.

Takahashi, T., Sutherland, S.C., Wanninkhof, R. et al. (2009). Climatological mean and decadal change in surface ocean pCO_2, and net sea-air CO_2 flux over the global oceans. *Deep Sea Research Part II 56* (8–10): 554–577. https://doi.org/10.1016/j.dsr2.2008.12.009.

Thomas, D.N. and Dieckmann, G.S. (2002). Antarctic Sea Ice – a habitat for extremophiles. *Science 295* (5555): 641–644. https://doi.org/10.1026/science.1063391.

Thompson, D.W.J., Solomon, S., Kushner, P.J. et al. (2011). Signatures of the Antarctic ozone hole in Southern Hemisphere surface climate change. *Nature Geoscience 4*: 741–749. https://doi.org/10.1038/NGEO1296.

Thompson, A.F. Stewart, A.L., Spence P. et al. (2018). The Antarctic Slope Current in a changing climate. *Reviews of Geophysics 56*: 741–770. https://doi.org/10.1029/2018RG000624.

Torstensson, A., Hedblom, M., Andersson, J. et al. (2013). Synergism between elevated pCO_2 and temperature on the Antarctic sea ice diatom *Nitzschia lecointei*. *Biogeosciences 10*: 6391–6401. https://doi.org/10.5194/bg-10-6391-2013.

Torstensson, A., Hedblom, M., Björk, M.M. et al. (2015). Long-term acclimation to elevated pCO_2 alters carbon metabolism and reduces growth in the Antarctic diatom *Nitzschia lecointei*. *Proceedings of the Royal Society B 282*: 20151513. https://doi.org/10.1098/rspb.2015.1513.

Trimborn, S., Brenneis, T., Sweet, E. et al. (2013). Sensitivity of Antarctic phytoplankton species to ocean acidification: growth, carbon acquisition, and species interaction. *Limnology and Oceanography 58* (3): 997–1007. https://doi.org/10.4319/Io.2013.58.3.0997.

Trimborn, S., Thoms, S., Brenneis, T. et al. (2017). Two Southern Ocean diatoms are more sensitive to ocean acidification and changes in irradiance than the prymnesiophyte *Phaeocystis antarctica*. *Physiologia Plantarum 160* (2): 155–170. https://doi.org/10.1111/ppl.12539.

Venables, H.J., Clarke, A. and Meredith, M.P. (2013). Wintertime controls on summer stratification and productivity at the western Antarctic Peninsula. *Limnology and Oceanography 58* (3): 1035–1047. https://doi.org/10.4319/Io.2013.58.3.1035.

Vernet, M., Martinson, D., Iannuzzi, R. et al. (2008). Primary production within the sea-ice zone west of the Antarctic Peninsula: I – sea ice, summer mixed layer, and irradiance. *Deep Sea Research II 55*: 2068–2085. https://doi.org/10.1016/j.dsr2.2008.05.021.

Wang, S., Bailey, D., Lindsay, K. et al. (2014). Impact of sea ice on the marine iron cycle and phytoplankton productivity, *Biogeosciences 11*: 4713–4731. https://doi.org/10.5194/g-11-4713-2014.

Wetzel, P., Winguth, A. and Maier-Reimer, E. (2005). Sea-to-air CO_2 flux from 1948 to 2003: a model study, *Global Biogeochemical Cycles 19* (2): GB2005. https://doi.org/10.1029/2004GB002339.

Whitworth, III, T. and Nowlin, Jr., W.D. (1987). Water masses and currents of the Southern Ocean at the Greenwich Meridian. *Journal of Geophysical Research 92* (C6): 6462–6476. https://doi.org/10.1029/JC092iC06p06462.

Williams, G.D., Bindoff, N.L., Marsland, S.J., Rintoul, S.R. (2008). Formation and export of dense shelf water from the Adélie Depression, East Antarctica, *Journal of Geophysical Research:* Oceans, *114* (C4): C04039. https://doi.org/10.1029/2007/C004346.

Wouters, B., Martin-Español, A., Helim, V. et al. (2015). Dynamic thinning of glaciers on the Southern Antarctic Peninsula. *Science 348* (6237): 899–903. https://doi.org/10.1126/science.aaa5727.

Xu, K., Fu, F.-X. and Hutchins, D.A. (2014). Comparative responses of two dominant Antarctic phytoplankton taxa to interactions between ocean acidification, warming, irradiance, and iron availability. *Limnology and Oceanography 59* (6): 1919–1931. https://doi.org/10.4319/Io.2014.59.6.1919.

Zondervan, I., Rost, B. and Riebesell, U. (2002). Effect of CO_2 concentration on the PIC/POC ratio in the coccolithophore *Emiliania huxleyi* grown under light-limiting conditions and different daylengths. *Journal of Experimental Marine Biology and Ecology 272* (1): 55–70. https://doi.org/10.1016/S0022 0981 (02) 00037–0.

10

Benthic Biome of the Southern Ocean: Present State of Knowledge and Future Perspectives

Moumita Bhowmik[1], Sumit Mandal[1]*, and Sarat C. Tripathy[2]

ABSTRACT

The Southern Ocean (SO) benthic biodiversity is a reflection of major geological events and climate change. This remote marine realm, dominated by glacial cycles and the Antarctic Circumpolar Current (ACC), is home to diverse benthic communities. The peculiar biotic and abiotic conditions prevailing in the SO have steered its benthic components towards key ecological adaptations. The disjunctive pattern witnessed in the distribution of the benthic fauna has resulted from vicariance during the separation of Gondwana. The high endemism found in the SO and Antarctic benthic assemblages have resulted from their evolution in long isolation. Moreover, the thermal gradient in water current acts as a determinant of the dispersal and connectivity of the species lineages. Owing to distinct trait modalities, the responses of SO benthic species to environmental changes differ considerably from those of species living in other ecosystems. The SO biome is prone to habitat alteration mediated by global climate change, which in turn has a cascading effect on the benthic biota, from the level of the genes to that of the community. It is therefore crucial to understand the benthic responses to climate change through the networks of biotic interactions. Our understanding of the benthic biodiversity of the SO and Antarctic region has improved immensely since the advent and application of modern molecular biological techniques. In addition, the Convention for Conservation of Antarctic Marine Living Resources has strengthened the conservation policies by designating and prioritizing important ecological zones as Marine Protected Areas and Vulnerable Marine Ecosystems. However, more integrated ecosystem studies and networking models are required to fill the knowledge gap and improve our understanding of the benthic biome of the SO and Antarctic regions. This review summarizes the current state of knowledge of the benthic biodiversity of the SO, appraises the evolutionary history that shaped the endemism of the SO benthos, and finally points out the potential threats of climate change on the biogeochemistry and ecosystem functioning in the SO.

10.1. INTRODUCTION

Antarctica was the last continent to be discovered on Earth and many of its parts, including the surrounding Southern Ocean (SO), are still largely unexplored (De Broyer and Koubby, 2014). The majority of it is underlain by deep sea, with a very small proportion of continental shelves. In the SO, the deep sea and abyssal

[1]*Marine Ecology Laboratory, Department of Life Sciences, Presidency University, 86/1, College Street, Kolkata, 700073, India*
[2]*ESSO-National Centre for Polar and Ocean Research, Ministry of Earth Sciences, Headland Sada, Vasco-da-Gama, Goa, 403804, India*
**Corresponding Author: Sumit Mandal (sumit.dbs@presiuniv.ac.in) ORCiD code: 0000-0002-2147-9466*

Systems Biogeochemistry of Major Marine Biomes, First Edition. Edited by Aninda Mazumdar and Wriddhiman Ghosh.

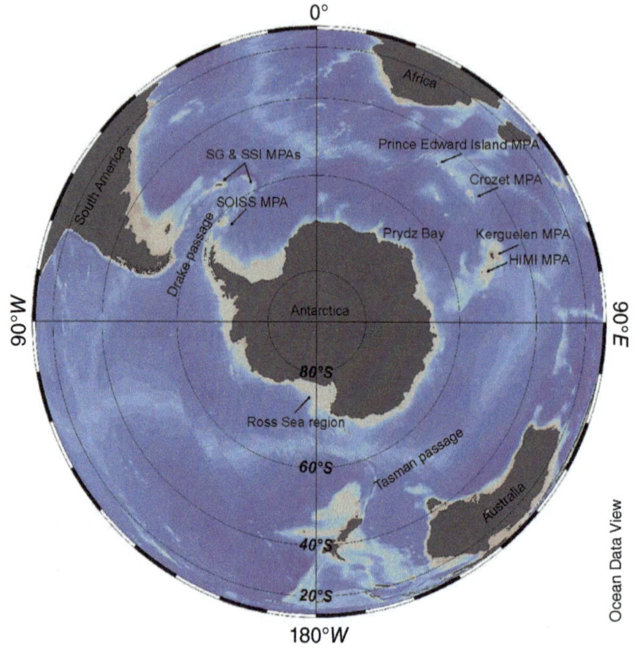

Figure 10.1 Map of Southern Ocean and Antarctica indicating the location of the Marine Protected Areas (MPAs). The map was taken from the open source web based utility C (https://odv.awi.de/) with permission from Ocean Data View.

area covers 34.8 million km^2 and 27.9 million km^2, respectively (Clark and Johnston, 2003), and include seven marine protected areas (MPA; Figure 10.1). It comprises three major ocean basins (Pacific, Indian, and Atlantic Ocean) that exceed depths of 3000 m, separated by submarine ridges, plateaus and chain of islands (Post et al., 2014). Water currents of the SO not only shapes the Antarctica and its marine ecosystems but also plays a crucial role in global ocean circulation (Griffiths, 2010). As the SO is spatially continuous around the globe, Antarctica and its coastal seas are largely insulated from the warmer lower latitudes by the enormous extent of winter sea ice present in the SO along with large ice shelves and grounded ice below the sea level (Figure 10.2). Turnover of water mass in the SO provides ventilation to the ocean realm by subducting surface waters, which transports atmospheric oxygen and CO_2 into the deep waters, and renewing the surface layers with upwelled nutrient-rich deep water (Post et al., 2014). The well-characterized circum-Antarctic oceanographic realm denotes the northernmost degree of cold surface water (Clarke et al., 2004). The Antarctic circumpolar current (ACC) is the principal circulation feature of the SO, which is stretched unbroken around the globe and serves as a unique global link to all other oceans. Antarctica is surrounded by cold deep waters accompanied by the strong ACC, steep gradients of temperature, and abundance and distribution of phytoplankton and zooplankton that collectively form a strong ecological barrier around Antarctica (Knox, 1994). The southern and northern boundaries of the SO have made it the most prominently demarcated marine ecosystem on Earth.

Benthos are the most biodiverse components of the ocean biome and they play crucial role in ecosystem functioning by rendering key processes such as bioturbation and nutrient exchange and carbon flux across sediment–water interfaces (Figure 10.3). They are also used as potential bio-indicators for several coastal marine habitats. The benthic communities of the SO are characterized by unique species assemblages and life history traits, but although it has been a century since the inception of benthic ecology research in Antarctica (Clarke, 2008), their functional characteristics in regulating ecosystem patterns and processes in the SO are poorly understood. The SO fronts are some of the most effective barriers of the global ocean, especially for organisms inhabiting distinct water depths and temperatures (Clarke et al., 2005; Gutt et al., 2015). These fronts also act as strong barriers to north–south water exchange and in so doing represent the most conspicuous discontinuity in marine biogeography. Certain areas that have undergone prolonged isolation usually have unique biotic composition and the SO is not an exception. Hence, it can be considered as a hotspot of species turnover, invasion, and local extinction (Griffiths et al., 2017). The deep-sea segment of the SO is specious and its distinguished habitats harbor more than 8200 nominal species with over 4100 benthic species (Clark and Johnston 2003; Griffiths, 2010; Halanych and Mahon, 2018). Shallow

(a) (b) (c) (d) (e) (f)

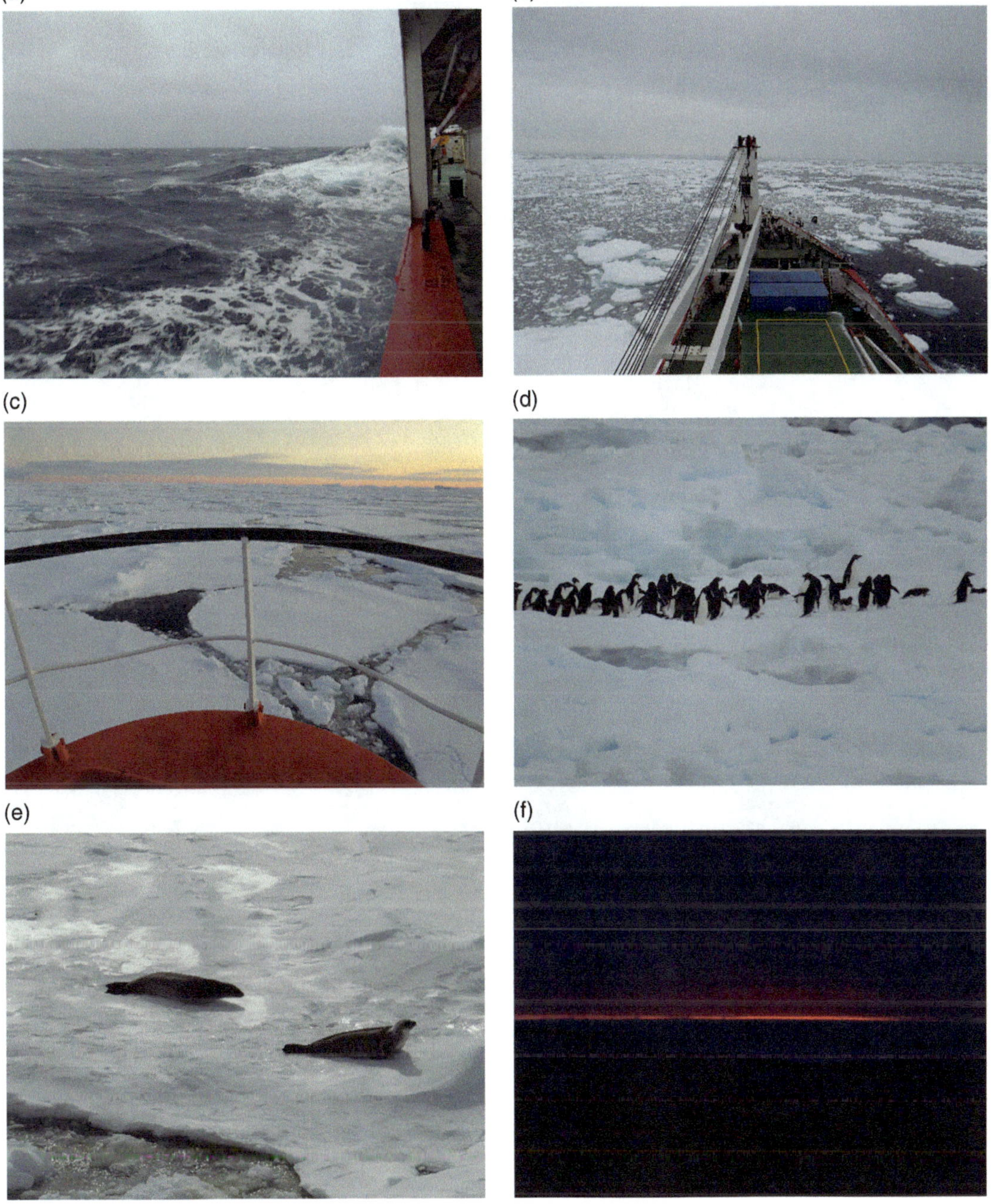

Figure 10.2 Surface physiography of the Southern Ocean (SO) and Antarctica. (a) Roaring 40s; (b,c) Floating sea ice of SO (~66°S); (d,e) Polar animals in the SO (~66°S); (f) Midnight sun in Antarctica. The pictures were taken by Sumit Mandal during the 11th Indian Expedition to SO, conducted by National Centre for Polar and Ocean Research (NCPOR) under the aegis of Ministry of Earth Sciences (MoES), Government of India (09.01.2020–08.03.2020). Not to scale. *Source:* Sumit Mandal

marine benthic species of the SO underwent adaptive radiation and thus show considerably high diversity with gigantism, longevity, slow growth, and late maturity (Brandt et al., 2007a). Arntz et al., (1994) delineated several factors such as temperature, extreme seasonality, and lack of terrestrial input, which shape the patterns and processes of the benthic ecosystems of the SO and Antarctic. Global warming mediated climate change, UV exposure, and ocean acidification have wreaked havoc on the biodiversity of this region, primarily because the native

Figure 10.3 Typical box core sediment sample collected fromthe Southern Ocean (SO) sea floor. (a,b) Box core collected benthic sample; (c,d) Close-up view of some macrobenthic organisms. The pictures were taken by Sumit Mandal during the 11th Indian Expedition to the SO, conducted by National Centre for Polar and Ocean Research (NCPOR) under the aegis of Ministry of Earth Sciences (MoES), Government of India (09.01.2020–08.03.2020). *Source:* Sumit Mandal

animals are unique and have less potential for resilience (De Broyer and Koubbi, 2014; Gutt et al., 2015). The history of deep-sea exploration in the SO dates back to 1872, starting from the expedition named "Challenger", and through the subsequent expeditions including ANDEEP I–III, considerable proportion of the SO benthic biodiversity has been revealed to date. Gradual development of institutions such as the Census of Marine Life Programme (CAML) and the Scientific Committee on Antarctic Research (SCAR) has further strengthened our knowledge on the SO biome. Furthermore, the Commission for the Conservation of Antarctic Marine Living Resources (CCAMLR) has taken extensive initiatives in conserving the SO ecosystem by adopting several MPAs and managing Vulnerable Marine Ecosystems (VME). The CCAMLR has set global benchmarks to operationalize ecosystem approaches using marine living resources. In this scenario, the present review synthesizes the knowledge available on the benthic biodiversity of the SO and the Antarctic coasts, and in so doing delineates its biogeographic patterns, dispersal potentials, and endemism. The implications of climate change and anomalous abiotic drivers (that prevail in the Anthropocene) on the benthic fauna and their habitat loss are also discussed.

10.2. BIOGEOCHEMISTRY OF THE SOUTHERN OCEAN

10.2.1 The High-Nutrient Low-Chlorophyll Zone and Productivity

The SO, the world's largest high-nutrient low-chlorophyll (HNLC) water body, plays a pivotal role in the biogeochemistry of the global ocean, and therefore climate regulation, by acting as a major sink for atmospheric carbon dioxide and enabling the redistribution of nutrients across the marine realm (Chisholm et al., 2001; Sigman et al., 2010). Various factors are responsible for the HNLC condition of the SO, among which lack of micronutrients (especially iron) is considered to be the foremost (De Baar et al., 1999), while the high grazing activity of zooplankton and limitation of sunlight and silicates also contribute to low phytoplankton production in the SO (Blain et al., 2007). Despite the HNLC characteristic, spatiotemporally variable phytoplankton blooms do occur in the surface waters of the SO (Moore et al., 1999; Moore and Doney, 2006; Tomalla et al., 2011; Sabu et al., 2014; Tripathy et al., 2014), particularly in regions where sources of biologically available trace nutrients (especially iron) are significant. It has been observed that, during the austral summers (December and January), the SO can contribute up to 60% of the global ocean's phytoplankton bloom. This is triggered mainly by iron inputs from Patagonia (dust advection) and regional physical processes such as upwelling and advection (Tripathy and Jena, 2019). Owing to geographical intricacies such as the presence of various frontal features, bottom topography, and varying advance and retreat of sea ice, the the SO cannot be taken as a single ecosystem, and is far more heterogeneous than previously considered (Cochlan, 2008). Accordingly, productivity across the the SO varies over space and time, with the frontal and coastal regions being generally more productive than the rest of the territories. Climate change-induced perturbations in the productivity and distribution of phytoplankton, zooplankton, and microbial and benthic species are likely to have an impact on overall ocean biogeochemistry (nutrient cycling, food web, carbon transfer and/or export, and benthic–pelagic coupling), including that of the SO, in the coming years (Henley et al., 2020).

10.2.2. Antarctic Circumpolar Current

The ACC is the largest ocean current encircling the Antarctica, thereby keeping the warm water mass away from the icy continent. The boundaries of the ACC are generally defined by zonal variations in certain water properties such as temperature and salinity (Gordon et al., 1977). The average temperature of ACC varies between −1 and 5°C depending on season and location. Average surface salinity decreases pole-ward, generally from 34.9 at 35°S to 34.7 at 65°S, with typical salinity values of 33.5–34.7 at 65°S (Orsi et al., 1995). The ACC connects the Pacific, Atlantic, and Indian oceans, and acts as the main avenue for the exchange of water mass, heat, nutrients, etc. among them. The eastward flow of the ACC is driven by strong westerly winds and is associated with several frontal zones such as the Sub-Antarctic Front (SAF), the Polar Front (PF), and the Southern ACC Front (SACC). The frontal regions are characterized as the zones of nutrients upwelling, so they support high productivity. Upwelling of subsurface water under the sea-ice brings considerable amounts of nutrients (macro and micro), enabling enhanced phytoplankton production. As the sea ice recedes and the water is exposed to more solar irradiance, epontic (attached under the ice) phytoplankton start growing during the first phase of the bloom followed by a succession of diatoms when the ice melts further south (Miller, 2004). Studies indicate that the ACC is usually dominated by diatoms with the sporadic presence of coccolithophores and silicoflagellates; furthermore, the territory to the north of the PF is dominated by smaller phytoplanktons (dinoflagellates and flagellates), whereas the area south of the PF is predominated by larger phytoplanktons such as the diatoms (Knox, 2007).

10.2.3. Carbon Immobilization and Benthic Carbon Flux

The SO is long known to play a critical role in global oceanic CO_2 sequestration. With the retreat of sea-ice around Antarctica (coastal SO), a huge 'blue carbon' (carbon that is captured as well as stored by marine system; Nellemann et al., 2009) sink is currently emerging in the form of living biomass. If effectively protected and promoted, the coastal SO with an annual blue-carbon fixation and sequestration capacity of 160 million tons could well become the largest natural negative feedback to the current trends of global warming (Gogarty et al., 2019). A recent study has highlighted that the benthic (bryozoans) carbon immobilization (net carbon accumulation per year) in the South Orkney Islands of the SO archipelago is the highest per species unit area than any other part of the polar oceans in an ice-free day (Barnes et al., 2016). The magnitude of carbon immobilization, in turn, is dependent on the duration of the *in situ* phytoplankton blooms (Barnes et al., 2016). This said, quantification of such carbon fluxes over large areas of the SO is a difficult proposition because of the inaccessibility and harsh environment of the region. In particular, very little accurate data is available for benthic carbon fluxes within the SO (Sachs et al., 2009). While benthic flux data can be obtained from porewater oxygen profiles determined at high resolution, and oxygen microprofiles *in situ* (at the seafloor), the *ex situ* flux estimations performed on sediment cores often suffer from temperature- and pressure-related artifacts leading to erroneous results. However, empirical relationship has been identified between benthic carbon fluxes and diatom species predominant in the sediment surface. Estimates based on remote sensing (using space-borne sensors) of phytoplankton biomass have further shown that benthic carbon fluxes in the Atlantic sector of the SO have significantly higher export efficiencies than what was anticipated previously (Sachs et al., 2009).

10.2.4. Retreat of Sea Ice and Benthic Blue Carbon

In the polar regions, the temporally varying extent of the sea ice is one of the principal factors associated with seasonal and multiyear climate variability (Bintanja et al., 2013). The impact of global warming on the polar regions is evident in the form of accelerated melting of ice in the Arctic and Antarctic. Nearly 75% of Antarctica's coastline is attached to continental ice sheets extending seaward into the SO (Rignot et al., 2013). Although in geological timescales the Antarctic ice-shelves have been relatively stable, decadal time-series reveals an unprecedented reduction in the sea ice in this region, which could be solely linked to climate change (Rintoul et al., 2018). Increased retreat/melting of sea-ice would lead to greater supply of iron in the coastal SO as well as beyond the PF, in addition to exposing the coastal SO water column to more solar irradiance. The coupled effect of iron availability and higher solar insulation is likely to increase the productivity of polar waters (Henley et al., 2020), thereby increasing the blue carbon sequestration capacity of the SO via benthic–pelagic coupling. It is also expected that with the retreat of sea ice the species diversity of phytoplanktons would experience a higher variability/uncertainty than the phytoplankton productivity; this in turn would change the established empirical relationship between dominant phytoplankton species (diatoms) in the water column and the benthic carbon fluxes. It has been noted that a significant portion of the biomass generated by phytoplankton blooms reaches the underlying sea floor within 1 week. The speed of sinking is largely dependent on the size of the phytoplankton, so it would change if the phytoplankton species native to a territory changed (Sachs et al., 2009), thereby altering the patterns of carbon flux to the higher trophic levels of the benthos. Blue carbon would emerge in areas of high biomass, making them vulnerable to harvesting and other anthropogenic activities. However, if the international community can protect those newly ice-freed areas from commercial exploitation climate change would be mitigated significantly (Gogarty et al., 2019).

10.3. BENTHIC ECOREGIONS AND BIODIVERSITY

Classification of regions characterized by homogeneous species composition is central to the delineation of biogeographic patterns within any major biome. A thorough understanding of the spatial distribution of species explains their responses to different environmental perturbations, so it is fundamental to the framing of conservation policies such as MPA demarcation and management (Douglas et al., 2014, Fabri-Ruiz et al., 2020). The rich benthic biodiversity of the SO varies considerably across the shelf, slope, and deep-sea territories. Whereas several benthic taxa are widespread, some are endemic and some are efficient in shifting their distribution ranges via northward or southward migration. Based on the biogeographic pattern, environmental parameters, and previous hierarchical classification framework, the SO can be broadly categorized into 23 benthic ecoregions and nine bathomes (0–4500 m+) (Last et al., 2010, Douglas et al., 2014). Over 85% of the SO seabed is situated at water depths >2000 m, and is often characterized by physiographic features such as seamounts, islands, and plateaus that act as extremely isolated habitats. Ecoregions on the Arctic shelf and slope are classified mainly based on species distribution. For bathomes, classification is solely based on the depth boundaries where rapid transitions in community composition take place. In this context, it is noteworthy that Fabri-Ruiz et al. (2020) defined 12 benthic ecoregions and provided a dynamic definition for MPAs based on a large pan-SO dataset documenting the distribution of echinoids alongside environmental drivers. Although, the boundaries of ecoregions within deeper oceans are based on the influence of environmental factors on species distribution, benthos that have a wide range of distribution (cosmopolitan species) are not properly distinguished through ecoregional classification. Similarly, for eurybathic taxa, the depth-based categorization can falsify their pattern of distribution.

The present day diversity of marine fauna in the SO is the result of climate change and crustal evolution of continents, which took place during the Cretaceous and Cenozoic era (Brandt, 2005). Clines of biodiversity and changes in the environmental gradients from the tropic to the pole are considered to be the most fundamental outlines of spatial heterogeneity in species distribution (Linse et al., 2006). The remarkable benthic biodiversity of the SO is the outcome of ecological differentiation and adaptive radiations along the tropic to pole axis of environmental variation. Intraspecific or interspecific competition for better utilization of resources followed by local extinction events drive the marine fauna towards key innovations and adaptations to the unique polar environment (Saucède et al., 2014). Three major sectors of the SO, namely (i) the Scotia Arc region, (ii) the Antarctic Peninsula, and (iii) the Atlantic sector are regarded as potential centers of diversification of marine life (Saucède et al., 2014). The first revision of the SO deep-sea benthic biodiversity (Brandt et al., 2007a) provided a comprehensive dataset on bathymetry and biogeography of benthic fauna sampled during ANDEEP expeditions between 2002 and 2005. The dataset encompassed a number of major taxa (Komokiacea, Porifera, Isopoda, Bivalvia, shelled Gastropoda, Scaphopoda and Polychaeta), and more than 90% of the species reported until then were either new to science or rare in terms of global occurrence (Brandt et al., 2007b).

10.3.1. Meiobenthos

Deep-sea meiobenthic assemblages are mostly dominated by foraminifera, nematodes, followed by harpacticoids and other minor phyla (Vincx et al. 1994). Foraminifera are comprised of calcareous, aglutinata, organic-walled taxa and most of the species are attached to hard substratum albeit, they are mostly dominated by either agglutinated or calcareous species (Brandt et al., 2007b; Gooday et al., 2014). Sessile individuals of agglutinated genera such as, Dendrophrya, Dendronina are mostly abundant around Scotia Sea and continental slope of Antarctic Peninsula and South Georgia (Earland 1933; 1934). The majority of them in the northwestern Weddell Sea are found either on dropstone or attached to biogenic substrates. They can also be found at all depths throughout Antarctica (Gooday et al., 2014). Free-living marine nematodes are the other key components of meiobenthic assemblages, and high species turnover rate and regional diversity of these animals are encountered in Antarctica (Brandt et al., 2007b). In the Drake Passage region, a decreasing density of nematodes and copepods has been reported with increasing water depth, and the phenomenon has been related to the rate of organic matter degradation by microbes and the length of euphotic zone (Gutzmann et al. 2004). Halanych and Mahon (2018) have documented 331 accepted species of marine nematodes from Antarctic waters. The number of harpacticoid species, however, is comparatively lower in the SO and most of the records are from shallow waters or deep-sea areas (George, 2014). Analysis of community DNA using metabarcoding approach has revealed an exceptionally high diversity of meiobenthic taxa in Antarctica, but because of the lack of precise taxonomic information understanding of their actual diversity and biogeography has remained obscure (Fonseca et al. 2017).

10.3.2. Macrobenthos

Among the macrobenthic taxa, isopods are well represented across water depths. The suborder Asellota contributes maximally to the overall diversity of isopods. Lack of proper taxonomic identification and investigation on cryptic species of isopods has limited our knowledge on their spatial distribution across the SO (Kaiser, 2014). For the order Cumacea, species from each of the eight families known worldwide are found in the SO, and the endemism rate here is greater (80%) than that in the Arctic (Mühlenhardt-Siegel, 2014). Members of Decapoda such as shrimps are predominantly benthic with few pelagic and symbiotic representatives in the SO (Basher and Costello, 2014); reptant decapods such as crabs and lobsters, on the other hand, are less explored in the context of the SO benthos. The water temperature and Circumpolar Deep Water (CDW) shape the distribution pattern of reptant decapods in this region. Among the 22 species known thus far, none were reported for a long time from areas with low water temperature (<0° C), apparently because cold water hinders Mg^{2+} regulation in the haemolymph (Griffiths et al., 2014; Frederich et al., 2000). Polar emergence of lithodids (Decapoda), therefore, is viewed as an instance of invasion (Griffiths et al., 2013, 2014).

Being the pioneer phylum in benthic communities, bryozoans are not only regarded as bioconstructors but are also one of the most successful taxa in the Antarctic shelf. They have a very narrow range of bathymetry, thus are sometimes used as indicators of biological and glacial history (Barnes and Hillenbrand, 2010). In terms of species richness, Polychaeta and Mollusca have significant representation in the SO macrofaunal community (Brandt et al., 2012), even though the proportion of polychaetes in the SO is apparently lower than that in the temperate latitudes (Schüller and Ebbe 2014). This said, DNA barcoding-based studies have revealed an increase in polychaete species with 10 new cryptic species and 10 additional morphospecies, thereby highlighting the potential richness of the hidden diversity of polychaetes in the SO and Antarctic waters (Brasier et al., 2016). Furthermore, strikingly different polychaete assemblages of syllides and polynoides with high levels of functional diversity have been reported from the Scotia and Amundsen Sea, respectively (Neal et al., 2018). To date, most of Antarctic deep-sea molluscan species belong to the macrofauna category with a few exceptions involving large-sized gastropod families such as Naticidae, Buccinoidae, Turridae, Bathydoridae, some Scaphopods, and Cephalopods (Brandt et al., 2007b). Like most of the SO benthic taxa, bivalves also exhibit the characteristic phenomenon of eurybathy, while, biogeographically, they have wide and narrow ranges of latitudinal and longitudinal distribution, respectively (Linse, 2014).

10.3.3. Megabenthos

Based on habitat or lifestyle, the deep-sea megabenthic fauna of the SO can be classified as sessile, errant benthic and errant benthopelagic (Brandt et al., 2007b). In terms of numerical abundance, species richness, and biomass contribution, Echinoderms, including representatives from all the five major classes (Asteroidea, Crinoidea, Echinoidea, Holothuroidea, and Ophuroidea), are the predominant errant megafauna in the Antarctic deep sea (Brandt et al., 2007b, 2012). Among all the groups present, Holothuroidea has the greatest diversity whereas Crinoidea has the least (Linse et al., 2007). For Asteroidea, representatives from all the seven orders are reported from Antarctic and sub-Antarctic waters. In addition 68% of the existing families can be found in the SO with a very low endemism rate at higher taxonomic level (Moreau et al., 2020) (Table 10.1). Few species of Echinoidea are strictly stenobenthic and most are endemic to the SO (Saucède et al., 2014). For the echinoids, genus richness decreases steadily southwards; however, changes in species richness vary greatly with increasing latitudes (Fabri-Ruiz et al., 2017; Saucède et al., 2014). Among the sessile megafauna, sponges, and anthozoans are important in terms of biomass and diversity, respectively. Among the polychaetes, the number of megabenthic species present in the SO is negligible. Pycnogonids, however, have high abundance and diversity in the SO, with individuals being particularly large-bodied (representing polar gigantism) as a result of the low metabolic rates driven by low *in situ* temperatures (Soler-Membrives et al. 2014).

10.3.4. Functional Types

Benthic assemblages in the SO and Antarctica – dominated by suspension feeders, deposit feeders, grazers and predators – are as functionally diverse and dynamic as they are structurally (taxonomically) (Kaiser et al., 2013). Gutt (2007) proposed a pioneering scheme of trophic guilds for classifying the shelf-inhabiting macrobenthic communities of Antarctica. The classification involved the functionality of benthic organisms (which is governed by their surrounding environmental parameters) and envisioned core communities dominated by sessile suspension feeders, and infauna and mobile epifauna. Subsequently, Gutt et al. (2013b) provided a detailed biogeographical coverage based on the previous classification scheme, and then further modified the previous classification scheme (Gutt et al., 2014) by including the mobile suspension feeders (e.g. Crinoides) into the category of SSFA-OTH (sessile suspension feeders with associated fauna) and excluding the mobile predators such as king crabs from SSFA-OTH and including them in MOIN (mobile deposit feeders, infauna and grazers). The interactional scheme based on advanced approaches, however, was envisaged to be more complex for the SO ecosystem. Moreover, functional groups can effectively surrogate the taxonomic composition and can also be used for benthic habitat categorization (Barnes and Sands, 2017).

10.4. EVOLUTIONARY SETTING AND UNIQUE TRAIT MODALITIES

The unique characteristics of the SO fauna originate in its survival of the Pliocene–Pleistocene glacial cycles (Allcock and Strugnell, 2012). This has shaped the characteristics of the SO fauna in such a way as to make it a model system for understanding how environmental factors influence the evolutionary history of the biota (Clarke et al., 2004; Barco et al., 2012). According to Dell (1972), the present hydrological conditions apparently initiated once the freezing process set in and the ice-shelf-dominated continental areas started producing cold bottom water in Antarctic. A total of 11 major evolutionary events have shaped the SO benthos. The most pronounced cooling event during the Eocene–Oligocene boundary changed the contemporary "greenhouse" world to an "ice-house" variant (Zachos et al., 2001), thereby mediating the extinction of several taxa and emergence of yet others. Of all the major mass extinction events, the one in the Cretaceous–Paleogene (K-Pg) boundary played the most profound role in reconstructing the modern benthos (Krug et al. 2009). Several biotic and abiotic factors played crucial roles in extinction selectivity and preferential recovery of certain taxa following the K-Pg event. Many opportunistic species, often referred to as "disaster taxa" proliferated rapidly after this ecological crisis, apparently owing to reduced competition and increased availability of vacated niches (Whittle et al., 2019). The opening of two marine gateways (namely the Drake Passage and the Tasman Passage, Figure 10.1) during the Cenozoic brought about major changes in oceanic circulation and exchange of marine fauna (Lawver et al., 2014). There are two global models for the origin and evolution of polar marine benthos: (i) the tropic model, which is applicable for shallow marine habitats, and (ii) the "thermohaline expressway" model for deep-sea habitats. The former model considers the tropics and poles as macroevolutionary sources and corresponding sinks, respectively. In contrast, the later model envisages exactly the opposite pathway for the deep-sea habitats. This theory considers the SO as the primary source of cold bottom water in the global ocean as well as the centre of origin of deep-sea organisms. It propounds that the gradual separation of continents and appearance of thermohaline circulation allowed all the

Table 10.1 Groupwise distribution of major benthic species with record of endemism, according to the the Register of Antarctic Marine Species (RAMS) database of World Register of Marine Species (WoRMS: http://www.marinespecies.org/rams/).

Group	Number of taxa	Number of species	Number of accepted species	Endemism	Remarks	Reference
Porifera	871	499	303	44% Species and 7% genus are endemic to SO	Half of the endemic genera and almost 70% endemic species of SO are shelf-restricted	Janussen and Downey (2014)
				Class level endemism: 69% in Hexactinellida, 53% in Calcarea and 38% in Desmospongiae		
Bryozoa	662	387	335	Five species are regionally endemic	Endemism is high at genus level and occasionally found at family level	Barnes and Griffiths (2008), Barnes and Downey (2014)
Gastropoda	1196	607	427	74% In Antarctic and 79% in SO	Species level endemism is higher compared with genus or family level	Griffiths et al. (2009), Schiaparelli and Linse (2014)
Bivalvia	434	185	127	57% Species level, 4.3% genus level in SO	Species level endemism is higher compared with genus or family level	Linse (2014), Linse et al. (2006)
Nematoda	791	459	373	Low endemism	Low endemism indicates that either they have radiated in abyssal environment a long time ago, or their high dispersal capability enables them to have wider distribution	Bik et al. (2010); Ingels et al. (2014)
Polychaeta	1399	773	546	28% From Ross sea endemic to SO, 40% from abyssal plain of Weddel sea and Scotia sea	Polychaetes of shelf and slope show higher endemism compared with abyssal polychaetes	Knox and Cameron (1998), Hilbig (2004), Schüller and Ebbe (2014)
Sipuncula	48	23	15	Absence of family, genera and species level endemism	They might have derived from deep water basin and colonized Antarctic shallow waters from deep sea	Clarke and Crame (1989), Salinas (2014)
Echiura	38	15	11	Absence of family, genera and species level endemism. Exception: *Ikeda* sp. (Echiura: Echiuridea)	They might have derived from deep water basin and colonized Antarctic shallow waters from deep Sea	Saiz Salinas (2014)
Pycnogonida	376	316	269	40% In Antarctic and 64% in SO	Antarctica acts a center of adaptive radiation and speciation for pycnogonids	Munilla and Soler-Membrives (2009); Soler-Membrives et al., (2014)

(Continued)

Table 10.1 (Continued)

Group	Number of taxa	Number of species	Number of accepted species	Endemism	Remarks	Reference
Amphipoda	1805	1105	954	83.6% Species level (benthic and bentho-pelagic species) at SO	High endemism in Amphipods reflects ACC-driven long isolation of SO as well as adaptive radiation in Antarctic waters	De Broyer and Jażdżewska (2014)
				66.6% Species level and 27.6% genus level in Antarctica		
Isopoda	793	524	406	87% of shelf and deep sea isopods are endemic to SO	High species level endemism indicates in situ radiation across depths Low endemism at family or genera level	Brandt (1992); Brandt et al. (2007a); Kaisar (2014)
Cumacea	145	101	91	80% In SO	Endemism is very high in SO compared with Arctic region	Mühlenhardt-Siegel (2011, 2014)
Asteroidea	640	405	295	23% Genera and 60% species endemic to SO	Endemism is lower at higher taxonomic level (order and family)	Moreau et al., (2020)
Crinoidea	146	80	42	31 Out of 43 recognized species are endemic to SO		Eléaume et al. (2014)
Echinoidea	183	96	80	68% Endemic to SO	Most of them are endemic, although they have small species richness in SO	Saucède et al. (2014)

ACC, Antarctic Circumpolar Current; SO, Southern Ocean.

shallow water Antarctic species to colonize the sea floor in previously inhospitable lower altitudes (Strugnell et al., 2008). This theory of polar submergence is clearly applicable for certain families of Antarctic isopods, foraminiferans, polychaetes, holothurians (elasipodid), and hexactinellid sponges (Crame, 2014). Furthermore, periodic extensions and retreat of Antarctic ice cap is thought to have played a pivotal role in the diversification of the continental shelf fauna, because this leads to the fragmentation of habitats, with allopatric populations on one hand, and on the other the opportunity for the previously isolated populations to mingle once again (Clarke and Crame, 1989, 1992).

Peculiar abiotic factors have directed the benthic fauna towards ecological adaptation via the development of new traits such as eurybathy, circumpolarity, prevalence of brooding, gigantism, slow growth rate, and prolonged longevity (Figure 10.4). The modern Antarctic benthic fauna is dominated by epifaunal suspension feeders as well as particle feeders associated with coarser substrates. Overall, Antarctic benthic organisms adapted with K-selected traits such as gigantism, slow growth rate, and prolonged longevity are the most common adapted features of polar benthic fauna (Knox, 2006). Several organisms that were free from predation pressure and calcium constraint were selected for gigantism (Arntz et al., 1992). Their reproductive strategies included prolonged gametogenesis, brooding, slow embryonic development, low fecundity, and seasonal reproduction. The most distinguishing feature of Antarctic benthic assemblage compared with that of the Arctic is the absence of shell-crushing predators. A great reduction in shell crushing (durophagous) predators during the onset of global cooling allowed animals to emerge that had possess thin and weakly calcified shells. Several gastropod taxa such as struthioloariids, ficids, and mitrids disappeared whereas, turrids *sensu lato* and buccinids underwent extensive radiation during this period (Crame, 1996). Poor calcification can also be a

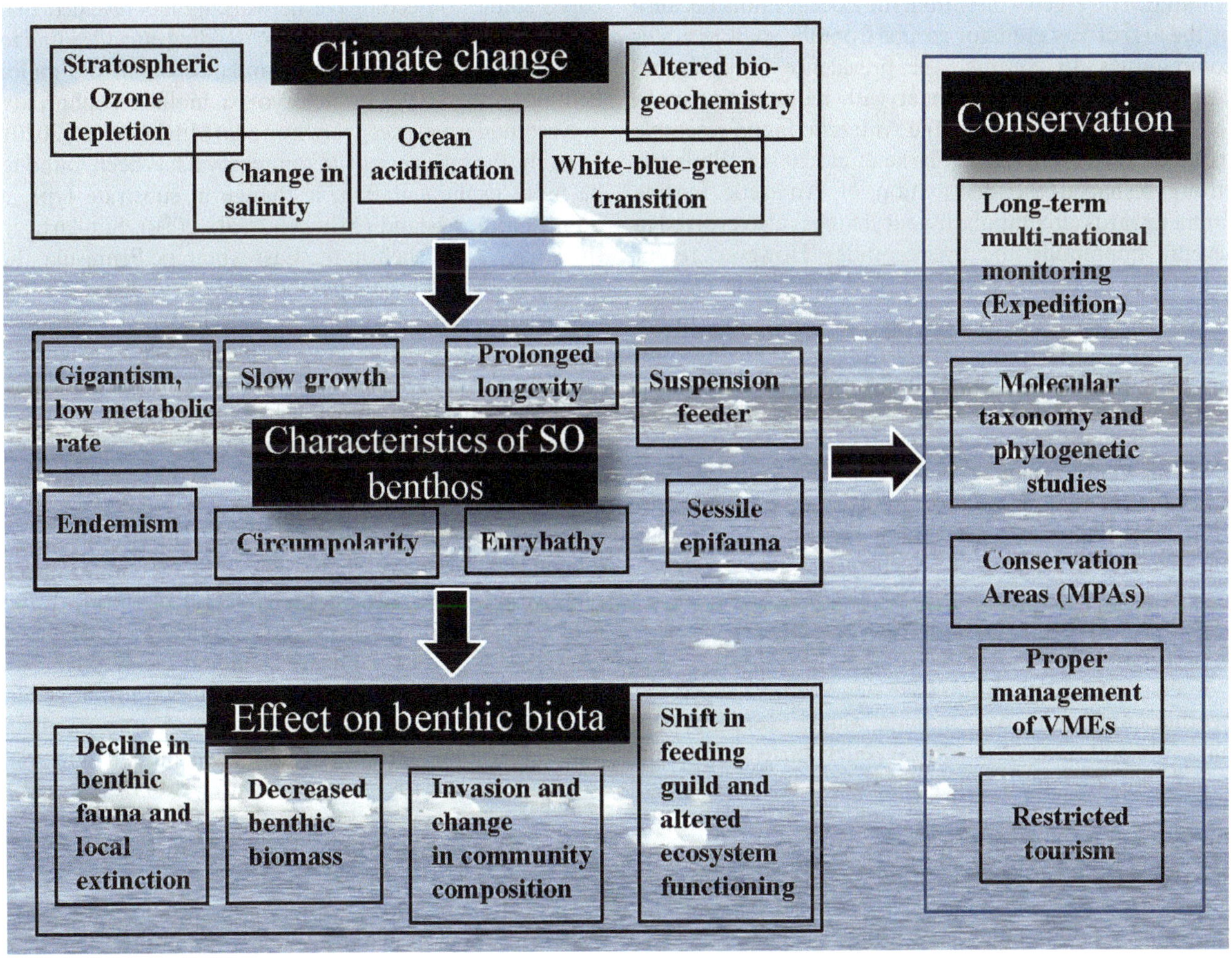

Figure 10.4 Schematic diagram showing effect of climate change on Southern Ocean benthos (with unique life history traits) and the possible conservation protocols.

consequence of the inverse relationship between water temperature and rate of calcium carbonate solubility (Fabry et al., 2009). Evidence for this can be found in modern SO gastropods that are similar to other shelled molluscs of the SO in having small size and thin, fragile shells (Schiaparelli and Linse, 2014). Moreover, the extinction of several decapods allowed the peracarid crustaceans to invade the empty niches (Brandt, 2000). These gradually evolved diverse morphological and physiological traits such as a wide range of feeding strategies and a highly specialized mating system (Kaiser, 2014). In addition, infaunal species started to deplete with the proliferation of epifaunal, suspension-feeder groups. Bryozoa are one of the most specious taxa in the SO benthos representing 412 species from Southern Polar Front (Barnes and Downey 2014). Crucial factors such as poorly sorted sediment granulometry and availability of dropstones have made the habitat preferable for the sessile taxa (Brandt, 2000; Brandt et al., 2012). Overall, the modern benthic fauna inhabiting the continental shelves of Antarctica diverged widely from the Eocene fauna because of the loss of key predator groups from the shallow water communities. In contrast, the prevalence of epifaunal suspension feeders in a habitat with reduced fine-grain terrigenous sediment made the Antarctic fauna resemble its Paleozoic counterpart (Clarke et al., 2004). While the strong reshaped the distribution of Antarctic marine fauna towards circumpolarity, eurybathy is also evolved in several amphipods and pycnogonides. However, recent discoveries of certain cryptic species (such as in *Glycera kerguelensis*) refute the concept of high endemism in the SO biome (Allcock and Strugnell, 2012). In conclusion, it is evident that the SO benthic biota has been reshaped several times through the geological time-scale, and the modern fauna is the product of previous complex events involving rapid warming and cooling, and mass extinctions. Precise and quantitative inferences on potential evolutionary links would require extensive molecular phylogenetic investigation of the benthos.

10.5. BIOTIC INTERACTION

Interspecific relationships have significantly shaped the benthic community of Antarctic waters ever since the physical conditions of its littoral zones started becoming comparatively stable (Gutt and Schickan, 1998). Different functional groups and their interactions with biotic and abiotic components also influence nutrient cycling and benthic carbon storage within the SO-Antarctic marine system (Barnes and Sands, 2017). Furthermore, the strong correlation between benthic–pelagic food webs and sea ice (including unpredictable changes in seasonality and properties) is thought to have significant effect on the entire SO biota (Gutt et al., 2020).

In this territory of the global ocean, predatory interactions between species are rare (Aronson and Blake, 2001), whereas symbiotic associations are widespread, especially in the kleptocommensal mode. The dynamics of Antarctic benthos has always been governed by species that provide biogenic substrate for others (Hedgpeth, 1971). The granulometry and pattern of the substrate play a determining role in the diversification of ecological traits. Owing to the scarcity of hard mineral substrates, the biotic counterparts provide new ecological niches for sessile organisms (Hétérier et al., 2008). Consequently, numerous epibionts and substrate species are found in Antarctic waters (Gutt and Schickan, 1998).

Most sponge species have wide-ranging environmental requirements, so have developed an exceptionally wide varieties of traits, including a variation in growth rate that affects their distribution pattern (Gutt and Schickan, 1998). Moreover, they dominate the biomass in most of the regions and hasten energy transfer in the sediment–water interface (Brasier et al., 2018). Recent Bayesian inference-based studies on ecological networking has revealed that removal of sponge species has a cascading effect leading to significant reduction in the abundance of all the major benthic groups (for example Bryozoa, molluscs, ophuroids, and echinoderms) except arthropods (Mitchell et al., 2020). Notably, the influence of taxon removal has been found to be more prominent than a change in substrate type at South Orkney Island (Mitchell et al., 2020). Similarly, in the Larsen A ice-shelf in the East Antarctic Peninsula, the disappearance of ascidians (the most obvious competitor) during 2007–2011 led to rapid decline of suspension-feeding ophiuroids and subsequent replacement with deposit feeders along with twofold increase in abundance of species that were previously rare. In addition, these unpredictable sea ice dynamics caused changes in food availability and thus hampered prey–predator interactions (Gutt et al., 2013b, 2020).

In the SO–Antarctic benthic biome, Bryozoans create habitat heterogeneity that allows changes in localized physical environments for better larval recruitment and greater ingress of dwellers and scavengers (Barnes and Downey, 2014). In the Lazarev Sea and the Weddell Sea, several important groups such as echinoderms, annelids, and molluscs are indirectly associated with bryozoan habitats (Gutt and Starmans, 1998), even though potential trophic relationships among them are yet to be resolved.

10.6. DISPERSAL AND ENDEMISM

Thatje (2012), after assembling all available molecular evidence, proposed that the genetic variability of Antarctic fauna has opened a new era of understanding of the theories of speciation, dispersal, and population connectivity. The geographic distribution and genetic connectivity between subpopulations of a species are correlated with successful dispersion from their natal population (Douglass et al., 2014). More precisely, the dispersal potential is directly correlated with the maintenance of homogeneous genetic population structure or in other words, haplotype diversity tends to increase when the dispersal distance decreases below the population distance. The ability of dispersion of early ontogenic stages played a crucial role in the evolution of Antarctic benthic assemblages (Thatje, 2012). The thermal gradient of PF acts as a potential biogeographical barrier for dispersal as well as limiting the organism's migration in and out of the SO (Halanych and Mahon, 2018). The abrupt heterogeneity in hydrological parameters between two sides of the PF prevents the entry of fauna from lower latitudes towards the SO (Griffiths et al., 2009a). Therefore, a prominent feature of the SO benthos is endemism (Clarke and Johnston, 2003; Griffiths, 2010). However, the surface and deep-sea currents act as powerful dispersal vectors in the SO and their intensity and direction have shaped the evolution of Antarctic marine life through time and space. The ACC and west-wind drift facilitate the dispersion of organisms in a clockwise direction around continents, which is evident in the biogeographic patterns of echinoids, pycnogonids, ascidians, and some of the bryozoan and molluscan taxa (David et al., 2005; Primo and Vázquez, 2007; Griffiths et al., 2009; Munilla and Soler Membrives, 2009). Consequently, ACC has promoted either long-distance connectivity between related species or diversification through allopatry (Barnes and Griffiths, 2008; Pearse et al., 2009; Janosik et al., 2011). In addition to hydrological parameters, taxon-specific reproductive strategy (brooder or broadcaster) plays a determining role in dispersal events. Organisms with a poor dispersal capability or lacking a dispersal phase (propagules) are often structured in haplogroups confined to narrow geographic ranges or become endemic (Rogers, 2007; Eléaume et al., 2014). Studies have revealed that a higher proportion of species in Cyclostomata are widely dispersed compared with gastropods because their reproduction is followed by cloning of zygotes before mass broadcast is released (Griffiths et al., 2009). Free-living nematodes lack active dispersal because of their limited mobility and conservative reproductive methods, albeit a small body size allows their shallow to deep water exchanges (Ingels et al., 2014). Furthermore, the discrepancy in their biogeographic pattern led to the conclusion of the "meiobenthic paradox" (Giere 2009). Foraminiferan populations exhibit a high degree of connectivity with distant parts of the SO, which are both geographically and bathymetrically separated (Brandt et al., 2014). Extensive studies on brooding paracarid crustaceans have portrayed their poor dispersal capability and they can therefore be used as biogeographical indicators. Similarly, their high endemism can be related to *in situ* radiation across water depths in the SO (Brandt 1992; Brandt et al., 2007a; De Broyer and Koubbi, 2014) (Table 10.1). Brandt et al., (2007b) recorded a higher percentage of species endemism in sponges and further investigation revealed that about 44% of the SO sponge species are endemic (Janussen and Downey, 2014). Furthermore, based on bathymetric data, authors have assigned the Antarctic shelf as a centre of sponge endemism in the SO. Limited research on cumaceans has revealed a high rate of endemism due to their low potentiality for active or passive dispersal; however, no significant level of faunal overlap exists in Cumacea (Mühlenhardt-Siegel, 2011, 2014). Endemism might differ between similar taxonomic groups with similar life history traits, as can be seen in the suborders of Bryozoa (Tilbrook and De Grave, 2005). More evidence can be found in two species of *Notocidaris* (Family: Ctenocidaridae) that differ in reproductive pattern (Gutt et al., 2011). One of them, *Notocidaris platyacantha*, is endemic to East Antarctic, whereas *Notocidaris mortenseni* is found across the continent (Saucède et al., 2014). These shelled gastropods show a high degree of species level endemism but little or no endemism at the generic or family level; this is because adaptive radiation in cold Antarctic environment can influence species or genera but not families (Linse et al., 2006; Griffiths et al., 2009). A complete absence of endemism has been recorded in Sipancula and Echiura (except *Ikeda* sp. Echiura: Echiuridea) (Table 10.1). Nevertheless, according to Halanych and Mahon (2018), species groups that are easier to identify have been reported to possess higher endemism compared with groups that are difficult to identify taxonomically. Moreover, most of the endemism records are based on morphology-based species identification, so the notion could change if molecular concepts were applied (Brasier et al., 2017)

Endemism in the SO is also defined as a result of the fauna's long history of evolution in isolation (Clarke et al., 2005). The onset of ACC and subsequent separation of Gondwana played crucial roles in thermal and biogeographical isolation, and vicariant speciation in associated marine fauna (Patarnello et al., 1996; Page and Linse, 2002; Lawver and Gahagan, 2003; Clarke et al., 2005). Therefore, the benthic composition of the SO is a combination of lineages with different life histories, where some existed before the formation of Antarctica and others have colonized relatively recently (Clarke and Crame, 1989; Verheye et al., 2017).

10.7. CLIMATE CHANGE AND BENTHOS

The SO ecosystem is highly susceptible to global climate change and anthropogenic perturbations (Ingels et al., 2012). The SO ventilates the global ocean by absorbing heat and anthropogenic CO_2 emissions (Turner et al., 2009). For example, the Ross Sea continental shelf is responsible for >25% of the total CO_2 uptake by the SO (Arrigo et al., 2008). The most notable climate-driven changes in the physical parameters of the SO include an increase in water temperature, acidification of the ocean, seasonality of the sea ice, and shift in PF (Figure 10.4). Like all other marine species, polar organisms also exhibit a range shift as a result of ocean warming, but have limited options to migrate (Griffiths et al., 2017). As the SO acts as a hotspot of species turnover, local extinction and invasion, an important question arises as to how the large proportion of endemic species would respond to this global climate change scenario in the near future (Beaugrand et al., 2015; Griffiths et al., 2017). Multivariate analysis on the georeferenced data of species occurrence and seafloor temperature suggested that 79% of the endemic species could experience loss of suitable habitat. The present thermal regime of the SO is associated with a high level of stenothermal fauna and their physiologies have been tuned to narrow temperature ranges. For this reason, elevation of seawater temperature will have severe consequences for the SO fauna and it is expected that more temperate species will invade from lower latitudes (Clarke, 1983; Berkman, 1991; Clarke and Crame, 1992). Amphipods and isopods, the most stenothermal organisms, would face several physiological challenges (in relation to respiration rate, growth rate, and feeding habit) and would most likely lose their migration potentiality and fail to disperse. In contrast, Antarctic post-metamorphic echinoides (opportunistic echinoides) would be able to acclimatize to the change-related to food availability (Ingels et al., 2012). Significant changes in water temperature and salinity in recent decades have been recorded in the Kerguelen plateau (Frenot et al., 1993). Species distribution models by Guillaumot et al. (2018) predicted that narrow-niche species (such as *Abatus cordatus*) will be unable to find a climatic refuge whereas species with a large distribution range (*Brisaster antarcticus*) will show greater potential resilience to future changes.

A rising sea temperature-mediated increase in phytoplankton productivity helps to create a prolonged growing season in the SO and Antarctica (Doney, 2006). Seasonal sea ice loss and subsequent phytoplankton bloom have a cascading effect on the trophic structure of the SO, especially on the megabenthic communities, which are dominated by glass sponges (Hexactinellida). This has been clearly witnessed in the climate change induced Antarctic sea ice collapse and the rapid expansion of glass sponges via a manifold increase in abundance and biomass. Moreover, their distribution pattern, which is skewed towards larger individuals, showed that recruitment in that particular area, after such catastrophic event, became highly active (Fillinger et al. 2013). Glacier retreat and disolution of the ice shelf allow light penetration with new phytoplankton blooms that ultimately fuel rapid benthic growth and increased sedimentation (Fillinger et al., 2013; Barnes, 2017; Rogers et al., 2019; Jansen et al., 2020). This phenomenon plays an inevitable role in blue carbon storage and sequestration. Thus, climate change mediated shift in sea colour "from white to blue to green" is a sign of increased carbon flow from the atmosphere to the benthic zone (Barnes, 2017). Furthermore, such changes in the magnitude of surface productivity is likely to alter the zooplankton seasonality pattern that eventually would lead to unpredictable shifts in bentho–pelagic coupling (Constable et al., 2014). According to recent investigations, the South Orkney Islands (SOI) has been hailed as the most important site for carbon immobilization by the benthos and distinguishable negative impacts on climate change (Barnes et al., 2016). Among all benthic animals, the high annual production of bryozoans plays a pivotal role in this biogeochemical process at SOI (Barnes et al., 2016). Furthermore, bryozoans are also regarded as the most useful taxon in biomonitoring environmental changes because of their high abundance in predicted zones of ocean warming and pH change in the SO (Barnes and Downey, 2014).

In addition to ocean warming, ocean acidification (OA) is going to be one of the most important threats to benthic habitats (Constable et al., 2014). The SO is prone to faster acidification because its low surface water temperature facilitates increased CO_2 solubility and greater upwelling of CO_2-rich bottom water (Guinotte and Fabry, 2008). Although a very limited amount of specific literature is available, tit has shown that OA will positively influence the bacterial community of the SO, whereas it will cause dissolution of skeleton, hinder reproductive success, and shorter life history stages in benthic organisms (Byrne, 2011; Ericson et al., 2012; Hancock et al., 2020). However, responses are most likely to be phylum specific: for example, as molluscs and echinoderms will be more at risk than crustaceans (Kroeker et al., 2010, 2013; Hancock et al., 2020). Most surprisingly, several taxa have shown some compensatory response to OA, such as production of thicker shells to prevent dissolution. In addition, in meiofaunal communities, unlike foraminiferans, nematodes have been more robust and showed species-specific responses to OA (Ingels et al., 2012; Gutt et al., 2020). Although the SO has been assumed as a single unit in terms of climate change, its responses are highly region specific (Rogers et al., 2019). Similarly, the effect on benthos would also differ across habitats on the continental self, slope, and deep sea (Brandt and Gutt, 2011). Cumulatively, benthic organisms of the SO are likely to respond positively as well as negatively towards climate change, or they may also completely change their contributions to ecosystem services (Griffiths et al., 2017; Hancock et al., 2020).

10.8. CONSERVATION AND FUTURE PERSPECTIVES

The SOI are considered globally significant conservation areas. Effective conservation and management strategies are highly dependent on the availability of sound scientific knowledge on the ecosystem in question. The SOI is a cold, remote area that partly remains covered by sea ice perennially. Owing to adverse geographical and climatic difficulties, benthic sampling is still a challenge in the SOI (Kaiser et al., 2013). However, recent advances in sampling technology, analytical methods, and database creation (for example Register of Antarctic Marine Species, ANTABIF) have significantly increased our knowledge of the benthic biodiversity of the SO. Taxonomy and systematics remain the fundamental approach in discovering the faunal composition of a major biome and determining their changes in relation to global processes (Costello et al., 2013). The advent of molecular taxonomy has significantly upgraded our understanding of cosmopolitan species, cryptic species, and invasion of benthic taxa in the SO and Antarctic realm. However, there are still gaps in our understanding of the dynamics of the ecosystem, and the mechanisms behind climate change mediated loss of biodiversity. Organizations such as CCAMLR and SCAR have made remarkable contributions in advancing our knowledge of the SO benthos and thereby conserving the ecosystem (Kaiser et al., 2013). For example, CCAMLR has declared seven MPAs (including the world's largest international MPA, the Ross Sea) spread over an ~12% area of the entire SO, of which, again, more than one-third encompasses no-take areas (Brooksid et al., 2020). The CCAMLR has also formed a risk management framework to identify VMEs within the SO. This framework uses biomass of VME taxa such as Cnidaria, Porifera and Bryozoa, which are vulnerable to destructive fishing activities, as indicators of ecosystem health: an area must contain a minimum 10 kg of VME taxa within the catch recovered from a single 1200 m segment of long-line gear; if it fails to do so fishing activities would be closed in that area until effective measures are taken by the commission (CCAMLR 2009b).

The impact and response of climate change is highly region specific in the SO. The South Orkney Islands Southern Shelf (SOISS), which is the first high-sea MPA, harbors approximately one-fifth of the entire SO fauna, and is mostly dominated by filter-feeding VME taxa (Brasier et al., 2018). The slow growth rate and recovery time of the VME are considered to be the early indicators of their vulnerability to disturbances and habitat destruction (Barnes and Souster, 2011). As a result of this, the South Georgia archipelago has been highly prioritized as a conservation zone (Barnes et al., 2011).

Despite the enormous efforts of the CAML, there are still knowledge gaps in the understanding of biogeographic patterns and ecosystem functioning within the SO. Xavier et al. (2016) has confirmed that, data on functionality and productivity of benthic organisms are still meager compared with their distributional information. Therefore, identification of important taxa, their biotic correlation with other taxa in ecosystem functioning and proper prioritization is a priority. However, there remains several methodological challenges in quantifying and structuring the benthic biome of the SO and adjacent areas. Long-term multinational expeditions or biomonitoring studies with advanced technological tools (that can analyze and include not only the distribution pattern but also life-history parameters) could be one of the ways forward. In addition, the use of molecular taxonomy tools (which can unravel the hidden diversities of this enormous biome along with the phylogenetic relationships of such cryptic diversities) should also be embraced (Figure 10.4). Next-generation technologies, improved methodologies, food-web forecasting systems, and integrated ecosystem studies with network modeling, and building comparative marine research programs between polar regions, would collectively help to assess and conserve the health of this unique frigid biome of the global ocean.

10.9. CONCLUSIONS

The greatest challenge in understanding and managing the benthic geobiology of the SO is the construction of the overall ecosystem framework. The enormous benthic biodiversity of the SO is an outcome of evolutionary events that shaped the spatial variation of the entire biota. This isolation and unique habitat heterogeneity have structured the benthic community in a very specialized way. Several unique trait modalities and life-history parameters have made the SO benthos vulnerable to the growing impact of climate change in the Anthropocene. Regionally, the impacts have been observed to propagate from producers to consumers through the food web, whereas the effects are more direct and prominent on the higher predators, range-restricted endemic species also extremely vulnerable to local extinction and invasion. Reconstruction of the overall trophic framework for the entire SO benthic biota and establishment of a comprehensive real-time database is key to the successful tracking and compensation of ecosystem perturbations caused by climatic change. In this context, special attention is warranted for the unexplored and underexplored SO regions so that the desired levels of knowledge can be attained before the effects of climate change become irreversible.

ACKNOWLEDGMENTS

We are thankful to the Editors of the book for the invitation to contribute a chapter. The authors express their gratitude to the Vice Chancellor, Presidency University for providing facilities and encouragement to carry out the above research work. The authors are also grateful to the Director, NCPOR, Ministry of Earth Sciences, Goa for his encouragement and giving permission for S.M. to participate in the 11th Indian Scientific Expedition to Southern Ocean. This work is supported by grants awarded to S.M. from NCPOR (NCPOR/2019/PACER-POP/BS-09 dated 05.07.2019) under the Pacer Outreach Program (POP) under the Polar Science and Cryosphere Programme (PACER) initiative. The authors are thankful to Volume Editor Wriddhiman Ghosh for language editing of the manuscript. The authors would also like to express their appreciation to the members of the Marine Ecology Laboratory for their help. This is NCPOR contribution number B-1/2021-22.

REFERENCES

Allcock, A.L. and Strugnell, J.M. (2012). Southern Ocean diversity: new paradigms from molecular ecology. *Trends in Ecology and Evolution 27* (9): 520–528. https://doi.org/10.1016/j.tree.2012.05.009

Arntz, W.E., Brey, T., Gerdes, D. et al. (1992). Patterns of life history and population dynamics of benthic invertebrates under the high Antarctic conditions of the Weddel Sea. *Marine Eutrophication and Population Dynamics 370*: 221–230.

Arntz, W.E., Brey, T. and Gallardo, V.A. (1994). Antarctic zoobenthos. *Oceanography and Marine Biology Annual Review 32*: 241–304.

Aronson, R.B. and Blake, D.B. (2001). Global climate change and the origin of modern benthic communities in antarctica. *American Zoologist 41* (1): 27–39. https://doi.org/10.1093/icb/41.1.27

Arrigo, K.R., van Dijken, G.L. and Bushinsky, S. (2008). Primary production in the Southern Ocean, 1997–2006. *Journal of Geophysical Research 113* (C08004). https://doi.org/10.1029/2007JC00455.

Barco, A., Schiaparelli, S., Houart, R. et al. (2012). Cenozoic evolution of Muricidae (Mollusca, Neogastropoda) in the Southern Ocean, with the description of a new subfamily. *Zoologica Scripta 41* (6): 596–616. doi:10.1111/j.1463-6409.2012.0055

Barnes, D.K.A. (2017). Polar zoobenthos blue carbon storage increases with sea ice losses, because across-shelf growth gains from longer algal blooms outweigh ice scour mortality in the shallows. *Global Change Biology 23* (12): 5083–5091. https://doi.org/10.1111/gcb.13772

Barnes, D.K.A. and Downey, R.V. (2014). Chapter 5.23. Bryozoa. In: *Biogeographic Atlas of the Southern Ocean* (eds. C. De Broyer, P. Koubbi, H.J. Griffiths H.J. et al.), 195–199. Cambridge: Scientific Committee on Antarctic Research.

Barnes, D.K.A. and Griffiths, H.J. (2008). Biodiversity and biogeography of southern temperate and polar bryozoans. *Global Ecology and Biogeography 17:* 84–99. https://doi.org/10.1111/j.1466-8238.2007.0034

Barnes, D.K.A. and Hillenbrand, C.D. (2010). Faunal evidence for a late quaternary trans-Antarctic seaway. *Global Change Biology 16* (12): 3297–3303. https://doi.org/10.1111/j.1365-2486.2010.0219

Barnes, D.K.A. and Sands, C.J. (2017). Functional group diversity is key to Southern Ocean benthic carbon pathways. *PLoS ONE 12* (6): e0179735. https://doi.org/10.1371/journal.pone.0179735

Barnes, D.K.A. and Souster, T. (2011). Reduced survival of Antarctic benthos linked to climate-induced iceberg scouring. *Nature Climate Change 1* (7): 365–368. https://doi.org/10.1038/nclimate1232

Barnes, D.K.A., Collins, M.A., Brickle, P. et al. (2011). The need to implement the Convention on Biological Diversity at the high latitude site, South Georgia. *Antarctic Science 23* (4): 323–331. https://doi.org/10.1017/S0954102011000253

Barnes, D.K.A, Ireland, L., Hogg, O.T. et al. (2016). Why is the South Orkney Island shelf (the world's first high seas marine protected area) a carbon immobilization hotspot? *Global Change Biology 22* (3): 1110–1120. https://doi.org/10.1111/gcb.13157

Basher, Z. and Costello, M.J. (2014). Chapter 5.22. Shrimps (Crustacea: Decapoda). In: *Biogeographic Atlas of the Southern Ocean* (eds. C. De Broyer, P. Koubbi, H.J. Griffiths H.J. et al.), 190–194. Cambridge: Scientific Committee on Antarctic Research.

Beaugrand, G., Edwards, M., Raybaud, V. et al. (2015). Future vulnerability of marine biodiversity compared with contemporary and past changes. *Nature Climate Change 5* (7): 695–701. https://doi.org/10.1038/nclimate2650

Berkman, P.A. (1991). Holocene meltwater variations recorded in Antarctic coastal marine benthic assemblages In: *Proceedings of the International Conference on the Role of Polar Regions in Global Change* (eds. G. Weller, C.L. Wilson and B.A.B. Severin), 440–449. Fairbanks: University of Alaska.

Bik, H., Thomas, W.K., Lunt, D. et al. (2010). Low endemism, continued deep-shallow interchanges and evidence for cosmopolitan distributions in free-living marine nematodes (order Enoplida). *BMC Evolutionary Biology 10* (1): 389.

Bintanja, R., van Oldenborgh, G.J., Drijfhout, S.S. et al. (2013). Important role for ocean warming and increased ice-shelf melt in Antarctic sea-ice expansion. *Nature Geoscience 6:* 376–379. DOI: 10.1038/NGEO1767.

Blain, S., Quéguiner, B., Armand, L. et al. (2007). Effect of natural iron fertilization on carbon sequestration in the Southern Ocean. *Nature 446*: 1070–1074. https://doi.org/10.1038/nature05700

Brandt, A., (1992). Origin of Antarctic Isopoda (Crustacea, Malacostraca). *Marine Biology, 113*: 415–423.

Brandt, A. (2000). Hypotheses on Southern Ocean peracarid evolution and radiation (Crustacea, Malacostraca). *Antarctic Science 12* (3): 269–275. https://doi.org/10.1017/s095410200000033x

Brandt, A. and Gutt, J. (2011): Biodiversity of a unique environment: the Southern Ocean benthos shaped and threatened by climate change/: In: *Biodiversity Hotspots* (eds. F. Zachos and J. Habel), 503–526. Berlin: Springer. https://doi.org/10.1007/978-3-642-20992-5_25.

Brandt, A., Gooday, A.J., Brandão, S.N., Brix, S., Brökeland, W., Cedhagen, T. et al. (2007a). First insights into the biodiversity and biogeography of the Southern Ocean deep sea. *Nature 447* (7142): 307–311. https://doi.org/10.1038/nature05827

Brandt, A., De Broyer, C., De Mesel, I., Ellingsen, K.E., Gooday, A.J., Hilbig, B. et al. (2007b). The biodiversity of the deep Southern Ocean benthos. *Philosophical Transactions of the Royal Society B: Biological Sciences 362* (1477): 39 66. https://doi.org/10.1098/rstb.2006.1952

Brandt, A., De Broyer, C., De Mesel, I. et al. (2012). Southern Ocean deep benthic biodiversity. In: *Antarctic Ecosystems: an Extreme Environment in a Changing World* (eds. A.D. Rogers, N.M. Johnston, E.J. Murphy et al.), 291–334. London: Philosophical Transactions of the Royal Society.

Brandt, A., Van de Putte, A. and Griffiths, H. (2014). Chapter 5.29. Southern Ocean benthic deep-sea biodiversity and biogeography. In: *Biogeographic Atlas of the Southern Ocean* (eds. C. De Broyer, P. Koubbi, H.J. Griffiths et al.), 234–239. Cambridge: Scientific Committee on Antarctic Research.

Brasier, M.J., Wiklund, H., Neal, L. et al. (2016). DNA barcoding uncovers cryptic diversity in 50% of deep-sea antarctic polychaetes. *Royal Society Open Science 3* (11): 1–19. https://doi.org/10.1098/rsos.160432

Brasier, M.J., Harle, J., Wiklund, H. et al. (2017) Distributional patterns of polychaetes across the west Antarctic based on DNA barcoding and particle tracking analyses. *Frontiers in Marine Science 4*: 356. doi.10.3389/fmars.2017.00356

Brasier, M.J., Grant, S.M., Trathan, P.N. et al. (2018). Benthic biodiversity in the South Orkney Islands Southern Shelf Marine Protected Area. *Biodiversity, 19* (1–2): 1–15. https://doi.org/10.1080/14888386.2018.1468821

Brooksid, C.M., Chown, S.L., Douglass, L.L. et al. (2020). Progress towards a representative network of Southern Ocean protected areas. *PLoS ONE 15* (4): (e0231361) 1–21. https://doi.org/10.1371/journal. pone.0231361

Byrne, M. (2011). Impact of ocean warming and ocean acidification on marine invertebrate life history stages: Vulnerabilities and potential for persistence in a changing ocean. In: *Oceanography and Marine Biology: An Annual Review,* Vol. *49* (eds. R.N. Gibson and R.J.A. Atkinson), 1–42. Taylor and Francis.

Chisholm, S.W., Falkowski, P.G. and Cullen, J.J. (2001). Discrediting ocean fertilization. *Science 294*: 309–310.

Clarke, A. and Johnston, N.M. (2003). Antarctic marine benthic diversity. In*: Oceanography and Marine Biology: an Annual Review*, Vol. *41* (eds. R.N. Gibson and R.J.A. Atkinson), 47–114. Taylor and Francis.

Clarke, A. (1983). Life in cold water: the physiological ecology of polar marine ectotherms. In: *Oceanography and Marine Biology. an Annual Review* Vol. *21* (ed. M. Barnes), 341–453, Aberdeen: Aberdeen University Press.

Clarke, A. (2008). Antarctic marine benthic diversity: patterns and processes. *Journal of Experimental Marine Biology and Ecology 366* (1–2): 48–55. doi:10.1016/j.jembe.2008.07.008

Clarke, A. and Crame, J.A. (1992). The Southern Ocean benthic fauna and climate change: a historical perspective. *Philosophical Transactions – Royal Society of London B 338* (1285): 299–309 https://doi.org/10.1098/rstb.1992.0150.

Clarke, A. and Crame, J.A. (1989). The origin of the Southern Ocean marine fauna. *Geological Society Special Publication 47:* 253–268. https://doi.org/10.1144/GSL. SP.1989.047.01.19

Clarke, A., Barnes, D.K.A. and Hodgson, D.A. (2005). How isolated is Antarctica? *Trends in Ecology and Evolution 20* (1): 1–3. https://doi.org/10.1016/j.tree.2004.10.004

Clarke, A.J., Aronson, R.B., Crame, J.A. et al. (2004). Evolution and diversity of the benthic fauna of the Southern Ocean continental shelf. *Antarctic Science, 16* (4): 559–68. doi: 10.1017/S095410200400232

Cochlan, W.P. (2008). Chapter. 12. Nitrogen uptake in the Southern Ocean. In: *Nitrogen in the Marine Environment* (eds. D.G. Capone, D.A. Bronk, M.R. Mulholland et al.), 569 596. San Diego, CA: Elsevier, Academic Press. doi:10.1016/B978–0-12–372522–6.00012–8.

Constable, A.J., Melbourne-Thomas, J., Corney, S.P. et al. (2014). Climate change and Southern Ocean ecosystems I: how changes in physical habitats directly affect marine biota. *Global Change Biology 20* (10): 3004–3025. https://doi.org/10.1111/gcb.12623

Costello, M.J., May, R.M. and Stork, N.E. (2013). Can we name earth's species before they go extinct? *Science 339* (6118): 413–416. https://doi.org/10.1126/science.1230318

Crame, J.A. 1996. Evolution of high-latitude molluscan faunas. In: *Origin and Evolutionary Radiation of the Mollusca* (ed. J.D. Taylor), 119–131. Oxford: Oxford University Press

Crame A. (2014). Chapter 3.1. Evolutionary Setting. In: *Biogeographic Atlas of the Southern Ocean* (eds. C. De Broyer, P. Koubbi H.J. Griffiths et al.), 32–35. Cambridge: Scientific Committee on Antarctic Research.

David, B., Choné, T., Moo, R. et al. (2005) *Antarctic Echinoidea, Synopses of the Antarctic Benthos*, Vol. *10*: (eds. J.W. Wägele and J. Sieg). Königstein: Koeltz Scientific Books.

De Baar, H.J.W. and Boyd, P.W. (1999): Chapter 4. The role of iron in plankton ecology and carbon dioxide transfer of the global oceans. In: *The Dynamic Ocean Carbon Cycle. a Midterm Synthesis of the Joint Global Ocean Flux Study. International Geosphere Biosphere Book Series* (eds. R.B. Hanson, H.W. Ducklow and J.G. Filed), 61–141., Cambridge: Cambridge University Press.

De Broyer, C. and Jażdżewska, A. (2014). Chapter 5.17. Biogeographic patterns of Southern Ocean benthic amphipods. In: *Biogeographic Atlas of the Southern Ocean* (eds. C. De Broyer, P. Koubbi, H.J. Griffiths et al.), 155–165. Cambridge: Scientific Committee on Antarctic Research.

De Broyer C. and Koubbi, P., (2014). Chapter 1.1. The biogeography of the Southern Ocean. In: *Biogeographic Atlas of the*

Southern Ocean (eds. C. De Broyer, P. Koubbi, H.J. Griffiths et al.), 2–9. Cambridge: Scientific Committee on Antarctic Research.

De Broyer C., Koubbi P., Griffiths H.J. et al. (eds.) (2014). *Biogeographic Atlas of the Southern Ocean,* XII. Cambridge: Scientific Committee on Antarctic Research.

De Broyer, C., Clarke, A., Koubbi, P. et al. (eds.) (2020). Register of Antarctic Marine Species. Accessed at http://www. marinespecies.org/rams on 2020-11-18

Dell, R.K. (1972). Antarctic benthos. *Advances in Marine Biology 10*: 1–216.

Doney, S.C. (2006). Oceanography: plankton in a warmer world. *Nature 444* (7120): 695–696. https://doi.org/10.1038/444695a

Douglass, L.L., Turner, J., Grantham, H.S. et al. (2014). A hierarchical classification of benthic biodiversity and assessment of protected areas in the Southern Ocean. *PLoS ONE 9* (7): 1–16. https://doi.org/10.1371/journal. pone.0100551

Douglas, L.L., Beaver D., Raymond B. et al. (2014). Chapter 10.1. Benthic Regional Classification. In: *Biogeographic Atlas of the Southern Ocean* (eds. C. De Broyer, P. Koubbi, H.J. Griffiths et al.), 414–417. Cambridge: Scientific Committee on Antarctic Research.

Earland, A. (1933). Foraminifera, Part II, South Georgia. *Discovery Reports 7*: 27–138.

Earland, A., (1934). Foraminifera. Part III. The Falklands sector of the Antarctic (excluding South Georgia). *Discovery Reports 10*: 1–208.

Eléaume, M., Hemery, L.G., Roux, M. et al. (2014). Chapter 5.25. Southern Ocean Crinoids. In: *Biogeographic Atlas of the Southern Ocean* (eds. C. De Broyer, P. Koubbi, H.J. Griffiths et al.), 208–212. Cambridge: Scientific Committee on Antarctic Research.

Ericson, J.A., Ho, M.A., Miskelly et al. (2012). Combined effects of two ocean change stressors, warming and acidification, on fertilization and early development of the Antarctic echinoid S*terechinus neumayeri*. *Polar Biology 35* (7): 1027–1034. https://doi.org/10.1007/s00300–011–1150–7

Fabri-Ruiz, S., Saucède, T., Danis, B. et al. (2017). Southern Ocean Echinoids database – an updated version of Antarctic, Sub-Antarctic and cold temperate echinoid database. *ZooKeys 697*: 1–20. https://doi.org/10.3897/zookeys.697.14746

Fabri-Ruiz, S., Danis, B., Navarro, N. et al. (2020). Benthic ecoregionalization based on echinoid fauna of the Southern Ocean supports current proposals of Antarctic Marine Protected Areas under IPCC scenarios of climate change. *Global Change Biology 26*: 2161–2180. https://doi.org/10.1111/gcb.14988

Fabry, V.J., McClintock, J.B., Mathis, J.T. et al. (2009). Ocean acidification at high latitudes: the Bellwether. *Oceanography 22* (4): 160–171. https://doi.org/10.5670/oceanog.2009.105

Fillinger, L., Janussen, D., Lundälv, T. et al. (2013). Rapid glass sponge expansion after climate-induced antarctic ice shelf collapse. *Current Biology 23* (14): 1330–1334. https://doi.org/10.1016/j.cub.2013.05.051

Fonseca, V.G., Sinniger, F., Gaspar, J.M. et al. (2017). Revealing higher than expected meiofaunal diversity in Antarctic sediments: a metabarcoding approach. *Scientific Reports 7*: 6094. https://doi.org/10.1038/s41598–017–06687-x

Frederich, M., Sartoris, F.J., Arntz, W.E. et al. (2000). Haemolymph Mg (2^+) regulation in decapod crustaceans : physiological correlates and ecological consequences in polar areas. *Journal of Experimental Biology 203*: 1383–1393.

Frenot, Y., Gloaguen, J.C., Picot, G. et al. (1993). *Azorella selago* Hook. used to estimate glacier fluctuations and climatic history in the Kerguelen Islands over the last two centuries. *Oecologia 95* (1): 140–144. https://doi.org/10.1007/BF00649517

George, K.H. (2014). Chapter 5.4 Southern Ocean Harpacticoida (crustacean: Copepoda). In: *Biogeographic Atlas of the Southern Ocean* (eds. C. De Broyer, P. Koubbi, H.J. Griffiths et al.), 88–93. Cambridge, Scientific Committee on Antarctic Research

Giere, O., (2009). *Meiobenthology. The Microscopic Motile Fauna of Aquatic Sediments*. Berlin: Springer-Verlag.

Gogarty, B., McGee, J., Barnes, D.K.A., et al (2019). Protecting Antarctic blue carbon: as marine ice retreats can the law fill the gap? *Climate Policy 20* (2): 149–162 DOI: 10.1080/14693062.2019.1694482

Gooday, A.J., Rothe, N., Bowser, S.S. et al. (2014). Chapter 5.2. Benthic Foraminifera. In: *Biogeographic Atlas of the Southern Ocean* (eds. C. De Broyer, P. Koubbi, H.J. Griffiths et al.), 74–82. Cambridge: Scientific Committee on Antarctic Research.

Gordon, A.L., Taylor, H.W. and Georgi, D.T. (1977). Antarctic oceanographic zonation. In: *Polar Oceans, Proceedings of the Polar Oceans Conference* 44–76. Arctic Institute of North America.

Griffiths, H.J. (2010). Antarctic marine biodiversity – what do we know about the distribution of life in the southern ocean? *PLoS ONE 5* (8): e1168. https://doi.org/10.1371/journal.pone.0011683

Griffiths, H.J., Barnes, D.K.A. and Linse, K. (2009). Towards a generalized biogeography of the Southern Ocean benthos. *Journal of Biogeography 36* (1): 162–177. https://doi.org/10.1111/j.1365–2699.2008.0197

Griffiths, H.J., Whittle, R.J., Roberts, S.J. et al. (2013). Antarctic crabs: invasion or endurance? *PLoS ONE 8* (7): e66981. https://doi.org/10.1371/journal. pone.0066981

Griffiths, H.J., Whittle, R.J., Roberts, S.J. et al. (2014). Chapter 5.21. Decapoda: Crabs and Lobsters. In: *Biogeographic Atlas of the Southern Ocean* (eds. C. De Broyer, P. Koubbi, H.J. Griffiths et al.), 185–189. Cambridge: Scientific Committee on Antarctic Research.

Griffiths, H.J., Meijers, A.J.S. and Bracegirdle, T.J. (2017). More losers than winners in a century of future Southern Ocean seafloor warming. *Nature Climate Change 7* (10): 749–754. https://doi.org/10.1038/nclimate3377

Guillaumot, C., Fabri-Ruiz, S., Martin, A. et al. (2018). Benthic species of the Kerguelen Plateau show contrasting distribution shifts in response to environmental changes. *Ecology and Evolution 8* (12): 6210–325. https://doi.org/10.1002/ece3.4091

Guinotte, J.M. and Fabry, V.J. (2008). Ocean acidification and its potential effects on marine ecosystems. *Annals of the New York Academy of Sciences 1134* (1): 320–342. https://doi.org/10.1196/annals.1439.013

Gutt, J. (2007). Antarctic macro-zoobenthic communities: a review and an ecological classification. *Antarctic Science 19* (2): 165–182. https://doi.org/10.1017/S0954102007000247

Gutt, J. and Schickan, T. (1998). Epibiotic relationships in the Antarctic benthos. *Antarctic Science 10* (4): 398–405. https://doi.org/10.1017/s0954102098000480

Gutt, J. and Starmans, A. (1998). Structure and biodiversity of megabenthos in the Weddell and Lazarev Seas (Antarctica): ecological role of physical parameters and biological interactions. *Polar Biology 20* (4): 229–247. https://doi.org/10.1007/s003000050300

Gutt, J., Barratt, I., Domack, E. et al. (2011). Biodiversity change after climate-induced ice-shelf collapse in the Antarctic. *Deep-Sea Research Part II: Topical Studies in Oceanography 58* (1–2): 74–83. https://doi.org/10.1016/j.dsr2.2010.05.024

Gutt, J., Cape, M., Dimmler, W. et al. (2013a). Shifts in Antarctic megabenthic structure after ice-shelf disintegration in the Larsen area east of the Antarctic Peninsula. *Polar Biology 36* (6): 895–906. https://doi.org/10.1007/s00300–013–1315–7

Gutt, J., Griffiths, H.J. and Jones, C.D. (2013b). Circumpolar overview and spatial heterogeneity of Antarctic macrobenthic communities. *Marine Biodiversity 43* (4): 481–487. https://doi.org/10.1007/s12526–013–0152–9

Gutt, J., Barnes, D.K.A. and Lockhart, S.J. (2014). Chapter 5.28. Classification and spatially explicit illustration of Antarctic macrobenthic assemblages: a feasibility study. In: *Biogeographic Atlas of the Southern Ocean* (eds. C. De Broyer, P. Koubbi, H.J. Griffiths et al.), 229–232 Cambridge: Scientific Committee on Antarctic Research.

Gutt, J., Bertler, N., Bracegirdle, T.J. et al. (2015). The Southern Ocean ecosystem under multiple climate change stresses – an integrated circumpolar assessment. *Global Change Biology 21* (4): 1434–1453. https://doi.org/10.1111/gcb.12794

Gutt, J., Isla, E., Xavier, J. C., Adams, B. J. et al. (2020). Antarctic ecosystems in transition – life between stresses and opportunities. *Biological Reviews 96* (3): 798–821. https://doi.org/10.1111/brv.12679

Gutzmann, E., Martínez Arbizu, P., Rose, A. et al. (2004). Meiofauna communities along an abyssal depth gradient in the Drake Passage. *Deep Sea Research Part II: Topical Studies in Oceanography 51* (14–16): 1617–1628. doi:10.1016/j.dsr2.2004.06.026.

Halanych, K.M. and Mahon, A.R. (2018). Challenging dogma concerning biogeographic patterns of antarctica and the Southern Ocean. *Annual Review of Ecology, Evolution and Systematics 49*: 355–378. https://doi.org/10.1146/annurev-ecolsys-121415–032139

Hancock, A.M., King, C.K., Stark, J.S. et al. (2020). Effects of ocean acidification on Antarctic marine organisms: a meta-analysis. *Ecology and Evolution 10* (10): 4495–4514. https://doi.org/10.1002/ece3.6205

Hedgpeth, J.W., (1971). Perspectives in benthic ecology in Antarctica. In: *Research in the Antarctic Publication No. 93* (ed. L.O. Quam,), 93–136. Washington, DC: American Association for the Advancement of Science.

Henley, S.F., Cavan, E.L., Fawcett, S.E. et al. (2020). Changing biogeochemistry of the southern ocean and its ecosystem implications. *Frontiers in Marine Science 7*: 1–31. doi:10.3389/fmars.2020.00581

Hétérier, V., David, B., De Ridder, C. et al. (2008). Ectosymbiosis is a critical factor in the local benthic biodiversity of the Antarctic deep sea. *Marine Ecology Progress Series 364:* 67–76. https://doi.org/10.3354/meps07487

Hilbig, B. (2004). Polychaetes of the deep Weddell and Scotia Seas – composition and zoogeographical links. *Deep-Sea Research II 51*: 1817–1825.

Ingels, J., Vanreusel, A., Brandt, A. et al. (2012). Possible effects of global environmental changes on Antarctic benthos: a synthesis across five major taxa. *Ecology and Evolution 2* (2): 453–485. https://doi.org/10.1002/ece3.96

Ingels, J., Hauquier, F., Raes, M. et al. (2014). Chapter 5.3. Antarctic free-living marine nematodes. In: *Biogeographic Atlas of the Southern Ocean* (eds. C. De Broyer, P. Koubbi, H.J. Griffiths et al.), 83–87. Cambridge: Scientific Committee on Antarctic Research.

Janosik, A.M., Mahon, A.R. and Halanych, K.M. (2011). Evolutionary history of Southern Ocean Odontaster sea star species (Odontasteridae; Asteroidea). *Polar Biology 34* (4): 575–586. https://doi.org/10.1007/s00300–010–0916–7

Jansen, J., Dunstan, P.K., Hill, N.A. et al. (2020). Integrated assessment of the spatial distribution and structural dynamics of deep benthic marine communities. *Ecological Applications 30* (3): e02065. https://doi.org/10.1002/eap.2065

Janussen, D. and Downey, R.V. (2014). Chapter 5.5. Porifera. In: *Biogeographic Atlas of the Southern Ocean* (eds. C. De Broyer, P. Koubbi, H.J. Griffiths et al.), 94–102. Cambridge: Scientific Committee on Antarctic Research.

Kaiser, S. (2014). Chapter 5.18. Antarctic and sub-Antarctic Isopod Crustaceans (Peracarida, Malacostraca). In: *Biogeographic Atlas of the Southern Ocean* (eds. C. De Broyer, P. Koubbi, H.J. Griffiths et al.), 166–172. Cambridge: Scientific Committee on Antarctic Research.

Kaiser, S., Brandão, S.N., Brix, S. et al. (2013). Review, concept and synthesis patterns, processes and vulnerability of Southern Ocean benthos: a decadal leap in knowledge and understanding. *Marine Biology 160* (9): 2295–2317. https://doi.org/10.1007/s00227–013–2232–6

Knox, G.A., (1994). *The Biology of the Southern Ocean*. Cambridge: Cambridge University Press.

Knox, G. (2006). *Biology of the Southern Ocean*. University of Canterbury.

Knox, G.A. (2007). *Biology of the Southern Ocean. CRC Marine Biology Series*. Boca Raton, FL: CRC Press.

Knox, G.A. and Cameron, D.B. (1998). The marine fauna of the Ross Sea: Polychaeta. *NIWA Biodiversity Memoir 108*: 1–125.

Kroeker, K.J., Kordas, R.L., Crim, R.N. et al. (2010). Meta-analysis reveals negative yet variable effects of ocean acidification on marine organisms. *Ecology Letters 13* (11): 1419–1434. https://doi.org/10.1111/j.1461–0248.2010.0151

Kroeker, K.J., Kordas, R.L., Crim, R. et al. (2013). Impacts of ocean acidification on marine organisms: quantifying sensitivities and interaction with warming. *Global Change Biology 19* (6): 1884–1896. https://doi.org/10.1111/gcb.12179

Krug, A.Z., Jablonski, D. and Valentine, J.W. (2009). Signature of the end-cretaceous mass extinction in the modern biota. *Science, 323*: 767–771.

Koubbi, P., De Broyer, C., Griffiths, H.J. et al. (2014). Chapter 12. Conclusions: Present and Future of Southern Ocean Biogeography. In: *Biogeographic Atlas of the Southern Ocean* (eds. C. De Broyer, P. Koubbi, H.J. Griffiths et al.), 470–475. Cambridge: Scientific Committee on Antarctic Research.

Last, P.R., Lyne, V.D., Williams, A. et al. (2010). A hierarchical framework for classifying seabed biodiversity with application to planning and managing Australia's marine biological resources. *Biological Conservation 143* (7): 1675–1686. https://doi.org/10.1016/j.biocon.2010.04.008

Lawver, L.A. and Gahagan, L.M. (2003). Evolution of cenozoic seaways in the circum-antarctic region. *Palaeogeography, Palaeoclimatology, Palaeoecology 198* (1): 11–37. https://doi.org/10.1016/S0031–0182 (03)00392–4

Lawver, L.A., Gahagam, L.M. and Dalziel, I.W.D. (2014). Chapter 3.2. Reconstructions of the southern ocean and antarctic regions. In: *Biogeographic Atlas of the Southern Ocean* (eds. C. De Broyer, P. Koubbi, H.J. Griffiths et al.), 36–42. Cambridge: Scientific Committee on Antarctic Research.

Linse, K. (2014). Chapter 5.11. Bivalvia. In: *Biogeographic Atlas of the Southern Ocean* (eds. C. De Broyer, P. Koubbi, H.J. Griffiths et al.), 126–128. Cambridge: Scientific Committee on Antarctic Research.

Linse, K., Griffiths, H.J., Barnes, D.K.A. et al. (2006). Biodiversity and biogeography of Antarctic and sub-Antarctic mollusca. *Deep-Sea Research Part II: Topical Studies in Oceanography 53* (8–10): 985–1008. https://doi.org/10.1016/j.dsr2.2006.05.003

Linse, K., Brandt, A., Bohn, J.M. et al. (2007). Macro- and megabenthic assemblages in the bathyal and abyssal Weddell Sea (Southern Ocean). *Deep-Sea Research Part II: Topical Studies in Oceanography 54* (16–17): 1848–1863. https://doi.org/10.1016/j.dsr2.2007.07.011

Miller, C.B. (2004). *Biological Oceanography*. Oxford: Blackwell Publishing.

Mitchell, E.G., Whittle, R.J. and Griffiths, H.J. (2020). Benthic ecosystem cascade effects in Antarctica using Bayesian network inference. *Communications Biology 3* (1): 1–7. https://doi.org/10.1038/s42003–020–01310–8

Moore, J.K. and Doney, S.C. (2006). Remote sensing observations of ocean physical and biological properties in the region of the Southern Ocean Iron Experiment (SOFeX). *Journal of Geophysical Research 111*: C06026. https://doi.org/10.1029/2005JC003289

Moore, J.K., Abbott, M.R., Richman, J.G. et al. (1999). SeaWiFS satellite ocean color data from the Southern Ocean. *Geophysical Research Letters 26* (10):1465–1468 DOI:10.1029/1999GL900242

Moreau, C., Jossart, Q., Danis, B. et al. (2020). The high diversity of Southern Ocean sea stars (Asteroidea) reveals original evolutionary pathways. *Progress in Oceanography 190*: 10247. https://doi.org/10.1016/j.pocean.2020.102472

Mühlenhardt-Siegel, U. (2011). Cumacean (Peracarida, Crustacea) endemism and faunal overlap in Antarctic deep-sea basins. *Deep-Sea Research Part II: Topical Studies in Oceanography 58* (1–2): 68–73. https://doi.org/10.1016/j.dsr2.2010.10.010

Mühlenhardt-Siegel, U., (2014). Chapter 5.20. Southern Ocean Cumacea. In: *Biogeographic Atlas of the Southern Ocean* (eds. C. De Broyer, P. Koubbi, H.J. Griffiths et al.), 181–184. Cambridge: Scientific Committee on Antarctic Research.

Munilla, T. and Soler Membrives, A. (2009). Check-list of the pycnogonids from antarctic and sub-antarctic waters: zoogeographic implications. *Antarctic Science 21* (2): 99–111. https://doi.org/10.1017/S095410200800151X

Neal, L., Linse, K., Brasier, M.J. et al. (2018). Comparative marine biodiversity and depth zonation in the Southern Ocean: evidence from a new large polychaete dataset from Scotia and Amundsen seas. *Marine Biodiversity 48* (1): 581–601. https://doi.org/10.1007/s12526–017–0735-y

Nellemann, C., Corcoran, E., Duarte, C. M. et al. (eds) (2009). *Blue Carbon. A Rapid Response Assessment*. United Nations Environment Programme, GRID-Arendal. www.grida.no

Orsi, A.H., Whitworth III, T. and Nowlin, Jr., W.D. (1995): On the meridional extent and fronts of the Antarctic Circumpolar Current. *Deep-Sea Research 42* (5): 641–673.

Page, T.J. and Linse, K. (2002). More evidence of speciation and dispersal across the Antarctic Polar front through molecular systematics of Southern Ocean Limatula (Bivalvia: Limidae). *Polar Biology 25* (11): 818–826. https://doi.org/10.1007/s00300–002–0414–7

Patarnello, T., Bargelloni, L., Varotto, V. et al. (1996). Krill evolution and the Antarctic ocean currents: evidence of vicariant speciation as inferred by molecular data. *Marine Biology 126* (4): 603–608. https://doi.org/10.1007/BF00351327

Pearse, J.S., Mooi, R., Lockhart, S.J. et al. (2009). Brooding and species diversity in the Southern Ocean: selection for brooders or speciation within brooding clades? In: *Smithsonian at the Poles: Contributions to International Polar Year Science*. (eds. I. Krupnik, M.A. Lang and S.E. Miller.), 181–196. IPY. https://doi.org/10.5479/si.097884601X.13

Post A.L., Meijers A.J.S., Fraser A.D. et al. (2014). Chapter 4. Environmental Setting. In: *Biogeographic Atlas of the Southern Ocean* (eds. C. De Broyer, P. Koubbi, H.J. Griffiths et al.), 46–64. Cambridge: Scientific Committee on Antarctic Research.

Primo, C. and Vázquez, E. (2007). Zoogeography of the Antarctic ascidian fauna in relation to the sub-Antarctic and South America. *Antarctic Science, 19* (3): 321–336. https://doi.org/10.1017/S0954102007000521

Rignot, E., Jacobs, S., Mouginot, J. et al. (2013). Ice-shelf melting around Antarctica. *Science 341* (6143): 266–270. doi:10.1126/science.1235798

Rintoul, S.R., Chown, S.L., DeConto, R.M., et al (2018). Choosing the future of Antarctica. *Nature 558*: 233–241. doi:10.1038/s41586–018–0173–4

Rogers, A.D. (2007). Evolution and biodiversity of Antarctic organisms: a molecular perspective. *Philosophical Transactions of the Royal Society B: Biological Sciences 362* (1488): 2191–2214. https://doi.org/10.1098/rstb.2006.1948

Rogers, A.D., Frinault, B.A.V., Barnes, D.K.A. et al. (2019). Antarctic futures: an assessment of climate-driven changes in ecosystem structure, function and service provisioning in the Southern Ocean. *Annual Review of Marine Science 12* (1): 87–120. https://doi.org/10.1146/annurev-marine-010419–011028

Sabu P., Anilkumar, N., George, J.V et al. (2014). The influence of air–sea–ice interactions on an anomalous phytoplankton bloom in the Indian Ocean sector of the Antarctic zone of the Southern Ocean during the austral summer. *Polar Science 8*: 370–384

Sachs, O., Sauter, E.J., Schlüter, M. et al. (2009). Benthic organic carbon flux and oxygen penetration reflect different plankton provinces in the Southern Ocean. *Reports on Polar and Marine Research Part I 56* (8): 1319–1335.

Saiz Salinas, J.I. (2014). Chapter 5.9. Sipuncula and Echiura. In: *Biogeographic Atlas of the Southern Ocean* (eds. C. De Broyer, P. Koubbi, H.J. Griffiths et al.), 117–121. Cambridge: Scientific Committee on Antarctic Research.

Saucède, T. Pierrat, B. and David, B. (2014). Chapter 5.26. Echinoids. In: *Biogeographic Atlas of the Southern Ocean* (eds. C. De Broyer, P. Koubbi, H.J. Griffiths et al.), 213–220. Cambridge: Scientific Committee on Antarctic Research.

Schiaprelli, S. (2014). Chapter 5.31. Biotic Interactions. In: *Biogeographic Atlas of the Southern Ocean* (eds. C. De Broyer, P. Koubbi, H.J. Griffiths et al.), 245–252. Cambridge: Scientific Committee on Antarctic Research.

Schiaparelli, S. and Linse, K. (2014). Chapter 5.10. Gastropoda. In: *Biogeographic Atlas of the Southern Ocean* (eds. C. De Broyer, P. Koubbi, H.J. Griffiths et al.), 122–125. Cambridge: Scientific Committee on Antarctic Research.

Schüller, M. and Ebbe, B. (2014). Chapter 5.13. Polychaetes. In: *Biogeographic Atlas of the Southern Ocean* (eds. C. De Broyer, P. Koubbi, H.J. Griffiths et al.), 134–137. Cambridge: Scientific Committee on Antarctic Research.

Sigman D.M., Hain, M.P. and Haug, G.H. (2010). The polar ocean and glacial cycles in atmospheric CO_2 concentration. *Nature 466* (7302): 47–55

Soler-Membrives, A., Munilla, T., Arango, P. et al. (2014). Chapter 5.14. Southern Ocean biogeographic patterns in Pycnogonida. In: *Biogeographic Atlas of the Southern Ocean* (eds. C. De Broyer, P. Koubbi, H.J. Griffiths et al.), 138–141. Cambridge: Scientific Committee on Antarctic Research.

Strugnell, J.M., Rogers, A.D., Prodöhl, P.A. et al. (2008). The thermohaline expressway: the Southern Ocean as a centre of origin for deep-sea octopuses. *Cladistics 24* (6): 853–860. https://doi.org/10.1111/j.1096-0031.2008.0023

Thatje, S. (2012). Effects of Capability for Dispersal on the Evolution of Diversity in Antarctic Benthos. *Integrative and Comparative Biology 52* (4): 470–482. https://doi.org/10.1093/icb/ics105

Thomalla S.J., Waldron, H.N., Lucas, M.I.,et al. (2011) Phytoplankton distribution and nitrogen dynamics in the southwest Indian subtropical gyre and Southern Ocean waters. *Ocean Science 7* (1):113–127.

Tilbrook, K.J. and De Grave, S. (2005). A biogeographical analysis of Indo-West Pacific cheilostome bryozoan faunas. In: *Bryozoan Studies A*. (eds. G.H.I. Moyano, J. Cancino and P. Wyse-Jackson), 341–35. London: Balkema.

Tripathy, S.C. and Jena, B. (2019). Iron-stimulated phytoplankton blooms in the Southern Ocean: a brief review. *Remote Sensing in Earth Systems Sciences 2* (1): 64–77. https://doi.org/10.1007/s41976-019-00012-y.

Tripathy S.C., Pavithran, S., Sabu, S. et al. (2014). Is phytoplankton productivity in the Indian Ocean sector of Southern Ocean affected by pigment packaging effect? *Current Science 107* (6): 1019–1026.

Turner, J., Comiso, J.C., Marshall, G.J. et al. (2009). Non-annular atmospheric circulation change induced by stratospheric ozone depletion and its role in the recent increase of Antarctic sea ice extent. *Geophysical Research Letters, 36* (8): 1–5. https://doi.org/10.1029/2009GL037524

Verheye, M.L., Backeljau, T. and d'Udekem d'Acoz, C. (2017). Locked in the icehouse: evolution of an endemic *Epimeria* (Amphipoda, Crustacea) species flock on the Antarctic shelf. *Molecular Phylogenetics and Evolution 114*: 14–33. https://doi.org/10.1016/j.ympev.2017.05.013

Vincx, M., Bett, B.J., Dinet, A. et al. (1994). *Meiobenthos of the Deep Northeast Atlantic. Advances in Marine Biology 30*: 1–88. doi:10.1016/s0065-2881(08)60061-9

Whittle, R.J., Witts, J.D., Bowman, V.C. et al. (2019). Nature and timing of biotic recovery in Antarctic benthic marine ecosystems following the Cretaceous–Palaeogene mass extinction. *Palaeontology 62* (6): 1–16. doi:10.1111/pala.12434

Xavier, J., Brandt, A., Ropert-Coudert, Y. et al. (2016). Future challenges in Southern Ocean Ecology research. *Frontiers in Marine Science 3* (94): 1–9. https://doi.org/10.3389/fmars.2016.00094.

Zachos, J., Pagani, H., Sloan, L. et al. (2001). Trends, rhythms and aberrations in global climate 65 Ma to present. *Science 292* (5517): 686–693. https://doi.org/10.1126/science.1059412

11

Biogeochemistry of the Antarctic Coasts: Implications for Biodiversity and Climate Change

Amrutha Karayakath[1], Jitendra K. Pattanaik[1]*, Khem C. Saini[2], Pushpendu Kundu[2], and Felix Bast[3]

ABSTRACT

Notwithstanding its largely lifeless appearance, Antarctica hosts considerable biodiversity and ecosystem (trophic-level) complexity. Antarctic soils, ice-covered areas, coastal regions, and the surrounding Southern Ocean contributes to this biodiversity. However, studies related to the number of species present, how and where they are located and their influence on the ecosystem processes are still scarce. Biodiversity and biogeochemical cycles often coincide with climate change studies. Photosynthetic carbon fixation by phytoplanktons in the Antarctic coasts influences the atmospheric CO_2 levels. On a geological timescale, such processes determine the climate. In this chapter, studies on Essential Ocean Variables and elemental cycles in Antarctica are reviewed to understand the biogeochemistry of the coast of Antarctica, and how these parameters and cycles influence the different coastal features of Antarctica. In the recent past, Antarctica has experienced a rapid rate of warming that in turn has threatened its biodiversity. The effects of climate change on biogeochemical parameters and their further consequences are therefore discussed here. The second half of this chapter deals with the prokaryotic as well as eukaryotic microbial diversities of Antarctica, and the implications they hold for climate change.

11.1. INTRODUCTION

Understanding life and its response to climatic changes at spatial scales is important for our existence. At a basic level, chemical elements are continuously exchanged between living organisms and the environment, following different biogeochemical pathways. Variations in elemental cycles can be interpreted as a response to perturbations that are happening in its immediate environment or as a result of global changes. Global warming, ocean acidification, eutrophication and acid rain are some of the indicators that show that cycling of elements have been perturbed at an enormous scale. Effects of these phenomena are hostile to the existence of life on Earth, so studies on these processes are important in climate-change-sensitive areas such as Antarctica. Antarctica is located at the southern pole and is perpetually covered with ice. It is the coldest and driest continent in the world and the Southern Ocean surrounds its landmass. Sea ice, coastal polynyas, fast ice, ice shelves and ice sheets are some of the major physical features observed in the Antarctic coast (Figure 11.1). The responses of these coastal features to climatic changes influence the biodiversity of the Antarctic coast. Impending changes in Antarctica's glacial ice mass would affect marine biogeochemistry and ecology as much as it would perturb global climate (Wassmann et al., 2011).

The glacial ice of Antarctica and Greenland cover approximately 10% of the total land area on Earth;

[1]Department of Geology, Central University of Punjab, Bathinda, 151401, India

[2]Centre for Biosciences, Central University of Punjab, Bathinda, 151401, India

[3]Department of Botany, Central University of Punjab, Bathinda, 151401, India

*Corresponding Author: Jitendra. K. Pattanaik (jitendra.bapi@gmail.com) ORCiD code: 0000–0002–3147–129X

Systems Biogeochemistry of Major Marine Biomes, First Edition. Edited by Aninda Mazumdar and Wriddhiman Ghosh.

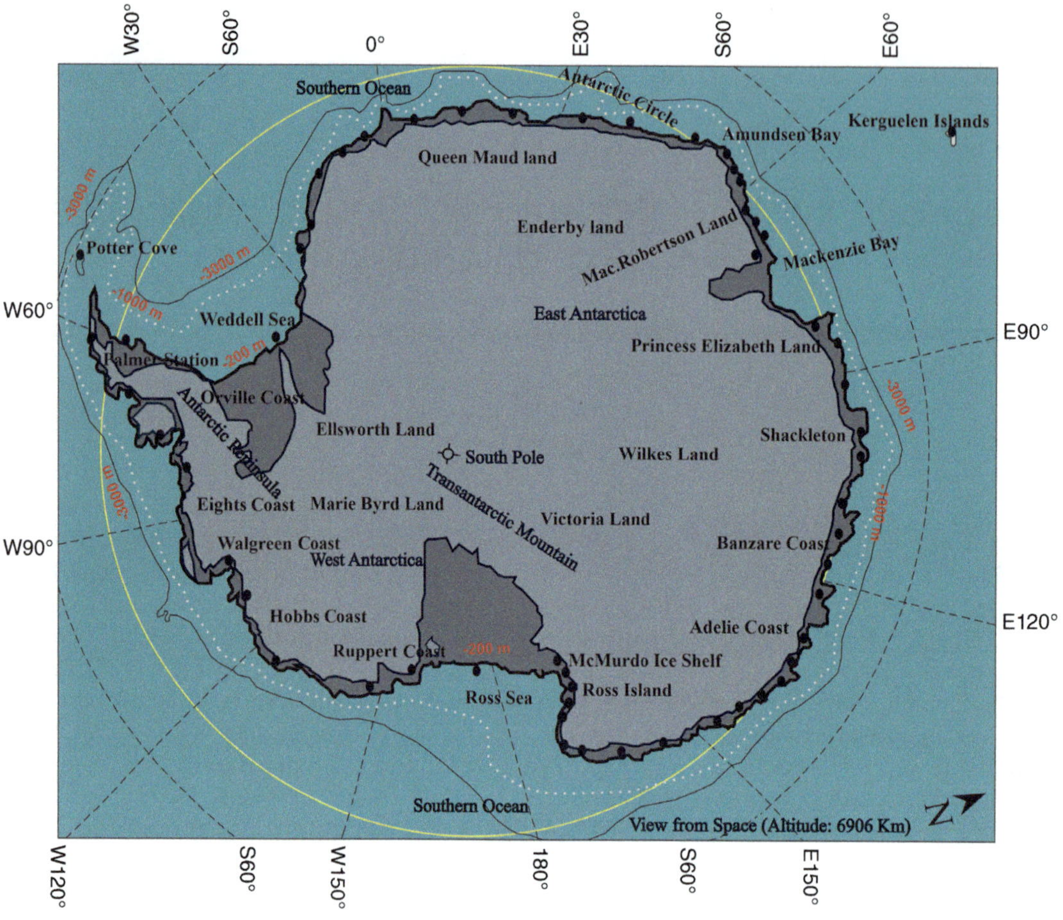

Figure 11.1 Map (modified from Google Maps, taken from altitude 6906 km) showing some of the locations studied in Antarctica. The yellow line indicates the Antarctic circle and black dots represent important land marks. The solid black line encircling Antarctica marks –3000 m bathymetry. White dotted lines mark the extent of sea ice (–1000 m), the dark gray tinted regions are the ice shelves (–200 m), and light gray colour shows the area covered by the continental ice sheet. Transantarctic Mountain separate East Antarctica from West Antarctica.

however, during the last ice age the maximum extent of land coverage by glaciers extended up to 32% (NSIDC, 2007). Approximately 90% of the Earth's ice mass is in Antarctica and for the past 40 million years, the continent has been partially covered by ice sheets (Ferrigno et al., 2002). The Lambert–Fisher glacier of Antarctica is the largest glacier in the world. At present, glaciers and their melting is important in the context of global warming. In addition to that, the cryosphere and hydrosphere have an inseparable correlation and hence glacier melting has an important role in global climate and ecosystem dynamics. Polar regions are rapidly affected by the climate change. For the past few years, the Antarctic Peninsula has experienced a rapid increase in ocean temperature and a decrease in sea ice cover, which in turn have affected the ocean stratification (see Legge et al., 2017 and references therein).

The Antarctic polar region functions as a major sink for global heat and carbon. In fact, half of the oceanic absorption of carbon (~1 petagram or billion tons) occurs in this zone. In addition, seasonal fluctuations of physical and chemical factors affect the span and duration of the ice mass in the Antarctic coast. This affects the biodiversity and cyclical gross primary productivity of the region. Ocean circulation dynamics influence the rate of CO_2 uptake from atmosphere and hence global carbon dynamics. Polar oceans are very sensitive to ocean acidification as their carbonate buffering capacity is comparatively lower (Shadwick et al., 2013). The pH, temperature, and salinity of the seawater determine the saturation level of $CaCO_3$. Ocean acidification lowers the level of $CaCO_3$ saturation in sea water (McNeil et al., 2004). Cold water itself has a lower level of $CaCO_3$ and hence the effect of ocean acidification is more pronounced in the Polar region (McNeil and Matear, 2008). This results in a growing concern about the existence of organisms that make structures from $CaCO_3$ because ocean acidification may influence their distribution with strong ecological impacts (Manno et al., 2012) and a threat to Antarctic biodiversity.

Biodiversity of Antarctica, both in the soil and ice-covered areas, is dependent on climatic factors such as global temperature variation, wind-driven changes in mixed layer depth, and differences in cloud cover on one hand, and the primary productivity of the region on the other (Montes-Hugo et al., 2009). Primary productivity in Antarctica is strongly controlled by the availability of micronutrients (especially iron), temperature, carbonate chemistry, light (seasonal variations in irradiance), and sea ice cover (Constable et al., 2014). Depending on the growth of primary producers, the consumers also thrive or decline in the ecosystem.

Energy supply to primary consumers in aquatic habitats situated at water depths <50 m is essentially rendered by phytoplanktons (Peck, 2018). The largest temporal source of iron for phytoplankton growth and physiological responses are not in the form of dust aerosols as originally thought, instead, iron is mobilized in water during contact with sediments and shelf break (Convey et al., 2014). Free water is also critical in freshwater environments, with additional important impacts from salinity, nutrient supply, oxygenation, and ice cover and extent (Convey et al., 2014). The ecological effects of reduced sea ice and increased levels of iceberg scour, reported mainly in the Antarctic Peninsula region, have been extensive in Antarctica (Peck, 2018). Antarctica features some of the strongest environmental gradients on the planet and therefore offers an ideal study ground for the importance of environmental variation to biodiversity (Convey *et al.*, 2014). First, terrestrial systems seem to be primarily abiotically driven, so that a direct connection between climate and biota should be created. Second, apart from two higher plant species which occur only in the northwestern Antarctic Peninsula, the vegetation consists entirely of cryptogamic organisms (Colesie et al., 2014). The Antarctic Peninsula is Antarctica's most vegetated region, although only 1.34% of the exposed surface has plant cover (Davey *et al.*, 2019).

The unevenness in the distribution of biodiversity indicates its connection with geographical, physical, chemical, biological, and climatic variables. Temperature, one of the important limiting factors of primary production, influences the phytoplankton growth rate, which is higher when associated with ice (Smetacek et al., 1992). The influence of photosynthetically active radiation (PAR) on biogeochemical cycles and associated changes in climate can be studied in polar region as the ice and snow cover act as a window for the underwater ecosystem, which varies locally and seasonally (Perovich et al., 1986). Ultraviolet (UV) radiations also affect the microbial mass as biological productivity decreases with increase in UV-B flux (Smith et al., 2000). Cowan et al. (2014) have reported that the microbial diversity in Antarctic soil mediate in the *in situ* carbon and nitrogen cycles. The diversity and abundance of microbes vary with the geographical location, temperature, and vegetation type present (Cowan et al., 2014). Even though Antarctica is rich in microorganisms, the resilient response of microorganisms to changes makes it difficult to predict the microbial response to climate changes (Signori et al., 2018). However, changes in the community composition of microbes in sediments reportedly affect carbon sequestration, so these can be used to link climate changes to carbon availability (Learman et al., 2016).

11.2. BIOGEOCHEMICAL CYCLES ON THE ANTARCTIC COAST

Nutrients are vital for living organisms to sustain life. They spend energy to absorb/extract chemicals from their environment and perform different biological activities (Begon et al., 2006). Nutrients exist in different sections of the biosphere: in the atmosphere, lithosphere, hydrosphere, and within the biota. Dead and living matters consist of elements in organic form whereas the other sections have inorganic forms of the elements (Begon et al., 2006). Elements are available to primary producers in the form of simple inorganic molecules or ions in the atmosphere or as dissolved ions in water for fixing energy and biochemical synthesis of organic materials (Lucas, 2001). These elements become incorporated into biomass as complex organic compounds. Elements such as C, N, P, H, and O are the basic building blocks of macromolecules such as carbohydrates, proteins, lipids, and nucleic acids. Anabolism uses these elements to create compounds essential for the organism. During catabolism the compounds are broken down to release energy. Metabolism correlates nutrient cycling with energy flux. In addition to the energy released, nutrients released into environment in organic and inorganic forms are available for the next cycle of energy production. An individual atom of a nutrient element may undergo repeated cycling (Moss and Balls, 1989). That is, the elements are recycled and used again but the molecules with which they are associated may change (Andreae et al., 1984). In this way, nutrients get trapped in living biomass, while decomposition of the biomass releases these elements back to the surrounding system/community. Weathering of rocks and dissolved minerals in rain also qualify as sources of supply (Bormann & Likens, 1967). In natural systems no element cycle occurs in isolation; instead, multiple cycles involving different physical or redox states of multiple elements operate simultaneously (Cole, 2013). Recycling of minerals is never comprehensive as some nutrients are transported from land to ocean by runoffs while some gaseous nutrients escape to the atmosphere.

Although geochemical cycles occur in the absence of living organisms, the latter's role in cycling nutrients in the biosphere is crucial (Waring and Schlesinger, 1985). Studies conducted in the coastal region of Antarctica using different biogeochemical parameters (Table 11.1) have suggested that seasonal variations, local processes, and atmospheric circulations influence the cycling of elements which is also modulated by the global climatic changes. Variations in different *in situ* parameters can be correlated with different chemical and physical processes. For example, a higher surface temperature accompanies when the breakout of sea ice occurs.

In Antarctic coasts, studies related to climate change start with the measurement of essential ocean variables (EOVs). More information about the EOVs can be obtained from the website of the Global Ocean Observing System (GOOS: www.goosocean.org/). These include, physical, biogeochemical, and biological and ecosystem variables that are essential for ocean observation at a global scale (Miloslavich et al., 2018). Routine measurement and analysis of these variables are used to understand the ocean dynamics. Past studies from the Antarctic coasts show that a wide range of methods have been used to measure different parameters associated with biogeochemistry. Remote sensing and traditional methods are also used to calculate/measure different biogeochemical parameters and their spatial and temporal variations. Physicochemical parameters such as salinity and temperature of the sea surface are measured using bucket thermometers (Shetye et al., 2017). An autoanalyzer based on standard colorimetric methods is used to measure nutrients such as silicate, phosphate, nitrate, and nitrite (Grasshoff et al., 1983). Coulometers can be used to determine the total carbon dioxide (TCO_2) content of seawater. A TOC-V-CSH analyzer is commonly used to measure total organic carbon (TOC) (Shetye et al., 2017). Different satellite-based data sets can also be used to calculate the variables such as the temperature of sea surface (SST), amount of PAR, the upwelling velocity of the region, chlorophyll *a* content, and sea-ice coverage. Understanding EOVs is the initial step in the study of the biogeochemistry of the Antarctic coasts, their biodiversity, and implications for global climate.

11.3. ANTARCTIC COASTAL FEATURES IN THE CONTEXT OF BIOGEOCHEMISTRY AND CLIMATE CHANGE

Proximity to both land and oceans and its nutrient resources make Antarctic coast considerably productive. Sea ice, coastal polynyas, fast ice, ice shelves, and ice sheets are the important physical features of Antarctic coastal ecosystem. While the majority of coastal biodiversity is associated with these features, the coastal biogeochemistry is also determined by the processes happening in these areas. Ice, as a moving entity in addition to its dynamic phase change, contributes to the geomorphology and biodiversity of the coast, where it is/was present in the form of sea ice, lake ice and/or glaciers (Viles and Spencer, 2014). Therefore, ice adds an extra dimension to the geological and ecological processes in Antarctic coast. Different components of the cryosphere such as snow, sea ice, and lake ice contribute/respond to local climate changes significantly. In contrast, the ice sheets, ice shelves, and glaciers, contribute/respond to global climate change and play significant roles in long-term changes, including the ice age cycles (Ravindra and Laluraj, 2012). Traditionally, Antarctica has been discussed in the context of global environmental change. However, it is important to explore the potential of this region to understand marine biogeochemistry and its feedbacks on climate and marine ecosystems (Shetye et al., 2017 and references therein). Sea ice, fast ice, coastal polynyas, ice shelves, and ice sheets (Figure 11.2) are important components of Antarctic coast and to understand the impact and feedback of climate change, it is essential to study these components.

11.3.1. Sea Ice

Sea ice cover acts as an interface between the atmosphere and the ocean controlling mass and energy transfer between the two realms. It has an important role in determining the radiative and thermodynamic properties of the sea surface (Massom and Stammerjohn, 2010). Sea ice reduces the pelagic productivity of the ocean by reducing the availability of sunlight necessary for primary production (Lizotte, 2001; Fitch and Moore, 2007). During the formation of sea ice, CO_2-rich brine water is flushed out from it (Rysgaard et al., 2007) and this increases the fresh water budget upon melting. In the Antarctic coastal area, this brine water rejection and fresh water release together influence the Antarctic Bottom Water (AABW) and further affect the global ocean circulation (Lubin and Massom, 2006). Therefore, sea ice is a vital component in shaping the thermohaline stratification of marine ecosystems and influences/controls the primary productivity and biogeochemistry of the ice-covered regions (Shetye et al., 2017).

Growing sea ice incorporates different constituents such as macronutrients, iron, organic matter, and sediments, which later get released into seawater when the ice melts. As it is a seasonal process, sea ice acts as a seasonal reservoir for some elements such as iron (Lannuzel et al., 2007). It reduces the iron limitation for phytoplankton growth (Lancelot et al., 2009) and is reflected as phytoplankton blooms (Arrigo et al., 2008a; Smith and Comiso, 2008). In the context of biological activity in the sea ice, it is important to note that in the Southern Ocean, ice algae contribute to 25% of overall primary production of ice-covered waters (Lizotte, 2001; Arrigo and Thomas, 2004). Sea ice continuously accumulates organic matter as a result of autotrophic and heterotrophic

Table 11.1 Summary of some of the studies carried out in the coastal region of Antarctica using different biogeochemical parameters are listed here.

Sl. No	Location and period of study	Biogeochemical parameters reported	Implications for global climate or/ and biodiversity	Reference
1	Eastern Weddell Sea, Antarctica, between 23 January and 23 February	Chl-a, DIC, PAR, pH, POC, PON, SST, SSS, TA, Oxygen concentration.Nutrients: Silicate, Phosphate, Nitrate	Algal blooms affect the nutrient dynamics and associated biogeochemical processes in sea ice habitat. The nutrient regeneration in sea ice ecosystems has no major impact by the heterotrophic oxidation of organic matter.	Gunther et al (1999)
2	Kerguelen Islands, Southern Ocean, austral spring 1995	Algal pigments, BSi, Chl-*a*, Dissolved iron, DMS, DMSP, PAR, SSS, SST.Nutrients : nitrate, ammonium and silicic acid	This highlights the role of a light mixing regime in controlling phytoplankton biomass and shows that region surrounding oceanic islands has high phytoplankton biomass	Blain et al (2001)
3	Fast ice in the Weddell sea, February 1998	Chl-a, pH, POC, PON, POP, stable carbon isotope compo-sition of DON, DOC, TA.Nutrients: phosphate, ammonium, nitrite, nitrate, and silicate	Discusses the Role of sea ice in the bentho-pelagic coupling and suggest that the stable carbon isotope composition of organic material can be used for ice algae tracer	Thomas et al (2001)
4	Palmer Station, Anvers Island, Western Antarctic Peninsula, 1995–1996	Precipitation, temperatures, UV-B and UV-A radiation	Reports the importance of temperature and water as environmental variables and shows that variation in UV radiation affects the ecosystem functioning	Convey et al (2002)
5	McMurdo dry valleys, East Antarctica, 1986–2000	Air temperature, PAR, relative humidity, soil temperature, solar flux, wind speed	Shows that the interaction between topography and air masses can greatly influence the microclimate.	Doran et al (2002)
6	37 Antarctic coastal polynya systems, 1997–2002	Chl-a	Reports the relationship between the annual primary production of a post-polynya and the size of its associated breeding colonies	Arrigo and Dijken (2003)
7	Northern and southern Victoria Land, Antarctica, December and January of 2003–2004	Ammonium and nitrate concentration, C and N isotope concentration, CO_2 flux, conductivity, gravimetric soil moisture, pH, soil temperature	Discusses the covariation in soil bacterial communities with soil biogeochemical properties and says that, across climate gradients and local landscapes, the influence of soil properties on habitat suitability vary	Barrett et al (2006)
8	Shallow-shelf habitats in Antarctica (<100 m depth)	$CaCO_3$ solubility, population of different species, SST	Global climate change affects the terrestrial, freshwater, and marine species by altering their geographic ranges, behaviours, and phenologies. Biological invasion is dependent on different physiological factors	Aronson et al (2007)

(*Continued*)

Table 11.1 (Continued)

Sl. No	Location and period of study	Biogeochemical parameters reported	Implications for global climate or/ and biodiversity	Reference
9	Victoria Land, 2005–2006	Al, Fe, OC, S, TN, w-eHg	Chemistry of cyanobacterial mats and sediments from Antarctic lakes can be used as a proxy to understand the effects of climatic and environmental changes on biogeochemical processes	Bargagli et al (2007)
10	Coastal area of Adélie Land, Spring 1999	Chl-a, CO_2, DIC, DMS, O_2, PCO_2, pH, SST, SSS,TA	Among three environments such as sea–ice brine, platelet ice-like layer, and underlying water showed different temporal variations of CO_2, O_2, and DMS. However, the temporal variations within each environment were similar.	Delille et al (2007)
11	Ross Sea, many annual cycles	FCO_2, pCO_2	Antarctic shelf waters are a strong sink for CO_2 because of its high biological productivity, intense winds, high ventilation rates, and extensive winter sea ice cover	Arrigo et al (2008b)
12	13 Coastal polynyas, 1992–2001	Air temperature, ice thickness, SSS, SST,wind speed	Antarctic coastal polynyas are important production house of Southern Ocean sea ice and the years 2000 and 2002 witnessed a significant decrease in the polynya area and the ice production.	Tamura et al (2008)
13	Marguerite Bay on the Antarctic Peninsula, February 2005 and March 2007	Seawater concentrations of bromoform ($CHBr_3$) and dibromomethane (CH_2Br_2)	Antarctic coasts are hot spots of Bromocarbon production and hence it has an influence on the chemistry of the Antarctic atmosphere	Hughes et al (2009)
14	Western Antarctic Peninsula, 1978–1986, 1998–2006	Chl-a	The phytoplankton dynamics along the WAP shelf is controlled by combination of atmosphere, ice and ocean processes	Montes-Hugo et al (2009)
15	Northern Marguerite Bay, Western Antarctic Peninsula, 1998–2012	Chl-a, conductivity, fluorescence macronutrients, PAR, pressure, temperature	Links the variability in phytoplankton bloom strength to changes in ice extent and winter mixed layer depths.	Venables et al (2013)
16	Antarctic ice sheet	DOC, Eh of the water, pO_2, POC TOC.Nutrients: N, P, C, Si and bioavailable Fe	Antarctic ice sheet has an important role in earth's carbon cycle and it acts as a mechanism for nutrients to be exported in runoff and icebergs	Wadham et al (2013)
17	Antarctic dry valleys	Carbon turnover, nitrogen turnover, PAR, temperature,	Microbial diversity in Antarctic soil drives key biogeochemical cycling processes such as the carbon and nitrogen cycles. Microbial abundance and diversity vary with vegetation type, sampling location and temperature.	Cowan et al (2014)

Table 11.1 (Continued)

Sl. No	Location and period of study	Biogeochemical parameters reported	Implications for global climate or/ and biodiversity	Reference
18	Arthur Harbor, Coastal Antarctic waters, 2012–2013	Biological oxygen saturation (ΔO_2/Ar), Chl-a, DIC, macronutrient concentration, NCP PAR, pCO_2, SSS, SST, TA	Variation in NCP is driven by factors such as physical mixing, grazing, and light availability. High frequency NCP variability is associated with the seasonal cycle. Discuss the role of climate in marine productivity	Tortell et al (2014)
19	Western Antarctica (covering the Ross Sea to the Weddell Sea), December 2013 to February 2014, November to December 2014	Ammonium concentration, $\delta^{13}C$ TOC, TN	Discusses the relationship between microbial communities and sediment biogeochemistry. Change in community composition modifies the carbon sequestration process and this can be used to link the climate change to carbon availability	Learman et al (2016)
20	McMurdo Ice Shelf (MIS) near Ross Island, Antarctica, 18 December 2012	Chl-a, DIC, DO, DOC, PC, pH, PN, SST, TOC. Nutrients: ammonium, phosphate, nitrate, and nitrite	Ice shelf thinning shifts the biogeochemistry of Antarctic coastal system and has a cascading effect on element cycles and biological processes	Vick-Majors et al (2016)
21	Western Antarctic Peninsula, 2009–2014 spring–summer seasons	Chl-a, dissolved organic matter,	Long term data collection on microbial community structure and function is important to understand the changes in transitions, mechanisms and implications of microbial community structure	Bowman et al (2017)
22	Ryder Bay, West Antarctic Peninsula, December 2010 to March 2014	Chl-a, DIC, FCO_2, fluorescence, pressure, salinity, TA, temperature, macronutrients	Carbon cycling is significantly affected by sea ice cover and changes in mixing pattern. Both DIC and TA exhibit a strong asymmetric seasonal cycle	Legge et al (2017)
23	Northern Marguerite Bay (West Antarctic Peninsula), 1997–2012	Chl-a, conductivity, fluorescence, pressure, temperature	Phytoplankton growth, biomass and composition in West Antarctic Peninsula is governed by sea ice cover and summer stratification. Seasonal changes in these factors are likely to have an effect on the entire Antarctic marine food web.	Rozema et al (2017)
24	Enderby Basin, Antarctic Coastal region, Indian Sector, from the austral summer until the onset of winter (January 2012, February 2010 and March 2009)	Air–sea CO_2 flux, Chl-a, PAR, PCO_2, pH, SSS, SST, TCO_2, TOC, upwelling velocity. Nutrients: silicate, phosphate, nitrate and nitrite	In Antarctic coast, the seasonal variation in biogeochemical factors is driven by sea ice and sea ice has a dual role in determining climate change	Shetye et al (2017)
25	Western Antarctic Peninsula	NA	The extra ordinary degree of spatial and temporal variability of climate in Western Antarctic peninsula is driven by sea ice variability, atmospheric interactions, glacial inputs, and exchange processes between open ocean and shelf waters	Hendry et al (2018)

(*Continued*)

Table 11.1 (Continued)

Sl. No	Location and period of study	Biogeochemical parameters reported	Implications for global climate or/ and biodiversity	Reference
26	Potter Cove (Antarctica), February and March 2015	Chl-a, DIC, fucoxanthin pigment concentration, oxygen concentration, phaeophytin pigment concentration, RQ, TC, TIC, TN, and TOC.Nutrients: phosphate, ammonium, nitrite, nitrate, and sulfate	Reports the secondary effects of glacial melting on benthic community and associated biogeochemical cycles. Shows that the benthic mineralization processes are also affected by this glacial melt	Hoffmann et al (2018)
27	Northern Antarctic Peninsula (NAP) Spring, Summer, and Autumn seasons from 2013 to 2015	Salinity, temperature	Temperature has an important role in shaping microbial diversity and community structure in NAP region	Signori et al (2018)
28	Barilari Bay, West Antarctic Peninsula, January 22 and 25 and February 22 and 24, 2010	Chl-a, conductivity, DO, $\delta^{18}O$, PAR, pCO_2, SSS, SST, temperature. Nutrients: phosphate, ammonium, nitrite, nitrate, and sulfate	Ice–ocean interaction impact the export of melt water and has played an important role in coastal biogeochemistry and carbon cycling. Ocean upwelling is the primary source of macronutrients	Cape et al (2019)
29	Western Antarctic Peninsula marine system	Salinity, temperature	Ecosystem restructuring owing to climate and sea ice changes can be demonstrated through ocean climate interactions and its biological and biogeochemical consequences	Henley et al (2019)

BSi, Biogenic silica; Chl-a, chlorophyll a; DIC, dissolved inorganic carbon; DMS, dimethyl sulphide; DMSP, dimethyl sulphonioproprionate; DO, dissolved oxygen; DOC, dissolved organic carbon; Eh, electronic potential; FCO_2, redox air–sea flux of CO_2; NA, not available; NCP, net community production; PAR, photosynthetically active radiation; PC, particulate carbon; PCO_2, partial pressure of CO_2; POC, particulate organic carbon; PON, particulate organic nitrogen; PON, particulate organic nitrogen; PN, particulate nitrogen; RQ, respiration quotient; SSS, sea surface salinity; SST, sea surface temperature; TA, total alkalinity; TCO_2, air–sea CO_2 flux; TIC, total inorganic carbon; TN, total nitrogen; TOC, total organic carbon; w-eHg, water-extracted Hg.

production. Hence wind and/or ocean current induced drift of sea ice redistributes this organic matter (Vancoppenolle et al., 2013). In Enderby basin, Shetye et al., (2017) reported a higher surface temperature in the month of February immediately after the break out of fast ice. From the Enderby Basin, the lowest salinity was reported during February and the highest during March. Melting of the glacial ice mass results in the dilution of surface water and causes low salinity, whereas freezing of the water during sea ice formation causes rejection/ concentration of salt to the surface water, resulting in higher salinity. Hence the fluctuation in salinity can be partly correlated with the freezing and thawing of sea ice.

Iron has an important role in the nutrient cycling. For example, nutrient uptake by diatoms is dependent on the Fe concentration of the environment. The Fe concentration in surface water is related to the melting of sea ice. Melting of sea ice releases the Fe trapped in it, which is maximum during February, as recorded in the The Enderby Basin (Shetye et al., 2017). This region recorded a high macronutrient concentration compared with open ocean (SiO_4 maximum during January, and NO_3 and PO_4 maxima during March) alongside pronounced seasonal cycling (Shetye et al., 2017). Furthermore, pH values recorded in the Enderby Basin were at a maximum during January and decreased further when freezing started in March (Shetye et al., 2017). Although sea ice appears to be a blockade to atmosphere–ocean gas exchange, it is rarely so in reality as the sea ice itself is permeable when warm (Golden et al., 1998). The presence of open waters such as polynyas within the ice also facilitates gaseous exchange (Morales Maqueda and Rahmstorf, 2002).

Sequestration of atmospheric CO_2 (Tréguer and Pondaven, 2002), ocean upwelling, outgassing, and ocean acidification are some of the processes that are dependent

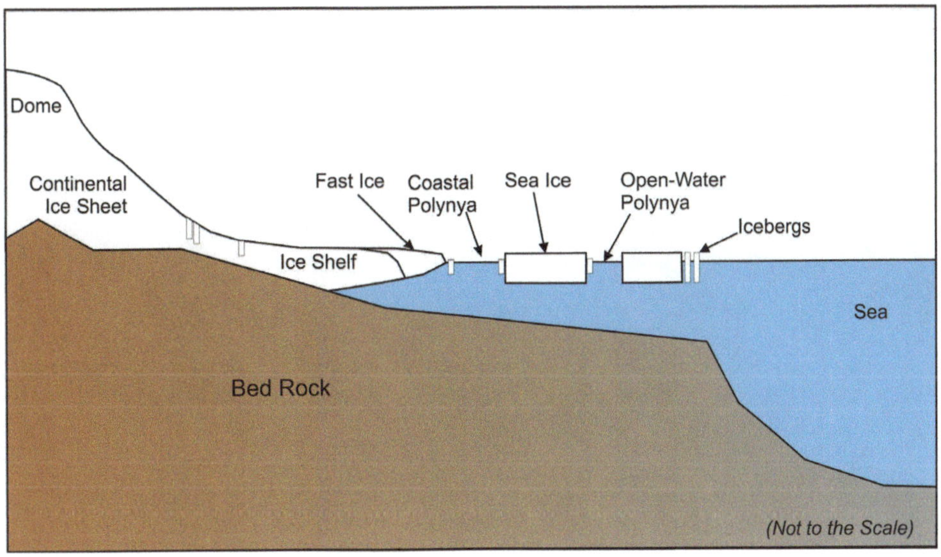

Figure 11.2 Schematic cross-section of the Antarctica Coast (not to the scale). The extent of continental ice sheet and ice shelf, relative position of fast ice, polynya, and sea ice are shown.

on the dynamics of sea ice (Fabry et al., 2009). Sea ice processes also contribute to other biogeochemical cycles such as those of carbon and sulfur. The ocean carbon cycle is driven by biological and solubility pumps. Fixation of inorganic carbon to organic carbon, $CaCO_3$ precipitation, and CO_2 release occurs through biological pumps (Sigman et al., 2010). Inorganic carbon transport during brine rejection associated with sea ice freezing plays a crucial role in marine carbon cycles (Vancoppenolle et al., 2013). Solubility pumps are associated with both the physical and chemical processes involving CO_2 dissolution and release (Vancoppenolle et al., 2013). Overall, the oceans absorb approximately one-quarter of all anthropogenic CO_2 emission (Sabine et al., 2004), thereby minimizing the effect of global warming. However, more absorption of CO_2 by ocean results in ocean acidification (Figure 11.3) and has a potential consequence for marine food webs (Doney et al., 2009).

Gaseous exchange between the atmosphere and ocean is affected by the seasonal variation in sea ice coverage and thickness. Studies of Antarctic coastal regions have shown that partial pressure of CO_2 (pCO_2), which drives CO_2 exchange between ocean and atmosphere, exhibits increasing trend from January to March; minimum TCO_2 was recorded during February-March. Important factors causing high pCO_2 during March could be the intense decay of organic matter (Pipko et al., 2002) and the ice cover preventing the release of CO_2 and other gases (Gordon et al., 1984; Klatt et al., 2002). In addition to this, variations in upwelling velocities of the Antarctic Circumpolar Current also contributes to the pCO_2 concentration, where high upwelling velocities ($0.3–0.4 \times 10^{-5}$ m s^{-1}) during the winter and lower velocities (0.1×10^{-5} m s^{-1}) in December/January results in a higher pCO_2 during March (Comiso et al., 1993). High latitude coastal regions of Antarctica are reported to be CO_2 sinks. For example, the Enderby Basin acts as a strong sink during January but transforms into a weak one during February–March (Shetye et al., 2017). Variations in pCO_2 in the Enderby Basin is biologically driven as it is too large to be developed from changes in temperature and salinity (Shetye et al., 2017). Longitude dependent variation in physicochemical parameters during January, February, and March shows the dynamic nature of coastal Antarctic waters (Louanchi et al., 2001) and the seasonal variation plays a pivotal role in regulating the biogeochemistry.

The extent and dynamics of sea ice have major impact on the marine ecosystem directly as well as indirectly. The effects occurring in the West Antarctic peninsula (WAP) is an example of this (Massom and Stammerjohn, 2010). Variability in sea ice is important in a biological context as it has either positive or negative effect on the different levels of the marine food web and primary production (Massom and Stammerjohn, 2010). Sea ices acts as a substrate for concentrated algal biomass and is a rich source of food for grazers such as Antarctic krill (*Euphausia superba*) when the water column lacks required nutrients (Smetacek et al., 1990; Quetin and Ross, 2009). Even the different life stages of Antarctic krill are related to the freezing and melting dynamics of the sea ice (Massom and Stammerjohn, 2010). Similarly, there is a positive correlation between the presence of ice krill (*Euphausia crystallorophias*) and the retreat of sea ice in the WAP

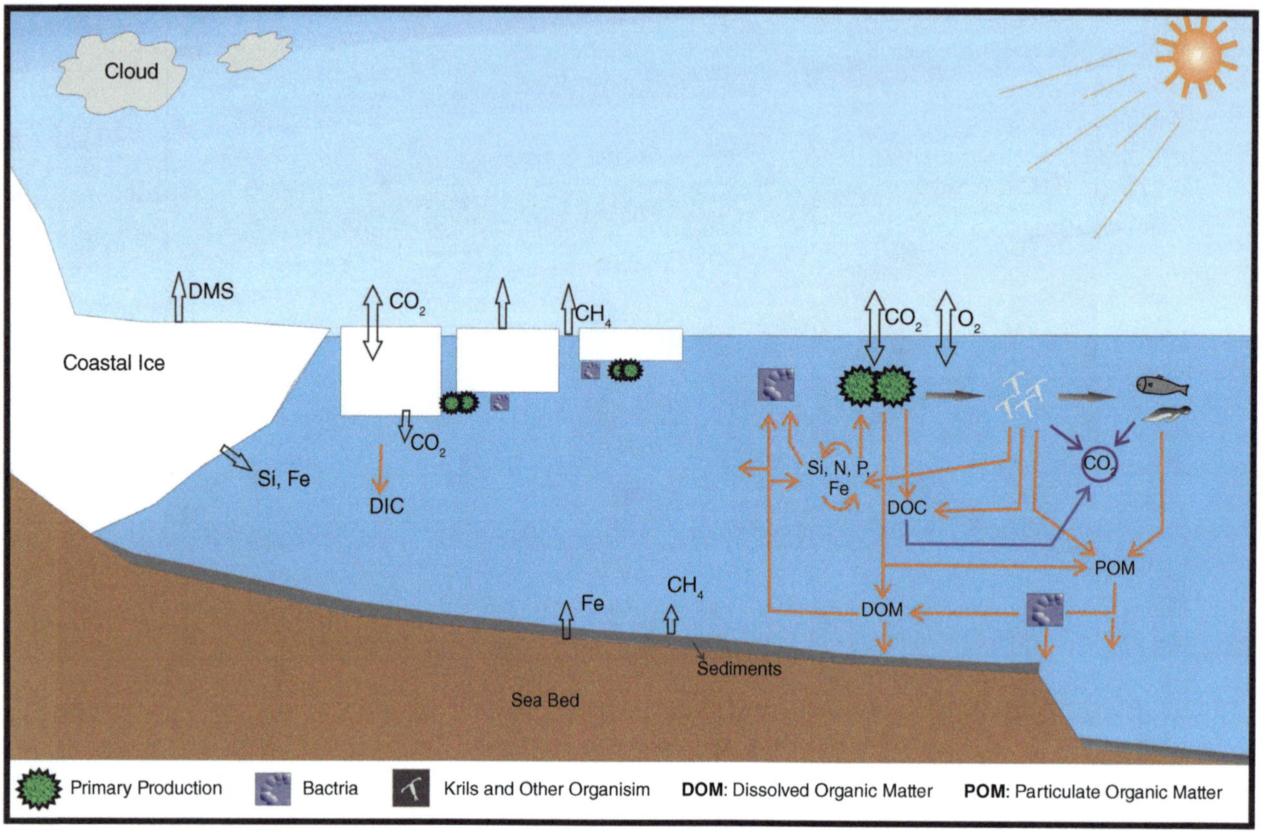

Figure 11.3 Biogeochemical cycle in the Antarctica coast (not to the scale) illustrating the food chain, nutrient cycle, CO_2 cycle and exchange of gases with the atmosphere. For a detailed discussion please refer to the text. DIC, dissolved inorganic carbon; DMS, dimethyl sulfide; DOC, dissolved organic carbon. (Modified from Grzymski et al., 2012, Vancoppenolle et al., 2013 and Cavan et al., 2019.)

region (Ross et al., 2008). Antarctic krills are present in the pelagic environment and are key members of the continental shelf ecosystem (Smith et al., 2006; Smetacek and Nicol, 2005). Furthermore, albeit sea ice acts as a barrier between water and air, it also serves as a habitat for many air-breathing predators (Tynan et al., 2010).

Sea ice processes contribute to an induced biological production of dimethyl sulfide (Figure 11.3). This cloud condensation nucleus has a potential cooling effect on the planet (Charlson et al., 1987; Watson and Liss, 1998). It is also an important osmotic regulator and cryoprotectant that contributes to the oceanic sulfur cycle (Tison et al., 2010). Significant ocean–atmosphere methane (CH_4) fluxes, and bromine oxide flux, are observed in sea ice zones (Vancoppenolle et al., 2013). Bromine components are very important in polar atmospheric chemistry (Simpson et al., 2007), while CH_4 is an important greenhouse gas that is incorporated into the water column and sea ice from anoxic sediments (Figure 11.3) (Shakhova et al., 2009). The degree to which these factors affect the biogeochemical system varies according to the areal extent and type of the sea ice, its concentration, dynamics, thickness distribution, and the degree of deformation (Massom and Stammerjohn, 2010). Complex interrelationships between these factors and their spatial and seasonal variations also contribute to this. Thomas et al. (2009) reported on the importance of sea ice in biogeochemical cycles of the Southern Ocean. This validates the role of sea ice as an ecological and biogeochemical driver (Massom and Stammerjohn, 2010) for not only coastal Antarctica but also areas well beyond the ice-cover (Turner et al., 2009).

11.3.2. Coastal Polynyas

Antarctic coastal polynyas, generally regarded as winter phenomena, are areas of reduced ice cover or open water within the ice pack (Arrigo and Van Dijiken, 2003). Coastal polynyas, although constituting only a small proportion of the overall sea ice zone, play globally important roles by producing sea ice and influencing AABW (Rintoul, 1998; Bindoff et al., 2000). Coastal polynyas are the areas of high sea ice production and salt flux (high salinity shelf water formation) and these areas have an important role in a variety of physical and biological processes (Vaz and Lennon, 1996; Bindoff et al., 2001; Buffoni et al., 2002). Surface waters associated with polynyas are the first among all polar marine systems to be exposed to spring solar radiation, and thereby experience enhanced levels of biological production during spring and summer (Mundy and Barber, 2001). The AABW has highest salinity in the Ross Sea and it is consistent with the observation that the Ross Ice Shelf polynya has the greatest amount of ice production (Orsi et al., 1999). Synoptic scale winds and strong katabatic scale winds control the formation of wind-driven latent heat polynyas and strong katabatic wind polynyas, respectively (Arrigo and Van Dijiken, 2003). Strong katabatic winds continuously replace the older ice with young fragile ice and the synoptic winds add upwelling or tidal forcing-driven heat component to the polynyas (Arrigo and Van Dijiken, 2003).

Polynya acts as an opening between deep ocean and atmosphere and it is the site of major phytoplankton blooms in spring (Arrigo and Van Dijken, 2003; Tremblay and Smith, 2007; Massom and Stammerjohn, 2010). The growth and abundance of diatoms, dinoflagellates and prymnesiophytes are higher within polynyas (Kopczyńska et al., 1995; Von Quillfeldt, 1997; Arrigo et al., 2000). This enhanced productivity along with bloom remnants and high fecal pellet production results in higher particle flux in the polynyas (Becquevort and Smith, 2001; Gowing et al., 2001; Cooper et al., 2002).

In areas such as WAP (Western Antarctic Peninsula), coastal polynyas are associated with high upwelling causing high productivity in the region (Ribic et al., 2008). These characteristics of this region have resulted in colonization by animals such as Emperor penguin (*Aptenodytes forsteri*) near recurring polynyas (Massom et al., 1998). On longer timescales the size and dynamics of coastal polynyas are controlled by climatic condition (Arrigo and Van Dijken, 2003). If the size of major polynyas decreases, this can have major impact on regional primary production, regional sea ice production and ocean fresh water budget and associated global thermohaline circulation (Rintoul, 2007).

11.3.3. Fast Ice

Fast ice forms around the ice sheet periphery, covering over a few kilometers to 200 km. It is often associated with icebergs that are grounded in waters <350–400 m deep and act as anchor points for fast ice formation (Giles et al., 2008; Massom et al., 2001). Variations in the patterns of fast ice formations and break up are sensitive to different atmospheric and ocean processes, hence they can act as indicators of climate change (Heil, 2006; Massom et al., 2009; Murphy et al., 1995).

Light availability and open water area are dependent on different wind and sea ice regimes. Fluctuations in this regime have a direct impact in determining phytoplankton biomass (Montes-Hugo et al., 2009). Fast ice acts as a breeding platform for Weddell seals (Thomas and DeMaster, 1983) and emperor penguins (Kirkwood and Robertson 1997; Burns and Kooyman, 2001) and because of its stability it is an important habitat for microorganisms (Lizotte, 2003; Deming, 2010) and sea ice algae (Satoh and Watanabe 1988; McMinn et al., 2000). While the emperor penguins are closely associated with the fast ice, the variability in pack ice also affects their colonies (Barbraud and Weimerskirch, 2001).

11.3.4. Ice Shelves

Once glaciers leave the land and float on the ocean surface, they are called ice shelves (Vick-Majors et al., 2016). Around 75% of Antarctica's coastlines are surrounded by ice shelves covering more than 1.561 million km^2 of coastal ocean in Antarctica (Rignot et al., 2013). The Ross Ice Shelf (RIS) accounts for one-third of the area of all Antarctic ice shelves and it is the most extensive ice shelf on earth (Fox et al. 1994; Depoorter et al., 2013). The RIS is an important part of Ross Sea (Figure 11.1), where AABW formation occurs and contributes to global thermohaline circulation (Jacobs et al., 1979; Budillon et al., 2011). Marine waters beneath ice shelves are reported to harbor active ecosystems. Even though thick ice prevents penetration of sunlight, the microbial ecosystem is supported by chemoautotrophic activity, advection of nutrients and biomass, and horizontal advection from adjacent open waters (Vick-Majors et al., 2016). However, many of the ice shelves surrounding Antarctica's coastline are under threat of collapse and thinning (Joughin et al., 2014; Paolo et al., 2016) and this loss might lead to imbalance in the biogeochemical cycle associated with the ice shelf region; for example, influencing a possible imbalance in carbon biogeochemistry at a local as well as global scale (Sigman et al., 2010).

11.3.5. Ice Sheets

The Antarctic Ice Sheet is a huge reservoir of organic carbon and microbial communities (Wadham et al., 2013) and is affected by the changes in atmospheric circulation, planetary albedo, and ice elevation/thickness (Pollard and DeConto, 2009). It also facilitates formation of bottom water beneath ice shelves (Flower and Kennett, 1994; Mackintosh et al., 2014) and influences oceanic temperature and circulation. The biologically active base of the ice sheet and the interchange of biogenic gases through ice sheet dynamics facilitate nutrient cycling in the surroundings (Wadham et al., 2013). The Antarctic ice sheet is the most extensive glacial microbial habitat on earth (Priscu et al., 2008). It has an important role in Southern Ocean primary productivity as it releases nutrients to melt waters, which are exported in runoff and icebergs to the Southern Ocean (Wadham et al., 2013 and references therein).

The East Antarctic Ice Sheet (EAIS) is the largest continental ice mass on Earth and about 98% of Antarctica is covered by thick ice sheets which constitute 90% of the world's permanent ice and is the main reservoir of fresh water (Wadham et al., 2013 and references therein). As the ice sheet is the largest continental ice mass, its response to climate change is of great importance. Different climate models (DeConto and Pollard, 2003; Huber et al., 2004; Pollard and DeConto, 2009) have suggested that the stability of Antarctic ice sheet is triggered by the atmospheric CO_2 concentration. Therefore, reconstructing the variability history of Antarctic ice sheets is essential in order to predict their stability in future years and to forecast their contribution to future sea level changes driven by climate change (Suganuma et al., 2014). Overall, the biogeochemistry of Antarctic coast is not a permanent/stable phenomenon behavior in any season. Depending on the location of the coast and type characteristics of sea ice, coastal polynyas, fast ice, ice shelves, and ice sheets, the biogeochemistry varies. Antarctic coastal features act as inhibitors or facilitators of the different transformations involved in the biogeochemical cycles. While the global climate change determines the fate of Antarctic coastal features, the biogeochemical cycles and biodiversity are directly affected by the regional response of coastal features to global/regional climate change.

11.4. INTERLINK BETWEEN BIOGEOCHEMICAL CYCLES AND CLIMATE CHANGE

The Earth's climate is reliant on diverse processes (physical, chemical, and biological) occurring in the biosphere (Denman et al., 2007). Therefore, changes or fluctuations in the biogeochemical cycles associated with the atmosphere, land, and ocean have direct and indirect effects on climate, and vice versa (Doney and Schimel, 2015; USGCRP, 2017). Biogeochemical cycles aid the transformation of energy and matter into usable forms for sustaining life and hence supports the functioning of the ecosystem (Brusseau, 2019). Biogeochemical cycles associated with the atmosphere, land, and ocean occur in tandem, where terrestrial biogeochemical cycles affect the concentration of greenhouse gases such as CO_2, CH_4, nitrous oxide (N_2O) and aerosol particles (USGCRP, 2017). These constituents determine the radiative properties of atmosphere and in doing so affect the entire biosphere. Human activities such as industrial emissions, fossil fuel burning, and deforestation have not only altered the earth's biosphere but also impacted upon the Earth's climate and biota (Doney et al., 2011).

Air mass circulation, cloud formation, precipitation, and fluctuations in relative humidity are some of the important atmospheric processes that can affect biogeochemical cycles. In addition to acting as transportation media for the distribution of various chemical components across the atmosphere, hydrosphere, and biosphere, these processes can also affect/control the genesis of various chemical constituents. This shows that variation in climatic pattern has a direct influence on the transfer and distribution of elements (Campbell et al., 2009). Biogeochemical cycles of chemical elements are important as they provide clues to the transformation of mobilizable chemicals that propagate through the environment as a result of their volatility or solubility (Andreae et al., 1984).

Climate change is considered to be a slow process; however, the effects of climate change on biogeochemical cycles vary in terms of timescale (Shaver et al., 2000). For example, biogeochemical processes involving variations in chemical components of soil owing to changes in desorption–absorption dynamics and microbial activities respond instantly to climate change. Feedback in intermediate time ranges involves changes in vegetation and an example of slow response, or long-term feedback, would be changes in soil genesis, and in soil organic matter content (Campbell et al., 2009).

Biogeochemistry and climate are interrelated as most of the chemical elements involved in the biogeochemical cycle have natural sources and sinks that respond to changes in climate (Doney and Schimel, 2015). Concentrations of different elements such as carbon and nitrogen affect the structure of plants on land and this in turn affects the amount and quality of chemical inputs into soils (USGCRP, 2017). Although greenhouse gases are an essential/natural part of the climate, increases in their concentrations have resulted in global warming and associated concerns. Although CO_2, as a greenhouse gas, is not as powerful as CH_4, the excess of anthropogenic CO_2 makes it the most significant gas and an important component of current global climate change (IPCC Working Group I, 2013). In the context of global warming it is therefore important to understand the global carbon cycle.

Global carbon cycle is a phenomenon involving carbon fluxes between a series of reservoirs (Ciais et al., 2013). The ocean is one of the most important reservoirs of carbon, with bicarbonates, carbonic acids and carbonate ions being its key inorganic forms. Carbon also exists in the form of dissolved organic carbon (Hansell et al., 2009). The ocean biological pump is an important component of the carbon cycle which determines atmospheric pCO_2 (Kwon et al., 2009). Particulate organic matter (POM) is consumed or decays within the mesopelagic zone (100–1000 m) of the water mass, releasing CO_2 to the atmosphere and leaving the remaining material to sink into the deep ocean (Martin et al., 1987). Depending on the efficiency of organic matter transfer along the depths, phytoplankton community structure changes (Weber et al., 2016). Atmospheric CO_2 is sensitive to this carbon transport; in addition, atmospheric CO_2-induced climate warming reduces ocean carbon storage as the diversity of subtropical communities increases in response to such changes (Weber et al., 2016). At the bottom of the ocean carbon is stored in the pellets and remnants of animal dead bodies. Either the zooplanktons and diatoms use this carbon or heterotrophic organisms feed on them. Fluctuations in the biogeochemical cycling of carbon are relevant to the understanding of global warming, which reduces CO_2 uptake by the ocean as warmer upper territories hold less CO_2 than the cold lower waters (Doney and Schimel, 2015). Anthropogenic emission of CO_2 directly affects ocean chemistry. Global warming can cause climatic alterations which in turn can affect the physical circulation of oceans and hence their biogeochemistry and biodiversity. Mixing of water masses, air–sea CO_2 flux, $CaCO_3$ precipitation/dissolution, photosynthesis, and respiration are the main factors that control dissolved inorganic carbon cycle (Legge et al., 2017).

Carbon cycling is central to climate change studies. Under natural and stable conditions every element is cycled in the biosphere (i.e. it reaches its sink and is ready for the next cycle). Anthropogenic activities have resulted in an excess concentration of certain elements in such a way that the sink is unable to accommodate them, thereby causing global warming, ocean acidification, acid rain, and/or eutrophication of water masses. Eutrophication occurs when an aquatic ecosystem receives excess amount of nutrients, causing excessive algal growth, called bloom. Algal bloom increases the biological oxygen demand of the ecosystem and reduces the dissolved oxygen present (Sullivan et al., 2010). When the level of greenhouse gases increases in the atmosphere, it causes its warming. An increase in sulfur dioxide and nitrate oxides in the atmosphere then react with water, forming compounds that reduce the pH of water and cause acid rain. This is a threat to organisms living in the ocean as well as the land. Ocean acidification occurs when the pH of the sea water decreases substantially, mainly owing to increased uptake of CO_2. Sea water becoming acidified due to an imbalance in biogeochemical cycles has direct and indirect effects on the flora and fauna as they find it difficult to exist at low pH levels. Therefore, depending on biodiversity and its seasonal variation, various biogeochemical cycles can be used as proxies to study global climate change.

11.5. FLORAL AND MICROBIAL DIVERSITY OF ANTARCTICA: DISTRIBUTION AND IMPLICATIONS FOR CLIMATE CHANGE

11.5.1. Diversity of Bacteria and Archaea

A huge diversity of microorganisms has been isolated from various habitats of Antarctica and identified using polyphasic taxonomic approaches (Shivaji and Reddy, 2010). Antarctic bacterial diversity has also been studied widely via culture-independent (environmental 16S rRNA gene sequence-based) methods. The various types of habitats explored include soil, water, cyanobacterial mats, sediments, subglacial outflow, sandstones, sediment core, ephemeral streams, ice core, geothermal vents, and those inhabited by penguins, and sponges (Shivaji, 2017). Different microbial communities involved in numerous ecological processes such as nutrient, carbon, phosphorus, nitrogen, and sulphur cycling are present in Antarctic snow and soil samples. Abundance of bacteria in different Antarctic habitats ranges from 0.2×10^7 to 0.6×10^{12} cells g^{-1} for ice (Carpenter et al., 2000), 0.2×10^2–10^7 cells ml^{-1} for water (Franzmann et al., 1990; Giudice et al., 2012), 8×10^6 to 2.4×10^7 cells g^{-1} for sediment (Stibal *et al.*, 2012), and 10^5–10^{10} cells g^{-1} for soil (Aislabie et al., 2009; Shivaji, 2017; Shivaji et al., 2019). Members of the genus *Sphingomonas* contribute heavily to the microbial biomass and nutrient recycling in the oligotrophic environment of coastal ice, and methylophilic genera such as *Methylobacillus*, *Methylophilus*, and *Methylovorus* are also present (Antony *et al.*, 2012). Bacterial diversity varies with Antarctic soil type: peninsular soils are dominated by Proteobacteria, west Antarctic soil by Actinobacteria, Transantarctic mountain soil by Bacteriodetes (Bottos *et al.*, 2014) and *Deinococcus–Thermus*, and Victoria Land soils by Actinobacteria (Shivaji, 2017). Based on polyphasic taxonomy, about 209 novel bacterial species have been described from different habitats of Antarctica (Bottos et al., 2014; Chattopadhyay et al., 2014; Reddy et al., 2016; Shivaji, 2017; Shivaji et al., 2019). The characteristic feature unifying all Antarctic bacteria is their psychrophilic adaptation for growth at zero/subzero to 30°C. Ji et al. (2017) reported that atmospheric trace gases serve as the primary energy sources for microbial communities of the Antarctic soil surface.

Antarctica's aquatic microbial diversity is limited as water constitutes only 0.4% of Antarctica's total seasonal ice-free part (Wilkins et al., 2013). Within this ice-free area bacterial communities are dominated by the phyla Proteobacteria (especially Alphaproteobacteria), followed by Bacteroidetes, Actinobacteria, Chloroflexi, Acidobacteria, Firmicutes, Gemmatimonadetes and Verrucomicrobia. The most common genera are *Arcicella*,

Brevundimonas, *Caulobacter*, *Duganella*, *Flavobacterium*, *Hymenobacter*, *Janthinobacterium*, *Pedobacter*, *Polaromonas*, *Prevotella*, *Sphingomonas*, *Rhodoferax*, and *Variovorax* (Wilkins et al., 2013).

Shivaji et al. (2004) reported diverse bacterial communities from Antarctica belonging to genera *Acidovorax*, *Brevundimonas*, *Burkholderia*, *Enterobacter*, *Flexibacter*, *Gemmatimonas*, *Microbacterium*, *Nevskia*, *Rhodococcus*, *Rhodopseudomonas*, *Serratia*, *Sphingomonas*, and *Xanthomonas*. Mikucki and Priscu, (2007) reported *Thiomicrospira*, *Betaproteobacteria*, *Deltaproteobacteria*, *Gammaproteobacteria*, and Bacteroidetes associated with Blood Falls, a Subglacial Outflow from the Taylor Glacier, Antarctica. Archer et al. (2017) reported Actinobacteria, Bacteriodetes, Chloroflexi, and Proteobacteria in sandstone and granite from the McMurdo Dry Valleys. Chaya et al. (2019) revealed high diversity of *Betaproteobacteria*, followed by Cyanobacteria, Bacteroidetes, and Firmicutes in the freshwater glacial lake Yukidori-Ike in Langhovde, East Antarctica. Sommers et al. (2018) reported the dominance of Bacteroidetes, followed by Proteobacteria, Cyanobacteria, Actinobacteria, Verrucomicrobia, Acidobacteria, Planctomycetes and Gemmatimonadetes in ice-lidded cryoconite holes on glaciers in the Taylor Valley. Pudasaini et al. (2017) revealed the bacterial communities of Browning Peninsula, Eastern Antarctica to be dominated by Actinobacteria, followed by Chloroflexi, Cyanobacteria and Proteobacteria (10%). Teixeira et al. (2010) reported the predominance of the genera *Bifidobacterium* (phylum Actinobacteria), *Arcobacter* (phylum Proteobacteria) and *Faecalibacterium* (phylum Firmicutes) in the rhizospheric soil associated with the plants *Deschampsia antarctica* Desv. (Poaceae) and *Colobanthus quitensis* (Kunth) Bartl (Caryophyllaceae) in the coastal areas of Admiralty Bay, maritime Antarctica. Peeters et al. (2012) showed that heterotrophic bacterial communities encompassing culturable members of Actinobacteria, Bacteroidetes, Proteobacteria, Firmicutes, and *Deinococcus–Thermus* dominated aquatic microbial mats in continental Antarctica and Antarctic Peninsula. González-Rocha et al. (2017) reported culturable diversity of *Flavobacterium* and *Pseudomonas* from Fildes Peninsula King George Island, Antarctica. Wang et al. (2017) reported that the predominant phyla in the intertidal sediments of Fildes Peninsula maritime Antarctic were Proteobacteria, Bacteroidetes, and Actinobacteria. Bowman et al. (2017) studied the community structure of bacteria based on ecophysiological data to form predictive models of microbial ecosystem functioning along the coasts of the WAP region. Signori et al. (2018) studied the seasonal and temporal variations in archaeal and bacterial diversities in the Northern Antarctic Peninsula during the spring, summer, and autumn seasons of the years 2013–2015. They concluded that the higher abundance of phytoplankton groups occurred during spring, and were characterized by the presence of the haptophyte *Phaeocystis* and the diatom *Corethron*, which in turn influenced the succession of heterotrophic bacterial communities. They found that microbial community structure varied significantly over time and with increase in sea surface temperature but not over space.

Only a few studies have reported the presence of archaea in Antarctic glacial environments. Archaea identified in the coastal Antarctic snowpack mainly belonged to two genera, the haloalkaliphilic *Natronococcus* sp. and halophilic *Halalkalicoccus* sp. (Antony *et al.*, 2016). Achberger et al. (2016) studied the microbial community structure of Subglacial Lake Whillans, West Antarctica, and discovered abundant chemolithoautotrophic bacteria alongside archaea that are capable of oxidizing reduced iron, sulfur, and nitrogen compounds. Saxton et al. (2016) investigated anaerobic oxidation of methane (AOM) in Lake Fryxell, Antarctica, and observed that AOM rates are highest in the deep waters where Atribacteria and Bathyarchaeota mediate AOM using humic acids as electron shuttles. A study by Monteiro et al. (2020) on ammonia-oxidizing archaea (AOA) and ammonia-oxidizing bacteria (AOB) showed that AOA exhibited enhanced adaptability to various geochemical changes compared with AOB in the McMurdo Dry Valleys of Antarctica. In contrast to bacterial diversity, archaeal diversity detected thus far on glacier surfaces is much lower (Antony *et al.*, 2016).

Little data is available to assess the direct and indirect impacts of increased temperature on Antarctica's microbial diversity. Although the immediate effect of elevated temperature on Antarctic microbial communities is unknown (Pearce, 2008), a study by Yergeau et al. (2007) has shown that the diversity of bacteria declines with increased climatic severity across a range of Antarctic terrestrial habitats. Furthermore, Yergeau et al. (2009) reported a decline in microbial diversity with increasing latitude from the Falklands Islands, South Orkney Islands (maritime Antarctic), and western Antarctic Peninsula. Rinnan et al. (2009) also studied soil bacterial communities over an Antarctic climate gradient and discovered that bacterial growth was lowest in the coldest regions and highest in Antarctica's warmest areas. Kim et al. (2018) performed open-top chamber (OTC) experiments in the Fildes Peninsula, maritime Antarctic, during the summer seasons of 2013–2015 and reported that warming increased the degradation rate of humic substances leading to increased carbon and nitrogen and enriched soil microbial communities. Dennis et al. (2019) revealed that increases in air temperature influence soil bacterial diversity in maritime Antarctica, while Van Gestel et al. (2020) estimated an approximately 25%

increase of bacterial abundance in Antarctic marine water and soil on account of a few degrees rise in temperature.

11.5.2. Diversity of Fungi

Fungi and yeasts are easily dispersed and colonized on various substrata, including desert, ice, soils, snow, sediments, rocks, permafrost, and vegetation such as bryophyte, lichens and flowering plants (Ruisi et al., 2007). The mycoflora of Antarctica is relatively diverse within the different climatic regions. A significant proportion of Antarctic microfungi is cosmopolitan, and most of them have been introduced to Antarctica by historic woods, soils, and ocean and wind currents (Ruisi et al., 2007; Duarte et al., 2019). The dispersal of fungi may also occur through vectors, such as dust, plant seeds, and birds (Margesin and Miteva, 2011). Antarctic fungi are adapted to low temperatures, low nutrient availability, repeated freeze and thawing cycles, osmotic stress, desiccation, and high UV radiation. Both culture-dependent and culture-independent methods have been used to assess Antarctic fungal diversity. Most of the major studies on Antarctic mycoflora, however, have used culture-dependent methods (Connell et al., 2018; Duarte et al., 2018, 2019).

The Antarctic mycoflora is dominated by the phylum Ascomycota, followed by Basidiomycota; Zygomycota, Chytridiomycota, Cryptomycota and Blastocladiomycota have also been reported from soils and ice-covered lakes (Cox et al., 2016; Rojas-Jimenez et al., 2017). Taxonomic identification of Antarctic yeasts has established the presence of different kinds of Basidiomycetes species, including *Candida humicola*, *Cryptococcus* sp. *Bulleria alba*, *Rhodotorula glutinis*, and *Rhodotorula rubra* (Goto *et al.*, 1969; Vishniac and Baharaeen, 1982; Shivaji and Prasad, 2009). Endemic fungi belonging to the class Eurotiomycetes have also been reported from Antarctica (Cox et al., 2016). Of all the species of fungi known from Antarctica 0.6% are water molds, while the rest are true fungi (Onofri et al., 2004; Ruisi et al., 2007). Baublis et al. (1991) described *Chrysosporium pannorum*, *Candida cferrii*, *Cephalosporium acremonium*, *Cladosporium elaturn*, *Phialophora lagerbergii*, and *Penicillium notaturn* from various microsites in Taylor Valley, Antarctica. Stchigel et al. (2003) isolated two new species of Ascomycetes, *Apiosordaria antarctica* and *Thielavia antarctica*, from soil and a lichen sample *Usnea* cf. *aurantioatra*, respectively, in King George Island, Antarctica (Stchigel et al., 2003). Rosa et al. (2010) described the diversity of endophytic fungi belonging to the genera *Aspergillus*, *Cadophora*, *Davidiella*, *Entrophospora*, *Fusarium*, *Geomyces*, *Gyoerffyella*, *Microdochium*, *Mycocentrospora*, and *Phaeosphaeria*, which grow on the surface of leaves of *C. quitensis* at Admiralty Bay, King George Island. Gonçalves et al. (2012) explored the diversity of filamentous fungi in different lakes of Antarctic Peninsula and identified 128 fungal isolates dominated by *Geomyces pannorum* and *Mortierella* sp. Alves et al. (2019) studied the diversity of rock fungi from the South Shetlands archipelago, maritime Antarctica, and reported the predominance of *Cladophialophora*, *Cladosporium*, *Cyphellophora*, *Eichleriella*, *Paracladophialophora*, and *Penicillium*. Hirose et al. (2016) investigated fungal diversity on different species of mosses in East Antarctica and found *Phomaherbarum*, *Alternaria*, *Phaeosphaeria*, and *Tetracladium* to be the predominant genera, while *Leotiomycetes*, *Pleosporales*, *Phaeosphaeria*, *Cosmospora*, and *Helotiales* were also present. Möller and Dreyfuss (1996) reported 14 non-lichenized fungi associated with Antarctic lichens, and the most abundant genera were *Phoma*, *Pseudogymnoascus*, *Chalara*, and *Phaeoseptonia*. While *Pseudogymnoascus* sp., *Thelebolus* sp., *Antarctomyces psychrotrophicus* and *Vishniacozyma victoriae* have been reported to be the parts of non-lichenized fungal diversity associated with *Usnea antarctica* and *Usnea aurantiaco-atra*, (Santiago et al., 2015; De Carvalho et al., 2019), mycelial fungi belonging to Leotiomycetes and Dothideomycetes have been found to predominate in natural and human-affected Antarctic soils from the surroundings of Russian research stations in maritime Antarctica (Kochkina et al., 2019).

The diversity of yeasts in Antarctica is believed to be low as the overall primary productivity of the region is below the levels required for luxurious yeast colonization. *Cryptococcus* species of yeasts can assimilate a wide range of carbon and nitrogen sources because of their repertoire of diverse enzymes. Members of *Aspergillus* and *Penicillium* are also capable of degrading a wide range of carbon substrates (Antony *et al.*, 2016). *Cryptococcus laurentii*, *Cryptococcus albidus*, and *Rhodotorula minuta* grow on Antarctic mosses while *Candida scottii* is the predominant soil yeast of this territory (Di Menna, 1960). Goto et al. (1969) isolated the basidiomycetous yeast *Moesziomyces antarcticus* (initially described as *Pseudozyma antarctica*), from 9 m deep sediments of Lake Vanda in Antarctica. Duarte et al. (2016) evaluated the diversity of yeasts in macroalgae and lichens in South Shetland Islands: while the dominant species in macroalgae were *Glaciozyma litorale*, *Metschnikowia australis*, *Mrakia* sp., and *Rhodotorula glacialis*, *Cryptococcus victoriae*, *Rhodotorula arctica*, *Rhodotorula laryngis*, *Mrakia* sp. and *Trichosporon* sp. predominated in lichens. Tsuji et al. (2017) isolated two new yeast species *Cystobasidium tubakii* and *Cystobasidium ongulense* from soil samples of East Ongul Island, East Antarctica. Ferreira et al. (2019) studied the diversity of epiphytic yeasts associated with the bryophytes *C. quitensis* and *D. antarctica* of Antarctic Peninsula, and found *Vishniacozyma victoriae*, *Mrakia gelida*, *Dioszegia antarctica*, and *Leucosporidium creatinivorum* to be the most abundant species. Most of the psychrophilic yeasts, however, were isolated from biofilms encompassing *Glaciozyma martinii*, *Glaciozyma antarctica*, *Mrakia frigida*, *M. gelida*, *Mrakia robertii*, *Phenoliferia glacialis* and *Phenoliferia psychrophenolica* (Ferreira *et al.*, 2019).

An increase in the diversity of yeast and fungi caused by climate change could have an adverse effect on Antarctic ecosystems. However, there are only a few reports evaluating the impact of global warming on Antarctic mycoflora. Newsham et al. (2016) used the 'space-for-time' substitution approach and observed that soil mycoflora is much greater in numbers in warmer habitats in maritime Antarctica. Selbmann et al. (2012) reported the extinction of an endemic Antarctic fungus *Cryomyces* sp. resulting from global warming.

11.5.3. Diversity of Lichens

Lichens are symbiotic organisms encompassing fungi (mycobionts) and green algae or cyanobacteria (photobionts). They can grow in extreme environments such as deserts, polar regions, and on exposed substrata such as rocks, stones, bark, soils, leaves, ice, ancient monuments, old buildings, and other man-made surfaces. Lichens are the pioneer species in ecological successions and act as bio-indicators of air pollution (Lücking, 1998; Conti and Cecchetti, 2001; Moore et al., 2011; Nayaka, 2014; Saini et al., 2019). According to Kirk et al. (2008), the lichenized fungi group consists of about 20,000 species worldwide. Dodge (1973) reported 429 species of lichens from Antarctica, while Convey (2006) summarized the biodiversity of Antarctic lichens under three biogeographic zones, with around 250 lichen species reported from the sub-Antarctic zone, 250 from the maritime zone, and 150 lichen species from the continental Antarctic zone. The wide diversity of lichens growing in Antarctica include members of *Acarospora*, *Buellia*, *Candellariella*, *Lecidea*, *Lecanora*, *Physcia*, *Rhizocarpon*, *Rinodina*, *Umbilicaria*, *Xanthoria*, *Turgidosculum*, and *Usnea* (Singh et al., 2018).

Upreti et al. (1999) reported 23 species (belonging to 16 genera) of lichens from the Maitri region of the Schirmacher Oasis, Antarctica. Seppelt et al. (2010) reported 29 species of lichens dominated by *Buellia*, *Caloplaca*, *Lecanora*, *Lecidea*, *Physcia*, *Rhizocarpon*, *Rhizoplaca*, *Umbilicaria*, and *Xanthoria* from Botany Bay, Granite Harbour, Ross Sea, Antarctica. Singh et al. (2015) reported 54 lichen species from the Schirmacher Oasis, Queen Maud Land, continental Antarctic, 48 from Victoria Land, Ross Sector, continental Antarctica, and 244 from Admiralty Bay, South Shetland Islands, and observed that 21 species were common to all these areas. Olech and Słaby (2016) studied changes in lichen diversity in the Lions Rump area, King George Island (1988–1990 vs. 2007–2008), and reported three species *Polyblastia gothica*, *Thelenella kerguelena*, and *Thelocarpon cyaneum* to be missing in the second expedition, in addition to a reduction in the number of sites for *Leptogium puberulum* and *Staurothelle gelida* and an increase in the range of pioneering species such as *Bacidia chrysocolla*, *Caloplaca johnstonii*, *Candelariell aurella*, and *Lecanora dispersa*. Singh and Nayaka (2017) reported 69 species of lichens from the Schirmacher Oasis and 25 species from Larsemann Hills, with microlichens as the predominant form in both the areas. Beck et al. (2019) reported the pioneer lichen *Placopsis* from maritime Antarctica. Laguna-Defior et al. (2016) studied *U. aurantiaco-atra* (fertile) and *U. antarctica* (sterile) from South Bay, Livingston Island. They concluded that both species had identical photosynthetic profiles in relation to temperature, optimum water content, and photosynthetic photon flux density. Domaschke et al. (2012) noted the low genetic diversity of *Cetraria aculeata* and its associated photobiont among Antarctic populations. Marečková et al. (2019) reported the effects of low temperature on photosynthetic performance of *Dermatocarpon polyphyllizum* from James Ross Island, Antarctica.

Sancho et al. (2007) investigated crustose lichens of Antarctica and observed that increased temperature and precipitation increase the growth rates of the lichens, and precipitation could be the most critical factor. Bokhorst et al. (2016) noted a 71% decline of dominant lichen *U. antarctica* owing to increased net winter respiration rates caused by higher temperatures and lower light, in OTC installations in maritime Antarctica; notably, however, the less dominant species did not show any significant response except for an increase in the species *Ochrolechia frigida*. The diversity of lichens decreases with increasing latitude across Antarctica (from ~350 lichen species in the northern part of the Antarctic Peninsula to ~120 species in the south, then to between 30 and 40 species at Mt. Kyffin and 12 species at the extreme latitude of 87°S; Green et al., 2011; Sancho et al., 2019). A study by Sancho et al. (2017) indicated the positive effects of increased annual temperatures on the Antarctic communities and reported that an increase of only 0.58°C in temperature results in a 26% increase in the growth rate of *U. antarctica*. Colesie et al. (2018) studied Antarctic lichens from Livingston Island, including *Stereocaulon alpinum*, *U. aurantiaco-atra* and *Placopsis contortuplicata*, and suggested that lichen species with broad ecological amplitude may be favored by the increase in temperature.

11.5.4. Diversity of Cyanobacteria

While endemic Antarctic cyanobacteria do occur, other non-endemic species, including tropical and Arctic species, are also commonly found. Native cyanobacteria are more widespread in Antarctica than their non-endemic counterparts and account for 67.9% of all Antarctic cyanobacteria (Taton et al., 2006). Jungblut et al. (2010) stated that Arctic cyanobacteria contribute more to the Antarctic cyanobacterial population than those from the other temperate regions. The probability of wind dispersion was theorized on the basis of similarities between Antarctic and Arctic cyanobacterial species revealed in the McMurdo Dry Valleys and East Antarctica. Cyanobacteria are commonly found as benthic communities in lakes across Antarctica, and their distribution appears to be more in the lower latitudes even though they are found in areas as far south as the 87°S latitude (Mills, 2018). In all benthic mats, *Leptolyngbya frigida*, and species of *Anagnostidis* and *Komarek* predominate; these organisms are also frequently found in soils and as epiphytes on mosses, thereby underscoring their high adaptability (Adams et al., 2006). Sabbe et al. (2003) revealed the presence of 33 cyanobacterial morphotypes, while Porazinska et al. (2004) reported the same species of *Chlorococcus*, *Chroococcus*, *Crinalium*, *Oscillatoria*, *Nostoc*, and *Spirulina* from streams and lakes, and five glaciers spanning the Taylor valley. Cyanobacteria are vital primary producers in many ecosystems, but an excessive abundance may well alter an ecosystem by adversely affecting other co-inhabitants (Mills, 2018).

11.5.5. Diversity of Seaweeds

An accurate estimate of the number of Antarctic macroalgae is difficult because of the limited availability of information; nevertheless, it is evident from the literature that West Antarctica harbors more macroalgal diversity than East Antarctica (Broyer and Koubbi, 2015). Skottsberg (1964) reported 96 species of macroalgae from Antarctica, while at least 119 species were recognized by Wiencke et al. (2002), and Hommersand et al. (2011) described five more species. Recently Pellizzari et al. (2017) reported two putative new species, *Prasiola* sp. and *Callophyllis* sp., from Antarctica. In a phylogenetic analysis, De Wever et al. (2009) showed that Antarctic freshwater isolates are affiliated mainly to the chlorophytan groups Chlorophyceae and Trebouxiophyceae and only one Antarctic strain (*Trichosarcina mucosa*, a filamentous green alga) belongs to Ulvophyceae. Although macroalgal diversity of Antarctica is low icompared with that of the world's temperate and tropical areas, it is comparable to that of the Arctic territory (Broyer and Koubbi, 2015; Wulff et al., 2009). Throughout the Antarctic marine flora, the degree of endemism is high, with Heterokontophyta and Rhodophyta being the groups most represented (Clayton, 1994) (Table 11.2). There has been one endemic order within the Phaeophyceae, the Ascoseirales, which consists of only one member, the kelp-like *Ascoseira mirabilis* Skottsberg. The order Desmarestiales originates in the southern hemisphere, and Antarctic members of this order share the same ecological niche as the order Laminariales of the northern hemisphere. Within the Phaeophyceae there are several endemic genera: *Himantothallus*, *Phaeurus*, and *Cystosphaera*. Among the Rhodophytes *Notophycus*, *Microrhinus*, *Hymenocladiopsis*, *Nereoginkgo*, *Antarcticothamnion*, *Gainia*, *Rhodokrambe*, *Leniea*, and *Austropugetia* are endemic, and among the Chlorophyta only *Lambia* is endemic. These species exist only in the Antarctic Peninsula (West Antarctica), whereas *Himantothallus grandifolius* and *Georgiella confluens* exist in both the western and eastern Antarctica; *Lola irregularis* occurs only in the eastern Antarctica (Broyer and Koubbi, 2015).

11.5.6. Diversity of Snow Algae

Psychrophilic algae (true snow algae) are a special community of freshwater microalgae that are adapted to snowy conditions, i.e. their optimum temperature of growth is around 0°C (Leya, 2013). Only a few species belonging to the genera *Chlamydomonas*, *Chloromonas*, and *Microglena* within the Chlorophyceae (Chlorophyta) and some *Desmotetra* spp. exhibit strong adaptation to such very low temperatures (Leya, 2013). Algal species found in grey, red and orange snows from the Windmill Islands area have been recorded by Ling and Seppelt (1998) and found to be distinct from those previously reported from the Arctic and northern alpine areas.

Table 11.2 Antarctic microalgae endemism (Based on Broyer and Koubbi, 2015; Hommersand *et al.*, 2009; Pellizzari *et al.*, 2017; Wiencke *et al.*, 2002).

	No. of species	No. of endemic species	% Endemic species
Rhodophyta	81	30	37
Heterokontophyta	27	12	44
Chlorophyte	18	4	22
Total	126	46	36.5

Further, some of these snow algae's closest relatives are soil or terrestrial algae found in warmer latitudes (Ling, 1996; Ling and Seppelt, 1998). *Mesotaenium berggrenii* has been found to be the predominant grey snow algae (Ling and Seppelt, 1990), while *Chloromonas rubroleosa* is the principal red snow algae (Ling and Seppelt 1993) in Antarctica. Green snow is caused by various species with large aggregations of vegetative cells, zoospores, gametes, planozygotes, zygospores, and akinetes (Ling, 1996). In Antarctica, because of the scarcity of land plants, snow algae contribute significantly to primary productivity. Their contribution to the productivity of terrestrial ecosystems in the Antarctic is likely to be greater than that in the Arctic and other alpine regions, because algal blooms in the latter regions tend to be patchy and close to other well-established vegetations (Davey *et al.*, 2019). Snow algae have been suggested to play a key role in climate change because they can minimize albedo ("bioalbedo") and thus increase the melting of snow and ice fields (Huovinen *et al.*, 2018). The number of red snow algae phylotypes found in both the polar regions is small, and the majority of them are unique to the Arctic or Antarctica. Segawa et al. (2018) reported that such red algal blooms in polar regions consisted primarily of cosmopolitan phylotypes but also contained endemic species distributed either in the Arctic or the Antarctic.

11.6. CLIMATE CHANGE AND ITS EFFECT ON ALGAL BIODIVERSITY

Ocean temperatures in the western Antarctic Peninsula has risen by an average of 1°C in the last half-century, and air temperatures have risen by ~0.25°C per decade since the 1950s (Schoenrock et al. 2016). Antarctica will not not only be impacted by the loss of stratospheric ozone and global warming but will also be threatened by ocean acidification because of human use of fossil fuels. As CO_2 reaches the ocean surface, pH and the concentration of carbonate ions decrease, while the concentrations of CO_2 and bicarbonate increase (Feely et al., 2004; Broyer and Koubbi, 2015). The pH, which was previously reported to range from 8.0 to 8.62, in western Antarctic surface waters is now as low as 7.932 units (Schoenrock et al. 2016). Global warming along the Antarctic Peninsula has resulted in a rise in the temperature of the season and a potentially significant rise in the length of the algal bloom season. Conversely, colder cycles with a shift in wind direction may maintain the current snow algae habitat. With habitat loss or when snow areas disappear entirely early in the summer, the ecosystems for that season may be totally lost (Davey et al., 2019). Temperature and changes in precipitation patterns have increased biological growth in lakes, mainly driven by the reduced length and extent of the lake ice cover and increased water column mixing (Convey, 2010). Increasing Antarctic temperatures will impact upon the biodiversity of cyanobacteria and other competing species as the rapid growth of one subspecies may outcompete other species in their ecological niche (Jungblut et al., 2010). The loss of stratospheric ozone over polar regions has contributed to a rise in UV-B radiation and may potentially have an impact on the distribution patterns of macroalgae (Zacher et al., 2009). Several studies have shown that UV radiation affects Antarctic macroalgae spores (most UV-susceptible stages). Therefore, increased UV radiation may produce changes in the depth distribution of these macroalgae during early succession (Campana et al., 2009; Broyer and Koubbi, 2015). The spatial distribution of seaweeds in the Antarctic is extremely restricted and is associated with environmental parameters. On a regional scale, the principal factors are the dynamics of ice formation and its appearance as floating sheets or icebergs, type of substrate, and the ocean current regime. The effects of temperature, salinity, pH, abrasion, photoperiod, irradiance, shading, UV rates, competition and grazing are also important (Zacher *et al.*, 2009; Mystikou *et al.*, 2014; Pellizzari *et al.*, 2017). During the last glaciation, some Antarctic species had expanded their distribution into the tropics. *Lamouroux confervoides*, *Desmarestia viridis*, and *Acrosiphonia arcta* had reached the equator only because the tropical water temperatures were low enough to allow the toughest stage of life history to move through the tropics. The large brown algae *Himantothallus grandifolius* and *Ascoseira mirabilis* are important structuring components of seaweed communities in the northern region of the Antarctic Peninsula and although they are not widely recognized in the Adelaide Island area (there is only one record of *H. grandifolius* from the Henkes Islands in 1975, and none from the area in 2010–2011), their presence as shelter-forming species would indicate a significant change in the phytobenthos (Mystikou et al., 2014). In addition to the obvious effect of temperature on the geographical distribution, there may be indirect effects mediated by milder and shorter periods of sea ice cover, particularly in the southern part of the Antarctic Peninsula (Smith and Stammerjohn, 2001; Stammerjohn et al., 2008; Broyer and Koubbi, 2015). The macro-algal populations will be able to spread to the south because of less strong ice abrasion and improved light conditions, and new ice-free areas will be colonized by seaweed as a result of glacial retreat (Broyer and Koubbi, 2015). To date, one non-native marine species *Ulva intestinalis* has been found to be thrive in Antarctic waters (south of 60°S). It was known elsewhere as an invasive species and potentially was transported by anthropogenic means to Antarctica (McCarthy et al., 2019). The role of microalgae in biogeochemical cycling is important. Species composition, density, cell size, and life history decide the depletion of organic matter and, based on these factors, C, N, and P are sequestered at various ratios. Drawing CO_2 from surface waters during photosynthesis and gradually transporting it to greater depths as organic C creates an air–sea CO_2 gradient that allows atmospheric CO_2 flow into the ocean. The efficiency of this biological C pump is partly determined by the availability of nutrients, which determines the size of the phytoplankton cell, the new production rate, and the essence of trophic interactions (Dunbar et al., 2003). Episodic release of Fe from the melting of sea ice is an important mechanism to control algal growth in Fe-deficient Antarctic waters, and is a primary cause of ice-edge algal blooms where the upper water column is sufficiently stable (Sedwick and DiTullio, 1997). In addition, differences in lipid composition and other metabolic modifications associated with adaptations of polar microalgae to altered niches may have complex consequences (Lyon and Mock, 2014). The ocean biological productivity, inferred from chlorophyll a concentration, has increased dramatically along the western shelf of Antarctic Peninsula. Summer time surface chlorophyll a concentration has decreased by 12% over the past 30 years along the WAP, with the highest equator ward decreases at 63°S and major rises in chlorophyll-a noted further south (Montes-Hugo et al., 2009). Low winter sea ice cover in the West Antarctic Peninsula contributes to decreased phytoplankton biomass and increased nanophytoplankton (Rozema et al., 2017). Severe fluctuation in the polar environments of the Antarctic Peninsula has caused a huge loss of flora and fauna in this region. The loss of algae population has caused a decline in krill that thrives on these algae, ultimately causing a decrease in penguin populations (Schoenrock, 2016).

11.7. SUMMARY

Temporal and spatial variations in Antarctic coastal features link the biogeochemistry and biodiversity of the coast to global climate change. Global climate change influences the physical and chemical characteristics of the coastal features, affecting the growth of associated biodiversity and biogeochemistry. Biogeochemical cycles that distribute elements within the biosphere have important roles in ecosystem function. The elements have natural source and their sinks respond to climatic changes. Fluctuations in biogeochemical cycles make the environment unfavorable for the organisms to survive, thereby affecting all the processes associated with the ecosystem. As Antarctica is extensively studied for global climate change, understanding its biogeochemical aspects is crucial. Different geochemical methods have been used to calculate the net primary production of Antarctic coasts, which is a primary factor in determining Antarctic biodiversity. Considering the harsh climatic conditions in the Antarctic, it is important to note that Antarctica is rich in biodiversity, likely with species yet to be discovered. Along the Antarctic coasts there exist a strong links between biogeochemical cycles and climate feedback, which will be highly active in future. To have a sustainable solution, continuous and long-term monitoring of the coasts is needed.

ACKNOWLEDGMENT

The authors thank the Vice-chancellor, Central University of Punjab, for his support on the completion of this work. We are also thankful to Director, NCPOR, Ministry of Earth Sciences, Goa, for support and help during the 36th Indian Scientific expedition to Antarctica. A.K. is grateful for INSPIRE fellowship (DST, New Delhi) towards her PhD. K.C.S. gratefully acknowledges Council of Scientific and Industrial Research (CSIR), New Delhi, India, for the financial support towards a PhD.

REFERENCES

Achberger, A., Christner, B., Michaud, A. et al. (2016). Microbial community structure of subglacial Lake Whillans, West Antarctica. *Frontiers in Microbiology 7*: 1457.

Adams, B., Bardgett, R., Ayres, E. et al. (2006). Diversity and distribution of Victoria Land biota, *Soil Biology and Biochemistry 38* (10): 3003–3018.

Aislabie, J., Jordan, S., Ayton, J. et al. (2009). Bacterial diversity associated with ornithogenic soil of the Ross sea region, Antarctica. *Canadian Journal of Microbiology 55* (1): 21–36.

Alves, I, Gonçalves, V, Oliveira, F, Schaefer, C, Rosa, C, and Rosa, L. (2019). The diversity, distribution, and pathogenic potential of cultivable fungi present in rocks from the south Shetlands archipelago, maritime Antarctica. *Extremophiles 23* (3): 327–336.

Andreae, M., Asami, T., Bertine, K. et al. (1984). Changing biogeochemical cycles. In: *Changing Metal Cycles and Human Health* (ed. J. O. Nriagu) 359–373. Springer, Berlin, Heidelberg.

Antony, R., Krishnan, K., Laluraj, C. et al. (2012). Diversity and physiology of culturable bacteria associated with a coastal Antarctic ice core. *Microbiological Research 167* (6). 372–380.

Antony, R., Sanyal, A., Kapse, N. et al. (2016). Microbial communities associated with Antarctic snow pack and their biogeochemical implications. *Microbiological Research 192*: 192–202.

Archer, S., De Los Ríos, A., Lee, K. et al. (2017). Endolithic microbial diversity in sandstone and granite from the McMurdo dry valleys, Antarctica. *Polar Biology 40* (5): 997–1006.

Aronson, R., Thatje, S., Clarke, A. et al. (2007). Climate change and invasibility of the Antarctic benthos. *Annual Review of Ecology, Evolution, and Systematics 38*: 129–154.

Arrigo, K. and Thomas, D. (2004). Large scale importance of sea ice biology in the Southern Ocean. *Antarctic Science 16* (4): 471–486.

Arrigo, K. and Van Dijken, G. (2003). Phytoplankton dynamics within 37 Antarctic coastal polynya systems. *Journal of Geophysical Research: Oceans 108* (C8): 3271.

Arrigo, K., DiTullio, G., Dunbar, R. et al. (2000). Phytoplankton taxonomic variability in nutrient utilization and primary production in the Ross Sea. *Journal of Geophysical Research: Oceans 105* (C4): 8827–8846.

Arrigo, K., van Dijken, G. and Bushinsky, S. (2008a). Primary production in the Southern Ocean, 1997–2006. *Journal of Geophysical Research: Oceans 113* (C8).

Arrigo, K, van Dijken, G. and Long, M. (2008b). Coastal Southern Ocean: a strong anthropogenic CO_2 sink. *Geophysical Research Letters 35* (21).

Barbraud, C. and Weimerskirch, H. (2001). Emperor penguins and climate change. *Nature 411* (6834): 183.

Bargagli, R. (2006). *Antarctic Ecosystems: Environmental Contamination Climate Change and Human Impact*, Vol. *175*. Springer Science and Business Media.

Bargagli, R., Monaci, F. and Bucci, C. (2007). Environmental biogeochemistry of mercury in Antarctic ecosystems. *Soil Biology and Biochemistry 39* (1): 352–360.

Barrett, J., Virginia, R., Wall, D., et al. (2006). Co-variation in soil biodiversity and biogeochemistry in northern and southern Victoria Land, Antarctica. *Antarctic Science 18* (4): 535–548.

Baublis, J., Wharton. Jr., R, and Volz, P. (1991). Diversity of micro-fungi in an Antarctic dry valley. *Journal of Basic Microbiology 31* (1): 1–12.

Beck, A., Bechteler, J., Casanova-Katny, A. et al. (2019). The pioneer lichen placopsis in maritime Antarctica: genetic diversity of their mycobionts and green algal symbionts, and their correlation with deglaciation time. *Symbiosis 79* (1): 1–24.

Becquevort, S. and Smith, Jr, W. (2001). Aggregation, sedimentation and biodegradability of phytoplankton-derived material during spring in the Ross Sea, Antarctica. *Deep Sea Research Part II: Topical Studies in Oceanography 48* (19–20): 4155–4178.

Begon, M., Townsend, C. and Harper, J. (2006). *Ecology: from Individuals to Ecosystems*. Oxford: Blackwell.

Bindoff, N., Rintoul, S. and Massom, R. (2000). Bottom water formation and polynyas in Adélie Land, Antarctica. *Papers and Proceedings of the Royal Society of Tasmania 133* (3): 51–56.

Bindoff, N., Williams, G. and Allison, I. (2001). Sea-ice growth and water-mass modification in the Mertz Glacier polynya, East Antarctica, during winter. *Annals of Glaciology 33*: 399–406.

Blain, S., Tréguer, P., Belviso, S.et al. (2001). A biogeochemical study of the island mass effect in the context of the iron hypothesis: Kerguelen Islands, Southern Ocean. *Deep Sea Research Part I: Oceanographic Research Papers 48* (1): 163–187

Bokhorst, S., Convey, P., Huiskes, A. et al. (2016). Usnea Antarctica, an important Antarctic lichen, is vulnerable to aspects of regional environmental change. *Polar Biology 39* (3): 511–521.

Bormann, F. and Likens, G. (1967). Nutrient cycling. *Science 155* (3761): 424–429.

Bottos, E., Woo, A., Zawar-Reza, P. et al. (2014). Airborne bacterial populations above desert soils of the McMurdo dry valleys, Antarctica. *Microbial Ecology 67* (1): 120–128.

Bowman, J., Amaral-Zettler, L., Rich, J. et al. (2017). Bacterial community segmentation facilitates the prediction of ecosystem function along the coast of the western Antarctic Peninsula. *The ISME Journal 11* (6): 1460–1471

Broyer, D., and Koubbi, C.J.C.S. (2015). Biogeographic atlas of the southern ocean. *CCAMLR Science 22*: 45–46.

Brusseau, M.L., Ramirez-Andreotta, M., Pepper, I.L. et al. (2019). Environmental impacts on human health and well-being. In: *Environmental and Pollution Science*, 477–499. Academic Press.

Budillon, G., Castagno, P., Aliani, S. et al. (2011). Thermohaline variability and Antarctic bottom water formation at the Ross Sea shelf break. *Deep Sea Research Part I: Oceanographic Research Papers 58* (10): 1002–1018.

Buffoni, G., Cappelletti, A. and Picco, P. (2002). An investigation of thermohaline circulation in Terra Nova Bay polynya. *Antarctic Science 14* (1): 83–92.

Burns, J. and Kooyman, G. (2001). Habitat use by Weddell seals and emperor penguins foraging in the Ross Sea, Antarctica. *American Zoologist 41* (1): 90–98.

Campana, G., Zacher, K., Fricke, A. et al. (2009). Drivers of colonization and succession in polar benthic macro-and microalgal communities. *Botanica Marina 52* (6): 655–667.

Campbell, J., Rustad, L., Boyer, E. et al. (2009). Consequences of climate change for biogeochemical cycling in forests of northeastern North America. *Canadian Journal of Forest Research 39* (2): 264–284.

Cape, M.R., Vernet, M., Pettit, E.C. et al. (2019). Circumpolar deep water impacts glacial meltwater export and coastal biogeochemical cycling along the West Antarctic Peninsula. *Frontiers in Marine Science 6*: 144.

Carpenter, E., Lin, S. and Capone, D. (2000). Bacterial activity in South Pole snow. *Applied Environmental Microbiology 66* (10): 4514–4517.

Cavan, E., Belcher, A., Atkinson, A. Campbell (2019). The importance of Antarctic krill in biogeochemical cycles. *Nature Communications 10* (1): 1–13.

Charlson, R., Lovelock, J., Andreae, M. et al. (1987). Oceanic phytoplankton, atmospheric sulphur, cloud albedo and climate. *Nature 326* (6114): 655.

Chaya, A., Kurosawa, N., Kawamata, A. et al. (2019). Community structures of bacteria, archaea, and eukaryotic microbes in the freshwater glacier lake yukidori-ike in langhovde, east Antarctica. *Diversity 11* (7): 105.

Ciais, P., Sabine, C., Bala, G. et al. (2013): Carbon and Other Biogeochemical Cycles. In: *Climate Change 2013: The Physical Science Basis. Contribution of Working Group I to the Fifth Assessment Report of the Intergovernmental Panel on Climate Change*. (eds. T.F. Stocker, D. Qin, G.-K. Plattner et al.), 465–570. Cambridge: Cambridge University Press.

Clayton, O. (1994). Evolution of the Antarctic marine benthic algal flora. *Journal of Phycology 30* (6): 897–904.

Cole, J. (2013). The carbon cycle, with a brief introduction to global biogeochemistry. In: *Fundamentals of Ecosystem Science* (ed. K. Weathers), 131–160. Oxford: Academic Press.

Colesie, C., Green, T., Türk, R. et al. (2014). Terrestrial biodiversity along the Ross sea coastline, Antarctica: lack of a latitudinal gradient and potential limits of bioclimatic modeling. *Polar Biology 37* (8): 1197–1208.

Colesie, C., Büdel, B., Hurry, V. et al. (2018). Can Antarctic lichens acclimatize to changes in temperature? *Global Change Biology 24* (3): 1123–1135.

Comiso, J., McClain, C., Sullivan, C. et al. (1993). Coastal Zone Color Scanner pigment concentrations in the Southern Ocean and relationships to geophysical surface features. *Journal of Geophysical Research: Oceans 98* (C2): 2419–2451.

Connell, L., Segee, B., Redman, R. et al. (2018). Biodiversity and abundance of cultured microfungi from the permanently ice-covered lake Fryxell, Antarctica. *Life 8* (3): 37.

Constable, A., Melbourne-Thomas, J., Corney, S. et al. (2014). Climate change and southern ocean ecosystems I: How changes in physical habitats directly affect marine biota. *Global Change Biology 20* (10): 3004–3025.

Conti, M. and Cecchetti, G. (2001). Biological monitoring: Lichens as bioindicators of air pollution assessment – a review. *Environmental Pollution 114* (3): 471–492.

Convey, P. (2006). Antarctic climate change and its influences on terrestrial ecosystems. In: *Trends in Antarctic Terrestrial and Limnetic Ecosystems* (eds. D.M. Bergstrom, P. Convey and A.H.L. Huiskes), 253–272. Dordrecht: Springer,.

Convey, P. (2010). Terrestrial biodiversity in Antarctica–recent advances and future challenges. *Polar Science 4* (2): 135–147.

Convey, P. and Smith, R. (2005). Responses of terrestrial Antarctic ecosystems to climate change. In: *Plants and Climate Change* (eds. J. Rozema, R. Aerts and H. Cornelissen), 1–12. Dordrecht: Springer.

Convey, P., Pugh, P., Jackson, C. et al. (2002). Response of Antarctic terrestrial microarthropods to long-term climate manipulations. *Ecology 83* (11): 3130–314

Convey, P. Chown, S. Clarke, A. et al. (2014). The spatial structure of Antarctic biodiversity. *Ecological monographs 84* (2): 203–244.

Cooper, L., Grebmeier, J., Larsen, I. et al. (2002). Seasonal variation in sedimentation of organic materials in the St. Lawrence Island polynya region, Bering Sea. *Marine Ecology Progress Series 226*: 13–26.

Cowan, D., Makhalanyane, T., Dennis, P. et al. (2014). Microbial ecology and biogeochemistry of continental Antarctic soils. *Frontiers in Microbiology 5*: 154

Cox, F., Newsham, K., Bol, R. et al. (2016). Not poles apart: Antarctic soil fungal communities show similarities to those of the distant arctic. *Ecology Letters 19* (5): 528–536.

Davey, M., Norman, L., Sterk, P. et al. (2019). Snow algae communities in Antarctica: metabolic and taxonomic composition. *New Phytologist 222* (3): 1242–1255.

De Carvalho, C., Santiago, I., Da Costa Coelho, L. et al. (2019). Fungi associated with plants and lichens of antarctica. In: *Fungi of Antarctica* (L. Rosa), 165–199. Dordrecht: Springer.

De Wever, A. Leliaert, F. Verleyen, E. et al. (2009). Hidden levels of phylodiversity in antarctic green algae: further evidence for the existence of glacial refugia. *Proceedings of the Royal Society B: Biological Sciences 276* (1673): 3591–3599.

DeConto, R. and Pollard, D. (2003). Rapid Cenozoic glaciation of Antarctica induced by declining atmospheric CO_2. *Nature 421* (6920): 245.

Delille, B., Jourdain, B., Borges, A. et al. (2007). Biogas. (CO_2, O_2, dimethyl sulfide) dynamics in spring Antarctic fast ice. *Limnology and Oceanography 52* (4): 1367–1379.

Deming, J. (2010). Sea ice bacteria and viruses. *Sea Ice 2*: 247–282.

Denman, K.L., Brasseur, G. Chidthaisong, A. et al. (2007). Couplings between changes in the climate system and biogeochemistry. In: *Climate Change 2007: The Physical Science Basis. Contribution of Working Group I to the Fourth Assessment Report of the Intergovernmental Panel on Climate Change* (eds. S. Solomon, D. Qin, M. Manning et al.), 499–587. Cambridge: Cambridge University Press.

Dennis, P., Newsham, K., Rushton, S. et al. (2019). Soil bacterial diversity is positively associated with air temperature in the maritime Antarctic. *Scientific Reports 9* (1): 1–11.

Depoorter, M., Bamber, J., Griggs, J. et al. (2013). Calving fluxes and basal melt rates of Antarctic ice shelves. *Nature 502* (7469): 89.

Di Menna, M. (1960). Yeasts from Antarctica. *Microbiology 23* (2): 295–300.

Dodge, C. (1973). *Lichen Flora of the Antarctic Continent and Adjacent Islands*: Canaan, NHPhoenix.

Domaschke, S., Fernandez-Mendoza, F.A., García, M. et al. (2012). Low genetic diversity in Antarctic populations of the lichen-forming ascomycete cetraria aculeata and its photobiont. *Polar Research 31* (1): 17353.

Doney, S.C. and Schimel, D.. (2015) Climate change and biogeochemical impacts. In: *eLS*. John Wiley and Sons, Ltd:. DOI:10.1002/9780470015902.a000324ub3

Doney, S., Balch, W., Fabry, V. et al. (2009). Ocean acidification: a critical emerging problem for the ocean sciences. *Oceanography 22* (4): 16–25.

Doney, S., Ruckelshaus, M., Duffy, J. et al. (2011). *Climate Change Impacts on Marine Ecosystems 4*: 11–37.

Doran, P.T., McKay, C.P., Clow, G.D. et al. (2002). Valley floor climate observations from the McMurdo Dry Valleys, Antarctica, 1986–2000. *Journal of Geophysical Research: Atmospheres 107* (D24): ACL-13.

Duarte, A.F., Passarini, M.Z., Delforno, T. et al. (2016). Yeasts from macroalgae and lichens that inhabit the South Shetland Islands, Antarctica. *Environmental Microbiology Reports 8* (5): 874–885.

Duarte, A.F., Barato, M., Nobre, F. et al. (2018). Production of cold-adapted enzymes by filamentous fungi from King George Island, Antarctica. *Polar Biology 41* (12): 2511–2521.

Duarte, A.F., Centurion, V. and Oliveira, V. (2019). Uncultivated fungi from Antarctica. In: *Fungi of Antarctica* (ed. L. H. Rosa), 19–41). Springer.

Dunbar, R., Arrigo, K., Lutz, M. et al. (2003). Non-redfield production and export of marine organic matter: a recurrent part of the annual cycle in the Ross Sea, antarctica. *Biogeochemistry of the Ross Sea 78*: 179–196.

Fabry, V., McClintock, J., Mathis, J. et al. (2009). Ocean acidification at high latitudes: the bellwether. *Oceanography 22* (4): 160–171.

Feely, R., Sabine, C., Lee, K. et al. (2004). Impact of anthropogenic CO_2 on the $CaCO_3$ system in the oceans. *Science 305* (5682): 362–366.

Ferrigno, J., Williams, R. and Thomson, J. (2002). *Coastal-Change and Glaciological Maps of the Antarctic Peninsula*. Reston, VA: US Geological Survey.

Ferreira, E.S., de Sousa, F.P., Rosa, L. et al. (2019). Taxonomy and richness of yeasts associated with angiosperms, bryophytes, and meltwater biofilms collected in the Antarctic Peninsula. *Extremophiles 23* (1): 151–159.

Fitch, D. and Moore, J. (2007). Wind speed influence on phytoplankton bloom dynamics in the Southern Ocean Marginal Ice Zone. *Journal of Geophysical Research: Oceans 112*: C08006).

Flower, B. and Kennett, J. (1994). The middle Miocene climatic transition: East Antarctic ice sheet development, deep ocean circulation and global carbon cycling. *Palaeogeography Palaeoclimatology Palaeoecology 108* (3–4): 537–555.

Fox, A., Paul, A. and Cooper, R. (1994). Measured properties of the Antarctic ice sheet derived from the SCAR Antarctic digital database. *Polar Record 30* (174): 201–206.

Franzmann, P., Deprez, P., Mcguire, A. et al. (1990). The heterotrophic, bacterial microbiota of Burton Lake, Antarctica. *Polar Biology 10* (4): 261–264.

Giles, A., Massom, R. and Lytle, V. (2008). Fast-ice distribution in East Antarctica during 1997 and 1999 determined using RADARSAT data. *Journal of Geophysical Research: Oceans 113*: C02S14.

Giudice, A., Caruso, C., Mangano, S. et al. (2012). Marine bacterioplankton diversity and community composition in an Antarctic coastal environment. *Microbial Ecology 63* (1): 210–223.

Gleick, P. (1996): Water resources. In: *Encyclopedia of Climate and Weather*, Vol. *2* (ed. S. Schneider), 817–823. New York: Oxford University Press

Golden, K., Ackley, S. and Lytle, V. (1998). The percolation phase transition in sea ice. *Science 282* (5397): 2238–2241.

Gonçalves, V., Vaz, A., Rosa, C. et al. (2012). Diversity and distribution of fungal communities in lakes of Antarctica. *FEMS Microbiology Ecology 82* (2): 459–471.

González-Rocha, G., Muñoz-Cartes, G., Canales-Aguirre, C. et al. (2017). Diversity structure of culturable bacteria isolated from the Fildes peninsula. (King George Island, Antarctica): a phylogenetic analysis perspective. *PLoS One 12* (6): 1–18.

Gordon, A., Chen, C.A. and Metcalf, W. (1984). Winter mixed layer entrainment of Weddell Deep Water. *Journal of Geophysical Research: Oceans 89* (C1): 637–640.

Goto, S., Sugiyama, J. and Iizuka, H. (1969). A taxonomic study of Antarctic yeasts. *Mycologia 61*, 748–774.

Gowing, M., Garrison, D., Kunze, H. et al. (2001). Biological components of Ross Sea short-term particle fluxes in the austral summer of 1995–1996. *Deep Sea Research Part I: Oceanographic Research Papers 48* (12): 2645–2671.

Grasshoff, K., Ehrhardt, M. and Kremling, K. (1983). *Methods of 660 Seawater Analysis 419*. Verlag Chemie.

Green, T., Sancho, L., Pintado, A. et al. (2011). Functional and spatial pressures on terrestrial vegetation in Antarctica forced by global warming. *Polar Biology 34* (11): 1643.

Grzymski, J., Riesenfeld, C., Williams, T. et al. (2012). A metagenomic assessment of winter and summer bacterioplankton from Antarctica Peninsula coastal surface waters. *The ISME Journal 6* (10): 1901–1915.

Günther, S., Gleitz, M. and Dieckmann, G. S. (1999). Biogeochemistry of Antarctic sea ice: a case study on platelet ice layers at Drescher Inlet, Weddell Sea. *Marine Ecology Progress Series 177*: 1–13.

Hansell, D., Carlson, C., Repeta, D. et al. (2009). Dissolved organic matter in the ocean: a controversy stimulates new insights. *Oceanography 22* (4): 202–211.

Hart, T. (1942). Phytoplankton periodicity in the Antarctic surface waters. *Discovery Reports 21*: 261–356.

Heil, P. (2006). Atmospheric conditions and fast ice at Davis, East Antarctica: a case study. *Journal of Geophysical Research: Oceans 111*:C05009.

Hendry, K., Meredith, M. and Ducklow, H. (2018). The marine system of the West Antarctic Peninsula: status and strategy for progress. *Philosophical Transactions of the Royal Society A: Mathematical, Physical and Engineering Sciences 376*: 1–6.

Henley, S., Schofield, O., Hendry, K. et al. (2019). Variability and change in the west Antarctic Peninsula marine system: research priorities and opportunities. *Progress in Oceanography, 173*: 208–237

Hirose, D., Hobara, S., Matsuoka, S. et al. (2016). Diversity and community assembly of moss-associated fungi in ice-free coastal outcrops of continental Antarctica. *Fungal Ecology 24*: 94–101.

Hoffmann, R., Pasotti, F., Vázquez, S. et al. (2018). Spatial variability of biogeochemistry in shallow coastal benthic communities of Potter Cove (Antarctica) and the impact of a melting glacier. *PLoS One 13* (12): e0207917.

Hommersand, M., Moe, R., Amsler, C. et al. (2009). Notes on the systematics and biogeographical relationships of Antarctic and sub-Antarctic rhodophyta with descriptions of four new genera and five new species. *Botanica Marina 52* (6): 509–534.

Huber, M., Brinkhuis, H., Stickley, C., et al. (2004). Eocene circulation of the Southern Ocean: was Antarctica kept warm by subtropical waters. *Paleoceanography 19* (4): 1–12.

Hughes, C., Chuck, A., Rossetti, H. et al. (2009). Seasonal cycle of seawater bromoform and dibromomethane concentrations in a coastal bay on the western Antarctic Peninsula. *Global Biogeochemical Cycles 23* (2): 1–13.

Huovinen, P., Ramírez, J., Gómez, O. et al. (2018). Remote sensing of albedo-reducing snow algae and impurities in the maritime Antarctica. *ISPRS Journal of Photogrammetry and Remote Sensing 146*: 507–517.

Jacobs, S., Gordon, A. and Ardai, J. (1979). Circulation and melting beneath the Ross Ice Shelf. *Science, 203* (4379): 439–443.

Ji, M., Greening, C., Vanwonterghem, I., Carere, C. et al. (2017). Atmospheric trace gases support primary production in Antarctic desert surface soil. *Nature 552* (7685): 400–403.

Joughin, I., Smith, B. and Medley, B. (2014). Marine ice sheet collapse potentially under way for the Thwaites Glacier Basin, West Antarctica. *Science 344* (6185): 735–738.

Jungblut, A., Lovejoy, C. and Vincent,I. (2010). Global distribution of cyanobacterial ecotypes in the cold biosphere. *The ISME Journal 4* (2): 191–202.

Kim, D., Park, H., Kim, J. et al. (2018). Passive warming effect on soil microbial community and humic substance degradation in maritime Antarctic region. *Journal of Basic Microbiology 58* (6): 513–522.

Kirk, P., Cannon, P., Minter, D. et al. (2008). *Dictionary of the Fungi*, 10e. Wallingford:. CABI Publishing, UK

Kirkwood, R. and Robertson, G. (1997). The foraging ecology of female emperor penguins in winter. *Ecological Monographs 67* (2): 155–176.

Klatt, O., Roether, W., Hoppema, M. et al. (2002). Repeated CFC sections at the Greenwich Meridian in the Weddell Sea. *Journal of Geophysical Research: Oceans 107* (C4): 5 1.

Kochkina, G., Ivanushkina, N., Lupachev, A.et al. (2019). Diversity of mycelial fungi in natural and human-affected Antarctic soils. *Polar Biology 42* (1): 47–64.

Kopczyńska, E., Goeyens, L., Semeneh, M. et al. (1995). Phytoplankton composition and cell carbon distribution in Prydz Bay, Antarctica: relation to organic particulate matter and its $\delta^{13}C$ values. *Journal of Plankton Research 17* (4): 685–707.

Kumar Chattopadhyay, M., Sathyanarayana Reddy, G., and Shivaji, S. (2014). Psychrophilic bacteria: Biodiversity, molecular basis of cold adaptation and biotechnological implications. *Current Biotechnology 3* (1): 100–116.

Kwon, E., Primeau, F. and Sarmiento, J. (2009). The impact of remineralization depth on the air–sea carbon balance. *Nature Geoscience 2* (9): 630.

Laguna-Defior, C., Pintado, A., Green, T. et al. (2016). Distributional and ecophysiological study on the Antarctic lichens species pair *Usnea antarctica/Usnea aurantiaco-atra*. *Polar Biology 39* (7): 1183–1195

Lancelot, C., Montety, A., Goosse, H. et al. (2009). Spatial distribution of the iron supply to phytoplankton in the Southern Ocean: a model study. *Biogeosciences 6* (12): 2861–2878.

Lannuzel, D., Schoemann, V., De Jong, J. et al. (2007). Distribution and biogeochemical behavior of iron in the East Antarctic sea ice. *Marine Chemistry 106* (1–2): 18–32.

Learman, D., Henson, M., Thrash, J. et al. (2016). Biogeochemical and microbial variation across 5500 km of Antarctic surface sediment implicates organic matter as a driver of benthic community structure. *Frontiers in Microbiology 7*: 284.

Legge, O., Bakker, D., Meredith, M. et al. (2017). The seasonal cycle of carbonate system processes in Ryder Bay, West Antarctic Peninsula. *Deep Sea Research Part II: Topical Studies in Oceanography 139*: 167–180.

Leya, T. (2013). Snow algae: adaptation strategies to survive on snow and ice. In: *Polyextremophiles. Cellular origin life in Extreme Habitats and Astrobiology* (eds. O.A. Seckbach and J. Stan-Lotter), 401–423). Dordrecht: Springer.

Ling, H.J.H. (1996). Snow algae of the windmill islands region, Antarctica. *Hydrobiologia 336* (1–3): 99–106.

Ling, H. U., and Seppelt, R. D. (1990). Snow algae of the windmill islands, continental Antarctica. *Mesotaenium berggrenii.* (Zygnematales, Chlorophyta) the alga of grey snow. *Antarctic Science 2*: 143–148.

Ling H.U. and Seppelt R.D. (1993). Snow algae of the windmill islands, continental Antarctica. 2. *Chloromonas rubroleosa* sp. nov. (Volvocales, Chlorophyta) an alga of red snow. *European Journal of Phycology 28*: 77–84.

Ling, H.U. and Seppelt, R.D. (1998). Snow algae of the Windmill Islands, continental Antarctica 3. *Chloromonas polyptera.* (Volvocales, Chlorophyta). *Polar Biology 20* (5): 320–324.

Lizotte, M. (2001). The contributions of sea ice algae to Antarctic marine primary production. *American Zoologist 41* (1): 57–73.

Lizotte, M. (2003). The microbiology of sea ice. In: *Sea Ice: an Introduction to its Physics Chemistry Biology and Geology* (eds. D.N. Thomas and G. S Dieckmann), 184–210. Oxford: Blackwell Publishing.

Louanchi, F., Ruiz-Pino, D., Jeandel, C.et al. (2001). Dissolved inorganic carbon, alkalinity, nutrient and oxygen seasonal and interannual variations at the Antarctic Ocean JGOFS-KERFIX site. *Deep Sea Research Part I: Oceanographic Research Papers 48* (7): 1581–1603.

Lubin, D. and Massom, R. (2006). *Polar Remote Sensing: Volume I: Atmosphere and Oceans*. Springer Science and Business Media.

Lucas, Y. (2001). The role of plants in controlling rates and products of weathering: importance of biological pumping. *Annual Review of Earth and Planetary Sciences 29* (1): 135–163.

Lücking, R. (1998). 'Plasticolous' lichens in a tropical rain forest at La Selva Biological Station, Costa Rica. *Lichenologist 30*: 287–290.

Lyon, B. and Mock, T.B. (2014). Polar microalgae: new approaches towards understanding adaptations to an extreme and changing environment. *Biology 3* (1): 56–80.

Mackintosh, A., Verleyen, E., O'Brien, P. et al. (2014). Retreat history of the East Antarctic Ice Sheet since the last glacial maximum. *Quaternary Science Reviews 100*: 10–30.

McMinn, A., Bleakley, N., Steinburner, K. et al. (2000). Effect of permanent sea ice cover and different nutrient regimes on the phytoplankton succession of fjords of the Vestfold Hills Oasis, eastern Antarctica. *Journal of Plankton Research 22* (2): 287–303.

Manno, C., Morata, N. and Primicerio, R. (2012). *Limacina retroversa*'s response to combined effects of ocean acidification and sea water freshening. *Estuarine, Coastal and Shelf Science 113*: 163–171.

Marečková, M., Barták, M. and Hájek, J. (2019). Temperature effects on photosynthetic performance of Antarctic lichen *Dermatocarpon polyphyllizum*: a chlorophyll fluorescence study. *Polar Biology 42* (4): 685–701.

Margesin, R. and Miteva, V. (2011). Diversity and ecology of psychrophilic microorganisms. *Research in Microbiology 162* (3): 346–361.

Martin, J., Knauer, G., Karl, D. et al. (1987). VERTEX: carbon cycling in the northeast Pacific. *Deep Sea Research Part A. Oceanographic Research Papers 34* (2): 267–285.

Massom, R. and Stammerjohn, S. (2010). Antarctic sea ice change and variability–physical and ecological implications. *Polar Science 4* (2): 149–186.

Massom, R., Harris, P., Michael, K. et al. (1998). The distribution and formative processes of latent-heat polynyas in East Antarctica. *Annals of Glaciology 27*:, 420–426.

Massom, R., Eicken, H., Hass, C. et al. (2001). Snow on Antarctic sea ice. *Reviews of Geophysics 39* (3): 413–445.

Massom, R., Hill, K., Barbraud, C. et al. (2009). Fast ice distribution in Adélie Land, East Antarctica: interannual variability and implications for emperor penguins *Aptenodytes forsteri. Marine Ecology Progress Series 374*: 243–257.

McCarthy, A., Peck, L., Hughes, K. et al. (2019). Antarctica: the final frontier for marine biological invasions. *Global Change Biology 25* (7): 2221–2241.

McNeil, B. and Matear, R. (2008). Southern Ocean acidification: a tipping point at 450-ppm atmospheric CO_2. *Proceedings of the National Academy of Sciences 105* (48): 18860–18864.

McNeil, B., Matear, R. and Barnes, D. (2004). Coral reef calcification and climate change: the effect of ocean warming. *Geophysical Research Letters 31* (22): L22309.

Mikucki, J, and Priscu, J. (2007). Bacterial diversity associated with blood falls, a subglacial outflow from the Taylor glacier, Antarctica. *Applied and Environmental Microbiology 73* (12): 4029–4039.

Miloslavich, P., Bax, N., Simmons, S. et al. (2018). Essential ocean variables for global sustained observations of biodiversity and ecosystem changes. *Global Change Biology 24* (6): 2416–2433.

Mills, F. (2018). *Green-blue algae in white wonderland – a review of current distributions and possible changes to cyanobacteria distribution caused by anthropogenic climate change in Antarctica.* University of Canterbury, New Zealand. (PCAS 20, 2017/2018).

Möller, C. and Dreyfuss, M. (1996). Microfungi from Antarctic lichens, mosses and vascular plants. *Mycologia 88* (6): 922–933.

Montes-Hugo, M., Doney, S., Ducklow, H. et al. (2009). Recent changes in phytoplankton communities associated with rapid regional climate change along the western Antarctic Peninsula. *Science 323* (5920): 1470–1473.

Monteiro, M.S., Baptista, M., Séneca, J. et al. (2020). Understanding the response of nitrifying communities to disturbance in the McMurdo dry valleys, Antarctica. *Microorganisms 8* (3): 404.

Moore, D., Robson, G. and Trinci, A. (2011). *21st Century Guidebook to Fungi* (with cd). Cambridge University Press.

Morales Maqueda, M. and Rahmstorf, S. (2002). Did Antarctic sea-ice expansion cause glacial CO_2 decline? *Geophysical Research Letters 29* (1): 11–1.

Moss, B. and Balls, H. (1989). Phytoplankton distribution in a floodplain lake and river system. II Seasonal changes in the phytoplankton communities and their control by hydrology and nutrient availability. *Journal of Plankton Research 11* (4): 839–839.

Mundy, C. and Barber, D. (2001). On the relationship between spatial patterns of sea-ice type and the mechanisms which create and maintain the North Water. (NOW) polynya. *Atmosphere–Ocean 39* (3): 327–341.

Murphy, E., Clarke, A., Symon, C. et al. (1995). Temporal variation in Antarctic sea-ice: analysis of a long term fast-ice record from the South Orkney Islands. *Deep Sea Research Part I: Oceanographic Research Papers 42* (7): 1045–1062.

Mystikou, A., Peters, A., Asensi, A. et al. (2014). Seaweed biodiversity in the south-western Antarctic Peninsula: surveying macroalgal community composition in the Adelaide island/ marguerite bay region over a 35-year time span. *Polar Biology 37* (11): 1607–1619.

National Snow and Ice Data Center (NSIDC) (2007). *World Glacier Inventory*. Boulder, CO: World Glacier Monitoring Service and National Snow and Ice Data Center/World Data Center for Glaciology.

Nayaka, S. (2014). Methods and techniques in collection, preservation and identification of lichens. *Plant Toxonomy and Biosystematics: Classical and Modern Methods* (eds. T.S. Rana, K.N. Nair and D.K. Upreti), 101–105. New Delhi: New India Publishing Agency.

Newsham, K., Hopkins, D., Carvalhais, L. et al. (2016). Relationship between soil fungal diversity and temperature in the maritime Antarctic. *Nature Climate Change 6* (2): 182–186.

Olech, M. and Słaby, A. (2016). Changes in the lichen biota of the lions rump area, King George island, Antarctica, over the last 20 years. *Polar Biology 39* (8): 1499–1503.

Onofri, S., Selbmann, L., Zucconi, L. et al. (2004). The mycota of continental Antarctica. *Terra Antartica Reports 11*: 37–42

Orsi, A., Johnson, G. and Bullister, J. (1999). Circulation, mixing, and production of Antarctic Bottom Water. *Progress in Oceanography 43* (1): 55–109.

Paolo, F., Fricker, H. and Padman, L. (2016). Constructing improved decadal records of Antarctic ice shelf height change from multiple satellite radar altimeters. *Remote Sensing of Environment 177*: 192–205.

Pearce, D. (2008). Climate change and the microbiology of the Antarctic Peninsula region. *Science Progress 91* (2): 203–217.

Peck, L. (2018). Antarctic marine biodiversity: adaptations, environments and responses to change. In: *Oceanography and Marine Biology*, Vol. *56* (eds. J. Hawkins, A. Dale, L. Firth et al.), 2–133. Taylor and Francis.

Peeters, K., Verleyen, E., Hodgson, D. et al. (2012). Heterotrophic bacterial diversity in aquatic microbial mat communities from Antarctica. *Polar Biology 35* (4): 543–554.

Pellizzari, F. Silva, M. Silva, E. et al. (2017). Diversity and spatial distribution of seaweeds in the South Shetland Islands, Antarctica: an updated database for environmental monitoring under climate change scenarios. *Polar Biology 40* (8): 1671–1685.

Perovich, D, Maykut, G. and Grenfell, T. (1986, August). Optical properties of ice and snow in the polar oceans. I: Observations. Crrel Report No: MP 2255. *Ocean Optics VIII 637*: 232–241.

Pipko, I., Semiletov, I., Tishchenko, P. et al. (2002). Carbonate chemistry dynamics in Bering Strait and the Chukchi Sea. *Progress in Oceanography 55* (1–2): 77–94.

Pollard, D. and DeConto, R. (2009). Modelling West Antarctic ice sheet growth and collapse through the past five million years. *Nature 458* (7236): 329.

Porazinska, D., Fountain, A., Nylen, T. et al. (2004). The biodiversity and biogeochemistry of cryoconite holes from McMurdo dry valley glaciers, Antarctica. *Arctic Antarctic and Alpine Research 36* (1): 84–91.

Priscu, J.C., Tulaczyk, S., Studinger, M. et al. (2008). Antarctic subglacial water: origin, evolution and ecology. In: *Polar lakes and rivers: limnology of Arctic and Antarctic Aquatic Ecosystems* (eds. W.F. Vincent and J. Laybourn-Parry), 119–135. Oxford: Oxford University Press.

Pudasaini, S., Wilson, J., Ji, M. et al. (2017). Microbial diversity of browning peninsula, eastern Antarctica revealed using molecular and cultivation methods. *Frontiers in Microbiology 8*: 591.

Quetin, L. and Ross, R. (2009). Life under Antarctic pack ice: a krill perspective. In: *Smithsonian at the Poles: Contributions to International Polar Year Science*. (eds. I. Krupnik, M.A. Lang and S.E. Miller), 285–298. Smithsonian Institution Scholarly Press

Ravindra, R. and Laluraj, C. (2012). Cryosphere research: Indian perspective. *Proceedings of the Indian National Science Academy 78* (3): 249–257.

Reddy, G., Chattopadhyay, M. and Shivaji, S. (2016). Biodiversity, adaptation and biotechnological importance of bacteria occurring in cold climates. In: *Biotechnology of Extremophiles* (ed. P.H. Rampelotto), 47–81. Springer.

Ribic, C., Chapman, E., Fraser, W. et al. (2008). Top predators in relation to bathymetry, ice and krill during austral winter in Marguerite Bay, Antarctica. *Deep Sea Research Part II: Topical Studies in Oceanography 55* (3–4): 485–499.

Rignot, E., Jacobs, S., Mouginot, J. and Scheuchl, B. (2013). Ice-shelf melting around Antarctica. *Science, 341* (6143): 266–270.

Rinnan, R., Rousk, J., Yergeau, E. et al. (2009). Temperature adaptation of soil bacterial communities along an Antarctic climate gradient: predicting responses to climate warming. *Global Change Biology 15* (11): 2615–2625.

Rintoul, S. (1998). On the origin and influence of Adelie Land bottom water. In: *Ocean, Ice and Atmosphere: Interactions at the Antarctic Continental Margin. Antarctic Research Series*, 75. (eds. S.S. Jacobs and R.F. Weiss), 151–172. Washington, DC: American Geophysical Union.

Rintoul, S. (2007). Rapid freshening of Antarctic Bottom Water formed in the Indian and Pacific oceans. *Geophysical Research Letters 34* (6): L06606.

Rojas-Jimenez, K., Wurzbacher, C., Bourne, E. et al. (2017). Early diverging lineages within Cryptomycota and Chytridiomycota dominate the fungal communities in ice-covered lakes of the McMurdo dry valleys, Antarctica. *Scientific Reports 7* (1): 1–11.

Rosa, L., Almeida Vieira, M. L., Santiago, I. et al. (2010). Endophytic fungi community associated with the dicotyledonous plant *Colobanthus quitensis*. (kunth) bartl. (Caryophyllaceae) in Antarctica. *FEMS Microbiology Ecology 73* (1): 178–189.

Ross, R., Quetin, L., Martinson, D. et al. (2008). Palmer LTER: Patterns of distribution of five dominant zooplankton species in the epipelagic zone west of the Antarctic Peninsula, 1993–2004. *Deep Sea Research Part II: Topical Studies in Oceanography 55* (18–19): 2086–2105.

Rozema, P., Venables, H., Van De Poll, W. et al. (2017). Interannual variability in phytoplankton biomass and species composition in northern Marguerite Bay. (West Antarctic Peninsula) is governed by both winter sea ice cover and summer stratification. *Limnology and Oceanography 62* (1): 235–252.

Ruisi, S., Barreca, D., Selbmann, L. et al. (2007). Fungi in Antarctica. *Reviews in Environmental Science and Biol Technology 6* (1–3): 127–141.

Rysgaard, S., Glud, R., Sejr, M. et al. (2007). Inorganic carbon transport during sea ice growth and decay: a carbon pump in polar seas. *Journal of Geophysical Research: Oceans 112* (C3): C03016.

Sabbe, K. Verleyen, E. Hodgson, D. et al. (2003). Benthic diatom flora of freshwater and saline lakes in the Larsemann hills and Rauer islands, east Antarctica. *Antarctic Science 15* (2): 227–248.

Sabine, C., Feely, R., Gruber, N. et al. (2004). The oceanic sink for anthropogenic CO_2. *Science 305* (5682): 367–371.

Saini, K, Nayaka, S., and Bast, F. (2019). Diversity of lichen photobionts: their coevolution and bioprospecting potential. In: *Microbial Diversity in Ecosystem Sustainability and Biotechnological Applications* (eds. T. Satyanarayana, S.K. Das and B.N. Johri), 307–323. Springer.

Sancho, L., Green, T. and Pintado, A. (2007). Slowest to fastest: Extreme range in lichen growth rates supports their use as an indicator of climate change in Antarctica. *Flora: Morphology, Distribution, Functional Ecology of Plants 202* (8): 667–673.

Sancho, L., Pintado, A., Navarro, F. et al. (2017). Recent warming and cooling in the Antarctic Peninsula region has rapid and large effects on lichen vegetation. *Scientific Reports 7* (1): 1–8.

Sancho, L., Pintado, A. and Green, T. (2019). Antarctic studies show lichens to be excellent biomonitors of climate change. *Diversity 11* (3): 42.

Santiago, I., Soares, M., Rosa, C. et al. (2015). Lichensphere: a protected natural microhabitat of the non-lichenised fungal communities living in extreme environments of Antarctica. *Extremophiles 19* (6): 1087–1097.

Satoh, H. and Watanabe, K. (1988). Primary productivity in the fast ice area near Syowa Station, Antarctica, during spring and summer 1983/84. *Journal of the Oceanographical Society of Japan 44* (6): 287–292.

Saxton, M., Samarkin, V., Schutte. et al. (2016). Biogeochemical and 16S rRNA gene sequence evidence supports a novel mode of anaerobic methanotrophy in permanently ice-covered Lake Fryxell, Antarctica. *Limnology and Oceanography 61* (S1): S119–S130.

Schoenrock, K., Schram, J.B., Amsler, C.D et al. (2016). Climate change confers a potential advantage to fleshy Antarctic crustose macroalgae over calcified species. *Journal of Experimental Marine Biology and Ecology 474*: 58–66.

Sedwick, P. and DiTullio, G. (1997). Regulation of algal blooms in antarctic shelf waters by the release of iron from melting sea ice. *Geophysical Research Letters 24* (20): 2515–2518.

Segawa, T. Matsuzaki, R. Takeuchi, N. et al. (2018). Bipolar dispersal of red-snow algae. *Nature Communications 9* (1): 1–8.

Selbmann, L., Isola, D., Fenice, M. et al. (2012). Potential extinction of Antarctic endemic fungal species as a consequence of global warming. *Science of the Total Environment 438*: 127–134.

Seppelt, R., Türk, R., Green, T. et al. (2010). Lichen and moss communities of Botany Bay, granite harbour, Ross Sea, Antarctica. *Antarctic Science 22* (6): 691–702.

Shadwick, E.H., Trull, T.W., Thomas, H. et al. (2013). Vulnerability of polar oceans to anthropogenic acidification: comparison of Arctic and Antarctic seasonal cycles. *Scientific Reports 3* (1): 1–7.

Shakhova, N., Sergienko, V. and Semiletov, I. (2009). Modern state of the role of the East Siberian Shelf in the methane cycle. *Herald of the Russian Academy of Sciences 79* (6): 507–518.

Shaver, G., Canadell, J., Chapin, F. et al. (2000). Global warming and terrestrial ecosystems: a conceptual framework for analysis: ecosystem responses to global warming will be complex and varied. Ecosystem warming experiments hold great potential for providing insights on ways terrestrial ecosystems will respond to upcoming decades of climate change. Documentation of initial conditions provides the context for understanding and predicting ecosystem responses. *BioScience 50* (10): 871–882.

Shetye, S., Jena, B. and Mohan, R. (2017). Dynamics of sea-ice biogeochemistry in the coastal Antarctica during transition from summer to winter. *Geoscience Frontiers 8* (3): 507–516.

Shivaji, S. (2017). Bacterial biodiversity, cold adaptation and biotechnological importance of bacteria occurring in Antarctica. *Proceedings of the Indian National Science Academy 83* (2): 327–352.

Shivaji, S. and Reddy, G. (2010). Bacterial biodiversity of Antarctica: conventional polyphasic and rRNA approaches. In: *Polar Microbiology: the Ecology Biodiversity and Bioremediation Potential of Microorganisms in Extremely Cold Environments*, (eds. A.K. Bej, J. Aislabie, and R.M. Atlas), 61–95. New York:CRC Press.

Shivaji, S., Chattopadhyay, M. and Reddy, G. (2019). Diversity of bacteria from Antarctica, arctic, Himalayan glaciers and stratosphere. *Proceedings of the Indian National Science Academy 85* (4): 909–923.

Shivaji, S. and Prasad, G. (2009). Antarctic yeasts: biodiversity and potential applications. In: *Yeast Biotechnology: Diversity and Applications* (eds. T. Satyanarayana and G. Kunze), 3–18). Dordrecht: Springer.

Shivaji, S., Reddy, G., Aduri, R. et al. (2004). Bacterial diversity of a soil sample from Schirmacher oasis, *Antarctica. Cellular and Molecular Biology 50* (5): 525–536.

Sigman, D., Hain, M. and Haug, G. (2010). The polar ocean and glacial cycles in atmospheric CO_2 concentration. *Nature 466* (7302): 47.

Signori, C., Pellizari, V., Enrich-Prast, A. et al. (2018). Spatiotemporal dynamics of marine bacterial and archaeal communities in surface waters off the northern Antarctic Peninsula. *Deep Sea Research Part II: Topical Studies in Oceanography 149*: 150–160.

Simpson, W., Glasow, R., Riedel, K. et al. (2007). Halogens and their role in polar boundary-layer ozone depletion. *Atmospheric Chemistry and Physics 7* (16): 4375–4418.

Singh, J. Singh, R. and Khare, R. (2018). Influence of climate change on Antarctic flora. *Polar Science 18*, 94–101.

Singh, S. and Nayaka, S. (2017). Contributions to the floral diversity of Schirmacher oasis and Larsemann hills, Antarctica. *Proceedings of the Indian National Science Academy 83* (2): 469–481.

Singh, S., Olech, M., Cannone, N. et al. (2015). Contrasting patterns in lichen diversity in the continental and maritime Antarctic. *Polar Science 9* (3): 311–318.

Skottsberg, (1964). Antarctic phycology. In: *Biologie Antarctique. 1st SCAR Symposium. Premier Symposium* (eds. C.R.M. Holdgate and J. Prévost), 147–154. Paris: Hermann.

Smetacek, V. and Nicol, S. (2005). Polar ocean ecosystems in a changing world. *Nature 437* (7057): 362.

Smetacek, V., Scharek, R. and Nöthig, E. (1990). Seasonal and regional variation in the pelagial and its relationship to the life history cycle of krill. In: *Antarctic Ecosystems* (eds. K.R. Kerry and G. Hempel), 103–114. Berlin, Heidelberg: Springer.

Smetacek, V., Scharek, R., Gordon, L. et al. (1992). Early spring phytoplankton blooms in ice platelet layers of the southern Weddell Sea, Antarctica. *Deep Sea Research Part A. Oceanographic Research Papers 39* (2): 153–168.

Smith, Jr., W, Ainley, D. and Cattaneo-Vietti, R. (2006). Trophic interactions within the Ross Sea continental shelf ecosystem. *Philosophical Transactions of the Royal Society B: Biological Sciences 362* (1477): 95–111.

Smith, J, Burritt, D. and Bannister, P. (2000). Shoot dry weight, chlorophyll and UV-B-absorbing compounds as indicators of a plant's sensitivity to UV-B radiation. *Annals of Botany 86* (6): 1057–1063.

Smith, R. and Stammerjohn, O. (2001). Variations of surface air temperature and sea-ice extent in the western Antarctic Peninsula region. *Annals of Glaciology 33*: 493–500.

Smith, W. and Comiso, J. (2008). Influence of sea ice on primary production in the Southern Ocean: a satellite perspective. *Journal of Geophysical Research: Oceans 113* (C5): C05S93.

Sommers, P., Darcy, J, Gendron, E. et al. (2018). Diversity patterns of microbial eukaryotes mirror those of bacteria in Antarctic cryoconite holes. *FEMS Microbiology Ecology 94* (1): fix167.

Stammerjohn, S., Martinson, D., Smith, R. et al. (2008). Trends in Antarctic annual sea ice retreat and advance and their relation to el Niño–southern oscillation and southern annular mode variability. *Journal of Geophysical Research: Oceans, 113* (C3): C03S90

Stchigel, A., Guarro, J. and Mac Cormack, W. (2003). *Apiosordaria antarctica* and *Thielavia antarctica*, two new Ascomycetes from Antarctica. *Mycologia 95* (6): 1218–1226.

Stibal, M., Hasan, F., Wadham, J. et al. (2012). Prokaryotic diversity in sediments beneath two polar glaciers with contrasting organic carbon substrates. *Extremophiles 16* (2): 255–265.

Stocker, T., Qin, D., Plattner, G. et al. (2013). IPCC, 2013: Summary for Policymakers in Climate Change 2013: the Physical Science Basis, Contribution of Working Group I to the Fifth Assessment Report of the Intergovernmental Panel on Climate Change. Cambridge University Press.

Suganuma, Y., Miura, H., Zondervan, A. et al. (2014). East Antarctic deglaciation and the link to global cooling during the Quaternary: evidence from glacial geomorphology and

10Be surface exposure dating of the Sør Rondane Mountains, Dronning Maud Land. *Quaternary Science Reviews 97*: 102–120.

Sullivan, A., Snyder, D. and Rounds, S. (2010). Controls on biochemical oxygen demand in the upper Klamath River, Oregon. *Chemical Geology 269* (1–2): 12–21.

Tamura, T., Ohshima, K. I. and Nihashi, S. (2008). Mapping of sea ice production for Antarctic coastal polynyas. *Geophysical Research Letters 35*(7): L07606.

Taton, A. Grubisic, S. Balthasart, P. et al. (2006). Biogeographical distribution and ecological ranges of benthic cyanobacteria in east Antarctic lakes. *FEMS Microbiology Ecology 57* (2): 272–289.

Teixeira, L., Peixoto, R., Cury, J. et al. (2010). Bacterial diversity in rhizosphere soil from Antarctic vascular plants of admiralty bay, maritime Antarctica. *The ISME Journal 4* (8): 989–1001.

Thomas, D., Kennedy, H., Kattner, G. et al. (2001). Biogeochemistry of platelet ice: its influence on particle flux under fast ice in the Weddell Sea, *Antarctica. Polar Biology 24* (7): 486–496.

Thomas, E., Dennis, P., Bracegirdle, T. et al. (2009). Ice core evidence for significant 100-year regional warming on the Antarctic Peninsula. *Geophysical Research Letters 36* (20): L20704

Thomas, J. and DeMaster, D. (1983). Diel haul-out patterns of Weddell seal. (*Leptonychotes weddelli*) females and their pups. *Canadian Journal of Zoology 61* (9): 2084–2086.

Tison, J., Brabant, F., Dumont, I. et al. (2010). High-resolution dimethyl sulfide and dimethyl sulfonio propionate time series profiles in decaying summer first-year sea ice at Ice Station Polarstern, western Weddell Sea, Antarctica. *Journal of Geophysical Research: Biogeosciences 115*: G04044.

Tortell, P., Asher, E., Ducklow, H. et al. (2014). Metabolic balance of coastal Antarctic waters revealed by autonomous pCO_2 and $\Delta O_2/A_r$ measurements. *Geophysical Research Letters 41* (19): 6803–681

Tréguer, P. and Jacques, G. (1992). Review dynamics of nutrients and phytoplankton, and fluxes of carbon, nitrogen and silicon in the Antarctic Ocean. In: *Weddell Sea Ecology*, pp. 149–162. Berlin, Heidelberg: Springer.

Tréguer, P. and Pondaven, P. (2002). Climatic changes and the cycles of carbon in the Southern Ocean: a step forward. (II). *Deep Sea Research 49*, 3103–3104.

Tremblay, J. and Smith Jr, W. (2007). Primary production and nutrient dynamics in polynyas. *Elsevier Oceanography Series, 74*, 239–269.

Tsuji, M., Tsujimoto, M., and Imura, S. (2017). *Cystobasidium tubakii* and *Cystobasidium ongulense*, new basidiomycetous yeast species isolated from East Ongul Island, *East Antarctica. Mycoscience 58* (2): 103–110.

Turner, J., Bindschadler, R., Convey, P. et al. (2009). *Antarctic Climate Change and the Environment*, 183–298. Cambridge: SCAR.

Tynan, C., Ainley, D. and Stirling, I. (2010). Sea ice: a critical habitat for polar marine mammals and birds. *Sea Ice* 395–424.

Upreti, D., Mukerji, K., Chamola, B. et al. (1999). *Studies on Antarctic Lichens: Biology of Lichens* (eds. K.G. Mukerji, B.P. Chamola, D.K. Upreti et al.), 333–342. New Delhi: Aravali Book International.

USGCRP (2017). *Climate Science Special Report: Fourth National Climate Assessment, Volume I* (eds. D.J. Wuebbles, D.W. Fahey, K.A. Hibbard et al) Washington, DC: US Global Change Research Program. doi:10.7930/J0J964

Vancoppenolle, M., Meiners, K., Michel, C. et al. (2013). Role of sea ice in global biogeochemical cycles: emerging views and challenges. *Quaternary Science Reviews 79*: 207–230.

Van Gestel, N., Ducklow, H. and Bååth, E. (2020). Comparing temperature sensitivity of bacterial growth in Antarctic marine water and soil. *Global Change Biology 26* (4): 2280–2291.

Vaz, R.N. and Lennon, G. (1996). Physical oceanography of the Prydz Bay region of Antarctic waters. *Deep Sea Research Part I: Oceanographic Research Papers 43* (5): 603–641.

Venables, H., Clarke, A. and Meredith, M. (2013). Wintertime controls on summer stratification and productivity at the western Antarctic Peninsula. *Limnology and Oceanography 58* (3): 1035–1047

Vick-Majors, T., Achberger, A., Santibáñez, P. et al. (2016). Biogeochemistry and microbial diversity in the marine cavity beneath the Mc Murdo Ice Shelf, Antarctica. *Limnology and Oceanography 61* (2): 572–586.

Viles, H. and Spencer, T. (2014). *Coastal Problems: Geomorphology Ecology and Society at the Coast*. Routledge.

Vishniac, H. and Baharaeen, S. (1982). Five new basidioblastomycetous yeast species segregated from *Cryptococcus vishniacii* emend. auct., an Antarctic yeast species comprising four new varieties. *International Journal of Systematic and Evolutionary Microbiology 32* (4): 437–445.

Von Quillfeldt, C. (1997). Distribution of diatoms in the Northeast Water polynya, Greenland. *Journal of Marine Systems 10* (1–4): 211–240.

Wadham, J, De'Ath, R., Monteiro, F. et al. (2013). The potential role of the Antarctic Ice Sheet in global biogeochemical cycles. *Earth and Environmental Science Transactions of the Royal Society of Edinburgh 104* (1): 55–67.

Wang, L., Liu, X., Yu, S. et al. (2017). Bacterial community structure in intertidal sediments of Fildes peninsula, maritime Antarctica. *Polar Biology 40* (2): 339–349.

Waring, R. and Schlesinger, W. (1985). *Forest Ecosystems. Analysis at Multiple Scales*. Orlando, FL: Academic Press.

Wassmann, P., Duarte, C.M., Agustí, S. et al. (2011). Footprints of climate change in the Arctic marine ecosystem. *Global Change Biology 17*: 1235e1249.

Watson, A. and Liss, P. (1998). Marine biological controls on climate via the carbon and sulphur geochemical cycles. *Philosophical Transactions of the Royal Society of London. Series B: Biological Sciences 353* (1365): 41–51.

Weber, T., Cram, J, Leung, S., et al. (2016). Deep ocean nutrients imply large latitudinal variation in particle transfer efficiency. *Proceedings of the National Academy of Sciences 113* (31): 8606–8611.

Wiencke, C. Clayton, M. and Wägele, J. (2002). *Antarctic Seaweeds*, Vol. *9*. ARG Gantner Verlag Ruggell.

Wilkins, D., Yau, S., Williams, T. et al. (2013). Key microbial drivers in Antarctic aquatic environments. *FEMS Microbiology Reviews 37* (3): 303–335.

Wulff, A., Iken, K., Quartino, M. et al. (2009). Biodiversity, biogeography and zonation of marine benthic micro-and macroalgae in the arctic and Antarctic. *Botanica Marina 52* (6): 491–507.

Yergeau, E., Newsham, K., Pearce, D. et al. (2007). Patterns of bacterial diversity across a range of Antarctic terrestrial habitats. *Environmental Microbiology 9* (11): 2670–2682.

Yergeau, E., Schoondermark-Stolk, S., Brodie, E. et al. (2009). Environmental microarray analyses of Antarctic soil microbial communities. *The ISME Journal 3* (3): 340–351.

Zacher, K. Rautenberger, R. Hanelt, D. et al. (2009). The abiotic environment of polar marine benthic algae. *Botanica Marina 52* (6): 483–490.

Biome IV
Extreme Environments

12

Geomicrobiology at a Physicochemical Limit for Life: Deep-sea Hypersaline Anoxic Basins

Nibendu Mondal*, Subhrangshu Mandal, and Wriddhiman Ghosh*

ABSTRACT

Despite their extreme physicochemical conditions, deep-sea hypersaline anoxic basins (DHABs), together with the overlying seawater–brine interfaces, constitute microbiologically diverse and biogeochemically active ecosystems. Extremely high salinity makes the brine waters of DHABs immiscible with the overlying oxic seawaters and the two water bodies are separated by suboxic interfaces. This salinity-based stratification creates pronounced chemoclines (haloclines) and pycnoclines within the systems. A comparison of all geochemical and microbiological data available for global DHAB systems showed that the seawater–brine interfaces are metabolically the most active segments of these physiographies. While sulfate reduction, methanogenesis and autotrophic carbon fixation are the only metabolic processes that have been identified thus far within the brine cores of only two DHABs of the Mediterranean Sea, the interfaces globally harbor diversified microbial communities having the potentials for sulfate reduction, sulfide oxidation, methanogenesis, aerobic/anaerobic methane oxidation, acetogenesis, carbon dioxide fixation aerobic/anaerobic ammonia oxidation, denitrification and/or nitrogen fixation. The overall geomicrobial framework of DHABs was considered in the context of comparator marine ecosystems and the emerging paradigm of ecosystem design was found to hold implications for our understanding of ancient microbiomes amidst the physicochemical extremes on a planetary scale.

12.1. INTRODUCTION

Deep-sea hypersaline anoxic basins (DHABs), also known as deep-sea hypersaline anoxic lakes (DHALs), are salt-accumulating aquatic systems located within fractured seabeds or depressions in various regions of the global ocean, below the usual seawater columns (generally a few kilometers beneath sea level). Whereas most brine lakes originate via dissolution of sub-seafloor salt deposits (ancient evaporite rocks) (Camerlenghi, 1990; Camerlenghi and McCoy, 1990; De Lange et al., 1990; Cita, 1991, 2006), alternative mechanisms such as dense shelf cascading, which transfers brine formed in shallow waters into deep waters as a result of density flow, have also been reported for brine formation in deep marine settings (Roveri et al., 2014). Deep-sea hypersaline anoxic lakes have 5–10 times higher salinity than normal seawater, and their hydrology and sedimentology are shaped by their protracted density-dependent separation from the overlying seawaters. In some locations, the brines of redissolved ancient geological evaporites with which the basins are filled, happen to be saturated with salts of divalent cations such as magnesium and calcium, and are therefore regarded as athalassohaline environments, i.e. having ionic compositions distinct from that of seawater, or other sea-derived or sea-like (thalassohaline) water bodies, which are rich in monovalent cations and subject to constant evaporation (Oren, 2006; Hallsworth et al., 2007; La Cono et al., 2011, 2019; Yakimov et al., 2013, 2015;

Department of Microbiology, Bose Institute, P-1/12 CIT Scheme VII M, Kolkata, 700054, India

**Corresponding Author: Nibendu Mondal (nibendumondal@gmail.com), Wriddhiman Ghosh (wriman@jcbose.ac.in) ORCiD code: 0000-0002-4922-7404*

Systems Biogeochemistry of Major Marine Biomes, First Edition. Edited by Aninda Mazumdar and Wriddhiman Ghosh.

Steinle et al., 2018). Because of their extremely high salinity, and therefore density, the brine waters of DHABs (specific gravity 1.1–1.2), in the absence of gravity flows, are immiscible with the overlying deep-sea water columns; consequently, the two water bodies are separated by chemoclines (gradients of chemical characteristics) and pycnoclines (gradients of water density), resulting in perennial density-stratification of the systems into anoxic brines below, the oxic marine water columns above and the suboxic interfaces (haloclines) in-between (De Lange et al., 1990; Hallsworth et al., 2007; La Cono et al., 2011, 2019; Wang et al., 2011, 2013; Yakimov et al., 2013, 2015; Steinle et al., 2018). Whereas in some deep-sea basins the brine pools are hundreds of meters deep from the surrounding seafloor, the seawater–brine interfaces, across global DHABs, are only a few meters thick (vertical expanse).

12.2. GEOGRAPHICAL AND GEOLOGICAL OVERVIEW OF DEEP-SEA HYPERSALINE ANOXIC BASINS

DHABs have thus far been discovered in four major territories of the marine realm: the Red Sea, the Mediterranean Sea, the Gulf of Mexico, and the Black Sea (Merlino et al., 2018; Figure 12.1). Key geomicrobial characteristics of the DHABs explored thus far are summarized in Tables 12.1 and 12.2, together with the most relevant publications considered for data collection. While maximum numbers of DHABs are located in the Red Sea, followed by the Mediterranean Sea and the Gulf of Mexico, only one hypersaline physiographic feature has been identified thus far in the Black Sea (Addy and Behrens, 1980; Williams and Lerche, 1987; Blum and Puchelt, 1991; Cita, 1991; Wallmann et al., 2002, 2006; Bohrmann et al., 2003; Aloisi et al., 2004; Ehrhardt et al., 2005; Cita, 2006; Pilcher and Blumstein, 2007; Feseker et al., 2009).

The brine lakes that are hosted in the deep seafloor depressions created due to tectonic activities (as in case of the Mediterranean DHABs) or salt withdrawal (as for the DHABs of the Red Sea or the Gulf of Mexico) involve dissolution (karstification) of underlying salt masses situated at shallow depths beneath the seafloor. In the halokinetic seafloors where DHABs have formed, salt welds, allochthons or other salt structures occur at such shallow sub-seafloor levels as to be able to dissolve and drain into the DHABs via faulted margins (seep zones) of the substratum (Camerlenghi, 1990). These phenomena perpetuate the DHAB brines, and in doing so stabilize the haloclines in the seawater–brine interfaces over geological time periods. Long-term physicochemical stability of the haloclines, in conjunction with the small amount of *in situ* oxygen (O_2) and relatively lower salinity of the interfaces (compared with the brines) gives rise to characteristic biodiversities, biogeochemistries, and sedimentologies in the water layers sitting atop the DHABs and impinging upon the upper rims of the salt-and-rock basin (Addy and Behrens, 1980; Pilcher and Blumstein, 2007). The physicochemical milieus of DHABs are characterized by complete darkness, lack of dissolved O_2, relatively elevated hydrostatic pressure and temperatures in the bottom-waters, and above all extreme salinity. The brine waters, in turn, contain typically higher quantities of ammonium, barium, bromine, borate, calcium, chlorine, magnesium, manganese, potassium, sodium, strontium, sulfate, and sulfide (compared with average seawater), albeit concentrations of these solutes vary widely across geographical and geological territories (De Lange et al., 1990). Water density based stratification is the most remarkable physiographic feature of DHAB systems. Salt concentration increases and oxygen concentration decreases eventually to zero in the oxic seawater to anoxic brine-bottom trajectory, through the suboxic interface. Temperature also increases along this trajectory: in some DHAB systems by only a few degrees Celsius whereas in others by tens of degrees Celsius (Williams and Lerche, 1987; Bregant et al., 1990).

12.3. DEEP-SEA HYPERSALINE ANOXIC BASINS AS PHYSICOCHEMICAL LIMITS OF THE EARTH'S BIOSPHERE

In situ combination of the above-mentioned extreme physicochemical conditions constraint the habitability of DHAB environments. While metabolically diverse microbial communities are known to live all along the chemoclines of these polyextreme habitats, the seawater–brine interfaces are relatively more biodiverse than the brine cores, and are therefore the main drivers of DHAB biogeochemistry (Merlino et al., 2018). The types and concentrations of salts and dissolved metals present in the brine determine their levels of water activity (A_w, which is 1 for pure water); this in turn determines how easily water becomes available to microbial cells for biochemical activities (Tapia et al., 2008). The chaotropic (order-breaking) or kosmotropic (order-making) nature of the ions present (including counter-ions and buffers), in conjunction with the pH and temperature of the aqueous milieu, entropically influences the structural integrity and activity of biomacromolecules (such as nucleic acids, proteins, lipids, etc.), and thereby the functionality of cell systems (Okur et al., 2017). The presence of high levels of chaotropic salts such as $MgCl_2$ (for example, in athalassohaline brines) reduce intracellular A_w and decrease the strength of hydrogen bonding and other electrostatic interactions in hydrated biomacromolecules, thereby

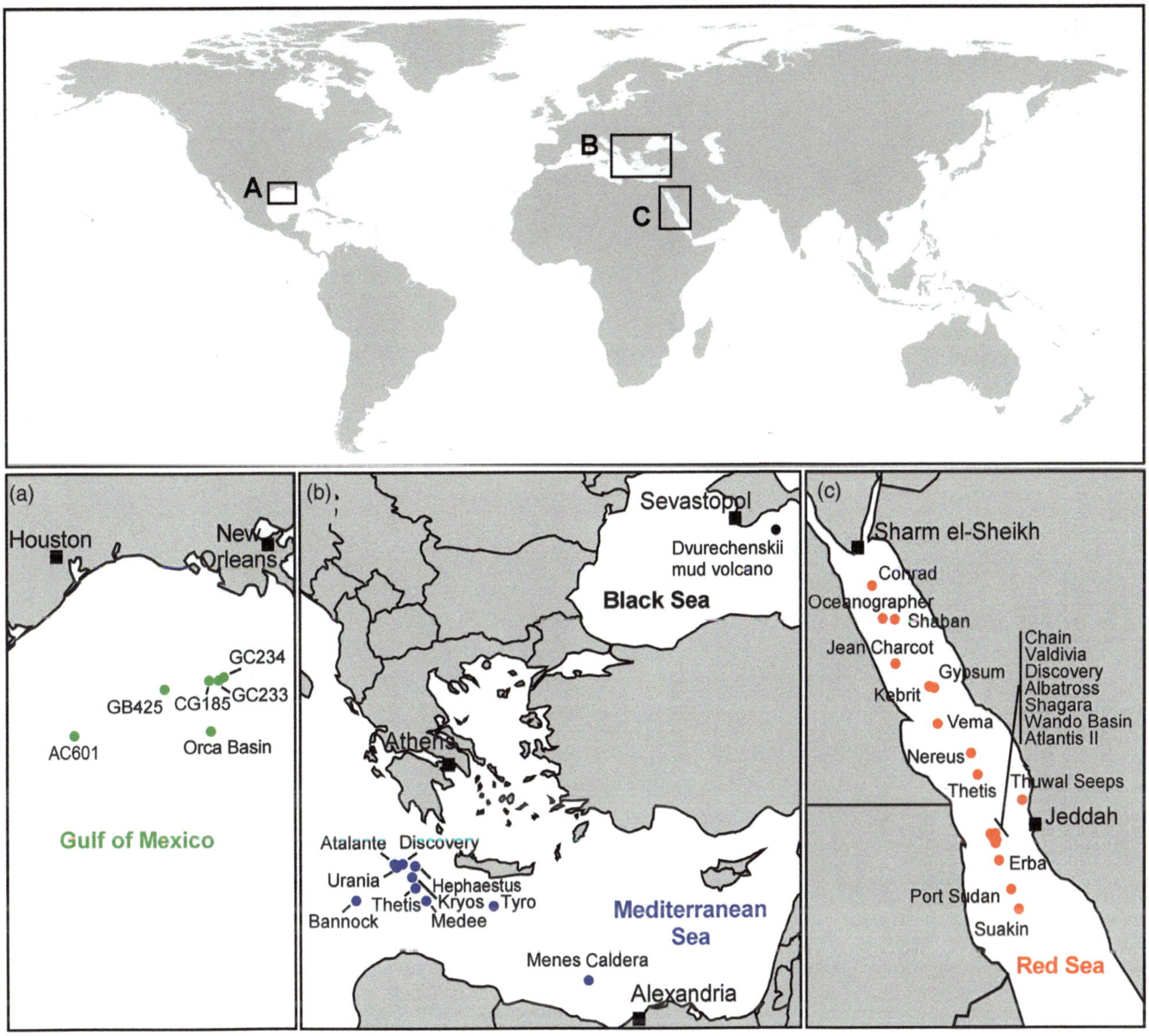

Figure 12.1 Locations of the well-studied DHAB systems of the global ocean: (a) the Gulf of Mexico (brines are indicated as green dots); (b) the Mediterranean Sea (brines are indicated as blue dots) and the Black Sea (brines indicated as a black dot); and (c) the Red Sea (brines are indicated as red dots). Merlino et al., 2018 (with kind permission from Oxford University Press); the locations of the DHABs AC601 (in the Gulf of Mexico) and Hephaestus (in the Mediterranean Sea) have been added to the adapted figure.

upsetting the latter's structures and functions (Koynova et al., 1997; Hallsworth et al., 2003). In contrast, the prevalence of kosmotropic salts such as NaCl in the aqueous milieu (as is the case in thalassohaline brines) potentially brings order to the biomacromolecules, mitigates against chaotropicity and render the ecosystem more biodiverse, as compared to athalassohaline environments that are not compensated by kosmotropes (Lee et al., 2018).

As lower A_w values constrain life biophysically, only the adapted few manage to survive at the extremes of physiological aridity (effectively xeric condition despite water being all around). Of all the DHAB environments of the global ocean, sites within Lake Kryos of the Mediterranean have been reported to contain microbial life at the lowest levels of *in situ* A_w. Steinle et al. (2018), based on the *in situ* presence of biogenic methane (CH_4) and empirically-determined *in situ* sulfate reduction and methane oxidation rates, reported carbon–sulfur cycling within the brine core of Kryos where A_w was calculated as 0.44 after accounting for $MgCl_2$ and $MgSO_4$ concentrations (Mg^{2+}, Cl^-, and SO_4^{2-} present at concentrations of 3936, 6545 and 144.2 mmol kg^{-1}, respectively), density

Table 12.1 Geographical, geological, and microbiological features of the well-studied DHAB systems of the global ocean.

Sl. no.	Brine nature	Name of the DHAB (reference)	Water-depth (mbsl)	Area (km^2), maximum thickness (m), volume (km^3)	Main mineral constituent	Geological age (Mya)	DHAB strata	Salinity ($g\ kg^{-1}$), A_W	Mg^{2+} (mM)	Na^+ (mM)	SO_4^{2-} (mM)	S^{2-} (mM)	O_2 (μM)	CH_4 (ppm)	*In situ* microbial processes detected thus far
							DHABs of the Mediterranean Sea								
1		Discovery[a]	>3580	~7.5, ~50, ~0.2	Bischofite ($MgCl_2.6H_2O$)	5.97–5.33	Brine pool	510, 0.382	5143	84	110.35	0.85	–	–	CO_2 fixation, methanogenesis, sulfate reduction.
							Seawater–brine interface	–, –	50–5050	–	–	–	–	–	CO_2 fixation, methanogenesis, sulfate reduction
2	Athalassohaline	Hephaestus[b]	>3373	~1.7, ~50, ~0.1	Bischofite ($MgCl_2.6H_2O$)	5.97–5.33	Brine pool	480, 0.395	4720	93	203	–	–	–	–
							Seawater-brine interface	–, –	70-4720	–	–	–	–	–	Acetogenesis, anaerobic ammonia oxidation, anaerobic methane oxidation, methanogenesis, sulfate reduction
3		Kryos[#,c]	>3344	~25, ~160, ~5	Bischofite ($MgCl_2.6H_2O$)	5.97–5.33	Brine pool	471, 0.399	3936* (4380*)	75* (125*)	144.2* (320*)	150[§]	–	20[§]	Methanogenesis, sulfate reduction
							Seawater–brine interface	330[†], 0.631[†]	83–665* (55–4380)	366-510*	31–61*	0–64[§]	0–182[§]	0.2–11[§]	Anaerobic methane oxidation, methanogenesis, sulfate reduction, sulfide oxidation
4		Bannock[d]	3350	–, –, –	–	5.97–5.33	Brine pool	–, –	644	4200	135.2	2.9	–	–	–
							Seawater–brine interface	–, –	–	–	–	–	–	–	Anaerobic ammonia oxidation, anaerobic methane oxidation, denitrification, methanogenesis, sulfate reduction
5	Thalassohaline	L'Atalante[e]	>3428	–, ~80, –	–	5.97–5.33	Brine pool	352, –	533	4670	323	2.9	–	–	–
							Seawater-brine interface	–, –	–	–	–	–	–	–	Anaerobic ammonia oxidation, CO_2 fixation, denitrification, methanogenesis, sulfate reduction
6		Medee[f]	>2900	110, ~180, 9	–	5.97–5.33	Brine pool	304–345, –	603–788	4022–4178	140.4–201	0.67–1.64	BDL	0.2–1.1	–
							Seawater-brine interface	–, –	–	–	–	BDL	BDL	–	CO_2 fixation, methanogenesis, sulfide oxidation
7		Thetis[g]	>3258	11, ~157, 0.7	Halite (NaCl)	5.97–5.33	Brine pool	348, –	604	4760	265	2.12	9.5	–	–
							Seawater–brine interface	–, –	–	–	–	–	–	–	Anaerobic methane oxidation, CO_2 fixation, denitrification, methanogenesis, nitrogen fixation, sulfate reduction, sulfide oxidation

8		Tyro[h]	–	–, –, –	–	5.97–5.33	Brine pool	–, –	71	5300	52.7	2.1	–	–	–
							Seawater–brine interface	–, –	–	–	–	–	–	–	–
9		Urania[i]	>3470	–, 200, –	–	5.97–5.33	Brine pool	–, –	315	3505	107	15	–	89.2	–
							Seawater–brine interface	240, –	–	–	–	–	–	–	Anaerobic ammonia oxidation, manganese oxidation, methanogenesis, sulfate reduction, sulfide oxidation
							DHABs of the Red Sea								
10		Atlantis II[j]	2062	–, ~120, –	Halite (NaCl)	14-13	Brine pool	255, –	29	4900	8.3	–	0.3	–	–
							Seawater–brine interface	–, –	62	700	32.5	BDL	2.2	–	Aerobic ammonia oxidation, aerobic methane oxidation, anaerobic methane oxidation, methanogenesis, nitrogen fixation, sulfate reduction.
11	Thalassohaline	Discovery Deep[j]	2141	–, 35, –	Halite (NaCl)	14-13	Brine pool	255, –	33	4700	8.4	–	0.3	–	–
							Seawater–brine interface	–, –	50	2300	26.7	BDL	0.3	0.035	Aerobic ammonia oxidation, CO_2 fixation, aerobic methane oxidation, anaerobic methane oxidation, methanogenesis, nitrogen fixation, sulfate reduction.
12		Erba Deep[k]	2392	–, 10, –	Halite (NaCl)	14–13	Brine pool	–, –	71	3200	41.6	–	0.3	–	–
							Seawater–brine interface	–, –	61	1500	38.6	BDL	0.3	0.01	Aerobic ammonia oxidation, methanogenesis, sulfate reduction.
13		Kebrit Deep[l]	1490	–, 3, –	Halite (NaCl)	14–13	Brine pool	–, –	119	4800	27.7	0.52	2.3	–	–
							Seawater–brine interface	–, –	107	3200	31.1	0.15	4.6	18.99	Aerobic ammonia oxidation, aerobic methane oxidation, anaerobic methane oxidation, methanogenesis, sulfate reduction
14		Nereus Deep[k]	2445	–, 12, –	Halite (NaCl)	14–13	Brine pool	–, –	67	3500	10.5	–	2.8	–	–
							Seawater–brine interface	–, –	67	2500	22.5	BDL	0.3	0.015	Aerobic ammonia oxidation, methanogenesis, sulfate reduction.

(*Continued*)

Table 12.1 (Continued)

Sl. no.	Brine nature	Name of the DHAB (reference)	Water-depth (mbsl)	Area (km^2), maximum thickness (m), volume (km^3)	Main mineral constituent	Geological age (Mya)	DHAB strata	Salinity ($g\ kg^{-1}$), A_W	Mg^{2+} (mM)	Na^+ (mM)	SO_4^{2-} (mM)	S^{2-} (mM)	O_2 (μM)	CH_4 (ppm)	*In situ* microbial processes detected thus far
							DHABs of the Gulf of Mexico								
15	Thalassohaline	AC601[m]	2334	–, –, –	–		Brine pool	–, –	–	–	–	–	–	–	–
							Seawater–brine interface	–, –	–	–	–	–	–	–	Anaerobic methane oxidation, sulfate reduction.
16		GB425[n]	600	–, –, –	Halite (NaCl)	175–145	Brine pool	–, –	–	–	–	–	–	–	–
							Seawater–brine interface	130, –	8.7	1790	<1	0.004	–	151440	Methanogenesis, sulfate reduction.
17		GC233[n]	650	0.00019, –, –	Halite (NaCl)	175–145	Brine pool	–, –	–	–	–	–	–	–	–
							Seawater–brine interface	121, –	97	1751	<1	0.002	–	8151	Acetogenesis, methanogenesis, sulfate reduction, sulfide oxidation
18		Orca Basin[o]	>2250	400, 200, –	–	175–145	Stratigraphy unclear	250, –	42	4240	20	0.025	–	–	Acetogenesis

A_W, water activity; mbsl, meters below sea level; –, data not available in the literature; BDL, below the detection limit of the experiment conducted.

*Data are in mmol (kg H_2O)$^{-1}$.

[#] Mg^{2+}, Na^+, SO_4^{2-}, S^{2-}, O_2 and CH_4 data for Lake Kryos have been taken from Steinle et al., 2018; Mg^{2+}, Na^+ and SO_4^{2-} concentration values that were also given for this DHAB by Yakimov et al., 2015 have been indicated in parenthesis.

[§] Data are in μmol (kg H_2O)$^{-1}$.

[†] Values were taken at an approximately two-thirds water-depth of the Kryos interface (Yakimov et al., 2015); the salinity value is in PSU (practical salinity units).

[a] Van der Wielen et al. (2005, 2006), Hallsworth et al. (2007), Roveri et al. (2014), Yakimov et al. (2015).

[b] Roveri et al. (2014), La Cono et al. (2019).

[c] Roveri et al. (2014), Yakimov et al. (2015), Steinle et al. (2018).

[d] De Lange et al. (1990), Daffonchio et al. (2006), Borin et al. (2013), Roveri et al. (2014).

[e] Van der Wielen et al. (2005), Yakimov et al. (2007, 2015), Borin et al. (2013), Roveri et al. (2014), Merlino et al. (2018).

[f] Yakimov et al. (2013), Roveri et al. (2014).

[g] Meyer and Kuever (2007), La Cono et al. (2011), Yakimov et al. (2013), Pachiadaki et al. (2014), Roveri et al. (2014).

[h] De Lange et al. (1990), Roveri et al. (2014), Merlino et al. (2018).

[i] Sass et al. (2001), Borin et al. (2009), La Cono et al. (2011), Roveri et al. (2014), Merlino et al. (2018).

[j] Antunes et al. (2011), Bougouffa et al (2013), Wang et al. (2013), Abdallah et al. (2014), Guan et al. (2015), Ngugi et al. (2015), Schmidt et al. (2015).

[k] Antunes et al. (2011), Guan et al. (2015), Ngugi et al. (2015), Schmidt et al. (2015).

[l] Antunes et al. (2011), Abdallah et al. (2014), Guan et al. (2015), Ngugi et al. (2015), Schmidt et al. (2015).

[m] Wankel et al. (2010), Merlino et al. (2018).

[n] Joye et al. (2005, 2009), Merlino et al. (2018).

[o] Van Cappellen et al. (1998), Nigro et al. (2016), Merlino et al. (2018).

Table 12.2 Main taxonomic groups detected corresponding to the individual microbial processes identified in the well-studied DHAB systems.

Sl. no.	Name of the DHAB (references)	DHAB strata	Microbiology technique used	Microbial processes and corresponding taxa										
				Sulfate reduction	Sulfide oxidation	Methanogenesis	Anaerobic methane oxidation	Aerobic methane oxidation	Acetogenesis	CO_2 fixation	Anaerobic ammonia oxidation	Aerobic ammonia oxidation	Denitrification	Nitrogen fixation
	DHABs of the Mediterranean Sea													
1	Discovery[a]	Brine	16S rRNA gene and partial genes sequencing	*Desulfobacteraceae*	–	*Methanohalophilus*	–	–	–	*Thiobacillus* sp.	–	–	–	–
		Seawater-brine interface		*Desulfobacteraceae*	–	*Methanohalophilus*	–	–	–	*Thiobacillus* sp.	–	–	–	–
2	Hephaestus[b]	Brine	16S rRNA gene and mRNA transcripts	–	–	–	–	–	–	–	–	–	–	–
		Seawater-brine interface		*Deltaproteobacteria*	–	Euryarchaeota	Euryarchaeota	–	*Acetothermia*	–	Thaumarchaeota	–	–	–
3	Kryos[c]	Brine	16S rRNA gene and mRNA transcripts	*Desulfovermiculus*	–	*Methanohalophilus*	–	–	–	–	–	–	–	–
		Seawater–brine interface		*Desulfovermiculus*	*Sulfurimonas*	*Methanohalophilus*	*Methylococcales*	–	–	–	–	–	–	–
4	Bannock[d]	Brine	16S rRNA gene and transcriptome	–	–	–	–	–	–	–	–	–	–	–
		Seawater-brine interface		*Desulfobacteriaceae*	–	IG	ANME1	–	–	–	*Planctomycetales*	–	IG	–
5	L'Atalante[e]	Brine	16S rRNA gene and functional gene analysis	–	–	–	–	–	–	–	–	–	–	–
		Seawater–brine interface		IG	–	MSBL1	–	–	–	*Thioclava*	*Planctomycetales*	–	IG	–
6	Medee[f]	Brine	16S rRNA gene and mRNA transcripts	–	–	–	–	–	–	–	–	–	–	–
		Seawater–brine interface		–	*Gammaproteobacteria*	MSBL1	–	–	–	Thaumarchaeota	–	–	–	–
7	Thetis[g]	Brine	16S rRNA gene and mRNA transcripts	–	–	–	–	–	–	–	–	–	–	–
		Seawater–brine interface		*Deltaproteobacteria*	*Gammaproteobacteria*	*Methanohalophilus*	ANME1	–	–	Thaumarchaeota	–	–	IG	IG
8	Tyro	Brine	–	–	–	–	–	–	–	–	–	–	–	–
		Seawater-brine interface	–	–	–	–	–	–	–	–	–	–	–	–
9	Urania[h]	Brine	16S rRNA gene	–	–	–	–	–	–	–	–	–	–	–
		Seawater–brine interface		*Deltaproteobacteria*	*Epsilonproteobacteria*	MSBL1	–	–	–	–	Crenarchaeota	–	–	–
	DHABs of the Red Sea													
10	Atlantis II[i]	Brine	16S rRNA gene and functional gene analysis	–	–	–	–	–	–	–	–	–	–	–
		Seawater–brine interface		*Desulfovibrio*	–	*Methanohalophilus*	ANME1	*Methylococcus*	–	–	–	*Nitrosopumilus*	–	Bradyrhizobium
11	Discovery Deep[j]	Brine	16S rRNA gene	–	–	–	–	–	–	–	–	–	–	–
		Seawater–brine interface		*Desulfovibrio*	–	Archaeoglobales	Archaeoglobales	*Methylococcus*	–	*Thermofilum*	–	*Nitrosopumilus*	–	Bradyrhizobium

(*Continued*)

Table 12.2 (Continued)

				Microbial processes and corresponding taxa										
Sl. no.	Name of the DHAB (references)	DHAB strata	Microbiology technique used	Sulfate reduction	Sulfide oxidation	Methanogenesis	Anaerobic methane oxidation	Aerobic methane oxidation	Acetogenesis	CO_2 fixation	Anaerobic ammonia oxidation	Aerobic ammonia oxidation	Denitrification	Nitrogen fixation
12	Erba Deep[j]	Brine	16S rRNA gene and functional gene analysis	–	–	–	–	–	–	–	–	–	–	–
		Seawater–brine interface		*Deltaproteobacteria*	–	*Methanohalophilus*	–	–	–	–	–	*Nitrosopumilus*	–	–
13	Kebrit Deep[k]	Brine	16S rRNA gene, functional gene analysis and isotope data	–	–	–	–	–	–	–	–	–	–	–
		Seawater–brine interface		*Deltaproteobacteria*	–	*Methanohalophilus*	IG	*Methylobacter*	–	–	–	*Nitrosopumilus*	–	–
14	Nereus Deep[j]	Brine	16S rRNA gene and functional gene analysis	–	–	–	–	–	–	–	–	–	–	–
		Seawater–brine interface		*Deltaproteobacteria*	–	*Methanohalophilus*	–	–	–	–	–	*Nitrosopumilus*	–	–
						DHABs of the Gulf of Mexico								
15	AC601[l]	Brine	–	–	–	–	–	–	–	–	–	–	–	–
		Seawater–brine interface	–	IG	–	–	IG	–	–	–	–	–	–	–
16	GB425[m]	Stratum not defined	16S rRNA gene and functional gene analysis	*Desulfobacterium*	–	*Methanolobus*	–	–	–	–	–	–	–	–
17	GC233[m]	Stratum not defined	16S rRNA gene and functional gene analysis	*Desulfobacterium*	*Epsilonproteobacteria*	*Methanolobus*	–	–	*Methanolobus*	–	–	–	–	–
18	Orca Basin[n]	Stratum not defined	16S rRNA gene and single cell genomics	–	–	–	–	–	KB1	–	–	–	–	–

–, Data not available in the literature; IG, metabolic process identified based on isotopic and/or other geochemical data.

[a] van der Wielen et al. (2005, 2006); Hallsworth et al. (2007).

[b] La Cono et al. (2019).

[c] Yakimov et al. (2015), Steinle et al. (2018).

[d] Daffonchio et al. (2006), Borin et al. (2013).

[e] Yakimov et al. (2007), Borin et al. (2013).

[f] Yakimov et al. (2013).

[g] Meyer and Kuever (2007), La Cono et al. (2011), Yakimov et al. (2013), Pachiadaki et al. (2014).

[h] Borin et al. (2009).

[i] Bougouffa et al. (2013), Wang et al. (2013), Abdallah et al. (2014), Guan et al. (2015), Ngugi et al. (2015), Schmidt et al. (2015).

[j] Guan et al. (2015), Ngugi et al. (2015), Schmidt et al. (2015).

[k] Antunes et al. (2011), Abdallah et al. (2014), Guan et al. (2015), Ngugi et al. (2015), Schmidt et al. (2015).

[l] Wankel et al. (2010).

[m] Joye et al. (2009).

[n] Nigro et al. (2016).

(1.31 kg dm^{-3}) and total dissolved solids (0.46 kg dm^{-3}). Before that, Yakimov et al. (2015), based on messenger RNA (mRNA) recovery data, had detected microbial activity along an approximately two-thirds water-depth of the seawater–brine interface of Kryos, characterized by a 2.27–3.03 M $MgCl_2$ concentration and 0.747–0.631 A_w. Whereas these water activity levels are considerably lower than the 0.72 cut-off at which microbial diversity is constrained in any other ecosystem (Stevenson et al., 2015; Lee et al., 2018), the exceptionally higher habitability of Lake Kryos, compared with other athalassohaline brines, is potentially attributed to the presence of relatively higher concentrations of sodium and sulfate ions that are kosmotropic (Yakimov et al., 2015). Similarly, the seawater–brine interface of Lake Hephaestus has also been found to harbor extremely halophilic microbial communities at an $MgCl_2$ concentration of 2.97 M, which corresponds to an A_w value of 0.653 (La Cono et al., 2019). In view of these peculiarities it is expected that further studies of biogeochemistry for DHAB systems would augment our understanding of the potential of life to colonize, adapt, and flourish in the biophysical extremities of not only the Earth's biosphere but also those of other extraterrestrial locations. Unfortunately, however, at this point of time, equivalent levels of geomicrobiological information are not available for all DHABs systems of the global ocean, as some are very well-studied whereas others are biogeochemically unexplored or poorly explored. This said, some members of the latter category do appear to have the potential to break new ground for the physicochemical limits of the Earth's biosphere. For example, the geophysically and geochemically idiosyncratic Dvurechenskii DHAB system is microbiologically unexplored (Sergeeva and Kopytina, 2014) and its geology suggests that it may harbor unique life strategies and biogeochemistries.

12.4. GEOMICROBIAL DYNAMICS IN DEEP-SEA HYPERSALINE ANOXIC BASINS ACROSS THE GLOBAL OCEAN

Most of the studies of microbial ecology available for different DHAB systems are based on diversity analyses that in turn have relied mostly on amplified 16S ribosomal RNA (rRNA) gene sequence analysis (Table 12.2). In contrast, information pertaining to *in situ* microbial processes and networking systems, based on functional metagenomic and metatranscriptomic data is limited. Of the distinct DHABs distributed across the global ocean, a number of Mediterranean systems are well-studied for their geomicrobiology, followed by a few in the Red Sea and in the Gulf of Mexico (Tables 12.1 and 12.2).

12.4.1. Deep-sea Hypersaline Anoxic Basins of the Mediterranean Sea

Within the Mediterranean Sea, brine lakes are situated mostly to the south of Crete, between Greece and the North African coast of Libya, at water-depths greater than 3000 m below sea level (mbsl) (Cita, 2006). These brines were generated from the dissolution of the evaporites produced during the Messinian salinity crisis, which occurred between 5.97 and 5.33 million years ago (Mya) in the Miocene period (Roveri et al., 2014). Dissolution of the Messinian evaporites, however, possibly occurred in more recent times (i.e. around 700–40000 years ago), when the ancient salt deposits were exhumed and exposed to the seawaters as a result of tectonic activities, and in the process dense brines flowed into the deep depressions (Fusi et al., 1996; Wallmann et al., 1997, 2002; Yakimov et al., 2007; La Cono et al., 2019). Furthermore, dense shelf cascading mechanism is recognized to have been active during the Messinian salinity crisis (a dramatic event characterized by the accumulation of large amounts of evaporites in the Mediterranean lows), playing crucial roles in the formation of DHABs in the eastern Mediterranean (Roveri et al., 2014).

Variations in the chemical composition of the DHABs of the Mediterranean Sea point also towards their origin in distinct geophysical events: while a majority of them (e.g. Bannock, L'Atalante, Medee, Thetis, Tyro, and Urania) are predominated by NaCl and so are thalassohaline (De Lange et al., 1990; Van der Wielen et al., 2005; La Cono et al., 2011), a few, such as Discovery, Hephaestus and Kryos, are athalassohaline and $MgCl_2$-dominated (Hallsworth et al., 2007; Yakimov et al., 2015; Steinle et al., 2018; La Cono et al., 2019) (Figure 12.1). The temperature of these brines is generally between 12°C and 19°C. Remarkably, however, Mediterranean DHABs appear to be mutually comparable in terms of their microbial community architecture. Although microbial communities are expected to be influenced significantly by the *in situ* composition and concentration of salts, other environmental factors such as dissolved O_2 concentration, organic matter content, and the presence/concentration of CH_4 and S^{2-} seem to be more crucial in determining microbiome structure/functions within DHAB systems.

Geomicrobiology of the Athalassohaline Brines of the Mediterranean Sea

The microbial ecologies of the athalassohaline brines of the Mediterranean Sea have been well-characterized, and can give insights into the potential biogeochemical cycles operating in these chaotrope-dominated extremities of the biosphere (Tables 12.1 and 12.2). Metagenome analysis for samples collected from the middle of the seawater–brine interface of Lake Hephaestus pointed to the potential presence of sulfate-reducing *Deltaproteobacteria*, anaerobically ammonium-oxidizing Marine Group I *Thaumarchaeota*, acetogenic KB1 group (*Acetothermia*) and, methanogenic and methanotrophic *Euryarchaeota* (La Cono et al., 2019). Possible *in situ* activities of these metabolic types, however, were corroborated only from the upper part of seawater–brine interface via detection of mRNA transcripts corresponding to ammonia monooxygenase (*amoA*), methylcoenzyme M reductase (*mcrA*), acetyl-CoA decarbonylase/synthase (*cdh*D), and dissimilatory sulfite reductase (*dsrAB*) genes (La Cono et al., 2019). Exploration of Hephaestus has further revealed that the dissolved organic matter content of this DHAB system includes thousands of species of small organic molecules, which have mean molecular weight <500 atomic mass units (La Cono et al., 2019). Small organic compounds that are apparently the products of microbial degradation of complex organic materials in the overlying water column also predominate in the dissolved organic matter content of Lake Kryos. In the Kryos brine, however, the chemical diversity of small organic molecules is lower than that in Hephaestus, although the prevalence of highly-oxygenated organic compounds (O/C ratio >0.6), reflective of biologically active environments, is higher. In the Hephaestus system, organic compounds having O/C ratios >0.6 are more prevalent in the seawater–brine interface. Comparative organic geochemistry thus indicated that the Hephaestus ecosystem is less influenced by oxygenated seawater than is the Kryos ecosystem (La Cono et al., 2019).

Other in-depth studies of biogeochemistry have also been conducted for Kryos. Here, *in situ* concentrations of SO_4^{2-}, H_2S, and CH_4 were all found to increase down the chemocline and reach very high levels at the brine core (Van der Wielen et al., 2005). The same trend was identified for the rate of *in situ* sulfate reduction. Notably, the rate of sulfate reduction in Kryos is approximately 5–20 times higher than those recorded in the other DHAB systems of the Mediterranean Sea (Pitzer, 1991; Van der Wielen et al., 2005). The only exception to this is Bannock, which has a sulfate reduction rate ~10 times higher than Kryos (Daffonchio et al., 2006; Steinle et al., 2018). Corroborating this, metatranscriptome analysis for the lower interface of Kryos identified mRNA transcripts corresponding to *dsrAB* genes which are signatures of sulfate-reducing activities (Yakimov et al., 2015). Sulfate reduction was also detected in the brine core of Kryos (Steinle et al., 2018). In this context it is further noteworthy that down-depth from the boundary of the upper and lower interfaces of this DHAB system, S^{2-} concentration decreases significantly, thereby indicating the potential presence of sulfide-oxidizing chemolithotrophic activities in the lower half of the seawater–brine interface (Yakimov et al., 2015). Carbon isotope ratios ($\delta^{13}C$) of the CH_4 present in the different strata of the Kryos chemocline could not resolve the relative contributions of thermogenic and biogenic processes to their origin. However, methanogenic activities, based on the presence of mRNA transcripts corresponding to *mcrA* genes, have been detected in the brine as well as the seawater–brine interface (Yakimov et al., 2015). *In situ* rates of methane oxidation, on the other hand, have been found to increase in the upper to lower interface trajectory; no methane oxidation apparently occurs within the brine core itself (Steinle et al., 2018).

Information on the biogeochemistry of Lake Discovery is derived largely from culture-independent (metaomics-based) microbiological data. For this system, methane and sulfide production was recorded in the seawater–brine interface as well as the anoxic hypersaline brine waters, while the *in situ* predominance of methanogens and sulfate-reducers were corroborated by 16S rRNA gene sequence-based (metataxonomic) diversity analysis of the *in situ* microbial community (Van der Wielen et al., 2005). Sequences corresponding to the autotrophic genes *cbbL* and *cbbM* which encode subunits of the enzyme ribulose-1,5-bisphosphate carboxylase/oxygenase (RuBisCO) were detected in the metagenomes isolated from the interface as well as the brine pool of the Discovery basin (Van der Wielen, 2006). The metabolically active status of the methanogenic and sulfate-reducing microbial populations were supported by the detection of mRNA transcripts corresponding to *mcrA* and *dsrAB* genes, albeit only in the interface and not in the brine water (Hallsworth et al., 2007).

Geomicrobiology of the Thalassohaline Brines of the Mediterranean Sea

A number of thalassohaline brines of the Mediterranean have also been well-studied for their microbial ecology and biogeochemistry, which do not differ greatly from those of their athalassohaline counterparts (Tables 12.1 and 12.2; Figure 12.1). In general, taxonomically diverse methanogens and sulfate-reducers are active in the anoxic layers of the seawater–brine interfaces of these DHAB systems, whereas methanotrophs and sulfur-oxidizers predominate the oxic strata of the interfaces.

Of all the Mediterranean DHABs, Lake Bannock has the highest sulfate reduction rate, which increases in this territory with the increase in salinity along the seawater to brine trajectory. The presence of sulfate-reducing microorganisms within the Bannock system was detected metataxonomically as well as metatranscriptomically throughout the anoxic zone of seawater–brine interface; methanogens and methanotrophs were identified in the deeper part of this anoxic interface where CH_4 and sulfate co-occur (Daffonchio et al., 2006). Analogous to the microbiome profile of Lake Bannock, exploration of the L'Atalante brine system via metataxonomic analysis of the *in situ* microbial diversity also detected sulfate-reducing, methanogenic, and heterotrophic microorganisms in the seawater–brine interface (Van der Wielen et al., 2005). At the same time, polymerase chain reaction (PCR) based detection of *cbbL* gene indicated the presence CO_2 fixation in the upper seawater–brine interface of L'Atalante (Yakimov et al., 2007). Furthermore, environmental 16S rRNA gene cloning and sequencing-based analyses indicated denitrification and anaerobic ammonium oxidation (anammox) to be the potential mechanisms of fixed nitrogen loss in the seawater–brine interface of both Bannock and L'Atlante ecosystems; detection of the hydrazine synthase (*hzsA*) gene reiterated the presence of anammox in the Bannock seawater–brine interface (Borin et al., 2013).

In other thalassohaline brines of the Mediterranean such as Lake Medee and Lake Thetis, mRNA transcripts corresponding to the *cbbL-1Ac* and *cbbL-1Aq* genes of the autotrophic marker protein RuBisCO, and the *aprA* gene, which encodes the alpha subunit of adenosine-5′-phosphosulfate (APS) reductase (and is regarded as a key enzyme for dissimilatory sulfate reduction on one hand and sulfur oxidation by certain photolithotrophs and chemolithotrophs on the other; Meyer and Kuever, 2007) have been detected in the oxygenated waters overlying the brines (La Cono et al., 2011; Yakimov et al., 2013). In both Lake Medee and Lake Thetis, active methanogenesis has been corroborated by recovering *mcrA* gene transcripts from the seawater–brine interfaces. Furthermore, in the interface of Lake Medee, organic geochemistry data suggested *in situ* acetogenic glycine–betaine fermentation, while sulfide oxidation was indicated by 16S rRNA gene sequence-based diversity analyses. In contrast, in the *Thetis* interface, a predominance of heterotrophic sulfate reduction and anaerobic methane oxidation was apparent from environmental 16S rRNA gene cloning and sequencing-based analyses (La Cono et al., 2011; Yakimov et al., 2013; Pachiadaki et al., 2014). In Thetis, recovery of transcripts corresponding to the nitrogenase enzyme complex (α and β chains of the protein as well as the ATPase component) corroborated nitrogen fixation in the lower half of the seawater–brine interface; similarly, transcripts of nitrous oxide and nitric oxide reductases across the interface indicated *in situ* denitrification (Pachiadaki et al., 2014).

The Urania DHAB system is remarkable among all the thalassohaline DHABs of the Mediterranean in having two sharp chemoclines: one beneath the oxic seawater (Interface 1) and another (Interface 2) between the two anoxic brine-layers having differential water density (Borin et al., 2009). Biogeochemical parameters such as sulfate reduction rate are higher in the upper part of Interface 1 compared with the other parts of this unique hydrography; sulfide oxidation, manganese oxidation and anammox are also active in this zone. However, the rate of methanogenesis is higher than sulfate reduction rate in the bottom of Interface 1 (Borin et al., 2009).

12.4.2. Deep-sea Hypersaline Anoxic Basins of the Red Sea

In the marine realm, maximum numbers of DHABs are situated within the deep axial part of the Red Sea (Pautot et al., 1984) (Figure 12.1), which was formed because of the splitting of the Arabian Peninsula from Horn of Africa owing to the movement of the Red Sea rift that started in the Eocene and is still continuing (Rose and Laking, 2008). During the Miocene epoch, the tectonic activity associated with these geodynamic movements reduced the connectivity of Red Sea with the global ocean (primarily via closure of the Bab Al Mandad Strait), and in doing so caused the accretion of thick evaporite rocks in the Red Sea basin floor (Searle and Ross, 1975). Through Middle Miocene (14-13 Mya), the basement became covered with evaporites or rift-fill sediments that apparently accumulated on transitional crust to begin with, but subsequently flowed (via fracture zones) into the axial (basaltic) zone of the Plio-Pleistocene oceanic crust, i.e. the geologically recent spreading seafloors of the Red Sea (Girdler and Whitmarsh, 1974; Ehrhardt et al., 2005). Thus, the flowing halite mixed with the axial hydrothermal circulation to give rise to several pools of dense hot brines and evaporite-enclosed deep basins along the deep axial part of the Red Sea (Feldens and Mitchell, 2015). However, in addition to these brines occurring along the deep axis of the Red Sea rift, a number of DHABs also occur in the shallower coastal margins, down the slope of the dipping sedimented seafloor (1.4–2.0 km below sea level) (Batang et al., 2012).

Red Sea DHABs, although all thalassohaline in nature (Table 12.1), can be geologically classified into two types (Schmidt et al., 2015; Merlino et al., 2018). The first type contains highly saline, low pH, and mildly warm water with low levels of heavy metals; examples of these include Kebrit Deep and Oceanographer. These DHABs have been formed from evaporite dissolution and sediment alteration; they do not involve volcanic rock interactions and are not enriched by hydrothermal fluid. In contrast, the second type of DHABs involve subsurface volcanic rock interaction and hydrothermal fluid input as they have been formed via redissolution of hydrothermal precipitates accreted from previous events of seafloor spreading (Bonatti 1985; Ehrhardt et al., 2005; Ehrhardt and Hübscher, 2015; Boschetti et al., 2018). Accordingly, DHABs such as Conrad, Nereus Deep and Shaban Deep located in the northern part of the Red Sea, and Albatros,

Atlantis II, Chain, Discovery Deep, Erba Deep, Port Sudan, Shagara, Suakin, Valdivia and Wando in the central region, have hot waters (often having temperatures ≥60° C) rich in metal ions such as iron, lithium, manganese and zinc.

Biogeochemical characteristics of the Red Sea DHABs are largely comparable to those of their Mediterranean counterparts (Guan et al., 2015). High variations in the density of water along the chemoclines cause entrapment of particulate matter and microbial debris in the seawater–brine interfaces of the Red Sea DHABs, which in turn leads to the increase of dissolved organic carbon and reduction of dissolved O_2 in the seawater to brine trajectory; consequently, the seawater–brine interfaces also have higher microbial activities than the brine cores (Ngugi et al., 2015).

Metagenomic exploration of the lower convective layer of the seawater–brine interface of Discovery Deep (*in situ* temperature 45°C) has revealed the presence of methanogens, anaerobic methane-oxidizers and phylogenetically diverse autotrophic archaea and bacteria. The autotrophic microorganisms are absent in the interface of Atlantis II (68°C), apparently owing to the extremely low CO_2 concentration of the fluid (Wang et al., 2013). Instead of autotrophs, however, heterotrophic bacteria (*Phyllobacterium*) capable of utilizing extracellular organic carbon are present in the Atlantis II interface, according to metagenomic data (Wang et al., 2013). Metagenomic data also indicated nitrogen fixation in the seawater–brine interfaces of both the DHAB systems (Wang et al., 2013). Furthermore, relative abundance of bacteria, compared with archaea, was higher in Atlantis II (~99.9%) than Discovery Deep (~49%) (Wang et al., 2011). While a decrease in SO_4^{2-} concentration down the Atlantis II and Discovery Deep water columns indicated microbial sulfate reduction and/or precipitation of anhydrites along both the interfaces (Bougouffa et al., 2013), sulfur isotope fractionation data from the seawater–brine interfaces of Kebrit Deep and Shaban Deep revealed sulfate reduction coupled with methane oxidation (Antunes et al., 2011). Furthermore, *dsrA* and *mcrA* gene sequences obtained from the seawater–brine interfaces of Kebrit Deep, Erba Deep, Nereus Deep, Discovery Deep as well as Atlantis II revealed enormous *in situ* diversities of sulfate-reducing bacteria and methanogenic archaea, respectively (Guan et al., 2015). Metagenomic data, 16S rRNA gene sequence analysis, and sequence analysis of particulate methane monooxygenase (*pmoA*) gene clone libraries provided evidence for the presence of aerobic methane oxidizers in the seawater–brine interfaces of Atlantis II, Discovery Deep, and Kebrit Deep (Abdallah et al., 2014). Metagenomic data for the seawater–brine interfaces of these five Red Sea DHABs have also revealed the nitrifying role of ammonia oxidizing archaea (AOA) in presence of low amounts of dissolved O_2; ammonia oxidizing bacteria (AOB), however, are conspicuously absent in these territories (Ngugi et al., 2015). In this context it is noteworthy that aerobic oxidation of ammonia has thus far been reported only from the seawater-brine interfaces of Red Sea DHABs (Figure 12.2).

12.4.3. Deep-sea Hypersaline Anoxic Basins of the Gulf of Mexico

Seismic events leading to the cracking and subsequent dissolution of evaporitic sheets and salt domes of the mid-Jurassic period (175–145 Mya) have given rise to thalassohaline DHABs in the Gulf of Mexico. Burial started during the Mesozoic, as documented by the first activity of salt diapirs, while Miocene depositions characterize the relatively recent geological history of the Gulf of Mexico basins (Bouma and Bryant, 1994; Roberts et al., 2007; Galloway, 2008; Cordes, et al., 2009; Merlino et al., 2018). The 400 km^2 Orca Basin located ~2.25 km below sea level, at the Walker Ridge and Green Canyon boundary (Pilcher and Blumstein, 2007), was the first DHAB to be discovered in the Gulf of Mexico (Shokes et al., 1977). The biogeochemistry of this huge thalassohaline brine system has been explored considerably over the last few decades, even though the exact stratigraphic location is not clear for all Orca Basin data-points. Single cell genomics and 16S rRNA gene sequence analysis has revealed the presence of acetogenic bacteria in this ecosystem (Nigro et al., 2016), carbon isotopic data have demonstrated the microbial control of carbon supply at the seawater–brine interface (Shah et al., 2013), and biogeochemical cycling of metals has been studied at the oxic–anoxic transition zone of the basin (Van Cappellen et al., 1998). In the Gulf of Mexico, other smaller brine pools or mud volcanoes such as NR-1/GC233, GC234, GB425, CG185 and AC601, characterized by relatively lower salinities, have also been discovered and well-studied for their geomicrobiology (Brooks et al., 1979; MacDonald et al., 1990; Joye et al., 2005, 2009). Of these, the brine pool GC233 and the mud volcano GB425 have been characterized for active *in situ* carbon–sulfur cycling. Both the systems are sulfidic and anoxic, and contain higher levels of dissolved inorganic carbon than the corresponding seawater columns. In GC233, the seawater–brine interface is dominated by sulfate-reducers, sulfide-oxidizers, acetogens and methanogens, according to PCR-based 16S rRNA and other gene sequence analyses. In contrast, the major metabolic types in the corresponding zone of the mud volcano GB425 are sulfate-reducers and methanogens. Notably, stable isotope data corroborated the biological origin of *in situ* methane of both GC233 and GB425 (Joye et al., 2009). Seawater being the source of sulfate in all deep-sea brines,

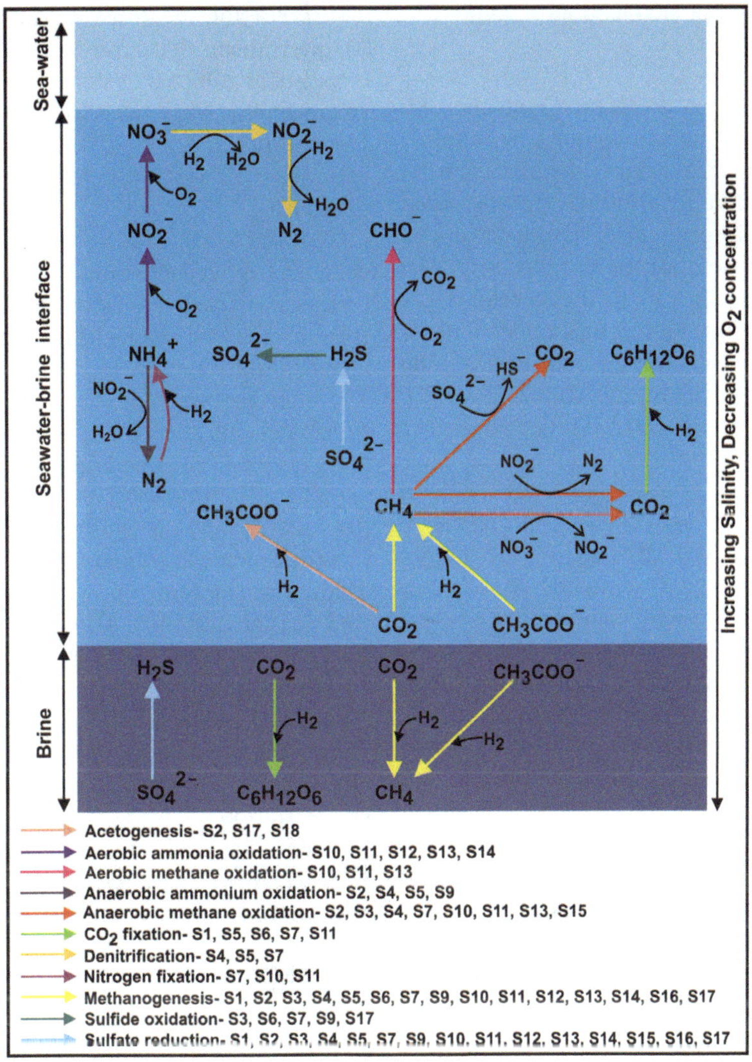

Figure 12.2 Schematic diagram mapping different metabolic processes of the carbon–sulfur–nitrogen cycle on to the different zones of a hypothetical model representing global DHAB systems. The different shades of blue represent the different physicochemical zones (seawater, seawater–brine interface and brine pool) that characterize a DHAB. The heights of the individual zones are neither in scale nor proportionate to the actual thickness of the zones encountered in different DHAB systems (this is because the water-depth, vertical expanse, and covered area of these physiographic features vary widely across the global ocean). Different arrow colors indicate the individual metabolic processes operating in situ (color codes are given in the lower panel of the figure). S1–S18 indicate the names of the different DHABs where the relevant processes have been identified: S1, Discovery; S2, Hephaestus; S3, Kryos; S4, Bannock; S5, L'Atalante; S6, Medee; S7, Thetis; S8, Tyro; S9, Urania; S10, Atlantis II; S11, Discovery Deep; S12, Erba Deep; S13, Kebrit Deep; S14, Nereus Deep; S15, AC601; S16, GB425; S17, GC233; S18, Orca basin.

its concentration in all these systems decreases down the water-depth. Sulfate reduction rate is 10 times higher in GC233 than in GB425; acetoclastic and hydrogenotrophic methanogenesis rates were 2 and 10 times higher in GB425 (compared with GC233), respectively (Joye et al., 2009). In the seawater–brine interface of another brine pool, AC601, anaerobic oxidation of methane and sulfate reduction have been detected based on *in situ* geochemical signatures (Wankel et al., 2010). Acetogenesis, which shares many enzymatic steps with methanogenesis (Drake et al., 2006), is apparently a characteristic of the microbial communities inhabiting the seawater–brine interfaces of the Gulf of Mexico DHABs (Joye et al., 2009; Nigro et al., 2016). Acetate produced by acetogenesis can be utilized as a substrate by methanogens as well as sulfate-reducers (Mand et al., 2014), thereby promoting syntrophic networking within the ecosystems.

12.4.4. The Solitary Deep-sea Hypersaline Anoxic Basin of the Black Sea

In the Black Sea, only the mud volcano Dvurechenskii has been studied thus far as the solitary hypersaline system. Geologically, Dvurechenskii is distinct from all other DHABs of the marine realm in having originated (in the Oligocene or Early Miocene epoch) from diagenetic processes and not the dissolution of evaporites (Aloisi et al., 2004). Physiographically also, Dvurechenskii is distinctive: it is a flat-topped structure which appears like a "mud pie" 800 m in diameter and 80 m in height. The temperature at Dvurechenskii's surface (up to 16.5°C), which is in contact with the sea water, is relatively higher than that of the overlying bottom-water (~9°C), suggesting advective transport of materials from the deep subsurface (Bohrmann et al., 2003). Furthermore, idiosyncratic to all other DHAB systems, the highly saline fluid of Dvurechenskii is enriched in barium, boron, iodine, lithium, strontium, and dissolved inorganic nitrogen (Aloisi et al., 2004). The sediment porewaters too are enriched in ammonium and chloride of deep subsurface origin, while there is also a potential flux of CH_4 from the sediment to the bottom-water, as apparent from the dissolved presence of the hydrocarbon in the water column (Bohrmann et al., 2003).

12.5. A UNIVERSAL BIOGEOCHEMICAL FRAMEWORK FOR ALL DEEP-SEA HYPERSALINE ANOXIC BASINS (AND OTHER HIGH-ENTROPY O_2-SCARCE MARINE MICROBIOMES?)

Deep-sea brine pools are filled with redissolved ancient geological evaporites. Those saturated with salts of divalent cations, such as magnesium and calcium, are known as athalassohaline brines, whereas the ones made of seawater-like fluids rich in monovalent cations are called thalassohaline brines (Oren, 2006; Hallsworth et al., 2007; La Cono et al., 2011, 2019; Yakimov et al., 2013, 2015; Steinle et al., 2018). Most of the well-explored DHABs are thalassohaline in nature, with elevated concentration of sodium ion, while only three out the 18 listed in Table 12.1 (namely, the Mediterranean Sea DHABs Discovery, Hephaestus and Kryos) are the athalassohaline and rich in $MgCl_2$. Compilation and comparison of all geochemical and microbiological data available for the 18 well-studied DHAB systems spread across the global ocean (Tables 12.1 and 12.2) revealed the following important facts.

• DHABs of the Gulf of Mexico are geologically most ancient, followed those of the Red Sea and then the Mediterranean.

• Except for the three athalassohaline brines of the Mediterranean that have bischofite as the main mineral constituent, DHAB systems generally are halite-based. Accordingly, Mg^{2+} largely exceeds Na^{+} in Discovery, Hephaestus, and Kryos, while the proportion of the two cations is the reverse in all other brines characterized thus far.

• For the Mediterranean systems, information on fluid chemistry is available mostly for the brine pools and very little is available for the seawater–brine interfaces; in contrast, for the Red Sea DHABs, more or less proportionate chemical information is available for both brine proper and interface, whereas for those situated in the Gulf of Mexico chemical data available are mostly for the interfaces (as for GB425 and GC233).

• 52.7–323 mM dissolved SO_4^{2-}, which is much higher than the 28 mM present-era average for global seawater (Canfield and Farquhar, 2009), is present in all the Mediterranean brine pools; on the other hand, within the Red Sea territory, 8.3–41.6 mM SO_4^{2-} has been detected in the brine pools and seawater–brine interfaces, with considerable enrichment reported for the interfaces of Atlantis II, Discovery Deep, Kebrit Deep and Nereus Deep. Among the DHABs of the Gulf of Mexico, only Orca Basin is known to have a substantive SO_4^{2-} concentration (20 mM).

• 0.85–15.0 mM dissolved S^{2-} is present in all the Mediterranean brine pools except Hephaestus, where this parameter has not been tested. In the brine cores and interfaces of Red Sea DHABs, except Kebrit Deep (where 0.52 mM and 0.15 mM S^{2-} is present in the brine proper and interface, respectively), S^{2-} is either untested or below the detection limit of the experiment conducted. The DHAB territories explored in the Gulf of Mexico, except AC601, contain 4–25 μM dissolved S^{2-}.

• In the Mediterranean Sea, 9.5 μM and up to 182 μmol (kg $H_2O)^{-1}$ dissolved O_2 have been detected, in the presence of 2.1 mM and up to 64 μmol (kg $H_2O)^{-1}$ dissolved S^{2-}, for the *Thetis* brine pool and *Kryos* interfaces, respectively. Across the brines and interfaces of the Red Sea DHABs, 0.3–4.6 μM O_2 is found in the presence of 0.01–18.99 ppm dissolved CH_4 in the interfaces of Discovery Deep, Erba Deep, Kebrit Deep and Nereus Deep, and 0.52 mM and 0.15 mM S^{2-} in the brine proper and interface of Kebrit Deep, respectively. While 89.2 ppm CH_4 has also been detected in the Urania brine pool in the Mediterranean, GB425 and GC233 interfaces in the Gulf of Mexico are reported to be saturated with 151440 and 8151 ppm CH_4, respectively.

Except for Tyro in the Mediterranean, all other DHAB territories of the marine realm have had at least some microbiological investigations carried out. So far as the brine proper is concerned, *in situ* microbial processes have thus far been detected only for the athalassohaline

Discovery and Kryos of the Mediterranean: methanogenesis and sulfate reduction are common to both, whereas CO_2 fixation is exclusive to Discovery. Apart from these two brine pools, the seawater–brine interface is the microbially (metabolically) active segment of all the DHAB physiographies. As this DHAB stratum is physicochemically less hostile to life (because of lower salinity and higher A_w), and nutrient-wise has more resources, it is intrinsically more biodiverse (supportive of microbial communities having potentials for greater numbers of metabolic processes) than the brine core (Tables 12.1 and 12.2; Figure 12.2). Metabolic processes that potentially operate in the brine cores and/or seawater–brine interfaces of global DHAB ecosystems are described in Table 12.3.

Of the 16 seawater–brine interfaces investigated for their microbiology across the Mediterranean, the Red Sea, and the Gulf of Mexico, plus the stratigraphically undefined territory of the Orca Basin in the Gulf of Mexico explored with similar objectives, potentials for sulfate reduction (15), methanogenesis (15), anaerobic methane oxidation (8), aerobic methane oxidation (3), sulfide oxidation (5), CO_2 fixation (5), aerobic ammonia oxidation (5), anaerobic ammonia oxidation (4), acetogenesis (3), nitrogen fixation (3), and/or denitrification (3) have been detected in the number of sites indicated within the parentheses. Notably, DHAB systems that do not have reports for some of the above-mentioned processes within their territories have either not been explored comprehensively for microbiology or were specifically investigated for select sets of microbial types and metabolic pathways.

According to the literature, nitrogen fixation is generally confined to the lower interfaces, while aerobic methane oxidation and aerobic ammonia oxidation are encountered exclusively at the upper interfaces, especially within the Red Sea DHAB systems (apparently, CH_4 produced in the brines and/or the lower interfaces diffuse up the water columns and are aerobically oxidized in the upper layers of the seawater–brine interfaces).

As particulate matter gets entrapped in the different layers of the water columns of brine systems as a result of differential salinity-driven densities of the layers, confined microbial populations tend to establish syntrophically networked sustainable communities in the different layers of the chemoclines (Ngugi et al., 2015). The current literature, as summarized in Tables 12.1 and 12.2, indicates that in a number of DHAB sites of the global ocean, microbial communities living in the seawater–brine interfaces have the potential metabolic diversities to constitute self-supporting ecosystems with sustainable energy flow and nutrient recycling pathways. All interface communities are sustained by the organic and inorganic nutrients received from the brine water columns as well as the overlying seawaters via intricate syntrophic networks and chemical diffusion. For example, methane oxidation and sulfide oxidation at the interfaces are apparently boosted by the substrates generated by methanogenesis

Table 12.3 Metabolic processes that potentially operate in the brine cores and/or seawater-brine interfaces of the well-studied DHAB systems of the global ocean.

Sl. no.	Name of the metabolic process	Chemical reaction(s)	Reference
1	Sulfate reduction	(a) $4H_2 + SO_4^{2-} + \; + H^+ \rightarrow 2HS^- + 4H_2O$	Muyzer and Stams (2008)
2	Sulfide oxidation	(a) $H_2S + 2O_2 \rightarrow 2SO_4^{2-} + 2H^+ + 8e^-$ (b) $HS^- + 1.6NO_3^- + 0.6H^+ \rightarrow 2SO_4^{2-} + 0.8N_2 + 0.8\,H_2O + 8e^-$	Roy and Trudinger (1970), Aminuddin and Nicholas (1973), Gevertz et al. (2000)
3	Methanogenesis	(a) $CH_3COO^- + H^+ \rightarrow 2CO_2 + CH_4$ (b) $4H_2 + CO_2 \rightarrow CH_4 + 2H_2O$	Thauer (1998)
4	Anaerobic methane oxidation	(a) $CH_4 + SO_4^{2-} \rightarrow HCO_3^{-+} HS^- + H_2O$ (b) $5CH_4 + 8NO_3^- + 8H^+ \rightarrow 5CO_2 + 4N_2 + 14H_2O$ (c) $3CH_4 + 8NO_2^- + 8H^+ \rightarrow 3CO_2 + 4N_2 + 10H_2O$	Cui et al. (2015)
5	Aerobic methane oxidation	(a) $CH_4 + O_2 + NADH \rightarrow CH_3OH + H_2O + NAD^+$	Zhu et al. (2016)
6	Acetogenesis	(a) $2HCO_3^- + 4H_2 \rightarrow CH_3COO^- + 4H_2O$	Ragsdale (2006)
7	CO_2 fixation	(a) $3CO_2 + 12e^- + 12H^+ + P_i \rightarrow C_2H_3O_2\text{-}CH_2OPO_4 + 4H_2O$	Swan et al. (2011)
8	Anaerobic ammonia oxidation	(a) $3NO_3^- + 5NH_4^+ \rightarrow 4N_2 + 9H_2O + 2H^+$	Dalsgaard et al. (2005)
9	Aerobic ammonia oxidation	(a) $NH_4^+ + O_2 + 2e^- + H^+ \rightarrow NH_2OH + H_2O$	Schmidt et al. (2002)
10	Denitrification	(a) $2NO_3^- + 10e^- + 12H^+ \rightarrow N_2 + 6H_2O$	Kuypers et al. (2011)
11	Nitrogen fixation	(a) $N_2 + 16ATP + 8e^- + 8H^+ \rightarrow 2NH_3 + H_2 + 16ADP + 16Pi$	Fisher and Newton (2002)

and sulfate reduction in the brine cores and, brine cores and seawaters, respectively; organic carbon flux from the seawaters is also a potential driver of microbiome functioning in the seawater–brine interfaces. In order to comprehensively illuminate microbiome functioning in these physicochemical extremities of the Earth's biosphere, holistic investigations of fine-resolution organic and inorganic biogeochemistry are needed not only for the individual strata of the global DHAB systems but also from the perspectives of the ecological continua, which constitute the functional frameworks of these systems.

Tandem operation of multiple H_2-requiring anaerobic processes of the carbon–sulfur cycle (such as sulfate reduction, methanogenesis, and acetogenesis) is the typical biogeochemical design for ecosystem functioning within DHAB territories. Remarkably, this is also true for a number of other anoxic marine microbiomes. For example, the structure/function of the microbial communities inhabiting the anoxic and extremely sulfidic hypersaline brines of the deep sea (>3500 mbsl) Urania basin in the eastern Mediterranean have marked resemblance to the microbiome and ecology of the upper sediment horizons underlying marine oxygen minimum zones (tens of centimeters below the OMZ seafloors). In the former habitat, methanogenesis (attributable to phylogenetically novel methanogens) not only co-occurs with sulfate reduction amidst high sulfate concentrations, but also outperform the latter process (Borin et al., 2009). Although there is no information on acetogenesis or anaerobic oxidation of methane in the Urania brine, anaerobic sulfur-oxidizing chemolithotrophs (mostly *Desulfobulbaceae* and *Epsilonproteobacteria* members) present *in situ* have been implicated in the conversion of a small fraction of the sulfate reduction-generated S^{2-} to SO_4^{2-}, via reduction of NO^{3-} and Mn^{4+} (chemolithoautotrophic manganese oxidation and anaerobic ammonium oxidation are said to regenerate these electron acceptors). In contrast, concurrent methanogenesis, sulfate reduction and/or AOM, amid high carbon export to the seafloor, has been reported from a 0–25 cmbsf (centimeters below sea floor) OMZ sediment horizon spread across water-depths of 70–1024 mbsls off the Peruvian margin (Treude et al., 2005; Maltby et al., 2016), and 0–45 cmbsf OMZ sediments of the eastern Arabian Sea located at water-depths of 530–580 mbsl (Bhattacharya et al., 2021). Exploration of 0–6 cmbsf sediments, under a water-depth of 87 m, off the central Chilean upwelling area, has revealed traces of methanogenesis dependent on non-competitive substrates, amid high dissimilatory sulfate reduction and prevalence of acetogens (Ferdelman et al., 1997). A high methane flux in the water column, attributable to sediment surface methanogenesis (potentially in the presence of porewater sulfate as well as nitrate), was reported from the heart of the Eastern Tropical North Pacific OMZ (350–650 mbsl); methanogenesis potential in that sediment surface was found to be reduced considerably below 2 cmbsf, probably owing to down-depth organic carbon depletion (Chronopoulou et al., 2017). In the hypersaline anoxic Kryos basin (located in the Eastern Mediterranean Sea, at >3330 mbsl), where $MgCl_2$ saturation within the brine leads to the lowering of *in situ* A_W to almost uninhabitable levels (~0.4), a combination of anaerobic carbon-sulfur metabolisms heightens in the "seawater–brine interface" to "heart of the brine" trajectory (Steinle et al., 2018). So, sulfate reduction activity and the build-up of hydrogen sulfide, ammonium, and biogenic methane ($\delta^{13}C_{CH4}$ value –47‰), are all higher within the Kryos brine. Although no anaerobic oxidation of methane has been detected within these anaerobic waters, high rates of methane oxidation, probably mediated by aerobic methanotrophs under micro-oxic conditions, have been recorded in the lower part of the seawater–brine interface (Steinle et al., 2018).

Interestingly, the O_2-scarce marine microbiome with maximum resemblance to DHAB systems is hosted in the high-temperature settings of the deep sea (~750 mbsl) hydrothermal system named Lost City Hydrothermal Field (LCHF). At LCHF, seawater infiltrating down to 500–1000 m inside the 100–400°C oceanic crust (through fissures) undergoes serpentinization reactions involving H_2O and dissolved HCO_3^- (or CO_2), and the olivine $[(Mg,Fe)_2SiO_4]$-rich crustal rock (Martin et al., 2008). This chemically modified fluid circulates through the crust and gets heated to ~200°C (Früh-Green et al., 2003; Proskurowski et al., 2006). Under the high heat of LCHF subsurface, the infiltrated seawater is reduced by crustal rock Fe^{2+}, yielding Fe^{3+} and H_2; this H_2, via *in situ* abiotic geochemistry, reduces the HCO_3^-/CO_2 dissolved in the water to yield various simple hydrocarbons such as formate, formaldehyde, methanol, methane, and acetate (Bach et al., 2006; Proskurowski et al., 2008). Eventually, 40–91°C highly alkaline effluent, rich in dissolved H_2, CH_4 and other low-molecular-mass hydrocarbons but extremely poor in CO_2, comes out of carbonate chimneys rising above the ultramafic seafloor (Kelley et al., 2001, 2005; Blackman et al., 2002; Proskurowski et al., 2006, 2008). Several lines of geological, biophysical, and biochemical evidences have predicted that these abiotic, mineral catalyzed, geochemical processes occurring in the mantle at LCHF were the primordial precursors of the enzyme- and cofactor-mediated energy-releasing biochemical reactions that are central to the metabolism of carbon and energy by extant methanogens and acetogens (Martin and Russell, 2006). This hypothesis is supported by the fact that the ultramafic foundation of the LCHF is geochemically similar to lavas that erupted into the primordial oceans, which purportedly hosted the first ecosystems on Earth (Kelley et al., 2005; Martin et al., 2008). Mirroring the putative origin of biochemistry from geochemistry, biological production of reduced organic compounds in the extant microbiome of LCHF also relies on reductive metabolisms that form CH_4, formate, and acetate using H_2 and CO_2 via reactions analogous to the serpentinization-derived H_2–CO_2 redox chemistry (Martin et al., 2008). Thus, methanogenesis by *Methanosarcinales*, sulfur and sulfate reduction by hyperthermophilic archaea, and bacteria belonging to the phylum *Firmicutes*, anaerobic oxidation of methane by organisms related to *Methanosarcinales*, as well as the ANME-1 and ANME-3 groups, form the metabolic framework of the LCHF microbiome, in tandem with a range of H_2-producing, organic matter fermentations by diverse bacteria and archaea (Martin et al., 2008).

The coexistence and tandem operation of sulfate reducers, methanogens and acetogens within a single biogeochemical system is idiosyncratic to normal ecological axioms as all four metabolisms employ the Acetyl-CoA pathway for either acetate synthesis or degradation, so their natural populations are expected to compete with each other for the common resource, H_2 (Drake et al., 2006). Although it may be presumed that abundance of H_2 in all the above marine environments makes this peculiar phenomenon possible, more studies of organic geochemistry and *in situ* community metabolisms are needed for these microbiomes in order to fully understand this peculiar and primordial ecosystem architecture. Major missing links in this information chain apparently pertain to organic geochemistry data on the flux/delivery and preservation/remineralization of labile organic matter within these systems, and community-level metabolic processes that can account for potential organic matter degradation under the prevailing micro-oxic and anoxic conditions to supply the simple fatty acids and reduced metabolites necessary to drive *in situ* anaerobic processes such as sulfate reduction and methanogenesis.

12.6. ASTROBIOLOGICAL IMPLICATIONS OF DEEP-SEA HYPERSALINE ANOXIC BASINS

Structures and functions of present-day DHAB microbiomes hold potential implications for our understanding of ancient ecosystems and metabolic strategies of early life on Earth (Dundas, 1998) and other extraterrestrial bodies (Mancinelli et al., 2004). The fact that geologically diverse O_2-stressed marine microbiomes portray a strikingly similar design of community structure and function implies that they are all retrospectively convergent in the geobiological evolution of the global ocean. The primordial ecosystem ancestral to these microbiomes may had well taken shape in a biogeochemical context that essentially featured the common parameters of these environments: i.e. acute O_2 stress, abundance of H_2 and simple C_1 compounds, and, above all, the surrounding regimen of a sulfatic sea. We hypothesize that the O_2-stressed ecosystem framework based on methanogenesis, methanotrophy, sulfate reduction, sulfur oxidation, and acetogenesis might had originated in a hydrothermal context (as conceived in the currently popular hypothesis for the origin of life on Earth), but subsequently proliferated to cooler marine environments in the course of evolution. Furthermore, studies envisaging plausible ecosystem functioning in planetary bodies such as Saturn's moon Enceladus and Jupiter's moon Europa have implicated sulfate, pyrite, and methane as potential bioenergetic molecules (McKinnon and Zolensky, 2003; Bouquet et al., 2015). It may therefore be that the geomicrobiology of DHABs has importance as an analog of the primordial marine ecosystem, on Earth as well as in comparator extraterrestrial locations.

REFERENCES

Abdallah, R.Z., Adel, M., Ouf, A. et al. (2014). Aerobic methanotrophic communities at the Red Sea brine-seawater interface. *Frontiers in Microbiology 5*: 487. https://doi.org/10.3389/fmicb.2014.00487

Addy, K.S., and Behrens, E.W. (1980). Time of accumulation of hypersaline anoxic brine in Orca basin (Gulf of Mexico). *Marine Geology 37*: 241–252. https://doi.org/10.1016/0025-3227(80):90104-8

Aloisi, G., Drews, M., Wallmann, K. et al. (2004). Fluid expulsion from the Dvurechenskii mud volcano (Black Sea): Part I. Fluid sources and relevance to Li, B, Sr, I and dissolved inorganic nitrogen cycles. *Earth and Planetary Science Letters 225*: 347–363. https://doi.org/10.1016/j.epsl.2004.07.006

Aminuddin, M., and Nicholas, D.J.D. (1973). Sulphide oxidation linked to the reduction of nitrate and nitrite in *Thiobacillus denitrificans*. *Biochimica et Biophysica Acta (BBA)-Bioenergetics 325*: 81–93. https://doi.org/1016/0005-2728(73)90153-9

Antunes, A., Ngugi, D.K., and Stingl, U. (2011). Microbiology of the Red Sea (and other) deep-sea anoxic brine lakes. *Environmental Microbiology Reports 3*: 416–433. https://doi.org/10.1111/j.1758-2229.2011.00264.x

Bach, W., Paulick, H., Garrido, C.J. et al. (2006). Unraveling the sequence of serpentinization reactions: petrography, mineral chemistry, and petrophysics of serpentinites from MAR 15 N (ODP Leg 209, Site 1274). *Geophysical Research Letters 33*: L13306. https://doi.org/10.1029/2006GL025681

Batang, Z.B., Papathanassiou, E., Al-Suwailem, A. et al. (2012). First discovery of a cold seep on the continental margin of the central Red Sea. *Journal of Marine Systems 94*: 247–253. https://doi.org/10.1016/j.jmarsys.2011.12.004

Bhattacharya, S., Mapder, T., Fernandes, S. et al. (2021). Sedimentation rate and organic matter dynamics shape microbiomes across a continental margin. *Biogeosciences 18*: 5203–5222. https://doi.org/10.5194/bg-18-5203-2021

Blackman, D.K., Karson, J.A., Kelley, D.S. et al. (2002). Geology of the Atlantis Massif (Mid-Atlantic Ridge, 30 N): implications for the evolution of an ultramafic oceanic core complex. *Marine Geophysical Researches 23*: 443–469. https://doi.org/10.1023/B:MARI.0000018232.14085.75

Blum, N., and Puchelt H. (1991). Sedimentary-hosted polymetallic massive sulphide deposits of the Kebrit and Shaban Deeps, Red Sea. *Mineralium Deposita 26*: 217–227. https://doi.org/10.1007/BF00209261

Bohrmann, G., Ivanov, M., Foucher, J.P. et al. (2003). Mud volcanoes and gas hydrates in the Black Sea: new data from Dvurechenskii and Odessa mud volcanoes. *Geo-Marine Letters 23*: 239–249. https://doi.org/10.1007/s00367-003-0157-7

Bonatti, E. (1985). Punctiform initiation of seafloor spreading in the Red Sea during transition from a continental to an oceanic rift. *Nature 316*: 33–37. https://doi.org/10.1038/316033a0

Borin, S., Brusetti, L., Mapelli, F. et al. (2009). Sulfur cycling and methanogenesis primarily drive microbial colonization of the highly sulfidic Urania deep hypersaline basin. *Proceedings of the National Academy of Sciences USA 106*: 9151–9156. https://doi.org/10.1073/pnas.0811984106

Borin, S., Mapelli, F., Rolli, E. et al. (2013). Anammox bacterial populations in deep marine hypersaline gradient systems. *Extremophiles 17*: 289–299. https://doi.org/10.1007/s00792-013-0516-x

Boschetti, T., Awaleh, M. and Barbieri, M. (2018). Waters from the Djiboutian Afar: a review of strontium isotopic composition and a comparison with Ethiopian waters and Red sea brines. *Water 10*: 1700. https://doi.org/10.3390/w10111700

Bougouffa, S., Yang, J. K., Lee, O.O. et al. (2013). Distinctive microbial community structure in highly stratified deep-sea brine water columns. *Applied and Environmental Microbiology 79*: 3425–3437. https://doi.org/10.1128/AEM.00254-13

Bouma, A.H., and Bryant, W.R. (1994). Physiographic features on the northern Gulf of Mexico continental slope. *Geo-Marine Letters 14*: 252–263. https://doi.org/10.1007/BF01274061

Bouquet, A., Mousis, O., Waite, J.H. et al. (2015). Possible evidence for a methane source in Enceladus' ocean. *Geophysical Research Letters 42*, 1334–1339. https://doi.org/10.1002/2014GL063013

Bregant, D., Catalano G., Civitarese G. et al. (1990). Some chemical characteristics of the brines in Bannock and Tyro Basins: salinity, sulphur compounds, Ca, F, pH, At, PO_4, SiO_2, NH_3. *Marine Chemistry 31*: 35–62. https://doi.org/10.1016/0304-4203(90)90030-G

Brooks, J.M., Bright, T.J., Bernard, B.B. et al. (1979). Chemical aspects of a brine pool at the East Flower Garden bank, northwestern Gulf of Mexico. *Limnology and Oceanography 24*: 735–745. https://doi.org/10.4319/lo.1979.24.4.0735

Camerlenghi, A. (1990). Anoxic basins of the eastern Mediterranean: geological framework. *Marine Chemistry 31*: 1–19. https://doi.org/10.1016/0304-4203(90)90028-B

Camerlenghi, A. and McCoy, F.W. (1990). Physiography and structure of Bacino Bannock (Eastern Mediterranean). *Geo-Marine Letters 31*: 21–33. https://doi.org/10.1007/BF02431018

Canfield, D.E. and Farquhar, J. (2009). Animal evolution, bioturbation, and the sulfate concentration of the oceans. *Proceedings of the National Academy of Sciences USA 106*: 8123–8127. https://doi.org/10.1073/pnas.0902037106

Chronopoulou, P.M., Shelley, F., Pritchard, W.J. et al. (2017). Origin and fate of methane in the Eastern Tropical North Pacific oxygen minimum zone. *The ISME Journal 11*: 1386–1399. https://doi.org/10.1038/ismej.2017.6

Cita, M.B. (1991). Anoxic basins of the eastern Mediterranean: an overview. *Paleoceanography 6*: 133–141. https://doi.org/10.1029/90PA01789

Cita, M.B. (2006). Exhumation of Messinian evaporites in the deep-sea and creation of deep anoxic brine-filled collapsed basins. *Sedimentary Geology 188*: 357–378. https://doi.org/10.1016/j.sedgeo.2006.03.013

Cordes, E.E., Bergquist, D.C. and Fisher, C.R. (2009). Macroecology of Gulf of Mexico cold seeps. *Annual Review of Marine Science 1*: 143–168. https://doi.org/10.1146/annurev.marine.010908.163912

Cui, M., Ma, A., Qi, H. et al. (2015). Anaerobic oxidation of methane: an "active" microbial process. *MicrobiologyOpen 4*: 1–11. https://doi.org/10.1002/mbo3.232

Daffonchio, D., Borin, S., Brusa, T. et al. (2006). Stratified prokaryote network in the oxic–anoxic transition of a deep-sea halocline. *Nature 440*: 203–207. https://doi.org/10.1038/nature04418

Dalsgaard, T., Thamdrup, B. and Canfield, D.E. (2005). Anaerobic ammonium oxidation (anammox) in the marine environment. *Research in Microbiology 156*: 457–464. https://doi.org/10.1016/j.resmic.2005.01.011

De Lange, G.J., Middelburg, J.J., Van der Weijden, C.H. et al. (1990). Composition of anoxic hypersaline brines in the Tyro and Bannock Basins, eastern Mediterranean. *Marine Chemistry 31*: 63–88. https://doi.org/10.1016/0304-4203(90)90031-7

Drake, H.L., Küsel, K., and Matthies, C. (2006). Acetogenic Prokaryotes. In: *The Prokaryotes* Vol. *2* (eds. M. Dworkin, S. Falkow, E. Rosenberg et al.), 354–420. New York: Springer. https://doi.org/10.1007/0-387-30742-7_13

Dundas, I. (1998). Was the environment for primordial life hypersaline? *Extremophiles 2*: 375–377. https://doi.org/10.1007/s007920050081

Ehrhardt, A. and Hübscher, C. (2015). The northern Red Sea in transition from rifting to drifting-lessons learned from ocean deeps. In: *The Red Sea* (eds. N.M.A. Rasul and I.C.F. Stewart), 99–121. Berlin, Heidelberg: Springer. https://doi.org/10.1007/978-3-662-45201-1_5

Ehrhardt, A., Hübscher, C. and Gajewski, D. (2005). Conrad Deep, Northern Red Sea: development of an early stage ocean deep within the axial depression. *Tectonophysics 411*: 19–40. https://doi.org/10.1016/j.tecto.2005.08.011

Feldens, P. and Mitchell, N.C. (2015). Salt flows in the central Red Sea. In: *The Red Sea* (eds. N.M.A. Rasul and I.C.F. Stewart), 205–218. (pp.). Berlin, Heidelberg: Springer. https://doi.org/10.1007/978-3-662-45201-1_12

Ferdelman, T.G., Lee, C., Pantoja, S. et al. (1997). Sulfate reduction and methanogenesis in a *Thioploca*-dominated sediment off the coast of Chile. *Geochimica et Cosmochimica Acta 61*: 3065–3079. https://doi.org/10.1016/S0016-7037(97)00158-0

Feseker, T., Pape, T., Wallmann, K. et al. (2009). The thermal structure of the Dvurechenskii mud volcano and its implications for gas hydrate stability and eruption dynamics. *Marine and Petroleum Geology 26*: 1812–1823. https://doi.org/10.1016/j.marpetgeo.2009.01.021

Fisher, K. and Newton, W.E. (2002). Nitrogen fixation – a general overview. In: *Nitrogen Fixation at the Millennium* (ed. G.J. Leigh), 1–3. Amsterdam: Elsevier. https://doi.org/10.1016/B978-044450965-9/50001-X

Früh-Green, G.L., Kelley, D.S., Bernasconi, S.M. (eds. N.M.A. Rasul and I.C.F. Stewart) et al. (2003). 30,000 years of hydrothermal activity at the Lost City vent field. *Science 301*: 495–498. https://doi.org/10.1126/science.1085582

Fusi, N., de Larderel, G.A., Borelu, A. et al. (1996). Marine geology of the Medriff Corridor, Mediterranean Ridge. *Island Arc 5*: 420–439. https://doi.org/10.1111/j.1440-1738.1996.tb00163.x

Galloway, W.E. (2008). Depositional evolution of the Gulf of Mexico sedimentary basin. *Sedimentary basins of the World 5*: 505–549. https://doi.org/10.1016/S1874-5997(08)00015-4

Gevertz, D., Telang, A.J., Voordouw, G. et al. (2000). Isolation and characterization of strains CVO and FWKO B, two novel nitrate-reducing, sulfide-oxidizing bacteria isolated from oil field brine. *Applied and Environmental Microbiology 66*: 2491–2501. https://doi.org/10.1128/AEM.66.6.2491-2501.2000

Girdler, R.W. and Whitmarsh, R.B. (1974). Miocene evaporites in Red Sea cores. Their relevance to the problems of the width and age of oceanic crust beneath the Red Sea. In: *Initial Reports of the Deep Sea Drilling Project*, Vol. *23*, 913–921. US Government Printing Office, Washington.

Guan, Y., Hikmawan, T., Antunes, A. et al. (2015). Diversity of methanogens and sulfate-reducing bacteria in the interfaces of five deep-sea anoxic brines of the Red Sea. *Research in Microbiology 166*: 688–699. https://doi.org/10.1016/j.resmic.2015.07.002

Hallsworth, J.E., Heim, S. and Timmis, K.N. (2003). Chaotropic solutes cause water stress in *Pseudomonas putida*. *Environmental Microbiology 5*: 1270–1280. https://doi.org/10.1111/j.1462-2920.2003.00478.x

Hallsworth, J.E., Yakimov, M.M., Golyshin, P.N. et al. (2007). Limits of life in $MgCl_2$-containing environments: chaotropicity defines the window. *Environmental Microbiology 9*: 801–813. https://doi.org/10.1111/j.1462-2920.2006.01212.x

Joye, S.B., MacDonald, I.R., Montoya, J.P. et al. (2005). Geophysical and geochemical signatures of Gulf of Mexico seafloor brines. *Biogeosciences 2*: 295–309. https://doi.org/10.5194/bg-2-295-2005

Joye, S.B., Samarkin, V.A., MacDonald, I.R. et al. (2009). Metabolic variability in seafloor brines revealed by carbon and sulphur dynamics. *Nature Geoscience 2*: 349–354. https://doi.org/10.1038/ngeo475

Kelley, D.S., Karson, J.A., Blackman, D.K. et al. (2001). An off-axis hydrothermal vent field near the Mid-Atlantic Ridge at 30°N. *Nature 412*: 145–149. https://doi.org/10.1038/35084000

Kelley, D.S., Karson, J.A., Früh-Green, G.L. et al. (2005). A serpentinite-hosted ecosystem: the Lost City hydrothermal field. *Science 307*: 1428–1434. https://doi.org/10.1126/science.1102556

Koynova, R., Brankov, J. and Tenchov, B. (1997). Modulation of lipid phase behavior by kosmotropic and chaotropic solutes: experiment and thermodynamic theory. *European Biophysics Journal 25*: 261–274. https://doi.org/10.1007/s002490050038

Kuypers, M.M., Marchant, H.K. and Kartal, B. (2018). The microbial nitrogen-cycling network. *Nature Reviews Microbiology 16*: 263–276. https://doi.org/10.1038/nrmicro.2018.9

La Cono, V., Smedile, F., Bortoluzzi, G. et al. (2011). Unveiling microbial life in new deep-sea hypersaline Lake Thetis. Part I: Prokaryotes and environmental settings. *Environmental Microbiology 13*: 2250–2268. https://doi.org/10.1111/j.1462-2920.2011.02478.x

La Cono, V., Bortoluzzi, G., Messina, E. et al. (2019). The discovery of Lake Hephaestus, the youngest athalassohaline deep-sea formation on Earth. *Scientific Reports 9*: 1679. https://doi.org/10.1038/s41598-018-38444-z

Lee, C.J.D., McMullan, P.E., O'Kane, C.J. et al. (2018). NaCl-saturated brines are thermodynamically moderate, rather than extreme, microbial habitats. *FEMS Microbiology Reviews 42*: 672–693. https://doi.org/10.1093/femsre/fuy026

MacDonald, I.R., Reilly, J.F., Guinasso, N.L. et al. (1990). Chemosynthetic mussels at a brine-filled pockmark in the northern Gulf of Mexico. *Science 248*: 1096–1099. https://doi.org/10.1126/science.248.4959.1096

Maltby, J., Sommer, S., Dale, A.W. et al. (2016). Microbial methanogenesis in the sulfate-reducing zone of surface sediments traversing the Peruvian margin. *Biogeosciences 13*: 283–299. https://doi.org/10.5194/bg-13-283-2016

Mancinelli, R.L., Fahlen, T.F., Landheim, R. et al. (2004). Brines and evaporites: analogs for Martian life. *Advances in Space Research 33*: 1244–1246. https://doi.org/10.1016/j.asr.2003.08.034

Mand, J., Park, H.S., Jack, T.R. et al. (2014). The role of acetogens in microbially influenced corrosion of steel. *Frontiers in Microbiology 5*: 268. https://doi.org/10.3389/fmicb.2014.00268

Martin, W. and Russell, M.J. (2006). On the origin of biochemistry at an alkaline hydrothermal vent. *Philosophical Transactions of the Royal Society B: Biological Sciences 362*: 1887–1926. https://doi.org/10.1098/rstb.2006.1881

Martin, W., Baross, J., Kelley, D. et al. (2008). Hydrothermal vents and the origin of life. *Nature Reviews Microbiology 6*: 805–814. https://doi.org/10.1038/nrmicro1991

McKinnon, W.B. and Zolensky, M.E. (2003). Sulfate content of Europa's ocean and shell: evolutionary considerations and some geological and astrobiological implications. *Astrobiology 3*: 879–897. https://doi.org/10.1089/153110703322736150

Merlino, G., Barozzi, A., Michoud, G. et al. (2018). Microbial ecology of deep-sea hypersaline anoxic basins. *FEMS Microbiology Ecology 94*: fiy085. https://doi.org/10.1093/femsec/fiy085

Meyer, B. and Kuever, J. (2007). Molecular analysis of the diversity of sulfate-reducing and sulfur-oxidizing prokaryotes in the environment, using *aprA* as functional marker gene. *Applied and Environmental Microbiology 73*: 7664–7679. https://doi.org/10.1128/AEM.01272-07

Muyzer, G. and Stams, A.J. (2008). The ecology and biotechnology of sulphate-reducing bacteria. *Nature Reviews Microbiology 6*: 441–454. https://doi.org/10.1038/nrmicro1892

Ngugi, D.K., Blom, J., Alam, I. et al. (2015). Comparative genomics reveals adaptations of a halotolerant thaumarchaeon in the interfaces of brine pools in the Red Sea. *The ISME Journal 9*: 396–411. https://doi.org/10.1038/ismej.2014.137

Nigro, L.M., Hyde, A.S., MacGregor, B.J. et al. (2016). Phylogeography, salinity adaptations and metabolic potential of the candidate division KB1 bacteria based on a partial single cell genome. *Frontiers in Microbiology 7*: 1266. https://doi.org/10.3389/fmicb.2016.01266

Okur, H.I., Hladílková, J., Rembert, K.B. et al. (2017). Beyond the Hofmeister series: ion specific effects on proteins and their biological functions. *Journal of Physical Chemistry B 121*: 1997–2014. https://doi.org/10.1021/acs.jpcb.6b10797

Oren, A. (2006). Life at High Salt Concentrations. In: et al. *Prokaryotes: a Handbook on the Biology of Bacteria*, Vol. *2* (eds. M. Dworkin, S. Falkow, E. Rosenberg et al.), 263–282. New York, USA: Springer. https://doi.org/10.1007/0-387-30742-7_9

Pachiadaki, M.G., Yakimov, M.M., LaCono, V. et al. (2014). Unveiling microbial activities along the halocline of Thetis, a deep-sea hypersaline anoxic basin. *The ISME Journal 8*: 2478–2489. https://doi.org/10.1038/ismej.2014.100

Pautot, G., Guennoc, P., Coutelle, A. et al. (1984). Discovery of a large brine deep in the northern Red Sea. *Nature 310*: 133–136. https://doi.org/10.1038/310133a0

Pilcher, R. S. and Blumstein, R. D. (2007). Brine volume and salt dissolution rates in Orca Basin, northeast Gulf of Mexico. *AAPG Bulletin 91*: 823–833. https://doi.org/10.1306/12180606049

Pitzer, K.S. (1991). Ion interaction approach: theory and data correlation. In: *Activity Coefficients in Electrolyte Solutions, Activity Coefficients in Electrolyte Solutions*, Vol. *2* (ed. K. S. Pitzer), 75–153. Boca Raton, FL: CRC Press.

Proskurowski, G., Lilley, M.D., Kelley, D.S. et al. (2006). Low temperature volatile production at the Lost City Hydrothermal Field, evidence from a hydrogen stable isotope geothermometer. *Chemical Geology 229*: 331–343. https://doi.org/10.1016/j.chemgeo.2005.11.005

Proskurowski, G., Lilley, M.D., Seewald, J.S. et al. (2008). Abiogenic hydrocarbon production at Lost City hydrothermal field. *Science 319*: 604–607. https://doi.org/10.1126/science.1151194

Ragsdale, S.W. (2006). Metals and their scaffolds to promote difficult enzymatic reactions. *Chemical Reviews 106*: 3317–3337. https://doi.org/10.1021/cr0503153

Roberts, H., Carney, R., Kupchik, M. et al. (2007). ALVIN explores the deep northern Gulf of Mexico slope. *Eos Transactions, American Geophysical Union 88*: 341–342. https://doi.org/10.1029/2007EO350001

Rose, P. and Laking, A. (2008). *Oceans: Exploring the Hidden Depths of the Underwater World*. Berkeley: University of California Press.

Roveri, M., Flecker, R., Krijgsman, W. et al. (2014). The Messinian Salinity Crisis: past and future of a great challenge for marine sciences. *Marine Geology 352*: 25–58. https://doi.org/10.1016/j.margeo.2014.02.002

Roy, A.B. and Trudinger, P.A. (1970). *The Biochemistry of Inorganic Compounds of Sulphur*. London: Cambridge University Press Aminuddin.

Sass, A.M., Sass, H., Coolen, M.J. et al. (2001). Microbial communities in the chemocline of a hypersaline deep-sea basin (Urania basin, Mediterranean Sea). *Applied and Environmental Microbiology 67*: 5392–5402. https://doi.org/10.1128/AEM.67.12.5392-5402.2001

Schmidt, M., Al-Farawati, R. and Botz, R. (2015). Geochemical classification of brine-filled Red Sea deeps. In: *The Red Sea* (eds. N. Rasul and I. Stewart), 219–233. Berlin, Heidelberg: Springer. https://doi.org/10.1007/978-3-662-45201-1_13

Schmidt, I., Sliekers, O., Schmid, M. et al. (2002). Aerobic and anaerobic ammonia oxidizing bacteria-competitors or natural partners? *FEMS Microbiology Ecology 39*: 175–181. https://doi.org/10.1111/j.1574-6941.2002.tb00920.x

Searle, R.C. and Ross, D.A. (1975). A Geophysical Study of the Red Sea Axial Trough between 20.5° and 22°N. *Geophysical Journal International 43*: 555–572. https://doi.org/10.1111/j.1365-246X.1975.tb00647.x

Sergeeva, N.G. and Kopytina, N.I. (2014). The first marine filamentous fungi discovered in the bottom sediments of the oxic/anoxic interface and in the bathyal zone of the Black Sea. *Turkish Journal of Fisheries and Aquatic Sciences 14*: 497–505. https://doi.org/10.4194/1303-2712-v14_2_21

Shah, S.R., Joye, S.B., Brandes, J.A. et al. (2013). Carbon isotopic evidence for microbial control of carbon supply to Orca Basin at the seawater-brine interface. *Biogeosciences 10*: 3175–3183. https://doi.org/10.5194/bg-10-3175-2013

Shokes, R.F., Trabant, P.K., Presley, B.J. et al. (1977). Anoxic, hypersaline basin in the northern Gulf of Mexico. *Science 196*: 1443–1446. https://doi.org/10.1126/science.196.4297.1443

Steinle, L., Knittel, K., Felber, N. et al. (2018). Life on the edge: active microbial communities in the Kryos $MgCl_2$-brine basin at very low water activity. *The ISME Journal 12*: 1414–1426. https://doi.org/10.1038/s41396-018-0107-z

Stevenson, A., Cray, J.A., Williams, J.P. et al. (2015). Is there a common water-activity limit for the three domains of life? *The ISME Journal 9*: 1333–1351. https://doi.org/10.1038/ismej.2014.219

Swan, B.K., Martinez-Garcia, M., Preston, C.M. et al. (2011). Potential for chemolithoautotrophy among ubiquitous bacteria lineages in the dark ocean. *Science 333*: 1296–1300. https://doi.org/ 10.1126/science.1203690

Tapia, M.S., Alzamora, S. M. and Chirife, J. (2008). Effects of water activity (a_w) on microbial stability: as a hurdle in food preservation. In: *Water Activity in Foods* (eds. G.V. Barbosa-Cánovas, A.J. Fontana, S.J. Schmidt et al.), 239–271. Hoboken, NJ: John Wiley and Sons. https://doi.org/10.1002/9780470376454.ch10

Thauer, R.K. (1998). Biochemistry of methanogenesis: a tribute to Marjory Stephenson: 1998 Marjory Stephenson prize lecture. *Microbiology 144*: 2377–2406. https://doi.org/10.1099/00221287-144-9-2377

Treude, T., Niggemann, J., Kallmeyer, J. et al. (2005). Anaerobic oxidation of methane and sulfate reduction along the Chilean continental margin. *Geochimica et Cosmochimica Acta 69*: 2767–2779. https://doi.org/10.1016/j.gca.2005.01.002

Van Cappellen, P., Viollier, E., Roychoudhury, A. et al. (1998). Biogeochemical cycles of manganese and iron at the oxic-anoxic transition of a stratified marine basin (Orca Basin, Gulf of Mexico). *Environmental Science & Technology 32*: 2931–2939. https://doi.org/10.1021/es980307m

Van Der Wielen, P.W. (2006). Diversity of ribulose-1, 5-bisphosphate carboxylase/oxygenase large-subunit genes in the $MgCl_2$-dominated deep hypersaline anoxic basin discovery. *FEMS Microbiology Letters 259*: 326–331. https://doi.org/10.1111/j.1574-6968.2006.00284.x

Van der Wielen, P.W., Bolhuis, H., Borin, S. et al. (2005). The enigma of prokaryotic life in deep hypersaline anoxic basins. *Science 307*: 121–123. https://doi.org/10.1126/science.1103569

Wallmann, K., Suess, E., Westbrook, G.H. et al. (1997). Salty brines on the Mediterranean sea floor. *Nature 387*: 31–32. https://doi.org/10.1038/387031a0

Wallmann, K., Aghib, F.S., Castradori, D. et al. (2002). Sedimentation and formation of secondary minerals in the hypersaline Discovery Basin, eastern Mediterranean. *Marine Geology 186*: 9–28. https://doi.org/10.1016/S0025-3227(02)00170-6

Wallmann, K., Drews, M., Aloisi, G. et al. (2006). Methane discharge into the Black Sea and the global ocean via fluid flow through submarine mud volcanoes. *Earth and Planetary Science Letters 248*: 545–560. https://doi.org/10.1016/j.epsl.2006.06.026

Wang, Y., Yang, J., Lee, O.O. et al. (2011). Hydrothermally generated aromatic compounds are consumed by bacteria colonizing in Atlantis II Deep of the Red Sea. *The ISME Journal 5*: 1652–1659. https://doi.org/10.1038/ismej.2011.42

Wang, Y., Cao, H., Zhang, G. et al. (2013). Autotrophic microbe metagenomes and metabolic pathways differentiate adjacent Red Sea brine pools. *Scientific Reports 3*: 1748. https://doi.org/10.1038/srep01748

Wankel, S.D., Joye, S.B., Samarkin, V.A. et al. (2010). New constraints on methane fluxes and rates of anaerobic methane oxidation in a Gulf of Mexico brine pool via *in situ* mass spectrometry. *Deep Sea Research Part II: Topical Studies in Oceanography 57*: 2022–2029. https://doi.org/10.1016/j.dsr2.2010.05.009

Williams, D. F. and Lerche, I. (1987). Salt domes, organic-rich source beds and reservoirs in intraslope basins of the Gulf Coast region. In: *Dynamical Geology of Salt and Related Structures* (eds. I. Lerche and J. J. O'Brien), 751–830. New York: Academic Press.

Yakimov, M.M., La Cono, V., Denaro, R. et al. (2007). Primary producing prokaryotic communities of brine, interface and seawater above the halocline of deep anoxic lake L'Atalante, Eastern Mediterranean Sea. *The ISME Journal 1*: 743–755. https://doi.org/10.1038/ismej.2007.83

Yakimov, M.M., La Cono, V., Slepak, V.Z. et al. (2013). Microbial life in the Lake Medee, the largest deep-sea salt-saturated formation. *Scientific Reports 3*: 3554. https://doi.org/10.1038/srep03554

Yakimov, M.M., La Cono, V., Spada, G.L. et al. (2015). Microbial community of the deep-sea brine Lake Kryos seawater-brine interface is active below the chaotropicity limit of life as revealed by recovery of mRNA. *Environmental Microbiology 17*: 364–382. https://doi.org/10.1111/1462-2920.12587

Zhu, J., Wang, Q., Yuan, M. et al. (2016). Microbiology and potential applications of aerobic methane oxidation coupled to denitrification (AME-D) process: a review. *Water Research 90*: 203–215. https://doi.org/10.1016/j.watres.2015.12.020

13

Ecology of Cold Seep Habitats

Mandar Nanajkar[1]*, Kalyan De[1], Aniket Desai[1,3], Sambhaji Mote[1], and Sabyasachi Sautya[2]

ABSTRACT

Cold seep ecosystems are fuelled by subsurface methane, hydrogen sulfide, and bicarbonate emission, which is exploited by diverse and endemic faunal communities. Although considered unconventional, at cold seep sites, chemosynthesis is the mechanism by which the bacterial community fixes carbon by utilizing hydrocarbon emissions. With discoveries of seep sites globally, including a new seep site in the Indian Ocean in 2018, it is imperative to understand its highly specialized faunal communities and their evolved functional attributes. Cold seeps have evolved unique metazoan lineages such as giant clams and long-living gutless polychaetes, which retain vital chemosymbiotic bacteria. With a very diverse invertebrate community, cold seeps share these endemics only with other such sites or hydrothermal vents and whale fall sites. Despite such sites being globally scattered thousands of kilometers apart, the communities therein display remarkable genetic affinity and connectivity. We discuss their survival, dispersal, and ecophysiology that has evolved to extract nutrition from hydrocarbons, resist toxic seepages, and regulate bacterial harvest. Seep communities have global implications in terms of climate change as they metabolize greenhouse gases at source; thus, their conservation is essential, with the potential of such sites as a commercial energy source. The seep organisms possess great scientific value as they can potentially synthesize biomedically important compounds. For the past four decades, workers have extensively studied cold seep ecosystems, their faunal attributes, and the mechanisms that govern these endemics: this review appraises the current research on seep ecosystems.

13.1. INTRODUCTION

The sun is the primary source of energy that transmits from one trophic level to another, starting with photosynthesis. However, in the last decade of the 19th century, it was discovered that apart from photosynthetically derived energy transfer, organic life exists with an alternative source. Winogradsky (1890) described some organisms as "Anorgoxydanten," which were certain bacteria that use energy from inorganic substances, and today we know it-as "chemosynthesis." Discovery of the hydrothermal vent (in the 1970s) and cold seep (in the 1980s) confirmed the work of Winogradsky (1890). Cold seeps are defined as a diffused or focused flow of hot and chemically modified seawater emerging from seafloor where microbial chemoautorophs thrive in a localized area of seabed (Van Dover 2000). Whereas hydrothermal vents are fissures on the seafloor from which geothermally heated water discharges. Hydrothermal vents are commonly found near volcanically active places, areas where tectonic plates are moving apart at spreading centers, ocean basins, and hotspots (Colín-García, María, 2016).

[1]CSIR-National Institute of Oceanography, Goa, India

[2]Regional Centre, CSIR-National Institute of Oceanography, Mumbai, India

[3]Academy of Scientific and Innovative Research, Dona Paula, Goa, 403004, India

*Corresponding Author: Mandar Nanajkar (mandar@nio.org): ORCiD code: 0000-0002-6608-7669; Kalyan De (kalyandeaqua@gmail.com): ORCiD code: 0000-0002-9722-2309; Aniket Desai (aniketdesai.4437@gmail.com): ORCiD code: 0000-0002-5650-7850; Sambhaji Mote (smote@nio.org): ORCiD code: 0000-0002-2238-4856; Sabyasachi Sautya (sautya@nio.org): ORCiD code: 0000-0003-2229-5834

Systems Biogeochemistry of Major Marine Biomes, First Edition. Edited by Aninda Mazumdar and Wriddhiman Ghosh.

Over the years, several deep-sea missions have discovered cold seeps in the world's oceans and have intensively investigated their physical, chemical, biological, and ecological properties (Foucher et al., 2009; Talukder, 2012). Chemosynthetic ecosystems such as seeps occur in the world's oceans and are aligned with geological formations such as mid-ocean ridges to seamounts to back-arc basins, margins, and trenches. These exhibit basalt pavement cracks with a diffuse flow of hot muds, brine lakes, mud volcanoes, carbonate pinnacles, and white and black smoker chimneys. These ecosystems can be categorized into hydrothermal vents and cold seeps, which can sustain unique life forms by chemosynthesis. Other than these two ecosystems, whale falls, and decaying wooden logs also support chemosynthetic communities (Smith et al., 1989).

Cold seeps and hydrothermal vents are maintained by subsurface hydrocarbon reservoirs wherein organic matter is progressively converted into hydrocarbons via microbial action. Hydrocarbons, for example, methane, are emitted on active and passive continental margins and subduction zones (Vigneron et al., 2013). Geological processes such as the dissociation of methane hydrates, salt diapirism, gravity compression, or plate subduction consistently supply hydrocarbon-rich fluids through faults and cracks to the system.

These seepages form a mosaic of habitats with a decreasing gradient of faunal activity from the center of fluid emission (Jørgensen and Boetius, 2007; Niu et al., 2017). This assortment of habitats and microhabitats is the combined function of temperature, salinity, pH, oxygen, carbon dioxide, methane, hydrogen sulfide, ammonia, hydrocarbon, metal contents, and other inorganic compounds (Fisher et al., 2007; Levin and Sibuet, 2012; Takai and Nakamura, 2010). Moreover, seeps are imperative pathways of biogeochemical cycling of elements (Boetius and Wenzh öfer, 2013), where reduced sulfur and methane rise from subsurface to the seabed and into the water column (Boetius and Wenzhöfer, 2013; Suess, 2014; Levin, 2005).

Studies on these chemosynthetic ecosystems were initiated with the discovery of hydrothermal vent (Corliss et al., 1979), four-decade ago in the deep sea, through towed camera systems and manned submersibles, followed by black smokers on the Eastern Pacific (Spiess et al., 1980), and cold seeps in the Gulf of Mexico (Paull et al., 1984). In the last few decades, technological advances in remotely operated vehicles (ROVs) and autonomous underwater vehicles (AUVs) with high-resolution seafloor mapping technologies have helped to improve our understanding immensely (Paull et al., 2015).

Continued exploration has discovered numerous deep-sea chemosynthetic ecosystems, indicating a similar distribution of fauna (Baker and German, 2004), with significant contributions from the arc and back-arc basins (Beaulieu et al., 2013). About 700 species from hydrothermal vents and 600 species from cold seeps (Ramirez-Llodra et al., 2005; German et al., 2011) have been reported at the closure of the Census of Marine Life program (Crist et al., 2010). Considering the frequency of new species discovered during each expedition, cold seeps offer a considerable potential to add many newer species. Every year between 2002 and 2010 about 25 megafaunal and macrofaunal species were added from chemosynthetic sites (Vanreusel et al., 2010). Advancement in molecular techniques such as next-generation sequencing (NGS) has added immensely to the inventory of organisms, especially microbes, as NGS is less susceptible to amplification bias, and yields a large set of genes that can be used for taxonomic identification and provide insights into metabolic functions (Stat et al., 2017). These advances have increased seep and vent diversity by orders of magnitude, concomitantly, adding to the understanding of ecosystem functions and community dynamics in extreme conditions.

Cold seeps demonstrate a highly productive ecosystem dependent on the microbial community for fixing chemical energy into organic carbon (Tunnicliffe et al., 2003). The seep environments act as a sink and a source of elements in the global biogeochemical cycles and heat budget (Levin et al., 2016). These systems generate elevated biomass compared with the surrounding deep-sea areas because of their ability to acquire carbon via chemosynthesis (Levin et al., 2016). Siboglinid tubeworms, large bivalves, and snails are typical inhabitants of the seeps and are infrastructure builders (Cordes et al., 2010). Shells of the dead organisms may persist for decades in these deep-sea habitats (Killingley et al., 1980), thus creating microhabitats and serving as a habitat for other fauna like crustaceans, polychaetes, and ophiuroids (Rodrigues, 2009; Kiel, 2010).

The majority of studies have established that bacterial communities fix the carbon heterotrophically, making it available for the metazoan community (Boetius et al., 2000; Briggs et al., 2012, Case et al., 2015). In addition, the bulk of literature on cold seeps uses stable isotopes that confirm the fixation of methane and sulfidic emissions that are made available to the dependent fauna in the form of carbon (Broecker and Peng, 1982; Levin and Michener, 2002; Becker et al., 2010). Chemosynthetically derived nutrition not only fulfills the demands of the seep community but disperses this by multiple pathways that spread horizontally as well as vertically in space. These pathways mediate the transfer of energy, resources, in- and out-migration, and gamete transfer as well as succession (Figure 13.1). Trophic flux at a seep site is largely governed by the symbiotic exchange where energy is synthesized by bacteria and utilized by higher invertebrates (Vanreusel et al., 2009;

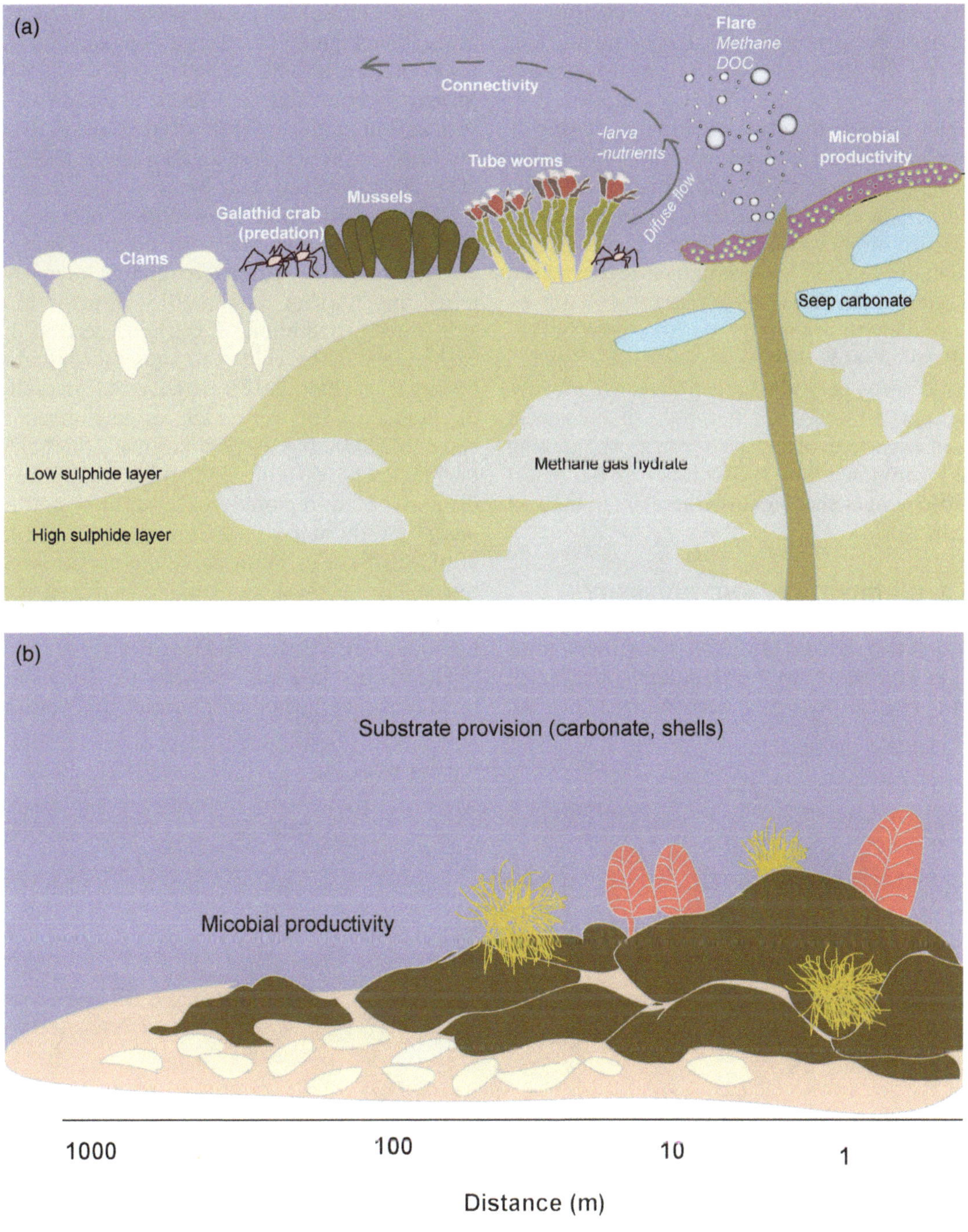

Figure 13.1 Schematic diagram showing ecosystem functioning at active (a) and non-active (b) cold seep areas.

Niemann et al., 2013). Studies have deciphered the chemical exchange and energetic transfer facilitated through specialized organs and novel physiological mechanisms (Southward and Newman, 1998).

The longevity of foundation metazoan communities at these sites is remarkable, and the associated faunal diversity that thrives has unique adaptations. The recent discovery of gas hydrate sites around the globe, such as in the Red Sea (Batanga et al., 2012) and in the Bay of Bengal (Mazumdar et al., 2019), has made it necessary to investigate these biological communities existing at new sites as well as their global connectivity. The most recently discovered seep site in the Indian shelf region is characterized by a high flux of hydrogen sulfide, supporting a high abundance of the chemosynthetic community (Mazumdar et al., 2019). This investigation revealed chemosymbiont bivalves (Vesicomidae, Mytilidae, Thyasiridae, and Solemyidae families), gastropods (Provannidae family), and polychaetes (Siboglinidae family) from seep-sites. In addition to the symbiont bearing community, Decapods, Ophiuroids, and Cirripeds were identified typically as nonchemosymbionts from the other cold seep regions (Mazumdar et al., 2019).

This chapter looks at the diversity and functioning of life at the seep ecosystem from a conventional or rather classical ecological perspective while keeping recently evolved concepts in mind. Admittedly, we do not aim to resolve the mysteries but explore worthy aspects from a different facet. Since Darwin, several theories have been postulated and synthesized, which are the basis of modern molecular studies today; few have been improvised, and some still under inconclusive debates. However, cold seeps defy all the conventional oceanographic concepts of seasonality, top-down ecosystem models, marine eco-toxicology, deep-sea deserts, making them ideal environments to revisit classical ecological concepts, which might shed light on these dark and cold corners of the world.

13.2. DISTRIBUTION AND DIVERSITY

Until the first two decades of cold seep discovery, most of the studies focused on the North Atlantic and North Pacific, but a few locations have been discovered in the southern hemisphere during recent times (e.g., Sellanes et al., 2004; Figure 13.2). Active seeps are found from shallow to deep environments (>6000 m; Sibuet and Olu, 2002). Moreover, a probabilistic estimate of chemosynthetic sites such as vents and seeps in the deep sea seems to be in many hundreds and even thousands (Beaulieu et al., 2013; Skarke et al., 2014; Beaulieu et al., 2015).

Methane-rich cold seeps support a range of organisms that have evolved with adaptation to extreme physicochemical settings to enable them to extract nutrition (Åström et al., 2018). Furthermore, rigid substrate by accretion of methane-derived authigenic carbonates at the cold seep sites boosts habitat complexity (Sen et al., 2018). However, the overall diversity of fauna is not very high at any given seep site (Levin, 2005), but is comparatively higher than the barren deep seafloor, thus contributing to the beta diversity at a regional scale, termed as "biological oases" of the deep sea (Cordes et al., 2010; Åström et al., 2018). Furthermore, niche partitioning carried out solely by the heterotrophic bacterial community increases biodiversity by dividing the area into microhabitats (Levin et al., 2012).

The major players of the seep ecosystems are vesicomyid clams, bathymodiolin mussels, and vestimentiferan

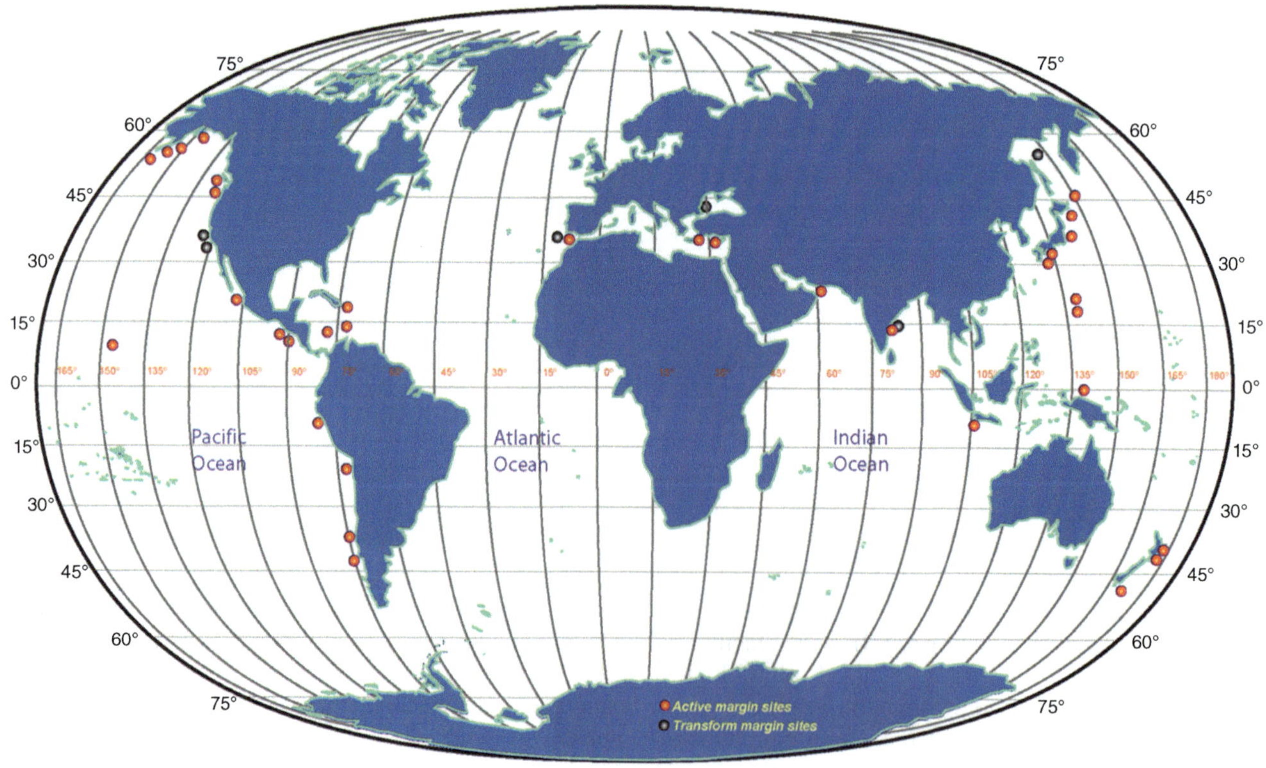

Figure 13.2 Global distribution of active and transform margin sites of the cold seep. The distribution map is based on modified after Suess (2014), and new sites from the Indian Exclusive Economic Zone (EEZ) from Mazumdar et al. (2019).

tubeworms (Kojima, 2002; Tunnicliffe et al., 2003; Levin, 2005; Baco et al., 2009). The associated fauna typically includes siboglinid, thyasirid, solemyid, lucinid bivalves, trochid gastropods, hymedesmid, and cladorhizid sponges, shrimp, amphipods, galathaeid crustaceans, and hesionid, ampharetid, polynoid, and dorvilleid polychaetes (Baco et al., 2009). One of the remarkable inhabitants occurring in the cold seeps is the *Lamellibrachia luymesi,* which grows slowly to a height of 3 m and has a life span of about 250 years (Fisher et al., 1997; Bergquist et al., 2000). Apart from typical macrofaunal communities (Levin et al., 2001, 2016), the other known example of biogenic habitat in the cold seeps is created by cold-water corals (Hovland and Thomsen, 1997; Cordes et al., 2009) and habitats formed by large sponges (Cordes et al., 2008; Quattrini et al., 2015).

Among the meiobenthic groups, foraminiferas have been studied from several seep sites, wherein they are abundant in the shallow seeps (Panieri, 2006) while numbers are low in deeper seeps. Several studies document the diversity of foraminiferans from active seep sites (Sen Gupta, 2007; Bernhard et al., 2010; Martin et al., 2010) and have also been suggested as paleoproxies for methane emission (Rathburn et al., 2003). By far, the most abundant meiofaunal group found in the cold seeps nematodes (Van Gaever et al., 2009; Zeppilli et al., 2011; Lampadariou et al., 2013). The influence of seep chemicals also impacts their distribution and may show reduced diversity and abundance while increasing the dominance of few resistant organisms such as *Halomonhystera* sp., *Sabatieria mortenseni,* and *Desmodora* sp. (Van Gaever et al., 2006). There was no single dominant species found across the seep sites implying that local settings and parameters have a significant role in structuring nematode community (Zeppilli et al., 2011). Generally, the next dominant species are the copepods, followed by several other groups (Lampadariou et al., 2013) such as kinorynchs (Robinson et al., 2004), ostracods, halacarids (Bright et al., 2010), isopods, and some polychaete species. The distribution of meiofauna is attributed to the nature of foundation species, *Caleptogena* sp. (Robinson et al., 2004) as well as microbial mats of *Beggiatoa,* which provides a niche to some of these smaller groups (Sommer et al., 2007).

13.3. BIOGEOCHEMICAL CYCLE OF COLD SEEP SEDIMENT

Cold seeps are considerable sources of carbon for the global carbon cycle (Judd and Hovland, 2007; Suess, 2014), and represent the largest source of methane on Earth (Kennicutt, 2017). Over three decades of cold seep discoveries have added many sites on the global continental margins (Hong et al., 2016), but the flux estimation for methane as well as its depth-integrated turnovers remain understudied (Egger et al., 2018; Luo et al., 2016; Feng et al., 2018). At a hotspot-scattered cold seep site, fluxes of CH_4 in the zone above the gas hydrate occurrence is estimated to be sizable (0.03 Gt year^{-1}), wherein annual methane emission to the water column was equivalent to 0.02 Gt C (Boetius and Wenzhöfer, 2013). The source of seep fluids has an spread of a few meters to many kilometers below the seabed (Suess, 2014), especially on the subductive oceanic plates.

Methanotrophic archaeal group and sulfate-reducing bacteria mediate the anaerobic oxidation of methane (AOM), the most critical biogeochemical process at the seep site (equation 1; Hoehler et al., 1994). Of the upward migration of dissolved methane in the overlying water mass, the majority of it is utilized at the sulfate–methane transition zone by AOM (Boetius et al., 2000; Luff et al., 2005), making it bioavailable.

$$(\mathrm{AOM}): \mathrm{CH_4} + \mathrm{SO_4}^{2-} \rightarrow \mathrm{HCO}^{3-} + \mathrm{HS}^- + \mathrm{H_2O} - 17\,\mathrm{kJ}\left(\mathrm{mol\,SO_4}^{-2}\right)^{-1} \quad (1)$$

Dissolved inorganic carbon (DIC) is produced by reaction (1), converting methane wherein authigenic carbonate precipitation partially removes it from solution (Boetius and Wenzhöfer, 2013). Although AOM consumes subsurface methane, a sizable quantity reaches the interface and mediates sulfur and carbon cycling in marine sediments (Orphan et al., 2001a; Egger et al., 2018). Cold seeps contribute a considerable amount of DIC and dissolved organic carbon (DOC) to the water column in the form of fossil carbon (Wang et al., 2001; Zachos et al., 2005; Snyder et al., 2007; Pohlman et al., 2011; Xu et al., 2018). Hence, seep-emitted carbon could increase acidification of the ocean as well as reduce the oxygen in seawater, ultimately contributing to global climate change (e.g. the Paleocene-Eocene thermal maximum; Archer et al., 2000; Zachos et al., 2005; Xu et al., 2018).

In addition, AOM couples with sulfate, nitrite, nitrate, Fe (III), and Mn (IV) playing a role in oxide reduction (equations 2–5; Boetius et al., 2000; Beal et al., 2009; Ettwig et al., 2010, 2016; Haroon et al., 2013). Electron transfer between methane-oxidizing and sulfate-reducing microorganisms has recently been reported by nanowire-dependent direct interspecies mechanism (McGlynn et al., 2015; Wegener et al., 2015). Interspecies electron transfer in sediments is also facilitated by conductive minerals such as magnetite and pyrite (Kato et al., 2012).

$$8NO_2^- + 3CH_4 + 8H^+ \rightarrow 3CO_2 + 4N_2 + 10H_2O^- 928\,kJ\left(mol\,NO^{2-}\right)^{-1} \quad (2)$$

$$8NO_3^- + 5CH_4 + 8H^+ \rightarrow 5CO_2 + 4N_2 + 14H_2O - 765\,kJ\left(mol\,NO^{3-}\right)^{-1} \quad (3)$$

$$4MnO_2 + CH_4 + 7H^+ \rightarrow HCO_3^- + 4Mn^{2+} + 5H_2O - 550\,kJ\left(mol\,MnO_2\right)^{-1} \quad (4)$$

$$8Fe(OH)_3 + CH_4 + 15H^+ \rightarrow HCO_3^- + 8Fe^{2+} + 21H_2O^- 270\,kJ\left(mol\,Fe(OH)_3\right)^{-1} \quad (5)$$

An array of interconnected processes along with diverse microbial mediated hydrocarbon oxidation plays an important role at the seepage site. Microbially mediated processes create and support high productivity (methane and hydrogen sulfide) and specialized biological diversity across trophic levels at cold seeps that are well-tuned to local biogeochemical regimes (Aharon et al., 1992; Sahling et al., 2002; Luff et al., 2005). Chemosynthetic bivalves/mussels take up methane and oxygen by diffusion as most of them harbor sulfide-oxidizing symbionts and methanotrophs/methanogens (Duperron et al., 2005). Therefore, most of the oxygen is rapidly consumed at or near the sediment–water interface by free-living and symbiotic microbes as well as metazoan respiration at the interface of sediment and water; thus, anaerobic metabolisms dominate at seeps.

The biomass accumulation at seep sites supported by phosphorus still requires an explanation (reviewed in Joye, 2020). Unlike, organic matter derived from terrestrial or pelagic marine processes with proportionate N and P the seeps fueled by hydrocarbon are deficient in N and almost without P. Deeply sourced brine fluids are known to be an essential source of N and P to the deep sea. This diffused P is taken up by sulfur-oxidizing microorganisms, especially *Beggiatoa* and *Thiomargarita* (giant sulfur-oxidizing bacteria) which concentrate and recycle the P (as polyphosphate), which is an essential process at seep sites (Jones et al., 2015).

The sediment geochemical profiles of hydrocarbon seeps (generally organic carbon-rich up to 10% by weight) sites significantly vary from the normal marine sediments (Joye et al., 2010). DIC and H_2S concentrations can reach up to or more than 15 mM (Joye, 2020). Depletion of sulfate prevalent within the upper layers of sediment (0–20 cm), and pore fluid CH_4 concentrations often found to be saturated to ~100 mM (Lapham et al., 2013). Hydrocarbon oxidation produces DIC and organic intermediates, but recycling of other nutrients is limited (Joye et al. 2004). Therefore, hydrocarbon seep sites are generally nutrient-poor, although organic matter is abundant, as the inadequate stocks of N and P are the main constraint to sustaining biological productivity (Joye et al., 2004). However, this chapter focuses on the ecology of the metazoan (multicellular) community that has evolved at cold seeps taking advantage of the chemotrophic microbiome while harmoniously establishing a functional ecosystem.

13.4. SYMBIOSIS

The cold seeps resemble the coral reef ecosystem (nonetheless, some seeps are known to harbor corals as well) in the sense of 'foundation species.' Foundation species are the structural and functional base of any habitat, solely carrying the responsibility of providing a niche for diverse faunal inhabitants of the ecosystem. Like corals, in the case of cold seeps, the clam/mussel beds and the polychaete gardens create a foundation for other associates to recruit and thrive, including macrofaunal mollusks, crustaceans, and other smaller metazoans such as nematodes, polychaetes and copepods, amphipods, isopods, and others. Another striking similarity is that the corals require a dinoflagellate symbiont for fixing carbon, while the cold seep megafauna require a bacterial symbiont. Here, the functional ecology is critically balanced on just a single interrelationship wherein both the ecosystems collapse entirely if they lose their obligate symbiont producers.

The unique congregation of organisms in and around the vents and cold seeps is exclusively dependent on the chemoautotrophic mode of nutrition (Cavanaugh et al., 1981; Roeselers and Newton, 2012). For example, organisms such as the vestimentiferan tubeworm *Riftia pachyptila* host symbiotic chemoautotrophic bacteria at high densities within their trophosomal tissue. These resident prokaryotic cells oxidize available sulfide and convert CO_2 to organic matter by reducing it, thus generating ATP via the Calvin–Benson Cycle (Cavanaugh et al., 1981).

Vesicomyiidae, Mytiliidae, Solemyidae, Thyasiridae, Lucinidae, Pogonaphora, and Porifera benefit from endosymbionts in terms of energy production by utilizing seepage sources (Southward, 1986; Sibuet and Olu, 1998). In the vestimentiferan tubeworms, the gut stores endosymbionts and these fulfill their metabolic requirements (Southward, 1989; Bright and Lallier, 2010). Vacelet et al., (1995) reported a symbiotic relationship between the sponge, *Cladrhiza* sp., and methanotrophic endosymbionts.

These endosymbionts show methanol dehydrogenase activity, whereas ATP sulfurylase nor adenosine-5′-phosphosulfate reductase activities were evident in the sponge, which indicates that the sponge derives nutrition directly from endosymbionts (Vacelet et al., 1995). Interestingly, multiple symbiotic bacterial species are used by several megafaunal, and macrofaunal species as many species are also reported in a single host (Lorion et al., 2012).

Cold seep chemosynthetic ecosystems are "bottom-up" ecosystems because the predatory pressure does not regulate the population dynamics at this habitat. Conventional ecosystems have a flow of nutrients and primary production is constant but constrained with minor fluctuations in a few parameters such as photosynthetically active radiation (PAR), nutrient ratios, and functional groups present. In contrast, the cold seeps are predominantly controlled by the quantum and ratios of elements released by geological activity. At the first trophic level, bacteria play the role of primary producers fixing lithogenic/geogenic sulfur and methane into organic matter. This organic matter is readily harvested by abundant metazoans that host these bacteria as symbionts. Other communities from different biomes that carry out such harvesting are social insects such as leaf cutter ants maintaining "fungal gardens", seed caching avifauna, and termites harboring gut bacteria to digest wood. However, this behavior of seep-derived bacterial 'gardening' is unique in a sense, as these are chemoautotrophic bacteria where the host metazoan requires refined evolutionary traits to resist the toxicants and nurture the microbes while siphoning nutrients for symbionts and deriving oxygen for itself. There are species of mat-forming bacteria common at the seep sites suggesting that chemoautotrophy by bacteria is not host-dependent, and the bacterial communities have independently evolved the chemosynthetic lifestyle. Therefore, it may be inferred that the cultivar has evolved a strategy to dock the bacteria and gain nutrition consistently. Another critical function by the host is maintaining a steady harvest with the management of yield for its growth and energy partitioning for reproduction. The holobiont metazoan is limited by subsurface resource supply, and fluctuations in it will result in dynamic standing stock of its crop (population of its bacterial symbiont).

13.5. SYMBIONTS AND ADAPTATIONS

Environmentally acquired symbionts come from a large and genetically diverse free-living bacterial population via a process known as "horizontal transmission". While in 'vertical transmission,' symbionts are also transferred by the host across generations via reproduction. Vertical transmission imposes genome limitations known as a 'transmission bottleneck' wherein female gametes are the carriers of symbiont to the next generation. With low population and low recombination, there is an accumulation of mild mutations (known as Muller's ratchet), which may aggravate and accelerate endosymbiont evolution (Moran, 1996). The genome of such symbiont becomes bulky as they retain genes necessary for a dual life as symbiont as well as competing in the community with other free-living prokaryotes (Bright and Bulgheresi, 2010).

Conversely, the eukaryotic hosts or the holobionts require a large cell volume to cater to the bacterial symbiont community. DNA selects for cell volume and growth rate by controlling nuclear volume (Cavalier-Smith, 1978). The "C-value enigma" states that the evolutionary complexity has no proportion with the quantity of DNA (Gregory, 2001). Thus the '*k*-selected' organisms need larger cells than '*r*-selected' ones, and that varying the total amount of DNA per cell is a way of controlling cell size. Based on the large size, steady reproduction, and delayed senescence of the seep foundation species, they can reasonably be categorized as '*k*-selected', although this categorization is relative (Adams 1880). Several combinations of factors may illustrate the lineages (Gregory, 2001) of the seep foundation holobionts. Nonetheless, cell size might not have any bearing on the size of the organism, although the possibility of congruence can be explored. Cell size, quantum of DNA, expressed portions, and co-evolutionary functions may provide some insights into their unique evolution. Eckelbarger and Watling (1995) revealed that the phylogenetic history of the species is responsible for rapid reproduction induced by abrupt input of organic matter. Traits to consume organic matter and increase the reproductive efficiency within short span suits opportunistic species in shallow habitats (Levin et al., 2016), while others that cannot cope with such conditions are found in the more stable habitat.

The meiobenthic groups, for example nematodes and copepods, have been known to undergo a process known as "chromatin diminution" wherein a significant portion (sometimes 90%) of the genome is hacked during differentiation of the somatic cell line (Wyngaard and Gregory, 2001). Physiological or phenotypic functionality is yet to be fully deciphered, but the hypothesis is that the DNA deletion may have a role in nongenic ejected DNA and not in differentiated somatic line but is more relevant in the germline or the early zygote (Cavalier-Smith, 1978, 1982; Gregory and Hebert, 1999). These traits may well be expressed in extreme ecosystems such as cold seeps, where the selection may favor avoiding deleterious mutations imposed by the ambient chemicals or biophysical environment.

From the perspective of the bacterial symbiont-bearing host, the community requires strategies to overcome any coevolutionary disadvantages. Few hypotheses have been proposed that might explain the genetic distinction as well as evolution of the genome as far as the host is concerned. The need to control the symbiont and manage its own requirements might have resulted in genome reduction to a specific allelic pool. This also depends on the life-history traits such as sedentary nature and reproductive patterns of a particular genus. Allozyme signatures depict that the genetic diversity of symbiont-bearing polychaetes such as *Alvinella* spp. is low (Denis et al., 1993; Vrijenhoek et al., 1993), compared with other opportunistic populations of the genera *Paralvinella* (Jollivet, 1995). The comparison with other non-endosymbiont seep communities such as the amphipod *Ventiella sulfuris* (France et al., 1992) and some limpets also revealed a distinction in genetic diversity.

In contrast, there are also reports of shallow water endosymbiont-bearing species of lucinids that display significant polymorphism (Dwiono et al., 1989). The other hypothesis based on ecophysiological performance states that species of *Alvinella* exhibit thermoregulation (Chevaldonné et al., 1991). Compared to these, the active amphipods and paralvinellids, which move continuously, experience a range of environmental contrasts and can evolve alleles to suit different environmental conditions with a heterozygous genotype (Koehn and Bayne, 1989). Another hypothesis that may explain the avoidance of a genetic bottleneck is displayed by paralvinellid species that hitchhike on the crabs of chemosynthetic habitats to move between habitats (Tunnicliffe and Jensen, 1987). This phoretic process may mediate the genetic exchange of populations while increasing polymorphism (Nevo et al., 1974).

The symbiotic bacterial community inside the seep invertebrate host is well documented, but there are very few records of other interactions such as parasitism (Zielinski et al., 2009). For example, the occurrence of disease-causing microbial parasites has been reported in the seep mussel, *Bathymodiolus* (Ward et al. 2004). Parasitism influences host life history, traits including phenotypic effects that control host reproduction, growth, longevity, and physiology (Dawkins, 1982). It is of particular interest to study the seep invertebrates because of the typical characteristics, such as gigantic size and increased life span, etc. For an altruistic strategy to evolve, there is a sort of compromise wherein some exchange is required, which results in an energetic disadvantage, but selection favors reproductive success. Indeed, engulfed organisms become organelles and interactions evolve while settling for stable strategies (Smith, 1982); apparently, seep interactions may portray a transition. This ancient ecosystem (Goedert et al., 1995) had enough time to produce cooperation for survival, although it could potentially produce deceiving interactions with species taking energetic advantage of the host. The primary question is why this habitat could not evolve several species with a lucrative and more cost-effective stable strategy of parasitism? Is it therefore possible that these co-evolved lineages are the relicts from past interactions that were negative or mildly harmful? Similarly, very little is known about diseases or syndromes from these ecosystems; perhaps we can expect more information in future investigations. Predation is known from the seep environment, although several species are nonobligate mobile predators from the surrounding benthic habitat, without much top-down control on the seep community.

One of the critical components of marine ecosystems that is not well studied are viruses, which have been known to play an essential role in deep-sea ecosystems (Ortmann and Suttle, 2005). Viruses tend to check the population of communities at all trophic levels in the marine food chain, including microbial diversity. Although the viral community at vent sites has been studied at a several sites (Juniper et al., 1998), their role in the seep communities is relatively understudied.

The giant tubeworm *R. pachyptila* harbors gammaproteobacterial endosymbionts Candidatus, *Endoriftia, Persephone,* which are well studied for their role in chemoautotrophy (Robidart et al., 2008). They acquire endosymbionts in the larval stage; however, Klose et al., (2015) suggested by use of an experimental set-up that symbionts escape from the host upon the death of the latter and migrate to the surface where they proliferate. Coelomic cells that harbor symbiotic bacteria known as trophosomes are not deterred by microbial fouling (Klose et al., 2016), while *Bathymodiolus* displays distinct hypertrophy with a manifold increase in gill surface area compared with their shallow-water relatives (Piquet et al., 2019), suggesting that they evolved forms to accommodate symbiont function. Various adaptive traits and co-evolutionary mechanisms explain the survival of seep organisms, one of which is 'apoptosis,' or cell death. Ciliated cells of metazoans are certainly the first to contact the surrounding fluid and thus come in close contact with a compound that is highly concentrated, that includes toxic sulfur and trace metals, which may trigger apoptosis (Circu and Aw, 2010). Cellular apoptosis might be an adaptive mechanism to avoid organismic death by sacrificing part of cells or organs, which may increase the longevity of the organism (Piquet et al., 2019). A higher metabolic rate produces lots of reactive oxygen species (ROS) which increases oxidative stress, eventually increasing apoptosis, and this may be the likely mechanism in *Mytilus* and *Bathymodiolus* species (Mone et al., 2014). Results from transcriptome studies and their subsequent interpretation directs one to the hypothesis that apoptosis was the primary process that helped mussels (initially *Bathymodiolus thermophilus*) to control the symbiotic bacterial population in the gills. These bacteria-filled bacteriocytes undergo apoptosis similarly in the tubeworm *R. pachyptila*, and this mediates the recovery of symbiont carbon and recycling of animal cells (Pflugfelder et al., 2009). Minimal apoptosis in bacteriocytes was observed, mainly where symbionts were most abundant, which could be explained by the endosymbiont-linked defense against apoptosis (Piquet et al., 2019). Gill specialization can be interpreted in the light of symbiosis, where enlarged gills require more robust ciliary activity and constant exposure to toxic compounds possibly leads to a high turnover of ciliated cells (Pflugfelder et al., 2009; Vigneron et al., 2014; Piquet et al., 2019).

13.6. GENETICS AND GLOBAL CONNECTIVITY

As Dobzhansky (1973) mentions, it only makes sense to discuss biology in an evolutionary context, and this notion is more pertinent while considering extreme environments. The functions of extremophilic organisms have solely evolved to exploit lithogenic chemosynthesis and can suitably be discussed and explained in an evolutionary context.

Isolation (e.g. Oceanic Islands) can vividly illustrate most of the evolutionary traits acquired by organisms (and its absences in mixed communities). The fundamentals of "insular isolation" (MacAurthur and Wilson, 1967) best suit the cold seep ecosystem, as these gaseous expulsions worldwide produce an extreme endemic community in a real sense. Cold seep and hydrothermal vent organisms mimic the life-history traits of island organisms (ironically, the seeps might be more ancient than the oceanic islands themselves). Typical characteristics include genetic isolation, endemism, gigantism, dwarfism, extinction, inbreeding depressions, heterosis, reduced predatory pressure, co-evolutionary interdependence, and various localized extreme adaptations, perhaps for reproduction or routine ecophysiology (Quammen, 2004). The only significant difference is that the oceanic island fauna has scope for the exchange of species with the mainland or the nearest similar habitat, thus are "dilute" compared with endemism in seeps. However, the extremophiles of the cold seep require a cold seep to survive with an obligate symbiotic bacterial community. This highly specialized adaptation principally makes them handicapped (much close to the 'handicap principle') because of overspecialization. Initially, this provides the perception that such an environment is mostly responsible for keeping these endemics on the edge of extinction but, as we explore further, their success is evident because of several other high-fitness strategies.

Primarily, a chemosynthetic habitat illustrates that the survival of unique and endemic fauna of the cold seep is a remarkable and rare phenomenon that has defied the dogma with a noteworthy similarity between and within the fauna of seep, vent, and whale fall sites. Nonetheless, lineages connect at the generic and family level with good genetic exchange globally (Bernardino et al., 2012).

The gene flow among Mytillidae (e.g. *B. thermophilus*) species possibly results from the dispersal of hardy pelagic larvae (Craddock et al., 1995b). Owing to the longevity of the larval phase, Mytillidae is well-represented at chemosynthetic sites that have been well separated geographically. Similarly, Southward et al. (1996) investigated the dispersal capabilities of vestimentiferan species (*Lamellibrachia barhami* and *Ridgela piscesae*), which showed excellent dispersal capabilities. Hecker's hypothesis states that seeps might serve as refuges for vent organisms (Craddock et al., 1995a). The study also suggests a distinct variability in mytilid species at vents and seep sites; thus, suitable studies are required to elucidate this. Genetic variation studies carried out on *Alviniconcha* spp. at two different locations, namely Lau and North Fiji basins in the southwest Pacific Ocean, revealed a higher degree of genetic differentiation within them. When a small number of individuals establish at a new site, they tend to have enormous genetic differences from the original population: a phenomenon termed as the "founder effect". Founder effects, along with abiotic environmental conditions, would act as a constraint in genetic exchange at the sites described (Denis et al., 1993).

Genetic exchange, diversity of gene pool, and the phenotypic expressions governed by the local environment, often prone to extinction, shape the populations of these habitats. Alvinellid polychaetes of the hydrothermal vent region do not show a reduction in polymorphism because of recurring population extinction (e.g. *Paralvinella grasslei*) but showed high genetic variability (Jollivet et al., 1995). The genetic data for *Calyptogena magnifica* reveals that despite producing nonplanktotrophic larvae, the species tend to be efficient dispersers. Their dispersal covers thousands of kilometers in the eastern Pacific and even equatorial currents failed to restrict them as the population was found to be conspecific at two different locations, located at the 18°S East Pacific Rise and at 9–10°N Galapagos Rift (Karl et al., 1996). Several models such as the "Steppingstone" and 'Island' model of Wright (1931, 1969) have been considered for a possible dispersal mode (Karl et al., 1996). The former fails to explain the conspecific nature found at vast distances as, with an increase in several steps taken by populations, there is a chance of a decrease in gene flow (Kimura and Wiess, 1964). In contrast, the Island model explains the dispersal pattern used by planktotrophic larvae, which are capable of floating and can cover vast distances with the help of surface currents. Nonplanktotrophic larvae possess better capabilities to mix in the water column and are long-lived with the ability to slow down or arrest development, thus they will be able to cover vast distances (Karl et al., 1996).

It is a fact that there is genetic traffic between close and near-distance seeps (Baco et al., 1999; Kojima et al., 2004; Audzijonyte et al., 2012). This again points to an extreme adaptation wherein larval longevity, as well as the settlement by seep organisms, can only besuccessfully sustained by establishing a genetic bridge and thus avoid localized extinction. Long-distance dispersal must be aligned with a more important trait of positive chemotaxis followed by recruitment at a favorable location thus increasing the efficiency of reproductive success. In the vast grouping of the spatial oceanic desert (any place other than the seep site), a pelagic larva has to apply a source apportionment model using cues in its vicinity to locate the seep site deep below. Otherwise, it may be assumed that there is a passive sporadic shower of larvae from the surface waters to the seafloor with only accidental survival if it falls over a seep site.

Moreover, it will be interesting to explore the global current vectors overlapping cold seeps sites to investigate the source and sink of this endemic gene pool and its interaction. Baco et al. (2016) noted that deep sea regions are vast voids of the homogeneous environment; thus, dispersal distances have to increase proportionately to the depths, and indeed several species have broad geographic ranges that encompass the entire ocean or more (Rex and Etter, 2010). On average, the deep-sea fauna displays a broader dispersal range compared with their shallow-water relatives, with a distance of up to 2000 km (Baco et al., 2016).

Similarity in faunal diversity and genetic links at whale fall, vents, and seeps has proved to be crucial in deciphering the distribution and affinity of fauna (Baco et al., 1999). Like land bridges for island allopatry, whale falls may play a pivotal role in larval bridges on a global scale. The fossil records of chemosynthetic communities mostly associated with whale fall (Squires et al., 1991) reveal that deep-sea association has existed since the early Oligocene (30–35 million years). Goedert et al. (1995) investigated fauna at eight whale fall sites (Makah and Pysht formations on the Olympic Peninsula, Washington State), which were represented by families of mytilid, thyasirid, and lucinid bivalves, but not the major species such as vesicomyids that are typical of reducing habitats. Unlike Baco et al. (1999), Goedert et al. (1995) opined that, considering the community distinction in their study, whale falls may not play a significant role in the genetic connectivity.

On this paradigm, the seep larvae face large voids of adverse ambiance (for their adults) if not settling on another seep site (or vent/whale fall). Because the number of whale fall distribution across the world oceans remains undetected, and its site abundance might be in question nonetheless, the numbers should be high considering the global population of whales (Smith et al., 1989). Furthermore, Smith et al. (1989) also calculated the mobility and probability of adult dispersal on the deep seafloor to reach a new chemosynthetic site, and it proves unlikely that many reach other specialized habitats. Estimates suggest that $3.9 \cdot 10^5$ whale carcasses sank every year before the advent of commercial whaling, thus making them six times more abundant compared with the present scenario (Jelmert and Oppen-Bernsten, 1996). There would have been frequent and spatial traffic of these endemics with seeps, vents, and well-distributed sporadic whale fall sites, but in recent times, the whale fall sites will have decreased because of industrial whaling.

It is far-reaching yet reasonable hypothesis that traditional whaling over two centuries might have influenced the distribution, diversity, and extinction of deep-sea chemosynthetic invertebrates globally (Butman et al., 1995). Smith et al. (1989) hypothesized that whale falls act as stepping stones and are important for the dispersal of the vent and seep species depending upon sulfide. Although this hypothesis is debatable (Tunnicliffe and Juniper 1990; Squires et al., 1991; Butman et al., 1995; Jelmert and Oppen-Bernsten, 1996; Tunnicliffe and Fowler, 1996), it still remains without a convincing alternative.

The connectivity and affinity of diverse fauna may imply that vents and whale falls might be ephemeral compared with seeps, and indeed, the seeps might be the site for "'ecological niche conservatism" wherein the original traits or functions are retained. Sympatry and hybridization will be rare within a given seep environment because of the cohesive nature of the habitat but predictably allopatry can be shared between any two seeps with a good chance of hybrids. It would be interesting to know what evolved first, the symbiosis with the prokaryotic community or the life strategy for long-distance dispersal.

13.7. CHEMOSYNTHESIS AND TROPHIC TRANSFER

Trophic interactions and transfer of chemical energy take place through the advection of particulate organic carbon inclusive of benthopelagic coupling and trophic transfer by predators and scavengers (Levin et al., 2016). Suspension-feeding chemoautotrophic serpulid worms and sea anemones rely on seep-derived particulate organic matter (Hessler et al., 1988). Some polychaete families (e.g. maldanids, onuphids, nephtyids, nereids and ampharetids), amphipods, tanaids, cnidarians, and nematodes display mixotrophic behavior by deriving nutrition chemosynthetically as well as photosynthetically in the form of organic matter occurring in sediment or overlying waters (Levin et al., 2010; Pape et al., 2011; Rodrigues et al., 2013).

Early studies to understand carbon source for those living in aphotic vent regions revealed that stable isotope values for carbon and nitrogen differ from those of other open ocean areas as $\delta^{13}C$ values were found to be lower (Rau, 1981). Subsequent studies carried out by several others have helped decipher the chemosynthetic organisms' relationship with the endosymbiont and origin of a local source of nutrition. Dissolved inorganic carbon in seawater acts as the primary source of carbon in tissue, whereas sulfides are a steady source of energy in chemosynthetic communities, where $\delta^{13}C$ values for vestimentiferans of vents are similar to those of seeps (Paull et al., 1985). The $\delta^{13}C$ values of carbonate sources from the shells of mussels and gastropods are trivial compared with those of abyssal dissolved inorganic carbon mostly used by molluscs. About 82% of carbon detected is almost similar to the $\delta^{14}C$ available in deep waters (Broecker and Peng, 1982). Later isotope studies of hydrocarbons carried out by Becker et al. (2010) on bivalves seeps authenticated the trends.

Carbon is made available and/or fixed in bathyal reducing environments through various processes, wherein methanotrophy and sulfur oxidation are the notable pathways of chemosynthesis in symbiotic organisms (Conway et al., 1994; Van Dover, 2000). The interaction between free-living or symbiotic sulfate reducers with chemoautotrophs is imperative (Boetious et al., 2000; Dubilier et al., 2001); thus, isotopic signatures reveal the nutritional mode of organisms inhabiting vent and seep habitat as there are numerous pathways of carbon fixation (Van Dover and Fry, 1989; Conway et al., 1994). Methane seeps are known to harbor subsurface chemosynthetic bacteria such as methane-oxidizers, sulfate-reducers (Orphan et al., 2001a, 2001b; Hinrichs et al., 2000). The abundant free-living chemosynthetic bacteria of filamentous nature form mats on the surface and subsurface, acting as primary producers for macrofaunal invertebrates such as dorivellids, amphaterids, nereids, and trichobranchids. The gastropods *Astyris permodestra* and *Provanna* sp. were found to be fuelled by mat microbes (Levin et al., 2003). Fluid expulsion at the seep region controls the distribution of megafauna and justifies the patchy distribution; these megafaunal species can act as "Flux Indicators" reflecting interactions among communities and sediment character (Sahling et al., 2002; Levin, 2005; Heyl et al., 2007). Sulfide concentrations present in sediment is a critical factor in controlling the settlement, distribution, and density of chemoautotrophic organisms, as higher concentrations of sulfide (ΣH_2S 10 cm >6.4 mol l^{-1}) inhibits the settlement or post-settlement survival of juvenile clams (Heyl et al., 2007). Predictably, the porewater chemistry might be responsible for the presence of mortal remains (shell beds) in addition to clam beds owing to the lack of evidence of predation or epidemic (Van Dover et al., 2003; Mills et al., 2003). The methane of biogenic origin fuels the methanotrophic symbiont, inferred from the $\delta^{13}C$ values obtained for *Bathymodiolus platiformis* (Barry et al., 2002; Fisher, 1990; Feng et al., 2015).

Bathymodiolin mussels of seeps are known indicators of fluid expulsion, the activity of seepage, coupling of sulfide, sulfate, and nitrogen cycling (Feng et al., 2015). Unlike studies carried out on the mode of nutrition adapted by organisms of a vent or seep habitats, little is known about their waste excretion and physiological adaptations. One such study was undertaken by Dattagupta et al. (2006) on *L. luymesi*, which adopted its meter-long roots to excrete additional ions (sulfate and hydrogen) unlike its distant cousin *R. pachyptila* which excretes via plume epithelium (Goffredi et al., 1999; Girguis et al., 2002; Dattagupta et al., 2006). A recently discovered species of encrusting seep sponge (*Pseudosuberites* spp.) from the Wairarapa site off New Zealand was found to be methanotrophic oriented and harboring diverse epibiotic macrofaunal communities (Thurber et al., 2010; Baco et al., 2010). Along with sponges, new genera of densely aggregating polychaetes from New Zealand were discovered with one spionid and two amphaterids (Sommer et al., 2010; Thurber et al., 2010).

13.8. IMPACT OF COLD SEEP HABITATS ON SURROUNDING OCEANS

The seep community has a systematic succession wherein there is a start, peak, and lag phase, and accordingly, the population and diversity changes (Bergquist et al., 2003; Cordes et al., 2010). There are well-documented fossilized records of extinct seep ecosystems due to the cessation of expulsion (Georgieva et al., 2019). In the course of succession, many non-seep organisms are attracted to this vicinity, take advantage, and join the seep-derived trophic transfer of energy. This increases the radii of seepage-derived organic transfer, laterally enriching the deprived seafloor communities in its surroundings as well as vertically (sometimes up to sea surface), making the water column productive (Burd and Thomson, 1994; Vereshchaka and Vinogradov, 1999; Cowen et al., 2001). An obligate seep community is also aligned with a facultative community that switches from geogenic to photosynthetically derived resource and vice versa. Studies have also shown that zooplankton takes advantage of plume-derived organic carbon entrained in the fluid just above seepage or plume site (Vereshchaka and Vinogradov, 1999; Cowen et al., 2001; Bennett et al., 2011).

Over centuries to millennia, the lag phase of succession experiences colonization of deep-water corals, which are slow-growing and long-lived species exploiting seep-derived energy, although they are sensitive to any small change in the environment (reviewed in Cordes et al., 2009). After the seepage ceases, cold-water corals at such sites utilize precipitated carbonate from methane oxidation (Cordes et al., 2008; Wheeler and Stadnitskaya, 2011). Symbiont-hosting invertebrates such as bamboo corals and other organisms are known to derive particulate organic material from sources of emissions with the help of chemoautotrophic microbes, for example in the Manus Basin (Erickson et al., 2009).

The benthic megafaunal community plays a significant role by consuming methane and sulfides, metabolizing these emissions before they reach the sea surface and thus prevent their contribution to climatic processes (Boetius and Suess, 2004; Sommer et al., 2006; Niemann et al., 2008). The influence of seep-derived elements may have more extensive biogeographic controls while a few elements such as iron might have global effects, whereas it is estimated that 15% of the total global methane emission is contributed by chemosynthetic ecosystems (Judd et al., 2002; Levin et al., 2016). The advection rate of cold seeps varies in space and time, ranging from few millimeters to 100 cm year^{-1}, with outliers of 3–6 m year^{-1} (Cordes et al., 2005; de Beer et al., 2006). Oxygen drawdown is a critical process at such sites as aerobic methane oxidation by methanotrophs ranges from 4 to 1450 mmol O_2 m^{-2} day^{-1} (Valentine, 2011; Boetius and Wenzhöfer, 2013). The function of the seep community from a broader perspective is essential, wherein they metabolize the methane, sulfides, and carbonates that are considered deleterious as far as the environment and the climate are concerned. Without these communities, there would have been more significant impacts of cumulative geogenic seepages globally. It may be said that the extended phenotypic effects (Dawkins, 1982) of the seep community have helped to reduce negative global impacts by metabolizing emissions.

13.9. SOCIETAL RELEVANCE OF COLD SEEPS

Owing to relatively constant and long-term emissions, slow-growing species dominate seep sites, with some species living up to 300 years. Understanding the mechanism by which these ecophysiological adaptations are acquired is related to cell death and links them to homeostasis, growth, and longevity. However, further insight into the functions of these habitats and their role in delayed senescence is required. Vestimentiferan species studied have high cell proliferation activities that are not comparable to other invertebrates and are close only to those of tumors and the wound-healing processes (Pflugfelder et al., 2009). These invertebrates may provide new traits or molecules useful in medical therapeutics and cancer biology. Conserving these unique traits is of importance as they have evolved over millions of years and may have application in conservation genetics, biomedical research, and other streams of biochemistry. These superorganisms may have significance in climate research and may help to mitigate human-mediated toxin release.

Organisms in extreme habitats evolve new biosynthetic pathways, generating unique biochemical compounds that could be beneficial to humankind and may have the potential for commercial synthesis (Davies-Coleman and Sunassee, 2012). There is a particular focus on obtaining materials such as biocatalysts, products for bioremediation, surfactants, and antifreeze from cold seeps (Bull et al., 2000). Several compounds generated by deep-sea corals and sponges are potentially useful for various pharmaceutical purposes such as cancer treatment and bone grafting (Foley et al., 2010; Maxwell, 2005). In recent times, the harnessing of chemosynthetic communities resulted in the discovery of enzymes such Fuelzyme (thermostable enzyme), Venuceane (utilized for DNA cloning, sequencing, and amplification), and Pyrolase (hydraulic fractionation), with great commercial applications (Krach, 2013). The β-lactamases produced by cold sediment bacteria are known to protect against the lethal effects of β-lactams (penicillin, cephalosporins, carbapenems, or monobactams) on cell wall synthesis (Song et al., 2005; Bush and Bradford, 2016; Bush, 2018).

Ethane is considered the second most copious constituent of natural gas after methane. Identification of an ethane-degrading anaerobe in cold seeps has extended our knowledge and understanding of previously unknown natural processes of these chemosynthetic organisms (Chen et al., 2019). Similarly, further knowledge and understanding of cold seeps could reveal many chemicals, which could be sources of secondary metabolites, to synthesize the natural products associated with their hosts and possibly yield future marine genetic resources and potentially valuable industrial processes (Marris, 2006; Molinsky et al., 2009). The unique and specific associations provide potential solutions for various technologies useful in carbon capture and storage (Minic and Thongbam, 2011; Pettit, 2011). Moreover, the accelerated development of "genome mining" or metagenomic approaches is allowing us to understand the functional diversity of microbes, which will lead to an improved approach towards metabolic mining of these communities (Barone et al., 2014).

13.10. ANTHROPOGENIC IMPACTS ON GLOBAL SEEP ECOSYSTEMS

The increasing human population and the rapid developments in technology, industry, and urbanization have created a desperate need for raw materials, energy, and mineral resources to increasing production. New avenues and resource reserves are explored for exploitation, making cold seeps and hot vents the sites to fulfill some of these demands. Methane hydrates are considered next-generation fuel sources due to the increasing feasibility to exploit these and efficiency with emerging technology (Chen et al., 2020; Yin and Linga, 2019). These ecosystems, which are distant and isolated from human activities, are now vulnerable because of a direct demand for energy and mineral resources (Ramirez-Llodra et al., 2011; Santos et al., 2012). The Nankai Trough is the world's first marine methane hydrate production site, and is thus under threat from drilling operations (Arata et al., 2011). It is also designated as an experimental site for understanding carbon sequestration by deep-sea burial and magmatic rocks (Goldberg et al., 2008), further exposing it to manipulation.

Fulfillment of human requirements is necessary, but while we do not fully understand the functions of these ecosystems in global biogeochemical cycles and their climatic implications, it is hard to conceive the dichotomy of exploiting minerals and energy from these systems. Seemingly "out of sight", the cold seep ecosystems on all the continental margins of the world are at risk because of bottom trawl fishing, mining, and oil and gas exploration. Furthermore, extraction of living resources, pollution, and waste disposal may have a significant impact on functionality and diversity (Baker et al., 2010). Inter-Ridge (2006) states that scientific activities have to be controlled by a set of standard guidelines, as stated under the "Statement of commitment to responsible research practices at deep-sea hydrothermal vents". According to Inter-Ridge (2006), 874 voyages between 1976 and 2012 were carried out to explore vents, but there is no data for cold seeps. Some consideration has to be given to the number of scientific expeditions and their detrimental effects on this fragile ecosystem (Godet et al., 2011). If we categorized these foundation species as per conventional IUCN criteria, most of the species would fall under the 'near extinct' or 'near threatened' category. Thus, scientific activities should have the highest ethics while abiding by national legislation as there is hardly any lawfulness for international waters.

The recognition of cold seeps as the most significant reserves of energy and endemic biodiversity should ignite a debate that will test the decision-making balance between long-term versus short-term gains. Understanding the seep community, its diversity, genetic connectivity, and its ecophysiology is essential not only to gain new insights into its biology but to obtain answers for many questions linked to chemosynthesis and, in general, ecology, evolutionary biology, and medicine. One has to be cautious and considerate while making decisions and formulating policies on future energy reserves as these sites also harbor rich organismic diversity – a natural capital that provides ecosystem goods and services worth billions. Knowing every entity, process, pathway, function, and functionally equivalent commercial worth is essential. Most of the other degraded ecosystems can possibly be restored to some extent, but chemosynthetic ecosystems will be impossible to restore, and thus, losing cold seeps will be irreversible in the real sense. Before thinking about the utilization of cold seep's resources, it is our ethical responsibility to weigh its sustainability to ensure a healthy planet for future generations. Once lost, unforeseen methods of resurrection ecology and "zombie biology" may have to be applied, but their repercussions are impossible to imagine.

ACKNOWLEDGMENT

The authors are grateful to the editors Dr. Aninda Mazumdar and Dr. Wriddhiman Ghosh, for their constructive suggestions. We are also thankful to two anonymous reviewers for the critical comments on the earlier version of the manuscript.

REFERENCES

Adams, P.A. (1880). Life history patterns in marine fishes and their consequences for fisheries management. *Fishery Bulletin 78* (1): 1–12.

Aharon, P., Graber, E.R. and Roberts, H.H. (1992). Dissolved carbon and 33–133–133–1 anomalies in the water column caused by hydrocarbon seeps on the northwestern Gulf of Mexico slope. *Geo-Marine Letters 12*: 33–40.

Arata, N., Nagakubo, S., Yamamoto, K. et al. (2011). Environmental impact assessment studies on Japan's methane hydrate RandD program. In: *Proceedings of the 7th International Conference on Gas Hydrates (ICGH 2011):* Edinburgh, UK.

Archer, D., Buffett, B. and Brovkin, V. (2000). Ocean methane hydrates as a slow tipping point in the global carbon cycle. *Proceedings of the National Academy of Sciences of the United States of America 2009*: 106, 20596–20601. https://doi.org/10.1073/pnas.0800885105

Åström, E.K.L., Carroll, M.L., Ambrose, Jr., W.G. et al. (2018). Methane cold seeps as biological oases in the high-Arctic deep sea. *Limnology and Oceanography, 68* (51): 5209–5231. https://doi.org/10.1002/lno.10732

Audzijonyte, A., Krylova, E.M., Sahling, H. et al. (2012). Molecular taxonomy reveals broad trans-oceanic distributions and high species diversity of deep-sea clams (Bivalvia: Vesicomyidae: Pliocardiinae) in chemosynthetic environments. *Systematics and Biodiversity 10*: 403–415. https://doi.org/10.1080/14772000.2012.744112

Baco, A., Smith, C., Peek, A. et al. (1999). The phylogenetic relationships of whale-fall vesicomyid clams based on mitochondrial COI DNA sequences. *Marine Ecology Progress Series 182*: 137–147. https://doi.org/10.3354/meps182137

Baco, A.R., Rowden, A.A., Levin, L.A. et al. (2009). Initial characterization of cold seep faunal communities on the New

Zealand margin. *Marine Geology 272* (1–4): 251–259. https://doi.org/10.1016/j.margeo.2009.06.015.

Baco, A.R., Rowden, A.A., Levin, L.A. et al. (2010). Initial characterization of cold seep faunal communities on the New Zealand Hikurangi margin. *Marine Geology 272*: 251–259. https://doi.org/10.1016/j.margeo.2009.06.015

Baco, A.R., Etter, R., Beerli, P. et al. (2016). A synthesis of genetic connectivity in deep-sea fauna and implications for marine reserve design. *Molecular Ecology 25* (14): 3276–3298. https://doi.org/10.1111/mec.13689

Baker, E.T. and German, C.R. (2004). On the global distribution of hydrothermal vent fields. In: *Mid-Ocean Ridges: Hydrothermal Interactions Between the Lithosphere and Oceans. Geophysical Monograph Series*, Vol. *148* (eds. C.R. German, J. Lin and L.M. Parson), 245–266. Washington, DC: American Geophysical Union.

Baker, M.C., Ramirez-Llodra, E.Z., Tyler, P.A. et al. (2010). Biogeography, ecology and vulnerability of chemosynthetic ecosystems in the deep sea. In: *Life in the World's Oceans* (ed. A.D. McIntyre), 161–182. Oxford: Wiley-Blackwell.

Barone, R., De Santi, C., Palma Esposito, F. et al. (2014). Marine metagenomics, a valuable tool for enzymes and bioactive compounds discovery. *Frontiers in Marine Science 1*: 38. https://doi.org/10.3389/fmars.2014.00038

Barry, J.P., Buck, K.R., Kochevar, R.K. et al. (2002). Methane-based symbiosis in a mussel, *Bathymodiolus platifrons*, from cold seeps in Sagami Bay, Japan. *Invertebrate Biology121*: 47–54. https://doi.org/10.1111/j.1744-7410.2002.tb00128.x

Batanga, Z.B., Papathanassiou, E., Al-Suwailema, A. et al. (2012). First discovery of a cold seep on the continental margin of the central Red Sea. *Journal of Marine Systems 94*: 247–253. https://doi.org/10.1016/j.jmarsys.2011.12.004

Beal, E.J., House, C.H. and Orphan, V.J. (2009). Manganese and iron-dependent marine methane oxidation. *Science 325*: 184–187.

Beaulieu, S.E., Baker, E.T., German, C.R. et al. (2013). An authoritative global database for active submarine hydrothermal vent fields: Global vent database. *Geochemistry Geophysics Geosystems 14*, 4892–4905. https://doi.org/10.1002/2013GC004998

Beaulieu, S.E., Baker, E.T. and German, C.R. (2015). Where are the undiscovered hydrothermal vents on oceanic spreading ridges? *Deep Sea Research II Topical Studies in Oceanography 121*: 202–212. https://doi.org/10.1016/j.dsr2.2015.05.001

Becker, E.L., Lee, R.W., Macko, S.A. et al. (2010) Stable carbon and nitrogen isotope compositions of hydrocarbon-seep bivalves on the Gulf of Mexico lower continental slope. *Deep Sea Research Part II: Tropical Studies. Oceanography*, *57*: 1957–1964. https://doi.org/10.1016/j.dsr2.2010.05.002

Bennett, S.A., Statham, P.J., Green, D.R.H. et al. (2011). Dissolved and particulate organic carbon export from seafloor hydrothermal venting: East Pacific Rise, 9°50'N. *Deep Sea Research Part I: Oceanography Research Papers 58*: 922–931. https://doi.org/10.1016/j.dsr.2011.06.010

Bergquist, D.C., Ward, T., Cordes, E.E. et al. (2003). Community structure of vestimentiferan generated habitat islands from the upper Louisiana slope cold seeps. *Journal of Experimental Marine Biology and Ecology 289*: 197–222. https://doi.org/10.1016/S0022-0981 (03)00046-7

Bergquist, D.C., Williams, F.M. and Fisher, C.R. (2000). Longevity record for deep-sea invertebrate. *Nature 403*, 400–500. https://doi.org/10.1038/35000647

Bernardino, A.F., Levin, L.A., Thurber, A.R. et al. (2012). Comparative composition, diversity and trophic ecology of sediment macrofauna at vents, seeps and organic falls. *PLoS ONE 7* (4): e33515.

Bernhard, J.M., Martin, J.B. and Rathburn, A.E. (2010). Combined carbonate carbon isotopic and cellular ultrastructural studies of individual benthic foraminifera: 2. Toward an understanding of apparent disequilibrium in hydrocarbon seeps. *Paleoceanography and Paleoclimatology 25*: PA4206. https://doi.org/10.1029/2009PA001846

Boetius, A. and Suess, E. (2004). Hydrate Ridge: a natural laboratory for the study of microbial life fueled by methane from near-surface gas hydrates. *Chemical Geology 205*: 291–310. https://doi.org/10.1016/j.chemgeo.2003.12.034

Boetius, A. and Wenzhöfer, F. (2013). Seafloor oxygen consumption fuelled by methane from cold seeps. *Nature Geoscience 6* (9): 725–734. https://doi.org/10.1038/ngeo1926

Boetius, A., Ravenschlag, K., Schubert, C.J. et al. (2000). A marine microbial consortium apparently mediating anaerobic oxidation of methane. *Nature 407*: 623–626. https://doi.org/10.1038/35036572

Briggs, B.R., Inagaki, F., Morono, Y. et al. (2012). Bacterial dominance in subseafloor sediments characterized by methane hydrates. *FEMS Microbiology Ecology 81*: 88–98. https://doi.org/10.1111/j.1574-6941.2012.01311.x

Bright, M. and Bulgheresi, S. (2010). A complex journey: transmission of microbial symbionts. *Nature Reviews Microbiology 8*: 218–230. https://doi.org/10.1038/nrmicro2262.

Bright, M. and Lallier, F. H. (2010). The biology of vestimentiferan tubeworms. *Oceanography and Marine Biology: an Annual Review 48*: 213–266.

Bright, M., Plumb, C., Riavitza, L.A. et al. (2010). Epizooic metazoan meiobenthos associated with tubeworm and mussel aggregations from cold seeps of the northern Gulf of Mexico. *Deep Sea Research II: Tropical Studies in Oceanography 57*: 1982–1989. https://doi.org/10.1016/j.dsr2.2010.05.003

Broecker, W.S. and Peng, T.H. (1982). *Tracers in the Sea*, 306–311. New York: Eldigio.

Bull, A.T., Ward, A.C. and Goodfellow, M. (2000). Search and discovery strategies for biotechnology: the paradigm shift. *Microbiology and Molecular Biology Reviews 64* (3): 573–606. https://doi.org/10.1128/mmbr.64.3.573-606.2000

Burd, B.J. and Thomson, R.E. (1994). Hydrothermal venting at Endeavor Ridge – effect on zooplankton biomass throughout the water column. *Deep Sea Research Part I: Oceanography Research Papers 41*: 1407–1423. https://doi.org/10.1016/0967-0637 (94)90105-8

Bush, K. (2018). Past and present perspectives on β-lactamases. *Antimicrobial Agents and Chemotherapy 62* (10): e01076–18. https://doi.org/10.1128/AAC.01076-18

Bush, K. and Bradford, P.A. (2016). β-Lactams and β-lactamase inhibitors: an overview. *Cold Spring Harbor Perspectives in Medicine 6* (8): a025247. https://doi.org/10.1101/cshperspect.a025247

Butman, C.A., Carlton, J.T. and Palumbi, S.R. (1995). Whaling effects on deep-sea biodiversity. *Conservation Biology 9*: 462–464. https://doi.org/10.1046/j.1523-1739.1995.9020462.x

Case, D.H., Pasulka, A.L., Marlow, J.J. et al. (2015). Methane seep carbonates host distinct, diverse and dynamic microbial assemblages. *mBio 6*: e01348–01315. https://doi.org/ 10.1128/mBio.01348–15

Cavalier-Smith T. (1978). Nuclear volume control by nucleo-skeletal DNA, selection for cell volume and cell growth rate and the solution of the DNA C-value paradox. *Journal of Cell Science 34*: 247–78.

Cavalier-Smith, T. (1982). Skeletal DNA and the evolution of genome size. *Annual Review of Biophysics and Bioengineering 11*: 273–302.

Cavanaugh, C.M., Gardiner, S.L., Jones, M.L. et al. (1981). Prokaryotic cells in the hydrothermal vent tube worm *Riftia pachyptila* Jones: possible chemoautotrophic symbionts. *Science 213*: 340–342. https://doi.org/10.1126/science.213.4505.340

Chen, L., Merey, S., Pecher, I. et al. (2020). A review analysis of gas hydrate tests: engineering progress and policy trend. *Environmental Geotechnics 1–17*. https://doi.org/10.1680/jenge.19.00208

Chevaldonné, P., Desbruyères, D. and Le Haître, M. (1991). Time-series of temperature from three deep-sea hydrothermal vent sites. *Deep Sea Research Part A. Oceanographic Research Papers 38* (11): 1417–1430.

Chen, S.C., Musat, N., Lechtenfeld, O.J. et al. (2019). Anaerobic oxidation of ethane by archaea from a marine hydrocarbon seep. *Nature 568* (7750): 108–111. https://doi.org/10.1038/s41586–019–1063–0

Circu, M.L. and Aw, T.Y. (2010). Reactive oxygen species, cellular redox systems and apoptosis. *Free Radical Biology and Medicine 48*: 749–762. https://doi.org/10.1016/j.freeradbiomed.2009.12.022

Colín-García, M. (2016). "Hydrothermal vents and prebiotic chemistry: a review". *Boletín de la Sociedad Geológica Mexicana 68* (3): 599–620. doi:10.18268/BSGM2016v68n3a13

Conway, N., Kennicutt, II, M. and Van Dover, C.L. (1994). Stable isotopes in the study of marine chemosynthetic based ecosystems. In: *Stable Isotopes in Ecology and Environmental Sciences* (eds. K. Lajtha, and R. Michener), 158–186. Oxford: Blackwell Scientific.

Cordes, E., Hourdez, S., Predmore, B. et al. (2005). Succession of hydrocarbon seep communities associated with the long-lived foundation species *Lamellibrachia luymesi. Marine Ecology Progress Series 305*: 17–29. https://doi.org/10.3354/meps305017

Cordes, E.E., McGinley, M.P., Podowski, E.L. et al. (2008). Coral communities of the deep Gulf of Mexico. *Deep Sea Research Part I: Oceanographic Research Papers 55* (6): 777–787. https://doi.org/10.1016/j.dsr.2008.03.005

Cordes, E.E., Bergquist, D.C. and Fisher, C.R. (2009). Macro-ecology of Gulf of Mexico cold seeps. *Annual Review of Marine Science 1* (1): 143–168. https://doi.org/10.1146/annurev.marine.010908.163912

Cordes, E.E., Cunha, M.R., Galéron, J. et al. (2010). The influence of geological, geochemical and biogenic habitat heterogeneity on seep biodiversity. *Marine Ecology 31*: 51–65. https://doi.org/10.1111/j.1439–0485.2009.00334.x

Corliss, J.B., Dymond, J., Gordon, L.I. et al. (1979). Submarine thermal springs on the Galápagos rift. *Science 203*: 1073–1083. https://doi.org/1126/science.203.4385.1073

Cowen, J.P., Bertram, M.A., Wakeham, S.G. et al. (2001). Ascending and descending particle flux from hydrothermal plumes at Endeavour Segment, Juan de Fuca Ridge. *Deep Sea Research Part I: Oceanography Research Papers 48*: 1093–1120. https://doi.org/10.1016/S0967–0637 (00)00070–4

Craddock, C., Hoeh, W.R., Gustafson, R.G. et al. (1995a). Evolutionary relationships among deep-sea mytilids (Bivalvia: Mytilidae) from hydrothermal vents and cold-water methane/sulfide seeps. *Marine Biology 121*: 477–485. https://doi.org/10.1007/BF00349456

Craddock, C., Hoeh, W.R., Lutz, R.A. et al. (1995b). Extensive gene flow in the deep-sea hydrothermal vent mytilid *Bathymodiolus thermophilus. Marine Biology 124*: 137–146. https://doi.org/10.1007/BF00349155

Crist, D.T., Ausubel, J. and Wagoner, P.E. (2010). *First Census of Marine Life 2010: Highlights of a Decade of Discovery. A Publication of the Census of Marine Life*. Washington, DC: Census of Marine Life International Secretariat.

Dattagupta, S., Miles, L.L., Barnabei, M.S. et al. (2006). The hydrocarbon seep tubeworm *Lamellibrachia luymesi* primarily eliminates sulfate and hydrogen ions across its roots to conserve energy and ensure sulfide supply. *Journal of Experimental Biology 209*: 3795–3895. https://doi.org/10.1242/jeb.02413

Davies-Coleman, M.T. and Sunassee, S.N. (2012). Marine bio-prospecting in southern Africa. In: *Drug Discovery in Africa – Impacts of Geneomics Natural Products Traditional Medicines Insights into Medicinal Chemistry and Technology Platforms in Pursuit of New Drugs* (eds. K. Chibale, M. Davies-Coleman and C. Masimirembwa), 193–209. Berlin: Springer-Verlag.

Dawkins, R. (1982). *The Extended Phenotype: the Long Reach of the Gene*. Oxford University Press.

de Beer, D., Sauter, E., Niemann, H. et al. (2006). *In situ* fluxes and zonation of microbial activity in surface sediments of the Håkon Mosby Mud Volcano. *Limnology and Oceanography 51* (3): 1315–1331. https://doi.org/10.4319/lo.2006.51.3.1315

Denis, F., Jollivet, D. and Moraga, D. (1993). Genetic separation of two allopatric populations of hydrothermal snails *Alvinoconcha* spp. (Gastropoda) from two Southwestern Pacific back-arc basins. *Biochemical Systematics and Ecology 21*: 431–440.

Dobzhansky, T. (1973). Nothing in biology makes sense except in the light of evolution. *The American Biology Teacher 35* (3): 125–129.

Dubilier, N., Mülders, C., Ferdelman, T. et al. (2001). Endosymbiotic sulphate reducing and sulphide oxidizing bacteria in an oligochaete worm. *Nature 411*: 298–302. https://doi.org/10.1038/35077067

Duperron, S., Nadalig, T., Caprais, J.C. et al. (2005). Dual symbiosis in a *Bathymodiolus* sp. mussel from a methane seep on the Gabon continental margin (Southeast Atlantic): 16S rRNA phylogeny and distribution of the symbionts in gills. *Applied and Environmental Microbiology 71*: 1694–1700.

Dwiono, S.A.P., Moraga, D., Le Pennec, M. et al. (1989). Genetic variability of the Lucinidae: *Loripes lucinalis, Lucinella divaricata* and *Lucinoma borealis minor* (Molluca: Bivalvia) from Brittany, France. *Biochemical Systematics and Ecology 17*: 463–468. https://doi.org/10.1016/0305–1978(89)90025–2

Eckelbarger, K.J. and Watling, L. (1995). Role of phylogenetic constraints in determining reproductive patterns in deep-sea invertebrates. *Invertebrate Biology 114*: 256–269. https://doi.org/10.2307/3226880

Egger, M., Riedinger, N., Mogollón, J.M. et al. (2018). Global diffusive fluxes of methane in marine sediments. *Nature Geoscience 11* (6): 421–425.

Erickson, K.L., Macko, S.A. and Van Dover, C.L. (2009). Evidence for a chemoautotrophically based foodweb at inactive hydrothermal vents (Manus Basin). *Deep Sea Research II 56*: 1577–1585. https://doi.org/10.1016/j.dsr2.2009.05.002

Ettwig, K.F., Butler, M.K., Le Paslier, D. et al. (2010). Nitrite-driven anaerobic methane oxidation by oxygenic bacteria. *Nature 464*: 543–548.

Ettwig, K.F., Zhu, B., Speth, D. et al. (2016). Archaea catalyze iron-dependent anaerobic oxidation of methane. *Proceedings of the National Academy of Sciences of the United States of America 113*: 12792–12796

Feng, D., Cheng, M., Kiel, S. et al. (2015). Using *Bathymodiolus* tissue stable carbon, nitrogen and sulphur isotopes to infer biogeochemical process at a cold seep in the South China Sea. *Deep Sea Research II: Topical Studies in Oceanography 4*: 52–59. https://doi.org/10.1016/j.dsr.2015.06.011

Feng, D., Qiu, J.W., Hu, Y. et al. (2018). Cold seep systems in the South China Sea: an overview. *Journal of Asian Earth Science 168*: 3–16.

Fisher, C.R. (1990). Chemoautotrophic and methanotrophic symbioses in marine invertebrates. *Reviews in Aquatic Science 2*: 399–436.

Fisher, C.R., Urcuyo, I.A., Simpkins, M.A. et al. (1997). Life in the slow lane: growth and longevity of cold seep vestimentiferans. *Marine Ecology 18*: 229–246. https://doi.org/10.1111/j.1439-0485.1997.tb00428.x

Fisher, C.R., Takai, K. and Le Bris, N. (2007). Hydrothermal vent ecosystem. *Oceanography 20* (1) 14–23. https://doi.org/10.5670/oceanog.2007.75

Foley, N.S., van Rensburg, T.M. and Armstrong, C.W. (2010) The ecological and economic value of cold-water coral ecosystems. *Ocean and Coastal Management 53* (7): 313–326. https://doi.org/10.1016/j.ocecoaman.2010.04.009

Foucher, J.P., Westbrook, G.K., Boetius, A. et al. (2009) Structure and drivers of cold seep ecosystems. *Oceanography 22* (1): 92–109. https://doi.org/10.5670/oceanog.2009.11.

France, S.C., Hessler, R.R. and Vrijenhoek, R.C. (1992). Genetic differentiation between spatially-disjunct populations of the deep-sea, hydrothermal vent-endemic amphipod *Ventiellasulfuris*. *Marine Biology 114*: 551–559.

Georgieva, M.N., Little, C.T.S., Watson, J.S. et al. (2019). Identification of fossil worm tubes from Phanerozoic hydrothermal vents and cold seeps. *Journal of Systematic Palaeontology 17* (4): 287–329. https://doi.org/10.1080/14772019.2017.1412362

German, C.R., Ramirez-Llodra, E., Baker, M.C. et al. (2011). Deep-water chemosynthetic ecosystem research during the census of marine life decade and beyond: a proposed deep-ocean road map. *PLoS ONE 6*: e23259. https://doi.org/10.1371/journal.pone.0023259

Girguis, P.R., Childress, J.J., Freytag, J.K. et al. (2002). Effects of metabolite uptake on proton-equivalent elimination by two species of deep-sea vestimentiferan tubeworm, *Riftia pachyptila* and *Lamellibrachia luymesi*: proton elimination is a necessary adaptation to sulfide-oxidizing chemoautotrophic symbionts. *Journal of Experimental Biology 205*: 3055–3066.

Godet, L., Zelnio, K.A. and Van Dover, C.L. (2011). Scientists as stakeholders in conservation of hydrothermal vents. *Conservation Biology 25*: 214–222. https://doi.org/10.1111/j.1523-1739.2010.01642.x

Goedert, J.L., Squires, R.L. and Barnes, L.G. (1995). Paleoecology of whale-fall habitats from deepwater Oligocene rocks, Olympic Peninsula, Washington State. *Palaeogeography Palaeoclimatology Palaeoecology 118*: 151–158. https://doi.org/10.1016/0031-0182 (94)00139-Y

Goffredi, S.K., Childress, J.J., Lallier, F.H. et al. (1999). The ionic composition of the hydrothermal vent tubeworm *Riftia pachyptila*: evidence for the elimination of SO_4^{2-} and H^+ and for a Cl^-/HCO_3^- shift. *Physiological and Biochemical Zoology 72*: 296–306. https://doi.org/10.1086/316664

Goldberg, D.S., Takahashi, T. and Slagle, A.L. (2008). Carbon dioxide sequestration in deep-sea basalt. *Proceedings of the National Academy of Sciences of the United States of America 105* (29): 9920–9925. https://doi.org/10.1073/pnas.0804397105

Gregory, T.R. and Hebert, P.D.N. (1999). The modulation of DNA content: Proximate causes and ultimate consequences. *Genome Research 9*: 317–324. https://doi.org/10.1101/gr.9.4.317

Gregory, T.R. (2001). Coincidence coevolution or causation? DNA content cell size and the C-value enigma. *Biological Reviews 76* (1): 65–101. https://doi.org/10.1017/S1464793100005595

Haroon, M.F., Hu, S., Shi, Y. et al. (2013). Anaerobic oxidation of methane coupled to nitrate reduction in a novel archaeal lineage. *Nature 500*: 567–570.

Hessler, R.R., Smithey, W.M., Boudrias, M.A. et al. (1988). Temporal change in megafauna at the Rose Garden hydrothermal vent (Galapagos Rift; eastern tropical Pacific). *Deep Sea Research Part I: Oceanography Research Papers 35*: 1681–1709. https://doi.org/10.1016/0198-0149 (88)90044-1

Heyl, T.P., Gilhooly, W.P., Chambers, R.M. et al. (2007). Characteristics of vesicomyid clams and their environment at the Blake Ridge cold seep, South Carolina, USA. *Marine Ecology Progress Series 339*: 169–184. https://doi.org/10.3354/meps339169

Hinrichs, K.U., Summons, R.E., Orphan, V. et al. (2000). Molecular and isotopic analysis of anaerobic methane-oxidizing communities in marine sediments. *Organic Geochemistry 31*: 1685–1701. https://doi.org/10.1016/S0146-6380 (00)00106-6

Hoehler, T.M., Alperin, M.J., Albert, D.B. et al. (1994). Field and laboratory studies of methane oxidation in an anoxic marine sediment: evidence for a methanogen-sulfate reducer consortium. *Global Biogeochemical Cycles 8*: 451–464.

Hong, W.L., Sauer, S., Panieri, G. et al. (2016). Removal of methane through hydrological, microbial and geochemical processes in the shallow sediments of pockmarks along eastern Vestnesa Ridge (Svalbard). *Limnology Oceanography 61*: S324–S343.

Hovland, M. and Thomsen, E. (1997). Cold-water corals – are they hydrocarbon seep related? *Marine Geology 137* (1–2): 159–164. https://doi.org/10.1016/S0025-3227 (96)00086-2

Inter-Ridge (2006). *Statement of Commitment to Responsible Research Practices at Deep-sea Hydrothermal Vents*. Kiel: Underwater Mining Institute.

Jelmert, A. and Oppen-Bernsten, D.O. (1996). Whaling and deep-sea biodiversity. *Conservation Biology 10*: 653–654. https://doi.org/10.1046/j.1523–1739.1996.10020653.x

Jollivet, D., Desbruykres, D., Ladrat, C. et al. (1995). Genetic differentiation of deep-sea hydrothermal vent alvinellid populations (Annelida: Polychaeta) along the East Pacific Rise. *Heredity 74*: 376–391.

Jones, D.S., Flood, B.E. and Bailey, J.E. (2015). Metatranscriptomic insights into polyphosphate metabolism in marine sediments. *The ISME Journal 10*: 1015–1019.

Jørgensen, B.B. and Boetius, A. (2007). Feast and famine – microbial life in the deep-sea bed. *Nature Reviews Microbiology 5*: 770–781. https://doi.org/10.1038/nrmicro1745

Joye, S.B. (2020). The geology and biogeochemistry of hydrocarbon seeps. *Annual Review of Earth and Planetary Sciences 48*: 205–231

Joye, S.B., Boetius, A., Orcutt, B.N. et al. (2004). The anaerobic oxidation of methane and sulfate reduction in sediments from Gulf of Mexico cold seeps. *Chemical Geology 205*: 219–238.

Joye, S.B., Bowles, M.W., Samarkin, V.A. et al. (2010). Biogeochemical signatures and microbial activity of different cold seep habitats along the Gulf of Mexico lower slope. *Deep Sea Research Part II 57*: 1990–2001.

Judd, A. and Hovland, M. (2007). *Seabed Fluid Flow, 408*. Cambridge: Cambridge University Press.

Judd, A., Jukes, V. and Leddra, M. (2002). MAGIC: a GIS database of marine gas seeps and seep indicators. *Geologiya I Geofizika 43* (7): 599–604.

Juniper, S.K., Bird, D.F., Summit, M. et al. (1998). Bacterial and viral abundances in hydrothermal event plumes over northern Gorda Ridge. *Deep Sea Research Part II 45*: 2739–2749. https://doi.org/10.1016/S0967–0645(98)00091–5

Karl, S.A., Schutz, S.J., Desbruykres, D. et al. (1996). Molecular analysis of gene flow in the hydrothermal-vent clam *Calyptogena magnifica*. *Molecular Marine Biology and Biotechnology 5*: 193–202.

Kato, S., Hashimoto, K. and Watanabe, K. (2012). Microbial interspecies electron transfer via electric currents through conductive minerals. *Proceedings of the National Academy of Sciences of the United States of America 109*: 10042–10046.

Kennicutt, M.C. (2017). Oil and gas seeps in the Gulf of Mexico. In: *Habitats and Biota of the Gulf of Mexico: Before the Deepwater Horizon Oil Spill* (ed. C. Ward), 275–358. New York: Springer.

Kiel, S. (eds). (2010). *The Vent and Seep Biota: Aspects from Microbes to Ecosystems*, Vol. *33*. Dordrecht: Springer. https://doi.org/10.1007/978-90-481-9572-5

Killingley, J.S., Berger, W.H., MacDonald, K.C. et al. (1980). 18O/16O variations in deep-sea carbonate shells from the Rise hydrothermal field. *Nature 287*: 218–221. https://doi.org/10.1038/287218a0

Kimura, M. and Weiss, W.H. (1964). The stepping stone model of genetic structure and the decrease of genetic correlation with distance. *Genetics 49*: 561–576.

Klose, J., Polz, M.F., Wagner, M. et al. (2015). Endosymbionts escape dead hydrothermal vent tubeworms to enrich the free-living population. *Proceedings of the National Academy of Sciences 112* (36): 11300–11305.

Klose, J., Aistleitner, K., Horn, M. et al. (2016). Trophosome of the deep-sea tubeworm *Riftia pachyptila* inhibits bacterial growth. *PLoS ONE 11* (1): e0146446. https://doi.org/10.1371/journal.pone.0146446

Koehn, R.K. and Bayne, B.L. (1989). Towards a physiological and genetical understanding of the energetics of the stress response. *Biological Journal of the Linnean Society 37*: 157–171. https://doi.org/10.1111/j.1095–8312.1989.tb02100.x

Kojima, S. (2002). Deep-sea chemoautosynthesis-based communities in the northwestern Pacific. *Journal of Oceanography 58*: 343–363. https://doi.org/10.1023/A:1015869927288

Kojima, S., Fukikura, K. and Okutani, T. (2004). Multiple trans-Pacific migrations of deep-sea vent/seep-endemic bivalves in the family Vesicomyidae. *Molecular Phylogenetics and Evolution 32*: 396–406. https://doi.org/10.1016/j.ympev.2004.02.016

Krach, M. (2013). EEZ seeps: exploration, exploitation and conservation EEZ seeps exploration, exploitation and conservation. UC San Diego, Center for Marine Biodiversity and Conservation. Retrieved from https://escholarship.org/uc/item/3qc625zw

Lampadariou, N., Kalogeropoulou, V., Sevastou, K. et al. (2013). Influence of chemosynthetic ecosystems on nematode community structure and biomass in the deep eastern Mediterranean Sea. *Biogeosciences 10* (8): 5381–5398. https://doi.org/10.5194/bg-10–5381–2013.

Lapham, L., Wilson, R., Riedel, M. et al. (2013). Temporal variability of *in situ* methane concentrations in gas hydrate bearing sediments near Bullseye Vent, Northern Cascadia Margin. *Geochemistry Geophysics Geosystems 14*: 2445–2459.

Levin, L., Ziebis, W., Mendoza, G. et al. (2003). Spatial heterogeneity of macrofauna at northern California methane seeps: influence of sulfide concentration and fluid flow. *Marine Ecology Progress Series 265*: 123–139. https://doi.org/10.3354/meps265123

Levin, L.A. (2005). Ecology of cold seep sediments: interactions of fauna with flow, chemistry and microbes. *Oceanography and Marine Biology – an Annual Review 43*: 1–46. https://doi.org/10.1201/9781420037449.ch1

Levin, L.A. and Michener, R. (2002). Isotopic evidence of chemosynthesis-based nutrition of macrobenthos: the lightness of being at Pacific methane seeps. *Limnology and Oceanography 47*: 1336–1345. https://doi.org/10.4319/lo.2002.47.5.1336

Levin, L.A. and Sibuet, M. (2012). Understanding continental margin biodiversity: a new imperative. *Annual Review of Marine Science 4*: 79–112. https://doi.org/10.1146/annurev-marine-120709–142714.

Levin, L.A., Etter, R.J., Rex, M.A. et al. (2001). Environmental Influences on regional deep-sea species diversity. *Annual Review of Ecology and Systematics 32* (1): 51–93. https://doi.org/10.1146/annurev.ecolsys.32.081501.114002

Levin, L.A., Mendoza, G.F., Gonzalez, J.P. et al. (2010). Diversity of bathyal macrofauna on the northeastern Pacific margin: the influence of methane seeps and oxygen minimum zones. *Marine Ecology 31* (1): 94–110.

Levin, L.A., Orphan, V.J., Rouse, G.W., et al. (2012). A hydrothermal seep on the Costa Rica margin: middle ground in a continuum of reducing ecosystems. *Proceedings of the Royal Society B: Biological Sciences 279* (1738): 2580–2588. https://doi.org/10.1098/rspb.2012.0205

Levin, L.A., Baco, A.R., Bowden, D.A., et al. (2016). Hydrothermal vents and methane seeps: rethinking the sphere of influence. *Frontiers in Marine Science 3*: 72. https://doi.org/10.3389/fmars.2016.00072

Lorion, J., Halary, S., Nascimento, J.et al. (2012) Evolutionary history of Idas sp. Med (Bivalvia: Mytilidae): a cold seep mussel bearing multiple symbionts. *CBM-Cahiers de Biologie Marine 53* (1): 77.

Luff, R., Greinert, J., Wallmann, K. et al. (2005). Simulation of long-term feedbacks from authigenic carbonate crust formation at cold vent sites. *Chemical Geology 216*: 157–174.

Luo, M., Dale, A.W., Haffert, L. et al. (2016). A quantitative assessment of methane cycling in Hikurangi Margin sediments (New Zealand) using geophysical imaging and biogeochemical modeling. *Geochemistry Geophysics Geosystems 17*: 4817–4835.

MacArthur, R. and Wilson, E.O. (1967). *The Theory of Island Biogeography, 215*. Princeton, NJ: Princeton University Press.

Marris, E. (2006). Drugs from the deep. *Nature 443* (7114): 904–905. https://doi.org/10.1038/443904a

Martin, R.A., Nesbitt, E.A. and Cambell, K.A. (2010). The effects of anaerobic methane oxidation on benthic foraminiferal assemblages and stable isotopes on the Hikurangi Margin of eastern New Zealand. *Marine Geology 272*: 270–284. https://doi.org/10.1016/j.margeo.2009.03.024

Maxwell, S. (2005). An aquatic pharmacy: the biomedical potential of the deep sea. *Journal of Marine Education 21* (4): 31–32.

Mazumdar, A., Dewangan, P., Peketi, A. et al. (2019). The first record of active methane (cold) seep ecosystem associated with shallow methane hydrate from the Indian EEZ. *Journal of Earth System Science 128*: 18. https://doi.org/10.1007/s12040-018-1044-y

McGlynn, S.E., Chadwick, G.L., Kempes, C.P. et al. (2015). Single cell activity reveals direct electron transfer in methanotrophic consortia. *Nature 526*: 531–534.

Mills, H.J., Hodges, C., Wilson, K. et al. (2003). Microbial diversity in sediments associated with surface-breaching gas hydrate mounds in the Gulf of Mexico. *FEMS Microbiology Ecology 46*: 39–52. https://doi.org/10.1016/S0168-6496 (03)00191-0.

Minic, Z. and Thongbam, P.D. (2011). The biological deep sea hydrothermal vent as a model to study carbon dioxide capturing enzymes. *Marine Drugs 9* (5): 719. https://doi.org/10.3390/MD9050719

Molinski, T.F., Dalisay, D.S., Lievens, S.L. et al. (2009). Drug development from marine natural products. *Nature Reviews Drug Discovery 8* (1): 69–85. https://doi.org/10.1038/nrd2487

Mone, Y., Monnin, D. and Kremer, N. (2014). The oxidative environment: a mediator of interspecies communication that drives symbiosis evolution. *Proceedings of the Royal Society B: Biological Sciences 281*: 20133112–20133112. https://doi.org/10.1098/rspb.2013.3112

Moran, N.A. (1996). Accelerated evolution and in endosymbiotic bacteria. *Proceedings of the National Academy of Sciences of the United States of America 93* (7): 2873–2878.

Nevo, E., Kim, Y.J., Shaw, C.R. et al. (1974). Genetic variation, selection and speciation in *Thomomys talpoides* pocket gophers. *Evolution 28*: 1–23.

Niemann, H. and Elvert, M. (2008). Diagnostic lipid biomarker and stable carbon isotope signatures of microbial communities mediating the anaerobic oxidation of methane with sulphate. *Organic Geochemistry 39*: 1668–1677. https://doi.org/10.1016/j.orggeochem.2007.11.003

Niemann, H., Linke, P., Knittel, K. et al. (2013). Methane-carbon flow into the benthic food web at cold seeps – a case study from the Costa Rica subduction zone. *PloS ONE 8* (10): e74894. https://doi.org/10.1371/journal.pone.0074894

Niu, M., Fan, X., Zhuang, G. et al. (2017). Methane-metabolizing microbial communities in sediments of the Haima cold seep area, northwest slope of the South China Sea. *FEMS Microbiology Ecology 93* (9). https://doi:10.1093/femsec/fix101

Orphan, V.J., Hinrichs, K.U., Ussler, 3rd, W. et al. (2001a). Comparative analysis of methane-oxidizing archaea and sulfate-reducing bacteria in anoxic marine sediments. *Applied and Environmental Microbiology 64*: 1922–1934.

Orphan, V.J., House, C.H., Hinrichs, K.U. et al. (2001b). Methane-consuming archaea revealed by directly coupled isotopic and phylogenetic analysis. *Science 293*: 484–487.

Ortmann, A. and Suttle, C. (2005). High abundances of viruses in a deep-sea hydrothermal vent system indicates viral mediated microbial mortality. *Deep Sea Research Part I: Oceanography Research Papers 52*: 1515–1527. https://doi.org/10.1016/j.dsr.2005.04.002

Panieri, G. (2006). The effect of shallow marine hydrothermal vent activity on benthic foraminifera (Aeolian arc, Tyrrhenian Sea). *Journal of Foraminiferal Research 36*: 1. https://doi.org/10.2113/36.1.3

Pape, E., Campinas Bezerra, T.N., Vanneste, H. et al. (2011). Community structure and feeding preference of nematodes associated with methane seepage at the Darwin mud volcano (Gulf of Cadiz). *Marine Ecology Progress Series 438*: 71–83. https://doi.org/10.3354/meps09278

Paull, C.K., Hecker, B., Commeau, R. et al. (1984). Biological communities at the Florida escarpment resemble hydrothermal vent taxa. *Science 226*: 965–967.

Paull, C.K, Jull, A.J.T., Tolin, L.J. et al. (1985). Stable isotope evidence for chemoautotrophy in an abyssal seep community. *Nature 317*: 709–711.

Paull, C.K., Caress, D.W., Thomas, H. et al. (2015). Seafloor geomorphic manifestations of gas venting and shallow subbottom gas hydrate occurrences. *Geosphere 11*: 491–513. https://doi.org/10.1130/GES01012.1

Pettit, R.K. (2011). Culturability and secondary metabolite diversity of extreme microbes: expanding contribution of deep sea and deep-sea vent microbes to natural product discovery. *Marine Biotechnology 13* (1): 1–11. https://doi.org/10.1007/s10126-010-9294-y

Pflugfelder, B., Cary, S.C. and Bright, M. (2009). Dynamics of cell proliferation and apoptosis reflect different life strategies in hydrothermal vent and cold seep vestimentiferan tubeworms. *Cell and Tissue Research 337* (1): 149–165. https://doi.org/10.1007/s00441-009-0811-0.

Piquet, B., Shillito, B., Lallier, F.H. et al. (2019). High rates of apoptosis visualized in the symbiont-bearing gills of deep-sea *Bathymodiolus* mussels. *PLoS ONE 14* (2): e0211499. https://doi.org/10.1371/journal.pone.0211499

Pohlman, J.W., Bauer, J.E., Waite, W.F. et al. (2011). Methane hydrate-bearing seeps as a source of aged dissolved organic carbon to the oceans. *Nature Geoscience 4*: 37–41.

Quammen, D. (2004). *The Song of the Dodo: Island Biogeography in an Age of Extinctions*. New York: Scribner.

Quattrini, A.M., Nizinski, M.S., Chaytor, J.D. et al. (2015). Exploration of the canyon-incised continental margin of the northeastern United States reveals dynamic habitats and diverse communities. *PLoS ONE 10* (10): e0139904. https://doi.org/10.1371/journal.pone.0139904

Ramirez-Llodra, E., Freitga, K., Blanco, M. et al. (2005). *ChEssBase: an Online Information System on Biodiversity and Biogeography of Deep-Sea Fauna from Chemosynthetic Ecosystems*. Version 2. World Wide Web Electronic Publications.

Ramirez-Llodra, E., Tyler, P.A., Baker, M.C. et al. (2011). Man and the last great wilderness: human impact on the deep sea. *PLoS ONE 6*: e22588. https://doi.org/10.1371/journal.pone.0022588.

Rathburn, A.E., Pérez, M.E., Martin, J.B. et al. (2003). Relationships between the distribution and stable isotopic composition of living benthic foraminifera and cold methane seep biogeochemistry in Monterey Bay, California. *Geochemistry Geophysics Geosystems 4* (12): 1106.

Rau, G.M. (1981). Hydrothermal vent clam and tubeworm 13C/"C. Further evidence of nonphotosynthetic food sources. *Science 213*: 338–340.

Rex, M.A. and Etter, R.J. (2010). *Deep-Sea Biodiversity: Pattern and Scale*. Cambridge, MA: Harvard University Press.

Robidart J.C., Bench S.R., Feldman R.A. et al. (2008). Metabolic versatility of the *Riftia pachyptila* endosymbiont revealed through metagenomics. *Environmental Microbiology 10* (3): 727–737.

Robinson, C.A., Bernard, J.M., Levin, L.A. et al. (2004). Surficial hydrocarbon seep infauna from the Blake Ridge (Atlantic Ocean, 2150 m) and the Gulf of Mexico (690–2240 m). PSZN: *Marine Ecology 25*: 313–336.

Rodrigues, C.F. (2009). *Macrofaunal Assemblages from Mud Volcanoes in the Gulf of Cadiz*. (Doctoral dissertation, 314 pp + annexes). Aveiro: Universida de de Aveiro.

Rodrigues, C.F., Hilario, A. and Cunha, M.R. (2013). Chemosymbiotic species from the Gulf of Cadiz (NE Atlantic): distribution, lifestyles and nutritional patterns. *Biogeosciences 10*: 2569–2581. https://doi.org/10.5194/bg-10–2569–2013

Roeselers, G. and Newton, I.L.G., (2012) On the evolutionary ecology of symbioses between chemosynthetic bacteria and bivalves. *Applied Microbiology Biotechnology 94* (1): 1–10. https://doi.org/10.1007/s00253–011–3819–9

Sahling, H., Rickert, D., Lee, R.W. et al. (2002). Macrofaunal community structure and sulfide flux at gas hydrate deposits from the Cascadia convergent margin, NE Pacific. *Marine Ecology Progress Series 231*: 121–138.

Santos, R.S., Morato, T. and Barriga, F.J.A.S. (2012). Increasing pressure at the bottom of the ocean. In: *Natural Resources Sustainability and Humanity: a Comprehensive View* (eds. A. Mendonça, A. Cunha and R. Chakrabarti), 69–81. Springer. https://doi.org/10.1007/978–94–007–1321–5_5

Sellanes, J., Quiroga, E. and Gallardo, V.A. (2004). First direct evidences of methane seepage and associated chemosynthetic communities in the bathyal zone off Chile. *Journal of the Marine Biological Association of the United Kingdom 84*: 1065–1066.

Sen Gupta, B.K. (2007). Attachment of Foraminifera to vestimentiferan tubeworms at cold seeps: refuge from seafloor hypoxia and sulfide toxicity. *Marine Micropaleontology 62* (1): 1–6.

Sen, A., Åström, E.K.L., Hong, W. et al. (2018). Geophysical and geochemical controls on the megafaunal community of a high Arctic cold seep. *Biogeosciences 15*: 4533–4559. https://doi.org/10.5194/bg-15–4533–2018

Sibuet, M. and Olu, K.R. (1998). Biogeography, biodiversity and fluid dependence of deep-sea cold-seep communities at active and passive margins. *Deep Sea Research Part II: Topical Studies in Oceanography 45*: 517–567. https://doi.org/10.1016/S0967–0645 (97)00074-X

Sibuet, M. and Olu, K.R. (2002). Cold seep communities on continental margins: structure and quantitative distribution relative to geological and fluid venting patterns. In: *Ocean Margin Systems* (eds. G. Wefer, D. Billett, D. Hebbeln et al.), 235–251. Berlin Heidelberg: Springer-Verlag.

Skarke, A., Ruppel, C., Kodis, M. et al. (2014). Widespread methane leakage from the seafloor on the northern US Atlantic margin. *Nature Geoscience 7*: 657–661. https://doi.org/10.1038/ngeo2232

Smith, C., Kukert, H., Wheatcroft, R. et al. (1989). Vent fauna on whale remains. *Nature 341*: 27–28. https://doi.org/10.1038/341027a0

Smith, M.J. (1982). *Evolution and the Theory of Games*. Cambridge University Press. https://doi.org/10.1017/CBO9780511806292

Snyder, G.T., Hiruta, A., Matsumoto, R. et al. (2007). Pore water profiles and authigenic mineralization in shallow marine sediments above the methane-charged system on Umitaka Spur, Japan Sea. *Deep Sea Research Part II 54*: 1216–1239.

Sommer, S., Pfannkuche, O., Linke, P. et al. (2006). Efficiency of the benthic filter: biological control of the emission of dissolved methane from sediments containing shallow gas hydrates at Hydrate Ridge. *Global Biogeochemical Cycles 20* (2): GB2019. https://doi.org/10.1029/2004GB002389

Sommer, S., Gutzmann, E. and Pfannkuche, O. (2007). Sediments hosting gas hydrates: oases for metazoan meiofauna. *Marine Ecology Progress Series 337*: 27–37.

Sommer, S., Linke, P., Pfannkuche, O. et al. (2010). Benthic respiration in a novel seep habitat dominated by dense beds of ampharetid polychaetes at the Hikurangi Margin (New Zealand). *Marine Geology 272*: 223–232.

Song, J.S., Jeon, J.H., Lee, J.H. et al. (2005). Molecular characterization of TEM-type beta-lactamases identified in cold-seep sediments of Edison Seamount (south of Lihir Island, Papua New Guinea). *Journal of Microbiology 43* (2): 172–178.

Southward, A.J. (1989). Animal communities fuelled by chemosynthesis: life at hydrothermal vents, cold seeps and in reducing sediments. *Journal of Zoology 217*: 705–709.

Southward, A.J. and Newman, W.A. (1998). Ectosymbiosis between filamentous sulphur bacteria and a stalked barnacle (*Scalpellomorpha*, Neolepadinae) from the Lau Back Arc Basin, Tonga. *Cahiers de Biologie Marine 39*: 259–262.

Southward, E.C. (1986). Development of the gut and segmentation of newly settled stages of *Ridgeia* (Vestimentifera): implications for relationship between Vestimentifera and Pogonophora. *Journal of the Marine Biological Association of the United Kingdom 68* (3): 465–487. https://doi.org/10.1017/S0025315400043344.

Southward, E.C., Tunnicliffe, V., Black, M.B. et al. (1996). Ocean-ridge segmentation and vent tubeworms (Vestimentifera) in the NE Pacific. In: *Tectonic Magmatic Hydrothermal and Biological Segmentation of Mid-ocean Ridges. Special Publication 118* (eds. C.J. MacLeod, P.A. Tyler and C.L. Walker), 211–224. London: Geological Society.

Spiess, F.N., MacDonald, K.C., Atwater, T. et al. (1980). East pacific rise: hot springs and geophysical experiments. *Science 207*: 1421–1433.

Squires, R.L., Goedert, J.L. and Barnes, L.G. (1991). Whale carcasses. *Nature 349*: 574.

Stat, M., Huggett, M.J., Bernasconi, R. et al. (2017). Ecosystem biomonitoring with eDNA: metabarcoding across the tree of life in a tropical marine environment. *Scientific Reports 7*: 12240. https://doi.org/10.1038/s41598–017–12501–5

Suess, E. (2014) Marine cold seeps and their manifestations: geological control, biogeochemical criteria and environmental conditions. *International Journal of Earth Science 103*: 1889–1916.

Takai, K. and Nakamura, K. (2010). Compositional, physiological and metabolic variability in microbial communities associated with geochemically diverse, deep-sea hydrothermal vent fluids. In: *Geomicrobiology: Molecular and Environmental Perspective* (eds. L.L. Barton, M. Mandl and A. Loy), 251–283. Dordrecht: Springer.

Talukder, A.R. (2012). Review of submarine cold seep plumbing systems: leakage to seepage and venting: seeps plumbing system. *Terra Nova 24*: 255–272.

Thurber, A.R., Kroger, K., Neira, C. et al. (2010). Stable isotope signatures and methane use by New Zealand cold seep benthos. *Marine Geology 272*: 260–269.

Tunnicliffe, V. and Fowler, M. R. (1996). Influence of sea-floor spreading on the global hydrothermal vent fauna. *Nature 379*: 531–533.

Tunnicliffe, V. and Jensen, R.G. (1987). Distribution and behaviour of the spider crab *Macroregonia macrochira* Sakai (Brachyura) around the hydrothermal vents of the northeast Pacific. *Canadian Journal of Zoology 65*: 2443–2449. https://doi.org/10.1139/z87–369

Tunnicliffe, V. and Juniper, S.K. (1990). Cosmopolitan underwater fauna. *Nature 344*: 300.

Tunnicliffe, V., Juniper, S.K. and Sibuet, M. (2003). Reducing environments of the deep seafloor. In: *Ecosystems of the Deep Oceans* (ed. P.A. Tyler), 81–110. Amsterdam, Elsevier.

Vacelet, J., Boury-Esnault, N., Fiala-Medioni, A. et al. (1995). A methanotrophic carnivorous sponge. *Nature 377*: 296.

Valentine, D.L. (2011). Emerging topics in marine methane biogeochemistry. *Annual Review of Marine Science 3* (1): 147–171. https://doi.org/10.1146/annurev-marine-120709–142734

Van Dover, C.L. (2000). *The Ecology of Deep-Sea Hydrothermal Vents*. Princeton, NJ: Princeton University Press.

Van Dover, C.L. and Fry, B. (1989). Stable isotope composition of hydrothermal vent organisms. *Marine Biology 102*: 257–263.

Van Dover, C.L., Aharon, P., Bernhard, J.M. et al. (2003). Blake Ridge methane seeps: characterization of a soft-sediment, chemosynthetically based ecosystem. *Deep Sea Research I: Oceanographic Research Papers 50*: 281–300.

Van Gaever, S., Moodley, L., de Beer, D. et al. (2006). Meiobenthos at the Arctic Hakon Mosby Mud Volcano, with a parental-caring nematode thriving in sulphide-rich sediments. *Marine Ecology Progress Series 321*: 143–155. https://doi.org/10.3354/meps321143.

Van Gaever, S., Galeron, J., Sibuet, M. et al. (2009). Deep-sea habitat heterogeneity influence on meiofaunal communities in the Gulf of Guinea. *Deep Sea Research Part II: Topical Studies in Oceanography 56* (23): 2259–2269. https://doi.org/10.1016/j.dsr2.2009.04.008.

Vanreusel, A., Andersen, A.C., Boetius, A. et al. (2009). Biodiversity of cold seep ecosystems along the European margins. *Oceanography 22*: 110–127.

Vanreusel, A., Fonseca, G., Danovaro, R. et al. (2010). The contribution of deep-sea macrohabitat heterogeneity to global nematode diversity. *Marine Ecology 31* (1): 6–20. https://doi.org/10.1111/j.1439–0485.2009.00352.x.

Vereshchaka, A. and Vinogradov, G. (1999). Visual observations of the vertical distribution of plankton throughout the water column above Broken Spur vent field, Mid-Atlantic Ridge. *Deep Sea Research Part I: Oceanography Research Papers 46*: 1615–1632. https://doi.org/10.1016/S0967–0637(99)00018–7

Vigneron, A., Cruaud, P., Pignet, P. et al. (2013). Archaeal and anaerobic methane oxidizer communities in the Sonora Margin cold seeps, Guaymas Basin (Gulf of California). *ISME Journal 7*: 1595–1608.

Vigneron, A., Masson, F., Vallier, A. et al. (2014). Insects recycle endosymbionts when the benefit is over. *Current Biology 24*: 2267–2273.

Vrijenhoek, R.C., Lutz, R.A., Craddock, C. et al. (1993). Gene flow and speciation in deep-sea hydrothermal vent organisms. In: *Evolution* 93(ed. F.M. Catzeflis), 484. Fourth Congress of the European Society for Evolutionary Biology, Montpellier, France.

Wang, X.C., Chen, R.F., Whelan, J. et al. (2001). Contribution of "Old" carbon from natural marine hydrocarbon seeps to sedimentary and dissolved organic carbon pools in the Gulf of Mexico. *Geophysical Research Letters 28*: 3313–3316.

Ward, ME., Shields, J.D., Van Dover, C.L. (2004). Parasitism in species of *Bathymodiolus* (Bivalvia: Mytilidae) mussels from deep-sea seep and hydrothermal vents. *Disease in Aquatic Organisms 62* (1–2): 1–16. doi: 10.3354/dao062001.

Wegener, G., Krukenberg, V., Riedel, D. et al. (2015). Intercellular wiring enables electron transfer between methanotrophic archaea and bacteria. *Nature 526*: 587–590.

Wheeler, A. J., and Stadnitskaia, A. (2011). Benthic deep-sea carbonates: reefs and seeps. *Developments in Sedimentology 63*: 397–455.

Winogradsky, S. (1890). Recherches sur les organisms de la nitrification. *Annales de l'Institut Pasteur 4*: 213–231.

Wright, S. (1931). Evolution in Mendelism populations. *Genetics 16*: 97–159.

Wright, S. (1969). *Evolution and the Genetics of Populations, Vol. 2: The Theory of Gene Frequencies*. Chicago: University of Chicago Press,.

Wyngaard, G.A. and Gregory, T.R. (2001). Temporal control of DNA replication and the adaptive value of chromatin diminution in Copepods. *Journal of Experimental Zoology 291*: 310–316.

Xu, C., Wu, N., Sun, Z. et al. (2018). Methane seepage inferred from pore water geochemistry in shallow sediments in the western slope of the Mid-Okinawa Trough. *Marine and Petroleum Geology 98*: 306–315.

Yin, Z. and Linga, P. (2019). Methane hydrates: a future clean energy resource. *Chinese Journal of Chemical Engineering 27* (9): 2026–2036. https://doi.org/10.1016/j.cjche.2019.01.005

Zachos, J.C., Röhl, U., Schellenberg, S.A. et al. (2005). Rapid acidification of the ocean during the Paleocene–Eocene thermal maximum. *Science 308*: 1611–1615.

Zeppilli, D., Mea, M., Corinaldesi, C. et al. (2011). Mud volcanoes in the Mediterranean Sea are hot spots of exclusive meiobenthic species. *Progress in Oceanography 91* (3): 260–272. https://doi.org/10.1016/j.pocean.2011.01.001.

Zielinski, F.U., Pernthaler, A., Duperron, S. et al. (2009). Widespread occurrence of an intranuclear bacterial parasite in vent and seep bathymodiolin mussels. *Environmental Microbiology 11* (5): 1150–1167.

14

Biogeochemical Characteristics of Hydrothermal Systems in the Indian Ocean

L. Surya Prakash[1,*], Sheryl Oliveira Fernandes[1,2], Baban Ingole[1], and John P. Kurian[1]

ABSTRACT

The Indian mid-oceanic ridge system is under explored for hydrothermal vent fields. Most vent fields are discovered along the slow to intermediate spreading Central Indian and ultra-slow spreading Southwest Indian ridges (CIR and SWIR). Detailed geological studies of Dodo, Solitaire, Edmond, Kairei, and Longqi vent fields and their fluids characteristics show that they are basalt-hosted systems. However, high hydrogen and methane concentrations in Kairei and Longqi fluids indicate fluid circulation through mafic–ultramafic lithologies. Abundant populations of (hyper)thermophilic hydrogenotrophic chemolithoautotrophs are associated with Dodo, Solitaire, and Kairei fluids. The Fe/Mn-oxidizing chemolithoautotrophs belonging to *Zetaproteobacteria* or various metal-tolerant genera within class Alphaproteobacteria and Gammaproteobacteria thrive on metal-rich Edmond and Longqui vent fluids. Therefore, the chemical composition of fluid and associated biosphere depend mainly on subsurface water–rock reactions and are independent of the spreading rate of the ridge. A very unusual animal community such as (i) vent crab *Austinograea rodriguezensis*, (ii) vent shrimps (*Rimicaris kairei*, and *Mirocaris indica*), (iii) deep-sea mussel (*Bathymodiolus marisindicus*), (iv) scaly-foot gastropods (*Chrysomallon squamiferum*, *Alviniconcha* sp.), and (v) barnacle (*Neolepas* sp.) are found at Indian Ocean vent fields. Most of the vent-specific fauna representing vent sites at the Central and Southwest Indian Ridges are found at Kairei field (> 24% of the total), making it biologically diverse and supporting the hypothesis that the Rodriguez Triple Junction (RTJ) could be an essential connecting point for dispersal of larvae to the nearby vent fields. Further, the Southwest–Southeast Indian Ridges could act as a corridor for dispersal of vent fauna between Atlantic, Indian and Pacific Oceans. Interestingly, similar chemosynthetic vent fauna (associated with Galatheidae, Neolepadidae, and Mitilidae families) have also been found at cold-seep methane hydrate systems in Indian continental margins. Detailed genomic studies of cold-seep community and its comparison with vent community and water column hydrothermal tracer-based studies are required to establish the ecosystem connectivity.

14.1. INTRODUCTION

14.1.1. Hydrothermal Circulation: Evolution of Fluids and Associated Chemosynthetic Ecosystem

Hydrothermal circulation plays a vital role in the cycling of elements, energy, and mass between the solid Earth and the Ocean (Lang, 2018; Urabe et al., 2015). It mainly

[1]*ESSO-National Centre for Polar and Ocean Research, Ministry of Earth Sciences, Headland Sada, Vasco-da-Gama, Goa, 403804, India*

[2]*Present address: School of Earth, Ocean and Atmospheric Sciences, Goa University, Taleigao Plateau Goa, 403206, India*

**Corresponding Author: Surya Prakash (surya@ncpor.res.in) ORCiD code: 0000-0003-0830-2145*

Systems Biogeochemistry of Major Marine Biomes, First Edition. Edited by Aninda Mazumdar and Wriddhiman Ghosh.

occurs at the fractured volcanic mid-ocean ridge system that extends over ~ 6 × 10^4 km, in which cold seawater percolates down through cracks in the oceanic crust and becomes heated (German and Von Damm, 2003). The chemical composition of seawater is modified during its reaction with the subsurface host rock. The modified seawater becomes hot and extraordinarily buoyant and rises back to the seafloor. These expelled vent fluids have high enrichments of dissolved gases, such as helium (^3He), hydrogen (H_2), methane (CH_4), carbon dioxide (CO_2), trace metals (e.g., Fe, Mn, Cu, Ni, Zn, Pb), and depletions such as Mg and SO_4 compared with ambient seawater. Most of the dissolved chemicals in hydrothermal fluids precipitate as (i) polymetallic-sulfides within cracks of host rocks as vein deposits, as mounds, and as thick columns named chimneys, and (ii) polymetallic oxyhydroxides and silicates upon mixing with the cold seawater (2 °C).

The chemical compositions of the fluids provide information on the depth of fluid circulation, heat source, pressure, and temperature (PT) conditions of the reactions they have experienced during their transit (residence time of fluids) through the crust. Generally, it is well understood that the percolated seawater is modified by the varying PT conditions, water–rock interactions, and various biological processes within the crust. Water–rock interaction processes begin during the downflow leg of seawater in the recharge zone, and they are indistinguishably connected (Figure 14.1). During fluid circulation, the seawater is modified and evolves as hydrothermal fluids, which are reducing because of the presence of H_2S. It is acidic and gains many metals (e.g. Fe^{2+} and Mn^{2+}) and gases (e.g. H_2, CO_2, CH_4, and ^3He) and trace elements (e.g. cobalt, nickel, copper, zinc, aluminum, silver, barium, tungsten, gold, lead, and rare earth elements, REE) by leaching them out from the rocks.

Chemical compounds spewed out through vent fluids support various chemolithoautotrophic microorganisms, which form the base of the food web for numerous animals associated with deep-sea hydrothermal ecosystems. The chemical composition of vent fluids can be altered by chemolithoautotrophic microbes during subsurface circulation. However, the biological influence on overall chemical fluxes, how widespread they are, and which elements are affected remain to be resolved. Recent studies (TAIGA concept: Trans-crustal Advection and *In-situ* biogeochemical processes of Global sub-seafloor Aquifers) have shed some light on the role of biology in sub-seafloor hydrothermal circulation (Urabe et al., 2015).

The buoyant high-temperature fluids expelled to the cold stratified oceanic water column from the vent lead to the formation of hydrothermal plumes (Figure 14.2). They are important for heat and energy as they can act as a driving force for mid-depth oceanic circulation (Helfrich and Speer, 1995), gross chemical fluxes (German et al., 1991), and for transporting both chemical compounds and larvae of many faunal species from one adjacent vent site to another (Van Dover et al., 2002). As the hydrothermal plume rises in the water column, it becomes progressively less buoyant and produces turbulence and eddies, which entrains seawater. In this process, the original vent fluid gets diluted, and it cannot rise beyond a finite maximum height of approximately 300–400 m above the seafloor (Figure 14.2). At this stage, the diluted hydrothermal fluid begins to disperse laterally, which is known as a non-buoyant plume or neutrally buoyant plume, and represents both directions of gene flow along the global mid-oceanic ridges and biogeochemical processes within. Potential sources of microorganisms in hydrothermal plumes primarily include communities from ambient seawater, vent chimneys, near-vent animal

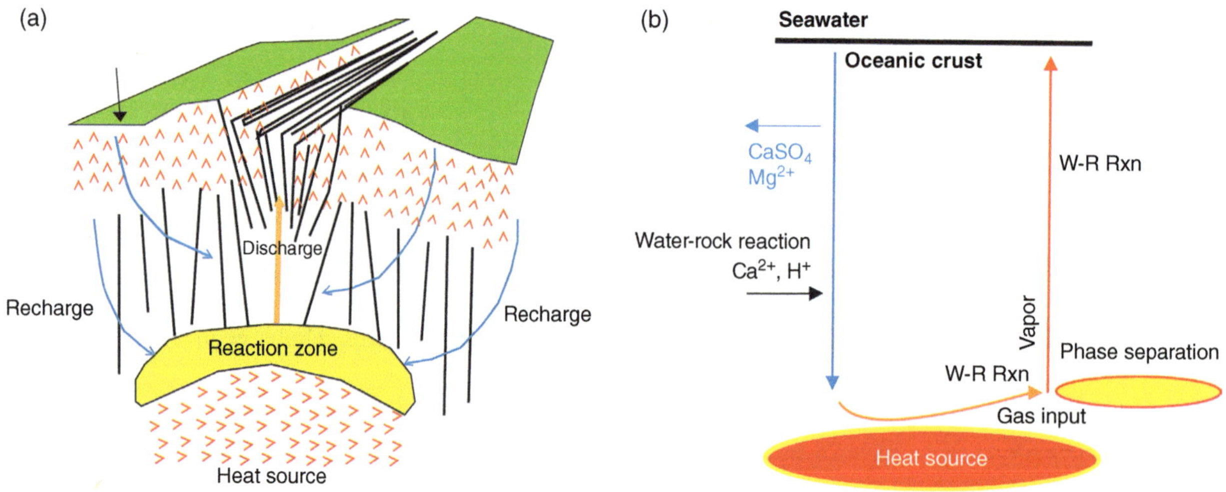

Figure 14.1 (a) Schematic illustration of three stages: recharge, reaction, and discharge of hydrothermal circulation (modified after German and Von Damm, 2003). (b) Schematic diagram shows the processes that control the chemical composition of hydrothermal vent fluid (after German and Von Damm, 2003). W-R RXN, water–rock reaction.

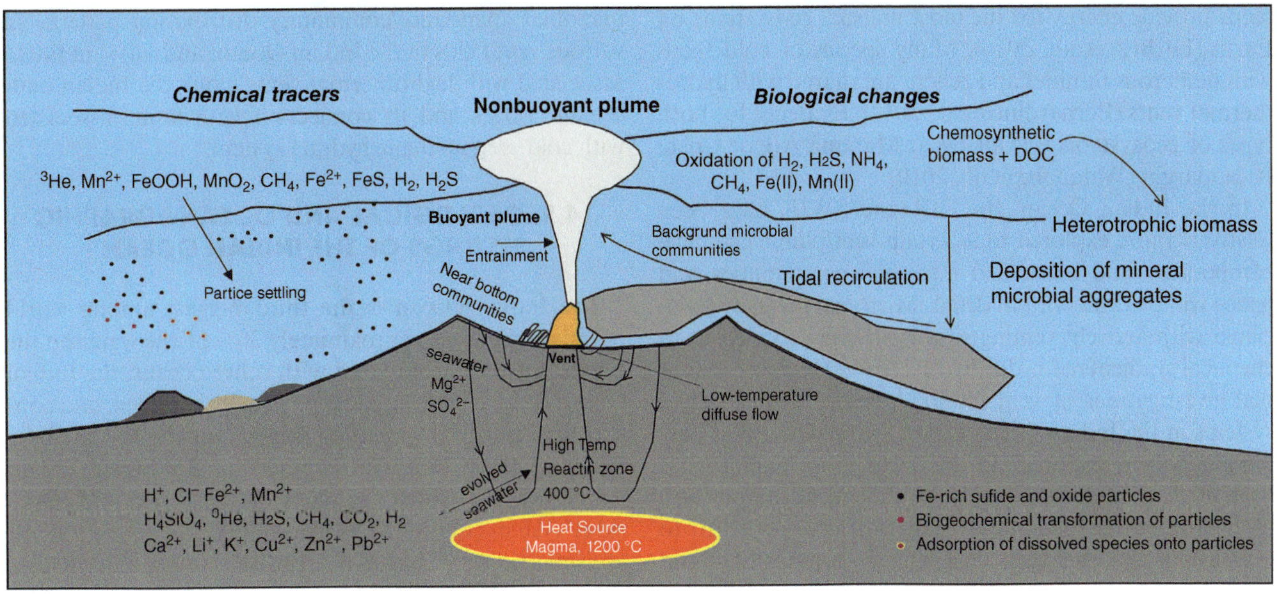

Figure 14.2 Schematic illustration of a hydrothermal plume in the deep water depicting dispersal of vent-derived chemical compounds, sources of plume microorganisms, and changes caused by them to hydrothermal components in ambient seawater. DOC, dissoved oranic carbon.

symbioses, subsurface environment, near-bottom waters, recirculation of aged plume or those growing within the plume (Figure 14.2; Dick et al., 2013). Nevertheless, plume chemical composition and microbes are distinct from those in the ambient water column (Winn et al., 1986; Naganuma et al., 1989; Juniper et al., 1998; Maruyama et al., 1998; German and Von Damm, 2003; Lam et al., 2004, 2008; Dick et al., 2009).

The vent fauna consumes the chemolithoautotrophic microbial mats that grow at hydrothermal vents. This is quite a contrast to the fauna associated with abyssal and seamounts sustained by the supply of phytodetritus from the water column (Van Dover 2000). Moreover, most of the chemosynthetic habitats occur discretely at geographically isolated sites. Therefore, the fauna associated with these ecosystems has unique survival strategies regarding reproduction and larval dispersal (Mullineaux et al., 2018). Alhough the Indian Ocean is considered the most biologically diverse, it is also under-sampled and ecologically less studied.

The diversity of vent specific communities has been described by many researchers (Haymon et al., 1993; Hashimoto et al., 2001; Van Dover et al., 2001; Nakamura et al., 2012; Beedessee et al., 2013; Chen et al., 2015; Copley et al., 2016; Watanabe et al., 2018; Zhou et al., 2018; Jang et al., 2020; Sun et al., 2020). However, most of the known hydrothermal vents sites are located in the Pacific and Atlantic Oceans, and very few active vent fields have been discovered in the Indian Ocean (Gamo et al., 2001; Hashimoto et al., 2001; Van Dover et al., 2001; Tao et al., 2011, 2012, 2014; Nakamura et al., 2012, 2013; Copley et al., 2016; Ji et al., 2017). Studies on the biodiversity and biogeography of Indian Ocean vent fauna were initiated only after discovering the Kairei vent field (Van Dover et al., 2001). Although the hydrothermal vents along the CIR and SWIR are located within a limited area, they are considered reasonably diverse in terms of biological communities associated with them (Nakamura and Takai, 2015b).

A chemosynthetic ecosystem similar to hydrothermal vents was also discovered recently on the eastern continental margins of India (Mazumdar et al., 2019). In 2009, chemosynthetic clams shells (*Calyptogena* sp.) associated with carbonates (methane derived) were reported in sediment cores collected from the Krishna–Godavari basin as evidence for paleo-cold seep activity (Mazumdar et al., 2009). Ten years later, in 2019, a shallow methane hydrate zone associated with the cold-seep ecosystem (chemosymbiont bivalves, polychaetes, and gastropods) was reported by Mazumdar et al. (2019). Bivalves comprised five genera from family *Provannidae* and family *Neritidae* represented by *Bathymodilus*, *Calyptogen Canchocele*, *Acharar* and *Pectin*, and Gastropoda. Polychaeta (Phylum: Annelida) was the second most dominant and was represented by four species (*Branchipolynoe seepensis*, *Sclerolinum* sp., *Glycera* sp., and *Serpulidae* sp.). The cold-seep system has similarities to the typical hydrothermal vent system at mid-oceanic ridges in which the ecosystem survives on chemical species such as H_2, H_2S, CH_4, Fe (II) (Fisher et al., 2007; Levin and Sibuet, 2012; Takai and Nakamura, 2010). The main difference is the temperature. The ecosystem at hydrothermal vents is surviving at a relatively higher temperature than the cold-seep methane system. Both hydrothermal vents and cold seeps provide energy for the most unusual ecosystems on Earth (Le Bris et al., 2016). Many species of cold seeps with numerous families and genera are shared with hydrothermal vents (Bernardino et al., 2012). Evidence for both types of ecosystems is available at Mariana Arc or Costa Rica margin (Watanabe et al., 2010).

In the Indian Ocean, the CIR and SWIR have been relatively more explored to ascertain vent/plume chemical components and associated microbial communities and hence will be dealt with in detail. Microbial diversity associated with actively venting sites is strongly connected to chemical gradients of hydrothermal fluids. However, microbial investigations of vent specific-fauna of hydrothermal systems in the Indian Ocean region are far fewer and they remain poorly explored. In this chapter, an attempt has been made to understand: (i) the geochemical composition of hydrothermal fluids in relation to source rock; (ii) to establish a link between geochemical composition and microbial abundance/community distribution patterns in various vent fields in the Indian Ocean; and (iii) vent fauna associated with hydrothermal vents along the Indian mid-oceanic ridges and its connection with fauna associated with cold-seep methane hydrate system.

14.2. GEOLOGICAL AND OCEANOGRAPHIC SETTINGS OF THE INDIAN OCEAN

The Indian Ocean is the third biggest among world oceans and covers approximately 20% of the water on the Earth's surface. Compared with other oceans, the Indian Ocean is the youngest, most physically complex, and constitutes four active spreading ridges, namely the Carlsberg Ridge (CR), CIR, SWIR, and the Southeast Indian Ridge (SEIR), which are parts of the world mid-ocean ridge system (Figure 14.3). Along with these, the longest aseismic ridges (Chagos–Laccadive, Ninety East Ridge,

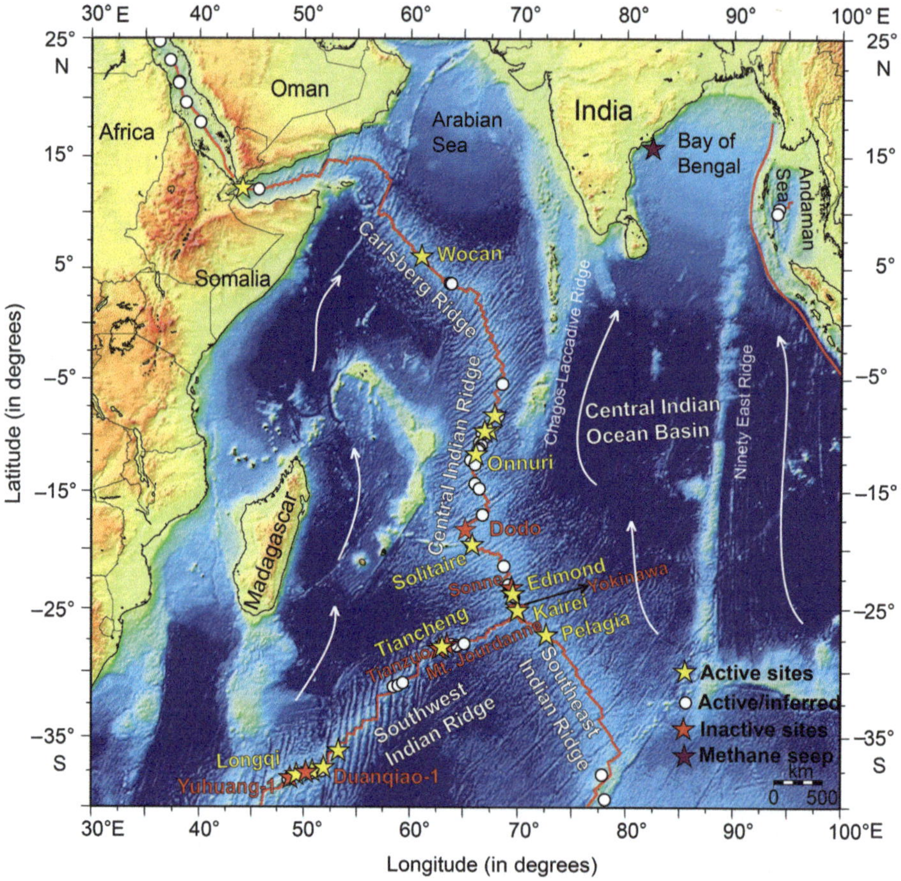

Figure 14.3 Geological setting of the Indian Ocean, which includes mid-oceanic ridges (Carlsberg Ridge, Central Indian Ridge, Southwest Indian Ridge, and Southeast Indian Ridge), aseismic ridges (Chagos–Laccadive and Ninety East ridge), and convergent plate boundary (Andaman back-arc basin). Active and inactive hydrothermal vent fields along the Indian mid-oceanic ridges are named and represented as yellow and blue stars. Active hydrothermal vents (unconfirmed) and plumes are shown as white circles. The cold seep methane hydrate zone is represented by a red star. The white line represents the general deep and bottom water flow in the Indian Ocean (redrawn from Mantyla and Reid, 1995; Piotrowski et al., 2009; Wilson et al., 2012).

Comorin Ridge, and Broken Ridge) and abyssal basins (Central Indian Ocean Basin, CIOB, Somali, Mascarene, Madagascar, and Mozambique basins); several plateaus (Agulhas, Kerguelen, Mascarene, Naturaliste, Wallaby, and Zenith) and seamounts are major structural features of the Indian Ocean.

The CR in the northwest Indian Ocean defines the plate boundary between the Indian and Somalia plates. The CR begins at the Owen fracture zone near 10°N and extends to the CIR near the equator with a northwest–southeast trend (Figure 14.3). The CR is a typical slow-spreading ridge with an average half-spreading rate of 11 to 16 mm/yr, a V-shaped rift valley, and a wide valley floor (Kamesh Raju et al., 2008). The Carlsberg Ridge is also characterized by rugged topography like the CIR, steep valley walls, and wide rift valley floor (13–26 km), which are characteristics of a slow-spreading ridge. Along the ridge axis mantle-derived peridotite was recovered from the inner valley floor (Ray et al., 2012), suggesting distinct variations in accretionary processes along the ridge.

The CIR, north of the Rodriguez Triple Junction (RTJ), is also a slow to intermediate spreading mid-ocean ridge in the Indian Ocean. The segmentation pattern of the 750 km long CIR between 3°S and 11°S (Figure 14.3) has been studied using multibeam bathymetry and magnetic data (Kamesh Raju et al., 2012b). The ridge has well-defined spreading segments and a wide rift valley floor 12–15 km wide, steep valley walls, rugged topography, and prominent volcanic constructional ridges on either side of the central rift valley (Kamesh Raju et al., 1997). These characteristics define the CIR to be a slow-spreading ridge. The spreading rates are slightly higher (average full spreading rates varying from 26 to 38 mm year^{-1}) than the slow-spreading CR. In the southern part of the CIR, the spreading rate is 47 mm year^{-1} (DeMets et al., 2010). Both the SWIR and SEIR meet the CIR at 25°S and form a ridge–ridge–ridge triple junction named the Rodriguez Triple Junction.

The SWIR extends from the RTJ at 25°S to the Bouvet Triple Junction (BTJ) at 0°E; 55°S, which is ~8000 km in length. With a full spreading rate of 14 mm year^{-1}, the SWIR is classified as an ultra-slow spreading ridge (Patriat and Segoufin, 1988; DeMets et al., 1994; Cannat et al., 2008). Absence of volcanic activity, insufficient melt supply, smooth topography, deep axial valley, and exposure of mantle-derived rocks in the axial valley and at some places with gabbro and basaltic rocks are characteristics of SWIR (Royer et al., 1988; Mendel et al., 1997; Mendel and Sauter, 1997; Sauter et al., 2001; Dick et al., 2003; Cannat et al., 2006, 2008). The SEIR extends from the RTJ at 25°S to the Macquarie Triple Junction (MTJ) at 162.76°E; 63°S and is approximately 6000 km long (Cochran and Sempéré, 1997). The SEIR is classified as an intermediate spreading ridge with a full spreading rate of 59–75 mm year^{-1} (Royer and Sandwell, 1989; Cochran and Sempéré, 1997; Small et al., 1999; DeMets et al., 2010).

The deep Indian Ocean is only ventilated from the south and flushes with the Antarctic Bottom Water (AABW) which is carried by the Antarctic Circum-polar Current (ACC), and North Atlantic Deep Water (NADW) carried by the Deep Western Boundary Current (DWBC) flow northward (Piotrowski et al., 2009). The Antarctic waters flow through Enderby Basin, on the west, and through Australian–Antarctic Basin (AAB), on the East (Vlastélic et al., 2001). The origin of water in the Enderby basin is mainly from the Weddell Sea, which flows through a gap in the SWIR and a gap between the Crozet and Kerguelen Plateaus. The same water spreads into the basin on the western side through the Crozet, Madagascar, Mascarene, and Somali basins and finally reaches the Arabian Sea. In contrast, water in the AAB originates from the Ross Sea, enters via the SEIR and spreads northwards into the Central Indian Basin. The general flow of bottom waters, such as AABW and NADW, in the western Indian Ocean, follows a south to north trend (Kumar and Li, 1996; Wilson et al., 2012).

14.3. HYDROTHERMAL VENT FIELDS IN THE INDIAN OCEAN

14.3.1. Chemical and Microbiological Aspects of Vent Fluids and Plume Waters

Indian Ocean ridges are poorly explored for hydrothermal vents when compared with the global mid-oceanic ridge system. Among the vent fields discovered, detailed studies on vent fluid chemical and microbiological characteristics in combination with geological setting are available for only five fields: Dodo, Solitaire, Edmond, and Solitaire along the CIR, and Longqi in the SWIR. Therefore, our discussion in this section will be focused mainly on these five fields.

Dodo

The Dodo field (18°20.1′S, 65°17.9′E; water depth of 2745 m) is located on the Dodo great lava plain where morphological characteristics of high-production basaltic melt are reported (Okino et al., 2015). It is similar to fast-spreading ridges such as the East Pacific Rise (EPR). Hydrothermal fluids from this field have high temperatures (356 °C) while the pH is 3.2. They are enriched in hydrogen (>2 nM) and chlorine (20%) compared with seawater (Table 14.1) and indicates sub-seafloor circulation through serpentinization of rocks and phase-separation (Kawagucci et al., 2008). However, the morphological features of these segments show only large-scale basaltic sheet-flow lava in this region

Table 14.1 Hydrothermal vent fluid compositions for different vent fields and seawater in the Indian Ocean.

Vent Field	Dodo	Solitaire	Edmond	Kairei	Longqi	Seawater
Location	18°20.1′S, 65°17.9′E	19°33.41′ S, 65°50.88′ E	23°52.68′S, 69°35.80′E	25°19.23′S, 70°02.42′E	49°39′E, 37°47′S	
Host Rock	Fresh Basalt (E-MORB)	Basalt (E-MORB)	Basalt (N-MORB)	Troctolite + D-MORB	Basalt	
Depth(m)	2745	2606	3270–3303	2420–2452	2765	
Temperature (°C)	356	307	382	359	365–379	2
pH (25°C)	3.27	4.40	3.13	3.36–3.50	3.58–3.68	7.8
Cl (mM)	684	489	927–950	637–642	612–614	542
Na (mM)	566	409	717–721	534–560	492–496	465
H_2 (mM)	2.70	0.43	0.12–0.14	2.97–3.89	0.16–0.20	<0.01
CH_4 (mM)	0.025	0.043	0.22–0.31	0.10–0.20	0.36–0.38	<0.01
CO_2 (mM)	4.6	16.2	5–13.7	3.94–10.1	–	–
H_2S (mM)	2.80	4.75	4.81	4.00	–	0
$^{3/4}He$ (R_{atm})	7.18	8.96	–	7.8	–	1.13
Mn (μM)	917	78	1430	840	1271–1392	<0.001
Fe (μM)	5408	60	13900	5400	10950–11660	<0.001

Data are compiled from: Van Dover et al. (2001), Gallant and Von Damm (2006). Kumagai et al. (2008), Nakamura et al. (2012), and Kawagucci et al. (2016) for CIR; and Ding et al. (2017), Ji et al. (2017), and Yue et al. (2019) for SWIR. MORB, mid-ocean ridge basalt; D-MORB, depleted.; E-MORB, enriched; N-MORB, normal.

(Okino et al., 2015), and ultramafic rocks are found over the Marie Celeste Fracture Zone (Hekinian, 1982), which is 50 km away from the vent field. Hence, it could be inferred that the hydrogen enrichment might have resulted from magmatic events and subsequent incipient fluid–rock interaction. An end-member concentration of Fe (5408 μM) and Mn (917 μM) indicates rapid sub-seafloor fluid cooling owing to a relatively shallow heat source and insufficient fluid–rock interaction (Kawagucci et al., 2016).

Hydrothermal fluid discharges and animal colonization at the Dodo field has ceased in the recent past (Kawagucci et al., 2016). The microbial cell density in the chimney (Tsukushi-1, Tsukushi-2, and Potsunen) habitats of the Dodo vent field have been reported to be similar in magnitude (10^7 cells g^{-1}; Kawagucci et al., 2016) to those from the Kairei hydrothermal field (KHF) (Takai et al., 2004a). Moreover, high H_2 concentrations in vent fluids, as in the KHF, support abundant populations of (hyper)thermophilic hydrogenotrophic chemolithoautotrophs such as methanogens and hydrogen- and sulfur-oxidizing bacteria that were closely related to the genera *Methanopyrus*, *Methanocaldococcus*, *Persephonella*, and *Sulfurimonas* (Kawagucci et al., 2016).

Solitaire

The Solitaire hydrothermal field lies within the Exclusive Economic Zone (EEZ) of the Republic of Mauritius and is located (19°33.413′S, 65° 50.888′E, at a water depth of 2606 m) several kilometers away from the ridge axis on the western flank of the CIR (Nakamura et al., 2012). The basalt-hosted field is ~50 × 50 m and has three major chimney sites – Toukon-3 (black smoker), Tenkoji, and Liger chimney (both with clear fluids) which are < 5 m high, with small mounds at their feet (Nakamura and Takai, 2015a). The hydrothermal fluids have a high temperature (307 °C) and are depleted in chlorine (10%) compared with seawater. This indicates phase-separation and segregation of a vapor-rich phase in discharging fluids (Nakamura et al., 2012; Nakamura and Takai, 2015b; Kawagucci et al., 2016). However, the hydrogen concentration of 0.46 mM (Nakamura and Takai, 2015b) is relatively low and comparable with typical hydrothermal fluids from mid-oceanic ridges (MOR (McCollom 2008). A relatively high pH of 4.8 (Table 14.1) indicates subseafloor mixing of hydrothermal fluids and seawater. The relatively low concentrations of Fe and Mn (60–78 μM) compared with other hydrothermal fluids of Dodo, Edmond, Kairei, and Longqi, indicates that subseafloor fluid cooling has been sufficiently long for fluid–mineral interaction (Kawagucci et al., 2016). The gas chemistry of hydrothermal fluids (CO_2, H_2 and CH_4), is comparable to the typical mid-oceanic ridge basalt-hosted system (Table 14.1). A high helium concentration (R_S/R_{atm} = 8.96; where R is the ratio of $^3He/^4He$ in sample and atmosphere) indicates the hotspot influence on the fluid chemistry, which is further supported by helium-rich basalts from this region (Füri et al., 2011). The Solitaire field is characterized by multiple diffuse flows from the talus throughout the hydrothermal field (Nakamura et al., 2012).

Permeable mineral structures and continued mixing of chemically reduced fluids and oxidized seawater provide favorable conditions that support the growth of endolithic microbial communities (Schrenk et al., 2003). Microbial cell density in the exterior surface of the Toukon-3 chimney fragment has revealed 2.1 × 10^4 cells g^{-1} whereas the inside part of the chimney harbored 2.4 × 10^3 cells g^{-1} (Kawagucci et al., 2016). Quantitative polymerase chain reaction (PCR) analysis of the 16S rRNA gene also revealed the highest bacterial abundance (10^5 copies g^{-1}) in the exterior surface of the Toukon-3 chimney but no detectable archaeal gene copy numbers. Members of Aquificales and Epsilonproteobacteria dominated the microbial community on the surface of the Toukon-3 chimney. Vent endemic Epsilonproteobacteria are responsible for actively oxidizing sulfur (Akerman et al., 2013). The Aquificales, with a microaerophilic and thermophilic bacterial lineage, and capable of oxidizing hydrogen, have been reported from deep-sea vents in the Mid-Atlantic (Reysenbach et al., 2000) and Indian Oceans (Reysenbach et al., 2002). Quantitative cultivation-based test revealed the existence of (hyper)thermophilic, hydrogenotrophic chemolithoautotrophs at the Toukon-3 chimney, which were most closely related to genera *Persephonella* and *Sulfurimonas*. Members of the genus *Persephonella* appear to be key players in carbon, sulfur, and nitrogen cycles in high-temperature habitats such as deep-sea vents (Mino et al., 2013). *Sulfurimonas* species have been isolated from sulfidic habitats (Inagaki et al., 2003) and can grow with nitrate or nitrite as electron acceptors and with a variety of sulfur species of different oxidation states or use hydrogen as electron donor (Labrenz et al., 2013). Genetic variation of 109 *Sulfurimonas* strains isolated from four geographically distant deep-sea hydrothermal vent regions (Okinawa trough, Mariana volcanic arc and trough, Central Indian Ridge, and the Mid-Atlantic Ridge) have also been studied based on 11 protein-coding genes. These included strains from the Solitaire vent field. High genetic variation corresponding to hydrothermal regions was observed, which indicated that allopatric speciation in the *Sulfurimonas* population was primarily influenced by geographical distance rather than the gas composition of vent fluid or habitat (Mino et al., 2017).

Edmond

The Edmond hydrothermal field is located (23°52.68′S, 69°35.80′E; water depth of 3290–3320 m) in an off-axis region of the northern end of the ridge segment (Van Dover et al., 2001) where typical abyssal hills are observed (Kumagai et al., 2008). The main area of hydrothermal venting in the Edmond hydrothermal field is spread to ~ 100 × 60 m, with black smoker complexes of up to 35 m high and 2 m diameter (Gallant and Von Damm, 2006). The most vigorous black-smoker venting is located in discrete clusters of large chimneys at the Nura Nura vent site (Kumagai et al., 2008). The field is dominated by old, disaggregated sulfide structures and massive sulfide taluses, which are suggestive of focused hydrothermal activity over several years (Van Dover et al., 2001). The fluid temperature of 382 °C (Table 14.1) is the highest ever measured in the Indian Ocean, with a pH range from 2.97 to 3.30 (Gallant and Von Damm, 2006; Kumagai et al., 2008). Chlorine concentrations are quite high (> 70%) when compared with seawater, indicating subseafloor phase separation along with brine condensation. However, hydrogen concentrations are relatively low and are comparable with fluids from EPR and Mid-Atlantic Ridge (MAR). Like the Solitaire hydrothermal field, extensive diffuse flows have been reported from the Edmond field (Nakamura and Takai, 2015a).

Orange-brown, iron oxyhydroxide sediments several centimeters thick have been found to accumulate in depressions, coating many sulfide structures and taluses in the Edmond field (Van Dover et al., 2001). Extensive microbial mats have also been found to cover the surface of sediments, which have been attributed to widespread sub-surface seawater entrainment and mixing processes (Van Dover et al., 2001). Rod-shaped microbes with s helical stalk-like filaments have been observed in microbial mats from a hollow inside a chimney structure from the Edmond field (Peng et al., 2007). *Mariprofundus ferrooxydans*, a neutrophilic, mesophilic, and iron-oxidizing chemolithoautotrophic marine bacterium affiliated to class *Zetaproteobacteria*, and isolated from iron-rich mats in hydrothermal fields of the Loihi Seamount (Emerson et al., 2007) has been shown to produce unique helical stalks consisting of organic compounds and iron oxides (Chan et al., 2011). It is likely that *Zetaproteobacteria* could also be dominating iron-rich mats (Kato et al., 2009) in the Edmond field. Bioprecipitation of Fe on the surface and in the interior of cells has been observed to occur along with microbial silicification in the mats (Peng et al., 2007). Scanning electron micrographs of mineralized crust samples have revealed abundant filamentous rod, and coccoidal forms encased in sulfur and sulfide mineral precipitates. These observations suggest that the biomineralization process by thermophilic chemolithoautorophs might be dependent on hydrothermal inputs (Peng et al., 2007) and may influence the architecture and sulfur cycling of sulfide spires in the vent field (Hoek et al., 2003).

Phylogenetic analysis of microbial communities associated with the mineralized crust coating a sulfide spire collected from the Edmond vent field revealed 26 bacterial and nine archaeal phylotypes (Hoek et al., 2003). More than 90% of the bacterial phylotypes were primarily affiliated with Epsilonproteobacteria. In contrast, the majority (97%) of the archaeal phylotypes grouped with the Deep-sea Hydrothermal Vent Euryarchaeotal Group (DHVEG) that appears to be endemic to vent ecosystems. The sulfur oxidation *soxB* gene of Epsilonproteobacteria from diffuse hydrothermal fluids suggests that *Sulfurimonas*-like species are responsible for actively oxidizing sulfur via the Sox pathway (Akerman et al., 2013). These autotrophic and mixotrophic sulfur-oxidizing bacteria (Sievert et al., 2008) can also use other electron acceptors such as nitrate, nitrite, and fumarate (Vetriani et al., 2004; Miller et al., 2007) and fix carbon via the reductive tricarboxylic acid (rTCA) cycle (Hügler et al., 2005). Archaea affiliated to DHVEG have been suggested to be obligate thermoacidophilic sulfur- or iron- reducing heterotrophs (Reysenbach et al., 2006). The thermophilic hydrogen-oxidizing *Hydrogenobacter thermolithotrophum* and hyperthermophilic sulfur-reducing *Thermococcales* have also been reported from sulfide spires from the Edmond vent field (Hoek et al., 2003).

Two primary lineages of Aquificales, *Persephonella marina* (Type strain HI) and *Persephonella* sp. nov. strain H3, were cultured in an enriched medium that was inoculated with ground sulfide samples obtained from the Edmond vent field (Reysenbach et al., 2002). These thermophilic, strictly chemolithoautotrophic, hydrogen-oxidizing microaerophiles have been reported from deep-sea hydrothermal vent sites in the Pacific Ocean (Takai et al., 2008; Mino et al., 2013). These bacteria grow at a temperature range between 55 and 80 °C and pH 4.7–7.5. They can use elemental sulfur, thiosulfate, or hydrogen as electron donors, while oxygen/nitrate is used as electron acceptors (Götz et al., 2002). Their occurrence at deep-sea vents globally has expanded our knowledge on their distribution. This knowledge reflects their preference to thrive in high-temperature, acidic fluids like those from the Edmond vent field.

Kairei

The Kairei hydrothermal field (KHF) is located (25°19.23′S, 70°02.42′E; water depth of 2415–2460 m) near the RTJ. Here, several hydrothermally active sites had been suggested to exist around the off-axis Hakuho knoll, which is ~ 6.4 km northeastward from the ridge axis of the CIR (Gamo et al., 1996). Several black smoker complexes with a maximum height of >10 m have been recognized in the 40 × 80 m field (Gamo et al., 2001), along with large areas of weathered sulfides and inactive chimneys (Nakamura and Takai, 2015a). The most vigorous black smoker discharges have been observed at the Kali vent site (Kumagai et al., 2008). The hydrothermal fluid temperature of 369 °C and pH range of 3.3–3.6 of the Kairei vent field are comparable with the Edmond field (Table 14.1). The chlorine concentrations are slightly higher (14%) compared with those of normal seawater. Based on basalt rocks and the chemical composition of fluids, the field was initially assumed to be a typical basalt-hosted hydrothermal system. Later studies have revealed enrichment of hydrogen (2.5–8.5 mM; Gamo et al., 2001; Van Dover et al., 2001; Gallant and Von Damm, 2006) and well-defined oceanic core complex structure with peridotite and gabbros. The field may have developed by seawater circulation through subsurface ultramafic rocks similar to hydrothermal fields along MAR (e.g. the Rainbow, Logatchev I and II, Ashaze I and II, and Nibelungen).

A microbial density of up to 2.3×10^6 cells l^{-1} has been detected in superheated Kali vent fluids (Takai et al., 2004). In the Fugen chimney vent orifice, which is ~ 50 m away from the Kali chimney (Takai et al., 2004), up to 10^7 cells g^{-1} wet weight were found. These values are much higher than those reported from high-temperature chimney fragments from the Solitaire hydrothermal fluids (Kawagucci et al., 2016). Takai et al. (2004) proposed that the Kairei ecosystem is sustained through primary production by hydrogenotrophic, hyperthermophilic methanogens, which utilize H_2 and CO_2 as primary energy and carbon sources. The thermophilic, methanogenic autotrophs *Methanocaldococcus indicus* (L'Haridon et al., 2003), *Thermodesulfatator indicus* (Moussard et al., 2004), *Methanotorris formicicus* and *Hydrogenimonas thermophila* (Takai et al., 2004) isolated from KHF were capable of using H_2 and CO_2 as sole energy and carbon sources. Subsequent studies using *Methanopyrus kandleri* strain 116, a dominant hyperthermophilic methanogen isolated from an *in situ* colonization system deployed in black smoker fluid of the KHF, were conducted under high temperatures (85–123 °C) and hydrostatic pressures (0.4–40 MPa) (Takai et al., 2008). Piezophilic growth of the strain significantly affected the stable carbon isotope fractionation of methanogenesis $-\delta^{13}C$ CH_4) from CO_2, shifting from −34‰ to −27‰ under conventional growth conditions to < −9.4‰ at 122 °C and 40 MPa. This indicated archaeal-mediated methanogenesis in hydrothermal systems wherein CO_2 is reduced to methane (CH_4) to produce cellular energy. An extremely high abundance of molecular hydrogen (2.5 mM) in KHF fluids could serve as the primary energy source for microbial methanogenesis. Moreover, significantly low $\delta^{13}C$-CH_4 in the vent fluids (−13.8 and −18.5‰) further supported the occurrence of the process wherein biogenic methane is produced from CO_2 (Takai et al., 2004).

In the KHF, active and inactive chimney structures host habitat-specific bacterial and archaeal communities (Han et al., 2018). Culture-independent molecular studies targeting the 16S rRNA gene revealed Epsilonproteobacteria like those belonging to order Campylobacterales, Nautiliales, and Aquificae to be the most abundant bacterial operational taxonomic units (OTUs) in active chimneys (Han et al., 2018). Epsilonproteobacteria, mainly related to various genera within the Campylobacterales, have also been reported from actively venting hydrothermal chimneys from the Manus Basin back-arc spreading center (Pjevac et al., 2018). Screening of chimney-associated microbial communities has revealed functional marker genes for hydrogen oxidation (*hupSCL*), sulfide oxidation (*sqr*), and Sox-pathway dependent sulfur oxidation (*soxB, soxAX, soxYZ*, and *soxCD*) (Pjevac et al., 2018), highlighting the importance of Epsilonproteobacteria in S cycling. Archaeal OTUs in active chimneys of KHF consisted of thermophiles showing taxonomic affinities to Archaeoglobales (utilize aromatic compounds as electron donors, and mediate Fe(III) reduction; Anderson et al., 2011; Manzella et al., 2015), Thermococcales (capable of producing Greigite, Fe_3S_4, nanocrystals on extracellular polymeric substances; Gorlas et al., 2018), Thermoprotei (use hydrogen as an energy source; Nakagawa and Takai, 2008) and Methanococcales (thermophilic H_2-consuming methanogens; Topçuoglu et al., 2016). The predominance of members of the Methanococcales in superheated hydrothermal emissions and chimney interiors, together with the other major microbial members of the Thermococcales at the KHF imply the existence of a hyperthermophilic subsurface lithoautotrophic microbial ecosystem (HyperSLiME) dominated by hyperthermophilic methanogens (Takai et al., 2004).

Chimney structures at the KHF are made up of sphalerite ((Zn,Fe)S), chalcopyrite ($CuFeS_2$), and pyrite (FeS_2) with minor barite ($BaSO_4$) and Fe oxides (Wang et al., 2012). Suzuki et al. (2004) found high cell density (10^7–10^8 cells g^{-1}) along with an abundance of magnetotactic bacteria such as *Magnetobacterium bavaricum*. They suggested that inactive chimneys provide a distinct habitat to microbes capable of thriving on reduced sulfide minerals. Characterization of bacterial communities in inactive chimney samples by Han et al. (2018) revealed that they were affiliated to *Desulfobulbus* and SAR324 (sulfate-reducing bacteria; Nakagawa et al., 2004), *Nitrosococcus* (ammonia-oxidizing bacteria; Ding et al., 2017), *Acidiferrobacter* (iron-oxidizing bacterium; Hallberg et al., 2011), Oceanospirillales (symbionts associated with hydrothermal vent gastropods, found colonizing sunken wood; Fagervold et al., 2012; Johnson et al., 2015), Alteromonadales (found on sunken wood, polycyclic aromatic hydrocarbon degraders; Fagervold et al., 2012; Sinha et al., 2019), and Thiotrichales (hydrothermal fauna symbionts; Crépeau et al., 2011). The archaeal communities from inactive chimneys mainly belonged to Woesearchaeota (formerly named DHVEG-6) and Thaumarchaeota (Han et al., 2018), which are involved in protein degradation and carbohydrate metabolism (Li et al., 2015).

Epifluorescence microscopic analysis of 4′,6-diamidino-2-phenylindole (DAPI)-stained microbial cells from ambient seawater adjacent to hydrothermal emission and in a rising plume was an order higher (10^9 cells l^{-1}) compared with normal deep seawater (Takai et al., 2004). However, flow cytometric analysis of plume samples from the KHF have revealed similar (1.77×10^7 cells L^{-1}), lower than (9.72×10^6 cells L^{-1}) or higher (2.20×10^7 cells L^{-1}) microbial abundance than at a distal reference station (0.9–1.4×10^7 cells L^{-1}) (Noguchi et al., 2015). Neutrally buoyant plumes are enriched in (trace) metals (Fe, Mn, Cu, Zn) and REEs (Haalboom et al., 2019). The fate of these substrates is influenced by microbially mediated biogeochemical reactions (Dick et al., 2009) and the binding of minerals onto organic material derived from vent biomass or *in situ* production through microbial activity (Breier et al., 2012). Microbial 16S rRNA phylotype composition showed that chemolithoautotrophs such as the SUP05 clade (*Thioglobus* sp., involved in sulfur oxidation; Marshall and Morris, 2013; Shah et al., 2019), *Methylobacter* and *Methylococcus* (methane oxidizers; Wartiainen et al., 2006; Kleiveland et al., 2012; Smith et al., 2018), and Epsilonproteobacteria thrived in KHF plumes (Noguchi et al., 2015). This finding corroborated earlier observations of Takai et al. (2004), who reported the highest proportion of Epsilonproteobacteria in KHF plume water. The KHF plume has an anomalously high H_2 concentration of up to 46.1 nmol L^{-1}. *Hydrogenimonas thermophila*, a novel thermophilic bacterium isolated from a black smoker vent orifice at the KHF, is capable of using molecular hydrogen as the sole energy source and carbon dioxide as the sole carbon source (Takai et al., 2004). Characterization of epsilonproteobacterial isolates from deep-sea hydrothermal vent systems has also revealed that they use H_2 or reduced sulfur compounds as an electron donor and O_2, nitrate, or elemental sulfur as an electron acceptor (Takai et al., 2003, 2004). Thus, an increase in SUP05 and epsilonproteobacterial phylotypes in the KHF plume is suggestive of microbially mediated H_2 consumption. The proportion of phylotypes belonging to heterotrophs such as *Alcanivorax* (mediating hydrocarbon and sulfur oxidation) were rare in hydrothermal fluid–seawater mixing zones near vents but increased in nonbuoyant plumes away from the KHF (Noguchi et al., 2015). However, the proportion of archaeal phylotypes associated with Marine Crenarchaeota Group I (MGI) were found to increase in ambient seawater adjacent to hydrothermal vents rather than in plumes (Takai et al., 2004), implying that, unlike Epsilonproteobacteria, they were not dependent on chemolithoautotrophic metabolism (Takai et al., 2003).

A considerable amount of dissolved organic compounds excreted by hydrothermal vent-associated faunal communities has been suggested to support the archaeal community. Epsilonproteobacteria have also been reported to colonize gill filaments of the hydrothermal vent gastropod *Alviniconcha* aff. *hessleri* from the KHF.

Based on compound-specific carbon isotopic analysis, inorganic carbon fixation in the epsilonproteobacterial endosymbiont was suggested to be based on the rTCA cycle (Suzuki et al., 2005). Many heterotrophic bacteria isolated from Kali chimney plumes have high multi-stressor tolerance to desiccation, peroxide exposure, UV, and gamma-ray irradiation (La Duc et al., 2007).

Longqi

The Longqi hydrothermal field (LHF) is located at 37°47′S, 49°39′E at a water depth of 2755–2785 m and is the first active vent field found on SWIR (Tao et al., 2012). The maximum temperature of hydrothermal fluids from the Longqi field is 379 °C (Table 14.1) with a pH of 3.21 (Ding et al., 2017; Ji et al., 2017; Yue et al., 2019). The field is known to be a basalt-hosted system (Tao et al., 2014). Several active vents have been located in a zone ~ 1000 m long, indicating that the area of hydrothermal activity in the Longqi field is very large (Tao et al., 2012, 2014). Hydrothermal fluids have a high concentration of chlorine when compared with ambient seawater, which is characteristic of fluid–rock interaction and phase-separation (Ji et al., 2017). High concentrations of dissolved gases and metals indicate circulation of hydrothermal fluids through a long reaction path of > 6 km involving both mafic and ultramafic lithologies (Tao et al., 2020). The hydrothermal fluids then ascend to the seafloor where they branch and expel from multiple vents (Ji et al., 2017). Endmember compositions of vents fluids from LHF arise from a single source and have shown no significant difference from those observed along the CIR (Table 14.1). This indicates that the spreading rate is not a key factor in controlling hydrothermal vent fluid chemistry (Ji et al., 2017).

Sulfide chimneys are formed as a product of high temperature (300–350 °C), hydrothermal activity, whereas Si-rich chimneys form at a fluid temperature of < 100 °C (Lei et al., 2017). The LHF is mainly constructed of chalcopyrite ($CuFeS_2$), kusachiite ($CuBi_2O_4$), pyrite (FeS_2), and manganese phosphate hydrate (Djurhuus et al., 2017). Detailed investigations have revealed that chimneys from LHF, with a fluid temperature of < 362 °C have pyrite and zinc sulfide as major components, while those with a temperature ≥ 379 °C are dominated by Bementite and Birnessite (Ding et al., 2017). In low-temperature hydrothermal chimneys on SWIR, microbial density is ~ 10^4–10^5 cells g^{-1} wet weight (Cao et al., 2014). In agreement with previous studies from hydrothermal systems elsewhere, microbial community structure analysis of chimney fragments from LHF has revealed greater bacterial diversity than that of Archaea (Ding et al., 2017) (Zhang et al., 2016). A distinct stratified pattern of microbial distribution in sulfide chimneys from SWIR has also been observed (Cao et al., 2014). *In situ* fluid temperature and gaseous components have been suggested to be the main components controlling microbial community structure (Ding et al., 2017; Lei et al., 2017). Sulfur-reducing Deltaproteobacteria dominate in the interior of the chimney (Sylvan et al., 2013), where they could be reducing sulfate to sulfide, resulting in the precipitation of metals (Cao et al., 2014). Lipid biomarkers have also supported the abundance of sulfate-reducing bacteria in sulfide chimneys (Lei et al., 2017), where they could coexist with methanogens (Sela-Adler et al., 2017) and compete for hydrogen (Hoehler et al., 1998). Epsilonproteobacteria, Gammaproteobacteria, Alphaproteobacteria, and Deltaproteobacteria and members of the phyla Bacteroidetes and Planctomycetes are the dominant bacterial lineages inhabiting the surface of sulfide chimneys (Cao et al., 2014; Ding et al., 2017). As observed in fluid/chimney samples collected from hydrothermal fields along the CIR (Kawagucci et al., 2016; Han et al., 2018) and elsewhere (Huber et al., 2010; Li et al., 2014), members within class Epsilonproteobacteria dominated the sulfur-oxidizing chemolithoautotrophic community. Sulfide-oxidizing autotrophs belonging to the genus *Sulfurovum* and the genus *Sulfurimonas* within Epsilonproteobacteria have been reported from high-temperature chimneys primarily composed of sulfide minerals (Lin et al., 2016). Sulfur-oxidizing bacteria could utilize sulfide as an energy source, leading to oxidization of sulfide to sulfate, followed by precipitation of sulfate in the outer chimney. Nitrates, metals, and hydrogen could function as electron donors or acceptors for sulfate-reducing and sulfur-oxidizing bacteria (Cao et al., 2014). Organotrophic Bacteroidetes grow as ectobionts on long, sheathed microbial filaments affiliated to the lithoautotrophic genus *Sulfurovum*. They utilize organic polymers produced by the host, thereby recycling organic matter in biofilms that grow on the surface of black smoker chimneys (Stokke et al., 2015). Members of Alphaproteobacteria, Deltaproteobacteria, and Gammaproteobacteria have also been reported to be dominant in inactive sulfide chimneys. They are thought to mediate diverse geochemical processes, including iron oxidation and nitrogen fixation ammonia oxidation, and denitrification (Li et al., 2017). Anaerobic metabolic abilities e.g., reduction of elemental sulfur to sulfide and carbohydrate fermentation are widely distributed among all significant Planctomycetes lineages, which they use for growth and survival (Elshahed et al., 2007).

Thaumarchaeota, Woesearchaeota, and Euryarchaeota were dominant archaeal lineages in LHF chimney fragments (Ding et al., 2017). Thaumarchaeota and Euryarchaeota have also been reported to colonize inactive vents (Zhang et al., 2016) and sediments (Wu et al., 2014; Wang et al., 2017) in hydrothermal fields. Thaumarchaeota, which are predominantly ammonia oxidizers (Wang et al., 2017), are abundant in Si-rich chimneys (Lei et al., 2017) and play a critical role in nitrogen cycling (Pester et al., 2011). Hyperthermophilic methanogens comprise most of the archaeal community in sulfide chimneys from the LHF (Lei et al., 2017). Euryarchaeota facilitates the anaerobic conversion of H_2, formate, acetate, methyl compounds, and simple alcohols to CH_4 and CO_2 (Nobu et al., 2016). Woesearchaeota can metabolize carbon and hydrogen under anaerobic conditions, which be might be associated with symbiotic and/or fermentation-based lifestyles (Castelle et al., 2015). Thus, the dominance of these major archaeal phyla in active chimneys is indicative of their role in biogeochemical cycling.

Hydrothermal plumes from LHF also have many bacterial and archaeal densities ranging from 10^8 to10^9 16S rRNA gene copies L^{-1}. These microbes are sourced from ambient deep seawater rather than from seafloor vent-derived niches (Dick et al., 2013; Li et al., 2016). Unlike at the CIR (Takai et al., 2004b), archaeal populations in hydrothermal plumes at the SWIR are spatially homogeneous and dominated by members of marine group I within the Crenarchaeota, which are involved in ammonia oxidation (Sintes et al., 2013). Conversely, bacterial communities in hydrothermal plumes are heterogenous in their distribution, consisting mainly of members within Alphaproteobacteria and Gammaproteobacteria (Li et al., 2016). Vent-specific genera (*Arcobacter*, *Caminibacter*, and *Sulfurimonas*) from the Epsilonproteobacteria and the SUP05 group from the Gammaproteobacteria have been reported from the LHF (Djurhuus et al., 2017). Epsilonproteobacteria are known to grow closer to the sulfide source (Patwardhan et al., 2018), but sequences from this class have not been reported from hydrothermal sediments in the vicinity of the Tiancheng vent field. Gammaproteobacteria are the most dominant microorganisms in these sediments, followed by Alphaproteobacteria and Deltaproteobacteria (Wu et al., 2014). Unlike *Epsilonproteobacteria*, metabolically versatile *Gammaproteobacteria* tend to become prevalent once the community is established (Patwardhan et al., 2018). Their abundance has been found to increase with distance from the vent area (Wang et al., 2015). Wu et al. (2014) found no overlap of bacterial OTUs (at 97% identity threshold) from hydrothermal sediments with those from previous studies on active and inactive chimney structures (Suzuki et al., 2004; Takai et al., 2004a), indicating different bacterial abundances within areas sampled in hydrothermal vent fields.

Few attempts have been made to isolate, identify and characterize bacterial strains from the SWIR. Exopolysaccharide producing novel strains such as *Alteromonas pelagimontana* sp. nov. and *Idiomarina* sp. 5.13 have been reported, and the latter strain possesses various genes for metal resistance (Sinha et al., 2017). Sequences of closely related known Mn (II)-oxidizing microorganisms (*Acinetobacter*, *Alteromonas*, *Erythrobacter*, *Halomonas*, *Marinobacter*, *Pseudoalteromonas*, *Pseudomonas*, and *Sulfilobacter*) have been retrieved in abundance from bacterial clone libraries of SWIR plume samples indicating a profusion of metal-resistant bacteria (Li et al., 2016). Gammaproteobacteria such as *Halomonas meridiana* and *Marinobacter algicola* isolated from SWIR waters can tolerate dissolved Mn concentrations up to two orders greater than in ambient seawater (Fernandes et al., 2018). They have the ability to detoxify elevated levels of Mn(II), converting it into an insoluble oxide which has similarities to the hydrothermal vein mineral Rhodochrosite. Exopolymeric material secreted by Mn(II) oxidizers acts as a nucleation site for the deposition of Mn oxide. Cell aggregates have been observed to be encrusted with microspheres (Figure 14.4). Scanning electron microscopic observations of low-temperature silica-rich deposits from the LHF have also revealed precipitation of silica spheres and Fe sulfides which is promoted and accelerated by the secretion of microbial extracellular polymer saccharide (Peng et al., 2011). A similar biomineralization style has been observed in Fe–Mn deposits from the Lau Basin (Fernandes et al., 2018). These findings revealed that bacterially mediated metal oxidation could mitigate excess levels of trace metals in hydrothermal ecosystems.

Autochthonous bacteria can contribute to a certain extent in the formation of mineral deposits along mid-oceanic ridges and have a potential for use in bioremediation and biotechnological applications (Fernandes et al., 2005, 2018). Several bacteria capable of degrading

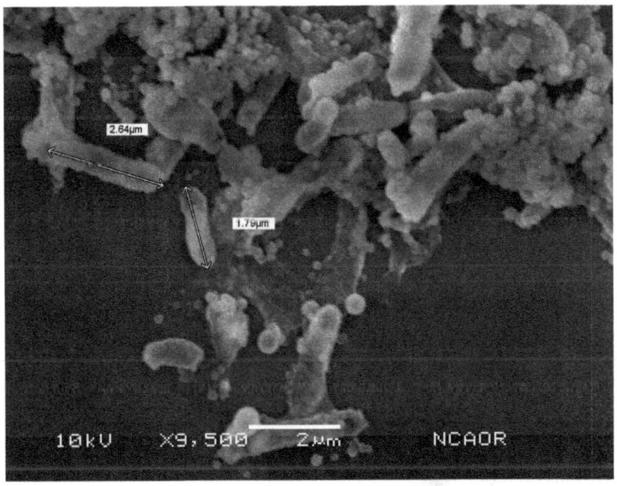

Figure 14.4 Scanning electron micrograph of Marinobacter algicola incubated in culture medium amended with 10 mmol l^{-1} Mn(II). The cells entrap Mn oxide in exopolymeric material and are encrusted with microspheres (modified after Fernandes et al., 2018) possibly formed from amorphous silica (Peng et al., 2011).

polycyclic aromatic hydrocarbons (PAH) and using them as a sole carbon source have been isolated from deep-sea waters (3946–4746 m) along the SWIR. Novel strains with PAH-degrading capabilities show close taxonomic affinities to *Alterierythrobacter*, *Citricella*, *Erythrobacter*, *Idiomarina*, *Lutibacterium*, *Maricaulis*, *Marinobacter*, *Martelella*, *Pseudidiomarina*, *Rhodobacter*, *Roseovarius*, *Salipiger*, *Sphingopyxis*, and *Stappia* spp. (Yuan et al., 2015). High-throughput DNA-based analysis of environmental samples has also shown that a considerable fraction of the microbial community along the Indian ridge system is composed of unclassified/unknown organisms (Kawagucci et al., 2016; Ding et al., 2017; Sinha et al., 2019). These findings provide an unprecedented opportunity to modify existing culturing techniques to harness pertinent microbes for bioprospecting.

The geochemical composition of hydrothermal fluids of the Dodo, Solitaire, Edmond, Kairei, and Longqi vent fields shows that basalts host all these systems. However, the Kairei fluids have high concentrations of hydrogen, and Longqi fluids have high concentrations of methane. These concentrations are quite unusual for basalt hosted system and indicates the possible fluid circulation through ultramafic rocks. Abundant populations of (hyper)thermophilic hydrogenotrophic chemolithoautotrophs are associated with Dodo, Solitaire, and Kairei fluids. In contrast, Fe/Mn-oxidizing chemolithoautotrophs belonging to *Zetaproteobacteria* or various metal-resistant genera within class Alphaproteobacteria and Gammaproteobacteria thrive on metal-rich Edmond and Longqui vent fluids. Therefore, it is evident that the fluid chemical composition and associated biology are primarily controlled by subsurface water-rock reactions. Detailed geological studies of these fields have supported the arguments. The Longqi field is located on the SWIR, whereas the other four fields, Dodo, Solitaire, Edmond, and Kairei, are located on the southern CIR. Geophysical studies have classified the SWIR as ultraslow-spreading and the southern CIR as intermediate-spreading ridges. Although the spreading rate of CIR and SWIR is different, hydrothermal systems discovered along these ridges are mostly basalt hosted. Therefore, it is perceived that the spreading rate has no role in the genesis of fluid chemical composition and associated biosphere.

14.3.2. Other Active Fields

Tiancheng

Tiancheng field is located at 63°32′E and 27°57′S; it is hosted by fractured basalts and is characterized by low-temperature (up to 13.2 °C) diffuse flow. Recent studies on trace metal in Tiancheng hydrothermal plumes (both buoyant and non-buoyant) show that the majority (75–100%) of the dissolved manganese is present in the soluble phase (< 1 kDa). Whereas more than 66% of dissolved Fe was removed in the buoyant stage of the plume, and 54–95% of the Fe is present as the colloidal phase in the buoyant plume (Fang and Wang, 2021). There is a widespread occurrence of hydrothermal fauna such as mussels, gastropods, crabs, shrimps, barnacles, and anemones on the seafloor (Chen et al., 2018). A new population of the scaly-foot snail *C. squamiferum* has been reported from the vent field. It has reddish-brown coloration from zinc sulfide deposits on the surface of its sclerites, unlike those at the Kairei field, which have black scales infused with iron sulfide nanoparticles. Scales lacking iron but still enriched in sulfur may be adapted to remove sulfur metabolites from their bodies (Sun et al., 2020).

Pelagia

Pelagia field is located at 26°S and 71°E of the SEIR (Han et al., 2018). The high-temperature fluids (332–256 °C) are highly enriched in dissolved Fe, Si, K, Li, Mn, and Zn, and apparently not affected by supercritical phase separation in the subsurface (Noowong et al., 2021). Chimneys are predominantly covered by swarms of hydrothermal vent endemic shrimps (*R. kairei*) and hydrothermal fluids have been observed to emanate from open fissures on the seafloor. Diffuse hydrothermal fluid and plume water samples in the Pelagia area harbor up to 107 microbial cells L^{-1} (Han et al., 2018). Active chimneys in the vent field appear to host habitat-specific microbial communities with low or no resemblance to known species. The bacterial community in the Pelagia active chimneys has been found to be dominated by OTUs classified as Epsilonproteobacteria, particularly the orders Nautiliales and Aquificae. These microbes can reduce sulfur to sulfide, thereby contributing to the formation of metal sulfides (Han et al., 2018). Dense colonies of endemic vent species such as mussels and other benthic communities have also been observed near the outlet of fissures.

Wocan

The Wocan Hydrothermal Field (WHF) was discovered in 2013 and has two hydrothermal sites (Wocan-1 and Wocan-2) on the axial volcanic ridge located at 6°22′N, 60°31′E on the slow-spreading CR (Wang et al., 2017). Iron- and copper-rich sulfides (Wang et al., 2017) along with the Fe–Si–Mn-oxyhydroxides (Popoola et al., 2019a) and metalliferous sediments (Popoola et al., 2019b) have been recovered from this field. The recovered hydrothermal precipitates can be classified into (i) Cu-rich chimneys, (ii) Cu-rich massive sulfides, (iii) Fe-rich massive sulfides, and (iv) silicified massive sulfides. Based on mineralogy and geochemistry of sulfide deposits, Wang et al. (2017) have inferred that Wocan-1 has undergone a cycle of heating with Cu-rich chimney growth and subsequent cooling, followed by late seafloor weathering, while Wocan-2 has undergone intermediate- to high-temperature mineralization followed by intense silicification of sulfides.

Onnuri

The active Onnuri vent field (OVF) was discovered at 11°24.9′S, 66°25.4′E at a water depth of 1990–2170 m. Disseminated sulfide assemblages with barite, altered rocks, Fe oxide crust, and vent organisms (hairy snails, mussels, barnacles, shrimps, and crabs) have been reported to be widely distributed around the field. Isolation of *Thermococcus indicus* sp. nov., a Fe(III)-reducing hyperthermophilic archaeon has suggested the presence and abundant occurrence of a thermophilic iron-reducing microbial population in the Onnuri vent field (Lim et al., 2020). Thin greenish layers of Cu-bearing secondary minerals were observed in breccia-type sulfides, which imply that: (i) the hydrothermal system might be affected by high-temperature fluids at least beneath the site, although plume data indicate low-temperature, diffuse venting is dominant; or (ii) high-temperature venting was present in the past, but the site is now waning (Kim et al., 2020).

Old City

The Old City hydrothermal field (OCHF; 27° 50.1′S; 64°35.1′E) along the SWIR (Sauter et al. 2013; Lecoeuvre et al., 2018) is a serpentinite-hosted hydrothermal site located ~3100 m below sea level. It is characterized by extremely slow and diffuse hydrothermal fluid flow. Carbonate–brucite chimneys at the OCHF are similar to those observed at the Lost City hydrothermal field (LCHF) along the mid-Atlantic ridge (Lecoeuvre et al., 2021). Metagenomic analysis of five chimney samples from OCHF has revealed significant taxonomic and metabolic heterogeneities among microbial populations (Lecoeuvre et al., 2021). Although formate- and CO-metabolizing microorganisms and sulfur-metabolizing bacteria are widespread, variable geochemical conditions with mixed but variable influences of reducing hydrothermal fluids has resulted in strong differences in microbial populations between chimneys. Nevertheless, similarities in bacterial and archaeal lineages were reported between the OCHF and LCHF chimney samples. The exteriors of less active or inactive chimneys at these hydrothermal fields are dominated by Proteobacteria that show metabolic capabilities for aerobic respiration, CO_2 fixation through the Calvin–Benson–Bassham cycle, and sulfur and nitrogen metabolism (Lecoeuvre et al., 2021).

Different sets of scientific information are available from different vent fields, and more detailed studies (chemical and biological properties of hydrothermal fluid along with geological background and vent organisms) are required to understand these systems.

14.3.3. Inactive Vent Fields

Sonne

The first hydrothermal field in the entire Indian Ocean was the SONNE field (Pluger, 1990), and the "MESO zone" at 23°21′S, 69°15′E is the area where massive sulfides were recovered in the MESO zone (Halbach et al., 1995). The MESO zone, which covers an area of about 0.6 km^2, is located at a water depth of 2850 m, and has sulfides that have formed over a period of at least 30,000 years. These sulfides are composed mainly of pyrite, marcasite, and chalcopyrite with traces of sphalerite, but they also contain minor amounts of amorphous silica (**Münch** et al., 1999). Chimneys show high concentrations of Cu and Fe _(>40 wt.%) followed by Zn (1 wt.%) and trace element contents such as Co up to 0.24 wt%. However, hydrothermal activity at the MESO zone ceased about 10 000 years ago (Lalou et al., 1998).

Yokoniwa

Yokoniwa hydrothermal field (YHF), which was identified in 2009, is an ultramafic-hosted system located on top of the Yokoniwa Rise (YR) and is very close to the Kairei field (Okino et al., 2015, Fujii et al., 2016). Hydrothermal sulfides and small dead chimneys were observed. The hydrothermal activity of the YHF was almost ended as indicated by weak venting of low-temperature fluids (<5°C) during submersible surveys. Serpentinized peridotite is extensively exposed along with small quantities of gabbro, dolerite, and basalt (Okino et al., 2015; Sato et al., 2015).

Mount Jourdanne

In 1998, the Mt. Jourdanne hydrothermal field was discovered at 27°51′S, 63°56′E and massive sulfide and chimney samples were collected (Münch et al., 2001). The sulfide samples recovered from Mount Jourdanne are classified as chimneys, mounds, and hydrothermal breccias (Nayak et al., 2014). The chimneys consist mainly of sphalerite and less chalcopyrite, set in matrix of late amorphous silica. In contrast, the mounds are dominated either by Fe sulfides (pyrite) or by a mixture of pyrite and chalcopyrite with less sphalerite, pyrrhotite, amorphous silica, and barite. The breccias consist of silicified basaltic material that is impregnated with sulfides and contain small chimney fragments within a matrix of low-temperature minerals such as sphalerite and pyrite. In addition, breccias also contain late-stage realgar, boulangerite, galena, Pb–As sulfosalts, and barite. Based on the different mineralogical associations, Nayak et al. (2014) have observed different thermal episodes ranging from black smoker mineralization conditions to cessation of the hydrothermal activity at Mt. Jourdanne.

Tianzuo

The Tianzuo hydrothermal field is an inactive sulfide field at 27°57′S, 63°32′E, which is hosted by ultramafic rocks and controlled by detachment fault (Tao et al., 2014, Chen et al., 2018). Polymetallic sulfide, reddish-brown iron hydroxide, polymetallic mud, and peridotite samples have been recovered from this field (Cao et al., 2021). No apparent mineralogical zonation was observed in the sulfides, which are dominated by marcasite and a small amount of pyrite (Py), opal, and covellite. Severe sulfide weathering alteration was found, and Fe oxyhydroxides were observed along the fissures. Cao et al. (2021) demonstrated that the composition of sulfides is controlled by multistage hydrothermal activity and the pervasive transform faults apart from the basement rock composition.

Yuhuang-1

Hydrothermal sulfides at Yuhuang-1 are the shallowest deposits (1400 m) on the SWIR at 49.26°E and 37.94°S (Han et al., 2010, Tao et al., 2014, Yu et al., 2021). The sulfide samples mainly consist of sphalerite, wurtzite, pyrite, marcasite, chalcopyrite, and secondary copper minerals and are classified as (i) Fe-rich sulfides, (ii) Zn-rich sulfides, and (iii) Cu-rich sulfides (Yu et al., 2021). The total volume of sulfide-rich mounds is calculated to be approximately 10.6×10^6 t. This includes at least ~7.5×10^5 tons of copper and zinc and ~18 tons of gold. The distribution and content of sulfide-rich deposits in the Yuhuang-1 hydrothermal field are composed of two areas ~ 500 m apart: the southwest sulfide area (SWS) and the northeast sulfide area (NES).

Duanqiao-1

Sulfides recovered from Duanqiao-1 hydrothermal field at 50°24′E and 37°39′S are mainly dominated by pyrite and sphalerite with a small amount of chalcopyrite and trace amounts of galena (Tao et al., 2012). Based on the detailed mineralogical studies of the sulfides, mineralization processes were divided into two main stages: (i) the medium–high-temperature stage and (ii) the medium–low-temperature stage (Zhang et al., 2018). Dating of the relict chimney and massive sulfides reveal four main episodes: (i) 68.9–84.3 thousand years, (ii) 43.9–48.4 thousand years, (iii) 25.3–34.8 thousand years and (iv) 0.7–17.3 thousand years, and about 84.3 (±0.5) thousand years ago; hydrothermal activity might have started and ceased about 0.737 (±0.023) thousand years ago (Yang et al., 2017). Sediments recovered from this field show mainly pelagic fractions (carbonate ooze) with minor clay contents, basaltic debris, Fe–Mn oxyhydroxides, and hydrothermal components (Liao et al., 2019). High concentrations of Cu, Pb, Zn, and Co and higher Cu/Fe and Zn/Fe ratios promote the precipitation of Cu sulfides first in the close proximity of the discharge area and followed by Cu, Zn, and Fe sulfides, and Fe-Mn oxyhydroxides with increasing distance from the source (Liao et al., 2019).

A total of 22 hydrothermal fields have been discovered so far along the Indian mid-oceanic ridge system. The majority of these fields are located along the SWIR and CIR. Among them, 10 fields are located along the SWIR and named Yuhuang-1 (49°16′ E/37°56′S), Longqi (49°39′E/37°47′S), Duanqiao-1 (50°24′E/37°39′S), 50°56′E carbonate field (50°56′E/37°37′S), 51°19′E field (51°19′E/37°27′S), 53°15′S field (53°15′E/36°60′S), Tiancheng (63°32′E/ 27°57′S), Tianzuo (63°31.8′E/27°57′S), Mount Jourdanne (63°56′E/27°51′S) and Old City (64°35.1É/27°50.1′S). Another 10 fields have been discovered along the CIR and named as Kairei (25°19.23′S/70°02.42′E), Yokoniwa (25°16′S/70°04′E), Edmond (23°52.68′S/69°35.80′E), Sonne (23°21′S/69°15′E), Solitaire (19°33.413′S/65°50.888′E), Dodo (18°20.1′S/65°17.9′E), Onnuri (11°24.9′S/66°25.4′E), 9°48.9′S field (9°48.9′S/66°40.6′E), 9°47.5′S field (9°47.5′S/ 66°41.9′E) and 8°10.1′S field (8°10.1′S/68°08.2′E). Only two fields have been discovered along the CR and SEIR and these have been named Wocan (6°22′N/60°31′E) and Pelagia (26°S/71°E).

14.3.4. Hydrothermal Plume Signatures in Water Column/Sediments/Ferromanganese Crusts

Several studies on water column tracers (both physical and chemical) in the CIR provide indirect evidence of hydrothermal activity in the region (Jean-Baptiste et al., 1992; Gamo et al., 1996, 2001; Kawagucci et al., 2008; Kumagai et al., 2008; Nakamura et al., 2009, 2012; Son et al., 2014; Ray et al., 2020). While some studies have focused on already discovered vent fields such as Dodo, Solitaire, Edmond, and Kairei, the others infer the existence of plume signatures in other unexplored areas. Very few reports are available on water column anomalies (German et al., 1998; Bach et al., 2002; Boulart et al., 2017) for active venting along with an inactive hydrothermal field (Tao et al., 2014; Han et al., 2018) in the SWIR and SEIR. To date, only one mega-event plume (CR2003; Murton et al., 2006) has been reported from the Carlsberg Ridge. This plume spread about 70 km (between 61°30E, 5°41′N and 60°33′E, 6°20′N) and additional small-scale events have been observed by Ray et al., (2008). In addition to this event plume, the first chronic plumes from two different locations have been reported at 3°42′N and 3°41.5′N (Ray et al., 2012, 2017). In subsequent years, a Chinese research team has recovered hydrothermal sulfides from this region (Tao et al., 2013).

Signatures of hydrothermal activity in ferromanganese crusts and sediments along the Indian ridge system and from other parts such as the back-arc basin (Andaman) and the abyssal basin (Central Indian Ocean Basin) have been reported by Nath et al., 1992, 1997; Rao et al., 1996; Kuhn et al., 1998, 2000; Kurian et al., 2008; Mascarenhas-Pereira and Nath, 2010; Kamesh Raju et al., 2012a; Surya Prakash et al., 2012; 2020; Kalangutkar et al., 2015; Li et al., 2016, 2017; Liao et al., 2018; Yu et al., 2018; Sinha et al., 2019; Agarwal et al., 2020.

14.3.5. Vent Fauna Associated with Hydrothermal Vents in the Indian Ocean

Studies on the biodiversity and biogeography of Indian Ocean vent fauna were initiated only after discovering the Kairei vent field (Van Dover et al., 2001). Even though hydrothermal vents along the CIR and SWIR are located within a limited area, they are considered to be reasonably diverse in terms of biological communities associated with them (Nakamura and Takai, 2015b). The diversity of vent-specific communities has been described by many researchers (Haymon et al., 1993; Hashimoto et al., 2001; Van Dover et al., 2001; Nakamura et al., 2012; Beedessee et al., 2013; Chen et al., 2015; Copley et al., 2016; Watanabe et al., 2018; Zhou et al., 2018; Jang et al., 2020; Sun et al., 2020). Based on published reports ~100 macrobenthic and megabenthic species/genera occur at the eight known active hydrothermal vent sites in the Indian Ocean (Table 14.2). The benthic fauna comprises 38 families belonging to eight phyla. The most dominant phylum in terms of the number of species/genera is Annelida, with 28 (29.2%) taxa belonging to eight families. Arthropoda is the second dominant phylum in order of faunal abundance, with 26 (27.1%) described/identified taxa belonging to nine families. Mollusca is the third most dominant phylum of vent fauna in the Indian Ocean (Table 14.3) and is represented by 25 (26.0%) species. Echinodermata is represented by three families (Chiridotidae, Prerasteridae, and Synaptidae). Cnidaria is represented by two families, whereas Nemertea and Platyhelminthes are represented with one family each. Chordata is represented by five families of ichthyofauna (fish). Of these, three families (Synaphobranchidae, Zoarcidae, and Macroudidae) were reported previously from the KHF (Hashimoto et al., 2001; Watanabe and Beedessee, 2015), and the families Ophidiidae and Notocanthidae were recently reported from the newly discovered Tiancheng hydrothermal vent field along the SWIR (Zhou et al., 2018; Sun et al., 2020). As shown in Table 14.3, most of the vent-dominant molluscs, crustacean, and annelid taxa are recorded from both CIR and SWIR vent fields. However, slightly higher benthic diversity (52%) has been reported from the four vent fields of the CIR systems (Figure 14.5). At the same time, with >35 confirmed benthic faunal taxa (genera/species), the KHF remains the most diverse among all the known active vent fields in the Indian Ocean. However, >30 vent specific benthic taxa have been reported from the Longqi vent field, making Longqi the most diverse and scientifically exciting vent field on the SWIR. Interestingly, most of the faunal taxon occurs concomitantly, suggesting a possible biogeographic and potential genetic connectivity between hydrothermal vent faunal communities along the CIR and SWIR (Sun et al., 2020). Although hydrothermal vent fauna of the Indian Ocean represents some of very unusual endemic species such as the Scaly-foot gastropod *C. squamiferum*, the biodiversity of this region is poorly understood compared with the well-studied Pacific and Atlantic vent systems (Sun et al., 2020). Nevertheless, analogous to the fauna of the other vents, a number of the species are believed to be endemic and are known to occur within the area of individual vents (Chen et al., 2015). In contrast, the vent crab *A. rodriguezensis*, vent shrimps *Rimicaris kairei*, and *M. indica*, the deep-sea mussel *B. marisindicus* and the Scaly-foot gastropods (*C. squamiferum* and *Alviniconcha* sp.) and barnacle (*Neolepas* sp.) are found at both the CIR and SWIR vent fields (Figure 14.6): they are among the most widely distributed vent species in the Indian Ocean (Nakamura et al., 2012; Copley et al., 2016; Zhou et al., 2018; Ryu et al., 2019; Jang et al. 2020; Kim et al., 2020; Sun et al., 2020). According to recent studies, the larvae of vent crab *A. rodriguezensis* can disperse among hydrothermal vent fields using surface currents (Beedessee et al., 2013). Moreover, phylogenetic and fossil analysis of hydrothermal vent stalked barnacles belonging to genus *Neolepas* is capable of long-distance migration (Watanabe et al., 2018). As these vent barnacles are widely distributed among different oceanic provinces and are the most conspicuous organisms in deep-sea hydrothermal vents (Herrera et al., 2015), they are considered excellent model organisms for studying the connectivity and biogeographic patterns in deep-sea chemosynthetic fauna.

Similarly, phylogenetic analysis of the vent shrimp *R. kairei* showed close lineages to the *Rimicaris exoculata* of the MOR vent field (Beedessee et al., 2013). Larvae of some vent species are known to survive away from the vent field for long periods (Herring and Dixon, 1998). Some shrimp larvae from Alvinocarididae family presumably feed on photosynthetic-based food, especially during their long dispersal periods (Beedessee et al., 2013). It is quite likely that some of the shrimp's species belonging to the family Alvinocarididae generally inhabit deep-sea hydrothermal vent areas, plausibly connected with cold-seep species present at the nearby cold-seep environments.

Table 14.2 Updated list of benthic fauna associated with Indian Ocean vent fields.

			Central Indian Ridge					Southwest Indian Ridge		
Phylum	Family	Species	Onnuri	Dodo	Solitaire	Emmond	Kairei	Tiancheng	Duanqiao	Longqi
Cnidaria	Actinostolidae	*Marianactis* sp.	+	+	+	+	+	+		
		Actinostolid sp. tiancheng						+		
		Actinostolid sp. sp. SWIR								+
	Cerianthidae	Cerianthidae sp.					+			
Arthropoda	Bythograeidae	*Austinograea rodriguezensis*	+	+	+	+	+	+		
	Kiwaidae	*Kiwa* sp. SWIR							+	+
	Alvinocarididae	*Rimicaris kairei*	+	+	+	+	+			+
		Chorocaris sp			+					+
		Mirocaris indica			+	+	+	+		+
		Alvinocaris solitaire			+					
	Galatheidae	*Munidopsis* sp.			+		+			
		Munidopsis sp. SWIR								+
		Galathidae indet						+		
	Ammotheidae	*Sericosura heteroscela*							+	+
		Sericosura bamberi							+	
		Sericosura n.sp SWIR							+	
	Dirivultidae	Dirivultidae indet						+		
		Aphotopointius limatulus	+							
		Aphotopointius Kiost	+							
		Aphotopointius muricatus	+							
		Benthoxynus constrictus	+							
		Stygiopontius spinifer	+							
		Stygiopontius horridus	+							
		Stygiopontius geminus	+							
		Stygiopontius quadripaxillifer	+							
	Schminkepinellidae	*Barathricola thermophilus*	+							
	Eolepadidae	*Neolepas marisindica*			+		+	+		
		Neolepas sp.						+	+	+
	Chionelasmatidae	*Eochionelasmus coreana*			+					
		Eochionelasmus sp.			+		+			

Mollusca	Mitilidae	*Bathymodiolus marisindicus*	+	+		+	+	+	+
		Gigantidas verijenhoeki	+						
	Lepetodrilidae	*Lepetodrilus* sp.		+		+			
		Lepetodrilus n. sp SWIR						+	+
		Pseudorimula sp. SWIR					+		+
		Pseudorimula sp.				+			
	Neolepetopsidae	*Eulepetopsis* sp.		+		+	+		
		Pseudorimula sp.					+		
	Peltospiridae (Scaly foot)	*Chrysomallon squamiferum*		+		+	+	+	+
		Gigantopelta aegis						+	+
		Dracogyra subfuscus							+
		Lirapex politus							+
	Phenacolepadidae	*Shinkailepas* sp.		+		+			
	Provannidae	*Alviniconcha* aff. *hessleri*		+	+	+			
		Alviniconcha cf. *marisindica*					+		
		Desbruyeresia cf. *marisindica*					+		
		Desbruyeresia sp.		+		+		+	
	Skeneidae	*Bruceiella* sp.				+			
	Raphitomidae	*Phymorhynchus* sp.		+	+	+			
		Phymorhynchus sp. *Tiancheng*					+		
		Phymorhynchus n. sp SWIR							+
	Simrothiellidae	*Helicoradomenia*? sp.				+			
	Sutulizonidae	*Sutulizona* sp.				+			
		Slit limpet sp. 1				+			
		Slit limpet sp. 2				+			
Annelida	Alvinellidae	Alvinellid gen. et sp.		+					
	Ampharetidae	*Amphisamytha* sp.				+			
		Amphisamytha n. sp. 1							+
		Amphisamytha n. sp. 2							+
	Amphinomidae	*Archinome* cf. *jasoni*		+		+			+
	Capitellidae	Capitellid sp. 1				+			
	Dorvilleidae	*Ophryotrocha* sp.				+			
		Cf. *Ophryotrocha* sp.					+		
		Ophryotrocha jiaolongi							+
	Hesionidae	Hesionid sp. 1				+			
		Hesionid sp. 2				+			
		Hesiolyra cf. *bergi*						+	+
		Hesionidae sp. Indet							+

(*Continued*)

Table 14.2 (Continued)

Phylum	Family	Species	Central Indian Ridge					Southwest Indian Ridge		
			Onnuri	Dodo	Solitaire	Emmond	Kairei	Tiancheng	Duanqiao	Longqi
	Polynoidae	*Branchinotogluma* sp.					+			
		Branchinotogluma bipapillata						+	+	+
		Branchipolynoe longqiensis								+
		Branchipolynoe sp.1			+	+	+			
		Branchipolynoe Sp. 2						+	+	+
		Harmothoe? sp.					+			
		Levensteiniella? sp					+			
		Polynoidae sp3 sensu						+		
		Peinaleopolynoe n. sp. Dragon							+	+
		Polynoidae n. gen.							+	
		Polynoidae sp. 1							+	+
		Polynoidae sp. 2								+
		Polynoidae sp. 3						+		
	Spionidae	*Prionospio* sp.					+			
		Laonice sp SWIR								+
		Prionospio cf. *unilamellata*								+
Echinodermada	Chiridotidae	*Apodacean* gen. et sp.			+					
		Chiridota sp.						+		
		Chiridota heheva								+
	Prerasteridae	Prerasteridae indet								
	Synaptidae	Gen. sp					+			
Nemertea		Unidentified			+		+			
Platyhelminthes		Unidentified			+		+			
Chordata	Synaphobranchidae						+			
	Zoarcidae						+			
	Ophidiidae	Ophidiidae indet						+		
		Ophidiidae sp						+		+
	Notacanthidae	cf. *Polyacanthonotus* sp.						+		
	Macroudidae	Macrouridae gen. et sp.			+			+		

Central Indian Ridge data from: Van Dover et al. (2001), Nakamura et al. (2012), Beedessee et al. (2013), Ryu et al. (2019), Jang et al. (2020), Kim et al., (2020), and other relevant reports. Southwest Indian Ridge (SWIR) data from: Copley et al., (2016), Zhou et al. (2018); Sun et al. (2020), and other relevant reports.

Table 14.3 Details of the hydrothermal vent fauna in the Indian Ocean based on described taxa.

Phyllum	No. of families	Percentage based on total known families	No. of taxa (species/ genera)	Percentage based on total known taxa
Cnidaria	2	5.1	4	4.1
Arthopoda	9	23.1	27	27.6
Mollusca	10	25.6	25	25.5
Annelida	8	20.5	29	29.6
Echinodermata	3	7.7	5	5.1
Nemertea	1	2.6	1	1.0
Platyhelminthes	1	2.6	1	1.0
Chordata	5	12.8	6	6.3
	39	100	98	100

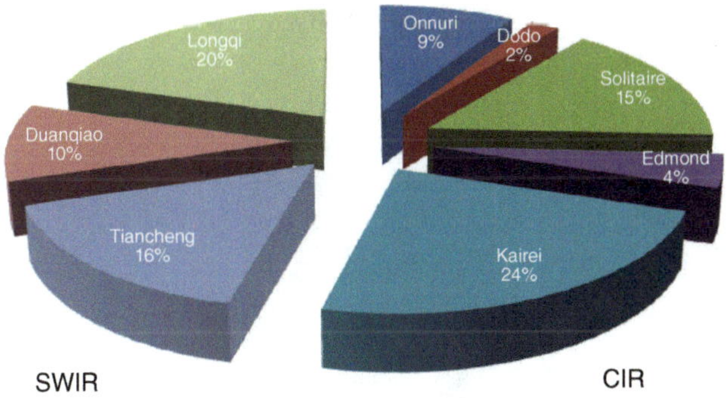

Figure 14.5 Benthic faunal richness (%) at each known vent field and comparison between CIR and SWIR vents fields. Values are based on described benthic species associated with hydrothermal vents in the Indian Ocean. Data source as shown in Table 14.2.

14.4. NATURE OF VENT-ASSOCIATED CHEMOAUTOTROPHS: ARE THEY LINKED WITH COLD-SEEP ECOSYSTEMS IN THE INDIAN OCEAN?

Mid-oceanic ridges are known to be the major source and extensive spread of hydrothermal helium, iron, and manganese in the Indian Ocean (Srinivasan et al., 2004; Nishioka et al., 2013; Vu and Sohrin, 2013; Takahata et al., 2018; Chinni et al., 2019). Recent studies on the biodiversity of the Indian Ocean hydrothermal vents from the Tiancheng field at SWIR show similarities with the Kairei and Edmond vent fields at CIR (Sun et al., 2020). Further, geochemical studies on ferromanganese crusts (Surya Prakash et al., 2020) along the Indian mid-oceanic ridges show along-ridge similarities and possible extension of deep water connectivity from south to north and follow the general bottom water flow (Antarctica Bottom Water and North Atlantic Deep Water; Kumar and Li, 1996; Wilson et al., 2012). Helium isotope studies in the Indian Ocean have identified four primary sources: (i) the mid-oceanic ridge, (ii) the Gulf of Aden, (iii) the convergent margin in the eastern Indian Ocean, and (iv) Indonesia through flow (Srinivasan et al., 2004). In addition, clockwise seawater circulation in the Northwestern Indian Ocean is responsible for the enrichment of hydrothermal helium (from the Gulf of Aden) in the Arabian Sea (Takahata et al., 2018). Similarly, we anticipate that anticlockwise seawater circulation in the eastern Indian Ocean may bring the volcanic inputs from the Indonesian Archipelago reported in the water column (Alibo and Nozaki, 2004). Therefore, in the eastern continental margin of India, the hydrothermal components (chemical and microbiological) can be expected from mid-oceanic ridges, the Andaman Sea, or Indonesia through flow.

Recent phylogeny, molecular, and fossil studies of newly described vent barnacle *Neolepas marisindica*, which is widely distributed in the Indian Ocean ridges, indicates that the genus *Neolepas* could have migrated to the Indian Ocean from the southern Pacific Ocean (Watanabe et al. 2018). Moreover, shrimp larvae from the Alvinocarididae family could presumably be feeding on photosynthesis-based food, especially during their long dispersal periods (Beedessee et al., 2013). Thus, it is quite likely that some of the species belonging to the family Galatheidae (crabs), Eolepadidae (barnacles), and Mitilidae (mussels) that generally inhabit deep-sea hydrothermal vent areas plausibly occur concurrently at nearby cold seep environments (Table 14.4) and hence are genetically interconnected. Detailed genomic studies of cold-seep communities along with detailed water column hydrothermal tracer-based (both chemical and microbiological) investigations are required to establish ecosystem connectivity between hydrothermal vents and cold-seep in the Indian Ocean.

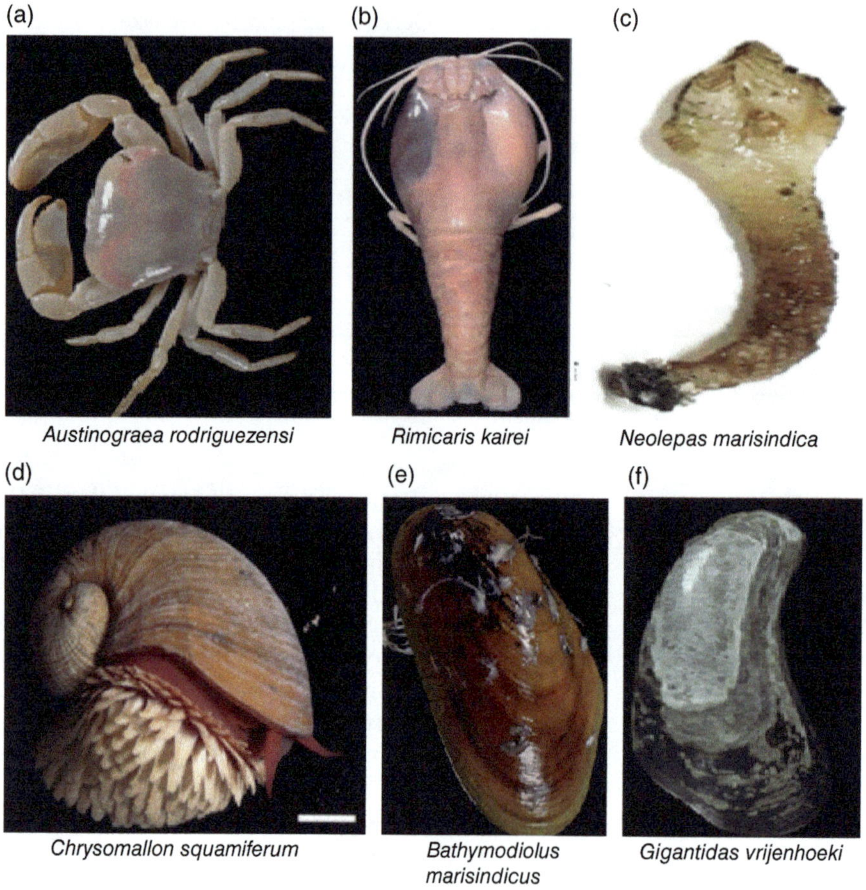

Figure 14.6 Vent fauna reported from the Central Indian Ridges (CIR) and Southwest Indian ridges (SWIR) vent fields. From: Sun et al. (2020) (a,e); Nakamura et al. (2012), (b,d); Ryu et al (2019), (c); Jang et al. (2020), (f).

Table 14.4 List of the closely matching hydrothermal vent and cold-seep fauna from the Indian Ocean[a].

		Seep fauna (KG Basin, Bay of Bengal)[a]	Vent fauna (Indian Ocean)
Phylum	Family	Species/genus	Species/genus
Arthropoda	Galatheidae	Galatheidae sp	
	Munidopsidae	Munidopsidae sp.	*Munidopsis* sp.
	Eolepadidae	*Neolepas* sp.	*Neolepas marisindica*
Mollusca	Bivalvia–Vesicomyidae	*Calyptogena magnifera*	
	Thuasiridae	*Conchocele* sp.	
	Solemyida	*Acharax* sp.	
	Pectinidae	*Pectin* sp.	
	Mitilidae	*Bathymodiolus* sp.	*Bathymodiolus marisindicus*
	Gastopoda–Provannidae	Provannidae	*Alviniconcha marisindica*
			Desbruyeresia marisindica
	Phenacolepadidae, Neritoidae	*Neritid* sp.	*Shinkailepas* sp.
	Lepetodrilidae		*Lepetodrilus*
Annelida	Polychaeta–Siboglinidae	*Sclerolinum* sp.	
	Glyceridae	*Glycera* sp.	
	Serpulidae	*Serpulidae* sp.	
	Polynoidae	Branchipolynoe seepensis	Branchipolynoe loggiensis
Echinodermada	Ophiuroidea	*Amphiodia* sp	

[a] Mazumdar et al. (2019). Central Indian Ridge data from: Van Dover et al. (2001), Nakamura et al. (2012), Beedessee et al. (2013), Ryu et al. (2019), Jang et al. (2020), and Kim et al. (2020, and other relevant reports. Southwest Indian Ridge data from Copley et al. (2016); Zhou et al. (2018), Sun et al. (2020) and other relevant reports.

14.5. CONCLUSION

The Indian mid-oceanic ridge system has varying spreading rates with ultraslow (Southwest), slow (Carlsberg), and intermediate (Southeast and Central Indian Ridges). This MOR hosts hydrothermal vents with diverse geochemical compositions, microbial biodiversity, and fauna. The CIR is relatively well explored compared with other ridges with nine hydrothermal fields discovered so far. The geochemical composition of fluids from Dodo, Solitaire, Edmond, Kairei, and Longqi vent fields indicates that they are basalt-hosted systems. However, only Kairei and Longqi fluids are enriched in hydrogen and methane because of mafic–ultramafic lithological influence. A high abundance of (hyper)thermophilic hydrogenotrophic chemolithoautotrophs has been observed to be associated with the Kairei fluids, which are enriched in hydrogen concentrations. In contrast, fluids from Edmond and Longqi are enriched in metal content. Extensive microbial mats mainly consisting of metal oxidizing chemolithoautotrophs showing taxonomic similarities to Zetaproteobacteria (as observed at Loihi seamount) have been recorded at the Edmond field. The geochemical composition of vent fluids and associated microbes inhabiting these extreme environments seems to be controlled mainly by subsurface water–rock interactions, and the spreading rate of the ridge has no significant role to play. Both the CIR and SWIR vent fields have similar vent fauna, for example, the crab *A. rodriguezensis*, vent shrimp *R. kairei*, and *M. indica*, the deep-sea mussel *B. marisindicus*, the scaly-foot gastropods (*C. squamiferum, Alviniconcha* sp.), and the barnacle (*Neolepas* sp.). Significant inter-ridge connectivity can thus be inferred. Further, chemosymbiont bearing bivalves, polychaetes, and gastropods have also been discovered in cold-seep methane hydrates in the eastern Indian continental shelf. The similarity in the fauna of hydrothermal vents and cold-seep in the Indian Ocean raises the question of whether the two ecosystems are interconnected. Genomic studies of cold-seep fauna along with the water column hydrothermal tracer studies (both chemical and microbial) may provide the link between these two geographically distant systems.

ACKNOWLEDGMENTS

The authors thank the Director, ESSO-National Centre for Polar and Ocean Research, Ministry of Earth Sciences, Goa, for his encouragement. The authors are also grateful to the Editors of *Systems Biogeochemistry of Major Marine Biomes* for their invitation to contribute this chapter and to two anonymous reviewers for a thorough review. This is NCPOR contribution number B-2/2021-22.

REFERENCES

Agarwal, D.K., Roy, P., Prakash, L.S. et al. (2020). Hydrothermal signatures in sediments from eastern Southwest Indian Ridge 63°E to 68°E. *Marine Chemistry 218*: 103732. https://doi.org/10.1016/j.marchem.2019.103732

Akerman, N.H., Butterfield, D.A. and Huber, J.A. (2013). Phylogenetic diversity and functional gene patterns of sulfur-oxidizing subseafloor *Epsilonproteobacteria* in diffuse hydrothermal vent fluids. *Frontiers in Microbiology 4*. https://doi.org/10.3389/fmicb.2013.00185

Alibo, D.S. and Nozaki, Y. (2004). Dissolved rare earth elements in the eastern Indian Ocean: chemical tracers of the water masses. *Deep Sea Research Part I: Oceanographic Research Papers 51* (4): 559–576.

Anderson, I., Risso, C., Holmes, D. et al. (2011). Complete genome sequence of *Ferroglobus placidus* AEDII12DO. *Standards in Genomic Sciences 5* (1): 50–60. https://doi.org/10.4056/sigs.2225018

Bach, W., Banerjee, N.R., Dick, H.J. et al. (2002). Discovery of ancient and active hydrothermal systems along the ultra-slow spreading Southwest Indian Ridge 10°–16°E. *Geochemistry, Geophysics, Geosystems 3* (7): 1–14.

Beedessee, G., Watanabe, H., Ogura, T. et al. (2013). High connectivity of animal populations in deep-sea hydrothermal vent fields in the central Indian ridge relevant to its geological setting. *PLoS ONE 8* (12): https://doi.org/10.1371/journal.pone.0081570

Bernardino, A.F., Levin, L.A., Thurber, A.R. et al. (2012). Comparative composition, diversity and trophic ecology of sediment macrofauna at vents, seeps and organic falls. *PLoS ONE 7* (4): e33515https://doi.org/10.1371/journal.pone.0033515

Boulart, C., Briais, A., Chavagnac, V. et al. (2017). Contrasted hydrothermal activity along the South-East Indian Ridge (130°E–140°E): From crustal to ultramafic circulation. *Geochemistry, Geophysics, Geosystems 18* (7): 2446–2458. https://doi.org/10.1002/2016GC006683

Breier, J.A., Toner, B.M., Fakra, S.C. et al. (2012). Sulfur, sulfides, oxides and organic matter aggregated in submarine hydrothermal plumes at 9°50′N East Pacific Rise. *Geochimica et Cosmochimica Acta 88*: 216–236. https://doi.org/10.1016/j.gca.2012.04.003

Cannat, M., Sauter, D., Mendel, V. et al. (2006). Modes of seafloor generation at a melt-poor ultraslow-spreading ridge. *Geology 34* (7): 605–608. https://doi.org/10.1130/G22486.1

Cannat, M., Sauter, D., Bezos, A. et al. (2008). Spreading rate, spreading obliquity and melt supply at the ultraslow spreading Southwest Indian Ridge. *Geochemistry, Geophysics, Geosystems 9* (4): Q04002. https://doi.org/10.1029/2007GC001676

Cao, H., Wang, Y., Lee, O.O. et al. (2014). Microbial sulfur cycle in two hydrothermal chimneys on the Southwest Indian Ridge. *MBio 5* (1). https://doi.org/10.1128/mBio.00980-13

Cao, H., Sun, Z., Jiang, Z. et al. (2021). Source origin and ore-controlling factors of hydrothermal sulfides from the Tianzuo hydrothermal field, *Southwest Indian Ridge. Ore Geology Reviews 134*: 104168

Castelle, C.J., Wrighton, K.C., Thomas, B.C. et al. (2015). Genomic expansion of domain archaea highlights roles for organisms from new phyla in anaerobic carbon cycling. *Current Biology 25* (6): 690–701. https://doi.org/10.1016/j.cub.2015.01.014

Chan, C.S., Fakra, S.C., Emerson, D. et al. (2011). Lithotrophic iron-oxidizing bacteria produce organic stalks to control mineral growth: Implications for biosignature formation. *ISME Journal 5* (4): 717–727. https://doi.org/10.1038/ismej.2010.173

Chen, C., Copley, J.T., Linse, K. et al. (2015). The heart of a dragon: 3D anatomical reconstruction of the "scaly-foot gastropod" (Mollusca: Gastropoda: *Neomphalina*) reveals its extraordinary circulatory system. *Frontiers in Zoology 12* (1). https://doi.org/10.1186/s12983-015-0105-1

Chen, J., Tao, C., Liang, J. et al. (2018). Newly discovered hydrothermal fields along the ultraslow-spreading Southwest Indian Ridge around 63°E. *Acta Oceanologica Sinica 37* (11): 61–67. https://doi.org/10.1007/s13131-018-1333-y

Chinni, V., Singh, S.K., Bhushan, R. et al. (2019). Spatial variability in dissolved iron concentrations in the marginal and open waters of the Indian Ocean. *Marine Chemistry 208*: 11–28. https://doi.org/10.1016/j.marchem.2018.11.007

Cochran, J.R. and Sempéré, J.-C. (1997). The Southeast Indian Ridge between 88°E and 118°E: gravity anomalies and crustal accretion at intermediate spreading rates. *Journal of Geophysical Research: Solid Earth 102* (B7): 15463–15487. https://doi.org/10.1029/97jb00511

Copley, J.T., Marsh, L., Glover, A.G. et al. (2016). Ecology and biogeography of megafauna and macrofauna at the first known deep-sea hydrothermal vents on the ultraslow-spreading Southwest Indian Ridge. *Scientific Reports 6*. https://doi.org/10.1038/srep39158

Crépeau, V., Cambon Bonavita, M.A., Lesongeur, F. et al. (2011). Diversity and function in microbial mats from the Lucky Strike hydrothermal vent field. *FEMS Microbiology Ecology 76* (3): 524–540. https://doi.org/10.1111/j.1574-6941.2011.0107

de Angelis, M.A., Lilley, M.D., Olson, E.J. et al. (1993). Methane oxidation in deep-sea hydrothermal plumes of the Endeavor Segment of the Juan de Fuca Ridge. *Deep-Sea Res 40*: 1169–1186.

DeMets, C., Gordon, R.G., Argus, D.F. et al. (1994). Effect of recent revisions to the geomagnetic reversal time scale on estimates of current plate motions. *Geophysical Research Letters.* https://doi.org/10.1029/94GL02118

DeMets, C., Gordon, R.G. and Argus, D.F. (2010). Geologically current plate motions. *Geophysical Journal International 181* (1): 1–80. https://doi.org/10.1111/j.1365-246X.2009.0449

Dick, G.J., Clement, B.G., Webb, S.M. et al. (2009). Enzymatic microbial Mn (II) oxidation and Mn biooxide production in the Guaymas Basin deep-sea hydrothermal plume. *Geochimica et Cosmochimica Acta 73* (21): 6517–6530. https://doi.org/10.1016/j.gca.2009.07.039

Dick, G.J., Anantharaman, K., Baker, B.J. et al. (2013). The microbiology of deep-sea hydrothermal vent plumes: Ecological and biogeographic linkages to seafloor and water column habitats. *Frontiers in Microbiology.* https://doi.org/10.3389/fmicb.2013.00124

Dick, H.J.B., Lin, J. and Schouten, H. (2003). An ultraslow-spreading class of ocean ridge. *Nature 426* (6965): 405–412. https://doi.org/10.1038/nature02128

Ding, J., Zhang, Y., Wang, H. et al. (2017). Microbial community structure of deep-sea hydrothermal vents on the ultraslow spreading southwest Indian ridge. *Frontiers in Microbiology.* https://doi.org/10.3389/fmicb.2017.01012

Djurhuus, A., Mikalsen, S.O., Giebel, H.A. et al. (2017). Cutting through the smoke: The diversity of microorganisms in deep-sea hydrothermal plumes. *Royal Society Open Science 4* (4). https://doi.org/10.1098/rsos.160829

Elshahed, M.S., Youssef, N.H., Luo, Q. et al. (2007). Phylogenetic and metabolic diversity of Planctomycetes from anaerobic, sulfide- and sulfur-rich Zodletone Spring, Oklahoma. *Applied and Environmental Microbiology 73* (15): 4707–4716. https://doi.org/10.1128/AEM.00591-07

Emerson, D., Rentz, J.A., Lilburn, T.G. et al. (2007). A novel lineage of proteobacteria involved in formation of marine Fe-oxidizing microbial mat communities. *PLoS ONE 2* (8). https://doi.org/10.1371/journal. pone.0000667

Fagervold, S.K., Galand, P.E., Zbinden, M. et al. (2012). Sunken woods on the ocean floor provide diverse specialized habitats for microorganisms. *FEMS Microbiology Ecology 82* (3): 616–628. https://doi.org/10.1111/j.1574-6941.2012.0143

Fang, Z. and Wang, W.X. (2021). Size speciation of dissolved trace metals in hydrothermal plumes on the Southwest Indian Ridge. *Science of the Total Environment 771*: 145367

Fernandes, S.O., Krishnan, K.P., Khedekar, V.D. et al. (2005). Manganese oxidation by bacterial isolates from the Indian Ridge System. *BioMetals 18* (5): 483–492. https://doi.org/10.1007/s10534-005-3000-y

Fernandes, S.O., Surya Prakash, L., Balan Binish, M. et al. (2018). Changes in morphology and metabolism enable Mn-oxidizing bacteria from mid-oceanic ridge environment to counter metal-induced stress. *Journal of Basic Microbiology 58* (5): 390–402. https://doi.org/10.1002/jobm.201700580

Fisher, C.R., Takai, K. and le Bris, N. (2007). Hydrothermal vent ecosystems. *Oceanography 20* (1): 14–23. https://doi.org/10.5670/oceanog.2007.75

Fujii, M., Okino, K., Sato, T. et al. (2016). Origin of magnetic highs at ultramafic hosted hydrothermal systems: Insights from the Yokoniwa site of Central Indian Ridge. *Earth and Planetary Science Letters 441*: 26–37.

Füri, E., Hilton, D.R., Murton, B.J. et al. (2011). Helium isotope variations between Réunion Island and the Central Indian Ridge (17°–21°S): New evidence for ridge-hot spot interaction. *Journal of Geophysical Research: Solid Earth 116* (2): 1–17. https://doi.org/10.1029/2010JB007609

Gallant, R.M. and Von Damm, K.L. (2006). Geochemical controls on hydrothermal fluids from the kairei and edmond vent fields, 23°–25°S, Central Indian Ridge. *Geochemistry, Geophysics, Geosystems 7* (6). https://doi.org/10.1029/2005GC001067

Gamo, T., Nakayama, E., Shitashima, K. et al. (1996). Hydrothermal plumes at the Rodriguez triple junction, Indian ridge. *Earth and Planetary Science Letters 142* (1–2): 261–270. https://doi.org/10.1016/0012-821x (96)00087-8

Gamo, T., Chiba, H., Yamanaka, T. et al. (2001). Chemical characteristics of newly discovered black smoker fluids and associated hydrothermal plumes at the Rodriguez Triple Junction, Central Indian Ridge. *Earth and Planetary Science Letters 193* (3–4): 371–379. https://doi.org/10.1016/S0012-821X (01)00511-8

German, C.R. and Von Damm, K.L. (2003). Hydrothermal Processes. In: *Treatise on Geochemistry*, Vol. 6–9 (eds. H.D.

Holland and K.K. Turekian), pp. 181–222. Elsevier. https://doi.org/10.1016/B0-08-043751-6/06109-0

German, C.R., Campbell, A.C. and Edmond, J.M. (1991). Hydrothermal scavenging at the Mid-Atlantic Ridge: Modification of trace element dissolved fluxes. *Earth and Planetary Science Letters 107* (1): 101–114. https://doi.org/10.1016/0012-821X (91)90047-L

German, C.R., Baker, E.T., Mevel, C. et al. (1998). Hydrothermal activity along the southwest Indian ridge. *Nature 395* (6701): 490–493.

Gorlas, A., Jacquemot, P., Guigner, J.M. et al. (2018). Greigite nanocrystals produced by hyperthermophilic archaea of Thermococcales order. *PLoS ONE 13* (8). https://doi.org/10.1371/journal. pone.0201549

Götz, D., Banta, A., Beveridge, T.J. et al. (2002). *Persephonella marina* gen. nov., sp. nov. and *Persephonella guaymasensis* sp. nov., two novel, thermophilic, hydrogen-oxidizing microaerophiles from deep-sea hydrothermal vents. *International Journal of Systematic and Evolutionary Microbiology 52* (4): 1349–1359. https://doi.org/10.1099/ijs.0.02126-0

Halbach, P., Blum, N., Plüger, W. et al. (1995). The Sonne Field – first massive sulfides in the Indian Ocean. *InterRidge News 4* (2): 12–15

Haalboom, S., Price, D., Mienis, F. et al. (2019). Successional patterns of (trace) metals and microorganisms in the Rainbow hydrothermal vent plume at the Mid-Atlantic Ridge. *Biogeosciences Discussions 1–44*. https://doi.org/10.5194/bg-2019-189

Hallberg, K.B., Hedrich, S. and Johnson, D.B. (2011). *Acidiferrobacter thiooxydans*, gen. nov. sp. nov.; an acidophilic, thermo-tolerant, facultatively anaerobic iron- and sulfur-oxidizer of the family Ectothiorhodospiraceae. *Extremophiles 15* (2): 271–279. https://doi.org/10.1007/s00792-011-0359-2

Han, X., Wu, G., Cui, R. et al (2010) Discovery of a hydrothermal sulfide deposit on the Southwest Indian Ridge at 49.2_E. In: *American Geophysical Union, Fall Meeting 2010*, Abstract OS21C–1531

Han, Y., Gonnella, G., Adam, N. et al. (2018). Hydrothermal chimneys host habitat-specific microbial communities: analogues for studying the possible impact of mining seafloor massive sulfide deposits. *Scientific Reports 8* (1). https://doi.org/10.1038/s41598-018-28613-5

Hashimoto, J., Ohta, S., Gamo, T. et al. (2001). First hydrothermal vent communities from the Indian Ocean discovered. *Zoological Science 18* (5): 717–721. https://doi.org/10.2108/zsj.18.717

Haymon, R.M., Fornari, D.J., Von Damm, K.L. et al. (1993). Volcanic eruption of the mid-ocean ridge along the East Pacific Rise crest at 9°45–52′N: Direct submersible observations of seafloor phenomena associated with an eruption event in April, 1991. *Earth and Planetary Science Letters 119* (1–2): 85–101. https://doi.org/10.1016/0012-821X (93)90008-W

Hekinian R. (1982). *Petrology of the Ocean Floor. Elsevier Oceanography Series, 33*. Amsterdam: Elsevier.

Helfrich K.R. and Speer K.G. (1995) Ocean hydrothermal circulation: mesoscale and basin-scale flow. *Geophysical Monographs (AGU) 91*: 347–356.

Herrera, S., Watanabe, H. and Shank, T.M. (2015). Evolutionary and biogeographical patterns of barnacles from deep-sea hydrothermal vents. *Molecular Ecology 24* (3): 673–689. https://doi.org/10.1111/mec.13054

Herring, P.J. and Dixon, D.R. (1998). Extensive deep-sea dispersal of postlarval shrimp from a hydrothermal vent. *Deep-Sea Research Part I: Oceanographic Research Papers 45* (12): 2105–2118. https://doi.org/10.1016/S0967-0637 (98)00050-8

Hoehler, T.M., Alperin, M.J., Albert, D.B. et al. (1998). Thermodynamic control on hydrogen concentrations in anoxic sediments. *Geochimica et Cosmochimica Acta 62* (10): 1745–1756. https://doi.org/10.1016/S0016-7037 (98)00106-9

Hoek, J., Banta, A., Hubler, F. et al. (2003). Microbial diversity of a sulphide spire located in the Edmond deep-sea hydrothermal vent field on the Central Indian Ridge. *Geobiology 1* (2): 119–127. https://doi.org/10.1046/j.1472-4669.2003.0001

Huber, J.A., Cantin, H.V., Huse, S.M. et al. (2010). Isolated communities of *Epsilonproteobacteria* in hydrothermal vent fluids of the Mariana Arc seamounts. *FEMS Microbiology Ecology 73* (3): 538–549. https://doi.org/10.1111/j.1574-6941.2010.0091

Hügler, M., Wirsen, C.O., Fuchs, G. et al. (2005). Evidence for autotrophic CO2 fixation via the reductive tricarboxylic acid cycle by members of the ε subdivision of proteobacteria. *Journal of Bacteriology 187* (9): 3020–3027. https://doi.org/10.1128/JB.187.9.3020-3027.2005

Inagaki, F., Takai, K., Kobayashi, H. et al. (2003). *Sulfurimonas autrotrophica* gen. nov., sp. nov., a novel sulfur-oxidizing E-proteobacterium isolated from hydrothermal sediments in the Mid-Okinawa Trough. *International Journal of Systematic and Evolutionary Microbiology 53* (6): 1801–1805. https://doi.org/10.1099/ijs.0.02682-0

Jang, S.J., Ho, P.T., Jun, S.Y. et al. (2020). A newly discovered *Gigantidas* bivalve mussel from the Onnuri Vent Field in the northern Central Indian Ridge. *Deep-Sea Research Part I: Oceanographic Research Papers 161*: 103299. https://doi.org/10.1016/j.dsr.2020.103299

Jean-Baptiste, P., Mantisi, F., Pauwells, H. et al. (1992). Hydrothermal 3He and manganese plumes at 19°29′S on the Central Indian Ridge. *Geophysical Research Letters 19* (17): 1787–1790. https://doi.org/10.1029/92GL00577

Ji, F., Zhou, H., Yang, Q. et al. (2017). Geochemistry of hydrothermal vent fluids and its implications for subsurface processes at the active Longqi hydrothermal field, Southwest Indian Ridge. *Deep-Sea Research Part I: Oceanographic Research Papers 122*: 41–47. https://doi.org/10.1016/j.dsr.2017.02.001

Johnson, S.B., Warén, A., Tunnicliffe, V. et al. (2015). Molecular taxonomy and naming of five cryptic species of *Alviniconcha* snails (Gastropoda: Abyssochrysoidea) from hydrothermal vents. *Systematics and Biodiversity13* (3): 278–295. https://doi.org/10.1080/14772000.2014.970673

Juniper, S.K., Bird, D.F., Summit, M. et al. (1998). Bacterial and viral abundances in hydrothermal event plumes over northern Gorda Ridge. *Deep-Sea Research Part II: Topical Studies in Oceanography 45* (12): 2739–2749. https://doi.org/10.1016/S0967-0645 (98)00091-5

Kalangutkar, N.G., Iyer, S.D., Mascarenhas-Pereira, M.B.L. et al. (2015). Hydrothermal signature in ferromanganese oxide coatings on pumice from the Central Indian Ocean Basin. *Geo-Marine Letters 35* (3): 221–235. https://doi.org/10.1007/s00367-015-0402-x

Kamesh Raju, K.A., Ramprasad, T. and Subrahmanyam, C. (1997). Geophysical investigations over a segment of the

Central Indian Ridge, *Indian Ocean. Geo-Marine Letters 17* (3): 195–201. https://doi.org/10.1007/s003670050026

Kamesh Raju, K.A., Chaubey, A.K., Amarnath, D. et al. (2008). Morphotectonics of the Carlsberg Ridge between 62°20′ and 66°20′E, northwest Indian Ocean. *Marine Geology 252* (3–4): 120–128. https://doi.org/10.1016/j.margeo.2008.03.016

Kamesh Raju, K.A., Ray, D., Mudholkar, A. et al. (2012a). Tectonic and volcanic implications of a cratered seamount off Nicobar Island andaman Sea. *Journal of Asian Earth Sciences 56*: 42–53. https://doi.org/10.1016/j.jseaes.2012.04.018

Kamesh Raju, K.A., Samudrala, K., Drolia, R.K. et al. (2012b). Segmentation and morphology of the Central Indian Ridge between 3°S and 11°S, Indian Ocean. *Tectonophysics 554–557*: 114–126. https://doi.org/10.1016/j.tecto.2012.06.001

Kato, S., Kobayashi, C., Kakegawa, T. et al. (2009). Microbial communities in iron-silica-rich microbial mats at deep-sea hydrothermal fields of the Southern Mariana Trough. *Environmental Microbiology 11* (8): 2094–2111. https://doi.org/10.1111/j.1462-2920.2009.0193

Kawagucci, S., Miyazaki, J., Noguchi, T. et al. (2016). Fluid chemistry in the Solitaire and Dodo hydrothermal fields of the Central Indian Ridge. *Geofluids 16* (5): 988–1005. https://doi.org/10.1111/gfl.12201

Kawagucci, Shinsuke, Okamura, K. et al. (2008). Methane, manganese and helium-3 in newly discovered hydrothermal plumes over the Central Indian Ridge, 18°–20°S. *Geochemistry, Geophysics, Geosystems 9* (10). https://doi.org/10.1029/2008GC002082

Kim, J., Son, S.K., Kim, D. et al. (2020). Discovery of active hydrothermal vent fields along the Central Indian Ridge, 8–12°S. *Geochemistry, Geophysics, Geosystems 21* (8): 0–3. https://doi.org/10.1029/2020GC009058

Kleiveland, C.R., Hult, L.T.O., Kuczkowska, K. et al. (2012). Draft genome sequence of the methane-oxidizing bacterium *Methylococcus capsulatus* (Texas). *Journal of Bacteriology 194* (23).. https://doi.org/10.1128/JB.01656-12

Kuhn, T., Bau, M., Blum, N. and Halbach, P. (1998). Origin of negative Ce anomalies in mixed hydrothermal-hydrogenetic Fe-Mn crusts from the Central Indian Ridge. *Earth and Planetary Science Letters 163* (1–4): 207–220. https://doi.org/10.1016/S0012-821X (98)00188-5

Kuhn, T., Burger, H., Castradori, D. et al. (2000). Volcanic and hydrothermal history of ridge segments near the Rodrigues Triple Junction (Central Indian Ocean) deduced from sediment geochemistry. *Marine Geology 169* (3–4): 391–409. https://doi.org/10.1016/S0025-3227 (00)00080-3

Kumagai, H., Nakamura, K., Toki, T. et al. (2008). Geological background of the Kairei and Edmond hydrothermal fields along the Central Indian Ridge: Implications of their vent fluids'distinct chemistry. In: *Geofluids* (Vol. *8*, pp. 239–251). https://doi.org/10.1111/j.1468-8123.2008.0022

Kumar, M.D. and Li, Y.H. (1996). Spreading of water masses and regeneration of silica and 226Ra in the Indian Ocean. *Deep-Sea Research Part II: Topical Studies in Oceanography 43* (1): 83–110. https://doi.org/10.1016/0967-0645 (95)00084-4

Kurian, S., Nath, B.N., Ramaswamy, V. et al. (2008). Possible detrital, diagenetic and hydrothermal sources for Holocene sediments of the Andaman backarc basin. *Marine Geology 247* (3–4): 178–193. https://doi.org/10.1016/j.margeo.2007.09.006

L'Haridon, S., Reysenbach, A.L., Banta, A. et al. (2003). *Methanocaldococcus indicus* sp. nov., a novel hyperthermophilic methanogen isolated from the Central Indian Ridge. *International Journal of Systematic and Evolutionary Microbiology 53* (6): 1931–1935. https://doi.org/10.1099/ijs.0.02700-0

La Duc, M.T., Benardini, J.N., Kempf, M.J. et al. (2007). Microbial diversity of Indian ocean hydrothermal vent plumes: microbes tolerant of desiccation. Peroxide exposure and ultraviolet and γ-irradiation. *Astrobiology 7*: 416–431. doi:10.1089/ast.2006.0060

Labrenz, M., Grote, J., Mammitzsch, K. et al. (2013). *Sulfurimonas gotlandica* sp. nov., a chemoautotrophic and psychrotolerant epsilonproteobacterium isolated from a pelagic redoxcline and an emended description of the genus *Sulfurimonas*. *International Journal of Systematic and Evolutionary Microbiology 63* (11): 4141–4148. https://doi.org/10.1099/ijs.0.048827-0

Lalou, C., Münch, U., Halbach, P. and Reyss, J.L. (1998). Radiochronological investigation of hydrothermal deposits from the MESO zone, *Central Indian Ridge. Marine Geology 149* (1–4): 243–254

Lam, P., Cowen, J.P. and Jones, R.D. (2004). Autotrophic ammonia oxidation in a deep-sea hydrothermal plume. *FEMS Microbiology Ecology 47* (2): 191–206. https://doi.org/10.1016/S0168-6496 (03)00256-3

Lam, P., Cowen, J.P., Popp, B.N. and Jones, R.D. (2008). Microbial ammonia oxidation and enhanced nitrogen cycling in the Endeavour hydrothermal plume. *Geochimica et Cosmochimica Acta 72* (9): 2268–2286. https://doi.org/10.1016/j.gca.2008.01.033

Lang, S.Q. (2018). Hydrothermal stamp on the oceans. *Nature Geoscience 11* (1): 10–12. https://doi.org/10.1038/s41561-017-0044-x

Le Bris N., Arnaud-Haond S., Beaulieu S. et al. (2016). Hydrothermal vents and cold seeps. In: *First Global Integrated Marine Assessment* (ed. United Nations), 853–862. Cambridge University Press.

Lecoeuvre A, Menez B, Lecourt L. et al. (2018). Microbial ecology linked to serpentinization processes and related carbonation at the Old City hydrothermal field. *Goldschmidt Abstracts 2018*: 1428.

Lecoeuvre, A., Ménez, B., Cannat, M. et al. (2021). Microbial ecology of the newly discovered serpentinite-hosted Old City hydrothermal field (southwest Indian ridge). *The ISME Journal 15*: 818–832. https://doi.org/10.1038/s41396-020-00816-7

Lei, J., Chu, F., Yu, X. et al. (2017). Lipid biomarkers reveal microbial communities in hydrothermal chimney structures from the 49.6°E Hydrothermal Vent Field at the Southwest Indian Ocean Ridge. *Geomicrobiology Journal 34* (6): 557–566. https://doi.org/10.1080/01490451.2016.1238979

Levin, L.A. and Sibuet, M. (2012). Understanding continental margin biodiversity: a new imperative. *Annual Review of Marine Science 4* (1): 79–112. https://doi.org/10.1146/annurev-marine-120709-142714

Li, J., Zhou, H., Fang, J., Wu, Z. et al. (2016). Microbial distribution in a hydrothermal plume of the Southwest Indian Ridge. *Geomicrobiology Journal 33* (5): 401–415. https://doi.org/10.1080/01490451.2015.1048393

Li, J., Cui, J., Yang, Q. et al. (2017). Oxidative weathering and microbial diversity of an inactive seafloor hydrothermal sulfide chimney. *Frontiers in Microbiology 8*. https://doi.org/10.3389/fmicb.2017.01378

Li, M., Toner, B.M., Baker, B.J. et al. (2014). Microbial iron uptake as a mechanism for dispersing iron from deep-sea hydrothermal vents. *Nature Communications 5*. https://doi.org/10.1038/ncomms4192

Li, M., Baker, B.J., Anantharaman, K. et al. (2015). Genomic and transcriptomic evidence for scavenging of diverse organic compounds by widespread deep-sea archaea. *Nature Communications 6*. https://doi.org/10.1038/ncomms9933

Liao, S., Tao, C., Li, H. et al. (2018). Surface sediment geochemistry and hydrothermal activity indicators in the Dragon Horn area on the Southwest Indian Ridge. *Marine Geology 398*: 22–34. https://doi.org/10.1016/j.margeo.2017.12.005

Liao, S., Tao, C., Dias, Á.A. et al. (2019). Surface sediment composition and distribution of hydrothermal derived elements at the Duanqiao-1 hydrothermal field, *Southwest Indian Ridge. Marine Geology 416*: 105975.

Lim, J.K., Kim, Y.J., Yang, JA. et al. (2020). *Thermococcus indicus* sp. nov., a Fe (III)-reducing hyperthermophilic archaeon isolated from the Onnuri Vent Field of the Central Indian Ocean ridge. *Journal of Microbiology 58* 260–267. https://doi.org/10.1007/s12275-020-9424-9

Lin, T.J., Ver Eecke, H.C., Breves, E.A. et al. (2016). Linkages between mineralogy, fluid chemistry and microbial communities within hydrothermal chimneys from the Endeavour Segment, Juan de Fuca Ridge. *Geochemistry, Geophysics, Geosystems 17* (2): 300–323. https://doi.org/10.1002/2015GC006091

Manzella, M.P., Holmes, D.E., Rocheleau, J.M. et al. (2015). The complete genome sequence and emendation of the hyperthermophilic, obligate iron-reducing archaeon "*Geoglobus ahangari*" strain 234T. *Standards in Genomic Sciences 10* (1). https://doi.org/10.1186/s40793-015-0035-8

Marshall, K.T. and Morris, R.M. (2013). Isolation of an aerobic sulfur oxidizer from the SUP05/Arctic96BD-19 clade. *ISME Journal 7* (2): 452–455. https://doi.org/10.1038/ismej.2012.78

Maruyama, A., Urabe, T., Ishibashi, J. et al. (1998). Global hydrothermal primary production rate estimated from the southern East Pacific Rise. *Cahiers de Biologie Marine 39* (3–4): 249–252. https://doi.org/10.21411/cbm. a.ec0e1360

Mascarenhas-Pereira, M.B.L. and Nath, B.N. (2010). Selective leaching studies of sediments from a seamount flank in the Central Indian Basin: Resolving hydrothermal, volcanogenic and terrigenous sources using major, trace and rare-earth elements. *Marine Chemistry 121* (1–4): 49–66. https://doi.org/10.1016/j.marchem.2010.03.004

Mazumdar, A., Dewangan, P., João, H.M. et al. (2009). Evidence of paleo-cold seep activity from the Bay of Bengal, offshore India. *Geochemistry, Geophysics, Geosystems 10* (6). https://doi.org/10.1029/2008GC002337

Mazumdar, A., Dewangan, P., Peketi, A. et al. (2019). The first record of active methane (cold) seep ecosystem associated with shallow methane hydrate from the Indian EEZ. *Journal of Earth System Science 128*: 18. https://doi.org/10.1007/s12040-018-1044-y

McCollom, T.M. (2008). Observational, experimental and theoretical constraints on carbon cycling in mid-ocean ridge hydrothermal systems. In: *Magma to Microbe: Modeling Hydrothermal Processes at Ocean Spreading Centers*, (eds. R.P. Lowell, J.S. Seewald, A. Metaxas et al.), 193–213. AGU.

Mendel, V. and Sauter, D. (1997). Seamount volcanism at the super slow-spreading Southwest Indian Ridge between 57°E and 70°E. *Geology 25* (2): 299–102. https://doi.org/10.1130/0091-7613 (1997)025<0099:SVATSS>2.O;2

Mendel, V., Sauter, D., Parson, L. et al. (1997). Segmentation and morphotectonic variations along a super slow-spreading center: the Southwest Indian Ridge (57°E–70° E). *Marine Geophysical Research.* https://doi.org/10.1023/A:1004232506333

Miller, A.J., Fan, X., Orsel, M. et al. (2007). Nitrate transport and signalling. *Journal of Experimental Botany 58*: 2297–2306). https://doi.org/10.1093/jxb/erm066

Mino, S., Makita, H., Toki, T. et al. (2013). Biogeography of *Persephonella* in deep-sea hydrothermal vents of the Western Pacific. *Frontiers in Microbiology 4*. https://doi.org/10.3389/fmicb.2013.00107

Mino, S., Nakagawa, S., Makita, H. et al. (2017). Endemicity of the cosmopolitan mesophilic chemolithoautotroph Sulfurimonas at deep-sea hydrothermal vents. *ISME Journal 11* (4): 909–919. https://doi.org/10.1038/ismej.2016.178

Moussard, H., L'Haridon, S., Tindall, B.J. et al. (2004). *Thermodesulfatator indicus* gen. nov., sp. nov., a novel thermophilic chemolithoautotrophic sulfate-reducing bacterium isolated from the Central Indian Ridge. *International Journal of Systematic and Evolutionary Microbiology 54* (1): 227–233. https://doi.org/10.1099/ijs.0.02669-0

Münch, U., Blum, N. and Halbach, P. (1999). Mineralogical and geochemical features of sulfide chimneys from the MESO zone, Central Indian Ridge. *Chemical Geology 155* (1–2): 29–44

Munch, U., Lalou, C., Halbach, P. et al. (2001). Relict hydrothermal events along the super-slow Southwest Indian spreading ridge near 63856 X E-mineralogy, chemistry and chronology of sulfide samples. *Chemical Geology 177*: 341–349.

Mullineaux, L.S., Metaxas, A., Beaulieu, S.E. et al. (2018). Exploring the ecology of deep-sea hydrothermal vents in a metacommunity framework. *Frontiers in Marine Science.* https://doi.org/10.3389/fmars.2018.00049

Murton, B.J., Baker, E.T., Sands, C.M. et al. (2006). Detection of an unusually large hydrothermal event plume above the slow-spreading Carlsberg Ridge: NW Indian Ocean. *Geophysical Research Letters 33* (10). https://doi.org/10.1029/2006GL026048

Naganuma, T., Otsuki, A. and Seki, H. (1989). Abundance and growth rate of bacterioplankton community in hydrothermal vent plumes of the North Fiji Basin. *Deep Sea Research Part A, Oceanographic Research Papers 36* (9): 1379–1390. https://doi.org/10.1016/0198-0149 (89)90089-7

Nakagawa, S. and Takai, K. (2008). Deep-sea vent chemoautotrophs: diversity, biochemistry and ecological significance. *FEMS Microbiology Ecology.* https://doi.org/10.1111/j.1574-6941.2008.0050

Nakagawa, T., Nakagawa, S., Inagaki, F. et al. (2004). Phylogenetic diversity of sulfate-reducing prokaryotes in active deep-sea hydrothermal vent chimney structures. *FEMS Microbiology Letters 232* (2): 145–152. https://doi.org/10.1016/S0378-1097 (04)00044-8

Nakamura, K. and Takai, K. (2015a). Geochemical constraints on potential biomass sustained by subseafloor water–rock interactions. In: *Subseafloor Biosphere Linked to Hydrothermal Systems: TAIGA Concept* (eds. J. Ishibashi, K. Okino, M. Sunamura), 11–30. Springer. https://doi.org/10.1007/978-4-431-54865-2_2

Nakamura, K. and Takai, K. (2015b). Indian Ocean hydrothermal systems: Seafloor hydrothermal activities, physical and chemical characteristics of hydrothermal fluids and vent-associated biological communities. In: *Subseafloor Biosphere Linked to Hydrothermal Systems: TAIGA Concept* (eds. J. Ishibashi, K. Okino and M. Sunamura), 147–161. Springer. https://doi.org/10.1007/978-4-431-54865-2_12

Nakamura, K., Morishita, T., Bach, W. et al. (2009). Serpentinized troctolites exposed near the Kairei Hydrothermal Field, Central Indian Ridge: insights into the origin of the Kairei hydrothermal fluid supporting a unique microbial ecosystem. *Earth and Planetary Science Letters 280* (1–4): 128–136. https://doi.org/10.1016/j.epsl.2009.01.024

Nakamura, K., Watanabe, H., Miyazaki, J. et al. (2012). Discovery of new hydrothermal activity and chemosynthetic fauna on the Central Indian ridge at 18°–20°S. *PLoS ONE 7* (3). https://doi.org/10.1371/journal. pone.0032965

Nakamura, K., Toki, T., Mochizuki, N. et al. (2013). Discovery of a new hydrothermal vent based on an underwater, high-resolution geophysical survey. *Deep-Sea Research Part I: Oceanographic Research Papers 74*: 1–10. https://doi.org/10.1016/j.dsr.2012.12.003

Nath, B.N., Balaram, V., Sudhakar, M. et al. (1992). Rare earth element geochemistry of ferromanganese deposits from the Indian Ocean. *Marine Chemistry 38* (3–4): 185–208. https://doi.org/10.1016/0304-4203 (92)90034-8

Nath, B.N., Plüger, W.L. and Roelandts, I. (1997). Geochemical constraints on the hydrothermal origin of ferromanganese encrustations from the Rodriguez Triple Junction, Indian Ocean. *Geological Society Special Publication 119*: 199–211. https://doi.org/10.1144/GSL. SP.1997.119.01.13

Nayak, B., Halbach, P., Pracejus, B. et al. (2014). Massive sulfides of Mount Jourdanne along the super-slow spreading Southwest Indian Ridge and their genesis. *Ore Geology Reviews 63*: 115–128.

Nishioka, J., Obata, H. and Tsumune, D. (2013). Evidence of an extensive spread of hydrothermal dissolved iron in the Indian Ocean. *Earth and Planetary Science Letters 361*: 26–33. https://doi.org/10.1016/j.epsl.2012.11.040

Nobu, M.K., Narihiro, T., Kuroda, K. et al. (2016). Chasing the elusive Euryarchaeota class WSA2: Genomes reveal a uniquely fastidious methyl-reducing methanogen. *ISME Journal 10* (10): 2478–2487. https://doi.org/10.1038/ismej.2016.33

Noguchi, T., Fukuba, T., Okamura, K. et al. (2015). Distribution and biogeochemical properties of hydrothermal plumes in the Rodriguez triple junction. In: *Subseafloor Biosphere Linked to Hydrothermal Systems: TAIGA Concept* (eds. J. Ishibashi, K. Okino and M. Sunamura), 195–204. Springer. https://doi.org/10.1007/978-4-431-54865-2_15

Noowong, A., Gomez-Saez, G.V., Hansen, C.T. et al. (2021). Imprint of Kairei and Pelagia deep-sea hydrothermal systems (Indian Ocean) on marine dissolved organic matter. *Organic Geochemistry 152*: 104141

Okino, K., Nakamura, K. and Sato, H. (2015). Tectonic background of four hydrothermal fields along the central Indian ridge. In: *Subseafloor Biosphere Linked to Hydrothermal Systems: TAIGA Concept* (eds. J. Ishibashi, K. Okino and M. Sunamura), 133–146. Springer. https://doi.org/10.1007/978-4-431-54865-2_11

Patriat, P. and Segoufin, J. (1988). Reconstruction of the Central Indian Ocean. *Tectonophysics 155* (1–4): 211–234. https://doi.org/10.1016/0040-1951 (88)90267-3

Patwardhan, S., Foustoukos, D.I., Giovannelli, D. et al. (2018). Ecological succession of sulfur-oxidizing epsilon- and gammaproteobacteria during colonization of a shallow-water gas vent. *Frontiers in Microbiology 9* (DEC). https://doi.org/10.3389/fmicb.2018.02970

Peng, X., Chen, S., Zhou, H. et al. (2011). Diversity of biogenic minerals in low-temperature Si-rich deposits from a newly discovered hydrothermal field on the ultraslow spreading Southwest Indian Ridge. *Journal of Geophysical Research: Biogeosciences 116* (3). https://doi.org/10.1029/2011JG001691

Peng, X.T., Zhou, H.Y., Yao, H.Q. et al. (2007). Microbe-related precipitation of iron and silica in the Edmond deep-sea hydrothermal vent field on the Central Indian Ridge. *Chinese Science Bulletin 52* (23): 3233–3238. https://doi.org/10.1007/s11434-007-0523-3

Pester, M., Schleper, C. and Wagner, M. (2011). The Thaumarchaeota: an emerging view of their phylogeny and ecophysiology. *Current Opinion in Microbiology 14* (3): 300–306. https://doi.org/10.1016/j.mib.2011.04.007

Piotrowski, A.M., Banakar, V.K., Scrivner, A.E. et al. (2009). Indian Ocean circulation and productivity during the last glacial cycle. *Earth and Planetary Science Letters 285* (1–2): 179–189. https://doi.org/10.1016/j.epsl.2009.06.007

Pjevac, P., Meier, D.V., Markert, S. et al. (2018). Metaproteogenomic profiling of microbial communities colonizing actively venting hydrothermal chimneys. *Frontiers in Microbiology 9*. https://doi.org/10.3389/fmicb.2018.00680

Pluger, W.L. (1990). Discovery of hydrothermal fields at the Central Indian Ridge. *Marine Mining 9* (1).

Popoola, S.O., Han, X., Wang, Y. et al. (2019a). Geochemical investigations of Fe-Si-Mn oxyhydroxides deposits in Wocan hydrothermal field on the slow-spreading Carlsberg ridge, Indian ocean: constraints on their types and origin. *Minerals 9* (1). https://doi.org/10.3390/min9010019

Popoola, S.O., Han, X., Wang, Y. et al. (2019b). Mineralogical and geochemical signatures of metalliferous sediments in wocan-1 and wocan-2 hydrothermal sites on the Carlsberg Ridge, Indian Ocean. *Minerals 9* (1). https://doi.org/10.3390/min9010026

Rao, P.S., Kamesh Raju, K.A., Ramprasad, T. et al. (1996). Evidence for hydrothermal activity in the Andaman Backarc Basin. *Current Science 70*, 379–384.

Ray, D., Mirza, I.H., Surya Prakash, L. et al. (2008). Water-column geochemical anomalies associated with the remnants of a mega plume: a case study after CR-2003 hydrothermal event in Carlsberg Ridge, *NW Indian Ocean. Current Science, 95* (3): 355–360.

Ray, D., Kamesh Raju, K.A., Baker, E.T. et al. (2012). Hydrothermal plumes over the Carlsberg Ridge, Indian Ocean. *Geochemistry, Geophysics, Geosystems 13* (1): Q01009. https://doi.org/10.1029/2011GC003888

Ray, D., Babu, E.V.S.S.K. and Prakash, L.S. (2017). Nature of suspended particles in hydrothermal plume at 3°40'N Carlsberg Ridge: A comparison with deep oceanic suspended matter. *Current Science 112* (1): 139–146. https://doi.org/10.18520/cs/v112/i01/139-146

Ray, D., KameshRaju, K.A., Srinivas Rao, A. et al. (2020). Elevated turbidity and dissolved manganese in deep water column near 10°47'S Central Indian Ridge: studies on hydrothermal activities. *Geo-Marine Letters 40*: 619–628. https://doi.org/10.1007/s00367-020-00657-5

Reysenbach, A.L., Banta, A.B., Boone, D.R. et al. (2000). Microbial essentials at hydrothermal vents. *Nature 404* (6780): 835. https://doi.org/10.1038/35009029

Reysenbach, A.L., Götz, D., Banta, A. et al. (2002). Expanding the distribution of the Aquificales to the deep-sea vents on Mid-Atlantic Ridge and Central Indian Ridge. *Cahiers de Biologie Marine 43* (3–4): 425–428.

Reysenbach, A.L., Liu, Y., Banta, A.B. et al. (2006). A ubiquitous thermoacidophilic archaeon from deep-sea hydrothermal vents. *Nature 442* (7101): 444–447. https://doi.org/10.1038/nature04921

Royer, J.Y. and Sandwell, D.T. (1989). Evolution of the eastern Indian Ocean since the Late Cretaceous: constraints from Geosat altimetry. *Journal of Geophysical Research 94* (B10). https://doi.org/10.1029/jb094ib10p13755

Royer, J.Y., Patriat, P., Bergh, H.W. et al. (1988). Evolution of the Southwest Indian Ridge from the Late Cretaceous (anomaly 34) to the Middle Eocene (anomaly 20). *Tectonophysics 155* (1–4): 235–260. https://doi.org/10.1016/0040-1951 (88)90268-5

Ryu, T., Woo, S. and Lee, N. (2019). The first reference transcriptome assembly of the stalked barnacle, Neolepas marisindica, from the Onnuri Vent Field on the Central Indian Ridge. *Marine Genomics 48*. https://doi.org/10.1016/j.margen.2019.04.004

Sato, H., Nakamura, K., Kumagai, H. et al. (2015). Petrology and geochemistry of mid-ocean ridge basalts from the Southern Central Indian Ridge. In: *Subseafloor Biosphere Linked to Hydrothermal Systems: TAIGA Concept* (eds. J. Ishibashi, K. Okino, M. Sunamura), 163–175). Springer.

Sauter, D., Patriat, P., Rommevaux-Jestin, C. et al. (2001). The Southwest Indian Ridge between 49°15'E and 57°E: focused accretion and magma redistribution. *Earth and Planetary Science Letters 192* (3): 303–317. https://doi.org/10.1016/S0012-821X (01)00455-1

Sauter, D., Cannat, M., Roumejon, S. et al. (2013). Continuous exhumation of mantle-derived rocks at the southwest Indian Ridge for 11 million years. *Nature Geoscience 6*: 314–20.

Schrenk, M.O., Kelley, D.S., Delaney, J.R. et al. (2003). Incidence and diversity of microorganisms within the walls of an active deep-sea sulfide chimney. *Applied and Environmental Microbiology 69* (6): 3580–3592.

Sela-Adler, M., Ronen, Z., Herut, B. et al. (2017). Co-existence of methanogenesis and sulfate reduction with common substrates in sulfate-rich estuarine sediments. *Frontiers in Microbiology 8*. https://doi.org/10.3389/fmicb.2017.00766

Shah, V., Zhao, X., Lundeen, R.A. et al. (2019). Morphological plasticity in a sulfur-oxidizing marine bacterium from the SUP05 clade enhances dark carbon fixation. *MBio 10* (3). https://doi.org/10.1128/mBio.00216-19

Sievert, S.M., Scott, K.M., Klotz, M.G., Chain, P.S.G., Hauser, L.J., Hemp, J. et al. (2008). Genome of the epsilonproteobacterial chemolithoautotroph *Sulfurimonas denitrificans*. *Applied and Environmental Microbiology 74* (4): 1145–1156. https://doi.org/10.1128/AEM.01844-07

Sinha, R.K., Krishnan, K.P. and John Kurian, P. (2017). Draft genome sequence of Idiomarina sp. strain 5.13, a highly stress-resistant bacterium isolated from the Southwest Indian Ridge. *Genome Announcements 5* (10). https://doi.org/10.1128/genomeA.01747-16

Sinha, R.K., Krishnan, K.P., Thomas, F.A. et al. (2019). Polyphasic approach revealed complex bacterial community structure and function in deep sea sediment of ultra-slow spreading Southwest Indian Ridge. *Ecological Indicators 96*, 40–51. https://doi.org/10.1016/j.ecolind.2018.08.063

Sintes, E., Bergauer, K., De Corte, D. et al. (2013). Archaeal amoA gene diversity points to distinct biogeography of ammonia-oxidizing Crenarchaeota in the ocean. *Environmental Microbiology 15* (5): 1647–1658. https://doi.org/10.1111/j.1462-2920.2012.0280

Small, C., Cochran, J.R., Sempéré, J.C. et al. (1999). The structure and segmentation of the Southeast Indian Ridge. *Marine Geology 161* (1): 1–12. https://doi.org/10.1016/S0025-3227(99)00051-1

Smith, G.J., Angle, J.C., Solden, L.M. et al. (2018). Members of the genus *Methylobacter* are inferred to account for the majority of aerobic methane oxidation in oxic soils from a freshwater wetland. *MBio 9* (6). https://doi.org/10.1128/mBio.00815-18

Son, J., Pak, S.J., Kim, J. et al. (2014). Tectonic and magmatic control of hydrothermal activity along the slow-spreading Central Indian Ridge, 8S-17S. *Geochemistry, Geophysics, Geosystems 15* (5): 2011–2020. https://doi.org/10.1002/2013GC005206

Srinivasan, A., Top, Z., Schlosser, P. et al. (2004). Mantle 3He distribution and deep circulation in the Indian Ocean. *Journal of Geophysical Research: Oceans 109* (C6).

Stokke, R., Dahle, H., Roalkvam, I. et al. (2015). Functional interactions among filamentous *Epsilonproteobacteria* and Bacteroidetes in a deep-sea hydrothermal vent biofilm. *Environmental Microbiology 17* (10): 4063–4077. https://doi.org/10.1111/1462-2920.12970

Sun, J., Zhou, Y., Chen, C. et al. (2020). Nearest vent, dearest friend: biodiversity of Tiancheng vent field reveals cross-ridge similarities in the Indian Ocean. *Royal Society Open Science 7* (3). https://doi.org/10.1098/rsos.200110

Surya Prakash, L., Ray, D., Paropkari, A.L. et al. (2012). Distribution of REEs and yttrium among major geochemical phases of marine Fe-Mn-oxides: Comparative study between hydrogenous and hydrothermal deposits. *Chemical Geology 312–313*: 127–137. https://doi.org/10.1016/j.chemgeo.2012.03.024

Surya Prakash, L., Ray, D., B.N.Nath., Satyanarayanan, M. et al. (2020). Anomalous phase association of REE in ferromanganese crusts from Indian mid-oceanic ridges: evidence for large scale dispersion of hydrothermal iron. *Chemical Geology 549*. https://doi.org/10.1016/j.chemgeo.2020.119679

Suzuki, Y., Inagaki, F., Takai, K. et al. (2004). Microbial diversity in inactive chimney structures from deep-sea hydrothermal systems. *Microbial Ecology 47*: 186–196). https://doi.org/10.1007/s00248-003-1014-y

Suzuki, Y., Sasaki, T., Suzuki, M. et al. (2005). Novel chemoautotrophic endosymbiosis between a member of the *Epsilonproteobacteria* and the hydrothermal-vent gastropod *Alviniconcha* aff. *hessleri* (Gastropoda: Provannidae) from the Indian Ocean. *Applied and Environmental Microbiology 71* (9): 5440–5450. https://doi.org/10.1128/AEM.71.9.5440-5450.2005

Sylvan, J.B., Sia, T.Y., Haddad, A.G. et al. (2013). Low temperature geomicrobiology follows host rock composition along a geochemical gradient in Lau Basin. *Frontiers in Microbiology 4*. https://doi.org/10.3389/fmicb.2013.00061

Takahata, N., Shirai, K., Ohmori, K. et al. (2018). Distribution of helium-3 plumes and deep-sea circulation in the central Indian Ocean. *Terrestrial, Atmospheric and Oceanic Sciences 29* (3): 331–340. https://doi.org/10.3319/TAO.2017.10.21.02

Takai, K. and Nakamura, K. (2010). Compositional, physiological and metabolic variability in microbial communities associated with geochemically diverse, deep-sea hydrothermal vent fluids. In: *Geomicrobiology: Molecular and Environmental Perspective*, 251–283. Springer. https://doi.org/10.1007/978-90-481-9204-5_12

Takai, K., Inagaki, F., Nakagawa, S. et al. (2003). Isolation and phylogenetic diversity of members of previously uncultivated ε-Proteobacteria in deep-sea hydrothermal fields. *FEMS Microbiology Letters 218* (1): 167–174. https://doi.org/10.1016/S0378-1097 (02)01147-3

Takai, K., Gamo, T., Tsunogai, U. et al. (2004). Geochemical and microbiological evidence for a hydrogen-based, hyperthermophilic subsurface lithoautotrophic microbial ecosystem (HyperSLiME) beneath an active deep-sea hydrothermal field. *Extremophiles 8* (4): 269–282. https://doi.org/10.1007/s00792-004-0386-3

Takai, K., Nealson, K.H. and Horikoshi, K. (2004a). Hydrogenimonas thermophila gen. nov., sp. nov., a novel thermophilic, hydrogen-oxidizing chemolithoautotroph within the E-Proteobacteria, isolated from a black smoker in a Central Indian Ridge hydrothermal field. *International Journal of Systematic and Evolutionary Microbiology 54* (1): 25–32. https://doi.org/10.1099/ijs.0.02787-0

Takai, K., Nealson, K.H. and Horikoshi, K. (2004b). *Methanotorris formicicus* sp. nov., a novel extremely thermophilic, methane-producing archaeon isolated from a black smoker chimney in the Central Indian Ridge. *International Journal of Systematic and Evolutionary Microbiology, 54* (4): 1095–1100. https://doi.org/10.1099/ijs.0.02887-0

Takai, K., Nunoura, T., Ishibashi, J.I. et al. (2008). Variability in the microbial communities and hydrothermal fluid chemistry at the newly discovered Mariner hydrothermal field, southern Lau Basin. *Journal of Geophysical Research: Biogeosciences 113* (2). https://doi.org/10.1029/2007JG000636

Tao, C.H., Li, H.M., Huang, W. et al. (2011). Mineralogical and geochemical features of sulfide chimneys from the 49°39′E hydrothermal field on the Southwest Indian Ridge and their geological inferences. *Chinese Science Bulletin 56* (26): 2828–2838. https://doi.org/10.1007/s11434-011-4619-4

Tao, C., Lin, J., Guo, S. et al. (2012). First active hydrothermal vents on an ultraslow-spreading center: Southwest Indian Ridge. *Geology 40* (1): 47–50. https://doi.org/10.1130/G32389.1

Tao, C., Wu, G., Deng, X. et al. (2013). New discovery of seafloor hydrothermal activity on the Indian Ocean Carlsberg Ridge and Southern North Atlantic Ridge-progress during the 26th Chinese COMRA cruise. *Acta Oceanologica Sinica 32* (8): 85–88. https://doi.org/10.1007/s13131-013-0345-x

Tao, C., Li, H., Jin, X. et al. (2014). Seafloor hydrothermal activity and polymetallic sulfide exploration on the southwest Indian ridge. *Chinese Science Bulletin 59* (19): 2266–2276. https://doi.org/10.1007/s11434-014-0182-0

Tao, C., Seyfried, W.E., Lowell, R.P. et al. (2020). Deep high-temperature hydrothermal circulation in a detachment faulting system on the ultra-slow spreading ridge. *Nature communications 11* (1): 1–9

Topçuoglu, B.D., Stewart, L.C., Morrison, H.G. et al. (2016). Hydrogen limitation and syntrophic growth among natural assemblages of thermophilic methanogens at deep-sea hydrothermal vents. *Frontiers in Microbiology 7*. https://doi.org/10.3389/fmicb.2016.01240

Urabe, T., Ishibashi, J.I., Sunamura, M. et al. (2015). Interdisciplinary studies: introduction of TAIGA concept. In: *Subseafloor Biosphere Linked to Hydrothermal Systems: TAIGA Concept* (eds. J. Ishibashi, K. Okino and M. Sunamura), 1–10. Springer. https://doi.org/10.1007/978-4-431-54865-2_1

Van Dover, C.L. (2000). *The Ecology of Deep-Sea Hydrothermal Vents*. Princeton: Princeton University Press.

Van Dover, C.L., Humphris, S.E., Fornari, D. et al. (2001). Biogeography and ecological setting of Indian Ocean hydrothermal vents. *Science 294* (5543): 818–823. https://doi.org/10.1126/science.1064574

Van Dover, C.L., German, C.R., Speer, K.G. et al. (2002). Evolution and biogeography of deep-sea vent and seep invertebrates. *Science 295* (5558): 1253–1257. https://doi.org/10.1126/science.1067361

Vetriani, C., Speck, M.D., Ellor, S.V. et al. (2004). *Thermovibrio ammonificans* sp. nov., a thermophilic, chemolithotrophic, nitrate-ammonifying bacterium from deep-sea hydrothermal vents. *International Journal of Systematic and Evolutionary Microbiology 54* (1): 175–181. https://doi.org/10.1099/ijs.0.02781-0

Vlastélic, I., Abouchami, W., Galer, S.J.G. et al. (2001). Geographic control on Pb isotope distribution and sources in Indian Ocean Fe-Mn deposits. *Geochimica et Cosmochimica Acta 65* (23): 4303–4319. https://doi.org/10.1016/S0016-7037(01)00713-X

Vu, T.D.H. and Sohrin, Y. (2013). Diverse stoichiometry of dissolved trace metals in the Indian Ocean. *Scientific Reports 3*: 1745. https://doi.org/10.1038/srep01745

Wang, H., Yang, Q., Ji, F. et al. (2012). The geochemical characteristics and Fe (II) oxidation kinetics of hydrothermal plumes at the Southwest Indian Ridge. *Marine Chemistry 134–135*: 29–35. https://doi.org/10.1016/j.marchem.2012.02.009

Wang, L., Cheung, M.K., Kwan, H.S. et al. (2015). Microbial diversity in shallow-water hydrothermal sediments of Kueishan Island, Taiwan as revealed by pyrosequencing. *Journal of Basic Microbiology 55* (11): 1308–1318. https://doi.org/10.1002/jobm.201400811

Wang, Y., Han, X., Petersen, S. et al. (2017). Mineralogy and trace element geochemistry of sulfide minerals from the Wocan Hydrothermal Field on the slow-spreading Carlsberg Ridge, Indian Ocean. *Ore Geology Reviews 84*, 1–19. https://doi.org/10.1016/j.oregeorev.2016.12.020

Wartiainen, I., Hestnes, A.G., McDonald, I.R. et al. (2006). Methylobacter tundripaludum sp. nov., a methane-oxidizing bacterium from Arctic wetland soil on the Svalbard islands, Norway (78°N). *International Journal of Systematic and Evolutionary Microbiology 56* (1): 109–113. https://doi.org/10.1099/ijs.0.63728-0

Watanabe, H. and Beedessee, G. (2015). Vent fauna on the central Indian ridge. In: *Subseafloor Biosphere Linked to Hydrothermal Systems: TAIGA Concept* (eds. J. Ishibashi, K. Okino and M. Sunamura), 205–212. Springer. https://doi.org/10.1007/978-4-431-54865-2_16

Watanabe, H., Fujikura, K., Kojima, S. et al. (2010). Japan: vents and seeps in close proximity. In: Topics in Geobiology (eds. N.H. Landman and P. Harries), 379–401. https://doi.org/10.1007/978-90-481-9572-5_12

Watanabe, H.K., Chen, C., Marie, D.P. et al. (2018). Phylogeography of hydrothermal vent stalked barnacles: a new species fills a gap in the Indian Ocean 'dispersal corridor' hypothesis. *Royal Society Open Science 5* (4). https://doi.org/10.1098/rsos.172408

Wilson, D.J., Piotrowski, A.M., Galy, A. et al. (2012). A boundary exchange influence on deglacial neodymium isotope records from the deep western Indian Ocean. *Earth and Planetary Science Letters 341–344*: 35–47. https://doi.org/10.1016/j.epsl.2012.06.009

Winn C.D., Karl D.M. and Massoth G.J. (1986). Microorganisms in deep-sea hydrothermal plumes. *Nature 320*: 744–746.

Winn, C.D., Cowen, J.P. and Karl, D.M. (1995). Microbes in deep-sea hydrothermal plumes. In: *The Microbiology of Deep-Sea Hydrothermal Vents* (ed. D. Karl), 255–273. Boca Raton, FL: CRC Press.

Winn, C.D., Karl, D.M. and Massoth, G.J. (1986). Microorganisms in deep-sea hydrothermal plumes. *Nature 320* (6064): 744–746. https://doi.org/10.1038/320744a0

Wu, Y., Cao, Y., Wang, C. et al. (2014). Microbial community structure and nitrogenase gene diversity of sediment from a deep-sea hydrothermal vent field on the Southwest Indian Ridge. *Acta Oceanologica Sinica 33* (10): 94–104. https://doi.org/10.1007/s13131-014-0544-0

Yang, W., Tao, C., Li, H. et al. (2017). 230 Th/238 U dating of hydrothermal sulfides from Duanqiao hydrothermal field, Southwest Indian Ridge. *Marine Geophysical Research 38* (1–2): 71–83.

Yu, Z., Li, H., Li, M. et al. (2018). Hydrothermal signature in the axial-sediments from the Carlsberg Ridge in the northwest Indian Ocean. *Journal of Marine Systems 180*: 173–181. https://doi.org/10.1016/j.jmarsys.2016.11.013

Yu, J., Tao, C., Liao, S. et al. (2021). Resource estimation of the sulfide-rich deposits of the Yuhuang-1 hydrothermal field on the ultraslow-spreading Southwest Indian Ridge. *Ore Geology Reviews 134*: 104169.

Yuan, J., Lai, Q., Sun, F. et al. (2015). The diversity of PAH-degrading bacteria in a deep-sea water column above the southwest Indian ridge. *Frontiers in Microbiology 6*. https://doi.org/10.3389/fmicb.2015.00853

Yue, X., Li, H., Ren, J. et al. (2019). Seafloor hydrothermal activity along mid-ocean ridge with strong melt supply: study from segment 27, southwest Indian ridge. *Scientific Reports, 9* (1). https://doi.org/10.1038/s41598-019-46299-1

Zhang, L., Kang, M., Xu, J. et al. (2016). Bacterial and archaeal communities in the deep-sea sediments of inactive hydrothermal vents in the Southwest India Ridge. *Scientific Reports 6*. https://doi.org/10.1038/srep25982

Zhang, B., Li, Z., Hou, Z. et al. (2018). Mineralogy and Chemistry of Sulfides from the Longqi and Duanqiao Hydrothermal Fields in the Southwest Indian Ridge. *Acta Geologica Sinica-English Edition 92* (5): 1798–1822.

Zhou, Y., Zhang, D., Zhang, R. et al. (2018). Characterization of vent fauna at three hydrothermal vent fields on the Southwest Indian Ridge: Implications for biogeography and interannual dynamics on ultraslow-spreading ridges. *Deep-Sea Research Part I: Oceanographic Research Papers 137*: 1–12. https://doi.org/10.1016/j.dsr.2018.05.001

INDEX

Systems Biogeochemistry of Major Marine Biomes, First Edition. Edited by Aninda Mazumdar and Wriddhiman Ghosh.

Printed in the USA/Agawam, MA
September 16, 2026

910661.051